Hesse–Meier–Zeeh

Spectroscopic Methods in Organic Chemistry

Prof. Dr. Stefan Bienz
Department of Chemistry
University of Zurich
Winterthurerstrasse 190
CH-8057 Zurich
Switzerland

Prof. Dr. Laurent Bigler
Department of Chemistry
University of Zurich
Winterthurerstrasse 190
CH-8057 Zurich
Switzerland

Dr. Thomas Fox
Department of Chemistry
University of Zurich
Winterthurerstrasse 190
CH-8057 Zurich
Switzerland

Prof. Dr. Herbert Meier
Department of Chemistry
Johannes-Gutenberg-University
Duesbergweg 18-14
D-55099 Mainz
Germany

3rd fully revised and
extended Edition

335 Figures, 132 Tables, 59 Schemes,
10 Graphical representations of full
analytical datasets

Georg Thieme Verlag

Stuttgart · New York

Library of Congress Cataloging-in-Publication Data is available from the publisher

This book is an extended authorized translation of the 9th German edition published and copyrighted 1979, 1984, 1987, 1991, 1995, 2002, 2005, 2012, 2016 by Georg Thieme Verlag, Stuttgart, Germany. Title of the German edition: Spektroskopische Methoden in der organischen Chemie.

© 2021. Thieme. All rights reserved,

Georg Thieme Verlag KG
Rüdigerstraße 14, 70469 Stuttgart, Germany
www.thieme-chemistry.com

Cover: © Thieme
Images: © Stefan Bienz and Laurent Bigler, Zurich, Switzerland.

Printed in Germany

by Firmengruppe Appl, Aprinta Druck GmbH, Senefelderstraße 3–11, 86650 Wemding

ISBN (print) 978-3-13-243408-0
ISBN (ePDF) 978-3-13-243410-3
ISBN (ePUB) 978-3-13-243411-0
DOI 10.1055/b000000049 1 2

Preface

Already more than 10 years have passed since the second edition of this textbook on spectroscopic methods appeared in 2008. Time has not been standing still, however, and organic analytics have again experienced a tremendous development. Not only did the instruments get refined, delivering better spectra and more data with lesser amounts of sample material, but also, hand in hand, computer technology made a big step forward, allowing us to handle and process efficiently the enormous amount of data arising with modern instruments and experiments—in particular with nuclear magnetic resonance (NMR) and mass spectrometry (MS) analyses.

The book *Spektroskopische Methoden in der organischen Chemie* by Manfred Hesse, Herbert Meier and Bernd Zeeh took account of these developments in the *German* version by two subsequent editions that appeared in 2012 and 2016. Even though the authorship has partly changed after more than 32 years, the textbook remained devoted to its original objective. It is still meant as a straightforward read and source of reference to complement lecture and laboratory courses devoted to structure elucidation and analytical characterization of organic compounds. It shall still offer enough information also to make it a reliable companion for Bachelor, Master, and PhD students, as well as for professionals in chemical teaching and research outside of universities.

The new *English* edition appears in a completely new guise. Not only has the layout been changed to a more lucid, modern, and colorful look, but also the content has been updated significantly, reflecting important developments in organic analytics. The major changes as compared with the previous edition are summarized shortly below.

In the *Ultraviolet/Visible Spectroscopy* (UV/Vis) chapter, attended by Herbert Meier (University of Mainz), the fundamentals of allowed and symmetry-forbidden electronic transitions are newly discussed by means of simple molecules. In addition, acknowledging the increasing importance of optoelectronic materials, the section about compounds with larger conjugated systems (aromatics, heteroaromatics, and open-chained oligomers) has been extended. More room is also given to determination of solvent polarities in the chapter on applications of UV/Vis spectroscopy.

Thomas Fox (University of Zurich) has taken over the authorship of the chapter of IR and Raman spectroscopy from Bernd Zeeh.

He particularly revised and complemented the parts describing the basics of these spectroscopic methods. For instance, the relationships between bond strengths and vibration frequencies or between molecule symmetries and resulting IR and Raman activities are newly presented. With regard to the instrumental part, the construction of IR and Raman spectrometers is discussed in more detail, giving special attention also to the laser technology that became increasingly important. A new section dedicated to the interpretation of spectra was added, considering more deeply vibration couplings, overtones, Fermi resonance, and combination and difference bands. Adaptions were also done to enhance the efficiency of the interpretation of IR spectra by use of already well-approved sample spectra. The important absorption bands are now directly linked to the tables of the characteristic group frequencies, which are completely revised and partly reorganized for even more effective use.

Major changes have been made in the NMR chapter, authored by Herbert Meier. Great emphasis is placed on modern one- and two-dimensional NMR techniques such as DEPT, APT, COSY, DQF-COSY, Ph-COSY, E-COSY, TOCSY, NOESY, ROESY, EXSY, HETCOR, HSQC, HSQC-TOCSY, HSQC-NOESY, HMBC, and INADEQUATE. The description of earlier methods, however, has not been forgone because their knowledge is required to be able to understand less recent publications. Since organic molecules are mainly constructed on the basis of carbon frameworks, ^{13}C NMR signals play a crucial role in the characterization of organic compounds. It is striking to observe, however, that respective signal assignments are omitted or erroneous in many publications. It thus became a special issue to deeply discuss electronic, steric, and anisotropic factors that affect the chemical shifts of ^{13}C signals in open-chain as well as cyclic compounds. Several respective tables have been added, and the graphical table with the compilation of ^{1}H and ^{13}C chemical shifts—displayed with compounds ordered according to substance classes—has been significantly extended and complemented with examples of more rare substance classes. In the NMR spectra shown, two opposing tendencies are taken into account: on the one hand there is the trend towards high field strengths; on the other hand, bench spectrometers with, e.g. 60 MHz became more important.

Stefan Bienz and Laurent Bigler (both University of Zurich) are the new authors of the MS chapter, replacing Manfred Hesse who died in 2011. Performance and user friendliness of mass spectrometers have tremendously advanced over the recent decades. New and sophisticated ion generation, ion separation, and ion detection methods, and many new accessories such as coupled separation and automatization modules, sample preparation kits, and evaluation software led to a fresh affection for the MS method and to a broad spread of MS instruments as routine and open-access equipment throughout many chemical facilities.

To account for these developments, the MS chapter was completely redesigned: the content was fully regrouped, freed of obsolete techniques and methods, and complemented with information to newest instrumental and methodical advances. Because the MS techniques became more and more different and facetted, a new chapter has been introduced, which gives guidance for selecting proper sample preparations and appropriate measuring procedures. The newest methods for the structural elucidation of small molecules up to biopolymers are introduced, based on accurate mass measurements (high-resolution MS, HR-MS) and collision-induced dissociation (CID). Although not in detail, fragmentation mechanisms and fragmentation patterns for CID processes are addressed along broad lines, which is relevant in connection with structural elucidations of unknown analytes, where information can often be gained from conclusions by analogy.

The completely new final chapter of the textbook, *Handling of Spectra and Analytical Data: Practical Examples*, is also authored by Stefan Bienz and Laurent Bigler. It shows by means of 10 real cases how analytical data are described and what kind of strategies could be followed to come up with these data to reasonable structural proposals. The compounds were measured without any exception with modern instruments, and the examples were chosen to demonstrate the information gain that can be acquired by the most commonly applied analytical methods, including two-dimensional NMR and HR-MS. As a supplement, a set of freely accessible exercises is provided online.

For the preparation of the new *English* edition, which embraces also the preparation of the two *German* editions of 2012 and 2016, we owe our gratitude to many colleagues. First of all, the marvelous groundwork of the two former authors, Manfred Hesse and Bernd Zeeh, is warmly acknowledged. Special thanks are also due to Heinz Berke and Ferdinand Wild (both University of Zurich), as well as Klaus Bergander (University of Münster) for their contributions to the IR/Raman chapter; Heinz Kolshorn, Johannes Liermann, and Ingrid Schermann (all University of Mainz) for their supports for the NMR chapter; and Urs Stalder, Armin Guggisberg, Yvonne Forster, Jrène Lehmann, and the students of the advanced chemical laboratory courses (all University of Zurich) for their inputs, analytical measurements, and samples that were used for the MS chapter, the practical examples in Chapter 5, and the electronic supplements.

Stefan Bienz
in the name of the authors

Contents

Download exercises and images now for free! In addition to the examples contained in the book, a series of exercises and all spectra and images from the book are available at http://www.thieme.de/HMZ.

1

Herbert Meier

UV/Vis Spectroscopy

1 UV/Vis Spectroscopy

1.1 Theoretical Introduction

1.1.1 Electromagnetic Waves and Electron Transitions in Molecules

Electromagnetic radiation is characterized by the **wavelength** λ or the **frequency** v. These values are connected with each other by the equation:

$$v \cdot \lambda = c$$

where c is the **velocity of light** (in vacuum $\approx 2.99 \cdot 10^{10}$ cm $\times$ s^{-1}). A quantum of light with frequency v has the *energy*

$$E = hv$$

Planck's constant h has the value $\approx 6.63 \cdot 10^{-34}$ J·s. The interaction of electromagnetic waves and molecules leads, in the case of absorption of ultraviolet (UV) and visible light (seldom near-infrared [NIR]), to the excitation of electrons, generally valence electrons. **Fig. 1.1** illustrates the relevant regions of the electromagnetic spectrum. The region of light visible to the human eye (Vis) is followed below $\lambda = 400$ nm by the UV region. Based on the different biological activities, a subdivision is made into UV-A (400-320 nm), UV-B (320-280 nm), and UV-C (280-10 nm). The IR region starts at 750 nm with the NIR. Sunlight consists on average of 44% IR, 52% Vis, and 4% UV-A. The portion of UV-B is in the $^o/_{oo}$ range. The terahertz (THz) region (100 μm $\leq \lambda \leq 1$ mm, 100 cm$^{-1} \geq \tilde{v} = 1/\lambda \geq 10$ cm^{-1}, $3 \cdot 10^{12}$ Hz $\geq v \geq 3 \times 10^{11}$ Hz) is located at the long-wavelength end of the IR. It is followed by the microwave region (1 mm $\leq \lambda < 30$ cm) and the radio wave region ($\lambda > 30$ cm). Spectroscopy of molecules is performed in all these wavelength regions. Apart from UV/Vis spectra in Chapter 1, IR spectra (Chapter 2) and in particular nuclear magnetic resonance (NMR) spectra (Chapter 3) are important for the structure determination of organic compounds. NMR is performed in magnetic fields by applying radio waves.

When visible light of a particular spectral color is absorbed, the human eye recognizes the **complementary color**:

Absorbed spectral color	Complementary color
Violet	Yellow-green
Blue	Yellow
Green-blue	Orange
Blue-green	Red
Green	Purple
Yellow-green	Violet
Yellow	Blue
Orange	Green-blue
Red	Blue-green
Purple	Green

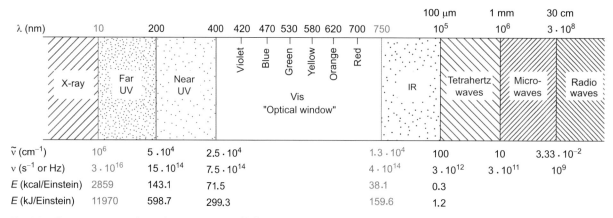

λ (nm)	10	200	400	420	470	530	580	620	700	750	10⁵ (100 μm)	10⁶ (1 mm)	3·10⁸ (30 cm)
$\tilde{v}$ (cm⁻¹)	10^6	$5 \cdot 10^4$	$2.5 \cdot 10^4$						$1.3 \cdot 10^4$		100	10	$3.33 \cdot 10^{-2}$
v (s⁻¹ or Hz)	$3 \cdot 10^{16}$	$15 \cdot 10^{14}$	$7.5 \cdot 10^{14}$						$4 \cdot 10^{14}$		$3 \cdot 10^{12}$	$3 \cdot 10^{11}$	10^9
E (kcal/Einstein)	2859	143.1	71.5						38.1		0.3		
E (kJ/Einstein)	11970	598.7	299.3						159.6		1.2		

Fig. 1.1 Electromagnetic radiation (1 Einstein = 1 mol light quanta).

In the past the wavelength was frequently given in Ångström, nowadays nanometer (nm) is usually used (1 nm = 10^{-9} m = 10^{-7} cm). Instead of giving frequencies in s^{-1}, it is customary to quote the **wavenumber** $\tilde{\nu}$ in cm^{-1}.

$$\tilde{\nu} = \frac{1}{\lambda} = \frac{\nu}{c}$$

If the energy is based on a quantum or an individual atomic or molecular process, the customary unit is the electron volt (eV). For a mole, i.e., 6.02×10^{23} quanta of light, the energy is given in kJ. Energy and wavenumber are directly proportional to each other. For conversions the following relationships are recommended:

$$1 \text{ eV} \cong 23 \text{ kcal·mol}^{-1} = 96.5 \text{ kJ·mol}^{-1} \cong 8{,}066 \text{ cm}^{-1}$$

$$1{,}000 \text{ cm}^{-1} \cong 12 \text{ kJ·mol}^{-1}$$

$$1 \text{ kJ·mol}^{-1} \cong 84 \text{ cm}^{-1}$$

If light with the appropriate frequency ν meets a molecule in the **ground state** ψ_0, it can be **absorbed** and will raise the molecule to an **electronically excited state** ψ_1. By **spontaneous emission** or by **stimulated emission**, caused by the light rays, the system can return to the ground state. The word "can" in these sentences expresses the **transition probability** of the two radiative processes, absorption and emission (**Fig. 1.2**).

The connection with the orbitals involved in the electronic transition is shown in **Fig. 1.3**. Taking Koopmans' theorem for granted, the energy of the highest occupied molecular orbital (**HOMO**) corresponds to the negative ionization potential (**IP**) and the energy of the lowest unoccupied molecular orbital (**LUMO**) to the negative electron affinity (**EA**). The orbital energies used here are related to **one-electron configurations**:

$$E(\text{LUMO}) - E(\text{HOMO}) = \text{IP} - \text{EA}.$$

According to **Fig. 1.3**, the following equation applies for the energy difference of the states:

$$E(S_1) - E(S_0) = h\nu = \text{IP} - \text{EA} - J + 2K.$$

The difference in energy between LUMO and HOMO is considerably greater than the excitation energy A for the transition

from the singlet ground state S_0 to the first excited singlet state S_1. The difference arises from the different electronic interactions (Coulomb term J, exchange term $2K$). The singlet–triplet splitting in this approximation is $2K$. Since $K > 0$ the lowest triplet state T_1 is always below S_1. Molecules, which have the same HOMO–LUMO gap, can have quite different excitation energies. The colorless anthracene represents a classical example, it has the same HOMO–LUMO energy difference as the blue azulene. As a further result of the configurational interaction, the HOMO–LUMO transition is not necessarily the lowest transition $S_0 \rightarrow S_1$ (cf. Fig. 1.21, p. 16). Modern **density functional theory** (**DFT**) calculations avoid the problem of additional terms for the configuration interaction and quote HOMO and LUMO levels, which correspond to those obtained, for example, from redox potentials determined by cyclic voltammetry. The measured spectrum can be directly compared with the energy of the electron transitions, when **time-dependent DFT** (**TD-DFT**) calculations are performed. Some explicit examples will be given in sections 1.3.3 and 1.3.4.

A measure for the transition probability is the **oscillator strength** f_{01}, a dimensionless quantity which classically represents the fraction of negative charge (electrons) that brings about the transition (by oscillation). The quantum mechanical equivalent of f is the vector of the **transition moment** M_{01}, which represents the change of the dipole moment during the transition. The **dipole strength** $D_{01} = |M_{01}|^2$ is directly proportional to f_{01}. If $D_{01} = |M_{01}| = f_{01} = 0$ then, even if the **resonance condition** $\Delta E = h\nu$ is fulfilled, no transition is possible. When the f value is small the term **forbidden transition** is used, if the f value is close to 1 the term **allowed transition** is used.

For diatomic or linear polyatomic molecules, as with atoms, **selection rules** for the allowed transitions between two different electronic states can be established based on the rule of the conservation of angular momentum. For other molecules, which constitute the overwhelming majority, these rules result in **transition exclusions**.

The spin exclusion rule states that the **total spin** S and the **multiplicity** $M = 2S + 1$ may not change during a transition, i.e.,

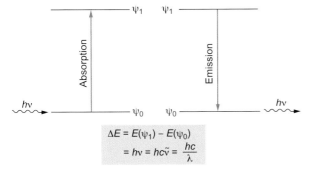

$$\Delta E = E(\psi_1) - E(\psi_0)$$
$$= h\nu = hc\tilde{\nu} = \frac{hc}{\lambda}$$

Fig. 1.2 Electronic transitions and radiative processes.

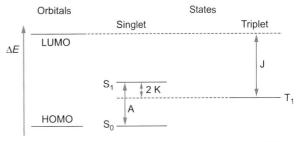

Fig. 1.3 Energy scheme for the electronic transition between HOMO and LUMO.

singlet states, for example, may undergo transitions to singlet states by absorption or emission, but not to triplet states. M_{01} can also be zero because of the symmetry of the orbitals (which are described by the wave functions φ_0 and φ_1 and represent the electronic part of the total wave functions ψ_0 and ψ_1). This is described as a **symmetry exclusion**. An easily comprehensible special case occurs for molecules containing a center of symmetry, whose wave functions are either symmetric (g, gerade = even) or antisymmetric (u, ungerade = odd). The symmetry exclusion states in such cases that electronic transitions between orbitals of the same **parity** are forbidden (parity exclusion, Laporte's rule):

allowed: g → u   forbidden: g ↛ g

u → g   u ↛ u

Movement of nuclei can reduce the symmetry, so that symmetry-forbidden transitions can, in fact, be seen. (An example of a vibrationally allowed transition is the long-wavelength absorption band of benzene; cf. p. 18.)

An important possible cause for the disappearance of the electronic transition moment is the so-called **overlap exclusion**, a special case of the symmetry exclusion. This takes effect when the two orbitals which are taking part in the electronic transition overlap poorly or not at all. That is quite clearly the case in an intermolecular **charge-transfer (CT) transition** where the electronic transition takes place from the donor to the acceptor molecule. There are also numerous intramolecular examples of the overlap exclusion (cf. **intramolecular charge transfer [ICT]** or the $n \to \pi^*$ transition of carbonyl compounds, p. 30 and p. 25, respectively).

If the possibilities of transitions between two orbitals of a molecule are worked out, it becomes apparent that exclusions become the rule and allowed transitions are the exceptions. However, forbidden transitions frequently occur, albeit with low transition probability, i.e., a low f value ($10^{-1} \geq f \geq 10^{-6}$). The spin exclusion rule is the most effective. Even spin-forbidden transitions can, however, be observed in cases of effective spin–orbit coupling (e.g., by heavy atoms) or in the presence of paramagnetic species.

If the molecule under investigation is considered in a Cartesian coordinate system whose axes are established, for example, with reference to the molecular axes, the vector M_{01} can be separated into its spatial components M_x, M_y, and M_z. For $M_{01} \neq 0$ at least one of the three components must be nonzero. When $M_x = M_y = 0$, and $M_z \neq 0$, the absorbed or emitted radiation is **polarized** in the z-direction. This optical anisotropy of the molecule cannot normally be observed, since the molecules are present in a random orientation. Polarization measurements are performed on single crystals or stretched plastic films.

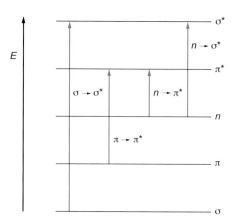

Fig. 1.4 Molecular orbitals and electron transitions.

A classification of the electronic transitions (bands) can be made from a knowledge of the molecular **orbitals** (MOs) involved. From occupied **bonding** σ- or π-orbitals or from **nonbonding** n-orbitals (lone pairs of electrons), an electron can be raised to an empty **antibonding** π^*- or σ^*- orbital. Correspondingly, the electron transitions (bands) are indicated as $\sigma \to \sigma^*$, $\pi \to \pi^*$, $n \to \pi^*$, $n \to \sigma^*$, etc. (**Fig. 1.4**).

Apart from this nomenclature based on a simplified MO description, there are several other conventions for the specification of electronic states and the possible transitions between them. Especially the last nomenclature (group theory) in **Table 1.1** is to be recommended.

Ethene (**1**) and formaldehyde (**2**) shall be used here as simple molecules for the illustration of allowed and forbidden electron transitions.

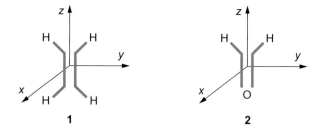

If the five σ bonds of ethene are considered to be located in the yz plane of a three-dimensional coordinate system, the p lobes of the π and π^* orbitals are aligned in the x-direction (**Fig. 1.5**). Ethene molecules have the following elements of symmetry in this arrangement:

- The x-, y-, and z-axes are twofold symmetry axes: $C_2(x)$, $C_2(y)$, $C_2(z)$.
- The origin of coordinates is an inversion center i.
- The xy-, xz-, and yz planes of the coordinate system are symmetry planes: $\sigma(xy)$, $\sigma(xz)$, $\sigma(yz)$.

Table 1.1 Nomenclature for electron transitions (absorptions)

System	Term symbol	State	Examples of electron transitions
enumerative	S_0	singlet ground state	$S_0 \rightarrow S_1$
	$S_1, S_2, S_3,...$	higher singlet states	$S_0 \rightarrow S_2$
	$T_1, T_2, T_3,...$	triplet states	$S_0 \rightarrow S_3$
according to Mulliken	N	ground state	$V \leftarrow N$
	Q, V, R	excited states	$Q \leftarrow N$
according to Platt	A	ground state	$B \leftarrow A$
	B, C, L	excited states	$C \leftarrow A$
			$L \leftarrow A$
according to Kasha	σ, π, n	orbital of origin	$\sigma \rightarrow \sigma^*$
			$\pi \rightarrow \pi^*$
	σ^*, π^*	orbitals of the excited electrons	$n \rightarrow \pi^*$
			$n \rightarrow \sigma^*$
group theory	symbols of the symmetry classes*		$^1A_2 \leftarrow {}^1A_1$
	A: sym. ⎤ related to rotation		$^1B_{1u} \leftarrow {}^1A_{1g}$
	B: antisym. ⎦ about the		$^1B_{2u} \leftarrow {}^1A_{1g}$
	rotional axis (axes)		$^1E_{1u} \leftarrow {}^1A_{1g}$
	C_n of maximum order		$^1A_1'' \leftarrow {}^1A_1'$
	E: doubly degenerate state		
	T: triply degenerate state		
	Indices		
	g: sym. ⎤ with respect to inversion		
	u: antisym. ⎦		
	1: sym. ⎤ with respect to C_2 axes which		
	2: antisym. ⎦ are perpendicular to C_n		
	': sym. ⎤ with respect to plane of		
	": antisym. ⎦ symmetry σ_h (perpendicular to C_n)		

* See a textbook on symmetry in chemistry.

Ethene (**1**) belongs to the point group D_{2h}, in short ethene (**1**) has D_{2h} symmetry. **Table 1.2** shows in the first column the irreducible representations of this point group and in the first row the above-mentioned symmetry elements (symmetry operations). The identity I is added because of group theoretical reasons. Each orbital of **1** must belong to a certain symmetry class (irreducible representation). Only **group orbitals** can be used and not, for example, single C–H **bond orbitals**. The **characters** +1 and –1 in **Table 1.2** express the symmetric or antisymmetric behavior toward the respective symmetry operation.

Fig. 1.5 and **Table 1.2** reveal that the π orbital of **1** belongs to the symmetry class b_{3u} and the π^* orbital to b_{2g}. Multiplication of

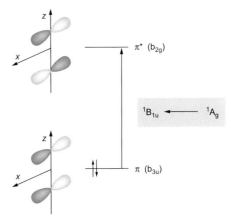

Fig. 1.5 $\pi \rightarrow \pi^*$ electron transition in ethene (**1**).

the characters for each electron reveals that the singlet ground state S_0 of **1** is an ^1A_g state ($b_{3u} \times b_{3u} = a_g$) and the lowest electronically excited singlet state S_1 is a $^1B_{1u}$ state ($b_{3u} \times b_{2g} = b_{1u}$).

The column on the right end of the character table (**Table 1.2**) contains the symmetry behavior of the x-, y-, and z-component of the **transition vector M**. The z-component corresponds to b_{1u} (B_{1u}). That means the $\pi \rightarrow \pi^*$ transition ($^1B_{1u} \leftarrow {}^1A_g$) is allowed and polarized in z-direction. It is observed in the far-UV at ~165 nm.

Formaldehyde (**2**), the second example, belongs to the point group C_{2v} with symmetry elements which can be seen in the drawing of the formula.

- Twofold symmetry-axis C_2 (z).
- Symmetry plane σ_v (xz).
- Symmetry plane σ_v (yz) (plane of the σ-bonds).

Table 1.3 shows the symmetry classes of the point group C_{2v}.

Table 1.2 Symmetry classes (irreducible representations) and characters of point group D_{2h}

			Symmetry elements						
D_{2h}	I	C_2 (z)	C_2 (y)	C_2 (x)	i	σ (xy)	σ (xz)	σ (yz)	
a_g	1	1	1	1	1	1	1	1	
b_{1g}	1	1	– 1	– 1	1	1	– 1	– 1	
b_{2g}	1	– 1	1	– 1	1	– 1	1	– 1	
b_{3g}	1	– 1	– 1	1	1	– 1	– 1	1	
a_u	1	1	1	1	– 1	– 1	– 1	– 1	
b_{1u}	1	1	– 1	– 1	– 1	– 1	1	1	z
b_{2u}	1	– 1	1	– 1	– 1	1	– 1	1	y
b_{3u}	1	– 1	– 1	1	– 1	1	1	– 1	x

UV/Vis-Spektren

Table 1.3 Symmetry classes (irreducible representations) and characters of the point group C_{2v}

C_{2v}	I	$C_2(z)$	$\sigma_v(xz)$	$\sigma_v'(yz)$	
a_1	$+1$	$+1$	$+1$	$+1$	z
a_2	$+1$	$+1$	-1	-1	
b_1	$+1$	-1	$+1$	-1	x
b_2	$+1$	-1	-1	$+1$	y

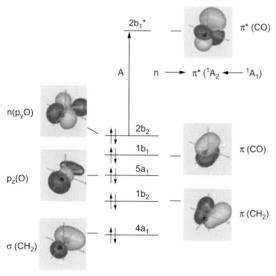

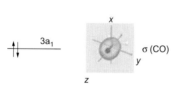

Fig. 1.6 Orbitals in the valence shell of formaldehyde (**2**) (energy increases from bottom to top). The labels of the group orbitals σ (CO), σ (CH$_2$), etc. correspond to the most important contributions. The $n \to \pi^*$ transition is the energy-lowest electron transition: $^1A_2 \leftarrow {}^1A_1$ (Table 1.3 reveals b$_2$ × b$_2$ = a$_1$, b$_2$ × b$_1$ = a$_2$). (S. Immel, Internet).

The group orbitals in the valence shell of formaldehyde (**2**) are depicted in **Fig. 1.6**. The symmetry classes of the orbitals can be seen from orbital pictures and formula **2** in the coordinate system. As already pointed out in the discussion of ethene (**1**),

it does not make sense to consider single C–H orbitals, since they do not belong to a symmetry class of the point group C_{2v}.

Fig. 1.6 does not contain the 1s orbitals of C and O, which have the symbols 1a$_1$ and 2a$_1$. The depicted orbitals 3a$_1$, 4a$_1$, 1b$_2$, 5a$_1$, 1b$_1$, and 2b$_2$ contain 6 × 2 electrons. The antibonding orbitals 2b$_1^*$, 6a$_1^*$, 3b$_2^*$, and 7a$_1^*$ are lying above the occupied orbitals. The $n \to \pi^*$ transition (2b$_1 \leftarrow$2b$_2$) is the energy-lowest transition, in which the nonbonding p$_y$ orbital of the oxygen atom delivers an electron to the π^* orbital of the CO double bond. The singlet ground state S$_0$ (an $^1A_1$ state) is thereby transformed to the S$_1$ state $^1A_2$ (b$_2$ × b$_1$ = a$_2$, **Table 1.3**). The column on the right side of **Table 1.3** reveals that none of the components x, y, and z of the transition vector has the characters of a$_2$. Accordingly, the $n \to \pi^*$ transition is forbidden. This statement corresponds to the overlap exclusion: $n(p_yO)$ and π^* are orthogonal to each other. A transition with a low intensity can be observed at ~300 nm.

In the $\pi \to \pi^*$ transition of **2**, an electron is raised from 1b$_1$ to 2b$_1$. This corresponds to the transition from the singlet ground state S$_0$ ($^1A_1$) to an electronically excited singlet state S$_n$ ($^1A_1$) (b$_1$ × b$_1$ = a$_1$). This transition is allowed and polarized in the z-direction (**Table 1.3**).

In general, $n \to \pi^*$ and $\pi \to \pi^*$ transitions are discussed for carbonyl compounds. *Ab initio* calculations reveal that the $\pi \to \pi^*$ transition of formaldehyde (**2**) has a high energy and should lie in the far-UV. Thus, a **Rydberg transition** seems to be the second transition, in which an $n(2b_2)$ electron is transferred to the 3s orbital of the next higher shell.

Considerations, which other electron transitions between the orbitals shown in **Fig. 1.6** are allowed or forbidden, are certainly a good exercise. However, it should be noted here that mixed transitions occur in many organic molecules. The discussion of allowed and forbidden electron transitions in other point groups exceeds the scope of this book. Textbooks on UV/Vis spectroscopy are compiled in the references on p. 40.

The statements in Section 1.1.1 apply to single-photon transitions. With the use of lasers **two-photon spectroscopy** has been developed. High photon densities allow the simultaneous absorption of two photons. This leads to altered selection rules; thus, for example, transitions between states of the same **parity** are allowed (g → g, u → u) and transitions between states of opposite parity are forbidden. The **degree of polarization** can also be determined in solution. Two-photon spectroscopy thus provides useful extra information in studies of electronically excited molecules.

At the end of this section, the photophysical processes of electron transitions are summarized in a modified Jablonski term scheme. From the ground state, which in general is a singlet state S$_0$, absorption leads to higher singlet states S$_1$, S$_2$, etc.

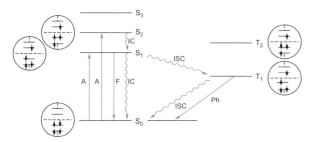

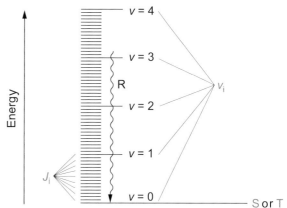

Fig. 1.7 Jablonski term diagram with a visual representation of the electronic configurations.
Radiative processes: →
A Absorption
F Fluorescence
Ph Phosphorescence
Nonradiative processes: ⤳
IC Internal conversion
ISC Intersystem crossing

Fig. 1.8 Schematic representation of the superimposition of electronic, vibrational, and rotational states; v_i vibrational quantum numbers, J_i rotational quantum numbers.

The return to S_0 from S_1, and more rarely from higher states S_n, can occur by the emission of radiation, known as **fluorescence**, or by nonradiative deactivation (**internal conversion**). Nonradiative spin-inversion processes (**intersystem crossing**) lead to triplet states T, which can return to S_0, either by emission of radiation known as **phosphorescence**—disregarding the spin exclusion—or by renewed intersystem crossing (**Fig. 1.7**).

"True" two-photon absorptions must be differentiated from processes in which two photons are absorbed one after the other. At high light intensities, populations of excited states can be attained which allow further excitation; for example, the process $S_0 \rightarrow S_1 \rightsquigarrow T_1$ can be followed by a triplet–triplet absorption $T_1 \rightarrow T_2$.

In contrast to atoms, the various electronic states of molecules have rather broad energies because of the added effect of vibrational and rotational levels. Each term in **Fig. 1.7** is therefore split into many energy terms, as shown schematically in **Fig. 1.8**. A specific energy level E_{tot} corresponds to a particular electronic, vibrational, and rotational state of the molecule.

To a first approximation the three energy components can be separated

$$E_{tot} = E_{electr.} + E_{vibr.} + E_{rot}$$

Born–Oppenheimer approximation

For an electronic transition it follows that

$$\Delta E_{tot.} = \Delta E_{electr.} + \Delta E_{vibr.} + \Delta E_{rot.}$$

The electronic part is always much larger than the vibrational part which, in turn, is much larger than the rotational part. The **relaxation** R (see **Fig. 1.8**) is an additional nonradiative deactivation within each electronic state. In addition to the monomolecular processes described here it should also be noted that **bimolecular photophysical processes** (energy transfer: **sensitization**, **quenching**) and **primary photochemical processes** can occur.

1.1.2 Light Absorption and the Spectrum

If a beam of light of intensity I_0 falls on a layer of homogeneous, isotropic material of thickness d, then apart from losses through reflection or diffraction it can be weakened by absorption. The intensity I of the emerging beam (transmission) is then given by

$$I = I_0 - I_{abs.}$$

The differential equation for the reduction of the intensity dI by an increment dx of the width of the absorbing layer is

$$dI = -\alpha \cdot I \, dx$$

and evaluation of the integral

$$\int_{I_0}^{I} \frac{dI}{I} = -\int_0^d \alpha \, dx$$

yields the function

$$I = I_0 \times e^{-\alpha d}$$

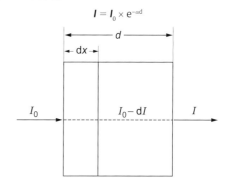

Here, α is a characteristic absorption coefficient for the medium. If consideration is restricted to dilute solutions, where only the solute, of concentration c, absorbs, then α can be replaced by $2.303 \cdot \varepsilon \cdot c$ to give

$$\ln \frac{I_0}{I} = 2.303 \cdot \varepsilon \cdot c \cdot d \text{ or } A = \log \frac{I_0}{I} = \varepsilon \cdot c \cdot d$$

Lambert–Beer law

The **absorbance** (extinction) A has no dimensions. The thickness d of the layer is measured in cm, the concentration c in $mol \cdot L^{-1}$. The molar **absorption coefficient** ε has the dimension $L \cdot mol^{-1} \cdot cm^{-1} = 1,000 \text{ cm}^2 \cdot mol^{-1} = cm^2 \cdot mmol^{-1}$; $L \cdot mol^{-1}$ is usually replaced by M^{-1}. This law, which was developed by Bouguer (1728), Lambert (1760), and Beer (1852), is valid for monochromatic light and dilute solutions ($c \leq 10^{-2} \text{ mol} \cdot L^{-1}$). The absorbance is, with a few exceptions, an additive property. For n absorbing species therefore it holds that

$$A_{\text{total}} = \log \frac{I_0}{I} = d \sum_{i=1}^{n} \varepsilon_i c_i$$

Special care is necessary when entering values for the concentrations of compounds which undergo a chemical change when dissolved, e.g., dissociation, dimerization.

If the absorbance is determined according to the **Lambert–Beer law** for all measured λ or $\tilde{\nu}$, and from that the substance-specific value ε, the absorption plot $\varepsilon(\tilde{\nu})$ or $\varepsilon(\lambda)$ can be obtained and thus the UV or UV/Vis spectrum. As a consequence of the width (in energy terms) of the electronic states it is a **band spectrum**. The individual bands are characterized by their properties of **position, intensity, shape**, and **fine structure**.

From **Fig. 1.9** it follows that the positions of the absorption bands depend on the nature of the electron transitions. For isolated chromophores, Table 1.4 (see p. 13) gives a guide. The position of the absorptions is, however, strongly influenced by

steric, inductive, and resonance effects—the latter being particularly strongly affected by inclusion of the chromophore in large conjugated systems (**Fig. 1.9**).

For certain chromophores the solvent also has a characteristic influence (see for example **Table 1.5** and **Fig. 1.34**).

A shift to longer wavelengths (red-shift) of a transition is called a **bathochromic effect**, and a shift to shorter wavelengths (blue shift) a **hypsochromic effect**.

The term **hyperchromic effect** is used to describe an increase in intensity. **Hypochromic** means the opposite, a decrease in intensity.

As described above, the transition moment $|M|$ or the oscillator strength f is a measure of the intensity of a transition. An alternative measurement for the intensity is the area S

$$S = \int_{(-\infty)}^{(+\infty)} \varepsilon d\tilde{\nu}$$

The relationship between f and S for a refractive index of $n \approx 1$ is given by

$$f = \frac{m \cdot c^2}{N_A \pi e^2} 10^3 (\ln 10) S$$

$$f \approx 4.32 \cdot 10^{-9} S$$

m mass of an electron
e charge of an electron
N_A Avogadro's constant
c velocity of light

S can often be determined by graphical integration or estimated very roughly from approximations such as

$$S = \varepsilon_{\text{max}} \cdot b$$

where b is the width of the band at half height (**Fig. 1.10**).

The higher the transition probability, the shorter is the radiation lifetime τ_0 of an excited state; τ_0 can be calculated from f and thus from S:

$$\tau_0 = \frac{c^3 m}{8\pi^2 \nu^2 e^2} \cdot \frac{1}{f}$$

As an approximation, τ_0 is given in seconds by

$$\tau_0 \approx \frac{1}{10^4 \cdot \varepsilon_{\text{max}}}$$

Usually the intensity of a band is judged simply by ε_{max}. The following assignments have become customary:

$\varepsilon \leq 10 \text{ M}^{-1} \text{ cm}^{-1}$ transition: forbidden
$10 < \varepsilon < 1,000$ weakly allowed
$1,000 < \varepsilon < 100,000$ allowed
$\varepsilon > 100,000$ strongly allowed.

Further important properties of absorption bands are shape and fine structure. Even if the different kinetic energies of individual

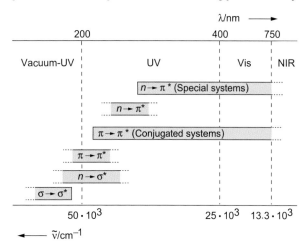

Fig. 1.9   Absorption regions of various electron transitions.

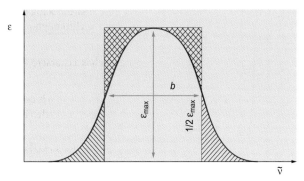

Fig. 1.10 True and approximated areas of an absorption band.

molecules are ignored, an electronic state does not have a uniform energy. Instead the overlying molecular vibrations and rotations must be taken into account, as described above. From Boltzmann statistics it will be appreciated that in the ground state S_0 the lowest vibrational level ($v = 0$) is almost exclusively populated.

For two states with an energy difference ΔE, the ratio of populations is given by

$$\frac{N_i}{N_j} = e^{-\Delta E/kT}$$

Since the Boltzmann constant $k = 1.38 \cdot 10^{-23}$ J K^{-1}, it can be seen that in the wavenumber scale $kT \approx 200$ cm^{-1} at room temperature. For a typical vibration in the IR region with $\tilde{v} = 1{,}000$ cm^{-1}, it holds that

$$\frac{N_i}{N_j} = e^{-1{,}000/200} = e^{-5} = 0.0067$$

The energetically higher vibrational level therefore has a population of less than 1%. Higher rotational levels on the other hand are appreciably populated. Thus, for rotations about single bonds ($\tilde{v} = 50$ cm^{-1}) the Boltzmann distribution at room temperature gives

$$\frac{N_i}{N_j} = e^{-50/200} = 0.78 = 44{:}56$$

The transition to S_1 leads to vibrational states with $v' = 0, 1, 2, 3 \ldots$ Because of very rapid relaxation to $v' = 0$ the fluorescence starts entirely from $v' = 0$ and leads to S_0 with $v = 0, 1, 2 \ldots$

Fig. 1.11 shows the situation schematically.

Spectra measured in solution do not show rotation lines—the electronic bands are composed of **vibrational bands**. The degree of structure observed in the absorptions depends on the substance. The vibrational fine structure is most likely to be seen in rigid molecules. In polyatomic molecules the vibrational levels lie very close together. Restricted rotation in solution and line broadening due to local inhomogeneities in the solvation result in unstructured bands. The measurement conditions can also

play an important role. **Fig. 1.12** shows for 1,2,4,5-tetrazine the reduction in structure with increasing interaction with the solvent and under the influence of temperature.

In line with the **Franck-Condon principle** the absorption probability is largest for a vertical transition from the energy

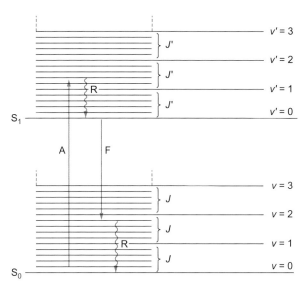

Fig. 1.11 Absorption and fluorescence as transitions between electronic, vibrational, and rotational levels.

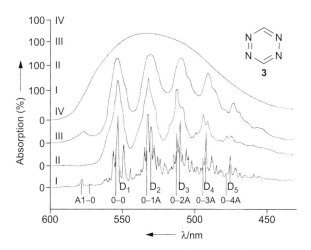

Fig. 1.12 Vibrational structure of the $n \rightarrow \pi^*$ absorption of 1,2,4,5-tetrazine (**3**) (from Mason SF. 1959, J. Chem. Soc., 1263).

I   Vapor spectrum at room temperature (with vibrational mode).
II Spectrum at 77 K in an isopentane/methylcyclohexane matrix.
III Spectrum in cyclohexane at room temperature.
IV The spectrum in water at room temperature.

The λ scale is referenced to I; II is shifted by 150 cm^{-1}; III by 250 cm^{-1} to higher wavenumbers; IV by 750 cm^{-1} to lower wavenumbers.

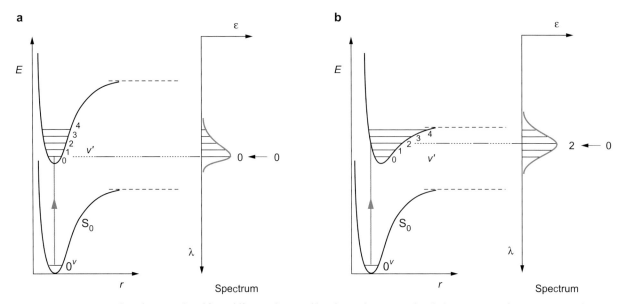

Fig. 1.13 Composition of an absorption band from different vibrational bands in a diatomic molecule (r, interatomic distance; E, energy).

a unsymmetric band with intense $0 \leftarrow 0$ transition.
b symmetric band with intense $2 \leftarrow 0$ transition.

hypersurface of the ground state into that of the electronically excited state, i.e., all molecular parameters (bond lengths and angles, configuration, conformation, solvation cage, etc.) remain unchanged during the transition.

From the expression for the vibrational component of the transition moment M, it can be seen that the transitions from the lowest vibrational level of the ground state ($v = 0$) to the various vibrational levels of the excited state ($v' = 0, 1, 2, \ldots$) do not have equal probability. Two extreme cases can be proposed for the superposition of the vibrational bands; these can be recognized from the shape of the absorption. In **Fig. 1.13** this is demonstrated for a diatomic molecule, using the simplification of the multidimensional energy hypersurface to the so-called Morse function $E_{pot} = f(r)$. The shape of the band is determined by whether the Morse function of the excited state is only vertically shifted (**Fig. 1.13a**) or is additionally shifted to other r values (**Fig. 1.13b**).

Analogue considerations are valid for the **emission. Fig. 1.14** shows the **absorption** A and the **fluorescence** spectrum F of anthracene (**4**). Fast **relaxation** R in the electronically excited state provokes that the fluorescence starts at the lowest vibrational level ($v' = 0$) of S_1. Absorption and fluorescence should ideally be mirror images which overlap in the $0 \rightleftharpoons 0$ transitions. However, the relaxed S_1 state can differ in its geometry and its solvent cage from the ground state S_0. Therefore, the $0 \rightleftharpoons 0$ transitions of absorption and fluorescence differ very often. The

difference of the corresponding wavenumbers is called **Stokes shift**. When the $0 \rightleftharpoons 0$ transitions are not visible the difference $\tilde{\nu}_{max}(A) - \tilde{\nu}_{max}(F)$ is often taken. The Stokes shift is large when the relaxation R causes a strong structural change. When the dipole moment of S_1 is much higher than that of S_0, polar solvents will provoke a large Stokes shift.

1.2 Sample Preparation and Measurement of Spectra

For analytical purposes, UV/Vis spectra are normally measured in solution. Optically pure solvents, available commercially, are used and allow transitions measured at concentrations of ~10^{-4} mol·L^{-1}. For the weak bands of forbidden transitions, the concentration must be increased appropriately. (As a guide the absorbance A should be ≈1. For a layer thickness of 1 cm— length of the wave path through a quartz cell—it follows from the Lambert–Beer law that $c \cdot \varepsilon \approx 1$. If $\varepsilon_{max} = 10^n$ M^{-1} cm^{-1} the measurement should therefore be made at a concentration of 10^{-n} mol·L^{-1}.)

Solvents with their own absorptions in the measurement region are unsuitable. The best transparency down to the vacuum-UV region is shown by perfluorinated alkanes such as **perfluorooctane**. Sufficiently transparent down to 195 nm (for $d = 1$ cm) are the saturated hydrocarbons **pentane**, **hexane**, **heptane**, or **cyclohexane**, and the polar solvents **water** and **acetonitrile**.

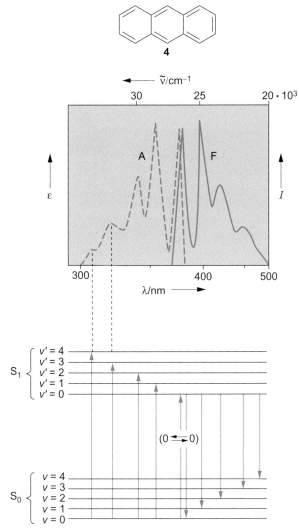

Fig. 1.14 Absorption band *A* and fluorescence band *F* of anthracene (**4**) as mirror images. The fluorescence intensity was adapted to $\varepsilon_{max} = 8{,}320\ M^{-1}\ cm^{-1}$ at $\lambda_{max} = 356$ nm. Assignment of vibrational quantum numbers: v (S_0) and v' (S_1).

Methanol, **ethanol**, and **diethyl ether** are useable down to ca. 210 nm. In order of increasing the lower measurement limit, then follow **dichloromethane** (220 nm), **chloroform** (240 nm), and **carbon tetrachloride** (250 nm). **Benzene**, **toluene**, and **tetrahydrofuran** are generally only applicable above 280 nm. An increase in the interaction between the compound being measured and the solvent leads to the loss of fine structure. It is therefore recommended to use nonpolar solvents wherever possible. The effect of solvent polarity on the position of absorption bands is discussed in **Table 1.5** and **Fig. 1.34** using the case of ketones as an example. In the customary double-beam

spectrometers, the cell with the solution to be measured is placed in one beam and a cell with the pure solvent in the other beam. The intensities are then compared over the whole spectral region. **Fig. 1.15** shows schematically the construction of a double-beam spectrometer.

Most instruments show the absorption *A* as a function of the wavelength λ. In contrast to *A*, the extinction coefficient ε has a specific value for a certain substance. It is therefore better to record a plot of ε against λ or even better against the wave number. Also $\tilde{\nu}$, unlike λ, is proportional to the energy. In the long-wavelength region, spectra which have a linear λ scale are expanded, in the short-wavelength region compressed. If strong and weak bands occur in the same spectrum, it is better to have log ε on the ordinate. **Fig. 1.16** shows a comparison of the four frequently used ways in which UV/Vis spectra are commonly displayed.

A special form of measurement is the recording of the fluorescence as a function of the wavelength of the excitation. The **excitation spectra** thus obtained are not always identical with the absorption spectra. Even with a very pure compound the participation of different rotational isomers can result in different spectra. Two-photon spectroscopy is often performed by this technique.

Some modern spectrometers permit the simultaneous measurement of several probes. Another modern technique concerns the determination of an ATEEM matrix by the simultaneous measurement of absorption, transmission, excitation, and emission.

1.3 Chromophores

1.3.1 Individual Chromophoric Groups and Their Interactions

As shown in Section 1.2, the position of an absorption band depends on the nature of the electronic transition involved. **Table 1.4** gives a list of the excitation energies of σ, π, and *n*-electrons in various isolated (**chromophoric**) **groups**. Although group orbitals have principally to be considered, bond orbitals are often used for the first characterization of an absorption.

Table 1.5 (p. 14) shows the influence of the solvent on the λ_{max} values of the $\pi \rightarrow \pi^*$ transition of acetophenone.

If a molecule has several π or *n* orbitals, which do not interact with each other, the spectrum will usually be the sum of absorptions assigned to the individual isolated chromophores. Steric effects, ring strain, etc. can lead to exceptions. Nonconjugated chromophores can also interact if they are near to each other, causing a shift or splitting of the bands (Davydov splitting). Where there are two identical chromophores there will often be two bands instead of the expected

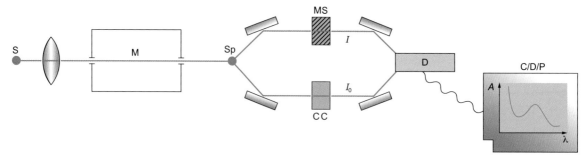

Fig. 1.15 Schematic diagram of a double-beam spectrometer.

S Radiation source (UV: hydrogen or deuterium lamp, Vis: tungsten–halogen lamp).
M (Double) monochromator using prisms and/or grating for spectral dispersion.
Sp Beam splitter (rotating mirror).
MS Measurement cell with solution.
CC Control cell with pure solvent.
D Detector (photoelectron multiplier, array of diodes).
C/D/P Computer/display/printer that records the transmission or absorption.

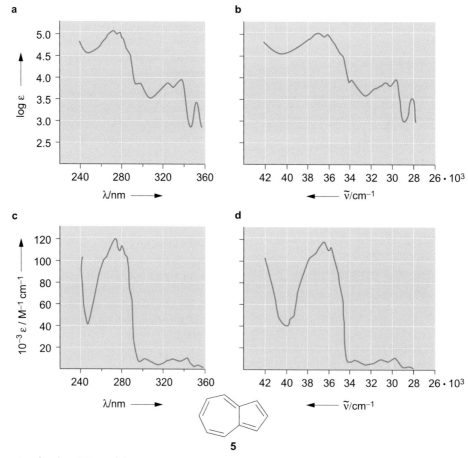

Fig. 1.16 UV spectra of azulene (**5**) in cyclohexane.
a $\log \varepsilon = f(\lambda)$
b $\log \varepsilon = f(\tilde{v})$
c $\varepsilon = f(\lambda)$
d $\varepsilon = f(\tilde{v})$
The blue color of azulene (**5**) arises from an absorption in the visible region of the spectrum, not shown in the above spectra.

UV/Vis-Spektren

Table 1.4 Absorptions of compounds bearing isolated (chromophoric) groups (lowest energy transitions)

Group	Transition	Example	$\lambda_{max}{}^a$ (nm)	$\varepsilon_{max}{}^a$ ($M^{-1} \cdot cm^{-1}$)
C—H	$\sigma \to \sigma^*$	CH_4	122	strong
C—C	$\sigma \to \sigma^*$	$H_3C—CH_3$	135	strong
—$\overline{O}$—	$n \to \sigma^*$	H_2O	167	1,500
	$n \to \sigma^*$	$H_3C—OH$	183	200
	$n \to \sigma^*$	$C_2H_5—O—C_2H_5$	189	2,000
—$\overline{S}$—	$n \to \sigma^*$	$H_3C—SH$	235	180
	$n \to \sigma^*$	$H_3C—S—CH_3$	228	620
	$n \to \sigma^*$	$C_2H_5—S—S—C_2H_5$	250	380
—$\overline{N}$—	$n \to \sigma^*$	NH_3	194	5,700
	$n \to \sigma^*$	$C_2H_5—NH_2$	210	800
	$n \to \sigma^*$	$C_2H_5—NH—C_2H_5$	193	3,000
	$n \to \sigma^*$	$(C_2H_5)_3N$	213	6,000
—Hal	$n \to \sigma^*$	$H_3C—Cl$	173	200
	$n \to \sigma^*$	$H_3C—Br$	204	260
	$n \to \sigma^*$	$H_3C—I$	258	380
	$n \to \sigma^*$	CHI_3	349	2,170
$\diagdown C=C \diagup$	$\pi \to \pi^*$	$H_2C=CH_2$	165	16,000
	$\pi \to \pi^*$	$C_2H_5—CH=CH—C_2H_5$		
		(E)	185	7,940
		(Z)	179	7,800
—C≡C—	$\pi \to \pi^*$	$HC≡CH$	173	6,000
	$\pi \to \pi^*$	$H—C≡C—C_2H_5$	172	2,500
$\diagdown C=\overline{O}$	$n \to \pi^*$	$H_3C—CH=O$	293	12
	$(\pi \to \pi^*)$	$H_3C—\overset{O}{\overset{\|}{C}}—CH_3$	187	950
	$n \to \pi^*$	$H_3C—\overset{O}{\overset{\|}{C}}—CH_3$	273	14
	$n \to \pi^*$	$H_3C—COOH$	204	41
$\diagdown C=\overline{S}$	$n \to \pi^*$	$H_3C—\overset{S}{\overset{\|}{C}}—CH_3$	460	weak
$\diagdown C=\overline{N}$—	$\pi \to \pi^*$	$H_3C—CH=N—OH$	190	8,000
	$n \to \pi^*$	$H_3C—CH=N—OH$	279	15
—$\overline{N}=\overline{N}$—	$n \to \pi^*$	$H_3C—N=N—CH_3$		
		(E)	353	240
		(Z)	368	weak
—$\overline{N}=\overline{O}$	$n \to \pi^*$	$(H_3C)_3C—NO$	300	100
		$(H_3C)_3C—NO$	665	20
—NO_2	$\pi \to \pi^*$	$H_3C—NO_2$	210	10,000
	$n \to \pi^*$		278	10
$\diagdown S=O$	$\pi \to \pi^*$	$H_3C—\overset{O}{\overset{\|}{S}}—CH_3$	216	2,950

^{a}The λ_{max} and the ε_{max} values depend to some extent on the solvent used. Measurements in the fat UV are generally made in the gas phase.

one, one of higher energy and one of lower energy than the energy of the isolated chromophore. This is shown by the examples of 1,4-pentadiene (**6**) and norbornadiene (**7**). The homoconjugation in **6** is hardly noticeable. The absorption band starts at 200 nm and extends (as for a monoolefin) into the vacuum-UV with a maximum at $\lambda = 178$ nm. The absorption of norbornadiene (**7**), on the other hand, starts at 270 nm, has a shoulder at 230 nm, and has a structured absorption between 226 and 199 nm with a maximum at 205 nm. The two nonconjugated double bonds in **7** therefore show a strong interaction, in contrast to **6**.

6 $\lambda_{max} = 178$ nm
$\varepsilon_{max} = 17,000$ $M^{-1} cm^{-1}$

7 $\lambda_{max} = 205$ nm
$\varepsilon_{max} = 2,100$ $M^{-1} cm^{-1}$

Conjugated chromophores are of special importance for UV/Vis spectroscopy. Classical examples are the polymethine dyes. As the conjugated system becomes larger, so the lowest energy $\pi \to \pi^*$ transition moves to longer wavelengths and becomes more intense; however, a convergence limit is reached for series of oligomers. A bathochromic effect and a hyperchromic effect are in general also observed when atoms or groups with n orbitals ($-\overline{O}H$, $-\overline{O}R$, $-\overline{N}H_2$, $-\overline{N}HR$, $-\overline{N}R_2$, $-\overline{S}H$, $-\overline{S}R$, $-\overline{Hal}$, etc.) are directly bound to a chromophoric group. In this context the term **auxochromic group** is used.

Interactions between several chromophores or chromophores and auxochromes will be extensively discussed in the following chapters. As explicit examples formaldehyde (**2**) and glyoxal (**8**) will be treated here.

The forbidden $n \to \pi^*$ transition of formaldehyde gives a band in the gas phase with extensive fine structure and a maximum at 303 nm.

2 $\lambda_{max}= 303$ nm
$\varepsilon_{max}= 18$ $M^{-1} cm^{-1}$

8 $\lambda_{max}= 450$ nm
$\varepsilon_{max}= 5$ $M^{-1} cm^{-1}$

Glyoxal (**8**) which, in contrast to the colorless formaldehyde (**2**), is yellowish-green in the gas phase, shows an absorption at 450 nm, shifted by some 150 nm. In the associated $n^+ \to \pi_3^*$ transition, neither the n nor the π orbital is comparable with the orbitals in formaldehyde. The two conjugated π-bonds in glyoxal are described by the bonding orbitals π_1 and π_2 and the two antibonding orbitals π_3^* and π_4^*; the latter are empty in the ground state. The two lone pairs of electrons (with p character) also interact and split into n^+ and n^-, with the symmetric combination $n+$ having the higher energy (**Fig. 1.17**, p. 14).

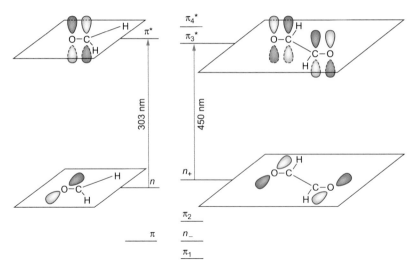

Fig. 1.17 Long-wavelength electronic transitions $S_0 \rightarrow S_1$ in formaldehyde (**2**) and s-*trans*-glyoxal (**8**).

Table 1.5 Absorption maxima ($\pi \rightarrow \pi^*$ transition) of acetophenone ($H_3C-CO-C_6H_5$) in different solvents

λ_{max} (nm)	Solvent
237	Cyclohexane
238	Dioxane
241	Dichloromethane
242	Ethanol
245	Water

1.3.2　Olefins and Polyenes

The $\pi \rightarrow \pi^*$ transition of ethylene lies in the vacuum-UV region with an intense band at $\lambda_{max} = 165$ nm ($\varepsilon_{max} = 16{,}000$ M^{-1} cm^{-1}). If a hydrogen atom is substituted by an auxochromic group, a bathochromic shift is observed. This results from the interaction of a lone pair of the auxochromic group with the π bond. From consideration of the resonance and inductive effects three new orbitals π_1 to π_3^* are predicted, as shown in **Fig. 1.18**. The shift to a longer wavelength of the absorption results from the reduction of the energy difference ΔE between the HOMO and the LUMO.

The introduction of alkyl groups also leads to a shift of the $\pi \rightarrow \pi^*$ absorption. This effect is frequently explained on the basis of hyperconjugation.

When two or more olefinic double bonds are conjugated, the average energy level of the π orbitals is indeed reduced by the mesomeric effect, but the energy difference between the HOMO and LUMO gets less with increasing chain length, as shown in **Figs. 1.19** and **1.20**, and **Table 1.6** (p. 15).

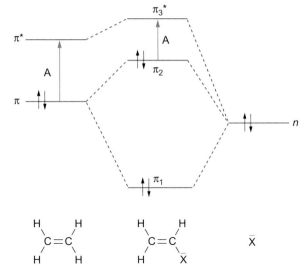

Fig. 1.18 Schematic energy diagram explaining the bathochromic shift of the $\pi \rightarrow \pi^*$ transition of ethylenes with auxochromic groups X.

Accurate calculations demonstrate, in agreement with the observations, that λ_{max} approaches a finite limiting value ($n \rightarrow \infty$). According to perturbation theory, different bond lengths (**Peierls distortion**) in the chain are the reason of this convergence. The limiting value $\lambda_{max} \approx 500$ nm should be reached for $n \approx 17$. The corresponding HOMO–LUMO gap of 2.48 eV is much higher than the band gap (0.56 eV) of solid all-*trans*-polyacetylene, which is a semiconductor.

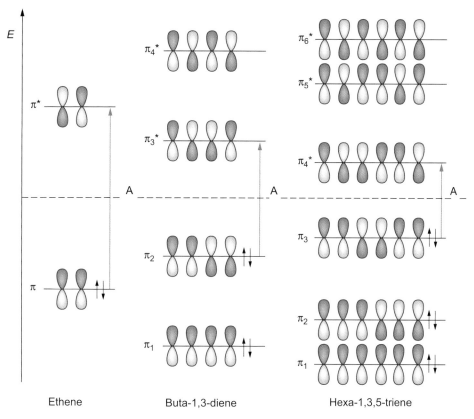

Fig. 1.19 HOMO–LUMO transitions in ethylene, 1,3-butadiene, and 1,3,5-hexatriene.

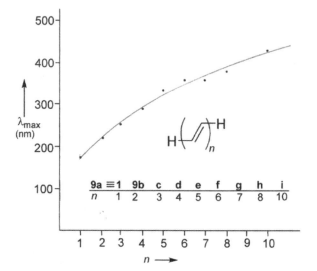

Fig. 1.20 Allowed long-wavelength transitions in *all-trans*-polyenes (**9a–i**).

Table 1.6 Allowed longest wavelength absorptions of conjugated *all-trans*-polyenes (λ_{max} in nm, ε_{max} in M^{-1} cm^{-1})

			$R—(CH=CH)_n—R$	
n	$R=CH_3$		$R=C_6H_5$	
	λ_{max}^a	ε_{max}	λ_{max}^b	ε_{max}
1	174	24,000	306	24,000
2	227	24,000	334	48,000
3	275	30,200	358	75,000
4	310	76,500	384	86,000
5	342	122,000	403	94,000
6	380	146,500	420	113,000

[a]Measured in petroleum ether or ether.
[b]Measured in benzene.

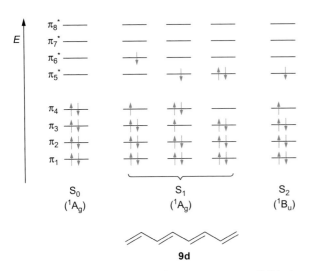

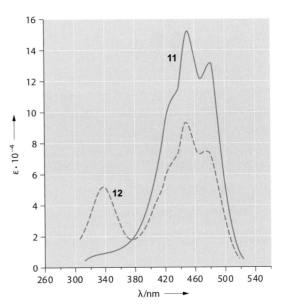

11 (all-*E*)-*β*-Carotene ————

12 (15*Z*)-*β*-Carotene - - - -

Fig. 1.21 π-Electron distribution of 1,3,5,7-octatetraene (**9d**) in the singlet states S_0, S_1, and S_2.

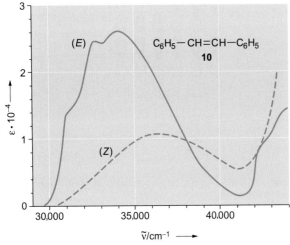

Fig. 1.22 UV spectrum of (*Z*)- and (*E*)-stilbene **10** at 295 K in methylpentane (from Dyck RH, McClure DS. 1962, J. Chem. Phys. **36**, 2336).

Fig. 1.23 Absorption spectra of β-carotenes of different configurations (**11, 12**).

Remarkably, the lowest excited state S_1 of linear *all-trans*-polyenes is not the optically allowed ^1B_u state, reached by the HOMO → LUMO transition, but an ^1A_g state, reached by a forbidden transition. Apart from HOMO − 1 → LUMO and HOMO → LUMO + 1 transitions a doubly excited configuration can make a considerable contribution to this state. This was first predicted by quantum mechanical calculations taking into account the interactions between configurations, and has been experimentally confirmed by two-photon spectroscopy. For the example of 1,3,5,7-octatetraene (**9d**), **Fig. 1.21** illustrates the electron distributions in S_0, S_1, and S_2.

The configuration of olefins also affects the position and intensity of absorptions. (*Z*)-Stilbene [(*Z*)-**10**] absorbs at slightly shorter wavelengths (higher wavenumbers) and less intensively than the (*E*)-isomer (*E*)-**10** (**Fig. 1.22**).

The (*Z*)- or (*E*)-configuration has particular influence on the higher energy transitions of polyolefins. The first overtone in β-carotene lies at 340 nm. In the *all*-(*E*)-configuration (**11**) it is symmetry forbidden (cf. parity rule). On inclusion of a (*Z*)-double bond, the symmetry is changed as shown for (15-*Z*)-β-carotenes (**12**). The electron transition at 340 nm is allowed and leads to the so-called (*Z*)-peak of the carotenes (**Fig. 1.23**).

Table 1.7 Long-wavelength UV absorptions of homoannular 1,3-dienes

Compound	λ_{max} (nm)	log ε	Solvent
 Cyclopentadiene	238	3.53	*n*-Hexane
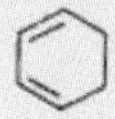 Cyclohexa-1,3-diene	256	3.90	*n*-Hexane
 Cyclohepta-1,3-diene	248	3.87	Isooctane
 Cycloocta-1,3-diene	228	3.75	Cyclohexane
1,3-Cyclododecadiene			
(*Z,Z*)-	226	3.98	*n*-Hexane
(*Z,E*)-	236	4.24	*n*-Hexane
(*E,E*)-	235	4.30	Isooctane

Table 1.8 UV absorption of dendralenes and radialenes

Dendralenes	 H $\sqsubset$ $\sqsupset_n$ H	λ_{max} (nm)
n = 3 3-Methylenepenta-1,4-diene	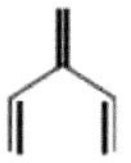	206, 231
n = 6 3,4,5,6-Tetramethyleneocta-1,7-diene	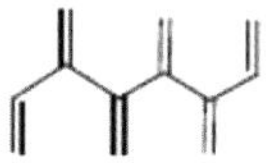	216

Radialenes	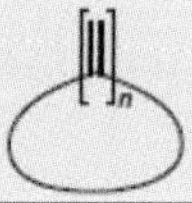	λ_{max} (nm)
n = 3 Trimethylenecyclopropane		213, 295
n = 6 Hexamethylenecyclohexane	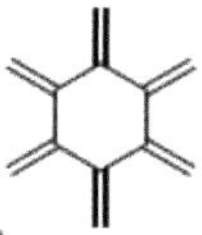	220

Table 1.9 Absorptions of annulenes

Compound	λ_{max} (nm)	log ε	Solvent	Colour of the solution	Character
 Cyclobutadiene	≈305	≈2.0			anti-aromatic
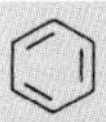 Benzene	262 208 189	2.41 3.90 4.74	Hexane	colour-less	aromatic
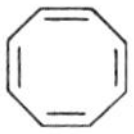 Cycloocta-tetraene	285	2.3	Chloro-form	yellow	non-aromatic
[10]Annulene	265 257	4.30 4.46	Methanol	yellow	non-aromatic
[14]Annulene	374 314	3.76 4.84	Iso-octane	red-brown	aromatic
[16]Annulene	440 282	2.82 4.91	Cyclo-hexane	red	anti-aromatic
[18]Annulene	764 456 379	2.10 4.45 5.5	Benzene	yellow-green	aromatic
[24]Annulene	530 375 360	3.23 5.29 5.26	Benzene	violet	(anti-aromatic)

Empirical rules for the absorption maxima of the long-wavelength $\pi \to \pi^*$ transitions of dienes and trienes were established by Woodward in 1942 and later and independently by Fieser and Scott. Because of many exceptions, these increment rules have lost their significance.

The series of (*Z,Z*)-1,3-cycloalkadienes in **Table 1.7** demonstrates the influence of steric factors on the 1,3-diene chromophore in carbocycles. Moreover, the effect of (*Z*)- and (*E*)-configurations is shown for 1,3-cyclododecadiene.

In contrast to the linear conjugated compounds, cross-conjugated systems do not exhibit a bathochromic shift, when the number of double bonds is increased. Typical dendralenes and radialenes are compiled in **Table 1.8**.

Table 1.9 contains the series of annulenes: **aromatic** $(4n + 2)$-π-electron systems, **antiaromatic** $4n$-π-electron systems, and nonplanar molecules with the so-called **nonaromatic** (olefinic) character. The similarity of the UV/Vis spectra of, for example, the aromatic [18]annulene and [6]annulene (=benzene) serves as an introduction to the next section.

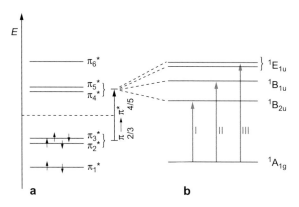

Fig. 1.24 (a) Energy scheme of the π-orbitals of benzene; (b) electronic excitations in benzene.

	λ_{max}	ε_{max}
I $^1B_{2u} \leftarrow {}^1A_{1g}$ according to Platt $^1L_b \leftarrow {}^1A$ according to Clar: α-band	256 nm,	204 M^{-1} cm^{-1}
II $^1B_{1u} \leftarrow {}^1A_{1g}$ according to Platt $^1L_a \leftarrow {}^1A$ according to Clar: p-band	203 nm,	7,400 M^{-1} cm^{-1}
III $^1E_{1u} \leftarrow {}^1A_{1g}$ according to Platt $^1B \leftarrow {}^1A$ according to Clar: β-band	184 nm,	60,000 M^{-1} cm^{-1}.

1.3.3 Benzene and Benzenoid Aromatics

In contrast to 1,3,5-hexatrienes (see p. 15) the π_2/π_3 and π_4^*/π_5^* orbitals of **benzene** form pairs of degenerate (i.e., equal energy) orbitals. As can be theoretically demonstrated, the four conceivable $\pi_{2/3} \rightarrow \pi_{4/5}^*$ transitions lead from the $^1A_{1g}$ ground state to the excited singlet states $^1B_{2u}$, $^1B_{1u}$, and $^1E_{1u}$. (The latter state is, as the symbol E implies, a degenerate state.) Because of the electron correlation the three excited states and therefore the three transitions are of different energy (**Fig. 1.24a, b**).

In the UV spectrum of benzene (**Fig. 1.25**), the highly structured α-band and the p-band correspond to symmetry forbidden transitions. The p-band, which appears as a shoulder, "borrows" intensity from the neighboring allowed transition (β-band). Because of the symmetry prohibition, there is no $0 \leftarrow 0$ transition in the α-band. The ν'_A vibration distorts the hexagonal symmetry and leads to the longest wavelength vibrational band. Further vibrational bands follow, separated by the frequency of the symmetric breathing vibration ν'_B (**Fig. 1.25**).

The introduction of a substituent reduces the symmetry of benzene, enlarges the chromophoric system, and changes the orbital energies and thus the absorptions, so that the p-band can overtake the α-band. The α-band, sometimes also called the B-band,

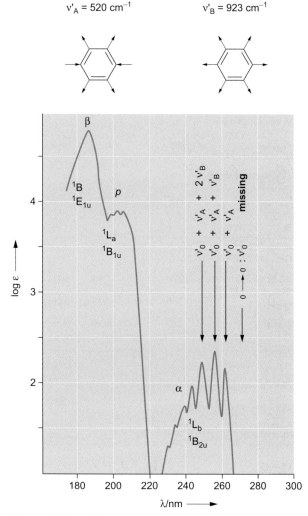

Fig. 1.25 Absorption spectrum of benzene.

gains intensity and often loses its fine structure; because of the reduction in symmetry its $0 \leftarrow 0$ transition becomes visible.

An overview of **monosubstituted benzenes** is given in **Table 1.10**.

The change in the spectra caused by the introduction of two or more substituents into the benzene ring, compared with the monosubstituted derivatives, is particularly marked in those cases where both an electron-withdrawing group and an electron-donating group are present (see **Table 1.11**). In these cases the increase in the size of the chromophore is linked to the possibility of an intramolecular charge transfer (**ICT**). p-Nitrophenol (**13**) is an example.

Table 1.10 UV absorptions of monosubstituted benzenes C_6H_5-R

Substituent R	Long wavelength, stronger transition		Long wavelength (forbidden) transition		Solvent
	λ_{max} (nm)	ε_{max} ($M^{-1}cm^{-1}$)	λ_{max} (nm)	ε_{max} ($M^{-1}cm^{-1}$)	
H	204	7,400	254	204	Water
	198	8,000	255	230	Cyclohexane
CH_3	207	9,300	260	300	Ethanol
C_2H_5	200	31,600	259	158	Ethanol
$CH(CH_3)_2$			251	250	Hexane
F			259	1,290	Ethanol
Cl	210	7,400	264	190	Water
Br	210	7,900	261	192	Water
I	207	7,000	257	700	Water
OH	211	6,200	270	1,450	Water
O^-	235	9,400	287	2,600	Water
OCH_3	217	6,400	269	1,480	Water
OC_6H_5	255	11,000	272	2,000	Water
			278	1,800	
NH_2	230	8,600	280	1,430	Water
NH^+_3	203	7,500	254	160	Water
$N(CH_3)_2$	251	12,900	293	1,590	Ethanol
NO_2			269	7,800	Water
$CH=CH_2$	244	12,000	282	450	Ethanol
$C\equiv CH$	236	12,500	278	650	Hexane
$C\equiv N$	224	13,000	271	1,000	Water
$CH=O$	242	14,000	280	1,400	Hexane
			330	60	
$CO-CH_3$	243	13,000	278	1,100	Ethanol
			319	50	
COOH	230	11,600	273	970	Water
COO^-	224	8,700	268	560	Water
SO_3H	213	7,800	263	290	Ethanol

Table 1.11 Long-wavelength absorptions λ_{max} (nm) of some *para*-disubstituted benzenes X^1-C_6H_4-X^2 (in water)

X^1	$X^2=H$		OH		NH_2		NO_2	
	λ_{max}	$\log \varepsilon$	λ_{max}	$\log \varepsilon$	λ_{max}	$\log \varepsilon$	λ_{max}	$\log \varepsilon$
H	254	2.31						
OH	270	3.16	293	3.43				
NH_2	280	3.16	294	3.30	315	3.30		
NO_2	269	3.89	310	4.00	375	4.20	267	4.16

13

The effect should be even stronger in the *p*-nitrophenolate anion than in *p*-nitrophenol itself (**Fig. 1.26a**). This is confirmed by **Fig. 1.26b** which, however, also shows that a similar effect occurs where there is *m*-substitution, and hence independently of the participation of quinonoid resonance structures.

The solvent can have a particularly strong effect in such cases, and even change the energetic order of the states. A good example of this is *p*-dimethylaminobenzonitrile (**14**).

14

The ICT can be stabilized by a twist about the CN single bond. The **TICT state** (twisted ICT) so formed has a large dipole moment (μ = 12 D) and is so energetically favored by polar solvents that it becomes the lowest electronically excited singlet state. Thus the fluorescence occurs from different singlet states in polar and nonpolar solvents. The dual fluorescence which is coupled with an **ICT** can also be explained by a solvent-induced pseudo-Jahn–Teller effect. The assumption of bond twisting is not necessary; the two singlet states must, however, have very similar energies.

Biphenyl (**15**) has in comparison to benzene a bathochromically shifted p-band (**Fig. 1.27**). Steric interaction between the *ortho*-standing H atoms of **15** causes a torsion of ~42° between the benzene ring planes. Increasing planarization in 9,10-dihydrophenanthrene (**16**) and 4,5,9,10-tetrahydropyrene (**17**) effects a further bathochromic shift of the absorption (**Fig. 1.27**).

An analogous argumentation is valid for the series of fluorene **18**, **19** and **20** for which the λ_{max} values 261, 247, and 235 nm are found for the allowed electron transition in ethanol.

18	19	20
λ_{max} = 261 nm	λ_{max} = 247 nm	λ_{max} = 235 nm

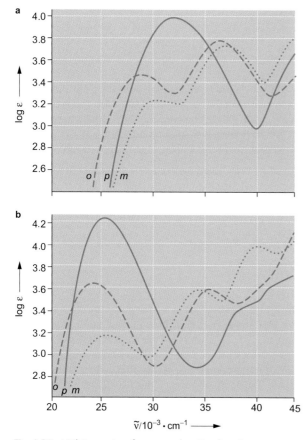

Fig. 1.26 UV/Vis spectra of *o*-, *m*-, and *p*-nitrophenol:
(a) in 10^{-2} molar hydrochloric acid,
(b) in 5×10^{-3} molar sodium hydroxide.

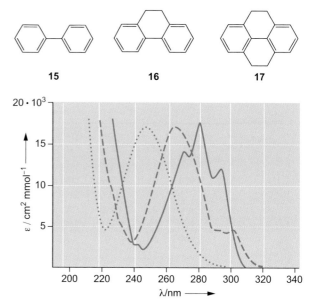

Fig. 1.27 UV spectra of biphenyl (**15** ⋯), 9,10-dihydrophenanthrene (**16** – – –), and 4,5,9,10-tetrahydropyrene (**17** —) in isooctane.

The prototype of polycyclic aromatic hydrocarbons (**PAHs**) consists exclusively of condensed benzene rings. **Linear anellation** as in anthracene or naphthacene and **angular anellation** as in phenanthrene, chrysene, etc. are subsumed as *kata-condensation*. *Peri*-**condensed** systems, such as pyrene or perylene, represent the second category of benzenoid aromatic hydrocarbons.

In the spectra of condensed benzenoid aromatics are many common features. The two highest occupied orbitals π_{n-1} and π_n and the two lowest empty orbitals π^*_{n+1} and π^*_{n+2} are no longer degenerate as they are in benzene. Four electron transitions I–IV are possible between them (**Fig. 1.28**). Due to the symmetrical arrangement of π and π^* orbitals related to the α-line (Coulomb integral) in the energy level scheme of alternating hydrocarbons, the transitions II and III are isoenergetic.

However, configuration interaction removes this degeneration. The topology of the condensed, benzenoid aromatics differentiates between a large (type a) and a small (type b) splitting of the orbital energies (**Fig. 1.28**). The α-band ($2 \le \log \varepsilon \le 3$) can be easily identified in the long-wavelength region of the absorption of type a systems, whereas it is engulfed in type b by the higher intensity of the p-band. The p-band corresponds to the HOMO → LUMO transition, that is polarized for acenes in the direction of the short axis. The β-band has an even higher intensity ($\log \varepsilon \approx 5$). The neighboring transition IV (β'-band) and still higher transitions play a minor role for the characterization of the UV/Vis spectra of benzenoid aromatics.

Fig. 1.29 shows the UV/Vis spectra of benzene, naphthalene, anthracene, naphthacene (tetracene), and phenanthrene. The α-band is visible in benzene, naphthalene, and phenanthrene; in anthracene and naphthacene it is superimposed by the intense p-band.

The intensity of the p-band remains more or less constant. The increase in the number of rings has no effect, because this

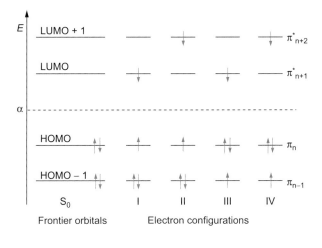

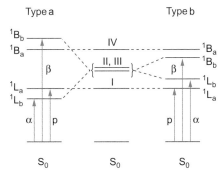

Fig. 1.28 Schematic representation of the excitation of electrons and the corresponding transitions in condensed benzenoid aromatics having strong (type a) or weak (type b) configuration interactions.

electronic transition is polarized parallel to the short axis. The bathochromic shift in the acene series leads to the members from tetracene onwards being colored:

Benzene, naphthalene, anthracene colorless

Tetracene (naphthacene) orange-yellow

Pentacene blue-violet

Hexacene dark green

If the anellation is nonlinear, characteristic changes occur in the spectra (e.g., anthracene and phenanthrene, **Fig. 1.29**). As well as the linear anellated tetracene the condensation of four benzene rings can lead to four angular systems, benz[*a*]anthracene, benzo[*c*]phenanthrene, chrysene, and triphenylene, and the *peri*-condensed system pyrene. Of these only naphthacene absorbs in the visible region; the others are colorless, but show colored fluorescence.

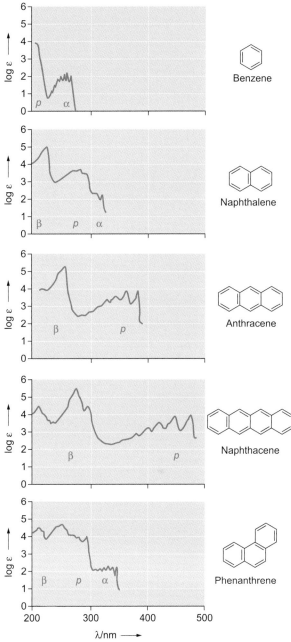

Fig. 1.29 UV/Vis spectra of condensed aromatic hydrocarbons (in heptane).

The diversity of structures in the series of *kata*-condensed aromatics increases fast with increasing numbers **n** of benzene rings:

Number of benzene rings **n**	General formula $C_{4n+2}H_{2n+4}$	Possible structure isomers
1	C_6H_6	1
2	$C_{10}H_8$	1
3	$C_{14}H_{10}$	2
4	$C_{18}H_{12}$	5
5	$C_{22}H_{14}$	17
6	$C_{26}H_{16}$	37
7	$C_{30}H_{18}$	123
8	$C_{34}H_{20}$	446
9	$C_{38}H_{22}$	1,689
10	$C_{42}H_{24}$	6,693

Table 1.12 Long-wavelength absorptions of benzenoid aromatics of the $C_{30}H_{18}$ series

Compound	λ_{max} [nm]/ (Solvent)	Color of crystal
Phenanthro[9.10-*b*]triphenylene	382(benzene)	colourless
Tribenzo[*a.c.j*]tetracene	423(benzene)	yellow
Dibenzo[*a.c*]pentacene	539(benzene)	red-violet
Benzo[*a*]hexacene	651(1-methyl- naphthalene)	blue-green
Heptacene	840(1-methyl- naphthalene)	black green

Many of these, only to a small part, known arenes are non-planar because of steric hindrance and show additionally stereoisomerism.

The number of isomers is still higher, when *peri*-condensed systems are included. (*Peri*-anellation has to be exercised with caution, because this anellation can not only lead to arenes with Kekulé structures but also to mono- and biradicals.)

Table 1.12 contains a selection of five compounds $C_{30}H_{18}$, which consist of seven *kata*-condensed benzene rings. When using a circle in the ring to show the existence of a complete π-electron sextet, one can see that their number decreases on going from the top to the bottom of the table. This corresponds to a red-shift of the absorption, which leads from the UV into the NIR region.

Monolayers of graphite are called graphenes. Graphene sheets can be very large and are insoluble in organic solvents. Their UV/Vis/NIR spectra can be only measured in thin films or in suspensions. **Fig. 1.30** presents the absorption of C216-nano-graphene (**21**) (p. 23), obtained in a selective synthesis. It has the area of a regular hexagon, but contains a hollow part in its center. Therefore, it can be regarded as an extended coronoid. The absorbance of **21** was measured in a suspension. Time-dependent density functional theory (TD-DFT) calculations provide the wavelengths of the p, β, and β' bands of this huge arene in an impressive manner.

Nonalternating PAHs are compounds which contain one or more odd-numbered rings. Whereas biphenylene (**22**) is an **alternating hydrocarbon**, fluoranthene (**23**) represents **nonalternating hydrocarbons**.

22
λ_{max} = 358 nm
ε = 8,710 $Lmol^{-1}cm^{-1}$
(Acetonitrile)

23
λ_{max} = 358 nm
ε = 8,630 $Lmol^{-1}cm^{-1}$
(Hexane)

The energy scheme of the π-orbitals of nonalternating hydrocarbons is nonsymmetrical relative to the α line (Coulomb integral).

The number of energetically close lying electronic transitions of PAHs increases with increasing numbers of π-electrons. At the end of this section, a comparison shall be drawn between the planar PAH $C_{60}H_{24}$ (**24**) and fullerene C_{60} (**25**), which has a spherical topology of 12 five-membered and 20 six-membered rings. Both compounds absorb in the entire range between 200 and 700 nm. Their maxima at longest wavelengths have a low intensity. Compound **24** is insoluble. It can only be measured in the solid state as a thin film.

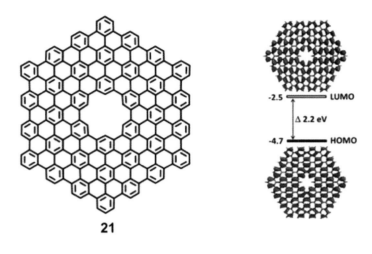

21

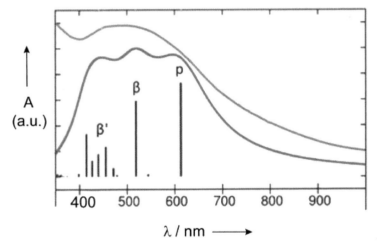

Fig. 1.30 UV/Vis/NIR spectrum of the C216-nanographene (**21**), extended coronoid: measurement in chlorobenzene suspension (*red curve*), TD-DFT calculation of the p-band (613 nm), β-band (519 nm), and the cluster of β'-bands (*black sticks*) and the corresponding simulation of the absorbance (*blue curve*) (from Müllen K et al. 2016, *J Am Chem Soc* **138**, 4322).

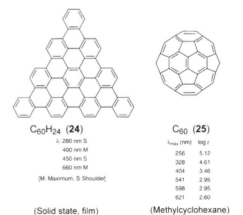

C60H24 (**24**)
λ: 280 nm S
  400 nm M
  450 nm S
  660 nm M
[M: Maximum, S: Shoulder]

(Solid state, film)

C60 (**25**)

λmax (nm)	log ε
256	5.12
328	4.61
404	3.46
541	2.95
598	2.95
821	2.60

(Methylcyclohexane)

1.3.4   Heteroaromatics (Hetarenes)

Furan (**26**), pyrrole (**27**), and thiophene (**28**) show in hexane almost structureless absorptions at the border to the far-UV.

	26	**27**	**28**
λmax (nm)	215	208	231
log ε	3.70	3.88	3.77

The spectra of five-ring hetarenes containing two or in some cases even three heteroatoms look similar. 1,2,3-Triazole (**29**) for example has in THF an absorption maximum at 215 nm, whereas 1,3,4-thiadiazole (**30**) has in cyclohexane a maximum at 305 nm.

	29	**30**	**31**
λ_{max} (nm)	215	305	251
log ε	3.64	4.33	3.30
	(THF)	(C$_6$H$_{12}$)	(C$_6$H$_{12}$)

The UV spectrum of pyridine (**31**) resembles the spectrum of benzene. However, the long-wavelength $\pi \to \pi^*$ transition is allowed. The $n \to \pi^*$ transition is located at ~270 nm below the foot of the $\pi\pi^*$ band. **Table 1.13** presents the long-wavelength absorption maxima of several azines. In particular, neighboring N-atoms lead to an interaction of their electron lone-pairs, which causes an energetical splitting to a higher and a lower lying n orbital. Hence, a bathochromic shift of the $n \to \pi^*$ transition occurs. The log ε values of the $\pi \to \pi^*$ transitions are between 3 and 4 and those of the $n \to \pi^*$ transition between 2 and 3.

The condensation of benzene rings to hetarenes leads to UV/Vis spectra, which are similar to the spectra of polycyclic arenes with the same number of rings. Going from pyrrole (**27**) to indole (**32**) and carbazole (**33**), a red-shift can be observed. An analogous effect is found for the series pyridine, quinoline, and acridine. The longest-wavelength maximum of quinoline and isoquinoline lies in cyclohexane at 314 and 317 nm, respectively, and for acridine even at 380 nm.

32 Indole		**33** Carbazole	
λ_{max}	log ε	λ_{max}	log ε
218	4.44	235	4.60
271	3.79	294	3.24
287	3.66	324	3.59
(Ethanol)		(Ethanol)	

Indigo (**34**) has a central CC double bond with *captodative*-substitution on both sides (**Fig. 1.31**). It represents a twofold cross-conjugated system whose long-wavelength absorption is predominantly due to a HOMO → LUMO electron transition. These frontier orbitals have π and π^* characters, respectively. The excitation S$_0$ → S$_1$ increases the electron density at the O atoms and decreases it at the N atoms. The long-wavelength absorption maximum is observed at 600 nm (acetonitrile). TD-DFT calculations give a similar value.

Table 1.13 Long-wavelength absorption maxima of azines in cyclohexane: λ_{max} (nm)/log ε

Pyridazine	Pyrimidine	Pyrazine
246/3.11	243/3.31	260/3.75
340/2.50	298/2.47	328/3.02

1,3,5-Triazine	1,2,4-Triazine[a]	1,2,4,5-Tetrazine
272/3.00	248/3.48	252/3.33
	374/2.60	320/1.42
		542/2.92

[a] Measurement in methanol

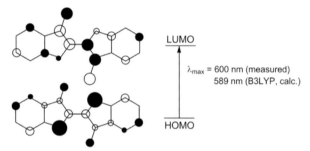

Fig. 1.31 HOMO → LUMO electron transition in indigo (**34**).

λ_{max} = 600 nm (measured)
589 nm (B3LYP, calc.)

34

Heterocycles can act as electron donors or electron acceptors in D-π–A systems. Compound **35** contains a carbazole unit as donor D and a terpyridine moiety as acceptor A. The HOMO of **35** is predominantly localized in the donor region, whereas the LUMO has a high electron density in the acceptor region (**Fig. 1.32a**).

35

a

b

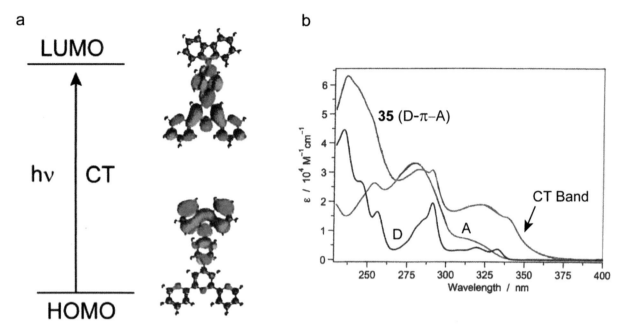

Fig. 1.32 **(a)** Electron density plot of the frontier orbitals of **35**; **(b)** UV spectrum of **35** (D–π–A) and its substructures carbazole (D) and phe-nylterpyridine (A) in acetonitrile (from Baschieri A, Sambre L, Gualandi I et al. 2013, *RSC. Adv.* **3**, 6507).

Thus, the HOMO → LUMO electron transition represents a typi-cal ICT. The CT band at the long-wavelength end of the absorp-tion spectrum (**Fig. 1.32b**) is not present in the absorption of the components.

1.3.5 Carbonyl Compounds

The carbonyl functional group contains σ-, π-, and n-electrons with s-character as well as n-electrons with p-character. This simple picture is based on the assumption of a nonhybridized oxygen atom. But even a detailed consideration of the deloca-lized group orbitals shows that the HOMO has largely the cha-racter of a p-orbital on oxygen (see **Fig. 1.6**). The excitation of an electron can occur into the antibonding π^* or σ^* orbital. For saturated aldehydes and ketones, the allowed $n \to \sigma^*$ and $\pi \to \pi^*$ transitions are in the vacuum-UV region. The forbidden $n(p) \to \pi^*$ transition lies in the region of 275 to 300 nm. The intensity of the $n \to \pi^*$ band is normally about $\varepsilon = 15$ to 30. (In β,γ-unsaturated ketones, it can, however, be increased by a factor of 10 to 100.)

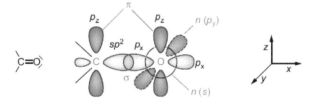

Auxochromes, such as OH, OR, NH_2, NHR, NR_2, Ha_l, etc., directly bonded to the carbonyl group, act as π-donors to increase the energy of the π^* orbital and as σ-acceptors to lower the n level. The result is a short-wavelength shift of the $n \to \pi^*$ transitions in **carboxylic acids** and their **derivatives (Table 1.14)**.

Conjugation of the carbonyl group with a (C–C) bond leads to a marked shift of the π-level; to a first approximation the n-orbital is unaffected (**Fig. 1.33**).

With increasing length of the conjugated chain in **enones**, the longest wavelength $\pi \to \pi^*$ transition moves further into the visible region, reaches the position of the $n \to \pi^*$ band, and hides it because of its considerably higher intensity, which also incre-ases strongly with increasing chain length (**Table 1.15**).

The λ_{max} values of the $n \to \pi^*$ transitions of α,β-unsaturated carbonyl compounds can be estimated from the extended Woodward rules (**Table 1.16**). However, these rules lost their significance, since major differences between calculated and experimental λ_{max} values were found for many examples. Five sufficiently exact and one unsatisfactory application of the increment system are compiled in **Table 1.17**.

As already mentioned in Section 1.3.1, certain absorptions are very solvent dependent. Such effects have been particularly carefully investigated for ketones. **Fig. 1.34** shows the example of benzophenone (**36**).

Table 1.14 $n \to \pi^*$ Transitions in saturated carbonyl compounds

Compound	λ_{max} (nm)	ε_{max} $(M^{-1}cm^{-1})$	Solvent
Acetaldehyde	293	12	Hexane
Acetone	279	15	Hexane
Acetyl chloride	235	53	Hexane
Acetic anhydride	225	50	Iso-octane
Acetamide	205	160	Methanol
Ethyl acetate	207	70	Petroleum ether
Acetic acid	204	41	Ethanol

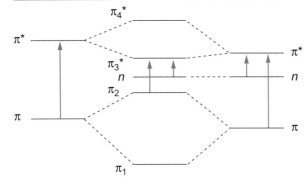

Fig. 1.33 Energy diagram of the electronic transitions in conjugated enones compared with alkenes and saturated carbonyl compounds

Table 1.15 Absorption maxima of the long-wavelength $\pi \to \pi^*$ transition in the vinylogous series $C_6H_5-(CH=CH)_n-CO-R$ (in methanol)

n	R=H		R=C₆H₅	
	λ_{max} (nm)	ε_{max} $(M^{-1}cm^{-1})$	λ_{max} (nm)	ε_{max} $(M^{-1}cm^{-1})$
0	244	12,000	254	20,000
1	285	25,000	305	25,000
2	323	43,000	342	39,000
3	355	54,000	373	46,000
4	382	51,000	400	60,000

The electronic states of benzophenone are lowered by solvation, hydrogen bonding in polar protic solvents being especially effective. The strongest effect is observed with the state of highest polarity, the π,π^* singlet state. Since the doubly occupied n-orbital on the oxygen atom is mostly responsible for the hydrogen bonding, the n,π^* singlet state of ketones has much poorer solvation properties (**Fig. 1.35**).

Similar solvent effects appear in certain heterocycles, azo compounds, nitroso compounds, thioketones, etc. However, the

Table 1.16 Increment system for calculating absorption maxima of α,β-unsaturated carbonyl compounds

$$\delta-C=C-C=C-C=O$$
$$\;\;\;\delta\;\;\gamma\;\;\beta\;\;\alpha\;\;\;X$$

(In methanol or ethanol)

Base values		
	X = H	207 nm
	X = Alkyl (or 6-membered ring)	215 nm
	X = OH, OAlkyl	193 nm

Increments

for each further conjugated (C=C) bond	+ 30 nm
for each exocyclic position of a (C=C) bond	+ 5 nm
for each homoannular diene component	+ 5 nm
for each substituent in	

	Position			
	α	β	γ	δ and higher
Alkyl (or ringrest)	10	12	18	18
Cl	15	12		
Br	25	30		
OH	35	30		50
O-Alkyl	35	30	17	31
O-Acyl	6	6	6	6
N(Alkyl)₂		95		

The base values are for measurements in alcohols. For other solvents the following solvent corrections must be applied

Water	+8 nm
Chloroform	−1 nm
Dioxane	−5 nm
Ether	−7 nm
Hexane	−11 nm
Cyclohexane	−11 nm

use of solvent dependence for characterizing the $n \to \pi^*$ and $\pi \to \pi^*$ transitions should be confined to aldehydes and ketones. Extreme solvatochromism such as, that of the zwitterionic pyridinium phenolates, is used to determine the polarity of solvents (p. 34–35)

The **quinones** represent special "enone" chromophores. As shown by a comparison of 1,4- and 1,2-benzoquinone (**37** and **38**), o-quinones absorb at longer wavelengths than the corresponding p-isomers (measurement in benzene):

1,4-Benzoquinone (yellow)
λ_{max} (nm) ε_{max} (M⁻¹cm⁻¹)
242   24,300 (allowed $\pi \to \pi^*$)
281   400 ($n \to \pi^*$)
434   20 (forbidden $n \to \pi^*$)

37

1,2-Benzoquinone (red)
λ_{max} (nm) ε_{max} (M⁻¹cm⁻¹)
390   3,020
610   20 (forbidden $n \to \pi^*$)

38

Table 1.17 Observed and calculated $\pi \to \pi^*$ absorptions of some enones (in ethanol)

Compound	Observed		Calculated
	λ_{max} (nm)	ε_{max} (cm^2· mmol^{-1})	λ_{max} (nm)
3-Penten-2-one H$_3$C–CH=CH–C(=O)–CH$_3$	224	9,750	215 + 12 = 227
1-Cyclohex-ene-1-carbaldehyde	231	13,180	207 + 10 + 12 = 229
1-Cyclohex-ene-1-carboxylic acid	217	10,230	193+10+12=215
Steroid type	241	–	215+10+12+5=242
Steroid type	388	–	215+2·30+5+ 39+12+3·18=385
4,6,6-Trimethyl-bicyclo[3.1.1]-hept-3-en-2-one	253	6,460	215+2·12=239

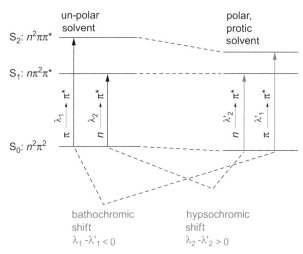

Fig. 1.35 The bathochromic and hypsochromic shifts of the $\pi \to \pi^*$ and $n \to \pi^*$ transitions of ketones on increasing the solvent polarity.

The reason for this is that the lowest π^*-orbital of the linearly conjugated *o*-quinone lies lower than that of the cross-conjugated *p*-quinone. Because of the interaction of the $n(p)$-orbitals of the two oxygen atoms, two $n \to \pi^*$ transitions are expected. In general they lie very close together. The $n \to \pi^*$ transitions are responsible for the yellow color of **37** and the red color of **38**.

The *p*- and *o*-quinonoid groups play a decisive role in many **organic dyestuffs**. The acid–base indicator phenolphthalein will be discussed as a typical example. The lactone form **39** contains solely isolated benzene rings and is therefore colorless. At pH = 8.4 the two phenolic protons are lost, forming the dianion **40**, which opens the lactone ring to give the red dye **41** ($\lambda_{max} = 552$ nm; $\varepsilon_{max} = 31,000$ M^{-1} cm^{-1}). With excess alkali the carbinol **42** is formed, a trianion in which the chromophoric **meriquinoid** group has disappeared again.

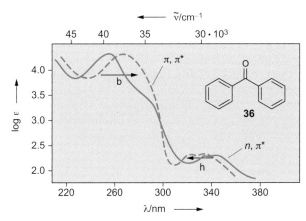

Fig. 1.34 Absorption spectra of benzophenone (**36**):

—in cyclohexane

---in ethanol

b bathochromic solvent effect (on increasing the solvent polarity)

h hypsochromic solvent effect (on increasing the solvent polarity).

39 colorless

40

41 deep red

42 colorless

1.3.6 Conjugated Oligomers and Polymers

In general, linear conjugated oligomers show a systematic bathochromic shift of the absorption at the longest wavelength with increasing number (n) of repeating units. A first example was already shown in **Fig. 1.20** for the series of polyenes.

In the case of **cyanines** and the related **polymethine dyes** with degenerate mesomeric resonance structures (see compounds **43a–d**), we observe for the lower members a more or less linear increase of λ_{max} with n (see following table). This rule is valid till for higher members the so-called cyanine limit is reached.

$$(H_3C)_2\overset{+}{N}=CH-(CH=CH)_n-N(CH_3)_2$$
$$\leftrightarrow (H_3C)_2N-(CH=CH)_n-CH=\overset{+}{N}(CH_3)_2$$

43	a	b	c	d
n	1	2	3	4
λ_{max} [nm]	309	409	510	612

$\lambda_{max}(n) - \lambda_{max}(n-1) \approx 100$ nm

In the case of nondegenerate systems, e.g., when the iminium group, $C=N^+(CH_3)_2$, in **43** is replaced by a formyl group, $CH=O$, to give a so-called merocyanine, as well as with completely different repeating units with aromatic or heteroaromatic building blocks, a convergent behavior of the energy and the wavelength of the absorption bands of lowest energy is established.

$$E(n) \rightarrow E_\infty \text{ and } \lambda(n) \rightarrow \lambda_\infty \text{ for } n \rightarrow \infty$$

At first one should check this convergence for the $0 \rightarrow 0$ transitions ($\lambda_{0,0}$) in the series of conjugated compounds; however, it is often also valid for the absorption maxima (λ_{max}). The absorption spectra of oligo(2,5-dipropoxyphenylenevinylene)s **44a-j** are shown as an example (**Fig. 1.36**).

When the energies $E(n)$ of the electronic transitions of **44a–j** are plotted against the reciprocal of the number of benzene rings, one obtains an apparently useful linear correlation. However, extrapolation to the polymer **44p** fails completely. On the other hand, when the E values of **44a – j** are based on an e function (dotted curve in **Fig. 1.36b**),

$$E(n) = E_\infty + (E_1 - E_\infty)\, e^{-a(n-1)}$$

the limiting value E_∞ for $n \rightarrow \infty$ corresponds to the measured value for the polymer **44p**. The difference $E_1 - E_\infty$ describes the conjugation effect; it affords the bathochromic shift between the first member and that with the "infinitely long" chain of the respective conjugated series. Furthermore, the **effective conjugation length** n_{ECL} indicates which oligomer reaches within $\lambda_\infty \pm 1$ nm (the error limit of a routine spectrophotometer) the limiting value. In the compound series **44**, this is the case for the undecamer **44i** according to calculation and measurement.

The synthesis of polymers is always accompanied by structural errors; E_∞ and n_{ECL} are important parameters for evaluating the length of defect-free segments in conjugated chains.

For extended chromophores the lowest energy electronic transitions may lie in the NIR region. For example, when a poly(phenylvinylene) system (PPV) is doped with an oxidant, an electron transfer leads to polymeric radical ions and doubly charged ions (polarons, bipolarons). The insulator **45** is thus transformed to the electrical semiconductor **46**.

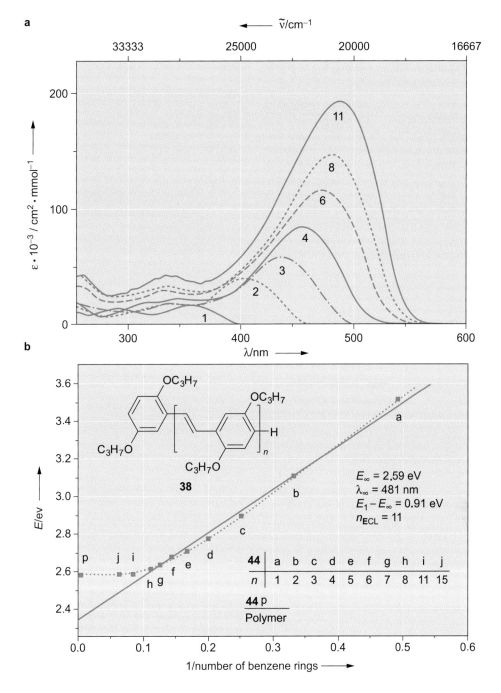

Fig. 1.36   **(a)** Long-wavelength absorption bands of the *all-E*-configured oligo(2,5-dipropoxyphenylene vinylene)s **44a–i** (*n* = 1, 2, 3, 4, 5, 6, 7, 8, and 11) in trichloromethane (from Stalmach U, Kolshorn H, Brehm I, Meier H. 1996, Liebigs Ann. Chem.1449). **(b)** Correlation of the transition energies *E*(*n*) of **44a–j** and the reciprocal of the number of benzene rings (from Meier H. 2005, Angew. Chem. Int. Ed. **44**, 2482).

UV/Vis-Spektren

On absorption measurements in solution one finds that, due to the doping, new bands occur beyond the absorption edge. The lowest energy transition can be shifted out of the visible wavelength region ($\lambda_{max} \approx 2{,}000$ nm); it can then only be detected by using a special spectrophotometer equipped for the NIR region.

In conjugated oligomers with terminal **donor–acceptor substitution (push–pull substitution)** D-π-A, the longest wavelength excitation is accompanied by an ICT (see also p. 4). If an electron moves from the donor to the acceptor region of the molecule, the energy of the electron interaction decreases. Such **CT bands** experience even larger red-shifts with increasing donor and acceptor strengths. **Table 1.18** illustrates this for the example of 4-dialkylamino-substituted *trans*-stilbenes **47a-f** ($n = 1$) with different acceptor groups A in the 4′-position. (The long, branched alkyl groups on the amino function serve for better solubilization in the case of larger numbers n of the repeating units.)

When the separation between the donor and acceptor substituents is increased, i.e., the conjugated chain ($n = 2, 3, 4, \ldots$) in **47** is lengthened, two opposing effects are observed. Extension of the conjugation corresponds to a **bathochromic effect**, whereas the declining influence of the ICT with increasing n exerts a **hypsochromic effect**. For the example of the oligo(1,4-phenylenevinylene)s **47** ($n = 1$–4), **Fig. 1.37** shows that for terminal substitution by a strong donor and a relatively weak acceptor (A = CN), like in the purely donor-substituted case (A = H) the extension of the conjugation dominates. The energy of the longest wavelength electron transition decreases with increasing n (red-shift). The opposite is the case in the series with the strong acceptor A = NO$_2$; on extension of the chromophore a hypsochromic effect is seen. For A = CHO, the two effects more or less cancel out, i.e., the length of the chromophore has hardly any influence on the longest wavelength absorption maximum. The convergence limit ($n \to \infty$) lies in all four cases at $\tilde{v}_\infty = 23.2 \times 10^3$ cm^{-1} ($\lambda_\infty = 430$ nm).

Similar results as for D-π-A series were found for D-π-A-π-D systems and for star-shaped compounds A(π-D)$_n$ or D-(πA)$_n$.

The rylenetetracarboxydiimides **48a–e** represent a series of oligomers, which have a laminar structure. A bathochromic shift of the intense absorption maxima can be seen (**Fig. 1.38**) for increasing numbers n of naphthalene segments. A convergence is not directly obvious, but the energies $\Delta E(n)$ of the electronic

Table 1.18 Long-wavelength absorption maxima of **47a-f** ($n = 1$) in CHCl$_3$

47	Acceptor group A	λ_{max} [nm]	Color
a	(H)	366	Colorless
b	CN	401	Yellow
c	CHO	423	Orange
d	NO$_2$	461	Red
e	CH = C(CN)$_2$	525	Dark red
f	C(CN)=C(CN)$_2$	670	Blue

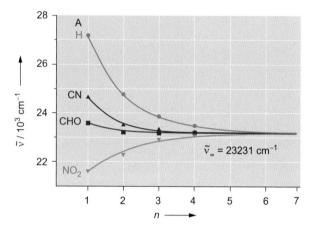

Fig. 1.37 Long-wavelength absorption maxima of the OPV series **47** ($n = 1$–4) bearing various acceptor groups A, measured in CHCl$_3$ (from Meier H, Gerold J, Kolshorn H, Mühling B. 2004 Chem. Eur. J. 360).

transitions correlate linearly with $1/L^2$. This result corresponds to the simple model of "electrons in a box." $\Delta E < 1$ eV can be expected for $1/L^2 \to 0$. An octarylene-tetracarboxydiimide ($n = 6$) with a substitution pattern, which differs somewhat from **48a–e**, is the highest known rylene to date. It is an almost colorless compound with a λ_{max} value of 1,066 nm, which corresponds to 1.16 eV.

In total, for conjugated oligomers there are the four major possibilities shown in **Fig. 1.39** for changes of the longest wavelength absorption with increasing number of repeating units n (extension of the conjugation). The bathochromic effect (case a) with convergence to λ_∞ is by far the most common; hypsochromic effects (case b) can occur for push–pull-substituted oligomers with strong donors and strong acceptors if the repeating unit contains aromatic building blocks.

A linear increase of λ_{max} with n (case c) up to a certain limit is typical for degenerate systems such as the cyanines **43** (p. 28). A "hyperlinear" increase of λ_{max} with n (case d) can be seen when the extension of the conjugation proceeds rather in two or more directions (area-like). Phenes and starphenes in the series of condensed arenes are examples for curve d.

UV/Vis-Spektren

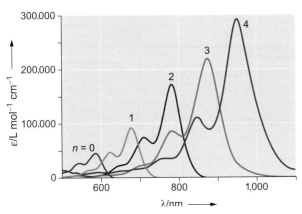

48a-e(n = 0–4)

L = N ←——→ N

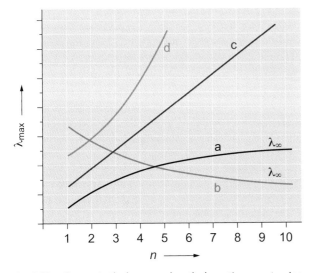

Fig. 1.38 Vis/NIR absorption spectra of rylenetetracarboxydiimides **48a–e** (n = 0–4) (from Herrmann A, Müllen K. 2006, Chem. Lett. **35**, 978).

Fig. 1.39 Changes in the long-wavelength absorption on extension of the conjugation: **(a)** bathochromic effect with convergence to λ_∞, **(b)** hypsochromic effect with convergence to λ_∞, **(c)** linear increase in λ_{max}, and **(d)** "hyperlinear" increase in λ_{max}.

Unfortunately, for cases c and d there are as yet no known really high oligomers.

1.3.7 Aggregated Molecules, Charge-Transfer Complexes

Due to electrostatic interactions, hydrogen bonding, or van der Waals forces, a **self-association** of molecules can occur. The electronic transitions of monomeric compounds are changed by self-association. New bands appear that are temperature- and concentration-dependent. A simple model starts with the association of two rod-like molecules, the transition moment **M** of which lies in the molecular axis. The angle α between the aggregation direction and the molecular axis then has decisive significance for the absorption (**Fig. 1.40**). At $\alpha = 0$ we speak of **J aggregates** (named after their discoverer Jelley) which lead to a bathochromic shift; at $\alpha = 90°$ we have **H aggregates** where the H expresses the hypsochromic shift.

In a two-molecule aggregate, the energy for the electron transition $h\nu$ is first changed by the van der Waals interaction W_1. The resulting $h\nu'$ is split into the two **Davidov components** $h\nu''$ and $h\nu'''$. The interaction energy W_2 valid for an exchange process in a two-molecule aggregate is proportional to the term $(1 - 3 \cos^2 \alpha)$ and thus becomes zero for the so-called *magic angle* $\alpha = 54.73°$. For $\alpha < 54.73°$ a bathochromic effect results $h\nu'' < h\nu'$ and for $\alpha > 54.73°$ a hypsochromic effect $h\nu'' > h\nu'$. This is illustrated in **Fig. 1.40** by the solid curve. For parallel transition moments **M**, it corresponds to the case **M** + **M**, i.e., the allowed transition. The dashed curve corresponds to the forbidden transition **M** − **M** = 0.

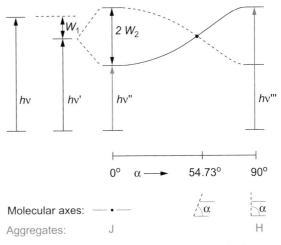

Fig. 1.40 H and J aggregates with hypsochromic or bathochromic shifts of absorption as compared with the absorption of the individual molecule.

H and J aggregates of more than two molecules obey an analogous set of rules. For transition moments that enclose an angle $\beta \neq 0$, both Davidov transitions are allowed and W_2 is proportional to the term $(\cos\beta - 3\cos^2\alpha)$. Aggregation in solution can usually be avoided by the choice of solvent and use of high dilutions; in solids, on the other hand, the interaction of chromophores is the rule.

Many detection reactions (color reactions) are based on the formation of a complex between a substance/class of substances and a detection reagent. Very common are electron **donor–acceptor** (EDA) **complexes**, also called **CT complexes** since the electron transition $S_0 \rightarrow S_1$ (**Fig. 1.41**) in these 1:1 complexes is accompanied by a partial transfer of charge from the donor to the acceptor.

The deep green quinhydrones such as **51**, discovered by F. Wöhler back in 1844, are typical CT complexes. The intensification of the color is caused by the π,π-interaction of the electron-rich hydroquinone **49** with the electron-poor 1,4-benzoquinone **50**. Hydrogen bonding reinforces the complex formation although it is not necessary as revealed by the corresponding complexes of hydroquinone dimethyl ether.

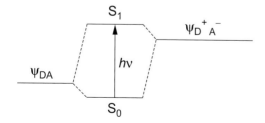

49	50	51
colorless	yellow	green

Further typical electron acceptors are tetranitromethane, tetracyanoethylene, 1,3,5-trinitrobenzene, picric acid, chloranil (2,3,5,6-tetrachloro-1,4-benzoquinone), 2,3,5,6-tetracyano-1,4-benzoquinone, 2,3-dichloro-5,6-dicyano-1,4-benzoquinone (DDQ), etc. Their colored EDA complexes with many unsaturated or aromatic compounds mostly exhibit broad, unstructured CT bands with ε values between 500 and 2,000 $cm^2\,mmol^{-1}$. The rates of formation can be examined by spectrophotometry to determine the equilibrium constants and ε values. The *bathochromic shift* relative to the absorption of the components depends on the donor and acceptor strengths.

Fig. 1.41 Electronic transitions in CT complexes.

Table 1.19 EDA complexes **52** from chloranil (2,3,5,6-tetrachloro-1,4-benzoquinone) and benzene or its methyl derivatives (measured in cyclohexane)

Compound	Number of CH$_3$ groups	λ_{max} [nm]
Benzene	0	346
Toluene	1	365
m-Xylene (1,3-dimethylbenzene)	2	391
Mesitylene (1,3,5-trimethylbenzene)	3	408
Durene (1,2,4,5-tetramethylbenzene)	4	452
Pentamethylbenzene	5	476
Hexamethylbenzene	6	505

Table 1.19 lists the positions of the CT bands of the complexes **52** of chloranil and benzene and its methyl derivatives. The donor strength increases continuously with the number of methyl groups; accordingly, the CT bands are shifted to larger wavelengths.

52

1.4 Applications of UV/Vis Spectroscopy

In combination with other spectroscopic methods, UV/Vis spectroscopy can be a valuable method of qualitative analysis and structure determination. New applications for UV/Vis spectroscopy have been opened up by the advances in the study of electronically excited states (photophysics and photochemistry and their utilization in materials science).

Quantitative analysis (colorimetry, photometry), photometric titration, and determination of equilibrium and dissociation constants are other important applications. UV/Vis spectroscopy is also a valuable method in the increasingly important area of trace analysis. An example of photometry is the determination of alcohol in blood. It involves the enzymatic dehydrogenation of ethanol to acetaldehyde. The hydrogen is taken up by NAD (nicotinamide adenine dinucleotide). This conversion can be very accurately assessed by absorption measurements. For chromatographic methods like HPLC (high-performance liquid chromatography), measurement of the UV absorption is the commonest detection method. In addition to fixed-wavelength photometers, photodiode arrays can also be used, thus allowing the measurement of a complete UV spectrum at each time point in the chromatogram.

UV/Vis spectroscopy enables a quantitative analysis of multicomponent mixtures (**chemometrics**), when the absorption coefficients $\varepsilon(\lambda)$ of the components are known. If for example the percentages of the components in a mixture of three compounds are unknown, three absorption measurements at different wavelengths λ_1, λ_2, and λ_3 are sufficient. Lambert–Beer law gives then an inhomogenous, linear system of three equations for the three concentrations c_1, c_2, and c_3.

In general, errors of the measurements and eventually systematic errors allow only an approximate solution and not an exact solution of the system of equations. On the basis of a process, known in mathematics as linear problem of averaged squares, it is advantageous to use an overdetermined system, that means to use m measurements (λ_1, λ_2, ..., λ_m) for n quested concentrations (c_1, c_2, ..., c_n) of n components ($m > n$). For the thickness $d = 1$ cm, the system of linear equation can be written in matrix notation:

$$\bar{A} = \bar{\bar{\varepsilon}}\,\bar{c}$$

$\bar{A}$ is the vector of the measured absorbances (constant terms) A_i:

$$\bar{A} = \begin{pmatrix} A_1 \\ A_2 \\ \vdots \\ A_m \end{pmatrix}$$

$\bar{\bar{\varepsilon}}$ represents the matrix of the coefficients, whereby ε_{ij} is the absorption coefficient of component j ($j = 1 - n$) at λ_i ($i = 1 - m$):

$$\bar{\bar{\varepsilon}} = \begin{pmatrix} \varepsilon_{11} & \varepsilon_{12} & \cdots & \varepsilon_{1n} \\ \varepsilon_{21} & \varepsilon_{22} & \cdots & \varepsilon_{2n} \\ \vdots & & & \\ \varepsilon_{m1} & \varepsilon_{m2} & \cdots & \varepsilon_{mn} \end{pmatrix}$$

and $\quad \bar{c} = \begin{pmatrix} c_1 \\ \vdots \\ c_n \end{pmatrix}$

represents the vector of n concentrations (solutions).

The system of equations becomes considerably simpler, when certain λ_i can be selected, so that discrete $\varepsilon_{ij} = 0$.

UV/Vis spectroscopy has a special role as an analytical tool in kinetic measurements. Whereas the measurement of spectra of slowly reacting systems has no problem, and can even be performed in the reaction flask using a light conductor system, special methods are required for fast reactions. The whole spectrum has to be measured as quickly as possible, and stored digitally. Optical multichannel analyzers are used. The measurement beam is shone onto a grating monochromator and then onto a two-dimensional array of photodiodes (diode array). The specific location of each diode corresponds to a particular wavenumber. The information from the individual channels gives the total spectrum (single scan ca. 100 milliseconds).

Even faster spectroscopy is possible with a flash laser apparatus. The excitation flash is followed in rapid succession by measurement flashes to establish the photochemically generated intermediates. In this way lifetimes of excited states in the ns and ps ranges can be measured. Recent advances have extended the measurable range to the femto- and attosecond range (1 fs $= 10^{-15}$ s, 1 as $= 10^{-18}$) so that even nonstationary states can be observed.

In the following sections, three simple applications of UV/Vis spectroscopy will be described. UV/Vis spectroscopy can be used for the **determination of the pK value** of a medium strong acid like 2,4-dinitrophenol (**53**) (**Fig. 1.42**).

The **dissociation equilibrium** is described by

$$\text{Acid} + H_2O \leftrightharpoons H_3O^+ + \text{Anion}^-$$

$$A_{\text{tot.}} = d(\varepsilon_s \cdot c_S + \varepsilon_a \cdot c_{A^-}) = d \cdot \varepsilon \cdot c.$$

The total absorbance with a formal ε and the weighed concentration $c = c_S + c_{A^-}$ is caused by the absorbing species S and A with extinction coefficients ε_s and ε_a. Rearrangement gives

$$\frac{c_S}{c_{A^-}} = \frac{\varepsilon - \varepsilon_a}{\varepsilon_s - \varepsilon} \qquad \varepsilon_s \neq \varepsilon$$

The terms ε_s and ε_a are obtained from absorption measurements on dilute solutions in a strongly acidic or alkaline medium, where the concentrations of A$^-$ and S are negligible. The determination of ε is best performed using buffer solutions with intermediate pH values.

Fig. 1.42 shows three such curves. They all cross in an **isosbestic point**. At its wavelength λ_i the two interconvertible species absorbing there, S and A$^-$, have the same ε value. From the known pH value of the buffer solution:

$$pK = -\log\frac{c_{H_3O^+} \cdot c_{A^-}}{c_S}$$

$$pK = pH + \log\frac{c_S}{c_{A^-}}$$

$$pK = pH + \log\frac{\varepsilon - \varepsilon_a}{\varepsilon_s - \varepsilon}$$

For the determination of the pK value, the values determined at different wavelengths should be averaged. For 2,4-dinitrophenol (**53**), this gives p$K = 4.10 \pm 0.04$.

The second example shows a **reaction spectrum** for the photofragmentation of the heterocyclic spirane **54**. In acetonitrile, cyclopentanone (**55**) is formed quantitatively with a quantum yield of 57%.

The irradiation is monochromatic at $\lambda = 365$ nm, which is close to the long-wavelength absorption maximum of **54** ($\lambda_{\text{max}} = 367$ nm,

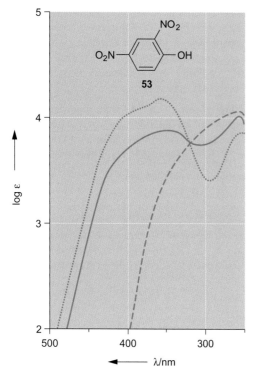

Fig. 1.42 pK value determination of 2,4-dinitrophenol (**53**) (from Flexer LA, Hammett LP, Dingwall A. J. Am. Chem. Soc. 1935; 57: 2103).
. . . . . . Solution of **53** in 0.1 molar sodium hydroxide.
———— Solution of **53** in 0.1 molar hydrochloric acid.
———— Solution of **53** in an acetate buffer of pH = 4.02.

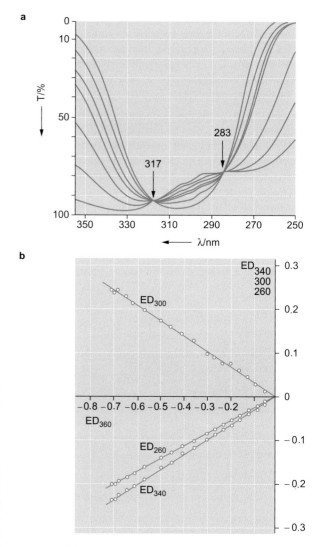

Fig. 1.43 **(a)** Reaction spectrum measured in % transmission for the photolysis **54** → **55** with monochromatic irradiation (λ = 365 nm) in acetonitrile; **(b)** corresponding extinction difference diagram (from Daniil D, Gauglitz G, Meier H. Photochem Photobiol 1977; **26**: 225).

log ε = 2.50). The reaction spectrum in **Fig. 1.43a** shows that this band is reduced in the course of the irradiation, i.e., the **transmission** there increases. At λ_1 = 317 nm this situation is reversed. In the range λ_1 = 317 > λ > 283 = λ_2, the transmission becomes less during the irradiation, since in this region the $n \rightarrow \pi^*$ transition of the product **55** builds up. At λ_2 there is a second reversal; λ_1 and λ_2 are the **isosbestic points** of this irreversible reaction. The appearance of isosbestic points shows the uniformity of the reaction. In particular, they rule out the possibility that the fragmentation takes place via an intermediate which increases in concentration and absorbs light. The uniformity of the reaction is often even more clearly shown by **extinction difference diagrams**. In such cases, $E(\lambda_m)$ must be a linear function of $E(\lambda_n)$; λ_m and λ_n can be any wavelengths in the absorption range. Instead of such **E diagrams** one can also construct an **ED diagram**.

Here the **extinction differences** $E(\lambda_m, t) - E(\lambda_m, t = 0)$ are plotted against the differences $E(\lambda_n, t) - E(\lambda_n, t = 0)$. In **Fig. 1.43b** ED ($\lambda$ = 340, 300, and 260) values are plotted as linear functions against ED (λ = 360). The reaction time is the parameter t. If

there are two or more independent (thermal or photochemical) reactions, the E or ED diagram is not linear.

The third example concerns the **determination of the polarity** of a medium by means of Reichardt's scale E_T (**30**). The so-called betaine-30 (**Fig. 1.44**), a pyridinium phenolate **56**, generates different colors in different solvents. The long-wavelength absorption maximum lies between 450 and 900 nm. The electronic excitation of **56** provokes an electron transfer from the phenolate ring to the pyridinium ring. Thus, the polarity of the dye molecule decreases on its way from the ground state S$_0$ to the excited state S$_1$. The energy of the corresponding **CT band**

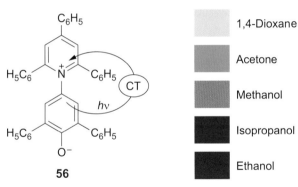

Fig. 1.44 The CT band of betaine-30 (**56**) generates different colors in different solvents.

1,4-Dioxane
Acetone
Methanol
Isopropanol
Ethanol

Table 1.20 E_T (**30**)-values of some solvents at 25 °C (calculated from the measured λ_{max}-values of betaine-30 (**56**))

Solvent	λ_{max} (nm)	E_T (30) (kcal · mol^{-1})
Carbondisulfide	872	32.8
Toluene	843	33.9
Diethylether	819	34.9
1,4-Dioxane	795	36.0
Ethylacetate	750	38.1
Trichloromethane	731	39.1
Dichloromethane	702	40.7
Acetone	677	42.2
Dimethylformamide	662	43.8
Dimethyl sulfoxide	634	45.1
Isopropanol	591	48.4
Propanol	564	50.7
Ethanol	550	51.9
Diethylene glycol	531	53.8
Methanol	515	55.4
Water [a]	453	63.1

[a] Very low solubility of **56** in water.

depends strongly on the polarity of the medium. Polar solvents lower the energy of the ground state S_0 much more than the energy of the S_1 state. That means polar solvents effect a hypsochromic shift of the CT band: **56** shows a **negative solvatochromism**.

The empirical polarity scale E_T (**30**) depends in a simple way on λ_{max}:

$$E_T (\mathbf{30}) = \frac{hc}{\lambda_{max}} N_A = \frac{28{,}591}{\lambda_{max}}$$

E_T (**30**) is obtained in kcal mol^{-1}, if Planck's constant h (6.63 × 10^{-34} J s), velocity of light (2.99 × 10^{10} cm s^{-1}), and λ_{max} (nm) are inserted. Due to the hyperbolic function, the statement can be made: the smaller the λ_{max} values are, the higher are the E_T (**30**) values, i.e., the higher is the polarity of the medium. The interpolation of E_T (**30**) values of two solvents is often nonlinear for mixtures of these solvents.

The color of solutions of **56** in particular solvents is always the complementary color in relation to the absorption range. Betaine-30 (**56**) has for example in acetone an absorption maximum at 677 nm (red light). Accordingly, the solution in acetone looks green (**Fig. 1.44**).

The E_T (**30**) values of some frequently used solvents at 25°C are compiled in **Table 1.20**. (E_T (**30**) values at higher temperatures are somewhat lower.) Betaine-30 is insoluble in apolar solvents such as *n*-hexane. Then a pyridinium phenolate can be used, which bears *tert*-butyl groups.

1.5 Derivative Spectroscopy

The recording of the first, second, or *n*th derivative of an absorption curve is an analytical aid, which has gained importance with the development of electronic differentiation. The mathematics of continuous, planar curves establishes the following relationships:

Absorption $A(\lambda)$	Maximum/ Minimum ↕	Inflection ↕
1st derivative $\frac{dA(\lambda)}{d\lambda}$	Zero crossing	Maximum/ Minimum ↕
2nd derivative $\frac{d^2A(\lambda)}{d\lambda^2}$		Zero crossing

Fig. 1.45 shows the long-wavelength part ($n \rightarrow \pi^*$ transition) of the UV spectrum of testosterone (**57**). The superimposition of the vibrational bands leads to a barely recognizable structure of the band. Above the first derivative is shown. The dashed line joins the absorption maximum with the zero crossing of the first derivative. The dotted lines join the inflection points in the left-hand part of the absorption curve with the extremes (maxima and minima) of the first derivative. The structure of the curve of the first derivative is much more obvious. The effect is even stronger in the curve of the second derivative, where zero crossings occur at the positions where extremes occur in the first derivative. Small changes in a spectrum, e.g., a shoulder, can be emphasized by derivative spectroscopy. The technique is also suited to the solution of difficult quantitative problems, e.g., in trace analysis, and in following the progress of reactions.

UV/Vis-Spektren

1. Der.

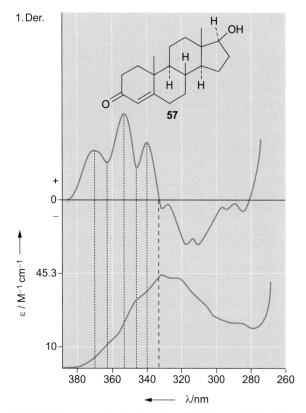

Fig. 1.45 Long-wavelength absorption of testosterone (**57**) in diethylene glycol dimethyl ether and the first derivative of the absorption curve (from Olson EC, Always CD. Anal. Chem. 1960;32: 370).

1.6   Chiroptical Methods

Chiroptical methods are optical measurements which depend on the **chirality** of the material under investigation. A substance is optically active if it rotates the plane of **linearly polarized light**. As can be seen in **Fig. 1.46**, this corresponds to the rotation of the vibrational direction of the electrical vector **E** of the light wave.

The optical rotation results either from a chiral crystal structure, as in quartz or cinnabar, or from the chirality of molecules (or ions). Of course, both may be involved. In camphor crystals,

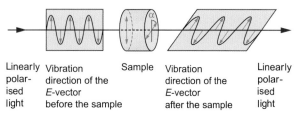

| Linearly polarised light | Vibration direction of the E-vector before the sample | Sample | Vibration direction of the E-vector after the sample | Linearly polarised light |

Fig. 1.46   Schematic representation of optical rotation.

for example, molecular and crystal effects overlap. A molecule (object) is chiral if it is not identical with its mirror image. This property requires that the molecule is asymmetric or only possesses symmetry elements which are **symmetry axes** C_n. **Planes of symmetry** σ or **rotation–reflection axes** S_n, including the **center of symmetry** $S_2 \equiv i$, must therefore not be present.

The **rotation angle** α measured with a **polarimeter** for a chiral compound in solution is given by the relationship:

$$\alpha = [\alpha]_\lambda^T \cdot l \cdot c$$

where α in degrees; l layer thickness in dm; c concentration in g mL^{-1}.

For the comparison of various optically active compounds, the rotation based on the molar mass M is often preferred:

$$[\phi]_\lambda^T = \frac{100\alpha}{l \cdot c} = \frac{[\alpha]_\lambda^T \cdot M}{100}$$

where ϕ, α in degrees; l layer thickness in cm; c concentration in mol· L^{-1}.

The **specific rotation** $[\alpha]_\lambda^T$ depends not only on the compound being measured, but also on the wavelength λ of the monochromatic radiation being employed and on the temperature T; α and $[\alpha]_\lambda^T$ have positive signs when the compound is **dextrorotatory**, i.e., when viewed against the light beam, **E** is rotated clockwise. The mirror-image isomer (enantiomer) is then **levorotatory** (counterclockwise) and has a **negative specific rotation** of the same magnitude.

The measured rotations can therefore be used for the determination of the enantiomeric purity. If, for example, the dextrorotatory form dominates in a mixture of enantiomers, then the following definitions apply:

Enantiomeric excess (%)

$$ee = E(+) - E(-)$$

Enantiomeric purity as the quotient

$$\frac{E(+) - E(-)}{E(+) + E(-)}$$

Optical purity

$$P = \frac{[\alpha]}{[\alpha]_{max}}$$

with $[\alpha]_{max}$ as the rotation of pure $E(+)$. The ratio $E(+)/E(-)$ is $1 + P/1 - P$, if the two enantiomers behave additively in the polarimetric measurement, otherwise optical purity and enantiomeric purity will not agree. Deviations from additivity are observed, for example, when there is association through hydrogen bonds.

Table 1.21 gives a collection of some specific rotations, measured in solution at 20°C with the sodium D line (589.3 nm).

The underlying chirality of **optically active compounds** is classified according to **chiral elements** (centers, axes, planes)—see a

Table 1.21 Specific rotations of some optically active compounds

Compound	Solvent	$[\alpha]_D^{20} \dfrac{\text{deg·cm}^2}{10\ g}$
R-Lactic acid (D-Lactic acid)	Water	−2.3
S-Alanine (L-Alanine)	Water	+2.7
S-Leucine (L-Leucine)	6 Molar hydrochloric acid	+15.1
	Water	−10.8
	3 Molar sodium hydroxide	+7.6
α-D-Glucose	Water	+112.2
β-D-Glucose	Water	+17.5
D-Glucose in solution equilibrium (mutarotation)	Water	+52.7
Sucrose	Water	+66.4
(1R,4R)-Camphor (D-Camphor)	Ethanol	+44.3
Cholesterol	Ether	−31.5
Vitamin D$_2$	Ethanol	+102.5
	Acetone	+82.6
	Chloroform	+52.0

Table 1.22 Specific rotation of [n]helicenes (**60**)

	Number of benzene rings n	Specific rotation $[\alpha]_D \dfrac{\text{deg.}\cdot\text{cm}^2}{10\ g}$
a	5	2,160
b	6	3,709
c	7	5,900
d	8	6,690
e	9	7,500
f	10	8,300
g	11	8,460
h	13	8,840

the absolute configuration: (+) corresponds to the (**P**)-helix and (−) to the (**M**)-helix.

60a–h (n = 5-11, 13)

textbook of stereochemistry. The commonest chiral element is the asymmetric carbon atom with four different ligands.

58

The H/D isotope effect is in principle sufficient for a measurable optical activity; thus 1-deuterio-1-phenylethane (**58**) has a specific rotation of 0.5 degree cm²/10 g.

Nevertheless, there are chiral compounds with $[\alpha] = 0$. An example is the enantiomerically pure 1-lauryl-2,3-dipalmityl glyceride (**59**). Although **59**, in contrast to the achiral 2-lauryl-1,3-dipalmityl glyceride, has no plane of symmetry, the difference between the 1-lauryl and the 3-palmityl group in relation to the chiral center at C-2 is too small to lead to an observable rotation! This is also the case for some chiral hydrocarbons.

59

Helicenes **60a–h** have extremely high specific rotations (**Table 1.22**). Their entire chromophore is responsible for their chirality. Moreover, their rotational direction is conformed to

For an understanding of optical rotation it is helpful to consider linearly polarized light as consisting of a **right-handed** and a **left-handed polarized wave** of the same amplitude and phase (**Fig. 1.47**). In an optically active medium, the two waves with opposite rotations have different velocities c (refractive indices n) and in the absorption region have different extinction coefficients ε as well. The case where $c_1 \neq c_r$ leads to a phase difference between the two waves and thus to a rotation of the **E** vector of the linearly polarized light which is reformed by combination of the two circularly polarized beams (**Fig. 1.47**).

The rotation angle depends on the wavelength used.

$$\alpha = \frac{180\,(n_1 - n_r)\,l}{\lambda_0}$$

where α rotation in degrees; l layer thickness; λ_0 vacuum wavelength in the same units of length; n_1, n_r refractive indices.

The **normal optical rotation dispersion** (ORD) $\alpha(\lambda)$ or $\phi(\lambda)$ is shown in **Fig. 1.48** for several steroids. Characteristic is the monotonous trend of the curves.

In the region of absorption bands, the normal ORD curve is overlaid with an S-shaped component to produce the so-called **anomalous ORD curve** (**Fig. 1.49**).

$|E_1| \neq |E_r|$ means that the **E** vectors of the two opposite circularly polarized light beams have different lengths after passing through the optically active medium due to differing absorption.

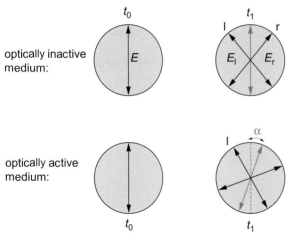

Fig. 1.47 Decomposition of the linearly polarized light beam into right- and left-handed circularly polarized beams. In an optically inactive medium $c_l = c_r$ and at any time point t the original vibration direction of the **E** vector is retained. In contrast, in an optically active medium where $c_r > c_l$ the vector **E**$_r$ of the right-handed wave has rotated at time t_1 by a larger angle than **E**$_l$. The resultant vibration direction shows a positive rotation α.

Their recombination produces an elliptical diagram (**Fig. 1.50**). The inclination of the ellipse is related to the optical rotation α.

If the end of the **E** vector runs through the ellipse in a clockwise direction, the **circular dichroism** (CD), $\Delta\varepsilon(\lambda)$, is said to be **positive**, if in a counterclockwise direction it is **negative**. Anomalous ORD and CD together make up the **Cotton effect**.

From the combination of positive or negative normal ORD curves with positive or negative Cotton effect, there are four types of anomalous ORD curves.

$\lambda_{Peak} < \lambda_{Trough}$ (**Fig. 1.49**) always implies a negative Cotton effect, $\lambda_P > \lambda_T$ a positive effect.

The extreme (maximum or minimum) of the CD curve lies at the same λ value as the crossing point of the anomalous and interpolated normal ORD curves (approximately the inflection point, **Fig. 1.49**). In simple cases this λ_{max} value corresponds roughly to the maximum of the normal UV/Vis absorption (see the steroids **62** and **63** in **Fig. 1.51**).

Instead of $\Delta\varepsilon(\lambda)$ the **molar ellipticity** $[\Theta]_M$ is often recorded as a function of the wavelength. The ellipticity Θ itself is defined as the angle the tangent of which is equal to the quotient of the smaller and larger half axes of the ellipse (**Fig. 1.50**). In analogy to the specific rotation, the **specific ellipticity** is defined as:

$$\Theta = [\Theta]_\lambda^T \cdot c \cdot l$$

c = concentration in g·mL^{-1}

l = layer thickness in dm.

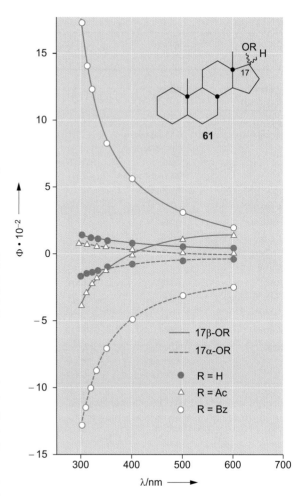

Fig. 1.48   Normal ORD curve of 5 α-androstanes **61** substituted at the C-17 position (from Jones PM, Klyne W. J. Chem. Soc. 1960; 871).

The **molar ellipticity** is then:

$$[\Theta]_M = \frac{\Theta \cdot M}{100 \cdot c \cdot l} = \frac{[\Theta]_\lambda^T \cdot M}{100}$$

M = molecular mass

A simple relationship between $\Delta\varepsilon = \varepsilon_l - \varepsilon_r$ and $[\Theta]_M$ can be derived. When c is given in mol·L^{-1}, l in cm, and ε in L·mol^{-1}·cm^{-1}, we obtain:

$$[\Theta]_M = 3,300\ \Delta\varepsilon$$

whereby the molar ellipticity has the dimensions deg·cm^2·dmol^{-1}.

The ellipticity is, like the optical rotation, suitable for determining enantiomeric purity. For the analytical interpretation of the Cotton effect with regard to structural information, there are several rules: theoretical, semiempirical, and purely empirical. One such is the octant rule for saturated ketones, which have an $n \rightarrow \pi^*$

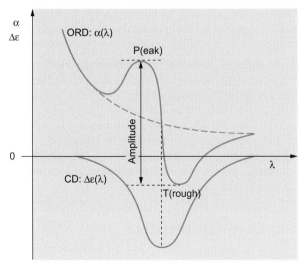

Fig. 1.49 Cotton effect—relationship between ORD and CD curves; in this example the CD is negative and the normal ORD (dashed line) positive.

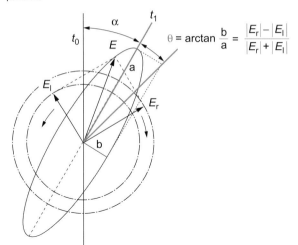

Fig. 1.50 Addition of the electric field vectors E_l and E_r after passage through an optically active medium with $n_l > n_r$ (i.e., $c_l < c_r$) in an absorption range $\varepsilon_l > \varepsilon_r$ (i.e., $|E_l| < |E_r|$).

transition at ~280 nm. The three nodal planes of the n and π^* orbitals divide the space into eight octants, which can be represented as the octants of a Cartesian x, y, z coordinate system. The four octants with positive y values are shown in **Fig. 1.52a**.

The xy plane corresponds to the σ-bonding plane of the carbonyl function and the carbonyl C atom is assumed to lie on the positive side of the y-axis. In the two octants shown in blue, the Cotton effect has a positive sign (looking from O to C, upper left and lower right), in the other two octants a negative sign. If a cyclohexanone structure is introduced into this coordinate system, as in **Fig. 1.52b**, then the two substituents at C-4 lie in the

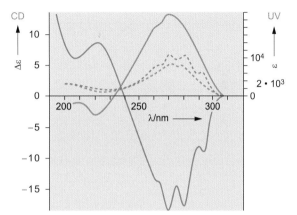

Fig. 1.51 CD curves and UV absorptions of ergosterol (**62**) and lumisterol (**63**).

yz plane and the *equatorial* substituents at C-2 and C-6 approximately in the xy plane; these make no contribution to the Cotton effect. Positive Cotton effects are produced by the *axial* C-2 substituents and by the *axial* and *equatorial* substituents at C-5; negative Cotton effects, on the other hand, by the *axial* C-6 and *axial* and *equatorial* substituents at C-3. As a matter of course, it must be remembered that only chiral cyclohexanone derivatives need be considered. For other types of compounds similar rules can be established. The reader should refer to the literature quoted in the bibliography. As just one more application, the determination of the secondary structure of polypeptides can be mentioned. For the example of the peptide constructed from L-(+)-lysine, **Fig. 1.53** shows the differentiation between the α-helix, β-pleated sheet, and random coil structures.

In general it can be established that polarimetry is useful for the determination of concentration or purity or, as in carbohydrate chemistry, for the study of rearrangement processes

a

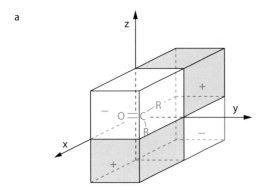

b

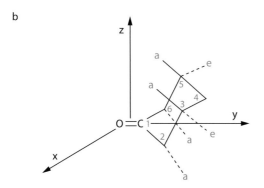

Fig. 1.52 Illustration of the octant rule for a saturated ketone (cyclohexanone derivative).

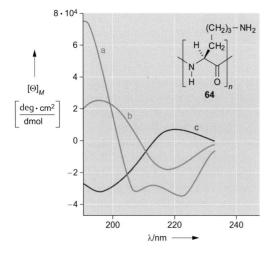

Fig. 1.53 Measurement of the molar ellipticity $|\theta|_M$ for the determination of the secondary structure of the peptide **64** (from Greenfield N, Fasman GD. Biochemistry 1969;8:4108).
(a) α-Helix
(b) β-Configuration (pleated sheet)
(c) Random coil
(The molar mass M refers to the building block of the biopolymer).

(mutarotation, inversion), whereas **ORD and CD spectra** provide valuable information about structure by characterizing absolute configurations, particularly in natural product chemistry.

At the end of this section it should be pointed out that substances in an external magnetic field are always optically active (**Faraday effect**). The vibrational plane of linearly polarized light, parallel to the magnetic lines of force, will be rotated by substances which are normally optically inactive. Thus magnetic optical rotatory dispersion (**MORD)** and magnetic CD (**MCD**) measurements can be a useful supplement to **ORD** and **CD** measurements. Literature references to further information on these techniques are given.

Supplementary Literature

UV/Vis spectroscopy

Books

Andrews DL. Applied Laser Spectroscopy, VCH, Weinheim; 1992

Bernath PF. Spectra of Atoms and Molecules. New York, NY: Oxford University Press; 2005

Clark BJ, Frost T, Russell MAUV. Spectroscopy. London: Chapman & Hall; 1993

Ewing GW. Instrumental Methods of Chemical Analysis. New York, NY: McGraw-Hill Book Comp; 1975

Fabian, J., Hartmann, H. Light Absorption of Organic Colorants. Berlin: Springer Verlag; 1980

Gauglitz G. Praxis der UV/Vis Spektroskopie. Tübingen: Attempto Verlag; 1983

Gauglitz G, Vo-Dinh T. Handbook of Spectroscopy. Weinheim: Wiley-VCH; 2003

Gorog S. Ultraviolet-Visible Spectrophotometry in Pharmaceutical Analysis. CRC, Boca Raton; 1955

Griffiths J. Colour and Constitution of Organic Molecules. New York, London: Academic Press; 1976

Jaffé HH, Orchin M. (1962), Theory and Applications of Ultraviolet Spectroscopy. New York, NY: Wiley; 1962

Klessinger M, Michl J. Lichtabsorption und Photochemie Organischer Moleküle. Weinheim: Verlag Chemie; 1989

Knowles C, Knowles A. Practical Absorption Spectrometry. London: Chapman & Hall; 1983

Maass DH. An Introduction to Ultraviolet Spectroscopy with Problems. In: Scheinmann F, ed. An Introduction to Spectroscopic Methods for the Identification of Organic Compounds. Vol. 2, New York, NY: Pergamon Press; 1973

Merkt F. Handbook of High-Resolution Spectroscopy. 3rd ed. New York, NY: J. Wiley & Sons; 2001

Murell JN. Elektronenspektren organischer Moleküle. Bibliographisches Institut 250/250a*, Mannheim; 1967

Olsen ED. Modern Optical Methods of Analysis. New York, NY: McGraw-Hill Book Comp.; 1975

Parikh VM. Absorption Spectroscopy of Organic Molecules. Reading: Addison-Wesley; 1974

Parker CA. Photoluminescence of Solutions. Amsterdam: Elsevier; 1968

Perkampus H-H. UV-Vis-Spektroskopie und ihre Anwendungen. Berlin: Springer; 1986

Perkampus H-H. Encyclopedia of Spectroscopy. Weinheim: VCH; 1995

Samson JA. Vacuum Ultraviolet Spectroscopy. San Diego: Academic Press; 2000

Schmidt W. Optische Spektroskopie. Weinheim: Verlag Chemie; 2005

Schmidt W, Redmond RW. Optical Spectroscopy in Life Sciences and Chemistry. Chichester: J. Wiley & Sons; 2003

Schulman SG. Molecular Luminescence Spectroscopy. New York, NY: Wiley; 1993

Sharma A, Schulman SG. Introduction to Fluorescence Spectroscopy. New York, NY: Wiley; 1999

Snatzke G. Elektronen-Spektroskopie, In: Korte F, ed. Methodicum Chimicum. Vol. 1/1. Stuttgart: Georg Thieme Verlag; 1973

Talsky G. Derivative Spectrophotometry of First and Higher Orders. Weinheim: VCH; 1994

Thompson CC. Ultraviolet-Visible Absorption Spectroscopy. Boston: Willard Grant Press; 1974

Thomas M. Ultraviolet and Visible Spectroscopy. Chichester: J. Wiley & Sons; 1996

Valeur B. Molecular Fluorescence. Weinheim: Wiley-VCH; 2001

Zollinger H. Color Chemistry. Weinheim: Wiley-VCH; 2003

Series

UV Spectrometry Group, Techniques in Visible and Ultraviolet Spectrometry, Chapman & Hall, London.

Data Collection, Spectral Catalogues

- Hershenson, HM. Ultraviolet and Visible Absorption Spectra. New York, NY: Academic Press.

- A.P.I. Research Project 44: Ultraviolet Spectral Data, Carnegie Institute and U.S. Bureau of Standards.

- Phillips, JP., Feuer, H., Thyagarajan, BS. (u.a.), Organic Electronic Spectral Data. New York, NY: Wiley.

- Pestemer, M., Correlation Tables for the Structural Determination of Organic Compounds by Ultraviolet Light Absorptiometry. Weinheim: Verlag Chemie.

- Lang, L., Absorption Spectra in the Ultraviolet and Visible Region. New York, NY: Academic Press.

- Perkampus H-H. UV-VIS Atlas of Organic Compounds. Weinheim VCH; 1992.

- UV-Atlas organischer Verbindungen. Weinheim: Verlag Chemie.

- Sadtler Standard Spectra (Ultraviolet). London: Heyden.

Chiroptical Methods

Monographs

Barron L. Molecular Light Scattering and Optical Activity. Cambridge: Cambridge University Press; 2004

Berova N. Polavarapu, PL. Nakanishi, K., Woody, R.W. Comprehensive Chiroptical Spectroscopy. Hoboken, NJ: J. Wiley & Sons; 2012

Caldwell DJ, Eyring H. The Theory of Optical Activity. New York, NY: Interscience; 1971

Charney E. Molecular Basis of Optical Activity: Optical Rotatory Dispersion and Circular Dichroism. New York, NY: Wiley; 1979

Crabbé P. An Introduction to the Chiroptical Methods in Chemistry. Mexico City: Syntex; 1971

Crabbé P. ORD and CD in Chemistry and Biochemistry –An Introduction. New York, London: Academic Press; 1972

Djerassi C. Optical Rotary Dispersion. New York, NY: McGraw-Hill Book Comp;1964

Fasman GD. Circular Dichroism and the Conformational Analysis of Biomolecules. New York, NY: Plenum; 1996

Harada N, Nakanishi K. Circular Dichroic Spectroscopy. New York, NY: University Science Books; 1983

Kobayashi N., Muranaka A., Mack J. Circular Dichroism and Magnetic Circular Dichroism Spectroscopy for Organic Chemists. London: RSC Publ.; 2012

Lightner D, Gurst JW. Organic Conformational Analysis and Stereochemistry from Circular Dichroism Spectroscopy. New York, NY: Wiley; 2002

Mason SF. Molecular Optical Activity and the Chiral Discriminiations. Cambridge: University Press; 1982

Michl J, Thulstrup EW. Spectroscopy with Polarized Light. Weinheim: VCH; 1986

Nakanishi K, Verova N, Woody RW. Circular Dichroism: Principles and Applications. Weinheim: VCH; 1994

Norden B, Rodger A, Dafforn T. Linear Dichroism and Circular Dichroism. London: RSC Publ., 2010

Olsen ED. Modern Optical Methods of Analysis. New York, NY: McGraw-Hill Book Comp; 1975

Purdie N, Brittain HG. Analytical Applications of Circular Dichroism. Amsterdam: Elsevier; 1994

Rodger A, Norden B. Circular Dichroism and Linear Dichroism. Oxford: Oxford Univ. Press; 1997

Snatzke G. Optical Rotary Dispersion and Circular Dichroism in Organic Chemistry. Heyden, Canada; 1967

Thulstrup EW. Aspects of the Linear and Magnetic Circular Dichroism of Planar Organic Molecules. Berlin: Springer; 1980

Velluz L, Legrand M, Grosjean M. Optical Circular Dichroism. Weinheim: Verlag Chemie; 1965

UV/Vis-Spektren

2

Thomas Fox

Infrared and Raman Spectra

2 Infrared and Raman Spectra

2.1 Introduction

Molecular vibrations and rotations can be excited by absorption of electromagnetic radiation in the infrared (IR) wavelength range. This adjoins the low-frequency edge of visible light and is also called *thermal radiation* as it can be sensed by the skin in the form of heat. There are two different ways to measure molecular vibrations:

- Directly as absorption in a classical IR spectrum.
- Indirectly as absorbed and *r*eemitted radiation by *Raman* spectroscopy.

Formerly, the position of IR absorption bands was expressed in *wavelength* units (λ, μm) of the absorbed light (1 μm = 10^{-3} mm = 10^4 Å). The absorptions of organic molecules cover a typical wavelength range between λ = 2.5 and 15 μm.

Nowadays instead of wavelengths, their reciprocal *wavenumber* (number of wave maxima per centimeter) is mostly used to express absorption frequencies:

$$\text{wavenumber } \tilde{\nu} = \frac{1}{\lambda} \, [\text{cm}^{-1}]$$

The following relationship applies for the conversion of wave**lengths** into wave**numbers**:

$$\text{wavenumber } \tilde{\nu} \, [\text{cm}^{-1}] = \frac{10^4}{\text{wavelength } \lambda \, [\mu\text{m}]}$$

Wavenumbers are especially useful because they are directly proportional to the frequency ν of the absorbed radiation and, therefore, to the excitation energy ΔE of the appropriate molecular vibration:

$$\lambda \cdot \nu = c$$
$$\nu = \frac{c}{\lambda} = c \cdot \tilde{\nu}$$
$$\Delta E = h \cdot \nu = \frac{h \cdot c}{\lambda} = h \cdot c \cdot \tilde{\nu}$$
$$\Delta E \sim \tilde{\nu}$$

- c speed of light ($3 \cdot 10^{10}$ cm·s^{-1})
- h Planck constant ($6.626 \cdot 10^{-34}$ J·s)
- ν frequency (Hz or s^{-1})
- λ wavelength (cm)
- $\tilde{\nu}$ wavenumber (cm^{-1})

Typical IR spectra cover a wavenumber range between 4,000 and 400 cm^{-1}.

The functional groups of organic molecules show characteristic vibrations which are related to appropriate absorption bands in a defined IR range. Such vibrations are usually limited to the location of a given functional group without significant interference by other parts of the molecule. This allows the identification of functional groups on the basis of the wavenumber of their absorption bands. This fact, together with a facile handling, makes IR spectroscopy the easiest, quickest, and highly reliable method of assigning a substance to a particular class of compounds. Usually one can check at once, whether an alcohol, amine, or ketone, or, for example, some aliphatic or aromatic compound is present. Frequencies, intensities, and shapes of absorption bands bear information about the presence and position of particular functional groups in a molecule. A huge and continuously increasing collection of IR reference spectra is available in the literature and computerized databases, facilitating the specification of compounds on the basis of their IR spectra.

2.2 Basic Principles

For an understanding of the basic processes which lead to an IR spectrum, the simple model of a classical harmonic oscillator can be regarded. If atoms are considered as point masses, the vibration of a diatomic molecule (e.g., HCl) can be described as shown in **Fig. 2.1**. The molecule consists of the masses m_1 and m_2 which are joined by an elastic spring (**a**). An enforced elongation of the equilibrium distance r_0 between the masses by an amount $x_1 + x_2$ (**b**) generates a restoring force K which lets the system vibrate about the equilibrium distance after release of the external stretching force.

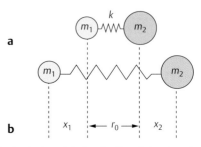

Fig. 2.1 Mechanical model of a vibrating diatomic molecule (stretching $\Delta r = x_1 + x_2$).

According to Hooke's law, the restoring force K is to a first approximation proportional to the stretching Δr:

$$K = -k \cdot \Delta r$$

The relation contains a negative sign since the restoring force K opposes the stretching force ($k \cdot \Delta r$). In the mechanical model, the factor k equals the elasticity constant of the spring. On the molecular level, it is the so-called *force constant* which directly correlates with the bond strength between the atoms and is measured in [N·cm^{-1}].

The potential energy $V(r)$ of the harmonic oscillator described above is a function of the distance r between the oscillating masses m_1 and m_2 and is quantized for *molecular* vibrations (**Fig. 2.2**).

$$V(\Delta r) = \frac{1}{2} k \cdot \Delta r^2 = 2\pi^2 \cdot \mu \cdot \nu_{osc}^2 \cdot \Delta r^2$$

V potential energy

Δr deflection $= x_1 + x_2$ (**Fig. 2.1**)

k force constant

ν_{osc} oscillator frequency

μ reduced mass $= \dfrac{m_1 \cdot m_2}{m_1 + m_2}$

From the above equation for the potential energy, $V(r)$ of a two-piece oscillator follows directly its vibration frequency:

$$\nu_{osc} = \frac{1}{2\pi} \sqrt{\frac{k}{\mu}}$$

The oscillation frequency ν_{osc} increases, therefore, with the force constant k or the bond strength. On the other hand, ν_{osc} decreases with an increasing reduced mass μ of the oscillator.

Conversion of the oscillator frequency ν_{osc} via substitution for the wavenumber and expression of the reduced mass μ in terms of atomic mass units yields a useful relation for the force constant k:

$$k = \frac{m_1 \cdot m_2}{m_1 + m_2} \left[\frac{\tilde{\nu}}{1303} \right]^2 [\text{N·cm}^{-1}]$$

This provides a direct correlation between the measured wavenumber of a diatomic molecule and the associated binding force k. Hydrochloric acid (H^{35}Cl) absorbs at a wavenumber of 2,887 cm^{-1}. This value, in conjunction with the atomic masses $m_1 = 1$ and $m_2 = 35$ yields a binding force of $k = 4.78$ N·cm^{-1} for the H–Cl bond.

The equation for a diatomic oscillator is not applicable for larger molecules, but for the latter there exist analogous relations between vibration frequency and binding force. For example, carbon–carbon bonds follow the order:

$$k_{C=C} > k_{C=C} > k_{C-C}$$

This allows a qualitative estimation of expected absorption frequencies (absorption band positions) of carbon–carbon vibrations in the IR spectrum.

For a more detailed description of *molecular* vibrations, the model of the *classical harmonic* oscillator is not suitable. First of all, it does not consider secondary effects which occur after absorption of higher amounts of energy resulting in larger vibrational amplitudes and, hence, in bond weakening. As a consequence, the frequencies of higher molecular vibration modes are smaller as predicted by the model of the *harmonic* oscillator. The more precise model of an *anharmonic* and *nonclassical* oscillator considers these circumstances. It exhibits asymmetric potential diagrams due to the weaker binding forces of stretched bonds ($r > r_0$), accompanied by smaller energy efforts for *further* bond expansion (**Fig. 2.3**). On the left side of the graph ($r < r_0$) the potential energy of the oscillator increases rapidly due to strong repulsive forces between the atom cores.

This model of a *nonclassical* oscillator also considers the absorption of *quantized* energy amounts leading to vibrational modes with discrete energies E, characterized by quantum numbers n. Up to room temperature, molecules basically perform vibrations

Infrared and Raman Spectra

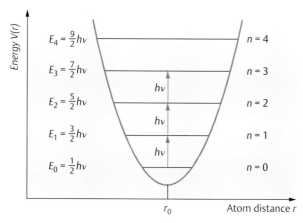

Fig. 2.2 Potential diagram of a molecular harmonic oscillator with discrete energy levels E_n.

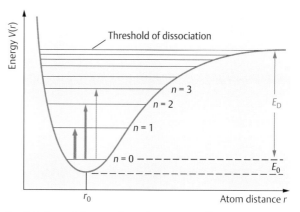

Fig. 2.3 Potential curve of an anharmonic oscillator; E_0, zero-point energy; E_D, dissociation energy (arrow thickness correlates to transition probability).

at the energetic ground state (zero-point energy $E_{n=0}$), which can be understood as oscillating deviations from the atom equilibrium distances (δr_0) according to Heisenberg's uncertainty principle.

The excitation energy of a vibration ΔE_{vib} is equal to the energy difference $E_{n+1} - E_n$ between two neighboring vibration states n and $n + 1$.

According to the Schrödinger equation, it results:

$$E_{vib} = h\nu_{osc} \left[n + \frac{1}{2}\right] = \frac{h}{2\pi} \sqrt{\frac{k}{\mu}} \left[n + \frac{1}{2}\right]$$

E_{vib} vibration energy
$\Delta E_{vib} = E_{n+1} - E_n = h\nu_{osc}$
n vibration quantum number; n = 0, 1, 2, ...
h Planck constant

The excitation of a bond vibration due to absorption of a light quantum lifts the molecule from a vibrational state n to the next higher state, $n + 1$. The energy difference of the involved vibration states equals the energy of the light quantum (resonance condition). Due to the anharmonicity of a molecular oscillator, the energy distances between neighbored vibration modes decrease with larger values of n and approach a continuum close to the edge of the bond dissociation energy.

Transitions between n = 0 and n = 1 are according to the excitation of *base vibrations*. However, a double quantum transition from n = 0 to n = 2 is called the *first overtone*, that one from n = 0 to n = 3 *second overtone*, etc., with frequencies always being slightly below the double, triple, etc., of the appropriate base vibration. The probabilities of double and multiple quantum transitions are very low with respect to single quantum transitions and decrease with increasing transition order. The intensities of the resulting absorption bands are accordingly weak but may significantly swell up for vibrations with distinct anharmonic behavior. This is to a large extent the case for vibrations between light atoms which exhibit especially large amplitudes and appropriate bond weakening.

Besides the resonance condition, which postulates equal energies of both absorbed light quantum and excited molecular vibration, there is another requirement for the excitation of vibrations: only those vibrations which induce periodical changes of molecular dipole moments can directly be excited by IR radiation. This, together with the resulting band intensities, depends on the symmetry of the molecule and its change during the vibration (see Section 2.15.2, Selection Rules).

2.3   Infrared Spectrometer

There are, in principle, two types of IR spectrometers featuring basically different data acquisition techniques: the traditional grating or prism (scanning) instruments, which are still in use,

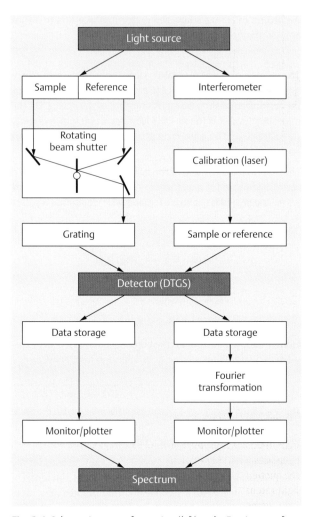

Fig. 2.4 Schematic setup of a grating (*left*) and a Fourier transform (*right*)-IR spectrometer.

and the more powerful Fourier transform (FT)-IR spectrometers (**Fig. 2.4**), which have largely replaced the former.

Both of them contain an IR light source whose radiation passes through the sample with loss of intensity due to excitation of molecular vibrations at appropriate wavelengths. In the classical IR spectrometer, the frequency-dependent intensity changes behind the sample are registered by a detector and directly stored as intensity-against-frequency plots. The FT-IR spectrometer, in contrast, acquires a primary *time*-dependent intensity interferogram which is converted into a *frequency*-dependent spectrum by subsequent *Fourier transformation* (**Fig. 2.5**).

IR sources have to comply with the characteristics of a Planck radiator, i.e., they need to emit light with a continuous frequency spectrum. According to Planck's radiation chart, the

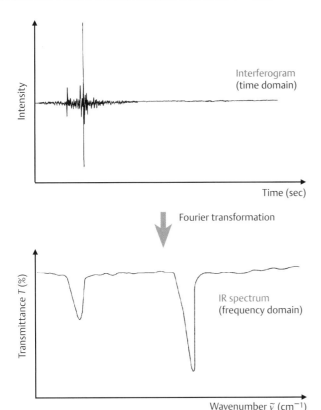

Fig. 2.5 From the time domain to the IR spectrum.

light intensity is proportional to the emitter temperature and decreases with the wavelength. As a consequence of the latter, the quotient of IR radiation is marginal with respect to the total light intensity. Therefore, the light source must function at as high a temperature as possible. Zirconium dioxide–based *Nernst rods* (working temperature 1,800 K) or, alternatively, more robust silicon carbide bars (*Globar*, 1,500 K) are mostly used.

The detector gathers the incoming radiation after passing through the sample and transforms it into electrical impulses. This is either a photodetector with the output voltage proportional to the light intensity or a thermal *DTGS* (***d**euterated **t**rigly-cine **s**ulfate*) detector based on the pyroelectric principle where the output signal is a function of temperature.

2.3.1 Classical (Scanning) Infrared Spectrometers

These instruments operate on the double beam principle: a beam splitter (*chopper*) divides the continuous radiation from the source into two beams of equal intensity. One of the beams is passed through the sample, and the other one serves as a control beam which passes a reference channel with an empty sample carrier. For gases the reference channel contains an

evacuated probe cell, for solutions a cell with the pure solvent, and for solid substances a pellet of the appropriate matrix material. For optical comparison, the sample and the reference beams are directed alternately onto a monochromator (prism or diffraction grating) by the aid of a fast-rotating mirror. The monochromator splits the alternating reference/sample beam into its spectral components, which are registered successively at the detector. Wavelengths which are not absorbed by the sample are registered at a constant intensity leading to a steady detector output voltage. In contrast, at wavelengths which are weakened in the sample channel due to absorption, the radiation intensity at the detector oscillates with the frequency of the rotating mirror. This induces an alternating voltage at the detector poles which is electronically amplified and registered as an absorption band. From the beginning of IR spectroscopy procedures, the detector impulses have been directly translated via a plotter onto a paper (*abscissa*: wavenumber, increasing from right to left; *ordinate*: transmittance in percentage). More contemporary IR instruments contain analog-to-digital converters which digitize the primary analog detector output for further computerized spectrum processing. The acquisition of a spectrum with a typical wavenumber range of 400 to 4,000 cm^{-1} takes about 10 minutes with this conventional method.

2.3.2 Fourier Transform Spectrometer

The FT technique represents a development in IR spectroscopy, combining the potentials of a modern computer and laser technology. It is now the established standard method and has almost completely superseded conventional IR spectrometers. Its advantage with respect to the *scanning* method is the tremendous time saving due to large sensitivity enhancement. This is achieved by simultaneous collection of all frequencies at the detector *at once* instead of time-consuming consecutive scanning of discrete wavelengths. This requires a totally different spectrometer design and the use of an interferometer which converts the original source light via (*constructive* and *destructive*) interference into light with a time-dependent oscillating intensity (interferogram, *time domain*, **Fig. 2.5**, top).

The central device of an FT-IR spectrometer is the Michelson interferometer (**Fig. 2.6**). After passing an aperture, the original light of the radiation source encounters a semitransparent splitting plate where it is split into two beams. One of them is first deflected by the plate before passing through it after reflection at a *fixed* mirror. The second beam passes the plate and is reflected by a *shifting* mirror (*time-dependent* position) which induces a *time-dependent* pathway difference of the beams after their recombination behind the plate. Thus, the continuous shifting of the drifting mirror generates a time-dependent mutual

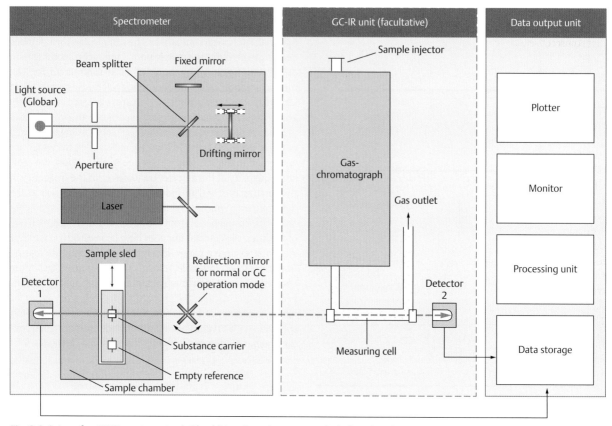

Fig. 2.6 Setup of an FT-IR spectrometer (with additional gas chromatography/infrared unit).

decrease and increase of the interfering light beams. The intensity pattern of the resulting interferogram is a function of the spectral composition of the light and the speed of the drifting mirror ($mm·s^{-1}$).

Such premodulated radiation undergoes characteristic changes during its passage through light-absorbing matter due to the excitation of molecular vibrations and the loss of appropriate frequency components. The output behind the sample is registered and then electronically subtracted from the interferogram of an empty reference sample. The result is converted into a band spectrum (*frequency domain*) with familiar appearance by aid of a mathematical operation, the *Fourier transformation* (**Fig. 2.5**).

Fourier transformation is a highly complex mathematical algorithm and required a good portion of measuring time at the beginning of FT-IR spectroscopy. Nowadays, data processing is straightforward and FT-IR measurements take just those few seconds which are necessary for the preparation of a time-dependent interferogram. The highly precise adjustments within a Michelson interferometer are handled with lasers which allow a mirror alignment on the micrometer scale for the generation of exactly defined light-pathway differences.

Compared to the conventional techniques, FT-IR spectroscopy offers three advantages:

1) *Considerable saving of time:* Since the total frequency range of the light source is registered simultaneously, the FT-IR acquisition time amounts only to a few seconds instead of several minutes for the scanning method (*multiplex* or *Fellgett advantage*).

2) *Better signal-to-noise ratio:* In contrast to the scanning technique where discrete wavelengths with low individual intensities are detected successively, the FT-IR detector is hit by the whole power of the light source all along (*Jacquinot advantage*).

3) *High wavenumber precision:* Monochromatic laser light with an exact known frequency is mixed with the output signal via a laser splitter. This allows a highly precise internal spectrum calibration (*Connes advantage*).

Furthermore, FT-IR spectrometers are single-beam instruments and this renders the error-prone use of separate sample and reference channels unnecessary. Instead, the substance and reference are placed separately into a sample carriage which forwards them consecutively into the beam of the Michelson interferometer.

2.4   Sample Preparation

IR spectra can be obtained for gaseous, liquid, and solid materials, in pure or diluted form. The choice of the method is based on the physical and chemical properties, as well as a possible air and moisture sensitivity of the sample. Also, some practical aspects may be decisive, like the efficiency and sensitivity of the measurement or the informative value of the resulting spectrum. In any case, it must be considered that both the position and the shape of IR absorptions (and Raman emissions, see Section 2.15, Raman Spectroscopy) sometimes depend significantly on the type of the sample. Due to different molecular interactions, for example, in the liquid or solid state, the appropriate IR spectra of one and the same compound can differ considerably. This applies particularly to functional groups which are part of bridging hydrogen bonds. The latter are also the reason for the band broadening effect of polar solvents, with corresponding negative influences on the spectral resolution.

There are two approaches to measure the absorption behavior of materials: via *transition* across the light-absorbing medium or via *reflection* at its surface. In spite of their different instrumental setups, both methods provide equivalent results, i.e., spectra, which represent the IR absorption property of a sample as a function of the light frequency. One notable advantage of the newer reflection method is the easy handling of the system, because it quite often allows instant measurements without special sample preparation (see Section 2.4.2, Reflection Measurements). Nevertheless, the classical transition technique is still used intensively, with regard to the manifold of scientific issues and different sample compositions.

2.4.1   Sample Preparation for Measurements in Transmission

a)   Measurement in the Gaseous Phase

Gases or liquids with a high vapor pressure are transferred into a tightly closable glass cuvette which is enclosed at both ends by IR-transparent plates. Due to the low density of the absorbing medium, its layer thickness amounts typically to a few centimeters. Nowadays, measurements in the gaseous phase are carried out particularly using IR spectrometers in combination with a gas-phase chromatograph (*gas chromatography/infrared [GC/IR] coupling*, **Fig. 2.6**). This setup allows gas chroma-

tographic separation of substance mixtures and instantaneous transfer of the isolated fractions with an inert gas stream into the IR measuring cell to register the individual spectrum of each sample component.

Measurements in the gaseous phase can also be favorable if disturbance of *inter*molecular interaction has to be minimized for improved spectral resolution. Small molecules show, for example, in the gaseous phase absorption bands with *rotational fine-structures* which are normally not resolvable in the condensed phase. This resolution enhancement is achieved at the cost of absorptions which are based on *inter*molecular interactions and are, therefore, weakened or even totally suppressed in the spectra of molecules in the gaseous phase.

b)   Measurement of Undissolved Liquids

In the easiest case, one drop of the pure liquid is placed between two compressed sodium chloride (NaCl) plates in the beam path of the spectrometer. NaCl is transparent between 4,000 and 667 cm^{-1} and, therefore, suitable for basic routine measurements. Depending on the absorption properties of the liquid, the distance between the NaCl plates and, correspondingly, the thickness of the liquid layer can be varied with the aid of spacers.

If the water content of the sample exceeds 2%, sample carriers with less water sensitivity such as calcium fluoride plates must be used alternatively. Cloudiness and bubbles in the liquid should always be avoided, because they give rise to internal scattering and reflections of the IR radiation, which are the main sources of disturbance in background absorption.

c)   Measurements in Solution

These are performed if band-broadening intermolecular interactions or extremely strong absorptions of the sample have to be reduced by dilution. For this purpose, a solution of 1 to 5% of the compound in a nonpolar solvent such as tetrachloromethane, chloroform, cyclohexane, or carbon disulfide is prepared and then measured in a sodium chloride cell against a reference containing the pure solvent. In exceptional cases aqueous solutions may also apply, which can be measured in cells made up of water-resistant materials such as calcium fluoride.

For a proper choice of the solvent, its self-absorption also has to be considered which should not overlap with the wavenumber range of interest (**Fig. 2.7**).

d)   IR Measurements of Solid Compounds

These are carried out either with a suspension of the powdered sample in oil or with pellets made of a pulverized mixture of the solid compound with an alkali halide (mostly potassium bromide [KBr]).

Suspensions in oil are mainly prepared for air- and moisture-sensitive substances. For this purpose, about 1 mg of the solid

Infrared and Raman Spectra

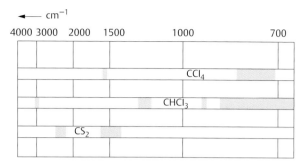

Fig. 2.7 Self-absorption of some common solvents (cf. **Fig. 2.15**, IR spectrum of CHCl₃).

compound is finely comminuted together with one drop of paraffin oil (*Nujol*) in an agate mortar. The thus obtained and preferably bubble-free paste is pressed between two sodium chloride plates. For the observation of (C−H) vibrations, the paraffin oil must be replaced by long-chained perchlorinated or perfluorinated carbon compounds which are commercially available for these spectroscopic purposes.

For the preparation of a KBr pellet, the substance is finely comminuted with a 10- to 100-fold quantity of KBr in an agate mortar and then compressed with the aid of a hydraulic press. This converts (*sinters*) the primarily pulverulent material into a glass-like pellet. Too coarse or too fine milling is unfavorable for the sintering process within the pellet, with the consequence of light scattering losses which are noticeable in terms of an increasing tilt of the baseline to the left (light scattering effects increase with the wavenumber).

The preparation of suspensions and pellets of a solid compound prior to its analysis in a spectrometer was formerly a standard procedure. Besides the preparative effort and the self-absorption of the embedding material, the use of KBr is additionally disadvantageous due to its hygroscopicity. The latter provokes bothersome OH bands above 3,000 cm⁻¹ since last traces of humidity can hardly be excluded for such kind of hygroscopic material. For these reasons, nowadays, KBr pellets are used only if there are no better alternatives on hand (see Section 2.4.2).

2.4.2   Reflection Measurements

Measurements at IR spectrometers which work on the principle of *attenuated total reflection* (ATR) do not require special sample preparation. The crystalline material is thoroughly pestled and a portion of it is placed on the sample carrier of the spectrometer. The material is tightly pressed onto the surface of the carrier plate with the aid of an integrated pressing mechanism to cause sintering, which minimizes reflection loss *inside* the sample.

Liquids need not be compressed and can readily be measured by putting one drop onto the carrier plate.

During the measurement the source beam is channeled in an *inclined direction* from the bottom of the transparent carrier plate toward the boundary layer between the carrier and the sample material. Some portion of the radiation penetrates the sample, where it undergoes weakening due to its excitation of molecular vibrations, which is accompanied by the absorption of appropriate spectral components. The hence attenuated reflection beam is subsequently registered according to the procedure for measurements in transmission and plotted as a usual absorption spectrum.

With the ATR technique, even strongly absorbing substances can be measured without dilution, because absorption takes place just in direct proximity of the sample boundary layer. Furthermore, the extent of absorption can be controlled by the inclination angle of the light beam.

2.4.3   Raman Measurements

In the case of Raman spectroscopy, the frequencies of excitation light and sample emission lines are mostly within the visible spectral range (see Section 2.15, Raman Spectroscopy). For this reason, *aqueous* solutions can also be prepared and measured without difficulty in *glass* cuvettes. Crystalline substances can also be measured in the pure form without further provisions and are directly placed on the sample carrier of the spectrometer, provided that they are not air-sensitive and need not be handled particularly under a protective gas atmosphere.

2.5   Infrared Spectrum

Fig 2.8 shows the IR spectrum of the paraffin *Nujol*. It is used as an embedding material for powders which have to be measured in suspension and causes absorption bands which are typical for hydrocarbons. These bands may overlap with the spectrum of the substance leading to difficulties in observing the appropriate bands of the sample. The absorption bands for the (C−C) chains are not the problem, because they are weak and appear only at increased layer thicknesses between 1,350 and 750 cm⁻¹. However, the (C−H) valence vibrations of *Nujol* are very intense and appear in the range of about 3,000 cm⁻¹. This makes *Nujol* less suitable for quantitative measurements with attention at this absorption range, due to the error-prone difference formation between the sample and comparison beam in the referencing channel of the spectrometer. In such cases, special perhalogenated polyethylenes without (C−H) bonds are used, which show self-absorption only in the fingerprint range below 1,300 cm⁻¹.

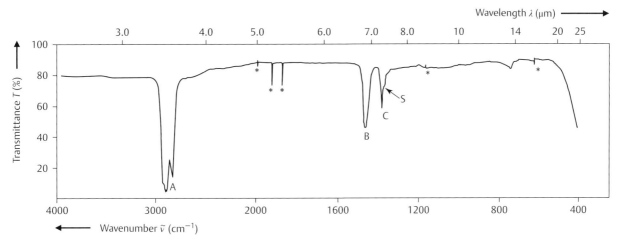

Fig. 2.8 IR spectrum of paraffin (*Nujol* film).

The ordinate of an IR spectrum represents the *transmission* (%T). This is the percentage of radiation which *passes* through the sample without absorption. As an intensity standard serves a comparison beam in the reference channel which passes through an appropriate medium without the sample substance. Instead of the transmission, also the absorption (%A) may be represented in the ordinate:

$$%T = 100 − %A$$

- A: Absorption band; at this wavenumber the CH_3 and CH_2 groups of the molecule absorb radiation energy under excitation of valence vibrations ν(C–H); cf. **Table 2.1**.
- B: Absorption band due to the excitation of asymmetric CH_3 and symmetric CH_2 deformation vibrations, $\delta_{as}(CH_3)$ and $\delta_s(CH_2)$; cf. Section 2.5.1 and **Table 2.1**.
- C: Absorption due to the excitation of symmetric CH_3 deformation vibrations $\delta_s(CH_3)$.
- S: "*Shoulder*" indicating an overlap with a neighboring band of similar wavenumber.
- *: The so-called "*spikes*"; occasionally occurring in the IR spectra of grating spectrometers; small and significantly sharp peaks due to hardware switching or electromagnetic disturbance.

The abscissa is calibrated in wavelengths (cm⁻¹) which are proportional to the energy of the absorbed light quanta. For the sparsely populated high-frequency range above 2,000 cm⁻¹ (left), an abscissa scale smaller than that for the *fingerprint* area is applied. The latter requires a better scale resolution due to the higher number of potential absorption bands. The primary wave*length* scale in λ (mm) is still met for spectra which have been recorded on older prism instruments. A scale in wavelengths is unfavorable since it does not correlate proportionally

with the absorbed energy. The consequences are deformed IR band shapes which are weakly resolved in the short-wave range.

2.5.1 Number and Types of Vibrations

The number of vibrations which may be adopted by a given molecule can readily be deduced.

A molecule consisting of *N* atoms possesses *3N* degrees of freedom, due to three independent space coordinates of each atom. However, the number of *vibrational* degrees of freedom is smaller since three molecular motions result in pure translation when all atoms move synchronously in one of the three orthogonally independent spatial directions. In the case of nonlinear molecules, another three degrees of freedom drop out due to synchronous rotation of all atoms about one of three orthogonally independent axes. The rotation of *linear* molecules about the molecular axis has no influence on the atom coordinates so that here just two rotational degrees of freedom have to be subtracted. Therefore, the **number *n* of vibrational degrees of freedom** for a molecule with *N* atoms can be calculated as:

$$n = 3N − 6 \text{ for nonlinear molecules}$$

$$n = 3N − 5 \text{ for linear molecules}$$

These are the so-called **normal** or **basis vibrations** of a molecule. The number of these fundamental vibrations which can *effectively* be excited by IR radiation is always smaller than the calculated value due to the application of *selection rules* which forbid the excitation of certain vibration patterns in the IR spectrum (see Section 2.15.2, Selection Rules). Besides the absorption bands due to *basis* vibrations, under certain circumstances additional bands of *overtones* occur, as well as *combination* and *difference* vibrations (see Section 2.5.2, Spectrum Interpretation).

Depending on the **type of vibration**, it can be distinguished between:

- **Valence** vibrations which go along with a change of bond lengths and
- **Deformation** vibrations being related to changes of bond angles.

A classification based on the **symmetry behavior** distinguishes between:

- **Symmetric** vibrations (index s) which take place under conservation of the symmetry of the involved groups or the entire molecule,
- **Asymmetric** vibrations (index as) which take place with the loss of one or more symmetry elements, as well as
- **Degenerate** vibrations, two or manifold, which deform the molecule in different spatial directions but with identical symmetry behavior and frequency.

The linear molecule CO_2, for example, performs **four** basis vibrations ($3N - 3 - 2$):

- **One** *symmetric* valence vibration (ν_s) which stretches and compresses both (C−O) bonds accordantly,
- **One** *asymmetric* valence vibration (ν_{as}) which stretches one (C−O) bond while it compresses the other one, and
- **Two** *degenerate* deformation vibrations (δ) which cause periodical changes of the (O−C−O) bond angle. These vibrations take place independently within two orthogonal planes which contain the molecular axis in their common line of intersection. All other deformations of the molecule can be described as a linear combination of the two basic deformation vibrations and have no influence on the degree of degeneration.

The bent water molecule (H_2O) adopts just **three** vibrational modes ($3N-3-3$), i.e., **two** (O−H) valence vibrations (ν_s, ν_{as}) and **one** deformation (δ).

$\nu_s(CO_2)$ $\nu_{as}(CO_2)$ $\delta(CO_2)$ $\delta(CO_2)$

$\nu_s(H_2O)$ $\nu_{as}(H_2O)$ $\delta(H_2O)$

For larger molecules, the number of allowed basis vibrations rises rapidly. Vibrations of molecular subunits are often just weakly coupled with those of the residual molecule. This is especially advantageous for the identification of functional groups which often give rise to characteristic subspectra with unambiguous absorption bands. The three-atomic methylene group ($-CH_2-$) performs the following localized vibrations:

$\nu_s(CH_2)$ $\nu_{as}(CH_2)$

+ = forward movement
− = backward movement

δ "bending" ρ "rocking" τ "twisting" ω "wagging"

$\delta(CH_2)$, *in plane* $\delta(CH_2)$, *out of plane*

For the assignment of the various vibration modes, the following symbols are used:

ν = *valence* or stretching vibration, e.g., $\nu_s(CH_2)$, $\nu_{as}(CH_2)$

δ = *deformation* vibration, e.g., $\delta_s(CH_3)$, $\delta_{as}(CH_3)$

ρ = "*rocking*" in the plane ("*in plane*")

γ = all vibrations outside of the plane ("*out of plane*") like

ω = "*wagging*," bent up and bent down

τ = "*twisting*," rotation about a torsion angle

2.5.2 Spectrum Interpretation

IR spectra can be divided into two areas:

- The range above 1,500 cm^{-1} is the domain of individual absorption bands which can readily be assigned on the basis of their wavenumbers via comparison with the values of appropriate tables. This yields a first overview of the presence or absence of potential functional groups. The valence vibrations of single bonds with hydrogen (such as C−H, O−H, and N−H) absorb at highest frequencies (left range of the acetone spectrum in **Fig. 2.9**), which is mainly a consequence of the small hydrogen mass (see Section 2.2, Basic Principles).
- The range below 1,500 cm^{-1} is less straightforward because it mostly contains a number of bands which cannot be assigned separately. In their entirety, in contrast, these absorptions appear like an unmistakable fingerprint of the molecule as a whole and, therefore, this area is named **fingerprint range**. It contains characteristic *skeletal vibrations* such as $\delta(CCH)$, $\delta(HCH)$, $\nu(C-O)$, or $\nu(C-N)$, as well as the low-frequency bands of many functional groups (see Section 2.11, Infrared Absorption in the *Fingerprint* Range). The analytical conclusiveness of these often crowded bands is rather low due to overlaps and **coupling** with other vibrations which results in sometimes delusive frequency shifts.

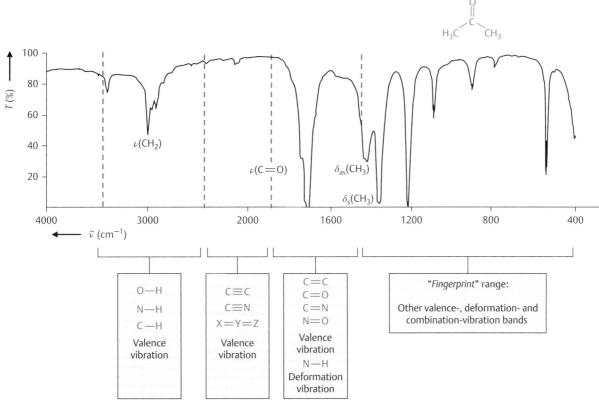

Fig. 2.9 Typical IR absorption areas (for details, see **Figs. 2.10–2.14** and **Tables 2.1–2.21**).

The fingerprint range may also contain bands due to **difference** and **combination** modes as well as **overtones** which are all basically weak. However, in exceptional cases they can be strong enough for potential misinterpretations. Therefore, these and other specific features of a spectrum shall be considered in more detail as follows.

Coupling

Coupling occurs between two vibrations which belong to mutually adjacent groups. The stronger the coupling, the smaller the frequency difference between the involved vibrations and the smaller the damping effect of the molecular parts in between. Then the vibrations enter a resonance state leading to an energy transfer from the lower frequency to the higher frequency vibration, i.e., the absorption band with the smaller wavenumber shifts to the right and the one with the higher wavenumber to the left.

Overtones

Overtones accompany anharmonic vibrations with large amplitudes. These lead to an overexpansion with a disproportionate weakening of the affected bond, making Hooke's

law less restrictive; besides the first vibrational state ($n = 1$), higher excitations ($n = 1, 2, 3, 4, …$) are now also allowed (see Section 2.2, Basic Principles). Also the *selection rules* (see Section 2.15.2, Selection Rules) are not applicable for overtones. The transition between the vibrational ground state ($n = 0$) and the second excitation state ($n = 2$) is called *first overtone*, that one between $n = 0$ and $n = 3$ *second overtone*, etc. The wavenumbers of overtones are slightly below the double, triple, or multiple values of the appropriate basis vibration, due to the over-proportionate bond stretching and weakening. Overtones exhibit basically weak intensities which decrease with the degree of excitation.

In the Raman spectrum of (*E*)-1,2-dichloroethene (**Fig. 2.35**), the two small bands at 1,630 and 1,680 cm⁻¹ are due to ν(C–Cl) overtones. The weak band in the acetone spectrum at around 3,400 cm⁻¹ (**Fig. 2.9**) is obviously the first overtone of the ν(C=O) vibration at 1,710 cm⁻¹.

Fermi Resonance

It is the result of the coupling between a basis vibration and an overtone, which is possible if certain symmetry conditions are fulfilled.

Fermi resonance increases the intensity and possibly the wavenumber of the involved overtone, making its unambiguous identification less straightforward. Sometimes the occurrence of overtones can also be of useful diagnostic relevance, as for example in the case of aldehyde groups. These can be unambiguously identified on the basis of characteristic double bands between 2,700 and 2,850 cm^{-1}, which originate from the coupling between ν(C–H) and 2·δ(C–C–H).

Combination and Difference Bands

These bands occur *under certain conditions* due to the absorption of light quanta, which excite two vibrations *at once* (*combination bands*), or transform an already excited vibration into another one at an elevated frequency (*difference bands*). The wavenumber of a combination band equals the sum of the simultaneously excited vibrations, and accordingly the wavenumber of a difference band equals the frequency difference of the involved vibrations. For an exemplary case of two basis vibrations at 1,100 and 500 cm^{-1}, the appropriate *difference* band would appear at 600 cm^{-1} and the *combination* band at 1,600 cm^{-1}, respectively. Combination and especially difference bands are mostly weak, but there are exceptions as shown in the IR spectrum of (*E*)-1,2-dichloroethene (**Fig. 2.34**), with a strong combination band occurring at 1,661 cm^{-1}, which is the sum of the two ν(C–Cl) basis vibrations (**Table 2.22**). With the sole knowledge of the IR spectrum, the absorption at 1,661 cm^{-1} cannot be recognized as a combination band. Due to symmetry restrictions (see Section 2.15.2, Selection Rules), just its asymmetric component ν_{as}(C–Cl) at 817 cm^{-1} (**Fig. 2.34**, band *D*) is observable in the IR spectrum, but *not* the symmetric one, ν_{s}(C–Cl). Only by means of a Raman spectrum the second component of the combination band can be discovered at 844 cm^{-1} (**Fig. 2.35**, band *D'*), which supports a correct assignment. Potential coupling with a third vibration (a variant of *Fermi resonance*) can thoroughly disguise a combination band and lead to some confusion, since its altered frequency will equal at the most by chance the sum of two basis vibrations.

The **wavenumbers of basis vibrations** (cm^{-1}) generally *decrease* with increasing atomic masses, as illustrated by the following examples:

Bond	$\tilde{\nu}$ (C–X) (cm^{-1})	Atom mass of X
C–H	$\approx$ 3,000	1
C–D	$\approx$ 2,100	2
C–C	$\approx$ 1,000	12
C–Cl	$\approx$ 700	35

Conversely, **frequencies** *increase* proportionally with the binding force between the involved atoms; hence, triple bonds tend to have higher absorption wavelengths than double and single bonds:

- C≡C wavenumbers (2,200 cm^{-1})
- C=C wavenumbers (1,640 cm^{-1})
- C–C wavenumbers (1,000 cm^{-1})

Deformation or bending vibrations are connected with the change of bond angles and occur at lower wavenumbers in the *fingerprint* area below 1,500 cm^{-1}. One exception is deformation vibrations with the participation of (–N–H) units which appear around 1,600 cm^{-1} (**Fig. 2.9**).

Besides the position or wavenumber of a band, its **shape** also yields valuable information. It can be more or less intense, sharp or broad, symmetric or asymmetric. For its **intensity**, the kind of vibration is decisive (see Section 2.15.2, Selection Rules). **Line broadening** is often an evidence for *inter*molecular interaction at a high concentration and particularly for the formation of bridging hydrogen bonds. Asymmetric shapes can be the result of band overlaps which are in tables often assigned as **shoulders**. Some functional groups are furthermore characterized by **double bands** which make them unambiguously distinguishable from other similar groups (e.g., aldehyde vs. ketone). Double bands often arise as a consequence of coupling between a basis vibration and an overtone or combination vibration at similar frequency.

A possible **strategy** for the interpretation of an IR spectrum may be the following:

First of all, one inspects the range above 1,500 cm^{-1} on the basis of **Figs. 2.10** to **2.13** to check if there is some evidence for particular structural elements or if structures can directly be excluded. A comparison of the *fingerprint* range with **Fig. 2.14** points out if there are typical bands which support, weaken, or exclude the possibility of a proposed structure. The tables of particular functional groups (**Tables 2.1–2.21**) are suitable for a closing refinement of the initial considerations.

A sole IR spectrum usually does not reflect the entire picture of vibrations, because a part of them can be subject to symmetry restrictions (see Section 2.15.2, Selection Rules) which prohibit the observation of appropriate absorptions in the IR spectrum. For this reason, the recording of a supplementary Raman spectrum is advisable. Furthermore, it is not always possible to assign *all* bands properly, because some absorptions may have other reasons than pure basis vibration (see Section 2.5.2). In Section 2.12, Examples of Infrared Spectra, the example spectra of some representative compounds are presented.

A suggested structure can readily be compared with IR spectra of similar compounds which are available as freeware on the Internet or in commercial databases (see Section 2.13, Information Technology Assisted Spectroscopy). In doing so, the

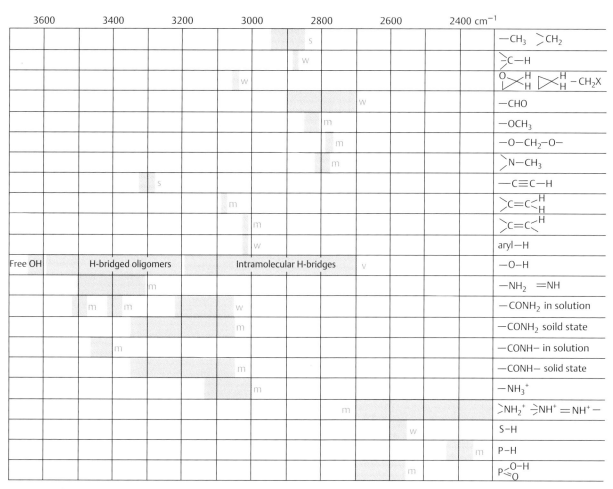

Fig. 2.10 Valence vibrations of hydrogen, ν(H–Y); intensity: **s**, strong; **m**, medium; **w**, weak; and **v**, variable.

2400	2300	2200	2100	2000	1900 cm⁻¹	
			w			—C≡CH
		v				—C≡C—
	v					—C≡N
	s					—N⁺≡N
		s				—S—C≡N
s						O=C=O
	s					—N=C=O
		s				—N=N=N
		s				—N=C=N—
		s				C=C=O
			s			—N=C=S
		s				C=N=N
			s			C=C=N—
				m		C=C=C

Fig. 2.11 Valence vibrations of triple bonds and cumulated double bonds; intensity: **s**, strong; **m**, medium; **w**, weak; and **v**, variable.

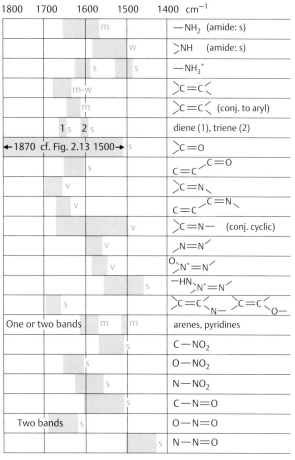

Fig. 2.12 Valence vibrations of double bonds ν(X=Y) and deformation vibrations δ(−N−H); intensity: **s**, strong; **m**, medium; **w**, weak; and **v**, variable.

appropriate recording conditions have to be considered because different sample media (e.g., *KBr/Nujol/pure*) and aggregate states have different influences on the spectrum (see Section 2.4, Sample Preparation). For a definite validation of the result, nuclear magnetic resonance (NMR) spectroscopy and mass spectrometry should be consulted whenever possible.

2.6 Characteristic Absorptions: An Overview

The example spectrum in **Fig. 2.9** is divided into four units which are depicted in the assignment overviews (**Figs. 2.10–2.14**). The important (C=O) absorptions appear typically between 1,800 and 1,500 cm^{-1} and are presented in **Fig. 2.13** and **Table 2.10a–o**.

The absorption ranges in **Figs. 2.10** to **2.14** are indicated with horizontal bars. Because *intensities* are also a useful assignment

criterion, these are labeled with the common notations **s** (strong), **m** (medium), **w** (weak), and **v** (variable).

In **Tables 2.1** to **2.21** (cf. *table overview*, below), the appropriate wavenumbers are listed. If not noted otherwise, the depicted bands are *strong*; apart from that apply the conventions described in **Figs. 2.10** to **2.14**.

Group	Table	Carbonyl compound	Table
single bonds			
C−H	2.1–2.3	aldehyde	2.10 f
O−H	2.4		
N−H	2.5, 2.6	amide	2.10 j
S−H	2.7		
P−H	2.7	carboxylate anion	2.10 i
X−D	2.7		
double bonds		carboxylic acid	2.10 h
C=O	2.10		
C=N	2.11	carboxylic acid anhydride	2.10 a
N=N	2.12		
C=C	2.13	carboxylic acid chloride	2.10 b
N=O	2.14		
cumulated double bonds		diacyl peroxide	2.10 c
C=C=C	2.9		
N=C=O	2.9	ester	2.10 d
X=Y=Z	2.9		
triple bonds		imide	2.10 l
C≡C	2.8		
X≡Y	2.8	ketone	2.10 g
aromatic systems			
	2.15	lactam	2.10 k
	2.16		
		lactone	2.10 e
fingerprint area			
S-derivatives	2.17	thioester and -acid	2.10 o
P-derivatives	2.18		
C−O and C−N	2.19	urea	2.10 m
N−O	2.14		
halogen compounds	2.20	urethane	2.10 n
inorganic ions	2.21		

2.7 Infrared Absorptions of Single Bonds with Hydrogen (cf. Tables 2.1–2.7)

2.7.1 (C−H) Absorption

Simple alkanes (paraffins) show comparatively plain spectra (see **Fig. 2.8**), because

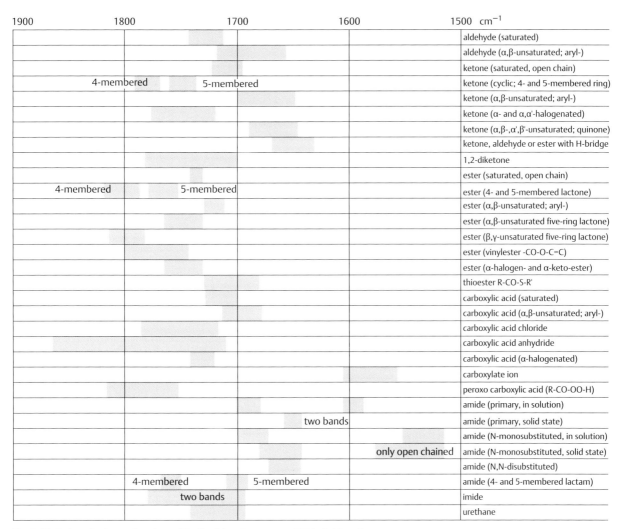

Fig. 2.13 Carbonyl valence vibrations ν(C=O); all bands are *strong* (**Table 2.10a–o**).

Infrared and Raman Spectra

- There are just two kinds of atoms and bonds (C−C and C−H).
- Some absorption bands have similar frequencies, leading to partial or total overlap.
- Many absorptions are weak or totally suppressed due to symmetry restrictions.

The appearance of (CH_2) vibration modes is exemplarily drafted in Section 2.5.1, Number and Types of Vibrations, whereas the absorption ranges of aliphatic, olefinic, and aromatic (C−H) vibrations are summarized in **Tables 2.1** to **2.3**. Since these groups are not affected by *bridging* hydrogen bonds, their band positions are not or just slightly influenced by the chemical environment or aggregate state.

Organic molecules contain generally (C−H) bonds so that these absorptions are diagnostically less valuable. The *absence* of a (C−H) band in the spectrum is certainly an unambiguous proof for the lack of this subunit within the examined compound. Unsaturated and aromatic (C−H) valence vibrations can readily be distinguished from (C−H) absorptions of saturated structures:

- *Saturated* C−H: wavenumber < 3,000 cm⁻¹
- *Unsaturated* C−H: wavenumber > 3,000 cm⁻¹

Additionally, the band intensities of unsaturated and aromatic (C−H) valence vibrations are significantly lower as compared to saturated (C−H) moieties. The *fingerprint* range of alkanes is shown in **Fig. 2.14**, whereas **Tables 2.1** to **2.3** show the band positions of select (C−H) vibrations.

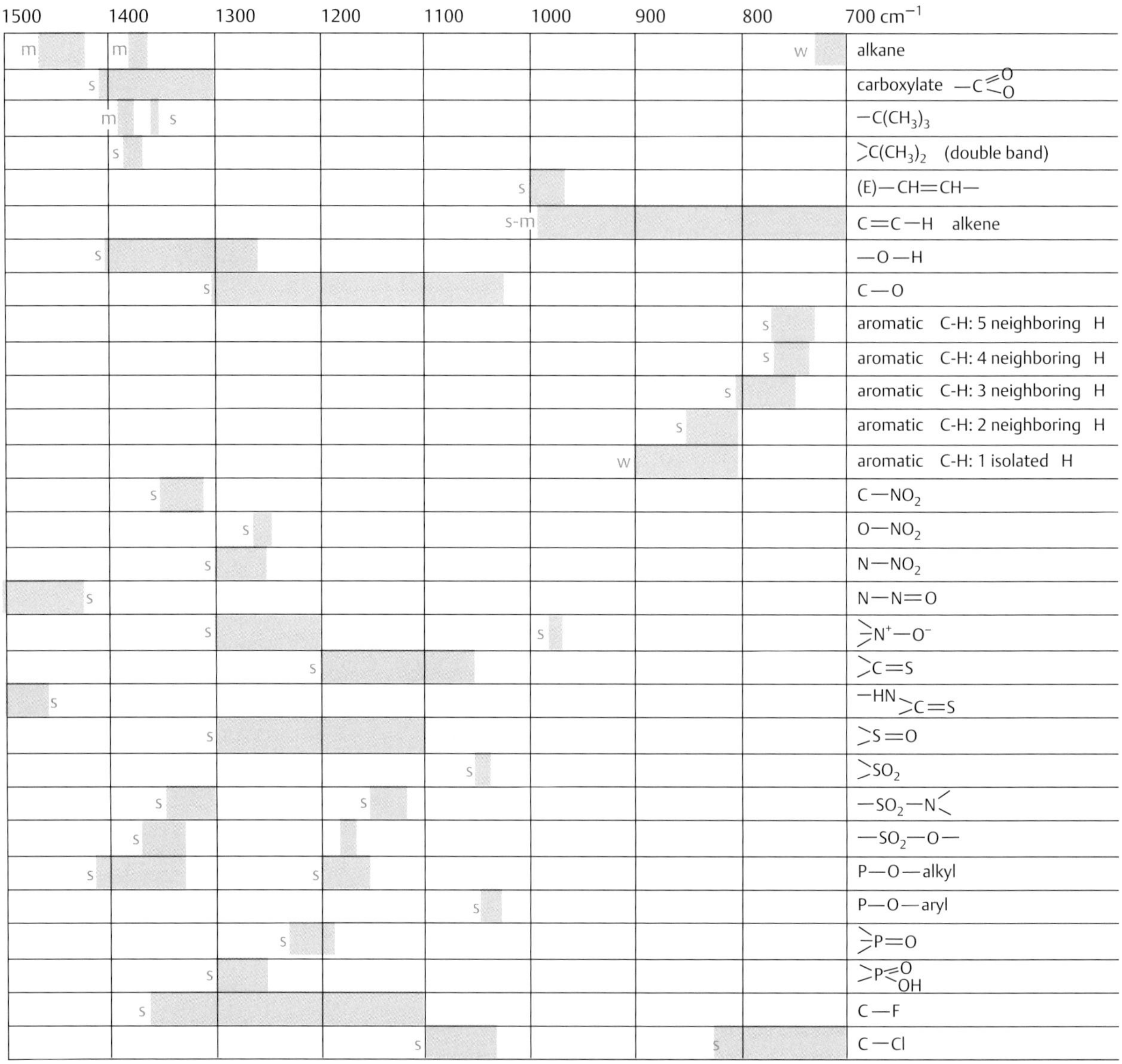

1500	1400	1300	1200	1100	1000	900	800	700 cm⁻¹	Assignment
m	m						w		alkane
	s								carboxylate $-C{\leq}{}^{O}_{O}$
	m	s							$-C(CH_3)_3$
	s								$>C(CH_3)_2$ (double band)
					s				(E)—CH=CH—
					s-m				C=C—H alkene
	s								—O—H
		s							C—O
							s		aromatic C-H: 5 neighboring H
							s		aromatic C-H: 4 neighboring H
						s			aromatic C-H: 3 neighboring H
					s				aromatic C-H: 2 neighboring H
					w				aromatic C-H: 1 isolated H
		s							$C-NO_2$
		s							$O-NO_2$
		s							$N-NO_2$
s									N—N=O
		s		s					$>N^+-O^-$
			s						$>C=S$
s									$-HN{>}C=S$
		s							$>S=O$
			s						$>SO_2$
		s	s						$-SO_2-N{<}$
		s							$-SO_2-O-$
	s		s						P—O—alkyl
			s						P—O—aryl
		s							$>P=O$
		s							$>P{\leq}{}^{O}_{OH}$
		s							C—F
			s				s		C—Cl

Fig. 2.14 Characteristic absorptions in the *fingerprint range*; intensity: **s**, strong; **m**, medium; and **w**, weak.

2.7.2 (O–H) and (N–H) Absorptions

The shape and position of (O–H) valence vibration bands are a convenient measure for the presence and strength of hydrogen bridges. The stronger the hydrogen bridge, the weaker and longer is the appropriate (O–H) bond. Consequently, the resulting (O–H) absorption band is shifted to a lower wavenumber and, at the same time it is broadened and exhibits more intensity. A *sharp* band for monomers with *isolated* (O–H) groups can be observed between 3,650 and 3,590 cm⁻¹ for gaseous phases and diluted solutions or if steric factors prevent the formation of bridging hydrogen bonds. In contrast, pure liquids, crystals, and normally concentrated solutions show *broad* bands of oligomer (O–H) composites between 3,600 and 3,200 cm⁻¹. The spectra of liquid phases can exhibit both kinds of absorption bands, depending on the polarity and concentration of the sample.

The influence of hydrogen bridges is also noticeable for some carbonyl groups (**Table 2.10**) which act as Lewis acceptors with accordingly lowered (C=O) valence vibration frequencies.

*Intra*molecular hydrogen bridges of nonchelate type (e.g., 1.2-diols) show one sharp band in the range of 3,570 to 3,450 cm⁻¹, which depends on the strength of the particular hydrogen

Table 2.1 (C−H) absorptions of alkanes

Group	Band	Notes
−CH$_3$	2980–2850	ν_s and ν_{as} valence vibrations;
	1470–1430 (m)	
	1390–1370 (m)	δ_{as} and δ_s deformation vibrations
−C(CH$_3$)$_3$	1395–1380 (m)	typical (*t*-butyl) double band;
	≈ 1365	δ_s deformation vibration
⟍C(CH$_3$)$_2$⁄	≈ 1380 (m)	nearly symmetric doublet
⟍CH$_2$⁄	2960–2850	ν(C−H), typically 2-3 bands; with halogen up to 3060;
	1470–1420 (m)	δ(C−H);
	≈ 720 (w)	(CH$_2$) rocking vibration
Cyclopropane	≈ 3100 (m)	ν_{as}(CH$_2$)
	≈ 3030 (w)	ν_s(CH$_2$)
−CH	2890–2800 (m)	with halogen also > 3000 cm^{-1}

Table 2.2 (C−H) absorptions of alkenes, alkynes, and aromatic rings

Group	Band	Notes
C≡C−H	≈ 3300	
⟍C=C⟋ H	3040–3010 (m)	
	840–790 (m)	
⟍C=C⟍ H / H	3095–3075 (m)	
	895–885	
H⟋C=C⟋ H ⟍	970–960	"out of plane" deformation; if in conjugation, e.g., with (C=O), shifted up to 990 cm^{-1}
H⟍C=C⟋ H /	730–675 (m)	
	995–985	
	940–900	
aromatic C−H	3100–3000 (m)	ν(C−H) of aromatic rings
	≈ 670	δ(C−H) of benzene for aromatic ν(C=C) and benzene substitution patterns see **Tables 2.15** and **2.16**

Table 2.3 Specific (C−H) absorptions of groups with O and N

Group	Band	Notes
epoxides (O triangle)	3060–2990	
aldehydes −CHO	2900–2700 (w)	two bands of similar intensity, one of them near 2720 cm^{-1}: Fermi resonance between ν(C−H) and δ(CHO)−overtone (**Fig. 2.28a**); very characteristic for aldehydes
−O−CH$_2$−O−	2790–2770 (m)	
−O−CH$_3$	2850–2810 (m)	
−O−CO−CH$_3$	1385–1365	often dominating bands of this spectral area
−CO−CH$_3$	1360–1355	
amines N−C−H	2820–2780 (m)	

concentration changes, as well as their respective absorption bands. **Intermolecular** hydrogen bridges, in contrast, weaken and break due to increasing dilution, with a diminishing and sharpening effect on the appropriate bands. With increasing dilution, the former (O−H) *oligomer* absorptions approach the appearance and wavenumber of bands of *isolated* (O−H) groups.

Spectra of solid samples which contain (O−H) units or crystal water show a broad and strong (O−H) absorption band between 3,400 and 3,200 cm^{-1}.

Table 2.4 (O−H) absorptions

Group	Band	Notes
water in solution	3710	
crystal water (solid state)	3600–3100 (w)	often accompanied by a weak band at 1,640–1,615 cm^{-1}
water traces in KBr pellets	3450 (v)	
free −OH	3650–3590 (v)	sharp; (O−H) valence vibration
−OH in H-bridges	3600–2500	mostly broad; can be sharp for intramolecular H-bridges; the lower the frequency, the higher the degree of aggregation
alcohols	3600–3200	
carboxylic acids	3200–2500 (v)	sometimes too broad to be recognized
R−O−H	1410–1070	δ(R−O−H) deformation vibration; for alcohols, see **Table 2.19**
−O−H....O=C	850–950 (m)	characteristical deformation vibration of H-bridges in carboxylic acids (cf. **Fig. 2.19**)

bridge. Intermolecular *dimerization* already induced some broadening but the bands are still distinguishable from those of higher aggregates. Dilution experiments can provide details about the kind and degree of a given oligomerization process. **Intramolecular** hydrogen bridges will not be affected by

Infrared and Raman Spectra

Carboxylic acids are mostly characterized by a broad and weakly resolved band series between 3,000 and 2,500 cm^{-1} (**Fig. 2.19**, stearic acid). The most prominent component at the highest frequency corresponds to (O–H) valence vibrations which are overlaid by (C–H) bands. The broad and sometimes characteristically serrated absorption pattern at lower wavenumbers down to 2,500 cm^{-1} is due to the overlap of combination vibrations and overtones. This feature is useful for the identification of carboxylic acids, all the more in conjunction with a carbonyl absorption band at an appropriate position (**Table 2.10**).

Absorption bands of (N–H) valence vibrations (**Table 2.5**) can sometimes be confused with those of (O–H) bonds. Since (N–H) groups show a weaker tendency toward hydrogen bridge formation, their absorptions are accordingly less broadened and shifted due to such effects. (N–H) bands are generally smaller compared to those of (O–H) groups and occur fairly below 3,600 cm^{-1}.

Additional weak bands in the domain of (N–H) and (O–H) vibrations above 3,200 cm^{-1} may be due to overtones of strong carbonyl absorptions (cf. spectrum in **Fig. 2.17**).

Table 2.5 (N–H) absorptions of amines, amides and imines

Group	Band	Notes
amine and imine $\diagdown$N—H$\equiv$N$\diagdown$H	3500–3300 (m)	primary amines with two bands for ν_s(N–H) and ν_{as}(N–H); sec. amine band usually weaker; pyrrole and indole bands sharp (**Fig. 2.30**)
	< 3250 (m)	ν(N–H) of amines with H-bridges; often accompanying the bands of unbridged amine (**Fig. 2.26**)
—NH$_2$	1650–1560 (m)	δ(–N–H), prim. amine and amide (**Fig. 2.26**)
$\diagdown$N—H	1580–1490 (w)	δ(–N–H) of sec. amines/amides; often too weak to be observed
unsubstituted amide —CO—N$\diagup^H_{\diagdown H}$	≈ 3500 (m) ≈ 3400 (m)	for solids ≈ 150 cm^{-1} lower; with H-bridges between 3300 and 3050 cm^{-1}
monosubstituted amide —CO—N$\diagup^H_{\diagdown R}$	3460–3250 (m)	lower for solids and H-bridges; extra band for lactams, solids and H-bridges (**Figs. 2.22** and **2.28a**)
	3250–3100 (m)	casually occurring combination band: ν(C=O) + δ(N–H), **Fig. 2.22**

Table 2.6 (N–H) absorptions of ammonium and iminium salts

Group	Band	Notes
—NH$_3^+$	3130–3030 (m)	solid state; broad ν(N–H)
	3060–3000 (m)	typical range for amino acids
	≈ 2500 (w-m)	broad bands; often observable
	≈ 2100 (w-m)	combinations and overtones
	≈ 1650 (w)	δ_{as}(N–H); rather Raman active
	≈ 1550 (m)	δ_s(N–H); cf. **Fig. 2.30**
$\diagup^+_{\diagdown}$NH$_2$ —$\overset{\mid}{\underset{\mid}{N}}H^+$ =N$^+\diagup_{\diagdown H}$	2700–2250 (m)	ν(N–H), broad; H-bridged units

Table 2.7 Other (X–H) and (X–D) absorptions

Group	Band	Notes
—S–H	2600–2550 (w)	weaker than ν(O–H); less influenced by H-bridges
$\diagdown$PH	2440–2350 (m)	sharp
$\overset{O}{\underset{\mid}{\overset{\|}{—P—OH}}}$	2700–2560 (m)	associated OH
X–D	value of X–H devided by 1.37	deuterium labeling useful for the assignment of (X–H) absorptions

The (N–H) stretching vibrations of amides are characterized by two absorption bands at wavenumbers between 3,500 and 3,070 cm^{-1}. Lactams show only one appropriate (N–H) absorption and can therefore easily be distinguished from acyclic amides. Amide (–N–H) deformation bands appear close to the appropriate carbonyl vibrations and account for the second typical amide absorption in this range (*amide I* and *amide II*, see **Table 2.10**).

2.8   Infrared Absorptions of Triple Bonds and Cumulated Double Bonds

Triple bonds (**Table 2.8**) and cumulated double bonds (**Table 2.9**) can readily be recognized because they absorb in a range which is nearly free of other noteworthy bands.

Table 2.8 Triple bonds X≡Y

Group	Band	Notes
—C≡C—H	2140–2100 (w)	
—C≡C—	2260–2150 (v)	for polyacetylenes often more bands at lowered wavelengths observable; IR-*inactiv*, if symmetrically substituted
—C≡N	2260–2200 (v)	ν(C≡N)-absorptions are stronger and shifted to lower wavelength, if conjugated; betimes very weak or absent (e.g., cyanhydrines)
isocyanide $-\overset{+}{N}\equiv\overset{-}{C}$	2165–2110	
nitrileoxide $-C\equiv\overset{+}{N}-\overset{-}{O}$	2300–2290	
diazonium salts $R-\overset{+}{N}\equiv N$	2270–2230	
thiocyanate R—S—C≡N	2175–2160 ≈ 2140	aromatic R aliphatic R

Table 2.9 Cumulated double bonds X=Y=Z

Group	Band	Notes
allene $\underset{/}{\overset{\backslash}{}}C=C=C\underset{\backslash}{\overset{/}{}}$	≈ 1950 (m)	two bands for terminal allenes, or if bound to electron- withdrawing groups (e.g., —COOH)
carbon dioxide O=C=O	2349 668	$\nu_{as}(CO_2)$ and $\delta(CO_2)$; can occur in connection with background measurements against air
isocyanate —N=C=O	2275–2250	very high intensity; position not influenced by conjugation
isothiocyanate —N=C=S	2140–1990	broad and very intense
azide $-\overset{+}{N}=\overset{-}{N}=\overset{-}{N}$	2160–2120	
carbodiimide —N=C=N—	2155–2130	very high intensity; if conjugated with aryl groups, split into an asymmetric doublet
ketene $\underset{/}{\overset{\backslash}{}}C=C=O$	≈ 2150	
diazoalkane $R_2C=\overset{+}{N}=\overset{-}{N}$	≈ 2100	

Table 2.9 (*Continued*)

Group	Band	Notes
diazoketone		
$-CO-CH=\overset{+}{N}=\overset{-}{N}$	2100–2090	
$-CO-CR=\overset{+}{N}=\overset{-}{N}$	2070–2060	
ketene imine $\underset{/}{\overset{\backslash}{}}C=C=N-$	≈ 2000	

The extraordinarily high double bond frequencies of X=Y=Z systems are supposed to be due to a strong coupling between the adjacent double bonds, going along with a large separation of the symmetric and asymmetric valence vibrations. This kind of coupling is possible if two groups of adequate symmetry and similar vibration frequency are close to each other. Other examples for such coupling are amid groups and the carboxylate ion (**Table 2.10i** and **j**).

2.9 Infrared Absorptions of Double Bonds C=O, C=N, C=C, N=N, and N=O

Carbonyl groups appear with prominent absorption bands in an area between 1,650 and 1,800 cm⁻¹, which is almost free of other vibrations (**Table 10a–o**). The band **intensities** are graduated as follows:

Carboxylic acid > ester > ketone ≈ aldehyde ≈ amide

Amide groups are complex vibration entities and may exhibit exceedingly varying band intensities.

Carbonyl groups are of particular interest due to their tendency toward **intra-** and **intermolecular interaction**, and the measured wavenumbers yield a lot of information about the chemical environment based on the following rules:

- The stronger the electron-withdrawing effect of *X* in a (R–CO–X) group, the higher is the carbonyl wavenumber.
- The (C=O) frequency of α,β-unsaturated carbonyl compounds is reduced by 15 to 40 cm⁻¹ compared to the saturated analog (except amides which are less affected).
- Further conjunction has negligible influence.
- Strained cyclic compounds are shifted to higher frequencies. This phenomenon serves as a remarkably reliable measure for the ring size, which allows us to distinguish unambiguously between four-, five-, and six-membered cyclic ketones, lactones, and lactams. Six-membered and larger rings show normal (C=O) frequencies which are comparable to those of their chain-like analogs.
- Hydrogen bridging with the participation of a carbonyl group decreases its frequency by 40 to 60 cm⁻¹. This applies especially to carboxylic acids, amides, enolized-β-oxocarbonyl compounds as well as o-hydroxy- and o-aminophenyl carbonyl compounds.

Infrared and Raman Spectra

▪ The total influence of several parts equals approximately the sum of the single effects.

The frequencies of carbonyl vibrations in the solid state are slightly lowered compared to those of appropriate diluted solutions.

The frequencies of *double-bond vibrations* increase with the **degree of substitution** around. In the case of **symmetric** substitution, the band intensities are weakened or even invisible in the IR spectrum, but in turn well observable in a Raman spectrum (Section 2.15.2, Selection Rules). For this reason, the IR absorptions of (*E*)-double bonds are commonly less pronounced than those of the appropriate (*Z*)-isomers. Additional consideration of associated (C–H) vibrations (**Table 2.3**) can supplement the structural information about an examined double bond.

Potential ring strains and the relative position with respect to a cyclic system also influence the frequency of double bond vibrations. An **exocyclic** double bond shows the same behavior as cyclic ketones: the frequency increases with decreasing ring size. A double bond **within** a cycle shows the opposite tendency: its frequency decreases with decreasing ring size. The changes in the related (C–H) valence vibrations are not so much significant; they increase slightly with increasing ring strain.

Table 2.10 Carbonyl Absorption C=O (all listed bands are strong)

Group	Band	Notes
2.10 (a) Carboxylic acid anhydride		
saturated	1850–1800 1790–1740	two bands, usually separeted by 60 cm⁻¹; the band at higher frequency is stronger for *acyclic* anhydrides, that one at lower frequency is stronger for *cyclic* anhydrides
aryl- and α,β-unsaturated	1830–1780 1770–1710	
saturated five-memberd ring	1870–1820 1800–1750	
all types	1300–1050	one or two strong bands due to (C–O) valence vibrations

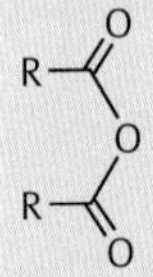

Table 2.10 (b) Carboxylic acid chloride (acyl chloride)

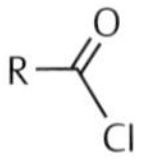

saturated	1815–1790
aryl- and α,β-unsaturated	1790–1750

Table 2.10 (c) Diacyl peroxide

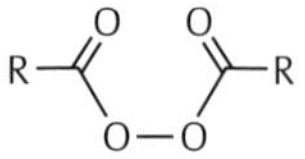

saturated	1820–1810 1800–1780
aryl- and α,β-unsaturated	1805–1780 1785–1755

Table 2.10 (d) Ester

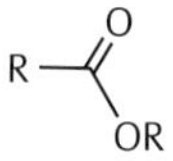 saturated	1750–1735	
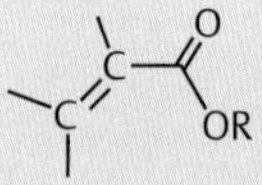 acrylic acid esters	1730–1710	
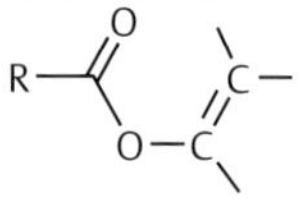 vinyl esters	1800–1750	the (C=C) stretching vibration is also shifted to higher frequency
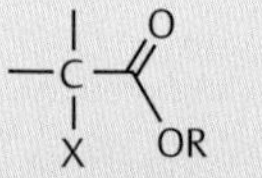 X: electronegative (e.g., halogen)	1770–1745	
α-keto-ester	1755–1740	
β-keto-ester	1750–1730	comparable with normal ester-wavelengths
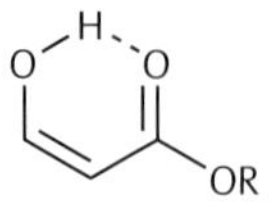 β-keto-ester in enol form	≈ 1650	(C=C) typically at 1630 cm⁻¹ (s)
all types of esters	1300–1050	one or two strong bands due to ν(C–O) stretching vibration

Table 2.10 (e) Lactone (without ring strain comparable to open-chained ester)

Structure	
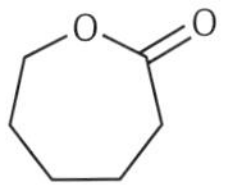	1730
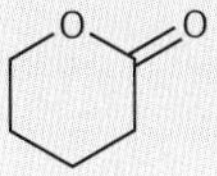	1750
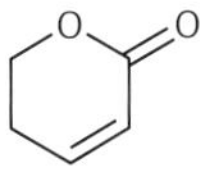	1720
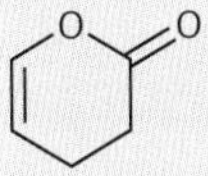	1760
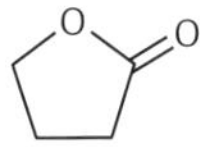	1775
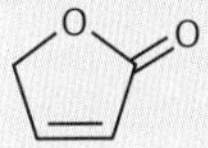	1770–1740
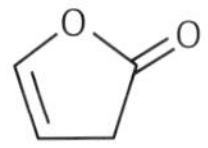	1800
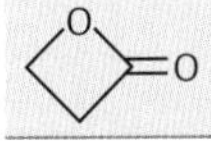	1840

Table 2.10 (f) Aldehyde

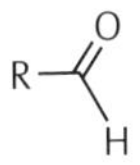

$\nu(C-H)$, cf. Table 2.3; $\delta(HCO)$ at 1390 cm^{-1} (w) without practical use.

for liquid films or solids the below values are reduced by 10-20 cm^{-1}.

measurements in gaseous phase lead to an increase (ca. +20 cm^{-1}).

saturated R	1740–1720	o-hydroxy and o-amino groups shift the wavenumbers toward 1655-1625 cm^{-1}, due to intra-molecular hydrogen bridges
aryl-CHO	1715–1695	
α,β-unsaturated R	1705–1680	
α,β- and γ,δ-unsaturated R	1680–1660	
β-ketoaldehydes in the enol form	1670–1645	low values due to hydrogen bonds of the chelate type

Table 2.10 (g) Ketone

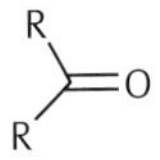

for liquid films or solids the below values are reduced by 10–20 cm^{-1}.

measurements in gaseous phase lead to an increase (ca. +20 cm^{-1}).

saturated R	1725–1705	
monoaryl	1700–1680	
o-hydroxy- and o-aminoaryl ketones	1655–1635	low wavenumbers due to intra-molecular H-bridges; sterical factors and other substituents influence the band position
α,β-unsaturated R	1685–1665	
α,β- and α',β'-unsaturated and diaryl	1670–1660	
cyclopropyl-	1705–1685	
four-membered ring ketones	≈ 1780	
five-membered ring ketones	1750–1740	conjugation with (C=C) bonds has similar effects as in the case of unsaturated open-chain ketones
six-membered and higher ring ketones	1725–1705	comparable with the values of saturated open-chain ketones
α-halogenated ketones	1745–1725	
α,α'-dihalogenated ketones	1765–1745	conformation dependent; highest wavenumbers for co-planar arrangement between halogens and (C=O) bond
1,2-diketones, s-trans, open-chain	1730–1710	ν_{as} (C=O)$_2$ ν_s IR-inactive (Raman-active)
1,2-diketones, s-cis, six-membered ring	1760 and 1730	
1,2-diketones, s-cis, five-memberd ring	1775 and 1760	
quinones	1690–1660	related ν(C=C) at 1600 cm^{-1} (s)
tropones	1650	in the presence of H-bridges lowered to around 1600 cm^{-1} (e.g., α-tropolones)

Infrared and Raman Spectra

Table 2.10 (h) Carboxylic acid

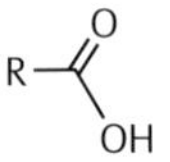

R—C(=O)OH

monomer (saturated R)	1760	*rarely* observed; sometimes found in solution, accompanying the lower frequent dimer
dimer (saturated R)	1725–1700	*mainly* observed; lowered frequencies due to H-bridges; in ether 1730 cm^{-1}
α,β-unsaturated carboxylic acids	1715–1690	
aryl-carboxylic acids	1700–1680	
α-halo-carboxylic acids	1740–1720	
additionally for all types	3000–2500	broad band of (O—H—O)—bridges which is overlain by shoulders due to over- and combination-tones, as well as by $\nu_{s,as}$ (CH$_n$)

Table 2.10 (i) Carboxylate anion

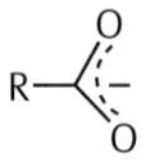

R—C(—O)(—O)⁻

most types	1610–1550 1420–1300	ν_s (COO) and ν_{as} (COO)
amino acids	≈ 1580	accompanied by $\delta(-NH_3^+)$ at ≈ 1550 cm^{-1}

Table 2.10 (j) Amide

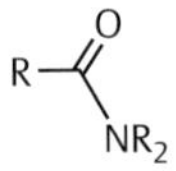

R—C(=O)NR$_2$

(see also **Tables 2.5** and **2.6** for (N—H) vibrations)

primary amides —CO—NH$_2$		
in solution solid state	≈ 1690 ≈ 1650	*amide I*: ν(CO)
in solution solid state	≈ 1600 ≈ 1640	*amide II*: δ(N—H); not for lactam. Usually less intense than *amide I*; in solid state potential overlap of *amide I* and *amide II*
N-monosubstituted amides —CO—NHR		

in solution solid state	1700–1670 1680–1630	*amide I* (cf. **Fig. 2.28**)
in solution solid state	1550–1510 1570–1515	*amide II*; just found for open-chained amides; usually less intense than *amide I*
N,N-disubstituted amides —CO—NRR	1670–1630	similar spectra in solution and solid state, due to missing hydrogen-bridges
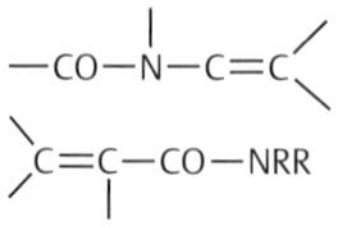 —CO—N—C=C C=C—CO—NRR		vicinal (C=C) bonds shift the wavenumber by + 15 cm^{-1}, possibly due to a dominating -I- *effect* of the double bond

Table 2.10 (k) Lactam

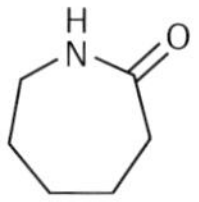 (7-membered ring)	1669	in the solid state shifted to lower wavelengths (see **Fig. 2.22**)
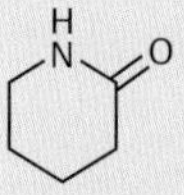 (6-membered ring)	1670	
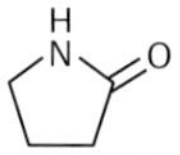 (5-membered ring)	1717	
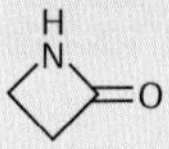 (4-membered ring)	1750	
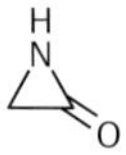 (3-membered ring)	1850	

Table 2.10 (l) Imide

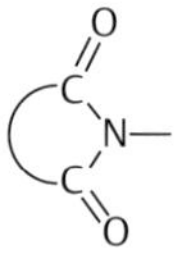

six-membered rings	≈ 1710 and 1700	shifted by + 15 cm^{-1} on conjugation with multiple bonds
five-membered ring	≈ 1770 and 1700	

Table 2.10 (m) Urea

N—C(=O)—N

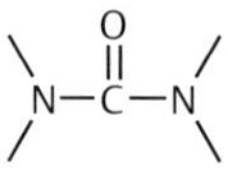	≈ 1660
six-membered rings	≈ 1640
five-membered ring	≈ 1720

Table 2.10 (n) Urethane

O ‖ R—O—C—N ⟨	1740–1690	an additional amide II band for δ(N—H) is visible, if at least one hydrogen is bound to nitrogen

Table 2.10 (o) Thioester and acid

CO—R—SH	1720
R—CO—SR	1690
R—CO—SAr	1710
Ar—CO—SR	1665
Ar—CO—SAr	1685

Table 2.11 Imines, Oximes, etc. $\rangle C{=}N\langle$

Group	Band	Notes
$\rangle C{=}N{-}H$	3400–3300 (m)	ν_s(N-H) vibration; lowered by hydrogen bridges
$\rangle C{=}N{-}$	1690–1640 (v)	difficult to identify due to large intensity variation and vicinity to the ν_s(C=C) vibration region
α,β-unsaturated	1660–1630 (v)	
conjugated cyclic systems	1730–1480 (v)	oxime bands usually very weak

Table 2.12 Azo compounds $\diagdown N{=}N\diagup$

Group	Band	Notes
$\diagdown N{=}N\diagup$	≈ 1575 (w)	very weak or inactive in IR; occasionally visible in Raman
$\diagdown \overset{+}{N}{=}N\diagup$ O⁻	≈ 1570	

Table 2.13 Alkenes $\rangle C{=}C\langle$

Group	Band	Notes
$\rangle C{=}C\langle$ non-conjugated	1680–1620 (v)	very weak, if symmetrically substituted
conjugated with aromatic rings	≈ 1625 (m)	more intensive than for non-conjugated systems
dienes, trienes etc.	1650 and 1600	sometimes overlapping; band at lower frequency usually more intensive
α,β-unsaturated carbonyl compounds	1640–1590	usually much weaker than corresponding (C=O) band
enol esters, enol ethers and enamines	1690–1650	

Table 2.14 Compounds with NO single and double bonds $N{-}O$ and $N{=}O$

Group	Band	Notes
nitro compounds —C—NO₂	≈ 1560 and ≈ 1350	$\nu_{s,as}$(NO₂), cf. **Fig. 2.27**; conjugation to multiple bonds lowers the values by ≈ 30 cm⁻¹
nitrates R—O—NO₂	1640–1620 1285–1270	symmetric and asymmetric stretching vibrations
nitramines $\rangle$N—NO₂	1630–1550 1300–1250	symmetric and asymmetric stretching vibrations
nitrites R—O—NO	1680–1650 1625–1610	each one band for s-cis and s-trans nitrite
nitroso compounds R—NO	1600–1500	
E-dimer	1290–1190	
Z-dimer	1425–1370	
nitrosamines $\rangle$N—NO	1460–1430	
N-oxides R₃N⁺—O⁻		pyridine-N-oxide in non-polar solvents at 1250 cm⁻¹; electron-withdrawing substituents shift the frequency to higher values and *vice versa*
aromatic	1300–1200	
aliphatic	970–950	
NO₃⁻	1410–1340	asymmetric stretching vibration; ν_s at ≈ 1055 cm⁻¹ Raman active

2.10 Infrared Absorption of Aromatic Compounds

Aromatic systems show several absorptions which are sufficiently characteristic of an unambiguous identification:

- 3,100 to 3,000 cm⁻¹: aryl–H valence vibration (**Table 2.3**)
- 2,000 to 1,600 cm⁻¹: several weak bands of overtones and combination vibrations
- 1,600 to 1,500 cm⁻¹: (C=C) valence vibrations; two or three bands of high significance for aromatics (**Table 2.15**). Polycyclic compounds and pyridines also show these absorptions
- 1,225 to 950 cm⁻¹: *fingerprint* bands of low diagnostic value
- 900 to 680 cm⁻¹: (C–C–H) deformations (*out of plane*); the number and position of the bands depend on the number of the neighboring hydrogen atoms at the ring and indicate its degree of substitution (**Table 2.16**).

Table 2.15 (C=C) absorptions of aromatic compounds

Group	Band	Notes
aromatic rings	1630–1570 (w-m)	
	1530–1440 (m)	often the strongest aromatic (C=C) absorption band
	1475–1400 (m)	
	2000–1650 (w)	overtones and combination vibrations of aromatic rings

Table 2.16 Substituted benzene rings ⬡—X$_n$ (CCH and ring deformation)

Group	Band	Notes
five neighboring H-atoms	770–730	monosubstituted benzene; cf. toluene, **Fig. 2.23**
	710–685	
four neighboring H-atoms	760–740	1,2-disubstitution; cf. o-xylene, **Fig. 2.24a**
three neighboring H-atoms	800–770	1,3-disubstitution (cf. **Fig. 2.24b**), 1,2,3-trisubstitution
two neighboring H-atoms	840–800	1,4-disubstitution (cf. **Fig. 2.24c**), 1,3,4-trisubstitution
isolated H-atom	900–800 (m)	1,3-disubstitution etc.; usually weak and of low analytical value

The diagnostic value of the bands at 2,000 to 1,600 cm⁻¹ and below 900 cm⁻¹ is lowered by the possible presence of other strong absorptions in this area: carbonyl groups may occur above 1,600 cm⁻¹ and carbon–halogen bonds below 900 cm⁻¹ (**Table 2.20**), leading to questionable band assignments which then have to be verified, e.g., by NMR spectroscopy.

Figs. 2.23 to **2.24c** showing the IR spectra of toluene and the three xylene isomers illustrate the influence of aromatic ring substitution on the appearance of the appropriate IR absorption pattern. In addition, the spectrum of tryptophane (**Fig. 2.30**) shows the characteristic bands for 1,2-disubstituted aromatic rings.

The values in **Table 2.16** roughly apply also to condensed ring systems and pyridine derivatives (**Fig. 2.29**). Strongly electron-withdrawing substituents shift the values generally toward higher frequency.

2.11 Infrared Absorption in the Fingerprint Range

Besides the *out-of-plane* vibrations of aromatics (**Table 2.16**), the *fingerprint* range also covers significant absorptions of groups containing an element of the third and higher periods (for sulfur and phosphorus, see **Tables 2.17** and **2.18**), as well

Table 2.17 Sulfur compounds S

Group	Band	Notes
S–H	2600–2550 (w)	ν(SH) much weaker than ν(OH); less influenced by H-bridges. Strong in the Raman spectrum
C=S	1200–1050	
$\overset{S}{\underset{}{\nwarrow}}$C–N$\overset{H}{\underset{}{\diagdown}}$	≈ 3400	ν(NH); in solids shifted down to 3150 cm⁻¹
	1550–1460	amide II: δ(NH)
	1300–1100	amide I: ν(CS)
S=O	1060–1040	
sulfones	1350–1310	
SO$_2$	1160–1120	
sulfonamides	1370–1330	
R–SO$_2$–N	1180–1160	
sulfonates	1420–1330	
R–SO$_2$–OR'	1200–1145	
sulfates	1440–1350	
RO–SO$_2$–OR'	1200–1145	

Table 2.18 Phosphorus compounds P

Group	Band	Notes
P–H	2400–2350	sharp
P-phenyl	≈ 1440	sharp
P–O–alkyl	1050–1030	
P–O–aryl	1240–1190	
P=O	1300–1250	
P–O–P	970–910	broad
$\overset{O}{\underset{}{\|}}$—P—OH	2700–2560	O–H in H-bridges
	1240–1180	ν(P–O)

as absorptions of single bonds such as C–O (**Table 2.19**) and C–halogen (**Table 2.20**), or those of inorganic ions (**Table 2.21**).

The typical absorptions of halogenated arenes above 1,000 cm⁻¹ (**Table 2.20**) are due to skeletons and *not* valence vibrations.

Table 2.19 Functional groups with (C—O) and (C—N) single bonds

Group	Band	Notes	
prim. alcohols	1140–1070 (m)	δ(C—O—H)	
	1080–990	ν(C—O)	
	980–870 (w)	ν(C—C)	*all* alcohols:
sec. alcohols	1150–1090 (m)	δ(C—O—H)	ν(OH) 3640–3620 (without H-bridges)
	1130–1090	ν(C—O)	
	880–790 (w)	ν(C—C)	
tert. alcohols	1220–1110	ν(C—O)	ν(OH) 3600 - 3200 (with H-bridges)
	760–690 (w)	ν(C—C)	
ether C—O—C ⟩C—O—C⟨	1150–1070 1275–1200 1075–1020	sometimes split, cf. **Fig. 2.18**	
epoxides	1270–1230 950–830 870–750		
ester	1330–1050	2 bands; the stronger of them at lower wavenumber (δ$_{asym}$)	
CH$_3$CO—O—CR	≈ 1240		
RCO—O—CH$_3$	≈ 1165	both ν$_{asym}$	
RC—N	1400–1000 (m)	ν(C—N), cf. **Fig. 2.26**; low analytical significance	

Table 2.20 Halogen compounds (C—Hal)

Group	Alkyl-Hal	Aryl-Hal
C—F	1365–1120	1270–1100
C—Cl	830–560	1100–1030 ν(skeleton)
C—Br	680–515	1075–1030 ν(skeleton)
C—I	640–450	1060–1050 ν(skeleton)

Table 2.21 Inorganic ions

Group	Band	Notes
ammonium	3300–3030	all bands are strong
cyanides, cyanates, thiocyanates	2200–2000	
carbonates	1450–1410	
sulfates	1130–1080	
nitrates	1380–1350	
nitrites	1250–1230	
phosphates	1100–1000	

2.12 Examples of Infrared Spectra

The spectra shown in **Figs. 2.15** to **2.30** illustrate the position, appearance, and relative intensity of the absorptions of typical representatives of some compound categories. They also demonstrate the relevance of the *fingerprint* region for a proper sample characterization.

Fig. 2.28a shows the spectrum of an *N-monosubstituted amide* with an additional aldehyde function (KBr pellet). The observed band diversity is typical for amides and is due to different degrees of molecular association. The broad shoulder at around 3,400 cm⁻¹ further indicates the presence of traces of water in the KBr pellet.

The IR spectrum of the same compound in *Nujol* (**Fig. 2.28b**) contains apparently no water, because it shows no appropriate (O—H) absorption above 3,300 cm⁻¹. Instead it is now partly superimposed by *Nujol* bands (N), especially in the important aldehyde (C—H) vibration area of the compound around 2,900 cm⁻¹.

Fig. 2.28c also shows an IR spectrum of this compound, now in *liquid solution* (in CHCl₃). The change from the solid (crystalline) to liquid (dilution) phase influences the shape and position of those absorption bands which are directly affected by *inter*molecular association. In solid amides, molecular aggregation is mainly accomplished by (N—H) hydrogen bridges which weaken the (N—H) and (C=O) bonds, thereby shifting the corresponding valence vibration bands toward lower frequency. Dilution breaks these intermolecular hydrogen bridges and restores the original (N—H) and (C=O) binding forces with an elevating effect on the frequency of the appropriate absorptions: the (N—H) valence vibrations are significantly shifted to the left (≈3,500 cm⁻¹) and the *amide* (C=O) absorption overlaps with the formerly distinguishable *aldehyde* (C=O) band (**Figs. 2.28a** and **c**; 1,670 and 1,690 cm⁻¹).

The solid state spectrum of this amide also shows *one* benzene absorption at 1,600 cm⁻¹ (**Fig. 2.28a**, KBr pellet), which in solution splits into *two* distinguishable bands (**Fig. 2.28c**). IR spectra of solutions are generally better resolved than those of solid compounds due to a reduced aggregation which also leads to a less number of bands. This is helpful for a straightforward interpretation, especially for absorption bands in the crowded *fingerprint* region.

Amino acids (**Fig. 2.30**) show absorptions of zwitterionic groups. The (N—H) band of the primary ammonium group ([–NH₃]⁺) at around 3,000 cm⁻¹ is overlaid by (C–H) stretching vibrations. At 2,500 and 2,100 cm⁻¹, two bands are appeared which are typical for the presence of amino acids and are due to overtones and combination vibrations. The double-bond range shows several bands of which at least one arises from the ionized carboxylic group.

Infrared and Raman Spectra

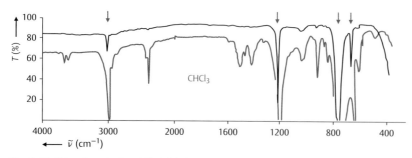

Fig. 2.15 Chloroform (as liquid film); black spectrum 9 μm layer, blue spectrum 100 μm. The figure shows the dependence of the absorption band strength on the layer thickness. Chloroform is a frequently used solvent with a self-absorption which limits the layer thickness to < 0.2 mm.

2995 cm⁻¹	Tab. 2.1: (C−H) valence vibration ν(CH)
1200 cm⁻¹	Tab. 2.1: (C−H) deformation vibration δ(CH)
760 cm⁻¹	Tab. 2.20: asymmetric (C−Cl) valence vibration ν_{as}(CCl)
670 cm⁻¹	Tab. 2.20: symmetric (C−Cl) valence vibration ν_{s}(CCl)

All other bands are combinations or overtones.

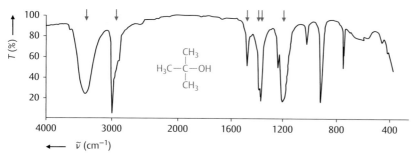

Fig. 2.16 *tert*-Butyl alcohol (as liquid film); alcohols can easily be identified by means of their strong (O−H) and (C−O) absorption bands.

≈ 3400 cm⁻¹	Tab. 2.19: (O−H) valence vibration ν(OH) of H-bridges; the higher frequent shoulder at
≈ 3600 cm⁻¹	indicates the presence of small amounts of non-associated (OH) groups
2975 cm⁻¹	Tab. 2.1: (C−H) valence vibration $\nu_{s,as}$(CH$_3$)
1470 cm⁻¹	Tab. 2.1: asymmetric (C−H) deformation vibration δ_{as}(CH$_3$)
1380 cm⁻¹	Tab. 2.1: characteristic double band for *t*-butyl groups
1365 cm⁻¹	Tab. 2.1: δ_{s}(C(CH$_3$)$_3$)
1200 cm⁻¹	Tab. 2.19: (C−O) valence vibration ν(C−O)

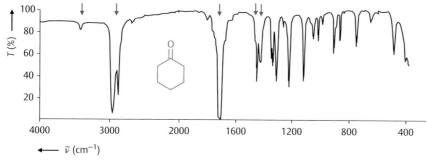

Fig. 2.17 Cyclohexanone (as film).

≈ 3400 cm⁻¹	Tab. 2.10g: overtone of the carbonyl group (2·1710 - Δ)
> 2960 cm⁻¹	Tab. 2.1: (C−H) valence vibrations $\nu_{s,as}$(CH$_2$)
1710 cm⁻¹	Tab. 2.10g: (C=O) valence vibration ν(CO), ketone
1450 cm⁻¹	Tab. 2.1: (C−H) deformation vibration δ(CH$_2$)
1420 cm⁻¹	(C−H) deformation vibration of (CH$_2$) group close to C=O

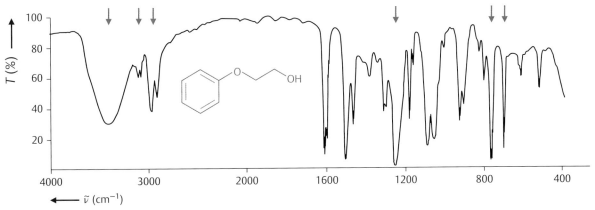

Fig. 2.18 2-Phenoxyethanol (as film).

≈ 3350 cm⁻¹	Tab. 2.19: (O−H) valence vibration ν(OH) of H-bridges
≈ 3100 cm⁻¹	Tab. 2.2: two valence vibrations ν(CH) of the benzene ring
2930 cm⁻¹	Tab. 2.1: (C−H) valence vibrations $\nu_{s,as}$(CH₂)
1250 cm⁻¹	Tab. 2.19: (C−O) valence vibration of ethers
760 cm⁻¹	Tab. 2.16: (C−C−H) deformation of monosubstituted aromates (five neighboring H-atoms)
695 cm⁻¹	Tab. 2.16: ring deformation of monosubsituted benzenes

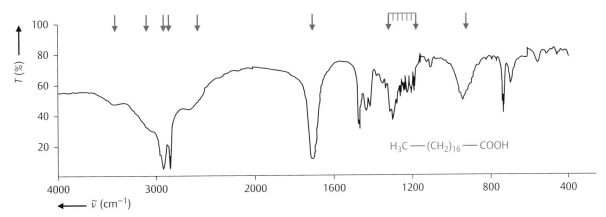

Fig. 2.19 Stearic acid (KBr pellet).

3450 cm⁻¹	Tab. 2.4: ν(O−H), water traces of KBr pellet
≈ 3100 cm⁻¹	Tab. 2.4: ν(O−H), H-bridges, carboxylic acid
2940 to 2880 cm⁻¹	Tab. 2.1: $\nu_{s,as}$(CH₂) and $\nu_{s,as}$(CH₃)
2600 cm⁻¹	characteristic broad shoulder of overlapping combination vibrations and overtones
1700 cm⁻¹	Tab. 2.10h: ν(C=O), carboxylic acid
1350 to 1190 cm⁻¹	characteristical "*progression bands*" of alkyl chains (> C₁₂) in the solid state; they originate from coupled *twisting* and *rocking* vibrations of (E)-oriented (CH₂) groups, τ(CH₂) and ρ(CH₂)
930 cm⁻¹	Tab. 2.4: δ(OH), H-bridges, carboxylic acid

Infrared and Raman Spectra

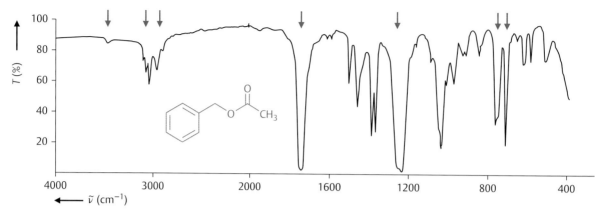

Fig. 2.20 Benzyl acetate (as film).

3450 cm⁻¹	overtone of ν(C=O)
3050 to 3020 cm⁻¹	Tab. 2.2: various valence vibrations ν(CH), benzene ring
2960 to 2880 cm⁻¹	Tab. 2.1: $\nu_{s,as}$(CH$_2$) and $\nu_{s,as}$(CH$_3$)
1740 cm⁻¹	Tab. 2.10d: ν(C=O), ester
1230 cm⁻¹	Tab. 2.10d: ν(C–O), ester
750 cm⁻¹	Tab. 2.16: δ(C–C–H), monosubstituted aromate
700 cm⁻¹	Tab. 2.16: ring deformation, monosubstituted aromate

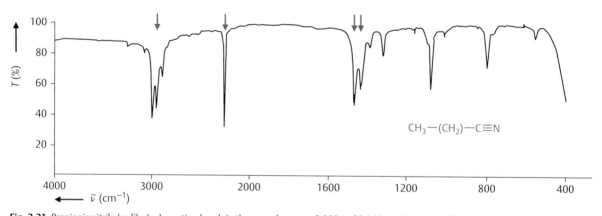

Fig. 2.21 Propionic nitrile (as film); absorption bands in the range between 2,300 and 2,000 cm⁻¹ are generally strong evidence for triple bonds.

2970 to 2860 cm⁻¹	Tab. 2.1: $\nu_{s,as}$(CH$_2$, CH$_3$), saturated alkyl groups
2250 cm⁻¹	Tab. 2.8: ν(CN), triple bond
1460 cm⁻¹	Tab. 2.1: δ(CH)
1430 cm⁻¹	Tab. 2.1: δ(CH) next to CN-group

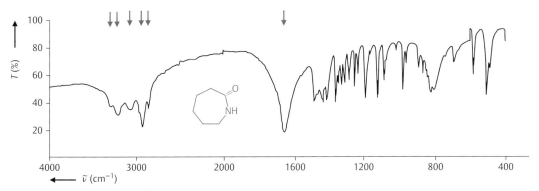

Fig. 2.22 ε-Caprolactam (KBr pellet).

3295 and 3210 cm⁻¹   Tab. 2.5: ν(N−H), N-monosubstituted amide with lactam extra band
            3100 cm⁻¹   Tab. 2.5, 2.10k: combination band δ(N−H) + ν(C=O)
   2950 to 2860 cm⁻¹   Tab. 2.1: $\nu_{s,as}$(CH₂)
            1660 cm⁻¹   Tab. 2.10k: ν(C=O), lactam

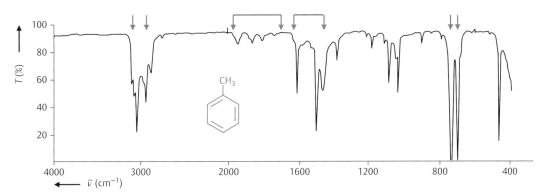

Fig. 2.23 Toluene (as film).

       > 3000 cm⁻¹   Tab. 2.2: aromatic ν(C−H)
       < 3000 cm⁻¹   Tab. 2.1: aliphatic ν(C−H)
1960 to 1700 cm⁻¹   Tab. 2.15: typical aromatic overtones and combination vibrations
1610 to 1450 cm⁻¹   Tab. 2.15: three typical aromatic ν(C=C) vibrations
  730 and 695 cm⁻¹   Tab. 2.16: δ(C−C−H) of monosubstituted aromatic rings (5 neighboring H-atoms)

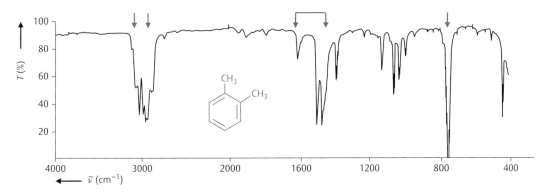

Fig. 2.24 (a) o-Xylene (as film).

       > 3000 cm⁻¹   Tab. 2.2: aromatic ν(C−H)
       < 3000 cm⁻¹   Tab. 2.1: aliphatic ν(C−H)
1610 to 1450 cm⁻¹   Tab. 2.15: three typical aromatic ν(C=C) vibrations
            740 cm⁻¹   Tab. 2.16: δ(C−C−H) of 1,2-disubstituted aromatic rings (4 neighboring H-atoms)

Infrared and Raman Spectra

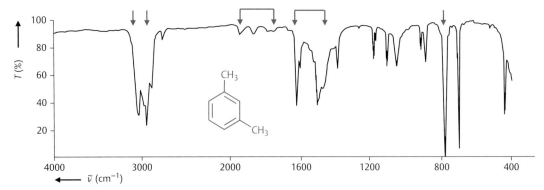

Fig. 2.24 (b) *m*-Xylene (as film).

> 3000 cm⁻¹   Tab. 2.2: aromatic ν(C−H)
< 3000 cm⁻¹   Tab. 2.1: aliphatic ν(C−H)
1940 to 1740 cm⁻¹   Tab. 2.15: typical aromatic overtones and combination vibrations
1610 to 1450 cm⁻¹   Tab. 2.15: three typical aromatic ν(C=C) vibrations
775 cm⁻¹   Tab. 2.16: δ(C−C−H) of 1,3-disubstituted aromatic rings (3 neighboring H-atoms)

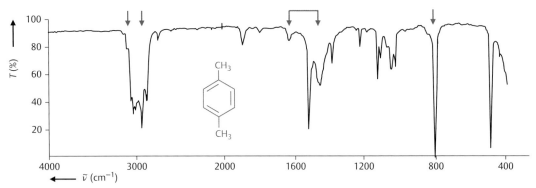

Fig. 2.24 (c) *p*-Xylene (as film).

> 3000 cm⁻¹   Tab. 2.2: aromatic ν(C−H)
< 3000 cm⁻¹   Tab. 2.1: aliphatic ν(C−H)
1610 to 1450 cm⁻¹   Tab. 2.15: three typical aromatic ν(C=C) vibrations
800 cm⁻¹   Tab. 2.16: δ(C−C−H) of 1,4-disubstituted aromatic rings (2 neighboring H-atoms)

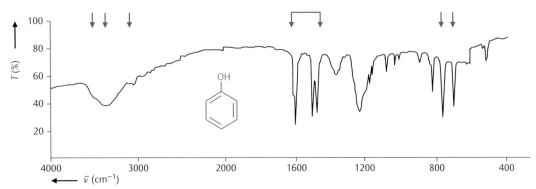

Fig. 2.25 Phenol (KBr pellet).

3500 cm⁻¹   Tab. 2.4: ν(OH) valence vibration of H-bridges in dimers
3360 cm⁻¹   Tab. 2.4: ν(OH) valence vibration of H-bridges in higher oligomers
3040 cm⁻¹   Tab. 2.2: aromatic ν(C−H)
1610 to 1450 cm⁻¹   Tab. 2.15: three typical aromatic ν(C=C) vibrations
755 and 690 cm⁻¹   Tab. 2.16: aromatic δ(C−C−H), 5 neighboring H-atoms

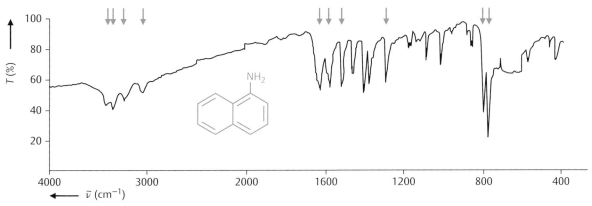

Fig. 2.26 1-Naphthylamine (KBr pellet).

3500 and 3420 cm⁻¹ Tab. 2.5: ν_s(N−H) and ν_{as}(N−H) of primary amine
　　　　3230 cm⁻¹ Tab. 2.5: ν(N−H) of H-bridged amine
　　　　3040 cm⁻¹ Tab. 2.2: aromatic ν(C−H)
　　　　1620 cm⁻¹ Tab. 2.5 δ(NH₂)
1570 and 1510 cm⁻¹ Tab. 2.15: aromatic ν(C=C)
　　　　1290 cm⁻¹ Tab. 2.19: ν(C−N)
　　　　795 cm⁻¹ aromatic δ(C−C−H), 3 neighboring H-atoms (**Table 2.16** applies approximately also to naphthalenes)
　　　　770 cm⁻¹ aromatic δ(C−C−H), 4 neighboring H-atoms

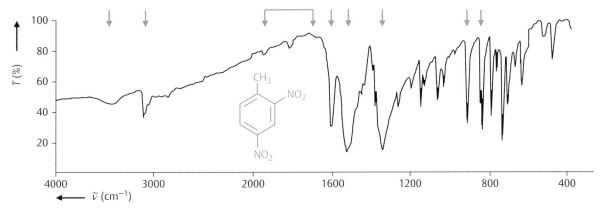

Fig. 2.27 2,4-Dinitrotoluene (KBr pellet).

　　　　3450 cm⁻¹ Tab. 2.4: water traces in KBr pellet
　　　　3100 cm⁻¹ Tab. 2.2: aromatic ν(C−H)
1940 to 1700 cm⁻¹ Tab. 2.15: typical aromatic overtones and combination vibrations
　　　　1605 cm⁻¹ Tab. 2.15: aromatic ν(C=C)
1520 and 1340 cm⁻¹ Tab. 2.14: ν_s and ν_{as}(NO₂)
　　　　915 cm⁻¹ Tab. 2.16: aromatic δ(C−C−H), isolated H (high-frequency shifted due to conjugated (NO₂) groups)
　　　　840 cm⁻¹ Tab. 2.16: aromatic δ(C−C−H), 2 neighboring H-atoms

2.13　Information Technology Assisted Spectroscopy

All modern IR spectrometers run in the *online* mode where the measurement device, computer hardware, and storage media form one composite unit. The software can be divided into five basic categories:

1. System operation software (setup and measurement control).
2. Spectrum processing software (Fourier transformation and spectrum visualization).
3. IR catalogues of fine chemicals, drugs, natural materials, etc., serve as a basis for integrated subroutines which compare the measured spectrum with the database and calculate appropriate correlation factors (*hit rates*).

Infrared and Raman Spectra

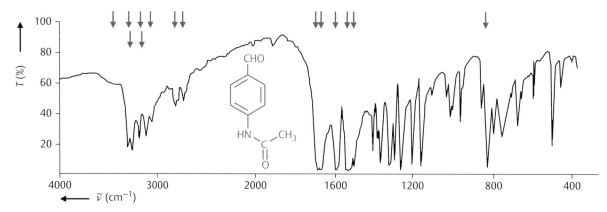

Fig. 2.28 (a) 4-(Acetylamino)-benzaldehyde (KBr pellet).

3450 cm⁻¹	Tab. 2.4: water traces in KBr pellet
3300 and 3260 cm⁻¹	Tab. 2.5: ν(N–H); monosubstituted amide with extra band for solid state
3190 and 3110 cm⁻¹	Tab. 2.5, 2.10j: ν(N–H) of H-bridged amide and probably [ν(C=O) + δ(N–H)] amide combination band
3060 cm⁻¹	Tab. 2.2: aromatic ν(C–H)
2810 cm⁻¹	Tab. 2.3: ν(C–H), aldehyde
2730 cm⁻¹	Tab. 2.3: δ(CHO)-overtone of aldehyde group, in Fermi-resonance with aldehyde ν(C–H)
1690 cm⁻¹	Tab. 2.10f: ν(C=O), aldehyde
1670 cm⁻¹	Tab. 2.10j: ν(C=O), N-monosubstituted amide (amide-I vibration)
1600 cm⁻¹	Tab. 2.15: aromatic ν(C=C)
1535 cm⁻¹	Tab. 2.10j: δ(N–H); amide-II vibration
1510 cm⁻¹	Tab. 2.15: aromatic ν(C=C)
835 cm⁻¹	Tab. 2.16: aromatic δ(C–C–H); p-disubstituted benzene with 2 neighboring H-atoms

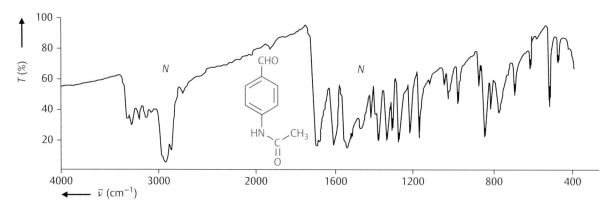

Fig. 2.28 (b) 4-(Acetylamino)-benzaldehyde (in *Nujol*, N); for band assignments, cf. **Fig. 2.28a.**

4. Spectrum interpretation software provides suggestions for possibly existing structural elements on the basis of the observed band patterns (*pattern recognition*) or which predicts the absorptions for given molecular subunits.

5. Software which combines IR, NMR, MS, and ultraviolet (UV) spectroscopy: it predicts NMR signals, MS fragmentation patterns, and UV bands for structural elements which have been recognized in the IR spectrum. It also supports the coordinated interpretation of spectra derived from these spectroscopic methods.

Software of types 1 and 2 is an integral component of each IR instrument. Also basic spectral catalogs and simple interpretation software are usually part of the standard equipment. More powerful versions of types 3 to 5 are commercially available from scientific software providers or spectrometer manufacturers.

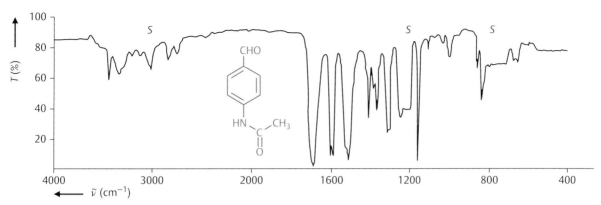

Fig. 2.28 (c) 4-(Acetylamino)-benzaldehyde (solvent CDCl$_3$, S); for band assignments, cf. **Fig. 2.28a**.

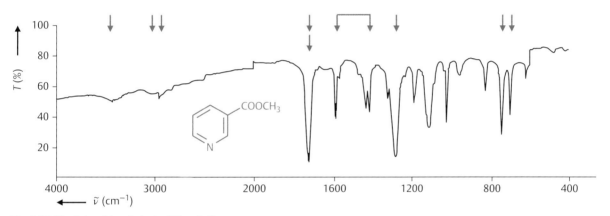

Fig. 2.29 Nicotinic acid methylester (KBr pellet).

3450 cm^{-1}	Tab. 2.4: water traces in KBr pellet
> 3000 cm^{-1}	Tab. 2.2: aromatic ν(C−H)
< 3000 cm^{-1}	Tab. 2.1: aliphatic ν(C−H)
1725 cm^{-1}	Tab. 2.10d: ν(C=O); ester
1725 cm^{-1}	Tab. 2.11: ν(C=N)
1590 to 1420 cm^{-1}	Tab. 2.15: three typical aromatic ν(C=C) vibrations
1290 cm^{-1}	Tab. 2.10d: ν(C−O); ester
745 and 705 cm^{-1}	Tab. 2.16: aromatic δ(C−C−H) of monosubstituted benzenes (Tab. 2.16 approximately valid also for pyridines)

2.14 Quantitative Infrared Spectroscopy

Similar to UV spectroscopy, IR spectra also allow quantitative conclusions to be made about the concentration of compounds in solution or in a mixture. The relationship between the amount of absorbed light and the concentration of an absorbing medium is described by the *Lambert–Beer law*:

$$\log \frac{I_0}{I} = \varepsilon \cdot c \cdot d = E_\lambda$$

At a given wavelength λ, the absorption is proportional to the concentration c and the thickness (known path length) d of the absorbing medium. The intensity ratio I_0/I of the light beam **before** and **after** its passage through the sample is an observable quantity. In the *Lambert–Beer law* it transforms into $\log(I_0/I)$, which is equal to the absorbance E_λ. The proportionality factor ε is the wavelength-dependent and substance-specific extinction coefficient. The aim of quantitative IR analysis is the determination of concentrations c on the basis of characteristic bands and their absorbance E_λ.

The *Lambert–Beer law* is strictly valid only for highly diluted solutions, because potential intermolecular interaction (e.g., aggregation) at higher concentrations may alter the coefficient

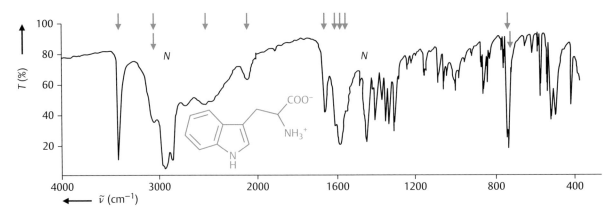

Fig. 2.30 D,L-tryptophan (in *Nujol*, N).

3400 cm^{-1}	Tab. 2.5: ν(N−H); indole
3030 cm^{-1}	Tab. 2.6: ν(N−H); (NH$_3^+$) group
3030 cm^{-1}	Tab. 2.2: aromatic ν(C−H)
2500 and 2100 cm^{-1}	Tab. 2.6: overtones and combinations; typical for (NH$_3^+$) ammonium salts
1665 cm^{-1}	Tab. 2.6: δ_{as}(NH) of (NH$_3^+$); exceptionally strong
1610 cm^{-1}	Tab. 2.15: aromatic ν(C=C)
1585 cm^{-1}	Tab. 2.10i: ν(C=O), amino acid zwitterion
1555 cm^{-1}	Tab. 2.6: δ_s(NH) of (NH$_3^+$)
755 and 745 cm^{-1}	Tab. 2.16: aromatic δ(C−C−H); indole

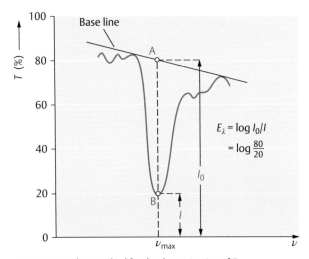

Fig. 2.31 Baseline method for the determination of E_λ.

$$E_\lambda = \log I_0/I$$
$$= \log \frac{80}{20}$$

ε and thus suspend the proportionality condition. In particular, pellets of solid compounds are not suitable for *accurate* quantitative IR measurements and provide at most *semiquantitative* results. A major problem in connection with intensity determination for solids is light reflection and scattering.

For quantitative IR analysis, an empirical **calibration curve** is created based on the intensities of a characteristic absorption band at various known concentrations. The obtained graph yields directly the unknown concentration c_x after measurement of the corresponding extinction E_λ (**Figs. 2.31** and **2.32**).

E_λ is determined by means of a *baseline algorithm* which makes use of an assisting tangent line at the outermost edges of the absorption curve as depicted in **Fig. 2.31**.

Quantitative IR analysis is used in quality control of plastics, pharmaceuticals, and plant protection products.

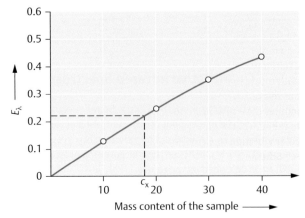

Fig. 2.32 Calibration curve.

2.15   Raman Spectroscopy

The Raman effect was theoretically predicted already in 1923 by **A. Smekal** and 5 years later experimentally confirmed by **C. V. Raman**. Its mechanism of molecular vibration excitation differs fundamentally from that of IR spectroscopy. Due to a different symmetry behavior (Section 2.15.2, Selection Rules), those vibrations can be visible in the Raman spectrum which are not observed in the IR spectrum, and vice versa. IR and Raman spectroscopies are thus mutually supplementary methods.

2.15.1   Excitation Mechanisms

Raman Effect

In contrast to IR spectroscopy, where a light quantum of proper energy excites a molecular vibration of the same frequency, in the case of Raman spectroscopy excitation is achieved by means of **monochromatic light** and follows an **indirect pathway** via the electron shell of the molecule. This process is feasible for vibrations which are accompanied by an *oscillating polarizability* (deformability) α of the electron shell. In the linear relation

$$\mu(t) = \alpha \cdot E(t)$$

α is the proportionality factor between the induced molecular dipole moment μ and the field strength E of the irradiated light. The collision with a photon transfers the electron shell of the molecule to a virtual swinging mode *without* electron-orbital rearrangement; the molecule remains energetically below HOMO-LUMO excitation (HOMO: highest occupied molecular orbital; LUMO: lowest unoccupied molecular orbital). The following relaxation into the electronic ground state can take place in various manners. Most probable is the reemission of a photon with an unchanged frequency, which amounts to *elastic* light scattering at the electron shell (**Rayleigh scattering**) without any effect on the molecular vibration mode.

Relaxation under emission of a light quantum with a *reduced* frequency equals an inelastic scattering process (**Stokes scattering**) after which the residual electron shell excitation can turn into molecular vibration. Also molecules which are at already excited vibration modes are capable of absorbing photons with the formation of electronically excited states. Simultaneous relaxation into the electronic and molecular ground states induces the emission of a light quantum with a frequency equal to the sum of those of the excitation light and the initial molecular vibration (*hyper*elastic impact or **anti-Stokes scattering**, see **Scheme 2.1, Raman Excitation**).

The predominant part of the irradiated light passes through the sample without interaction and has no influence on the Raman spectrum. One of 10,000 light quanta is scattered elastically and only contributes to the **Rayleigh** line, which has to be filtered out. In contrast, the relevant **Stokes-** and **anti-Stokes** (*Raman*)

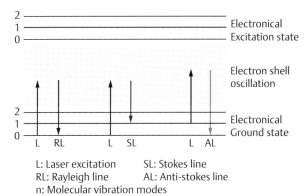

Scheme 2.1 Raman excitation.

L: Laser excitation   SL: Stokes line
RL: Rayleigh line   AL: Anti-stokes line
n: Molecular vibration modes

lines which correlate with molecular vibration frequencies are, however, comparatively weak and appear with quantum yields of only about 10^{-8}. Therefore and with respect to the excitation mechanism, **laser** emitters are used as light sources for Raman spectrometers as they provide extremely intense and strictly monochromatic light. Apart from special applications like *resonance Raman* measurements (see section 2.15.1), Raman spectra are usually excited in the *visible* or *near-IR* frequency range, in order to avoid undesired fluorescence background radiation.

Resonance Raman Effect

Photons of appropriate energy induce electronic *HOMO–LUMO* transitions which have a strong **intensifying effect** on the resulting emission lines. These underlie a considerable reabsorption and are superimposed by intense fluorescence background radiation disturbance (**Scheme 2.2**). Under normal conditions, these negative effects predominate the benefit of increased signal intensities as long as no additional corrective techniques are applied.

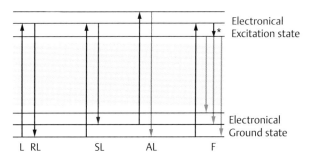

L: Laser excitation   SL: Stokes line
RL: Rayleigh line   AL: Anti-Stokes line

F: Fluorescence after vibrational relaxation*

Scheme 2.2 Resonance Raman excitation.

Infrared and Raman Spectra

The spectral bandwidth of resonance Raman excitation is significantly reduced since it initiates mainly vibrations at those parts of the molecule which are affected by the primary *HOMO–LUMO* transition. This circumstance is no downright disadvantage. It is, in contrast, often exploited for special applications like **selective** Raman measurements (Section **2.15.4**, Applications). In practice it has to be kept in mind that light irradiation in the energetic range of electronic (*HOMO–LUMO*) transitions is accompanied by the potential danger of photolytic decomposition of the sample.

Hyper-Raman Effect

Extremely high excitation light intensities above a threshold limit cause disproportionately strong deformations of the electron shell. The induced molecular dipole moment μ is no longer proportional to the field strength E and can just be represented by a power series:

$$\mu = \alpha E + 1/(2!) \cdot \beta E^2 + 1/(3!) \cdot \gamma E^3 + \cdots$$

In this relation also **multiple quantum** impacts are taken into account, which occur with a non-negligible probability under these conditions. If the power series contains the first two terms, it describes a double-quantum transition where two photons *synchronously* generate a virtual swinging state of the electron shell. The *hyperpolarizability* β correlates with the square of the field strength of the excitation radiation and gains significance only at extremely high quantum densities. *Hyper*-Raman lines appear at frequencies equal to those of the appropriate normal lines **plus** the excitation frequency (**Scheme 2.3**). Due to their diminutive probabilities, double-quantum transitions have much lower intensities than normal (*linear*) Raman emissions. The probabilities of higher multiple-quantum transitions (degree > 2) are vanishingly small and can be neglected.

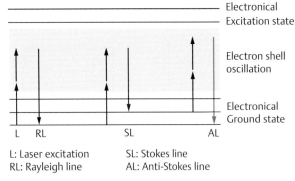

L: Laser excitation   SL: Stokes line
RL: Rayleigh line   AL: Anti-Stokes line

Scheme 2.3 Hyper-Raman excitation.

2.15.2 Selection Rules

A molecular vibration is **IR**-*active* only if it is accompanied by a change in the molecular dipole moment. In the case of *point-symmetric* molecules with an inversion center, this applies to **uneven** (*u-type*) vibrations that cause molecule deformations with the loss of the symmetry center. **Even** *g-type* vibrations have no influence on the symmetry center of point-symmetric molecules and are thus IR-forbidden but, instead, **Raman**-*active*. The alternative IR and Raman activities of molecular vibrations are classified via the *mutual exclusion rule*: **u-type** vibrations are **IR-active** and Raman-forbidden; **g-type** vibrations are, on the other hand, **Raman-active** and IR-forbidden.

This simple rule is inherently valid just for molecules with a symmetry center and shall be illustrated by means of the linear carbon dioxide molecule whose polarizability α and dipole moment μ are plotted against ν_s and ν_{as} in **Fig. 2.33**.

During the symmetric valence vibration ν_s with the amplitudes a and b (*g-type*), the dipole moment remains unchanged. This vibration is thus IR-inactive and causes no absorption band. In contrast, ν_s induces an oscillation of the polarizability of the electron shell (minimum during compression a, maximum during stretching b), making this vibration Raman active. The asymmetric valence vibration ν_{as} (*u-type*) with its amplitudes c and d shows opposite Raman and IR activities. Due to contemporaneous compression and elongation of the left and right part of the molecule, the overall expansion of the electron shell remains almost constant so that this vibration is not observable in the Raman spectrum. In the IR spectrum, in contrast, ν_{as} causes an absorption band which is associated with an oscillating dipole moment pointing toward opposite directions during the deflections c and d (**Fig. 2.33**).

Figs. 2.34 and **2.35** show the IR and Raman spectra of the point-symmetric planar (*E*)-1,2-dichloroethene. They reveal a complementary picture: the absorptions of *u-type* vibrations appear in the IR spectrum, whereas the emission bands of *g-type* vibrations are collected in the Raman spectrum. In **Table 2.22** the IR bands and Raman lines are listed and assigned to the appropriate vibrations.

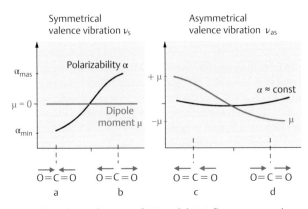

Fig. 2.33 Valence vibrations of CO_2 and their influence on α and μ.

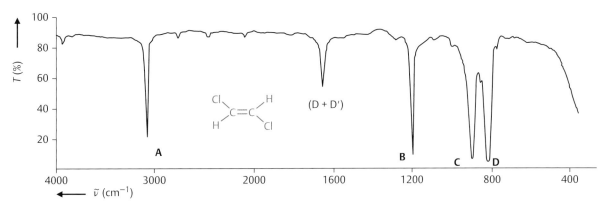

Fig. 2.34 *IR* spectrum of (*E*)-dichloroethene (band assignment, cf. **Table 2.22**).

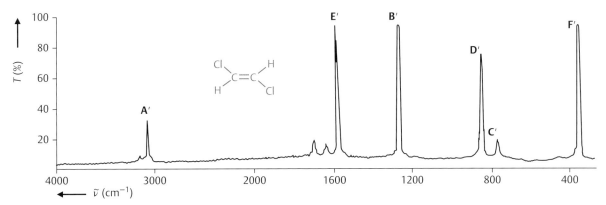

Fig. 2.35 *Raman* spectrum of (*E*)-dichloroethene (band assignment, cf. **Table 2.22**).

Although the *mutual exclusion rule* is not applicable to vibrations of molecules *without* symmetry centers, IR and Raman activities can be estimated according to potential changes of the polarizability and dipole moment during a given vibration. In case of doubt, systematic considerations by means of the *group theory* are indispensable, i.e., first the symmetry (*point group*) of the given molecule is determined on the basis of existing symmetry elements (rotation axes, mirror planes, and symmetry center). The obtained symmetry leads to an appropriate correlation table (*character table*) where the types and number of potential vibrations are listed.

For special Raman applications like *depolarization* measurements (see Section 2.15.4, Applications), further differentiation between **total symmetric** and **non-total symmetric** vibrations is essential. A vibration is named *total symmetric* if **all** symmetry elements remain unaffected. This may not be confused with *g-type* vibrations for which the only criterion is the preservation of the symmetry center of the molecule.

The symmetric properties of a molecule depend on its chemical environment and on whether it is incorporated into a crystal lattice or located in a liquid or gaseous medium. Such circumstances can have influence on the IR and Raman activity of particular vibrations and affect the appearance of an appropriate spectrum. A band can, for example, be strong for a sample in the solid state and nonetheless be weak for the same compound in the gaseous phase; in gases, molecules are essentially present as monomer units which are generally characterized by higher symmetries compared to larger molecular assemblies. Higher symmetries are directly associated with tighter symmetry restrictions which can in turn cause intensity attenuation of particular bands.

2.15.3 Raman Spectrometer

The technical layout of a classical Raman spectrometer resembles a grating IR instrument. The main difference is represented by the radiation source in the form of a laser which supplies highly intense and strictly monochromatic light. Furthermore, no reference channel is necessary since *emission* lines are registered at the detector (**Fig. 2.36**).

Table 2.22 Assignment of the IR and Raman bands of (E)-dichloroethene (cf. **Figs. 2.34** and **2.35**)

Kind of vibration	Asymmetric (IR-active)	Band (cm⁻¹)	Symmetric (Raman active)	Band (cm⁻¹)
ν(C−H)		A (3090)		A' (3070)
ν(C−Cl)		D (817)		D' (844)
δ(HCCH)		B (1200)		B' (1270)
γ(HCCH)		C (895)		C' (760)
ν(C=C)	–	–		E' (1576)
δ(Cl−CC−Cl)		< 300 cm⁻¹		F' (350)

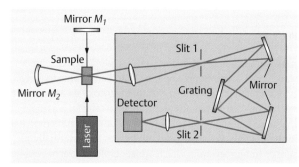

Fig. 2.36 Schematic setup of a classical Raman spectrometer (*blue lines*: scattered light).

The original laser light is scattered at the sample and additionally reflected at focusing mirrors (M_1 and M_2) for intensity enhancement. In many cases, the given compound is simply positioned on an easily accessible carrier plate like in a microscope. The alignment of the sample is not crucial as the arising scattering light spreads in a spherical manner. The intense Rayleigh light at laser wavelength is to be filtered out directly behind the sample to avoid superposition with the much weaker Raman lines and an overload of the detector. The remaining light enters the spectrometer perpendicularly to the primary laser beam, and is directed toward the entrance slit (S_1).

A grating splits the scattered radiation into its spectral components which are registered by a detector behind the exit slit (S_2).

Typically red or green gas lasers (He/Ne or Ar), tunable dye lasers (e.g., Rhodamine 6G), or YAG (Y/Al garnet) solid-state lasers are used as light sources. For colorless samples that mainly absorb light in the ultraviolet range, visible light emitting gas lasers are used in order to minimize self-absorption and fluorescence stray light. For samples which absorb in the visible range or which contain fluorescent impurities, IR YAG lasers (> 1,000 nm) are advisable, although the resulting IR scattering is comparatively weak (scattering decreases with λ^4). For resonance Raman measurements, variable laser frequencies are requisite. These are made available by tunable dye lasers which are, for fluorescence suppression, adjusted at a frequency slightly *below* the electronic transition of interest.

In the first Raman instruments, photographic plates were used as detectors which required a light exposure lasting for hours. These have later been substituted by photo-multipliers (photocathodes) which discharge electron quantities in proportion to incoming light intensities. Also, this technique is nowadays obsolete, because photo-multipliers are **one-channel detectors** which require a rotatable grating for the successive registration of separate emission lines. Latest devices are **multiple-channel detectors** (CCD [*charge-coupled device*]) which contain a large number (> 10⁶) of microscopic and individually connected semiconductors on their surface. A CCD is comparable to a video camera which records the whole of the split spectral components *simultaneously* behind the grating. Accordingly, the registered frequency-dependent intensities are transferred onto the data storage unit.

Raman spectra can also be obtained by the aid of the FT technique (**FT-Raman** or **FT-R**; cf. *FT-IR*, Section 2.3.2, Fourier Transform Spectrometers). The corresponding sensitivity enhancement particularly facilitates the detection of the inherently weak IR scattering radiation of YAG lasers. Often no explicit FT-Raman instrument is required for this purpose since appropriately equipped FT-IR systems are designed for a facile conversion into a FT-Raman spectrometer, and a reverse one (*FT-IR/R* instruments). The effort for such reconstruction is limited to the substitution of the IR Nernst rod for a Raman laser radiator and, respectively, the exchange of the detector. Apart from that, the technical layouts of FT-Raman and FT-IR spectrometers are practically identical (**Fig. 2.6**).

2.15.4 Applications

At its early stage, Raman spectroscopy could not at once be applied as a routine spectroscopic method. The main handicap was its low sensitivity compared to the classical IR spectroscopy, which required the use of sunlight as the radiation source, resulting in the need of accordingly elaborate accessories for its utilization. Also, after the introduction of more effective low-pressure mercury lamps, Raman spectroscopy was for a long time at best considered as a useful supplementary method.

By now, highly efficient laser radiators are commonly used as light sources, making Raman spectroscopy a well-established and independent spectroscopic method far beyond its original supplementary character. Its scope of applications is not limited to cases for which IR spectroscopy is less suitable like the characterization of **less polar** materials or **aqueous** solutions. Instead, the use of Raman spectroscopy extends to areas, such as examinations of **single crystals** and **polymers**, as well as the excitation of particularly **low-frequency** vibrations. It is also beneficial in the context of **homoatomic** vibrations like those of C−C rings and skeletons. On the other hand, the strong and characteristic IR bands of polar groups like $C=O$ and $X-H$ are just weakly represented in a Raman spectrum. Besides the registration of classical Raman emission spectra for assignment purposes or the determination of binding forces, there are further applications which are based on particular physical effects. These shall be presented in the following, even though they exceed the limits of routine methods and may require additional equipment.

Resonance Raman Spectroscopy

The **enhanced sensitivity** of *resonance* Raman spectroscopy facilitates the measurement of *micro*molar aqueous solutions, which are, for example, encountered in biology or biochemistry.

The relaxation of electronic excitation states toward the ground state often leads to a reinforced appearance of higher order overtones. These are useful for the calculation of **potential curves** and **anharmonicities** of appropriate molecular vibrations which in turn allows quantitative conclusions about binding situations (**Fig. 2.3**).

If a molecule contains two or more well separated chromophoric groups with different absorption frequencies, these can be excited selectively. As a consequence, only the Raman lines of the directly involved functional group make an appearance. A successive excitation of individual chromophores results in a series of nonoverlain subspectra, which facilitates the **line assignment** of large molecules. Accordingly works selective observation of low-concentrated molecules in Raman-active solvents. In any case, the UV/vis spectrum of the given compound must be known, in order to adjust the Raman laser to the frequency of the appropriate electronic transition. For this reason, tunable dye lasers with adequate wavelength ranges are necessary for these selective techniques.

Additional arrangements are required to avoid **fluorescence background radiation** and the **reabsorption** of emission lines which might annihilate the achievements of sensitivity enhancement. Line reabsorption can be minimized by appropriate experimental layout like tangential irradiation along the surface of a stretched sample. Fluorescence can sufficiently be suppressed by use of *time-coordinated* detection techniques. These take advantage of the delayed appearance of fluorescence stray radiation which can be handled by appropriate detector shutter times. Alternatively, disturbing fluorescence can be eliminated *spectrally* if its frequency is quite beyond the relevant Raman lines (**Scheme 2.2**).

Hyper-Raman Spectroscopy

Due to their particular excitation mechanism (**Section 2.15.1**), hyper-Raman vibrations are not subjected to the *Raman* selection rules but rather to those of *IR* excitation (**Section 2.15.2**). The main feature of hyper-Raman spectroscopy is its potential for the excitation of vibrations which are **neither Raman nor IR active**. This is the case for **amorphous solids** such as **glasses** or **polymers** and for readily polarizable compounds with a large dipole moment. The latter can for example consist of two reversely polarized functional groups which are connected with each other by an extensive conjugate π-system.

The application of this method is, however, severely limited due to its comparatively low sensitivity and the excessive thermal impact on the sample material which demands especially designed spectrometers. The required light intensities are typically generated by the so-called *giant pulses* of IR YAG-lasers which are properly focused onto a microscopically small part of the sample. Overheating is prevented by concurrent sample rotation and external cooling. Also, the detector has to meet extraordinary requirements since the output of hyper-Raman radiation is orders of magnitude less intense than the already extremely weak Raman emission.

Depolarization

Besides the position, intensity, or shape of a Raman emission line, its degree of depolarization ρ can also be helpful for the **assignment**. Depending on the kind of excited vibration, a resulting emission line can be polarized completely, partly, or no more. ρ represents the ratio of Raman emissions which oscillate in the original polarization plane of the excitation wave (intensity $I_{\parallel}$) and in the perpendicular direction (intensity $I_{\perp}$), respectively:

$$\rho = I_{\perp}/I_{\parallel}.$$

For optical *ani*sotropic molecules, the polarizability α is direction-dependent, since its spatial components α_x, α_y, and α_z are different. Therefore, the electrical field components E_x, E_y, and E_z of the excitation light take different parts in the induced molecular dipole moment μ:

$$\mu = \alpha_x \cdot E_x + \alpha_y \cdot E_y + \alpha_z \cdot E_z$$

Infrared and Raman Spectra

As a consequence, the polarization plane of the resulting Raman scattering $\alpha \times E$ is inclined, in comparison with the laser excitation light E (**Scheme 2.4a** and **b**).

For optically isotropic compounds, the polarizability α is not direction-dependent because its components α_x, α_y, and α_z are equal and, thus, the polarization plane of the induced dipole moment $\alpha \times E$ and the corresponding Raman radiation will remain parallel with respect to the primary laser radiation (**Scheme 2.4c**).

For liquid and gaseous phases, there exists a simple relationship between the depolarization ρ of a Raman line and the kind of the associated molecular vibration:

$\rho = 0$: the compound is optically isotropic and the excited vibration total symmetric,

$0 < \rho < \frac{3}{4}$: the compound is optically *anisotropic* and the excited vibration total symmetric,

$\rho = \frac{3}{4}$: the excited vibration is *not* total symmetric.

Also for powders consisting of statistically oriented microcrystals, these rules are basically applicable, as long as they are embedded into an optically isotropic matrix of similar refractivity to avoid reflection and refraction loss (see Section 2.4, Sample Preparation, *Suspensions* and *Pellets*).

Laser light is by nature linearly polarized and is, therefore, predestinated for depolarization measurements. The degree of depolarization can simply be determined on the basis of Raman spectra which are recorded with a polarization filter in the beam path in front of the detector. The aperture direction of the filter is successively set in *parallel* and *perpendicular* orientations with respect to the laser polarization plane, resulting in two Raman spectra which contain emission lines with the intensities $I_{\parallel}$ and $I_{\perp}$, respectively.

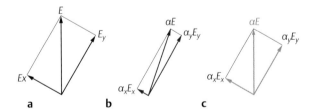

Scheme 2.4 Depolarization within a two-dimensional plane.

a The components E_x and E_y of a laser excitation field E (laser irradiation in z-direction).

b $\alpha_y > \alpha_x$: the component $\alpha_y \cdot E_y$ of the induced dipole moment $\alpha \cdot E$ is disproportionately high, tilting the latter in the y-direction.

c $\alpha_y = \alpha_x$: the components $\alpha_x \cdot E_x$ and $\alpha_y \cdot E_y$ contribute proportionately to $\alpha \cdot E$, which, therefore, oscillates in the direction of the excitation field E.

Fig. 2.37 shows three Raman spectra of nearly isotropic tetrachloromethane (CCl_4) in the low-frequency area, measured *without* polarization filter (**a**), as well as *with* filter in *parallel* (**b**), and *perpendicular* (**c**) aperture alignments, relative to the polarization plane of the laser.

The emission line at 458 cm^{-1} is due to the total-symmetric valence vibration (ν_s) which at any time maintains the geometry of the molecule. The two lines at 313 and 218 cm^{-1} correspond to deformation vibrations (δ_s and δ_{as}) which are connected to changes in bond angles and, thus, to changes in the molecular symmetry. A visual comparison of the signal intensities already allowed an unambiguous differentiation between the symmetric valence vibration (ν_s) and the two deformations (δ_s and δ_{as}).

In the spectrum, measured *without* polarization filter (**Fig. 2.37a**), all signals have comparable intensities.

The measurement with a *parallelly* aligned polarization filter with respect to the laser polarization plane (**Fig. 2.37b**) evokes one dominating high-frequency band (left) and two suppressed low-frequency bands (right). The intense band corresponds to the total-symmetric valence vibration (ν_s) of the CCl_4 molecule which induces no significant inclination between the polarization planes of the laser and the resulting Raman line which, therefore, passes the polarization filter without intensity loss. The two smaller lines at the right are in contrast due to the *non*-total symmetric deformation vibrations: the induced Raman emissions have considerable components in the perpendicular direction with respect to the laser and filter polarization planes and are, thus, to a large extent removed from the spectrum.

An opposite picture arises from a *perpendicular* setting of the laser and filter polarization planes (**Fig. 2.37c**): the two emission lines (right) of the *non*-total symmetric deformations δ_s and δ_{as} now display a comparatively high intensity since their polarization planes are more efficiently declined from the laser plane toward the filter aperture. Therefore, they rather pass the filter, in contrast to the ν_s emission (left) which retains its perpendicular polarization with respect to the filter and gets mostly wiped out of the spectrum.

2.15.5 Comparison of Infrared and Raman

Both IR and Raman spectroscopies are generally suitable for the examination of solid, liquid, and gaseous samples. They are complementary methods which follow different selection rules for vibration excitation and are characterized by features which are summarized below.

Characteristics of IR spectroscopy:

▪ Strong bands for ionic bonds,

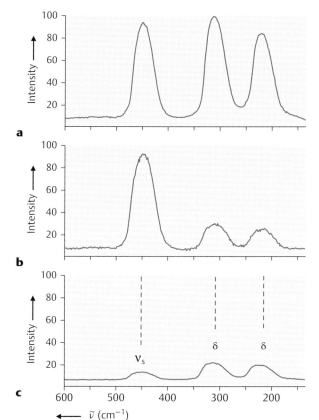

Fig. 2.37 Three Raman spectra of CCl$_4$ (laser frequency: 514 nm): **(a)** without polarization filter; **(b)** with polarization filter, *parallel* aperture and laser polarization; **(c)** with polarization filter, *crossed* aperture and laser polarization.

- Deformation vibrations often more intense compared to Raman,
- Allows quantitative determination of concentrations.

But:

- Not suitable for aqueous samples (self-absorption of water and glass cells),
- Limited observability of low-frequency (< 200 cm^{-1}) vibrations,
- Solid state bands often broader compared to Raman.

Characteristics of Raman spectroscopy:

- Convenient for stretching vibrations of covalent bonds and less polar groups,
- Usually clearly arranged spectra with less and sharper lines compared to IR,
- Allows depolarization experiments,
- Enables selective excitation of different parts of a molecule (resonance Raman),
- Requires just smallest amounts of sample material due to advanced laser technology,

- Aqueous samples unproblematic,
- Air-sensitive substances easily measured in glass cells,
- No reference channel needed.

But:

- Careful choice of the laser source necessary to avoid fluorescence and sample photolysis.
- Concentration measurements were not straightforward (Lambert–Beer *absorption* law was not applicable, also reabsorption of emission lines and laser-dependent line intensities).

Literature

Christy AA, Ozaki Y, Gregoriou VG. Modern Fourier Transform: Infrared Spectroscopy. Amsterdam: Elsevier; 2001

Colthup NB, Daly LH, Wiberley SE. Introduction to Infrared and Raman Spectroscopy. New York, London: Academic Press; 2012

Griffiths PR, De Haseth JA, Winefordner JD. Fourier Transform Infrared Spectrometry (Chemical Analysis: A Series of Monographs on Analytical Chem). Hoboken: Wiley & Sons; 2007

Guenzler H, Gremlich HU. IR Spectroscopy: An Introduction. Weinheim: Wiley-VCH; 2002

Harris DC, Bertolucci MD. Symmetry and Spectroscopy. New York, NY: Dover Publications; 1989

Jaffé HH, Orchin M. Symmetry in Chemistry. Huntington, New York: Krieger Publishing Company; 1997

Mackenzie M. Advances in Applied Fourier Transform Infrared Spectroscopy. New York, NY: Wiley; 1988

Nakamoto K. Infrared and Raman Spectra of Inorganic and Coordination Compounds. New Jersey: Wiley; 2009

Nishikida K, Nishio E, Hannah R. Selected Applications of Modern FT-IR Techniques. Tokyo: Kodansha Ltd.; 1995

Schrader B, Ed. Infrared and Raman Spectroscopy. Methods and Applications. Weinheim: Wiley-VCH; 1995

Smith BC. Fundamentals of Fourier Transform Infrared Spectroscopy. Boca Raton: CRC Press; 2011

Stuart BH. Infrared Spectroscopy: Fundamentals and Applications. Chichester: Wiley; 2004

Thompson JM. Infrared Spectroscopy. Singapore: Pan Stanford Publishing; 2018

Raman Spectroscopy

Baranska H, Lobudzinska A, Terpinski T. Laser Raman Spectroscopy. New York, NY: Wiley; 1983

Demtröder W. Laser Spectroscopy 1: Basic Principles. Berlin: Springer; 2014

Demtröder W. Laser Spectroscopy 2: Experimental Techniques. Berlin: Springer; 2015

Ferraro JR, Nakamoto K. Introductory Raman Spectroscopy. San Diego, London: Academic Press; 1994

Gupta VP. Molecular and Laser Spectroscopy: Advances and Applications. Amsterdam: Elsevier; 2018

Hendra P, Jones C, Warnes G. Fourier Transform Raman Spectroscopy. Chichester: Ellis Horwood; 1991

Long DA. The Raman Effect. Chichester: Wiley & Sons; 2002

McCreery RL. Raman Spectroscopy for Chemical Analysis. New York, Toronto, Weinheim: Wiley; 2000

Smith E, Dent G. Modern Raman Spectroscopy: A Practical Approach. Chichester: Wiley; 2005

Particular Techniques

Burns DA, Ciurczak EW. Handbook of Near-Infrared Analysis. Boca Raton: CRC Press; 2008

Fayer MD. Ultrafast Infrared and Raman Spectroscopy. New York, Basel: Marcel Dekker; 2001

Guozhen W. Raman Spectroscopy: An Intensity Approach. Singapore: World Scientific Publishing; 2017

Hasegawa T. Quantitative Infrared Spectroscopy for Understanding of a Condensed Matter. Tokyo: Springer; 2017

Herres W. Capillary Gas Chromatography - Fourier Transform Infrared Spectroscopy. New York, Heidelberg, Basel: Hüthig Verlag; 1987

Milosevic M. Internal Reflection and ATR Spectroscopy. Hoboken: Wiley & Sons; 2012

White R. Chromatography/Fourier Transform Infrared Spectroscopy and its Applications. New York, NY: Marcel Dekker; 1990

Databases and Application

Bertie JE, Keefe CD, Jones RN. Tables of Intensities for the Calibration of Infrared Spectroscopic Measurements in the Liquid Phase. Hoboken: Wiley; 1995

Cozzolino D. Infrared Spectroscopy: Theory, Developments and Applications. New York, NY: Nova Science Publishers; 2014

Krishnan K, Ferraro JR. Practical Fourier Transform Infrared Spectroscopy: Industrial and Laboratory Chemical Analysis. San Diego, CA: Academic Press; 1990

Kuptsov AH, Zhizhin GN. Handbook of Fourier Transform Raman and Infrared Spectra of Polymers. Amsterdam: Elsevier; 2011

Larkin P. Infrared and Raman Spectroscopy: Principles and Spectral Interpretation. Amsterdam: Elsevier, San Diego, CA: Oxford; 2011

Moore E. Fourier Transform Infrared Spectroscopy (FTIR): Methods, Analysis and Research Insights. New York, NY: Nova Science Publishers; 2016

Nyquist RA. Interpreting Infrared, Raman, and Nuclear Magnetic Resonance Spectra. San Diego, CA: Academic Press; 2001

Pouchert CJ. The Aldrich Library of FT-IR Spectra. Oxford: Wiley-Blackwell; 2009

Smith B. Infrared Spectral Interpretation: A Systematic Approach. Boca Raton: CRC Press; 1999

Socrates G. Infrared and Raman Characteristic Group Frequencies. Chichester: Wiley & Sons; 2004

Wartewig S. IR and Raman Spectroscopy: Fundamental Processing. Weinheim: Wiley-VCH; 2003

Workman J. Practical Guide and Spectral Atlas for Interpretive Near-Infrared Spectroscopy. Boca Raton: CRC Press; 2012

Herbert Meier

3 Nuclear Magnetic Resonance Spectroscopy

3 Nuclear Magnetic Resonance Spectroscopy

3.1 Physical Principles

3.1.1 The Resonance Phenomenon

Most atomic nuclei have an angular momentum P (**nuclear spin**) and therefore a **magnetic moment** $\mu = \gamma P$. The **gyromagnetic ratio** γ is a characteristic constant for each individual nuclear type. From quantum theory

$$P = \sqrt{I(I+1)} \cdot \frac{h}{2\pi}$$

and

$$\mu = \gamma \cdot \sqrt{I(I+1)} \cdot \frac{h}{2\pi}$$

where I is the **nuclear angular momentum quantum number** or **nuclear spin quantum number** of the particular atomic nucleus and can have integer or half-integer values (**Table 3.1**).

$$I = 0, 1/2, 1, 3/2, 2, 5/2, 3, \ldots$$

In a **homogeneous, static magnetic field** B_0, the angular momentum vector P can take up specific selected angles to the B_0 vector (**quantization of direction**). In these positions, the components of P in the direction of the field are given by

$$P_B = m \cdot \frac{h}{2\pi}$$

For the **orientational** or **magnetic quantum number** m, the allowed values are

$$m = +I, I-1, I-2, \ldots, -I+1, -I$$

The eigenstates ($2I + 1$, in total), so-called nuclear Zeeman levels, have different energies, given by

$$E_m = -\mu_B \cdot B_0 = -\gamma \cdot P_B \cdot B_0$$
$$= -\gamma \cdot m \cdot \frac{h}{2\pi} \cdot B_0$$
$$(m = +I, \ldots, -I)$$

For the hydrogen nucleus, the proton, $I = 1/2$, and therefore $m = \pm 1/2$.

The resultant energy level scheme is shown in **Fig. 3.1**. In the lower energy state, μ precesses with the Larmor frequency $\nu_0 = |\gamma| B_0/2\pi$ about B_0 and conversely in the higher energy state about $-B_0$. (If E_m is defined with a positive sign, then the magnetic quantum numbers m in **Fig. 3.1** must be exchanged.)

In thermal equilibrium, the ^1H nuclei are subject to **Boltzmann distribution**. Since the energy difference

$$\Delta E = \gamma \cdot \frac{h}{2\pi} \cdot B_0$$

is very small compared to the average thermal energy, the lower energy state is only very slightly more populated.

The relationship of the populations is given by

$$\frac{N_{(m=-1/2)}}{N_{(m=+1/2)}} = e^{-\frac{\Delta E}{kT}} .$$

Irradiation with quanta of energy of ΔE induces spin inversion. Because of the difference in populations, absorption A dominates[*]. The **resonance condition** is defined by the relationship:

$$h\nu = \Delta E = \gamma \cdot \frac{h}{2\pi} \cdot B_0$$

For a field strength of 2.35 T, the resonance frequency for protons $\nu = f(B_0)$ is close to 100 MHz, corresponding to radio waves with $\lambda = 3$ m. Commercial nuclear magnetic resonance (NMR) spectrometers have a proton frequency of up to 1,000 MHz, corresponding to a magnetic field of 23.49 T[**]. The production of 1.1 and 1.2 GHz NMR spectrometers is announced. If the resonance condition is fulfilled, the absorption would very rapidly eliminate the population difference between the two nuclear Zeeman levels; the system would then be described as being saturated, unless the reverse process, **relaxation**, takes place to a sufficient extent.

The energy released by the transition of a nucleus from a higher to a lower energy state can be absorbed by the environment in the form of heat (**spin–lattice relaxation**). This process takes place with a rate constant $1/T_1$. T_1 is called the **longitudinal relaxation time** because the process alters the magnetization in the direction of the field. The transverse magnetization also varies with time because of the interaction of the magnetic moments with each other (**spin–spin relaxation**). This process is characterized by the **transverse relaxation time** T_2.

---

[*] Nuclear magnetic emission is also possible, as in the CIDNP effect (chemically induced dynamic nuclear polarization). For details, see NMR texts quoted in the bibliography.

[**] For comparison, the magnetic field of Earth is $(50 \pm 25)10^{-6}$ T, (0.002 MHz = 2 kHz).

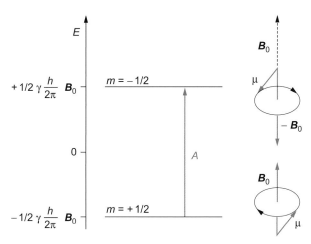

Fig. 3.1 Energy levels of protons in a magnetic field B_0.

As shown earlier, a magnetic moment $\mu \neq 0$ is a precondition for the NMR experiment. (The only nuclei with magnetic moment $\mu = 0$ are the e,e-nuclei with even mass and atomic numbers.)

Furthermore, it is of advantage if $I = 1/2$, since nuclei with larger spin quantum numbers additionally possess an electric **nuclear quadrupole moment**, which produces complications in the spectra (signal broadening).

In organic chemistry, the most important nuclei are ^{1}H, ^{7}Li, ^{11}B, ^{13}C, ^{15}N, ^{17}O, ^{19}F, ^{29}Si, ^{31}P, and ^{77}Se. For ^{13}C, ^{15}N, ^{29}Si, and ^{77}Se, the low natural abundance is disadvantageous (**Table 3.1**).

In addition, many nuclei of metal atoms have $I = 1/2$, for example ^{109}Ag, ^{111}Cd, ^{113}Cd, ^{115}Sn, ^{117}Sn, ^{119}Sn, ^{123}Te, ^{125}Te, ^{187}Os, ^{195}Pt, ^{199}Hg, ^{203}Tl, ^{205}Tl, and ^{207}Pb. The absolute sensitivity of these nuclei is often low taking into account their natural abundance. However, there are exceptions: ^{117}Sn, ^{119}Sn, ^{125}Te, ^{195}Pt, ^{207}Pb, and in particular ^{203}Tl and ^{205}Tl have a higher sensitivity than ^{13}C. ^{205}Tl has a natural abundance of 70.5% and a magnetic moment of $\mu = 1.638\mu_N$ and therefore an absolute sensitivity that is 800 times higher than that of ^{13}C and only seven times lower than that of ^{1}H.

A large magnetic moment for the nucleus to be studied is advantageous, since signal intensity is proportional to the third power of the magnetic moment. **Table 3.1** shows the important properties of those nuclei which are of relevance to organic chemistry.

The energy level scheme (**Fig. 3.1**) needs to be changed for nuclei with $I > 1/2$ in accordance with the equation for E_m. Since $\Delta m = 1$, however, the resonance condition remains the same. At a constant field B_0, the resonance frequencies ν of the various nuclei are directly related to the γ values. Since $\gamma(^1\text{H})/\gamma(^{13}\text{C}) = 2.675/0.673 = 3.975$, it follows that a ^{1}H resonance frequency of, e.g., 400 MHz corresponds to a ^{13}C resonance frequency of $400/3.975 \approx 100.6$ MHz.

In the following sections (3.1.2–3.1.5), the important properties of resonance signals will be discussed:

- Their **position** (resonance frequency; Section 3.1.2, Chemical Shift, p. 88)

Table 3.1 Properties of nuclei of relevance to NMR spectroscopy of organic compounds

Isotope	Spin quantum number	Gyromagnetic ratio γ [10^7 rad/Ts]	Magnetic moment μ [in units of μ_N]	Natural abundance (%)	Relative sensitivity per nucleus	Absolute sensitivity taking account of natural abundance	Resonance frequency ν_0 (MHz) at a field of 2.3488 T
^{1}H	1/2	26.752	2.793	99.985	1.000	1.000	100.000
^{2}H≡D	1	4.107	0.857[b]	0.015	0.010	$1.45 \cdot 10^{-6}$	15.351
^{6}Li	1	3.937	0.822[b]	7.42	0.009	$6.31 \cdot 10^{-4}$	14.716
^{7}Li	3/2	10.396	3.256[b]	92.58	0.294	0.27	38.862
^{10}B	3	2.875	1.801[b]	19.6	0.020	$3.90 \cdot 10^{-3}$	10.747
^{11}B	3/2	8.584	2.688[b]	80.4	0.165	0.13	32.084
^{13}C	1/2	6.728	0.702	1.10	0.016	$1.76 \cdot 10^{-4}$	25.144
^{14}N	1	1.934	0.404[b]	99.634	0.001	$1.01 \cdot 10^{-3}$	7.224
^{15}N	1/2[a]	−2.712	0.283	0.366	0.001	$3.85 \cdot 10^{-6}$	10.133
^{17}O	5/2[a]	−3.628	1.893[b]	0.038	0.029	$1.08 \cdot 10^{-5}$	13.557
^{19}F	1/2	25.181	2.627	100.0	0.833	0.833	94.077
^{29}Si	1/2[a]	−5.319	0.555	4.67	0.008	$3.69 \cdot 10^{-4}$	19.865
^{31}P	1/2	10.841	1.132	100.0	0.066	0.066	40.481
^{33}S	3/2	2.053	0.643[b]	0.76	0.003	$1.72 \cdot 10^{-5}$	7.670
^{77}Se	1/2	5.101	0.532	7.6	0.007	$5.25 \cdot 10^{-4}$	19.067

[a] In these cases $\gamma < 0$, i.e. the magnetic moment and nuclear spin point in opposite directions
[b] These nuclei also have an electric quadrupole moment

▪ Their **fine structure** (Section 3.1.3, Spin–Spin Coupling, p. 89)

▪ Their **linewidth** (p. 97).

▪ Their **intensity** (p. 98).

NMR spectra can be calculated relatively accurately using density functional theory. However, the time required is large, even for simple molecules.

3.1.2   Chemical Shift

The exact resonance frequency of a particular nuclear type depends in a characteristic fashion on the environment of the nucleus. The **effective strength of the magnetic field** at the nucleus differs from B_0 by the induced field:

$$B_{\text{eff}} = B_0 - \sigma B_0.$$

Including the dimensionless **shielding constant** σ in the resonance condition leads to

$$v = \frac{\gamma}{2\pi} B_0 (1 - \sigma)$$

The stronger a nucleus is shielded, i.e., the larger σ is, the smaller B_{eff} becomes. This means that, for a constant frequency, the applied field B_0 must be stronger in order to bring the nucleus to resonance. Similar considerations show that at constant B_0 field, v decreases as the shielding increases.

Because of the relationship $v = f(B_0)$, the position of the NMR absorption cannot be given by an absolute scale of values of v or B_0. Instead, the signal position is related to that of a **reference compound**. For ^{1}H and ^{13}C NMR spectroscopy, **tetramethylsilane** [**TMS**, $\text{Si}(\text{CH}_3)_4$] is the usual standard. At an observation frequency v, the difference of the positions of the signals of the observed nucleus X and TMS is given by

$$\Delta B = B(\mathbf{X}) - B(\mathbf{TMS})$$

and similarly for a frequency scale in Hz

$$\Delta v = v(X) - v(\text{TMS}) = \frac{\gamma}{2\pi} \cdot \Delta B$$

To specify the position of the signal, the **chemical shift** δ of the nucleus X is defined by

$$\delta(X) = 10^6 \frac{\Delta v}{v} \quad \text{(with } \delta(\text{TMS}) = 0 \text{ for } {}^1\text{H and } {}^{13}\text{C)}$$

δ is a dimensionless quantity, independent of the measurement frequency or the magnetic field strength, characteristic of the observed nucleus in its environment. (In ^{1}H NMR, the τ scale was originally popular, $\tau = 10 - \delta$.) Since Δv is very small compared to v, the factor 10^6 has been introduced and δ is quoted in **ppm** (**parts per million**); ppm is not a dimension, but is conventionally added to the δ value. The range of the δ scale for ^{1}H NMR is about 12; for ^{13}C NMR about 220 ppm. If extreme values are considered, the ranges increase to 40 or 350 ppm, respectively.

The exact resonance frequencies vary about v_0 (**Table 3.1**) in the ppm range.

Fig. 3.2 shows an example of the ^{1}H NMR and the ^{13}C NMR spectrum of acetic acid (**1**). In each case two absorptions are observed: the H- and C-atoms of the carboxy group appearing at lower field. These atoms (nuclei) are therefore less shielded than the methyl protons or the methyl carbon atoms, respectively. The calculation of the δ values follows from the equation given earlier for the chemical shifts.

Let us take the methyl signal in the ^{1}H NMR spectrum as an example. At an observation frequency of 200 MHz it appears 420 Hz to low field of the TMS signal. This corresponds to 2.10 ppm. Following current convention, one writes $\delta = 2.10$ or $\delta = 2.10$ ppm.

$$\delta_H(\text{CH}_3) = 10^6 \frac{420}{200 \cdot 10^6} = 2.10$$

δ values are positive in the direction of increasing resonance frequency.

The sensitivity of the **chemical shift** to changes in the environment of the measured nucleus is of great importance for the determination of the structures of organic compounds. The **shielding constant** σ, which determines the resonance position, is made up of three terms:

$$\sigma = \sigma_{\text{dia}} + \sigma_{\text{para}} + \sigma'$$

The diamagnetic term σ_{dia} corresponds to the opposing field induced by the external field in the electron cloud surrounding the nucleus. Electrons near the nucleus shield more than distant ones. The paramagnetic term σ_{para} corresponds to the excitation of p-electrons in the field and has an opposite effect to the diamagnetic shielding. Since for hydrogen only s-orbitals are present, only σ_{dia} is important. For higher nuclei such as ^{13}C, the paramagnetic term dominates. The term σ' describes the influence of neighboring groups, which can decrease or increase the field at the nucleus. Finally, σ also depends on intermolecular effects, which can be included by the addition of an extra term σ_{Medium}.

The width of the region of chemical shifts depends on the kind of nucleus. Apart from extreme cases, the following ranges can be assumed:

▪ ^{1}H ~ 12 ppm

▪ ^{13}C ~ 220 ppm

▪ ^{19}F ~ 400 ppm

▪ ^{31}P ~ 1,000 ppm

▪ ^{15}N ~ 900 ppm.

Other nuclei can have a significantly broader range of **chemical shifts**. The range of ^{119}Sn amounts to 6,000 ppm, for example. Broader the range, smaller the differences can become apparent in the chemical environment of the nuclei studied.

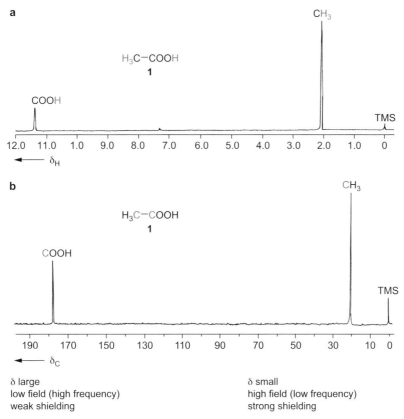

Fig. 3.2 (a) ^{1}H NMR spectrum of acetic acid (**1**) in CDCl$_3$. **(b)** ^{13}C NMR spectrum of acetic acid (**1**) in CDCl$_3$ (^{1}H broadband decoupled, i.e., without the effects of the (^{13}C, ^{1}H) couplings, see Section 3.1.3).

3.1.3 Spin–Spin Coupling

The signals observed in NMR spectra often show fine structure. Depending on the number of components of each signal, they are referred to as singlets, doublets, triplets, quadruplets, etc., or in general multiplets. The cause of this fine structure is the interaction with neighboring nuclei which possess a magnetic moment. This **spin–spin coupling** occurs between nuclei of the same type (**homonuclear**) and between nuclei of different type (**heteronuclear**) and means that the orientation of the spin of nucleus A influences the local field at the coupling nucleus X and vice versa. For two nuclei A and X, both having nuclear spin 1/2, there exist in principle four energy levels, corresponding to the four possible orientations of the two nuclei. Without spin–spin interaction ($J = 0$), each of A and X shows two absorptions of the same energy (**Fig. 3.3**, middle). This degeneracy is lifted by the **coupling** J. J is defined as having a positive sign if the energy levels of spin states, where both nuclei have the same spin orientation with respect to the external field, are raised. Spin states with opposite orientations are decreased in energy by the same amount. The reverse

situation applies when $J < 0$. Both cases cause the splitting of the A and X signals into doublets (**Fig. 3.3**).

The magnitude of the coupling is given by the **coupling constant** J, which in this case can be measured directly from the separation of the X lines or from the identical separation of the A lines. For proton–proton coupling, the values lie between about –20 and +20 Hz. With other nuclei, much larger values can occur. Thus in acetylene, the ^{13}C,^{13}C coupling is 171.5 Hz and the ^{13}C,^{1}H coupling is 250 Hz.

Particularly important is the fact that the coupling constant is independent of the external field $\boldsymbol{B}_0$. Two lines in a spectrum can either be singlet signals of two uncoupled nuclei with different chemical shifts or a doublet, arising from one nucleus, which forms an AX system with another coupling nucleus. A distinction between these two cases is easily made by measuring two spectra at different frequencies (see Section 3.3.8, p. 145). If the separation remains the same, the cause is coupling; if the separation (in Hz) increases with increasing observation frequency, the lines are two singlets of different chemical shifts. (On the δ scale, the separation is independent of the observation frequency ν_0.)

a

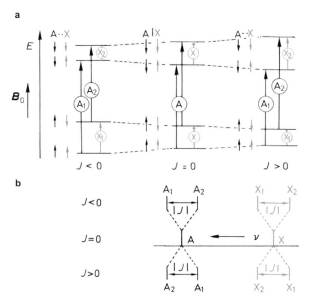

Fig. 3.3 (a) The four possible spin combinations and energy levels of a two-spin system ($m = \pm 1/2$) and the corresponding NMR transitions for $J \lessgtr 0$: ↑ A resonances, ↑ X resonances. **(b)** Stick spectra of the cases where $J < 0$, $J = 0$, and $J > 0$.

The coupling between two nuclei A and X, in an isotropic liquid phase, takes place in general through the bonds within the molecule (**scalar coupling**).

$$-\overset{|}{\underset{|}{^{13}C}}\,{^{1}\!H}$$

If there is **one** intermediate bond, the coupling is referred to as a **direct coupling** ^{1}J, e.g. $^{1}H{-}^{19}F$, $^{1}H{-}^{13}C$, etc. This term should not be confused with the **direct dipole–dipole coupling**, a through-space effect that appears preferentially in orientated phases (liquid crystalline states, solids). In isotropic fluids, these nonscalar couplings are averaged to zero by the thermal motion of the molecules.

With two or three intermediate bonds, the coupling is referred to as a **geminal coupling** ^{2}J or a **vicinal coupling** ^{3}J, respectively, for example:

$$^{2}J: \quad {^{1}H}-\overset{|}{\underset{|}{C}}-{^{1}H} \qquad {^{1}H}-\overset{|}{\underset{|}{C}}-{^{13}\!C}-$$

$$^{3}J: \quad {^{1}H}-\overset{|}{C}={\underset{|}{C}}-{^{1}H} \qquad {^{1}H}-C\equiv C-{^{13}\!C}-$$

With increasing numbers of bonds between A and X, $J(H,H)$ generally decreases. In the case of couplings with heavier nuclei, there is often a nonuniform decrease of $|J|$ with the number of bonds, instead there may be an intermediate maximum. For the

detection of **long-range couplings** ^{n}J, the resolving power of the spectrometer is of critical importance.

The complexity of the coupling pattern increases with the number of coupling nuclei. If the chemical shifts δ_i of all nuclei i in a molecule are known, as well as the coupling constants $^{n}J_{i,j}$ for all the possible pairs of nuclei, the NMR spectrum can be calculated (simulated). Conversely, the δ and J values can be directly determined from simple spectra.

In order to treat the coupling phenomena in more complicated systems than the AX case described earlier, it is necessary to have a general systematic **nomenclature for spin systems**.

n **isochronous nuclei**, i.e., n nuclei, which have the same chemical shift, either by coincidence or as a result of their **chemical equivalence** (cf. Section 3.2, p. 100), form an system. If there is in addition a set of m nuclei, again isochronous within the set, the spin system is described as A_nB_m, A_nM_m, or A_nX_m, depending on whether the resonance frequency v_B of the second set differs slightly, moderately, or strongly from v_A. This notation has an important restriction, namely, that for any nuclear combination A_iB_j (or A_iM_j or A_iX_j, as appropriate) ($i = 1, ..., n; j = 1, ..., m$), the coupling is the same.

Isochronous nuclei A_i, which only have **one** spin–spin interaction with nuclei of a neighboring group, are said to be **magnetically equivalent**. The same applies to the nuclei B, M, or X. (Since the spin–spin coupling is a reciprocal property, it is not possible for the A spins of an A_nB_m system to be magnetically equivalent and the B spins to not be.) Where there are more than two sets of spins, e.g., in the system $A_nB_mX_i$, the definition of magnetic equivalence requires that there is only a single value of J_{AB}, J_{AX}, and J_{BX}, each. **Isochronosity** is a necessary, but not sufficient, condition for magnetic equivalence, while magnetic equivalence is a sufficient, but not necessary, condition for isochronosity. For a better understanding, this can be demonstrated by two examples. Difluoromethane (**2**) has two isochronous H nuclei and two isochronous F nuclei. Each coupling $^{2}J(H,F)$ is the same. Therefore, the two H nuclei and the two F nuclei are chemically and magnetically equivalent as pairs . CH_2F_2 has an A_2X_2 spin system. In 1,1-difluoroethylene (**3**), the H nuclei are also chemically equivalent, as are the F nuclei, but there are two different couplings $^{3}J(H,F)$—as seen from a single hydrogen, (Z)-F and (E)-F couple differently. Isochronous nuclei which are not magnetically equivalent are indicated by a prime. 1,1-Difluoroethylene (**3**) has an AA′XX′ spin system.

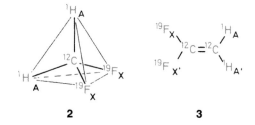

2   **3**

For each molecule, the ^1H and ^{19}F spectra show one half of the total spectrum; in the ^1H spectrum the A part and in the ^{19}F spectrum the X part. If one considers the case where the spin-free ^{12}C nuclei are replaced by ^{13}C nuclei, one would obtain in the coupled ^{13}C spectrum for difluoromethane (**2**) the M part of an A_2MX_2 system and for 1,1-difluoroethylene (**3**) the MN part of an AA'MNXX' system.

For the interpretation of the coupling patterns, it is important to note that **spin–spin coupling between magnetically equivalent nuclei has no effect on the spectrum,** although such nuclei do in fact have a coupling.

A single set of nuclei, as appearing in, for example, the ^1H spectrum of methane, ethane, ethylene, acetylene, or benzene, gives a singlet absorption. (The coupling between ^1H and ^{13}C is not usually observed in routine ^1H NMR spectra, since the natural ^{13}C content is small: 1.1%.) In these examples, the protons are isochronous on account of their chemical equivalence, but even if the isochronosity is coincidental, no splitting is observed.

An example is given in **Fig. 3.4**. Methyl 3-cyanopropionate (**4**) shows a single, unsplit signal at $\delta = 2.68$ for the two chemically inequivalent methylene groups.

Spin systems of the A_nX_m or A_nM_m type with two sets of magnetically equivalent nuclei are easy to interpret, as long as $|v_A - v_M|$ is at least a factor of ≈ 10 greater than $J_{A,M}$. The number of lines, the so-called **multiplicity of the signal**, is then

for A: $m \cdot 2I_X + 1$;

for X: $n \cdot 2I_A + 1$.

I_X and I_A are the spins of the nuclei X and A. When $I_X = I_A = 1/2$, a **first-order spectrum** is obtained with

$(m + 1)$ lines in the A part and

$(n + 1)$ lines in the X part.

Consider as an example the ^1H spectrum of bromoethane (**5**; **Fig. 3.5**). It has an A_3X_2 spin system. The local field at the position of the three chemically and magnetically equivalent methyl protons is affected by the nuclear spin of the two methylene

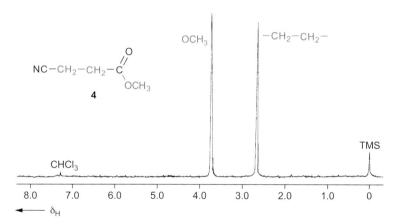

Fig. 3.4 ^1H NMR spectrum (60 MHz) of methyl 3-cyanopropionate (4) in $CDCl_3$.

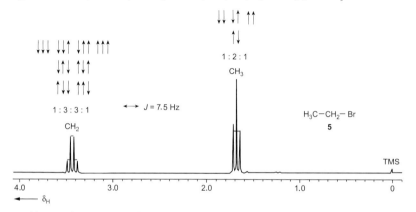

Fig. 3.5 ^1H NMR spectrum of bromoethane (**5**) in $CDCl_3$. (The splitting of the methyl signal into a triplet and the methylene signal into a quadruplet is explained by the spin orientations of the coupling protons in the neighboring groups.)

Nuclear Magnetic Resonance Spectroscopy

protons. These can be both parallel, both antiparallel, or one parallel, one antiparallel to the external field. There are four resultant energy levels, of which the two with opposing spins are degenerate. Because of the equal population probabilities for the individual spin states, the coupling from the methylene protons results in a triplet signal for the methyl protons, with an intensity distribution of 1:2:1. The chemical shift is given by the **center of gravity** of the signal: 1.67 ppm. In an exactly analogous way, the local field at the position of the methylene protons is influenced by spin–spin coupling from the methyl protons. For the three protons of the CH_3 group, there are eight spin combinations. The one of lowest energy has a total spin $m = 3/2$, the one of highest energy $m = -3/2$; in between are three degenerate states with $m = +1/2$ and three degenerate states with $m = -1/2$. The methylene group therefore produces from the coupling with the methyl group a quartet with the intensity distribution 1:3:3:1 (**Fig. 3.5**). The center is at $\delta = 3.43$ ppm. The separation of the lines in the triplet and the quartet is equal to the coupling constant J.

Table 3.2 shows the splitting patterns for the signal of a nucleus (or group of magnetically equivalent nuclei) in first-order spectra, as they vary with the number of coupling partner nuclei.

The **chemical shifts** of the nuclei in an A_nX_m spin system are given by the centers of the multiplets. The coupling constant J_{AM} can be directly measured in Hz from the separation of any two neighboring lines in either the A part or the X part of the spectrum (cf. **Fig. 3.5**). As a further example, the ^{13}C spectra of chloroform compared to deuteriochloroform and of dichloromethane compared to $CDHCl_2$ and CD_2Cl_2 will be considered. For $CHCl_3$ a doublet is obtained; for $CDCl_3$ a triplet (**Fig. 3.6**). Because of the isotope effect, there is a small difference in the chemical

shifts. The intensity ratios are 1:1 for $CHCl_3$ and 1:1:1 for $CDCl_3$. These ratios can be directly taken from **Table 3.2**, bearing in mind that protons have a spin of 1/2 and deuterons a spin of 1. Very noticeable is the difference in the coupling constants $J(C,H)$ and $J(C,D)$. The $J(C,H)$ coupling is larger by a factor that is very nearly equal to the ratio of the gyromagnetic ratios: $\gamma_H/\gamma_D \approx 6.5$.

In the coupled ^{13}C spectrum, CH_2Cl_2 gives a 1:2:1 triplet $[\delta(^{13}C) = 53.8, \, ^1J(C,H) = 177.6 \text{ Hz}]$, CD_2Cl_2 a 1:2:3:2:1 quintet, and $CHDCl_2$ a doublet of triplets with six lines of equal intensity. The rule is:

If a nucleus A or a set of magnetically equivalent nuclei A_n couples with two sets of neighboring nuclei M_m and X_l, the multiplicity of the signal of A is the product of the multiplicities caused by M and X, i.e., for spin 1/2 nuclei: $(m + 1) \cdot (l + 1)$.

The doublet in the A part of an AM spectrum for example becomes a doublet of doublets from additional AX coupling. If by coincidence $J_{AM} = J_{AX}$, then two of the four lines fall together, and a 1:2:1 triplet is obtained (**Fig. 3.7**).

Generally, the $(m + 1) \cdot (l + 1)$ lines of the A part of an $A_nM_mX_l$ spin system form a $(m + l + 1)$ multiplet if $J_{AM} = J_{AX}$. The A nuclei behave as if they were "seeing" $(m + l)$ magnetically equivalent neighboring nuclei. Consider for example the isopropyl and n-propyl groups in the structurally isomeric nitropropanes **6** and **7**.

$$H_3C - \underset{\underset{\textstyle NO_2}{|}}{CH} - CH_3 \qquad\qquad H_3C - \underset{\textstyle \gamma}{CH_2} - \underset{\textstyle \beta}{CH_2} - \underset{\textstyle \alpha}{NO_2}$$

$$\textbf{6} \qquad\qquad\qquad\qquad \textbf{7}$$

2-Nitropropane (**6**) forms an A_6X spin system. The six methyl protons are chemically and magnetically equivalent. The coupling with the methine proton splits their signal into a doublet.

Table 3.2 Coupling patterns in first-order spectra caused by spin–spin interactions

Number of coupling neighboring nuclei with spin		Number of lines (Signal multiplicity)	Relative intensities[a]
$I = 1/2$	$I = 1$		
0		1 (singlet)	1
1		2 (doublet)	1:1
2		3 (triplet)	1:2:1
3		4 (quartet, quadruplet)	1:3:3:1
4		5 (quintet, quintuplet)	1:4:6:4:1
5		6 (sextet)	1:5:10:10:5:1
6		7 (septet)	1:6:15:20:15:6:1
	0	1 (singlet)	1
	1	3 (triplet)	1:1:1
	2	5 (quintet, quintuplet)	1:2:3:2:1
	3	7 (septet)	1:3:6:7:6:3:1

[a] When $I = 1/2$ the relative intensities are the binomial coefficients which can be calculated from Pascal`s triangle

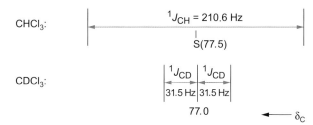

Fig. 3.6 ^{13}C stick spectra of CHCl$_3$ and CDCl$_3$ [taking into account the (C–H) and (C–D) coupling].

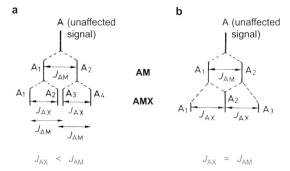

Fig. 3.7 Coupling pattern of the A part of an AMX spectrum.

The methine proton itself appears as a septet at lower field (**Fig. 3.8**).

The protons in 1-nitropropane (**7**) form an A$_3$M$_2$X$_2$ spin system, assuming magnetic equivalence of the protons within each methylene group (but see Section 3.2.2, p. 100). The methyl protons A and the α-CH$_2$ protons X each have the two protons of the β-methylene group as neighbors. For the A and X signals,

a triplet is therefore expected, and for the M protons a dodecet [12 = (3 + 1)·(2 + 1)]. Since the coupling constants $^3J_{AM}$ and $^3J_{AX}$ however are in practice of equal magnitude, a sextet is observed for the β-methylene group of **7** (see **Fig. 3.9**).

The simplest second-order spectrum is produced by the AB system. As in the AX case, there are four lines (**Fig. 3.10**).

If a molecule has two homonuclear sets of nuclei A$_n$B$_m$, where the quotient $|v_A - v_B|/J_{AB}$ is less than 10, then the rules for **first-order** spectra lose their validity. In **Fig. 3.9**, one can already see that the intensities of the lines in the two triplets are no longer exactly in the ratio 1:2:1. The lines that lie nearer to the signal of the coupling partners (β-CH$_2$ group) are more intense than those that lie further away. This phenomenon is referred to as the **roof effect**. Since

$$\frac{[V_A - V_M]}{J_{AM}} < \frac{[V_X - V_M]}{J_{XM}}$$

the roof effect is more marked in the methyl triplet than in the methylene triplet. For the same reason, the overriding roof effect in the β-methylene group is toward the methyl triplet. In complicated spectra, the roof effect can help to identify the coupling partners. **Second (higher) order** spectra occur also with $\Delta v/J > 10$ in those cases where sets of nuclei are present that are chemically equivalent but magnetically nonequivalent (e.g., AA′XX′). In general, spectra can be classified as **zero order** (only singlet signals), **first order**, and **higher order**.

The simplest second-order spectrum is produced by the AB system. As in the AX case, there are four lines (**Fig. 3.10**).

The spectrum shows the same symmetry about the center 1/2 $(v_A + v_B)$ as the AX system, and the separation of the two A or

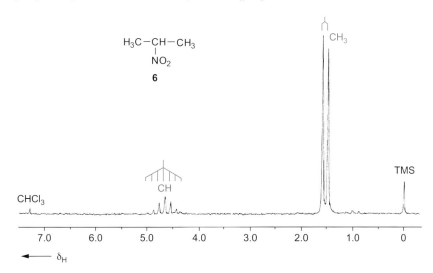

Fig. 3.8 ^{1}H NMR spectrum of 2-nitropropane (**6**) in CDCl$_3$.

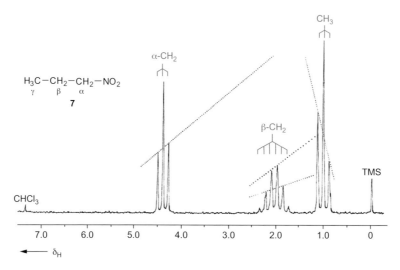

Fig. 3.9 ^{1}H NMR spectrum of 1-nitropropane (**7**) in CDCl$_3$.

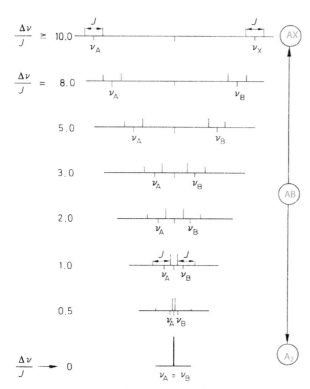

Fig. 3.10 Stick spectra of an AB system with constant coupling J_{AB} and varying ratio $\Delta v/J$.

B lines is also equal to the coupling constant J_{AB}. However, the intensities of the lines are fundamentally different for AX and AB systems. As **Fig. 3.10** shows, the appearance of the spectrum depends on the ration $\Delta v/J$. For Δv near 0, the appearance of the

spectrum approaches the singlet of an A_2 system. The opposite limit, when Δv is large, approaches an AX spectrum.

The analysis of an AB spectrum will be demonstrated with the example of (*E*)-cinnamic acid (**8**; **Fig. 3.11a**). The olefinic protons form an AB system (long-range coupling to the carboxyl proton or the phenyl protons is not observed at this resolution). Since

$$v_1 - v_2 = v_3 - v_4 = 16 \text{ Hz},$$

it follows immediately that $J_{AB} = 16$ Hz.

Since

$$V_1 - V_3 = V_2 - V_4 = \sqrt{(V_A - V_B)^2 + J^2} = 82 \text{ HZ},$$

it follows that

$$v_A - v_B = 80 \text{ Hz (at 60 MHz)}.$$

From the known position of the center point of the spectrum,

$$\tfrac{1}{2}(v_A + v_B) = \tfrac{1}{2}(v_1 + v_4) = \tfrac{1}{2}(v_2 + v_3)$$

the positions of v_A and v_B can be calculated. On the δ scale, the values obtained are $\delta_A = 7.82$ and $\delta_B = 6.47$. If the same compound is measured at 300 MHz (**Fig. 3.11b**), the AB system becomes an AX system and δ_A and δ_B are at the centers of the doublet signals. The small differences $\Delta\delta_A = 0.02$ and $\Delta\delta_B = 0.03$ ppm between **Figs. 3.11a** and **3.11b** are caused by different concentrations. The assignment of these two signals to the two different olefinic protons cannot be made from the spectrum alone; it requires the use of a system of chemical shift increments (see p. 144) or comparison with similar compounds or more sophisticated measurements, which will be discussed later.

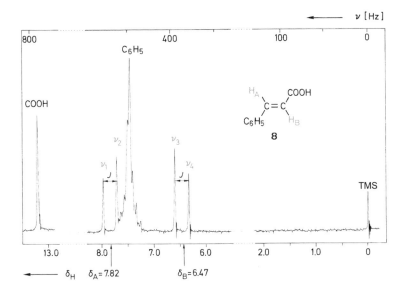

Fig. 3.11 (a) 60-MHz ^{1}H NMR spectrum of (*E*)-cinnamic acid (**8**) in CDCl$_3$ (v_1 = 478 Hz, v_2 = 462 Hz, v_3 = 396 Hz, v_4 = 380 Hz, v_A = 469 Hz, v_B = 388 Hz).

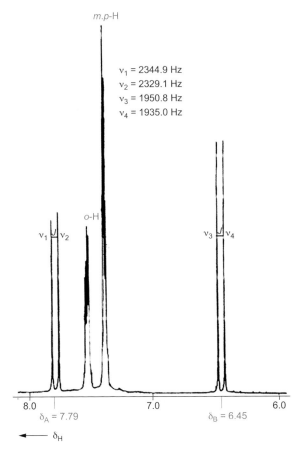

v_1 = 2344.9 Hz
v_2 = 2329.1 Hz
v_3 = 1950.8 Hz
v_4 = 1935.0 Hz

Fig. 3.11 (b) 300-MHz ^{1}H NMR spectrum of (*E*)-cinnamic acid (**8**) in CDCl$_3$ (the broad singlet of the COOH group [δ = 12.7 ppm] is not shown).

Of the possible three-spin systems A$_3$, A$_2$X, A$_2$M, A$_2$B, AMX, ABX, and ABC, the A$_2$B, ABX, and ABC cases cannot be treated according to first-order rules. The maximum possible number of lines in these three systems for nuclei with I = 1/2 is 9, 14, and 15, respectively, though some of these are often weak transitions that do not appear in routine spectra. (Spin systems AXX′ or ABB′ only occur when two chemically nonequivalent nuclei X or B are coincidentally isochronous, i.e., have the same chemical shift.)

For the exact analysis of three or more spin systems, the reader should refer to NMR textbooks. In cases that are only weakly second order, it is often possible to make an approximate analysis on the basis of first-order rules. An example is the ^{1}H NMR spectrum of phenyloxirane (styrene oxide **9**, **Fig. 3.12**). It has the ABM system. If it is treated as an AMX case, the parameters obtained (chemical shifts and coupling constants) deviate only insignificantly from those obtained by an exact analysis as an ABX case. (The analysis of an ABX case is described with reference to a heteronuclear example in Section 3.6.2, p. 266.) **Fig. 3.12** shows the 400 MHz ^{1}H NMR spectrum of phenyloxirane (**9**). The oxirane part represents an AMX spin pattern.

As shown on p. 93, an AMX system with three different coupling constants has four lines for each nucleus A, M, and X. (In an ideal case, all 12 lines should have the same intensity.) The coupling constants can be obtained directly from the frequencies of the individual lines (see **Figs. 3.7** and **3.12**).

Of the **higher spin systems**, only the following symmetrical **four-spin systems** will be considered further (**Table 3.3**).

For a better understanding of these spin systems, the following examples (**10–12**) have been selected and arranged so that the chemical shift difference decreases from left to right.

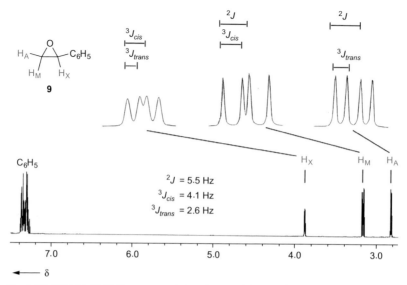

Fig. 3.12 400-MHz ^{1}H NMR spectrum of phenyloxirane (**9**) in CDCl$_3$.

Table 3.3 Symmetrical four-spin systems with two sets of nuclei

System	chemical shifts		Couplings	Transitions, max. number of lines [a]
First order				
A$_2$X$_2$	ν_A	ν_X	J_{AX}	6
Higher order				
AA'XX'	ν_A	ν_X	$J_{AA'}, J_{AX}$, $J_{AX'}, J_{XX'}$,	20
A$_2$B$_2$	ν_A	ν_B	J_{AB}	16 (18)
AA'BB'	ν_A	ν_B	$J_{AA'}, J_{AB}$, $J_{AB'}, J_{BB'}$	24 (28)

[a] for spins with $I = 1/2$

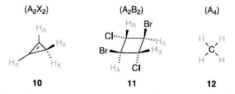

10   11   12

1,2-Disubstituted ethanes such as 3-nitropropionitrile (**13**) deserve special attention. While the examples cyclopropene (**10**), methane (**12**), thiophene (**14**), and ethene (**15**) are "rigid" molecules, rotation about the C–C bond must be considered for 1,2-disubstituted ethanes. This results in the chemical equivalence of the two protons labeled A as well as the two protons labeled X. The rotation does not however necessarily lead to

magnetic equivalence, i.e., $^3J_{AX}$ and $^3J_{AX'}$ can be different (see Section 3.2.2, p. 100).

(AA'XX')   (AA'BB')   (AA'A"A''')

13   14   15

Whether (**13**) is an AA'XX' or an AA'BB' system then depends on the difference $|\nu_A - \nu_X|$.

Fig. 3.13 shows the AA'**BB'** spectrum of o-dichlorobenzene (**16**) under conditions of high resolution; 24 lines can be identified, thus allowing all the parameters to be determined.

In many cases of four-spin systems, the spectra obtained under routine conditions show remarkably few lines. An example is the ^{1}H spectrum of furan (**17**; **Fig. 3.14**), an AA'XX' spin system. The interpretation of the two triplets as arising from an A$_2$X$_2$ system with a single value for J_{AX} would be incorrect. An exact analysis gives $J_{AX} = J_{A'X'} = 1.8$ and $J_{AX'} = J_{A'X} = 0.8$ Hz.

To conclude this section, it should be realized that the splitting of an NMR signal depends not only on the resolution of the instrument, but also on the measurement frequency (proportional to the field strength). For example, the quotient $\Delta\nu/J$, which determines whether or not spectra are first order, increases on going from 60 MHz (1.41 T) to 360 MHz (8.45 T) by a factor of 6; this means that a spectrum that is first order at 360 MHz can be second order at 60 MHz and therefore look quite different! A spin system with chemically equivalent but magnetically nonequivalent nuclei will give a second-order spectrum irrespective

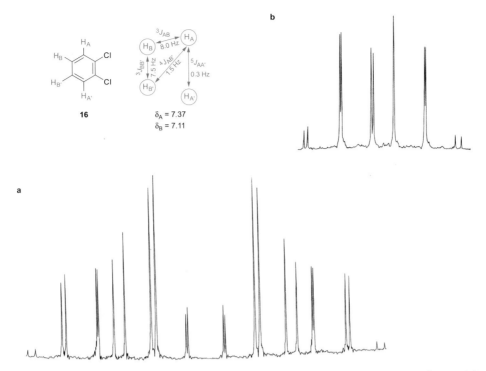

Fig. 3.13 High-resolution ¹H NMR spectrum of *o*-dichlorobenzene (**16**). (**a**) AA′BB′ system with 24 lines symmetrically spaced about the center, measured at 90 MHz; (**b**) MM′ part of the AA′MM′ spin system, measured at 400 MHz.

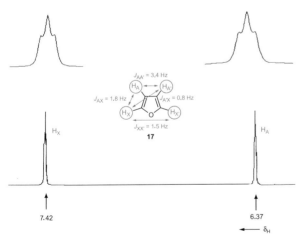

Fig. 3.14 ¹H NMR spectrum of furan (**17**) in CDCl₃ (AA′XX′ system, seemingly consisting of two triplets).

of the field (see **Fig. 3.13**), although the appearance of the spectrum can change considerably as the field is changed.

3.1.4 Linewidths

Fig. 3.15 shows the typical shape of an NMR signal. The **linewidth *b*** measured at half-height (often called $\mathbf{b}_{1/2}$) is

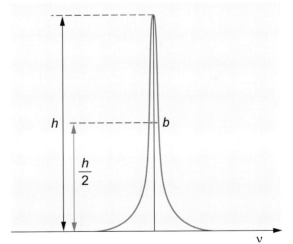

Fig. 3.15 Shape of an NMR signal; Lorentzian curve, *h* height, *b* linewidth.

considerably greater than the "natural linewidth" predicted by the Heisenberg uncertainty principle. *b* depends not only on the inhomogeneity of the magnetic field but also on long-range couplings and on the **relaxation times** T_1 and T_2 of the nucleus giving rise to the signal (see Section 3.1.1, p. 86).

Nuclei with an electric quadrupole moment, such as ^{14}N, or paramagnetic compounds present in the sample reduce the spin–lattice relaxation time T_1 and therefore broaden the lines. (Paramagnetic compounds themselves are poorly suited to NMR measurements for this reason.) In an analogous fashion, a reduction of the spin–spin relaxation time T_2, caused perhaps by an increase in the viscosity of the solution, also increases the linewidth. A special effect is the **line broadening** caused by **exchange phenomena**. In these cases, a distinction must be made between **intermolecular** and **intramolecular processes**. An example of the former is **proton transfer** in carboxylic acids, alcohols, or amines. If the ^1H spectrum of aqueous methanol is measured at 0°C, two OH signals are observed if the water concentration is less than 5%. At higher water concentrations, the proton transfer is accelerated. At first, a broadening of the OH signals is observed, and finally in the region where exchange is fast, a single OH signal is observed at an averaged chemical shift. As the lines broaden, so the coupling of the OH protons with the protons of the methyl groups disappears. In pure methanol, this latter effect can be produced by raising the temperature from 0°C to +10°C. (Acceleration of proton exchange on warming.)

Intramolecular exchange phenomena arise from **flexibility of molecules** (rotation, inversion, etc.) or chemical reactions (rapid rearrangements, valence isomerism, etc.). Examples will be discussed in Section 3.2.2 (p. 100). As a general rule, two nuclei, which exchange their chemical environments, will give two separate signals, v_A and v_B, if the exchange is slow on the NMR time scale. This is the case if the **lifetime** τ in the two states satisfies the condition $\tau \cdot |v_A - v_B| \gg 1$.

Conversely, if $\tau \cdot |v_A - v_B| \ll 1$, a single, averaged signal is obtained. In the intermediate region, the **coalescence** of the two signals is observed (see p. 108). In this region, the line shape is very dependent on τ. Since τ is a function of temperature, the spectra in this region are highly temperature-dependent. Determination of the coalescence point, or more precisely a **line-shape analysis**, allows the thermodynamic parameters of such processes to be calculated.

3.1.5   Intensity

The **area under the absorption curve** of an NMR signal is a measure of the intensity of the transition. The integration is drawn by the spectrometer as a **step curve**.

In ^1H spectra, the intensity measured by the height of the steps is proportional to the number of ^1H nuclei in the molecule which are responsible for the signal (see ethyl *p*-toluenesulfonate, **18**; **Fig. 3.16**).

In the quantitative analysis of mixtures, the number of chemically equivalent protons responsible for each signal must be taken into account. If the area F_A corresponds to n_A protons of substance

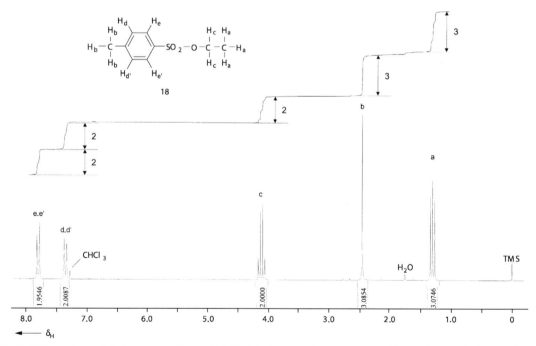

Fig. 3.16 ^1H NMR spectrum of ethyl *p*-toluenesulfonate (**18**; 60 MHz, CCl_4) with integration curve (the numbers under the signals are values of the integration magnitudes).

A (as determined by the structural formula) and the area F_B similarly corresponds to n_B protons of substance B, then the molar concentrations c in the solutions being measured are given by

$$\frac{c_A}{c_B} = \frac{F_A \cdot n_B}{F_B \cdot n_A}.$$

In routine ^{13}C spectra, exact quantitative conclusions cannot be drawn from the intensities of the signals. The intensity of a signal is proportional to the effective difference in populations between the energy levels giving rise to the signals, and therefore depends critically on the relaxation times. The T_1 relaxation times of ^{13}C nuclei lie in the region from 10^{-1} to 3×10^2 s. Protons bound directly to the relevant ^{13}C nucleus are most effective at reducing T_1. Quaternary ^{13}C atoms show the largest T_1 values, and consequently their signals are the weakest. Furthermore, molecular size and molecular motions have considerable influence on the longitudinal relaxation. This is clearly shown by the examples of toluene (**19**), styrene (**20**), and polystyrene (**21**).

The T_1 values, given in seconds, show marked differences between large and small molecules and between hydrogen-bearing

and quaternary C-atoms. The anisotropy of the molecular motions also plays a role. Thus, the *para*-carbons, which lie on the preferred axis of rotation, have the lowest T_1 values. (In special cases, this effect can be used for signal assignments.)

The T_1 relaxation times of protons are often shorter than those of the corresponding ^{13}C nuclei. The following values were found for cyclohexane and benzene at room temperature:

	T_1 (^{1}H)	T_1 (^{13}C)
C_6H_{12}	0.8 s	19.7 s
C_6H_6	2.2 s	26.8 s

The time interval between two successive pulses (cf. p. 119) is generally too short in ^{13}C NMR to allow relaxation of the spin system to the equilibrium state.

Fig. 3.17 shows the ^{13}C spectrum of 3-methylphenol (*m*-cresol; **22**). The seven ^{13}C signals have quite different intensities.

Irradiation at or near the NMR frequency of a nucleus leads to perturbation of the relaxation of neighboring nuclei. This leads to a change in the intensities of the signals of these neighboring nuclei (**nuclear Overhauser effect, NOE**). In ^{1}H NMR, this only occurs in double resonance experiments (see p. 155f). In ^{13}C NMR, however, routine spectra are measured with ^{1}H broadband decoupling. This removes the splitting caused by the coupling to the protons—the spectrum consists of individual singlet peaks (see **Fig. 3.2b**, p. 89, or **Fig. 3.17**). The heteronuclear NOE caused by the decoupling produces increases in intensity of

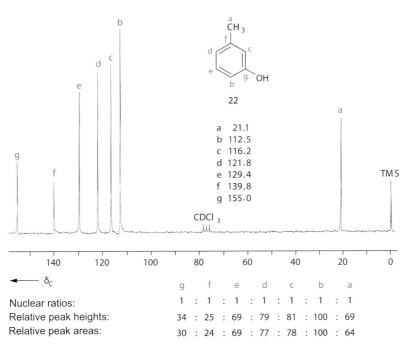

Nuclear ratios:

g		f		e		d		c		b		a
1	:	1	:	1	:	1	:	1	:	1	:	1

Relative peak heights:

| 34 | : | 25 | : | 69 | : | 79 | : | 81 | : | 100 | : | 69 |

Relative peak areas:

| 30 | : | 24 | : | 69 | : | 77 | : | 78 | : | 100 | : | 64 |

Fig. 3.17 ^{13}C NMR spectrum of *m*-cresol (**22**; in CDCl$_3$, ^{1}H broadband decoupled) with an evaluation of the signal intensities.

up to 200%. **Differential relaxation and differential NOEs are therefore the reason for the deviations of measured from theoretical intensity ratios of ^{13}C NMR signals.** Furthermore, the large influence of the measurement conditions on peak intensities must also be taken into account. It is however possible to avoid these disadvantages of ^{13}C NMR spectroscopy by removing the effects and thus obtaining spectra which can be integrated if necessary (see Section "Integration of NMR spectra," p. 206).

3.2 NMR Spectra and Molecular Structure

3.2.1 Molecules with "Rigid" Atomic Positions

The number of NMR signals which appear in a spectrum is dependent on the symmetry of the molecule under investigation. **Two nuclei in a molecule are chemically equivalent if they can be transposed by a symmetry element present in the molecule or if they become identical in the time average as a result of a fast (intramolecular) exchange process.** For a thorough understanding, a few examples with a rigid carbon skeleton will be considered. **Table 3.4** gives the number of expected ^{13}C and ^{1}H NMR signals of a few selected structures of various symmetries (point groups). The higher the symmetry of the molecule, the fewer the number of signals. In buckminsterfullerene C_{60} (point group I_h), for example all 60 carbon atoms have the same chemical shift ($\delta = 143.2$).

Fullerene C_{70} is less symmetric (point group D_{5h}) and has five δ values (150.7, 148.0, 147.4, 145.4, and 130.9). Still higher fullerenes have even more ^{13}C NMR signals. C_{76} (point group D_2) for example gives 19 δ values.

The ways in which ^{13}C and ^{1}H NMR spectroscopy usefully complement each other is demonstrated through the example of some disubstituted benzenes in **Table 3.5**.

3.2.2 Intramolecular Motion

As was stated at the beginning of this section, nuclei that are not related by a symmetry element can be chemically equivalent if they become identical as a result of an intramolecular motion. For example, the three protons of a freely rotating methyl group are chemically equivalent. Alanine (**23**) possesses no symmetry element, and therefore certainly no C_3 axis through the methyl carbon atom; nevertheless the three protons H_A are identical because of the rotation of the methyl group. The A_3X system of the four protons bound to carbon atoms gives in D_2O a doublet at $\delta = 1.48$ and a quartet at $\delta = 3.78$.

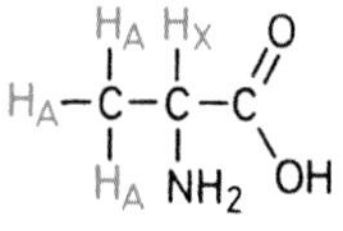

23

In the same way, the tertiary butyl group of compound **24** shows a single ^{1}H signal (singlet) and two ^{13}C signals for the quaternary and the three primary C atoms. Deviations from this behavior occur when the rotation of the t-butyl group is restricted. There are only a few examples of the freezing out of a methyl rotation; thus 9-methyltriptycene-1,4-quinone (**25**) shows anisochronous methyl protons at −141°C.

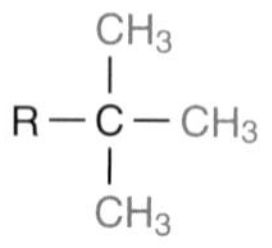

24 (R = OH, Cl, 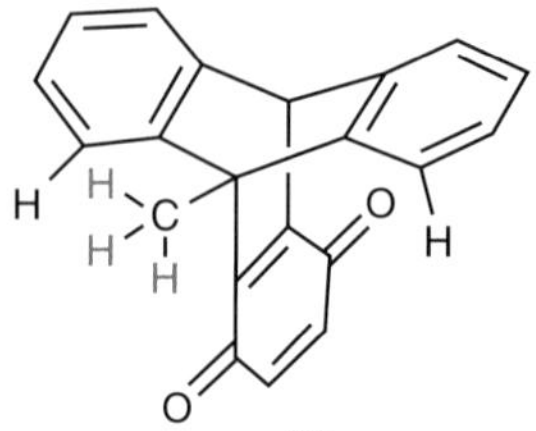 ...)   **25**

For CX_2 groups (X = H, CH_3, etc.), the situation is more complicated. Rotation does not always average the differences in the chemical environments of the two protons of a methylene group. Consider for example the following structure:

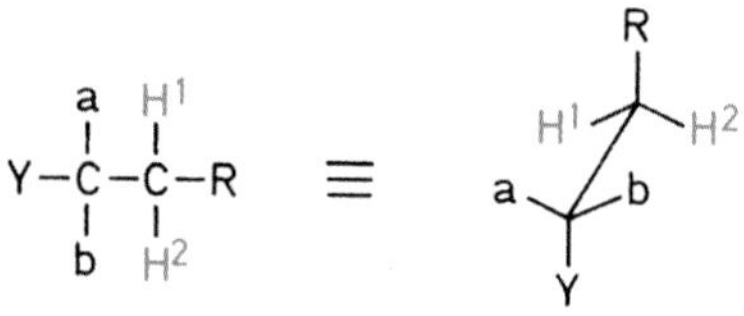

For Y = a or Y = b, there are conformations with a plane of symmetry:

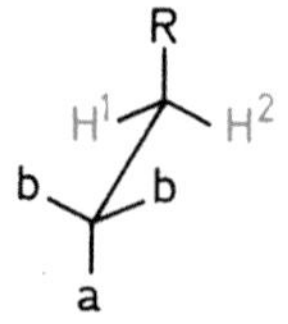

If the CH_2R group rotates, the chemical environments of the two methylene protons become different. The change for H^1 on rotation in a particular direction is exactly the same as the change for H^2 on rotation in the opposite direction. Independent of the population of the various rotational conformers, H^1 and H^2 are therefore chemically equivalent, and give a singlet if there is no coupling. If $Y \neq a$ or b, this is not possible. H^1 and H^2 are not identical in any of the conformers and form an AB system. Such protons are described as being **diastereotopic**. This is also true for the case where $Y = CH_2R$. The presence of neighboring **chiral** or **prochiral** groups can therefore lead

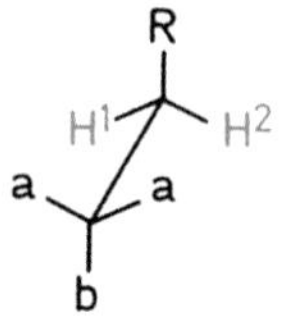

Table 3.4 Chemical equivalence of ^1H and ^{13}C nuclei in selected structures of various symmetries (point groups)

Structure	Point group	Applicable symmetry elements	Groups of chemically equivalent carbon nuclei	Groups of chemically equivalent hydrogen nuclei
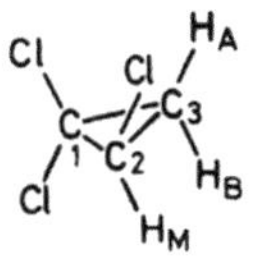 1,1,2-Trichlorocyclo-propane	C_1	–	C–1 C–2 C–3	H_A H_B H_M (ABM)
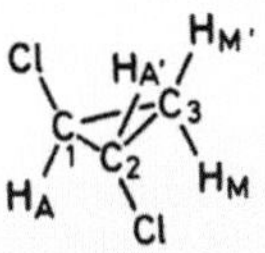 trans-1,2-Dichlorocyclo-propane	C_2	C_2	C–1, C–2 C–3	H_A, $H_{A'}$ H_M, $H_{M'}$ (AA'MM')
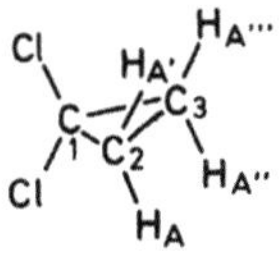 1,1,-Dichlorocyclo-propane	C_{2v}	2σ (σ, C_2)	C–1 C–2, C–3	H_A, $H_{A'}$, $H_{A''}$, $H_{A'''}$ Singlet
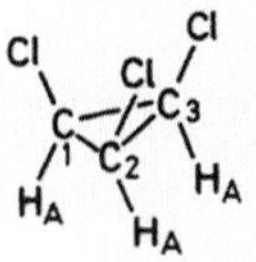 all-cis-1,2,3-Trichloro-cyclopropane	C_{3v}	C_3 (3σ)	C–1, C–2, C–3	H_A, H_A, H_A Singlet
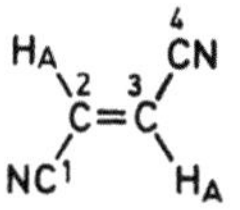 Fumaric acid dinitrile	C_{2h}	$i \equiv S_2$ (C_2)	C–1, C–4 C–2, C–3	H_A, H_A Singlet
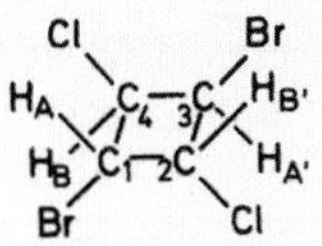 r-1,t-3-Dibromo-c-2,t-4-dichlorocyclobutane	C_i	$i \equiv S_2$	C–1, C–3 C–2, C–4	H_A, $H_{A'}$ H_B, $H_{B'}$ (AA'BB')

Table 3.4 Continued

Structure	Point group	Applicable symmetry elements	Groups of chemically equivalent carbon nuclei	Groups of chemically equivalent hydrogen nuclei
1,2,6-Trichlorobicyclo-[2,2,2]octa-2,5-diene-8,8-dicarboxylic acid	C_s	σ	C–1 C–2, C–6 C–3, C–5 C–4 C–7 C–8 C–9, C–10	H_A H_X, H_X (AX_2) H_B, H_B Singlet H_Y, H_Y Singlet Long range couplings ignored
Allene	D_{2d}	$2\,\sigma, S_4$	C–1, C–3 C–2	H_A, $H_{A'}$, $H_{A''}$, $H_{A'''}$, (Singlet)
Naphthalene	D_{2h}	$2\,\sigma$ $(3\,C_2)$	C–1, C–4, C–5, C–8 C–2, C–3, C–6, C–7 C–4a, C–8a	H_A, $H_{A'}$, $H_{A''}$, $H_{A'''}$, H_B, $H_{B'}$, $H_{B''}$, $H_{B'''}$ (AA′A″A‴BB′B″B‴)
1,3,5-Trifluorobenzene	D_{3h}	C_3 $(3\,\sigma, 3\,C_2)$	C–1, C–3, C–5 C–2, C–4, C–6	H_A, $H_{A'}$, $H_{A''}$, (A-part of AA′A″XX′X″)

to the nonequivalence of the protons of a methylene group. (If however two methylene protons can be interchanged by a symmetry operation which applies to the whole molecule, then chiral or prochiral groups do not affect the chemical equivalence!) The same considerations apply to the X signals of a CX_2 group, e.g., for the ^{13}C and ^1H signals of the two methyl groups of an isopropyl group.

In general, the following possibilities exist for two X groups in a molecule: **homotopic, enantiotopic**, and **diastereotopic**. These three cases are distinguished by considering the substitution of each of the groups in turn by a hypothetical achiral group T, not already present in the molecule, and comparing the molecules X|T and T|X so formed.

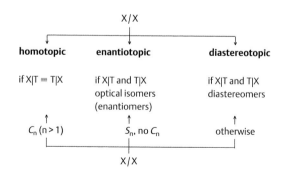

The distinctions are more easily made with the help of symmetry operators. Homotopic nuclei (groups) are interconverted by C_n axes ($n = 2, 3, ...$).

Table 3.5 Number of signals and spin systems in the NMR spectra of disubstituted benzenes

Arrangement		^{13}C signals (no coupling)	^1H signals (spin system)
Same substituents			
o	3	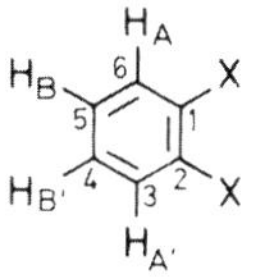 $\begin{cases} C_1 = C_2 \\ C_3 = C_6 \\ C_4 = C_5 \end{cases}$	AA'BB'
m	4	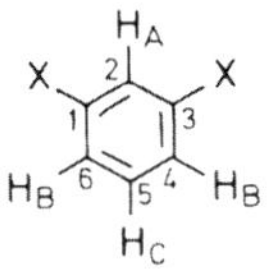 $\begin{cases} C_1 = C_3 \\ C_2 \\ C_4 = C_6 \\ C_5 \end{cases}$	AB_2C
p	2	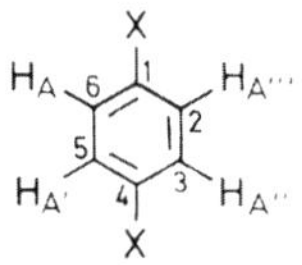 $\begin{cases} C_1 = C_4 = \\ C_2 = C_3 = \\ C_5 = C_6 \end{cases}$	AA'A''A''' Singlet
Different substituents			
o		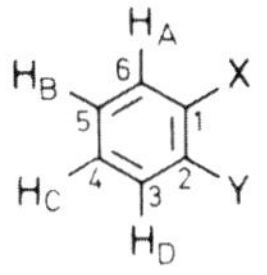 6 different carbons	ABCD
m		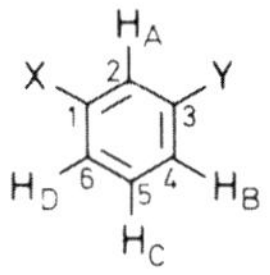 6 different carbons	ABCD
p	4	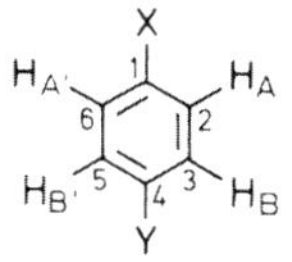 $\begin{cases} C_1 \\ C_2 = C_6 \\ C_3 = C_5 \\ C_4 \end{cases}$	AA'BB'

Since enantiotopic groups are interconverted by rotation–reflection axes S_n (including $S_1 \equiv \sigma$ and $S_2 \equiv i$), they can only occur in achiral molecules. Intramolecular motions or fast isomerizations can enhance the effective symmetry. Pseudo-symmetry elements appear, which should be included in the topicity discussion.

- **Homotopic groups are chemically equivalent and always lead to a single signal for each nuclear type.**

- **Enantiotopic groups give isochronous signals in achiral or racemic media,** anisochronosity may be observed in a chiral medium.
- **Diastereotopic groups are chemically nonequivalent and can only be isochronous by coincidence.**

For a better understanding of these situations, a few examples will be considered. While the two methylene protons in glycine (**26**) and the three methyl protons in alanine (**23**) are chemically equivalent, the two diastereotopic protons in phenylalanine (**27**) form the AB part of an ABC system.

In valine (**28**) and leucine (**29**), the asymmetric carbon atom, a chiral center, causes the methyl groups to be nonequivalent in the ^1H and ^{13}C spectra. As the distance between the isopropyl group and the chiral center increases, the differences in chemical shift decrease. Thus, for cholesterol (**30**), the ^{13}C spectrum (100 MHz) does indeed show two separate methyl signals for the isopropyl group, but the ^1H spectrum (400 MHz) does not distinguish between them.

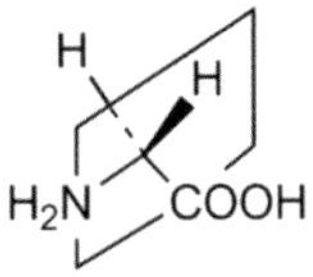

2,4-Diaminoglutaric acid (**31**) exists as two enantiomers and an achiral *meso*-form. The two chiral molecules possess a C_2 axis

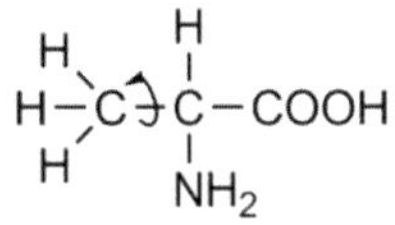

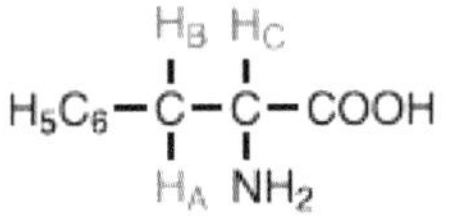

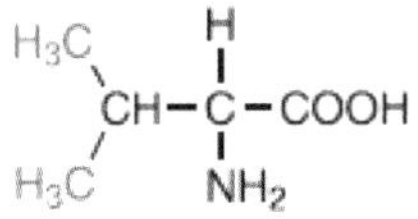

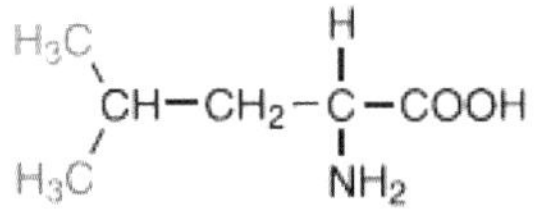

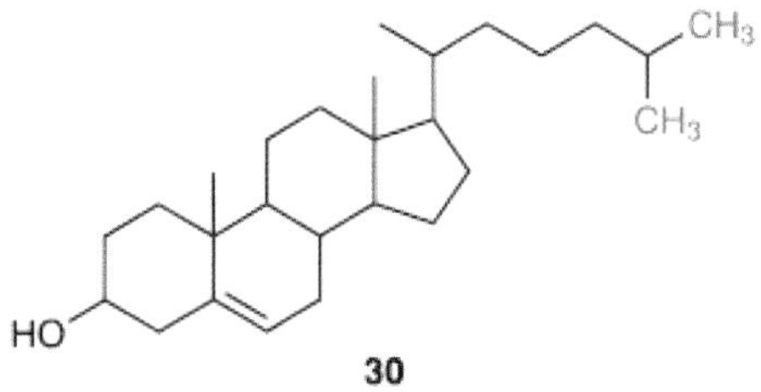

Nuclear Magnetic Resonance Spectroscopy

in the conformation drawn, and the two methylene protons are therefore chemically equivalent (homotopic).

31

In contrast, in the *meso*-form, the CH$_2$ group forms the AB part of an ABC$_2$ system.

Finally, a prochiral case will be considered. In glycerol (**32**) or in citric acid (**33**), for example, the two CH$_2$ groups are enantiotopic. The two H atoms within a methylene group are however diastereotopic (AA'BB'C or AB system, respectively).

32

33

Easy and nice to distinguish are homotopic (h), enantiotopic (e), and diastereotopic (d) nuclei (groups) in the diethyl acetals of symmetrical (**34**) and unsymmetrical carbonyl compounds (**35**).

34　(R = H, CH$_3$, ...)

35　(R^1 ≠ R^2)

The unhindered rotation of phenyl groups fundamentally leads to the equivalence of the two *o*- and of the two *m*-protons. This also applies in the presence of chiral centers. The spectrum of compound **36** is shown as an example (**Fig. 3.18**).

The two rigidly bound protons on the oxirane ring form an AB system with its center at δ = 3.32. The rotating CH$_2$Cl group bound to the chiral center similarly gives an AB system (center at δ = 4.00). Rotation of the *p*-nitrophenyl group by 180° reproduces the same structure; the relevant protons and C-atoms are identical in the time average. In the ^{13}C spectrum, there are four signals for aromatic carbon atoms, and in the ^{1}H spectrum an AA'BB' pattern (**Fig. 3.18**).

The nonequivalence of the two protons in a methylene group attached to a chiral or suitable prochiral center is also of relevance for amines. In a "rigid" compound of type **37**, the two protons of the methylene group are diastereotopic. This is also true for the case where R^2 = CH$_2$R^1. As a result of the rapid

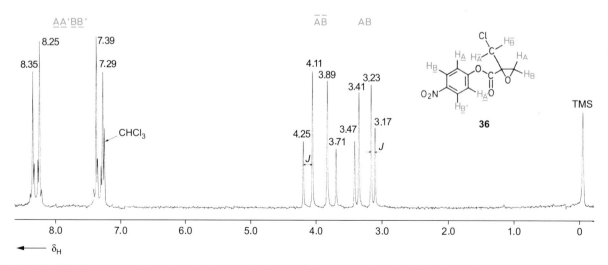

Fig. 3.18 ^{1}H NMR spectrum of the *p*-nitrophenyl ester of 1-chloromethyloxirane-1-carboxylic acid (**36**) in CDCl$_3$.

inversion at the nitrogen atom, they become chemically equivalent (enantiotopic).

Since the inversion can only occur in the free amine, and not in the protonated form, the process becomes slower with decreasing pH value. From the rate of exchange of the methyl protons, the rate constant for the inversion can be calculated, even at room temperature.

The inversion can be slowed down by including the nitrogen atom in a ring. A typical example is 1-ethylaziridine (**38**), the ring protons of which give an AA'BB' system at room temperature. Only above 100°C does the inversion at nitrogen lead to chemical equivalence (AA'A"A'''). (The two methylene protons of the ethyl group are equivalent even without the inversion!)

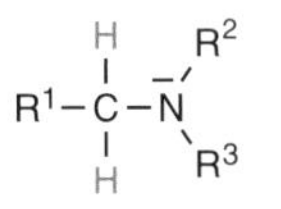

37 (R^1, R^2, R$^3 \neq$ H; R$^2 \neq$ R^3) 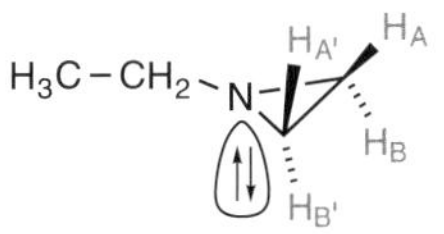 **38**

Nitrogen inversion is not possible in Tröger's base **39**, which has bridgehead nitrogen atoms. The two CH$_2$ groups in the eight-membered ring give an AB pattern in the proton spectrum (δ = 4.08 and 4.63), whereas the N–CH$_2$–N protons give a singlet (δ = 4.28). The compound is chiral, but has a C$_2$ axis; its racemization in acidic media can only be explained by a bond-breaking mechanism.

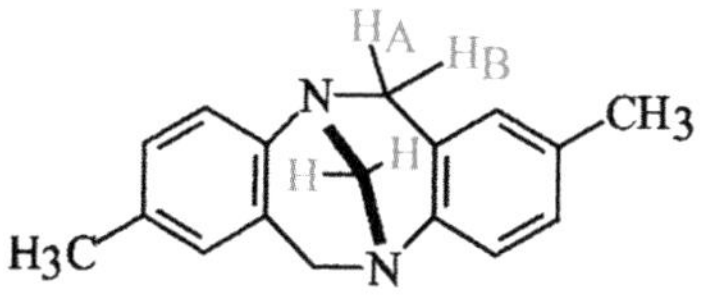

39

In the head-to-tail polymerization of vinyl monomers RCH = CH$_2$ or RR'C = CH$_2$, a chain is formed in which every second carbon atom is a chiral center. There are three distinguishable types of **tacticity**.

Isotactic: sequence of identical configurations.

Syndiotactic: regular alternation of configuration.

Atactic: random (statistical) sequence.

Shown below is an atactic polymer chain (**40**) which contains the **i**sotactic and **s**yndiotactic **diads D** and **triads T** and the **h**eterotactic triads **T**. In the syndiotactic triads, the methylene protons are homotopic from a consideration of the local symmetry. They are therefore observed in a syndiotactic polymer as a singlet. In an isotactic polymer, on the other hand, they form an AB system. In both cases, the groups R as well as R' are chemically

equivalent, each. For atactic polymers, it is usually sufficient to consider triads for R and R', whereas for the methylene protons the local symmetry must be determined at the tetrad level. The spectrum is therefore composed of elements arising from iso- and syndiotactic triads and their "connecting links," the heterotactic triads. The region of the *geminal* protons in particular can become very complex. In these cases, an evaluation of the ^{13}C NMR spectra should be preferred.

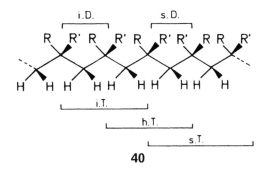

40

Following the description of **chemical equivalence** of the two protons of a methylene group, we will now consider their **magnetic equivalence**. While ethyl compounds **41** without steric hindrance form A$_3$B$_2$, A$_3$M$_2$, or A$_3$X$_2$ systems depending on the chemical shift difference between the CH$_3$ and CH$_2$ groups, it is at first sight surprising that 1,2-disubstituted ethanes **42** form AA'BB', AA'MM', or AA'XX' systems.

X—CH$_2$—CH$_3$ X—CH$_2$—CH$_2$—Y

41 (X $\neq$ H) **42** (X $\neq$ Y $\neq$ H)

Let us consider the three energetically preferred "staggered" conformations **42a**, **42b**, and **42c** with relative populations p_1, p_2, and p_3 ($p_1 + p_2 + p_3$ = 1).

In **42a** the substituents X and Y are *anti* to each other. The molecule has a plane of symmetry. The protons form an AA'BB' system with two chemical shifts v_A and v_B and four coupling constants. The two enantiomeric conformations **42b** and **42c** have identical ^{1}H NMR spectra (in an achiral medium) corresponding to an ABCD system (four chemical shifts, six coupling constants). Moreover $p_2 = p_3$, since **42b** and **42c** have the same energy.

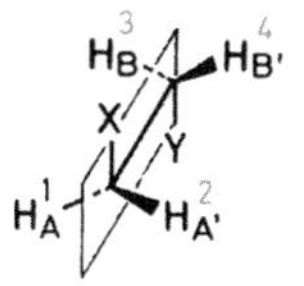

42a (p$_1$) 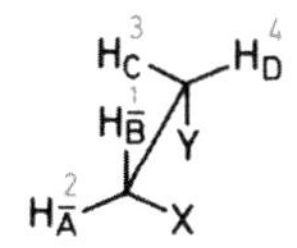   **42b** (p$_2$) 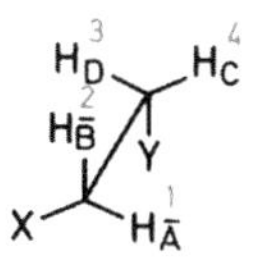   **42c** (p$_3$ = p$_2$)

Nuclear Magnetic Resonance Spectroscopy

If rotation is fast, **42** shows a spectrum with averaged shift and coupling parameters (the numbers in blue identifying that the H atoms are retained during the rotation). The averaged chemical shift of H^1, for example, is given by

$$v_1 = p_1 v_A + p_2 v_{\bar{B}} + p_2 v_{\bar{A}}$$

and for H^2 by

$$v_2 = p_1 v_{A'} + p_2 v_{\bar{A}} + p_2 v_{\bar{B}} .$$

Since $v_{A'} = v_A$ (plane of symmetry), $v_1 = v_2$. Similarly $v_3 = v_4$. The question now is whether **42** should be classified as an A$_2$B$_2$ spin system with **one** coupling constant $^3J_{AB}$ or as an AA'BB' system with $^3J_{AB} \neq {^3J_{AB'}}$. To decide this question, we must compare the coupling between H^1 and H^3 with that between H^1 and H^4.

$$^3J_{1,3} = p_1 J_{AB} + p_2 J_{\bar{BC}} + p_2 J_{\bar{AD}}$$
$$^3J_{1,4} = p_1 J_{AB'} + p_2 J_{\bar{BD}} + p_2 J_{\bar{AC}}$$

In contrast to the chemical shifts, the *vicinal* coupling constants do not necessarily average to the same value on rotation. $^3J_{1,3}$ can be different from $^3J_{1,4}$; this means that **42** must be described as an AA'BB' system. A nice example is the spectrum of 1-bromo-2-chloroethane (**43**) reproduced in **Fig. 3.19**. The measurements at 60 and 400 MHz differ strongly. However, both spectra are of higher order. $^3J_{1,3}$ and $^3J_{1,4}$ can however be so similar that a simpler ^{1}H spectrum is obtained. An example is 3-chloropropionitrile (**44; Fig. 3.20**). One should however take care not to assume from the 60 MHz spectrum that **44** has an A$_2$M$_2$ spin system.

The situation of longer chains R–(CH$_2$)$_n$–R is more complicated than it may appear on first sight. At least for a sufficiently high resonance frequency ($v_0 \geq 400$ MHz), a ^{1}H NMR spectrum is obtained, which can be interpreted according to first order. A study of the diesters (R=CO$_2$CH$_3$), for example, exposes only singlet signals (zero order) for $n = 1$ and 2, for $n = 3$ (at 400 MHz) a spectrum of first order (triplet at $\delta = 2.39$ and quintuplet at $\delta = 1.95$ ppm) is obtained, whereas a spectrum of higher order (A$_2$A$_2'$ X$_2$X$_2'$, $\delta_A = 1.67$, $\delta_X = 2.34$ ppm) results for $n = 4$. The decisive feature is $^3J(A,A') \neq 0$ for the chemically but not magnetically equivalent protons 3-H (A$_2$) and 4-H (A$_2'$). This is even valid for the case in which all vicinal coupling constants ^3J have the same size. Such ^{1}H NMR spectra of higher order are typical for symmetrically substituted chains R–(CH$_2$)$_n$–R with even numbers $n = 4, 6 \ldots$ In fact, butane ($n = 4$, R = H) gives two complex multiplet signals at the resonance frequences $v_0 \geq 500$ MHz.

In ring systems, it is necessary to consider the temperature-dependent **ring inversion** before deciding on the chemical equivalence of the various nuclei. The ^{1}H NMR spectrum of cyclohexane, for example, shows a singlet signal at $\delta = 1.43$ at room temperature. *Axial* and *equatorial* protons are equivalent as a result of the rapid inversion. On cooling, this process becomes slower and is eventually frozen out. In monosubstituted

cyclohexanes **45**, this leads to the appearance of isomers with the substituent in *axial* and *equatorial* positions. An example is discussed in Section 3.4.2 (p. 177). Similar considerations apply to the ^{13}C spectrum of 7,7-dimethylcycloheptatriene (**46**). The methyl C atoms are only chemically equivalent at temperatures where the ring inversion is rapid.

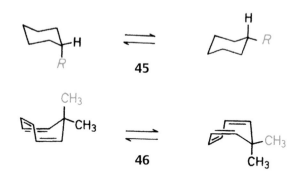

Chemical **and** magnetic equivalence shall be discussed with reference to morpholine (**47**) as an example.

This system can be considered as a further example of the systems found in 1,2-disubstituted ethanes **42** earlier. In the chair form, **47** has an ABCD spin system if transannular couplings are ignored. The rapid ring inversion at room temperature converts this into an AA'MM' system (**Fig. 3.21**), i.e., the two protons of each methylene group become chemically equivalent, but not magnetically equivalent.

In the cases discussed so far, the rapid ring inversion at room temperature is slowed down by cooling, which can be studied via the measurement of low-temperature spectra. Dibenzocyclooctyne (**48**) exists at room temperature in the chiral C$_2$ configuration. The aliphatic protons form an AA'BB' system (**Fig. 3.22**). On warming, the ring inversion becomes apparent. The signals first become broader, combine at the **coalescence temperature** (112°C), and at 145°C in the region of rapid ring inversion form a singlet. H$_a$ and H$_b$ exchange so quickly that a single, averaged signal is obtained in the NMR spectrum. The coupling of the protons does no longer affect the spectrum. The ring inversion in this case is equivalent to the racemization of the compound.

Further, very interesting examples where NMR is of use in the study of the flexibility of rings are the annulenes (see p. 125).

Alongside "sterically" hindered rotation about σ-bonds and "pseudorotation" in rings, another important example is the **rotation about bonds with partial double bond character**.

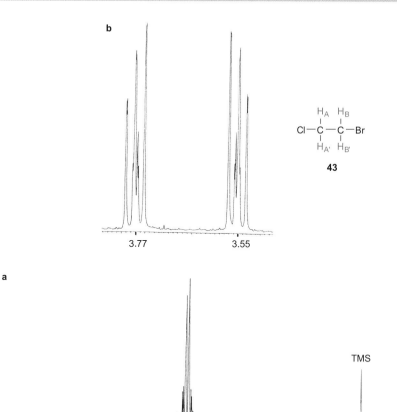

Fig. 3.19 ¹H NMR spectra of 1-bromo-2-chloroethane (**43**) in CDCl₃. (**a**) 60 MHz measurement; (**b**) 400 MHz measurement.

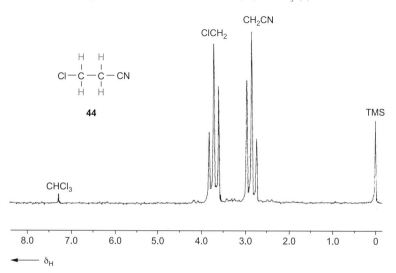

Nuclear Magnetic Resonance Spectroscopy

Fig. 3.20 ¹H NMR spectrum of 3-chloropro-pionitrile (**44**) in CDCl₃.

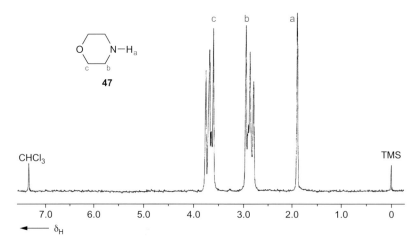

Fig. 3.21 ¹H NMR spectrum of morpholine (**47**) in CDCl₃ at room temperature.

Typical cases are groups such as:

$$D-Z\overset{A}{\diagup} \quad \longleftrightarrow \quad D^+=Z\overset{A^-}{\diagup}$$

D = NR₂, OR; Z = CR, N; A = O, S, (NR)

$$D-CH=CH-Z\overset{A}{\diagup} \quad \longleftrightarrow \quad D^+=CH-CH=Z\overset{A^-}{\diagup}$$

D = NR₂, (OR); ZA = CO−R, NO₂, etc.

An electron acceptor (A = O, S) and an electron donor D are bound to a central atom (Z = C, N). As a result of the participation of the dipolar resonance form, the D⋯Z bond (N⋯C, C⋯C, O⋯C, C⋯N, N⋯N, and O⋯N) shows restricted rotation.

Push–pull substituted alkenes represent vinylogous systems, which show restricted rotations about the D⋯CH and CH⋯ZA bonds.

At room temperature, the two methyl groups of dimethylformamide (**49; Figs. 3.23 and 3.24**) and dimethylnitrosamine (**50**), for example, are already chemically nonequivalent.

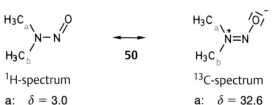

50

¹H-spectrum	¹³C-spectrum
a: δ = 3.0	a: δ = 32.6
b: δ = 3.8	b: δ = 40.5

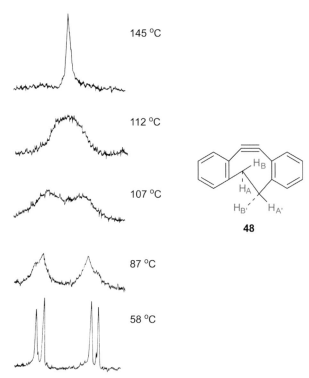

Fig. 3.22 Temperature-dependent 90 MHz ¹H NMR spectra of **48** in deuteriobromoform (from Meier H., Gugel H., Kolshorn H. 1976, Z. Naturforsch. B, **31**, 1270).

Amides with only one substituent on the N atom (e.g., *N*-ethylacetamide, **51**) exist predominately or exclusively in the form shown, with the substituent and the carbonyl O atom in (*Z*) arrangement.

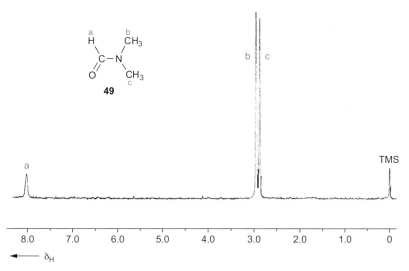

Fig. 3.23 ¹H NMR spectrum of dimethylformamide (**49**) at room temperature (above 120°C, a singlet is obtained for the methyl groups).

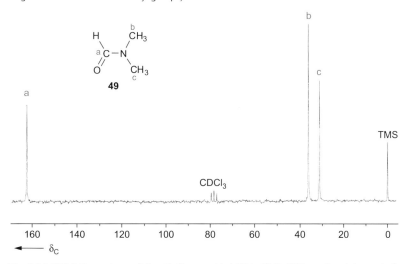

Fig. 3.24 ¹³C NMR spectrum of dimethylformamide (**49**) in $CDCl_3$ (¹H broadband decoupled).

51

¹H spectrum	¹³C spectrum
a: $\delta = 1.1$	a: $\delta = 14.6$
b: $\delta = 3.2$	b: $\delta = 33.7$
c: $\delta = 8.2$	CO: $\delta = 169.5$
d: $\delta = 2.0$	d: $\delta = 22.5$

Hydroxamic acids exist predominantly in the *s-cis* configuration. The *Z/E* ratio of propionhydroxamic acid (**52**) amounts to 9:1 in dimethyl sulfoxide (DMSO).

(*Z*)-**52**

(*E*)-**52**

9:1

In enamines too, the (C–N) rotation can be frozen out at room temperature, as the example of 3-dimethylamino-1,2-dihydropentalene (**53**) shows:

53

Position	$\delta\,(^1H)$	$\delta\,(^{13}C)$
1	2.91	22.9
2	3.13	38.7
3	–	163.6
3a	–	122.4
4	6.21	107.3
5	6.68	129.4
6	5.91	106.3
6a	–	147.5
CH_3	3.20/3.32	41.1/41.5

Compound **53** resembles the push–pull substituted alkenes. (*E*)-1-Dimethylamino-2-nitroethene (**54**) shows two different signals for the methyl groups.

54

δ (DMSO):   1-H 8.30, 2-H 6.80, CH_3 3.20, CH_3 2.85

The temperature-dependent behavior of (*E*)-4-dimethylaminobut-3-en-2-one (**55**) is still more interesting. A broad singlet ($\delta = 2.82$) for six protons is obtained for the $N(CH_3)_2$ group at room temperature. On cooling, this signal splits into two singlets of equal intensity ($\delta = 2.68$ and 2.98). Further lowering of the temperature leads to a splitting of the singlet obtained for the acetyl group ($\delta = 1.99$) into two singlets of different intensities. Two frozen rotamers exist at these low temperatures, whereby the *s-cis* conformer **55b** predominates.

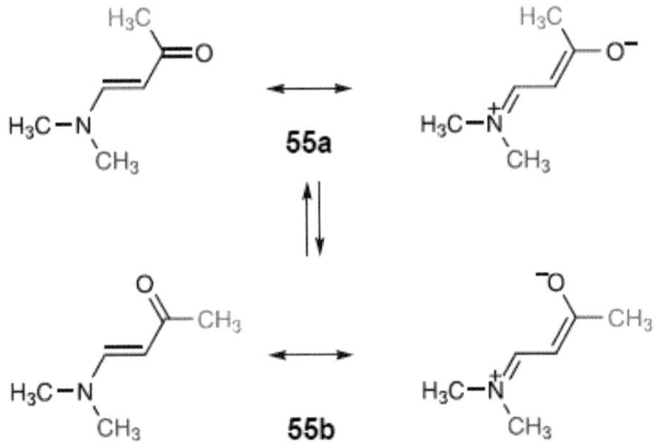

55a

55b

3.2.3   Chemical Exchange Processes

Alongside internal molecular motions inter- and intramolecular chemical exchange processes are also important for the determination of equivalence of nuclei.

Rearrangement reactions are mostly so slow on the NMR timescale that both isomers can be observed in the spectrum. The same often applies to **tautomerism**, as shown by the example of acetylacetone (**56**; **Fig. 3.25a, b**).

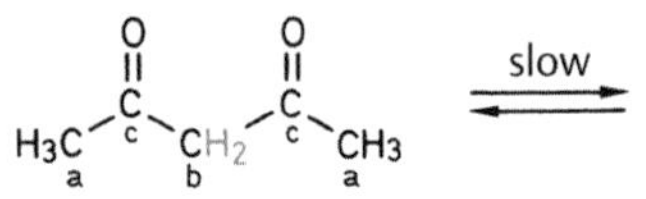

56a

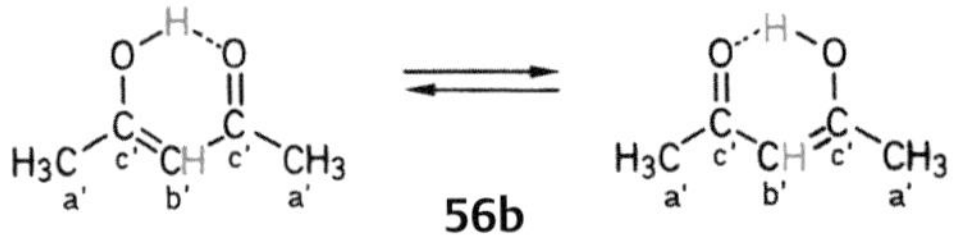

56b

At room temperature, in $CDCl_3$ the keto form **56a** is present in about 14% concentration and the enol form **56b** in about 86%. Only on raising the temperature does the reversible prototropy between C and O become rapid enough to result in the observation of averaged signals. From the observed spectra, it can further be seen that **56a** and **56b** both possess a similar symmetry element which causes the chemical equivalence of the methyl groups and the carbonyl C atoms. There are two possible explanations for this. The acidic proton could shift between the two O atoms (with simultaneous shift of the double bonds) so quickly that the enol form *de facto* appears to be symmetrical. The other possibility is a change of the coordination position of the proton on the mesomeric β-diketonate. The temperature independence of the ^1H, ^{13}C, and ^{17}O spectra of the enol form (even in unsymmetrical cases) supports the resonance model, in which the dynamic phenomenon relates only to the proton and not to the chain.

A similar question arises with tropolone (**57**). In the ^1H spectrum, it shows an AA′BB′C spin pattern. If the mobile proton is replaced by a methyl group (2-methoxycycloheptatrienone, **58**) which cannot migrate, an ABCDE spectrum is obtained. Tautomeric proton rearrangements between heteroatoms are often rapid on the NMR time scale; the number of NMR signals then corresponds to the pseudosymmetry. Further examples are imidazole, pyrazole, benzimidazole, and benzotriazole (cf. p. 274 and 278).

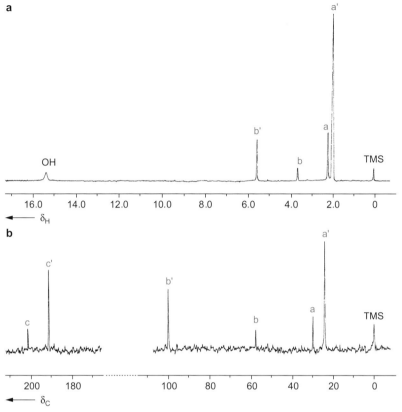

Fig. 3.25 NMR spectra of acetylacetone (**56**) in CDCl₃ at room temperature. (**a**) ¹H NMR spectrum; (**b**) ¹³C NMR spectrum (¹H broadband decoupled).

Porphyrin (**59**) shows at room temperature a fast proton exchange between the pyrrole rings. Therefore, the ¹H NMR spectrum of **59** contains only three singlet signals (**Fig. 3.26a**) and the ¹³C NMR spectrum contains three chemical shifts (**Fig. 3.26b**). The bigger the difference $\Delta\delta$ of exchanging nuclei is, the more difficult is the identification of the broad, averaged signals. This becomes obvious for the ¹³C NMR signal of β-C and even more for the signal of α-C. The *meso* carbon atoms (*m*-C) have the same δ-value in both tautomers **59**. Therefore, the intensity of the *m*-C signal is high.

A tautomeric equilibrium can lie so far on the side of one tautomer that the other tautomer is below the detection limit. This is, for example, the case for benzoin (**60**) and pyridoin (**61**). Routine measurements show only the hydroxy ketone structure of **60** and the enediol structure of **61**.

Nuclear Magnetic Resonance Spectroscopy

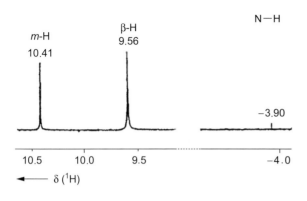

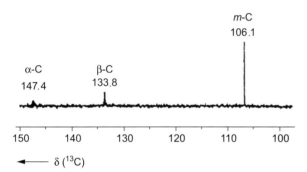

Fig. 3.26 NMR spectra of porphyrin (**59**) in [D₈]THF at room temperature: (**a**) ¹H NMR; (**b**) ¹³C NMR.

The measurement of tautomers is naturally not confined to equilibrium situations. The proportion of vinyl alcohol (**63**) in acetaldehyde (**62**), for example, is under the detectable limit for NMR spectroscopy. It can be measured after selective formation of the metastable species if the rearrangement is sufficiently slow [δ(¹³C) and δ(¹H)]:

H₃C—C 30.8 199.9 H 2.20 9.80 **62** ⇌ 88.0 C=C 149.0 3.82 H OH H H 4.18 6.45 **63**

A fast proton exchange between OH (or NH groups) disenables the coupling to neighboring nuclei (cf. p. 152). Lowering of the temperature or the concentration decelerates the intermolecular proton exchange. At room temperature, solvents have this effect if they act as **hydrogen-bond acceptors**. Anhydrous acetone or particularly DMSO is used for this purpose.

The OH signal of ethanol is in CDCl₃ a singlet at $\delta \approx 2.6$; in dry CD₃SOCD₃, however, a triplet at $\delta \approx 4.3$ ppm. The quadruplet q

of the CH₂ group of ethanol becomes a doublet of quadruplets dq at this change of solvents.

In the presence of D₂O, OH protons are subjected to a H/D exchange and an HDO signal can be measured. **The addition of D₂O to the solution studied is frequently applied for the identification of OH signals.**

Intramolecular hydrogen bonds can be annulled by DMSO too. Salicylaldehyde (**64**) serves here as an example.

[structure **64**]

The formation of hydrogen bonds is also discernible in the ¹³C NMR spectroscopy. Notably striking are the differences of the δ (¹³C) values of (Z)- and (E)-enols of 1,3-dicarbonyl compounds **65**. The δ values of C-1 and C-3 of the unusual (E)-isomer (E)-**65** are strongly up-field shifted and δ(C-2) is down-field shifted compared to the normal (Z)-isomer (Z)-**65**. The proton resonance of the OH group in the chelate form (Z)-**65** can exhibit a down-field shift of 5 to 10 ppm.

[structures Z and E]

65 (R = Aryl)

Besides reversible proton shifts, **tautomerism**, reversible electron shifts (**valence tautomerism**) are of especial interest. The growth of interest in this type of reaction in recent years would have been unthinkable without NMR spectroscopy. Bullvalene (**66**) will be described as a classic example. It undergoes a degenerate Cope rearrangement between 10!/3 identical isomers. At 120°C, this process is so fast that a sharp singlet is obtained for all 10 protons and for all 10 carbons. The term **fluctuating structure** is used to describe such situations. Below −60°C, the region of slow exchange is reached. Four distinct ¹³C signals are observed, in line with the symmetry of the rigid structure. (In the ¹H spectrum, some signals are coincidentally isochronous.) The coalescence region is at room temperature. At 15°C, the ¹H spectrum shows a very broad band; the ¹³C signals are lost in the noise (**Figs. 3.27 and 3.28**).

[structures **66**]

The activation energy for this valence tautomerism is ca. 49 kJ·mol^{-1}. (The rearrangement also takes place in the solid state. There, however, a reorientation of the molecule is also necessary, in order to retain the position of the molecule in the lattice; the activation energy in the solid is 63 kJ·mol^{-1}.)

Fluctuating systems must be clearly distinguished from **resonance systems**. Let us compare a monosubstituted cyclooctatetraene (**67**) and a monosubstituted benzene (**68**):

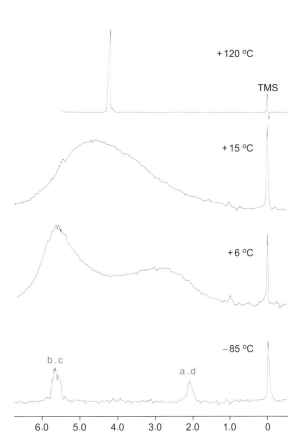

At room temperature, the double bonds in cyclooctatetraene oscillate. There is a simultaneous inversion of the tub-shaped ring. Both processes are fast on the NMR time scale, so that instead of eight different ring C-atoms and seven different H-atoms only five ^{13}C signals and four ^{1}H signals are observed

$$(H_A = H_{\bar{A}}, H_B = H_{\bar{B}}, H_C = H_{\bar{C}})$$

In the benzene derivative **68**, the protons *ortho-* to the substituent R are chemically equivalent, as are the *meta*-protons. The same applies to the corresponding C atoms. This is an example of the **static** phenomenon of resonance in these planar molecules and not a **dynamic** process as in the case of cyclooctatetraene. On cooling **67**, the movement of the double bonds, and independently also the ring inversion, becomes slower and eventually freezes out.

In **68**, there is no such temperature dependence of the NMR spectra. This difference is clarified by the energy diagrams in **Fig. 3.29**.

As well as H and C, heteroatoms can of course also be involved in exchange processes. An example is the furoxane **69**.

In ^{1}H NMR the ABCD spin system of benzofuroxan changes to an AA′BB′ spin system on warming; in the ^{13}C NMR the six signals reduce to three.

Fig. 3.27 Temperature-dependent ^{1}H NMR spectra of bullvalene (**66**) in CS$_2$: at −85°C the signals of the olefinic protons b and c (6H) are observed at low field and the signals of the three protons a on the three-membered ring and the bridgehead proton d at high field (from Schröder G. et al. 1965, Angew. Chem. **77**, 774).

Intra- and intermolecular exchange processes also play an important role in the NMR spectroscopy of organometallic compounds and metal complexes.

2,4-Dimethyl-2,3-pentadiene (tetramethylallene), for example, forms an Fe(CO)$_4$ complex **70**, in which all four methyl groups are equivalent at room temperature. In the η^2-complex, the iron atom changes its π-ligands rapidly. At low temperature, however, a "frozen out" structure with reduced symmetry is observed. The 2:1:1 distribution of the methyl groups is a consequence of the iron atoms being attached to only one double bond.

70

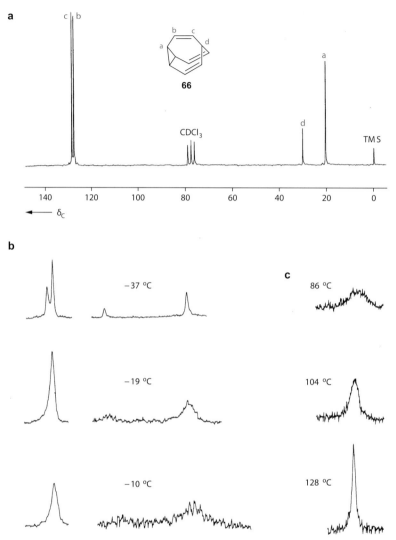

Fig. 3.28 Temperature-dependent ^{13}C NMR spectra of bullvalene **(66): (a)** spectrum at –62°C in $CDCl_3$, broadband decoupled; **(b)** spectra in the region of slow exchange (–37 . . . –10°C); **(c)** spectra in the region of fast exchange (+86 . . . +128°C) (from Günther H., Ullmen J. 1974, Tetrahedron **30**, 3781).

For many organometallic compounds, the temperature dependence of the NMR spectra is a consequence of association and dissociation processes.

Trimethylaluminum (**71**), for example, shows a single methyl signal at room temperature, which splits below –40°C into two signals. This is an example of a monomer–dimer equilibrium, which is so slow at low temperatures that a distinction can be made between terminal and bridging methyl groups. The monomer concentration is below the limit of sensitivity.

At the end of this section, some mention will be made of the application of the temperature dependence of NMR spectra in kinetic studies. In the processes mentioned, the intramolecular motion (rotations, inversions, etc.) and the intermolecular reactions are **equilibria** between two or more **conformers, tautomers,**

a

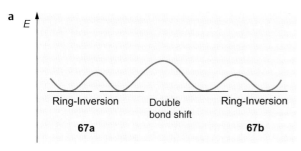

b

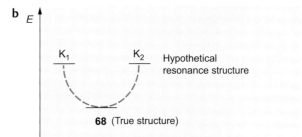

Fig. 3.29 (a) Schematic energy diagram for the cyclooctatetraene derivative **67** with valence isomerism between the structures **67a** and **67b** and ring inversion. (From experiments with the sixfold deuterated compound with R = C(CH₃)₂OH, values of $\Delta G^{\#}$ = 61.5 kJ·mol⁻¹ for the ring inversion and $\Delta G^{\#}$ = 71.6 kJ·mol⁻¹ for the double-bond shift (both at –2°C) were determined.) **(b)** Schematic energy diagram for the benzene derivative (**68**) and its two hypothetical Kekulé structures.

or **valence isomers**. If we take the simple case of a reversible exchange process between A and B with first-order kinetics:

$$A \underset{k'}{\overset{k}{\rightleftharpoons}} B$$

The relative populations are n_A and n_B ($n_A + n_B = 1$). If the energy contents of A and B are different, then the distribution will not be 1:1, but a temperature-dependent equilibrium:

$$\frac{n_B}{n_A} = e^{-\frac{\Delta G}{RT}}$$

ΔG difference in Gibb's free energy
R gas constant
T absolute temperature

The **Eyring equation** gives the rate constant k:

$$k = \frac{RT}{N_a \cdot h} e^{-\frac{\Delta G^{\#}}{RT}} = \frac{k_B T}{h} = e^{-\frac{\Delta G^{\#}}{RT}}$$

$\Delta G^{\#}$ free energy of activation
N_a Avogadro's number
h Planck's constant
k_B Boltzmann's constant

If the exchange A ⇌ B is slow, the signals of A and B are observed separately in the NMR spectrum; if it is fast however, only an averaged signal is observed for the exchanging nuclei. (The terms "slow" and "fast" are defined by comparison to k or $\tau = 1/k$ as described in Section 3.1.4.)

Fig. 3.30 shows the simple case where A and B have singlet signals of equal intensity which then merge on fast exchange to a twice as intensive signal v_m. If the effect of temperature on the shifts v_a and v_b can be ignored, then

$$v_m = \frac{v_A + v_B}{2}$$

Between the regions of fast and slow exchange, broad signals appear. The curve labeled T_c is the coalescence situation. T_c is

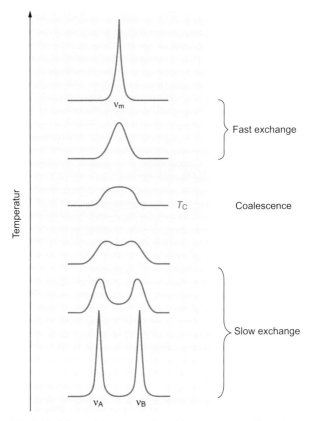

Fig. 3.30 Schematic representation of the temperature-dependent NMR spectra for a process A ⇌ B with no coupling between the exchanging nuclei.

called the **coalescence temperature**. k at coalescence is given approximately by

$$k_{T_c} = \frac{\pi}{\sqrt{2}} |v_A - v_B|$$

therefore $k_{T_c} \approx 2.22 \Delta v$.

Nuclear Magnetic Resonance Spectroscopy

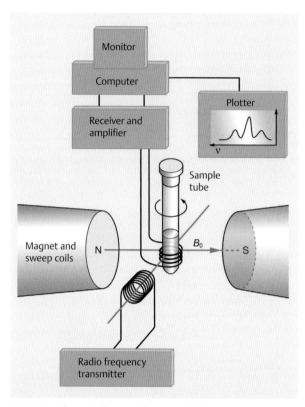

Fig. 3.31 Schematic diagram of a simple NMR spectrometer.

Inserting this relation into the Eyring equation results in

$$\frac{\pi}{\sqrt{2}} \, |v_A - v_B| = \frac{RT_c}{N_a \cdot h} \, e^{-\frac{\Delta G^{\neq}}{RT_c}}$$

or

$$\Delta G^{\neq} = RT_c \cdot \ln \frac{RT_c \sqrt{2}}{\pi \cdot N_a \cdot h \, |v_A - v_B|} \, .$$

If T_c is measured in K and the shifts v in Hz, then the free energy of activation is given (in kJ) by

$$\Delta G^{\neq} = 19.1 \cdot 10^{-3} \cdot T_c \, (9.97 + \log T_c - \log |v_A - v_B|).$$

Considerably more accurate than this approximate solution based on the coalescence temperature is **line shape analysis** (for details, see the NMR literature quoted in the bibliography).

If one assumes that

$$|v_A - v_B| \lesssim \begin{array}{l} 150 \text{ Hz} \ \ (\text{in } ^1\text{H NMR}) \\ 300 \text{ Hz} \ \ (\text{in } ^{13}\text{C NMR}) \end{array}$$

then this method can be used to study the reversible interconversion of states which have a **lifetime** $\tau = 1/k$ between 10^{-1} and 10^{-3} s; in certain cases, τ values one or two orders of magnitude smaller can be determined. This method can be extended to the case of coupling nuclei (see, for example, **Fig. 3.22**, p. 108).

Fig. 3.32 1,000-MHz NMR spectrometer (reproduced with kind permission of the Bruker Company).

In such cases,

$$k_{T_c} \approx 2.22 \sqrt{\Delta v^2 + 6J_{AB}^2} \, .$$

Since Δv increases with the measurement frequency, the coalescence temperature must also increase with $\boldsymbol{B}_0$. If in one molecule there are several pairs of nuclei with different Δv, then their T_c values must also vary. Any specification of the **coalescence temperature** must therefore always be accompanied by the measurement frequency and the relevant pair of nuclei A/B.

The kinetics of slower reactions can naturally also be determined by NMR. Concentrations of components being formed or disappearing can be determined by integration of the signals of components as they appear or disappear.

By application of the **Gibbs–Helmholtz equation**,

$$\Delta G^{\neq} = \Delta H^{\neq} - T\Delta S^{\neq}$$

the enthalpy $\Delta H^{\neq}$ and entropy of activation $\Delta S^{\neq}$ can be determined from k. By taking logarithms of both sides of the Eyring equation,

$$\log \frac{k}{T} = 10.32 - \frac{\Delta H^{\neq}}{19.1} \cdot \frac{1}{T} + \frac{\Delta S^{\neq}}{19.1}$$

and plotting log k/T against $1/T$, $\Delta H^{\neq}$ is derived from the slope and $\Delta S^{\neq}$ from the intercept of the straight line obtained. Obviously, it is advantageous to have as many values of k and T as possible.

The **Arrhenius equation** for the rate constant k contains the **Arrhenius activation energy** E_a, which can be obtained from the linear plot $\ln k = f(1/T)$:

$$k = A \times e^{-E_a/RT} \text{ or } \ln k = \frac{-E_a}{R} \times \frac{1}{T} + \ln A$$

The simple connection $E_a = \Delta H^{\#} + RT$ explains why the values of E_a and $\Delta H^{\#}$ are very similar. Their difference $(E_a - \Delta H^{\#})$ is often within the limits of error, since RT amounts to 0.6 kcal ≈ 2.5 kJ at room temperature.

3.3 ¹H NMR Spectroscopy

3.3.1 Sample Preparation and Measurement of ¹H NMR Spectra

NMR spectra for analytical purposes are normally measured in solution. A concentrated but not viscous solution is made in a proton-free solvent. Alongside CCl_4 and CS_2, a variety of deuterated **solvents** are commercially available (**Table 3.6**). **The most commonly used is $CDCl_3$.** Since the degree of deuteration is always somewhat less than 100%, weak signals from the solvent must always be expected. The δ values of these solvent signals are given in **Table 3.6**. While the $CHCl_3$ impurity in $CDCl_3$ (typically 0.2%) gives a singlet signal at $\delta = 7.24$, solvents with CD_3 groups give a quintet because of the coupling to deuterium ($I = 1$) in CHD_2 groups (cf. **Table 3.2**).

Many solvents also often have a detectable concentration of water, which gives rise to an H_2O or HOD signal (cf. **Table 3.6**). Trichloromethane (chloroform) contains HOD and DCl. Therefore, it is important to purify $CDCl_3$ before it is used for the measurement of compounds sensitive to water or acid.

The choice of solvent has some effect on the measured shifts. In the case of overlapping signals, the **solvent shift** can be useful. C_6D_6 is particularly suitable because of its high magnetic anisotropy (see, for example, **Fig. 3.48**).

[D_6]DMSO slows the proton exchange of OH groups and is recommended as a solvent if the coupling patterns of OH protons are to be observed (see p. 132). Deuterated solvents serve for the permanence of the magnetic field B_0. The D resonance of the solvent is used (**D lock**) to avoid temporal drifts of the magnetic field. Measurements in undeuterated solvents must be run by applying an external lock.

As a **reference substance** for establishing the zero-point of the δ scale, TMS is used, either added directly to the solution (**internal standard**) or in a capillary tube inside the measurement tube (**external standard**). If an external standard is used, the δ values for the chemical shifts must be corrected:

$$\delta_{corr.} = \delta_{meas.} + 6.67 \cdot 10^5 \cdot \pi \cdot [\chi_v(\text{Standard}) - \chi_v(\text{Sample})]$$

χ_v volume susceptibility

The use of TMS is unnecessary, if, for example, the residual $CHCl_3$ signal in $CDCl_3$ can be directly used.

Measurements are normally made at room temperature. **Low-** or **high-temperature spectra** are important for the study of intramolecular mobility (rotations, inversions, etc.) and for kinetic studies of reactions.

Fig. 3.31 (p. 116) gives a schematic representation of the construction of a simple NMR spectrometer. The magnetic field, produced by a permanent magnet or an electromagnet, should be as homogeneous as possible. It splits the nuclear Zeeman levels (see Section 3.1.1, p. 86) by an amount proportional to B_0. The sample is kept in a tube, which—for other than the so-called inverse measurement—is rotated about its vertical axis. This averages out horizontal field inhomogeneities.

The nuclei are excited by a radiofrequency (RF) transmitter of high stability. At resonance the induced magnetization created by the spin inversion results in a current in the receiver coil, which is perpendicular to both the magnetic field and the transmitter coil. Instead of the second coil, a bridge circuit can be used. The amplified signal is recorded on an x,y-plotter, which draws the spectrum.

There are two possible ways of achieving resonance (see Section 3.1.1). Either the frequency v can be varied at constant field B_0 (**frequency sweep**) or the field B can be varied at constant frequency (**field sweep**). In both methods the individual signals are observed consecutively by constant variation of v or B; this is described as a **continuous wave** (**CW**) technique.

To improve the **signal-to-noise (S/N) ratio** when measuring dilute solutions, many spectra can be measured consecutively, stored in a small computer, and the resultant spectra averaged (the so-called time-averaging, CAT-method, computer averaged transients). The random noise tends to cancel out, so the S/N ratio improves by a factor of $\sqrt{n}$, where n is the number of individual scans. The time required for this procedure limits its usefulness.

The **CW technique** is easy to understand and is therefore described here because of didactic reasons. Nowadays, modern NMR spectrometers use the **pulse Fourier transform** (**PFT**) **technique** and work with a superconducting magnet cooled by liquid helium (**solenoid**).

NMR spectrometers with a ¹H frequency $v_0 = 300$ MHz (7.05 T) are normally sufficient for routine ¹H and ¹³C NMR spectroscopy. If higher spectral dispersion or S/N ratio are required,

Nuclear Magnetic Resonance Spectroscopy

Table 3.6 Solvents for ^1H NMR spectroscopy

Solvent	^1H NMR shift δ	$H_2O/$ HDO δ	M.P.* (°C)	B.P.$_{760}$* (°C)
Carbon tetrachloride (CCl_4)	-		– 23	77
Carbon disulfide (CS_2)	-		– 112 T	46
Hexachloro-1,3-butadiene (C_4Cl_6)	-		– 21	215 H
Dichlorodifluoromethane (CCl_2F_2)	-		– 160 T	–30
[D_1] Chloroform ($CDCl_3$)	**7.24**	**1.5**	**– 64**	**61**
[D_4] Methanol (CD_3OD)	3.35	4.9	– 98 T	64
	4.78			
[D_6] Acetone (CD_3COCD_3)	2.04	2.8	– 95 T	56
[D_6] Benzene (C_6D_6)	7.27	0.4	6	80
[D_{12}] Cyclohexane (C_6D_{12})	1.42		7	81
[D_8] Toluene ($C_6D_5CD_3$)	2.30	0.4	– 95 T	111
	7.19			
[D_5] Nitrobenzene	7.50		6	211 H
($C_6D_5NO_2$)	7.67			
	8.11			
[D_2] Dichloromethane (CD_2Cl_2)	5.32	1.5	– 97 T	40
[D_1] Bromoform ($CDBr_3$)	6.83		8	150 H
[D_2] 1,1,2,2-Tetrachloroethane	6.00		–44	146 H
($C_2D_2Cl_4$)				
[D_3] Acetonitrile (CD_3CN)	1.93	2.1	– 45	82
[D_{10}] Diethylether ($C_4D_{10}O$)	1.07		– 116 T	35
	3.34			
[D_8] Tetrahydrofuran	1.73	2.4	– 108 T	66
(C_4D_8O)	3.58			
[D_8] Dioxane ($C_4D_8O_2$)	3.58		12	102
[D_6] Dimethylsulfoxide	2.49	3.3	19	189 H
(CD_3SOCD_3)				
[D_5] Pyridine (C_5D_5N)	7.19	5.0	– 42	115
	7.55			
	8.71			
[D_2] Water (D_2O)	4.65	4.8	0	100
[D_4] Acetic acid	2.03	11.6	17	118
(CD_3COOD)	11.53			
[D_1] Trifluoroacetic acid (CF_3COOD)	11.5		– 15	72
[D_{18}] Hexamethyl-	2.53		7	233 H,C
phosphoric acid triamide (HMPT) ($[(CD_3)_2N]_3PO$)				

* Refers to the undeuteriated compound
T suitable for low temperature measurements
H suitable for high temperature measurements
C highly carcinogenic

instruments with 400 to 900 MHz (9.39–21.14 T) or even 1,000 MHz = 1 GHz (23.49 T) are available. However, it should be pointed out that the resolution of multiplets of complex spin systems is not better when a higher field is used. On the contrary, one has to realize that the linewidth measured at 800 MHz is broader (due to field inhomogeneities) than that measured in a 400 MHz spectrometer.

Extremely high field strengths (above 20 T) are sometimes needed in biochemical studies or in medical chemistry. **Fig. 3.32** (p. 116) shows the first 1,000 MHz spectrometer fabricated in 2009. The development of 1.1 and 1.2 GHz NMR spectrometers is in progress. On the other hand, there is a trend towards low-field **NMR** spectrometers which can be used on the bench.

In contrast to the **CW technique**, all nuclei of a particular type, e.g., all protons in a sample, are excited simultaneously by an intense RF pulse. As explained in Section 3.1.1 (p. 86), a consequence of the Boltzmann distribution is that in an external field $\boldsymbol{B}_0$ more nuclei precess about the direction of $\boldsymbol{B}_0$ than about the opposite direction. The vector sum of the magnetic moments of these excess nuclei results in an **equilibrium magnetization $\boldsymbol{M}_0$** in the direction of $\boldsymbol{B}_0$ (**longitudinal magnetization**). When $\boldsymbol{M}_0$ has the direction z and the RF pulse the direction x, the magnetization vector is rotated by an angle θ, which is proportional to the power and the period t_p of the pulse ($\boldsymbol{M}_0 \rightarrow \boldsymbol{M}$). If the **pulse width t_p** is selected in the ms range so that the **pulse angle** $\theta = 90°$, then the maximum **transversal magnetization $|\boldsymbol{M}_y| = |\boldsymbol{M}_0|$** is obtained. $\boldsymbol{M}_y$ induces a signal in the receiver coil. **Fig. 3.33** illustrates the situation for a 45° and a 90° pulse.

At the end of the RF pulse, i.e., after some ms, the vector M returns to the equilibrium position. To understand this relaxation better, a coordinate $\bar{x}, \bar{y}, z$ system can be conceived that rotates with the Larmor frequency (cf. p. 86) as M does. The transverse magnetization $M_y = M_{\bar{y}}$ depends then only on the time. If a 90° pulse is assumed, the M_z component increases from 0 to M_0 and $M_{\bar{y}}$ becomes 0. The definition

$$\frac{dM_z}{dt} = \frac{M_0 - M_z}{T_1}$$

contains the **longitudinal relaxation time** T_1 (cf. p. 86), which is also called spin–lattice relaxation time. During the return to the equilibrium, the excess energy is delivered as heat to the "lattice." The T_1 times of protons are short and do not play an important role for the ¹H NMR measurements. This statement does not hold for ¹³C NMR measurements.

The **transverse magnetization** is defined in an analogous way:

$$\frac{dM_{\bar{y}}}{dt} = -\frac{M_{\bar{y}}}{T_2^*}$$

The rate constant $1/T_2^*$ contains $1/T_2$ and an inevitable contribution of field inhomogeneities ($T_2^* \leq T_2 \leq 2T_1$). T_2 is also called spin–spin relaxation time (see p. 86), because an energy exchange among the individual protons occurs. Integration of the upper differential equations gives:

$$M_z(t) = M_0 \left(1 - e^{-\frac{t}{T_1}}\right)$$

$$M_{\bar{y}}(t) = M_{\bar{y}}(0)\, e^{-\frac{t}{T_2^*}}$$

The actual **free induction decay** (FID) is however more complicated. The RF pulse has a basic frequency v_0. The short period t_p of the RF pulse (some ms) causes a polychromatic frequency band ($v_0 \pm 1/t_p$) according to Heisenberg's uncertainty principle. The resonance of all nuclei (all ¹H or all other nuclei suitable for v_0) is excited. On the basis of Δv, a modulation is formed, which is expressed by a cosinus term (damped vibration).

$$M_{\bar{y}}(t) = M_{\bar{y}}(0)\, e^{-\frac{t}{T_2^*}} \cos(2\pi \Delta v t)$$

The upper part of **Fig. 3.34** shows the attenuated cos function in black and its envelope (blue exponential function). A mathematical operation, called **Fourier transformation**, leads from the time domain into the frequency domain. The interferogram (FID) provides thereby the curve shown on the lower part of **Fig. 3.34**. This curve represents a **Lorentz curve** at the frequency $v_i = v_0 - \Delta v_i$. The same curve can be obtained in a CW spectrum. The **half-width $b_{1/2}$** of the Lorentz curve is determined by the relaxation time T_2^*. A slow spin–spin relaxation, i.e., a light damping of the FID, leads to a sharp resonance line, which

Nuclear Magnetic Resonance Spectroscopy

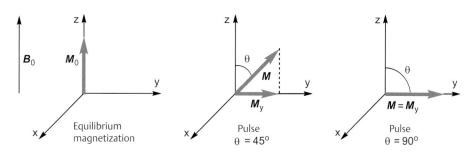

Fig. 3.33 Change of the direction of the magnetization vector by RF pulses.

a

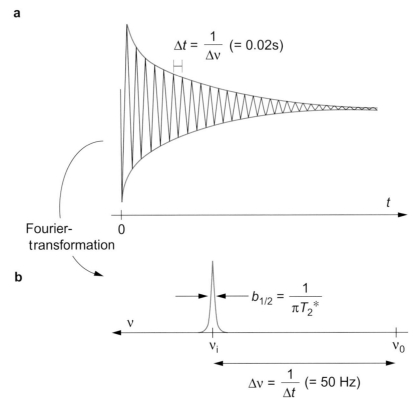

$$\Delta t = \frac{1}{\Delta \nu} \ (= 0.02\text{s})$$

t

Fourier-transformation   0

b

$$b_{1/2} = \frac{1}{\pi T_2^*}$$

ν

ν_i ν_0

$$\Delta \nu = \frac{1}{\Delta t} \ (= 50 \text{ Hz})$$

Fig. 3.34 PFT-NMR signal of a set of identical nuclei after 1 pulse (Δt = 20 ms, $\Delta \nu$ = 50 Hz). **(a)** FID (free induction decay) in the time scale; **(b)** Lorentz curve of a single resonance in the frequency scale.

has a small half-width (T_2^* = 0.32s, for example, corresponds to $b_{1/2}$ = 1.0 Hz).

n resonance lines correspond to n superimposed FIDs per scan. **Fig. 3.35** demonstrates the time flow of a PFT measurement. Immediately after the pulse with a **pulse width** t_p in the ms range, an acquisition time A_t follows (in the range of seconds). Ahead of the next pulse, a **relaxation interval** R_t will be inserted, which guarantees an extensive relaxation ($R_t \approx 10T_2^* \leq 5T_1$). A small pulse angle of, for example, 30° can shorten the relaxation interval to about $3T_1$ or less. The FID of 8 to 32 **scans** is accumulated for a routine ^{1}H NMR measurement. Very diluted probes or an end-group determination of polymers require for example n = 256. Less sensitive nuclei such as ^{13}C normally need a few thousand scans, before the Fourier transformation will be performed.

The **S/N ratio** improves by a factor $\sqrt{n}$. In addition, it can be further improved by a mathematical process, in which the FID is multiplied by an exponential function e^{-ct}. This **sensitivity enhancement** is on the account of the **resolution**, because the attenuation of the FID causes a line broadening.

The application of the function e^{+ct} has the opposite effect. In particular, an $e^{c_1t - c_2t^2}$ ($c_1, c_2 > 0$) multiplication of the FID proves its value

for the resolution enhancement. The c_2t^2 term in the exponent corresponds to a Lorentz-Gauss transformation. The resolution is improved at the expense of the S/N ratio. The distinct resonance lines become sharper (smaller half-width $b_{1/2}$ and more separate lines can be seen in complex spin patterns. An exaggerated application of the Gauss manipulation causes negative amplitudes in the signal patterns. Nowadays, so-called TRAF functions are used. They improve the resolution and do not degrade the sensitivity.

Fig. 3.36 depicts the ^{1}H NMR spectrum of trichloromethane (chloroform) measured with a high sensitivity in [D$_6$]acetone. A lot of disturbing signals become visible.

The rotation of the sample tube can lead to spinning side bands, which lie symmetrically on either side of the main signal. Their distance from the main signal increases with increasing rotational frequency, and their intensity becomes so small that confusion with real signals can be eliminated.

Further signals symmetrically placed about the main signal are the so-called ^{13}C satellites. They result from coupling to ^{13}C nuclei. Since the natural abundance of ^{13}C is only 1.1%, they are only observed for the strongest signals in routine ^{1}H NMR spectra (but see p. 155).

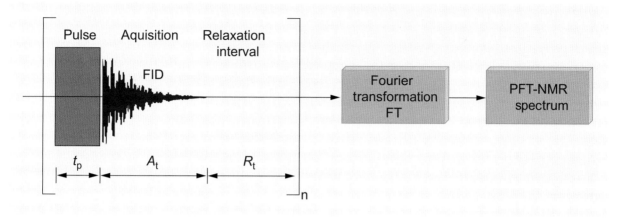

Fig. 3.35 Scheme for the time flow of a one-dimensional PFT spectroscopy.

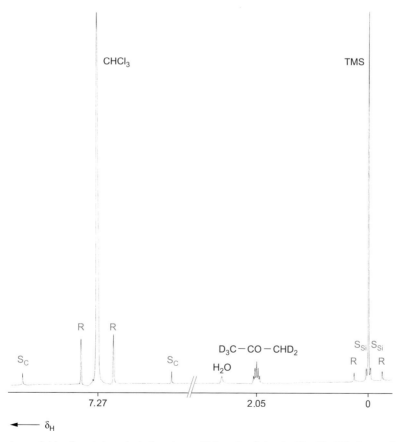

Fig. 3.36 ¹H NMR spectrum of chloroform in hexadeuterioacetone with "spurious" signals: CD₃–CO–CHD₂ (incomplete deuteration), H₂O (water content). R: spinning sidebands (because of very slow rotation), S꜀: ¹³C satellites (¹³C,¹H coupling in CHCl₃), S_Si: ²⁹Si satellites (²⁹Si,¹H coupling in TMS).

As in every spectroscopic study for structure determinations, attention should be paid in NMR on introduced impurities (traces of undeuterated solvents, softeners of plastics, grease from joints, etc.). In particular, suspended (ferromagnetic) particles or dissolved paramagnetic compounds can reduce the **high resolution**.

The instruction to avoid paramagnetic impurities does not mean that NMR spectra of **paramagnetic compounds** cannot be obtained. The coupling of the magnetic moment of an unpaired electron to the magnetic moments of the measured nuclei causes a fast relaxation of the nuclei. Broad, unresolved NMR signals are the consequence. Moreover, the shielding of the nuclei is altered so dramatically that chemical shifts (δ values) are obtained, which differ strongly from those measured for diamagnetic compounds.

The methyl groups of the nitroxyl radical TEMPO (2,2,6,6-tetramethylpiperidine-N-oxyl, **72**) give a broad resonance signal $\delta\,(^1H)$ at −16.51, while it is found at $\delta = 1.06$ for the corresponding hydroxylamine **73**. The difference of the $\delta\,(^{13}C)$ values is still bigger.

72   **73**

3.3.2 ^1H Chemical Shifts

The normal range (scale) of proton chemical shifts is approximately 12 ppm broad. If extreme values are considered, it is extended to about 50 ppm.

$$\delta\,(^1H)$$
$$20\text{---------}12\text{------}0\text{---------}-30\ \text{ppm}$$

Ignoring media effects, the chemical shift of a CH proton is determined by four factors:

- Hybridization of C
- Distribution of electron density (charge)
- Anisotropy effects
- Steric effects.

In 6-methylhept-4-en-1-yne (**74**),

74

the sequence

$$\delta(-CH_3) < \delta\,(\equiv CH) < \delta\,(-\overset{|}{C}H) < \delta\,(-\overset{|}{C}H_2) < \delta\,(=CH)$$

can be observed.

In general, one finds in hydrocarbons:

$$\delta\,(-\overset{|}{\underset{|}{C}}-H) \approx \delta\,(\equiv C-H) < \delta\,(=\overset{|}{C}-H)$$

The electron shells of a nucleus and its direct neighbors shield the nucleus from the external field $\boldsymbol{B}_0$ (see Section 3.1.2, p. 88). The electron density and therefore the shielding constant σ are affected by inductive and mesomeric effects. Consider firstly element–hydrogen bonds X–H. With increasing electronegativity of X, the shielding of the proton is reduced and the signal is shifted to a lower field. The ^1H shifts of ethanol (**75**) and ethanethiol (**76**) are given as examples. Under comparable measurement conditions, the OH protons absorb at a lower field compared to the SH protons.

$$H_3C-CH_2-OH \qquad\qquad H_3C-CH_2-SH$$

$\delta:$ 1.24   3.71   2.56   1.30   2.44   1.46

75 (in CCl_4)   **76** (in CCl_4)

While a comparison of the shifts of the α-protons of ethanol and ethanethiol shows a similar effect, the electronegativity of X is no longer dominant for the shifts of the β-protons. **Table 3.7** demonstrates how the influence of substituents (Hal, OR, NR$_2$, and NO$_2$) decreases in a n-butyl chain.

All these substituents cause down-field shifts, which are large in the α-position, noticeable in the β-position, small in the γ-position, and negligible in the δ-position. As **Table 3.8** shows, the down-field shift is markedly enhanced by the presence of additional electronegative substituents.

In contrast, protons bound to electropositive central atoms absorb at a very high field. In metal complexes, for example iron tetracarbonyl hydride (**77**), other influences must also be considered. In the literature, δ values down to −30 are found for such compounds.

$$HFe(CO)_4 \qquad \delta = -10.5$$
77

Table 3.7 ^1H NMR shifts of 1-substituted n-butane in $CDCl_3$

X	$-CH_2$	CH_2	CH_2	CH_3
H	0.91	1.30	1.30	0.91
Cl	3.42	1.69	1.41	0.92
I	3.20	1.80	1.41	0.93
OH	3.63	1.55	1.39	0.94
OCH$_3$	3.37	1.55	1.37	0.92
OCOCH$_3$	4.05	1.59	1.40	0.95
SH	2.53	1.60	1.41	0.91
NH$_2$	2.69	1.42	1.35	0.92
N(CH$_3$)$_2$	2.24	1.42	1.35	0.92
NHCOCH$_3$	3.20	1.48	1.35	0.93
NO$_2$	4.42	2.05	1.43	0.99

Table 3.8 ¹H shifts of the halogenated methanes (δ values)

	CH$_3$X	CH$_2$X$_2$	CHX$_3$
X = F	4.27	5.45	6.49
X = Cl	3.06	5.30	7.24
X = Br	2.69	4.94	6.83
X = I	2.15	3.90	4.91

Electron-donor groups D and electron acceptor groups A on CC double or triple bonds cause inductive and mesomeric effects which change the electron density and consequently the chemical shifts. The following cases for electronic effects have to be discussed:

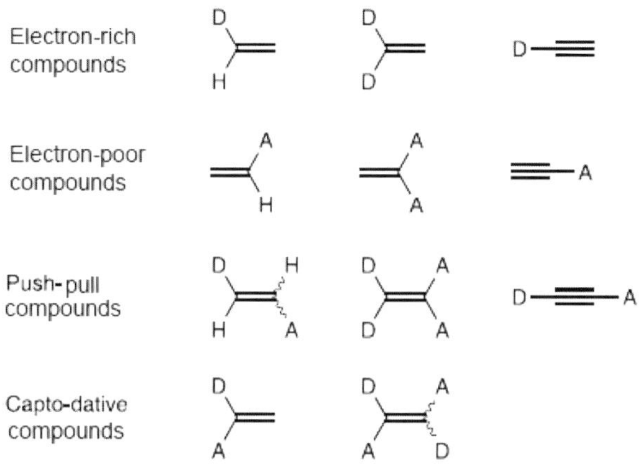

Electron-rich compounds

Electron-poor compounds

Push-pull compounds

Capto-dative compounds

In acrolein (**78**) and ethyl vinyl ether (**79**), the influence of an acceptor group and a donor group, respectively, becomes obvious (δ values):

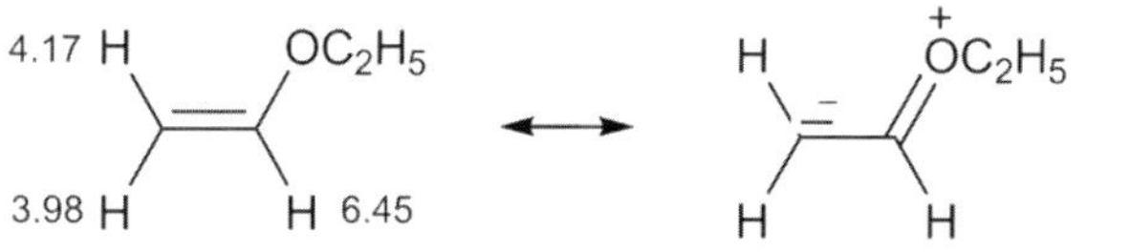

Compared to ethylene (**81**), two acceptor groups are present in the diester **80** (2-methylenemalonic acid dimethyl ester) and two donor groups in the ketene acetal **82** (1,1-dimethoxyethene).

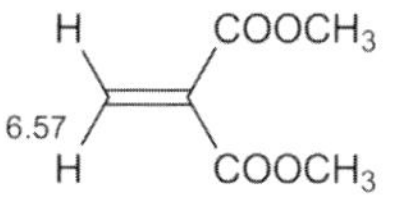

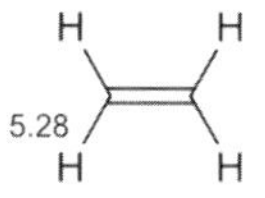

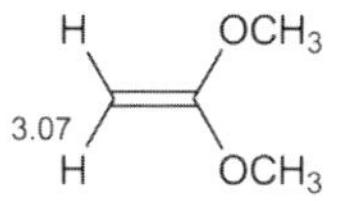

Similar effects exist for the series of alkynes **83–85**.

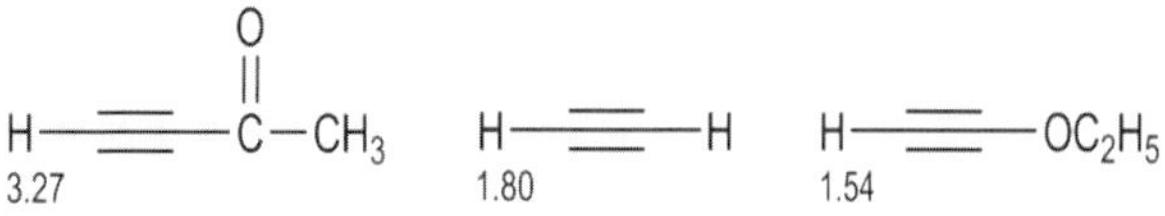

A drastic difference in the ¹H chemical shifts is observed for the push–pull system **86** ((E)-3-ethoxyacrylonitrile) and the capto-dative system **87** (2-ethoxyacrylonitrile). The olefinic protons exhibit $\Delta\delta = 2.55$ for **86** and $\Delta\delta = 0.10$ for **87**.

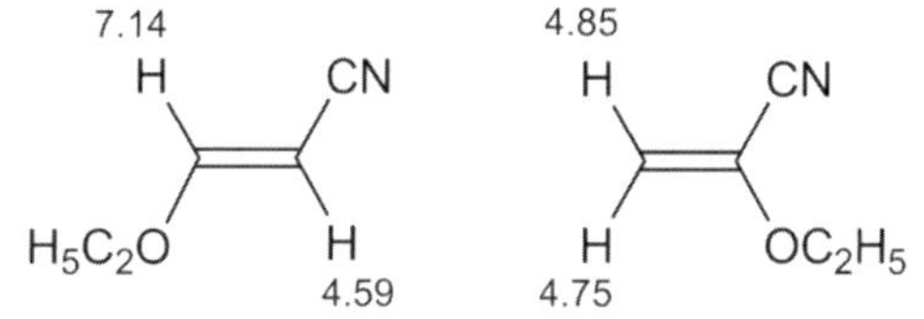

Changes of charge density caused by formation of cations (**89**, **91**, and **92**) or anions (**93**, **95**, and **97**) often show marked effects on the ¹H chemical shifts:

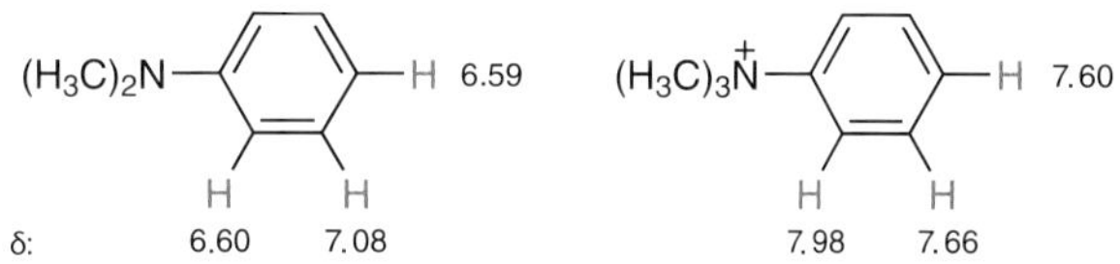

δ: 6.60   7.08
88

δ: 7.98   7.66
89

H$_3$C$-$CH$_2$$-CH_3$
δ: 0.91   1.33   0.91
90

H$_3$C$-\overset{+}{\text{C}}$H$-$CH$_3$
δ: 5.06   13.50   5.06
91

$[$CH$_2$$\doteq$CH$\doteq$CH$_2$$]^+$
δ: 8.97   9.54   8.97
92

$[$CH$_2$$\doteq$CH$\doteq$CH$_2$$]^-$
δ: 2.46   6.28   2.46
93

H$_3$C$-$CH$_2$$-CH_2$$-$COOH
δ: 0.97   1.68   2.34
94

H$_3$C$-$CH$_2$$-CH_2$$-COO^-$
δ: 0.90   1.56   2.16
95

δ: 8.07   6.90
96

δ: 7.78   6.04
97

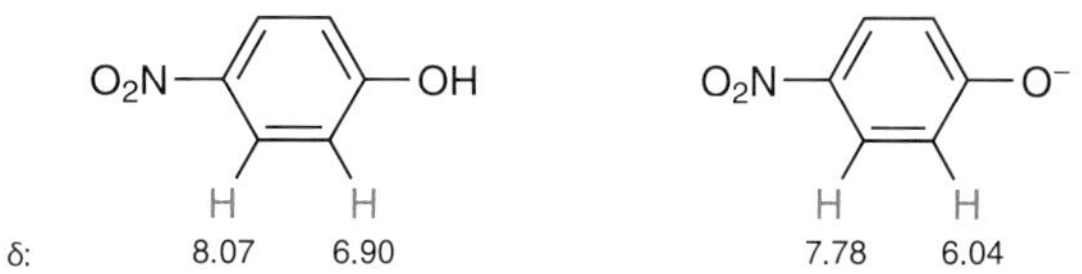

After the electronic effects, **anisotropy effects** shall be discussed.

Chemical bonds are in general magnetically anisotropic, i.e., the susceptibility χ depends on the spatial orientation. Double and triple bonds, three-membered rings, and cyclic conjugated systems show particularly strong magnetic anisotropy. **Fig. 3.37** demonstrates the effect of using cones of anisotropy. In the positive areas, marked in green-blue, the shielding is large. Protons in these regions are shifted to higher field (smaller δ values). Protons in the negative regions absorb at low field (large δ values) because of the reduced shielding.

Protons bound to olefinic C atoms have δ values between ca. 4 and 8, i.e., at much lower field than similar protons bound to saturated C atoms. This is due to the anisotropy as well as the change in hybridization. Aldehyde protons absorb between about 9.3 and 10.7. Here the effect of the electronegativity of the O atom adds to those of hybridization and anisotropy. Acetylenic H atoms should be more deshielded than olefinic H atoms because of the polarity of the C$-$H bond. The anisotropy effect of the C$\equiv$C bond however causes an opposite effect: the signals of protons on sp carbons are found between $\delta = 1.8$ and 3.2. If methyl groups are considered instead of H atoms, then bonding to C=C or C=O bonds also causes a down-field shift. The methyl protons therefore lie in the negative region of the cone of anisotropy (**Fig. 3.37**). Methyl groups bound to sp carbons also show down-field shifts because of the small angle of the anisotropy cone. These shift effects are illustrated by the δ values of the following compounds **90**, **98-100**, and **101-103**.

In norbornane (**101**), the chemical shift of the bridgehead protons lies as expected at a relatively high field. The introduction of one or two (C=C) bonds shifts them to successively larger δ values. A similar effect is seen for the *syn*-protons of the methylene bridge; the signal of the *anti*-proton in norbornene (**102**) however is shifted to a higher field. Similar differences are seen for the shifts of the *exo*- and *endo*-protons. Finally, it is noteworthy that the olefinic protons in norbornadiene (**103**) absorb at unusually high δ values. From these examples, it can be appreciated that the anisotropy effects of individual bonds or structural elements often combine in quite complicated ways. The schematic representation of anisotropy effects in **Fig. 3.37** can only be taken as a rough guide.

The **ring current model** for the explanation of anisotropy effects in cyclic conjugated π-systems has proved to be particularly fruitful, despite certain theoretical problems. As shown in **Fig. 3.38** for a benzene ring, one can imagine that the magnetic field causes a "ring current" of the electrons in an aromatic ring. The opposing field thus induced is described by the lines of force represented by the dashed lines. In the positive zone above, below, and inside the benzene ring, the B_0 field is reduced, i.e., the shielding increased. The signals of protons lying in these zones experience an up-field shift. The signals of protons in the negative zone outside the ring are shifted to low field.

The ^{1}H shifts of the following examples (**104-107**, **19**, **108**, and **109**) show these effects very clearly.

Propane **90**   Propene **98**

Acetaldehyde **99**   Propyne **100**

Norbornane **101**   Norbornene **102**   Norbornadiene **103**

104   **105**   **106**   **107**   **19**

108   **109**

In the annulene series, the ring current is often used as a qualitative criterion for aromaticity. The model proposed for benzene in **Fig. 3.38** applies to planar, cyclic conjugated $(4n + 2)\pi$ electron systems. The "external" protons are deshielded and the "internal" protons shielded. In contrast to this **"diamagnetic" ring current** of the **"diatropic" compounds** (**aromatics**), cyclic $(4n)\pi$ electron systems, **"paratropic" compounds** (**antiaromatics**) show a **"paramagnetic" ring current**. Shielding and deshielding are then exactly reversed. This does *not* mean that the direction of the induced field is reversed; instead, antiaromatics have a small HOMO–LUMO energy difference based on Jahn–Teller splitting. In the

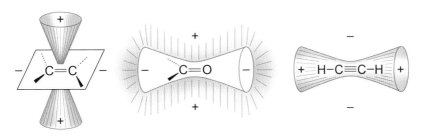

Fig. 3.37 Magnetic anisotropy of C=C, C=O, and C≡C bonds.

magnetic field B_0, this leads to a mixing of wave functions of the excited states with the ground state, increasing the paramagnetic part of the shielding constant (cf. p. 88). The ring current effect is an easily determined physical criterion of **aromaticity**. An illustration is given by the **aromatic (diatropic)** [18]annulene (**110**) and the **antiaromatic (paratropic)** [16]annulene (**111**).

As the temperature increases, the mobility of these annulenes increases to such an extent that the internal and external protons exchange their positions and only an averaged signal is observed. While [18]annulene shows separate signals for the two kinds of protons at room temperature, the much reduced activation energy in [16]annulene results in a singlet.

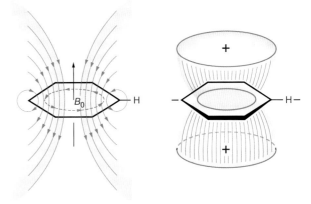

Fig. 3.38 Ring current model for aromatic systems (example: benzene).

In bay regions as, for example, in phenanthrene (**113**) and in fjord regions as in benzo[c]phenanthrene (**114**), steric effects and anisotropy effects are superimposed. The protons in these regions exhibit large δ values.

110	12 Hₐ: δ = 9.28
−70 °C	6 Hᵢ: δ = −2.99
+110 °C	18 H: δ = 5.45

111	12 Hₐ: δ = 5.2
−120 °C	4 Hᵢ: δ = 10.32
+30 °C	16 H: δ = 6.70

The ¹H NMR signals of the aromatic 1,6-methano[10]annulene (**112**) are at about 7 ppm for the ring protons and at δ = -0.5 ppm for the protons on the methano bridge.

Compound **114** can be regarded as [4]helicene. The higher helicenes, starting with [6]helicene have δ values between 6 and 7 ppm for certain protons on the terminal benzene rings. An up-field shift operates in this region in accordance with the ring current model (**Fig. 3.38**).

Kekulene (**115**) represents an extreme example for a downfield shift of aromatic protons. The inner protons show their resonance at δ = 10.45. The two perimeters of **115** have 18 and 30 π centers, respectively. This corresponds to the aromaticity rule of (4n + 2) conjugated π centers. However, the δ values of the inner and outer protons reveal a benzenoid and not an annulenoid aromaticity (cf. the 18-π perimeter of **110**, whose inner protons have a δ value of -2.99 ppm at low temperature).

If the ring current model is extended to condensed aromatics, the estimates of the chemical shifts must be summed over all the rings involved.

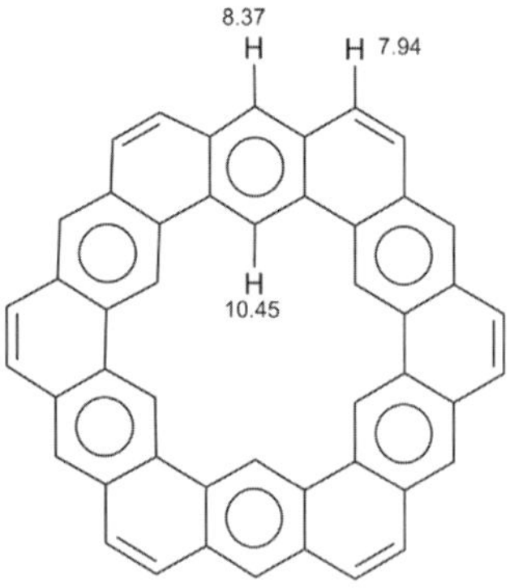

8.37
H   H 7.94

H
10.45

115

Dinaphtho[1,2-*a*:2′,1′-*j*]anthracene (**116**) is a further condensed benzenoid hydrocarbon which gives proton signals at very low field.

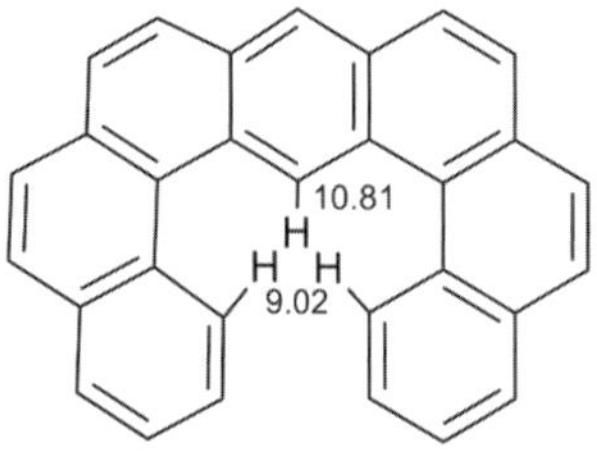

10.81
H
H H
9.02

116

The ring current concept as a model for the anisotropy effect can be extended to heteroaromatics [cf., for example, 1,2,3,6-tetrahydropyridine (**117**) and pyridine (**118**)].

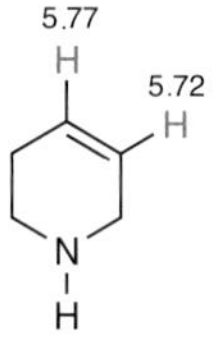

5.77
H   5.72
       H

7.75
H   7.38
       H
       8.59
       H

117 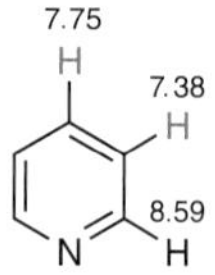   **118**

Another good example is coproporphyrin (**119**).

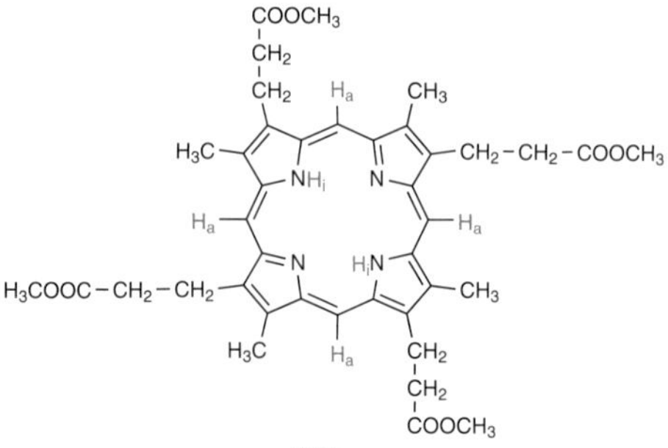

COOCH₃
CH₂
CH₂   Hₐ   CH₃
H₃C
        NHᵢ   N   CH₂–CH₂–COOCH₃
Hₐ   Hₐ
        N   HᵢN
H₃COOC–CH₂–CH₂   CH₃
H₃C   Hₐ   CH₂
             CH₂
             COOCH₃

119

The NH protons exchange their positions on the four nitrogen atoms so quickly that all the pyrrole rings are equivalent. Only two methyl signals are observed, one being due to the ester groups. The signal of the NH protons is at $\delta = -4$ and is shifted by 11 ppm compared to the NH signal of pyrrole. The protons H_a of the methine bridges absorb at $\delta = 10$, i.e. at very low field.

The ring current model applied to the periphery of the porphyrin framework gives a simple explanation for these effects, the NHa protons being internal and the methine protons being external.

Recently prepared subporphins represent a further nice example for the ring current. The boron complex of tribenzosubporphin (**120**) belongs to this category. The nonplanar compound has a C_3 axis. The central network of 14π electrons leads to low-field signals for the *meso*-H and the benzene ring protons and to a high-field signal at $\delta = -2.60$ ppm for the axial OH proton. The ^{11}B resonance is located at $\delta = -14.6$ ppm in relation to $BF_3 \times (OC_2H_5)_2$ as an external standard. Analogous subporphins, which do not have condensed benzene rings, show a similar diatropic ring current.

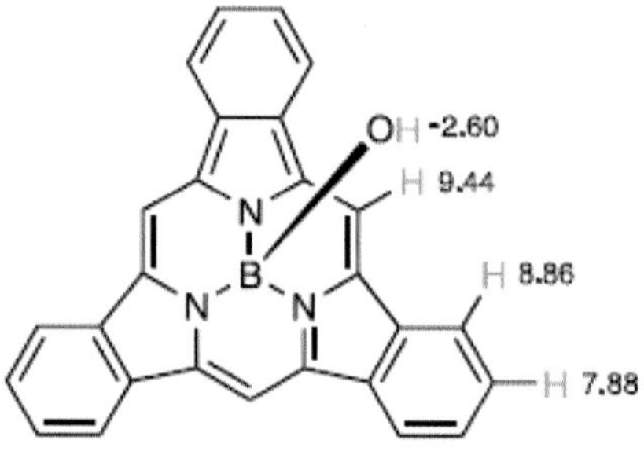

OH -2.60
H 9.44
H 8.86
H 7.88

120

The shifts of the quasiaromatic ions **121** to **125** show, in addition to the effects of anisotropy, a marked influence of the electron density caused by the charge.

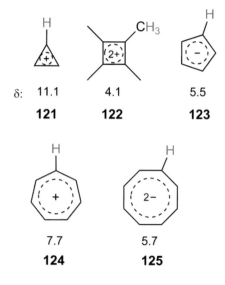

H   CH₃   H

δ: 11.1   4.1   5.5

         121   **122**   **123**

H   H

         7.7   5.7

         124   **125**

Table 3.9 ¹H chemical shifts of cycloalkanes and cyclic ethers (δ values in CDCl$_3$ or CCl$_4$)

	δ		δ		
			α	β	γ
Cyclopropane	0.22	Oxirane	2.54	–	–
Cyclobutane	1.96	Oxetane	4.73	2.72	–
Cyclopentane	1.51	Tetrahydrofuran	3.75	1.85	–
Cyclohexane	1.44	Tetrahydropyran	3.56	1.58	1.58
Cyclodecane	1.51				
Cyclododecane	1.34				

The dianion of [16]annulene shows a signal for the four internal protons at very high field ($\delta = -8.17$) and for the 12 external protons at low field. This contrast to the uncharged compound is typical for the transition from $4n$ to $(4n + 2)\pi$ electron systems.

Anisotropy effects are important for a number of other, not necessarily conjugated, ring systems. Cyclopropane ring will be mentioned here, protons on which are strongly shielded (**Table 3.9**). The effect is also shown by heterocyclic three-membered rings.

Cyclic π-electron systems, which are neither aromatic nor anti-aromatic because of their strong deviation from planarity, show shifts which confirm their analogy to open chain alkenes. Thus, cyclooctatetraene (**126**) shows a singlet at $\delta = 5.69$. Here it must be realized that this signal is averaged by two processes which are both fast on the NMR time scale: ring inversion and double bond shifts (degenerate valence isomerism, **automerization**) (see p. 115).

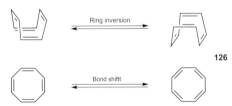

126

Strong van der Waals interactions between two protons or between a proton and a neighboring group lead to deformation of the electron clouds. The **steric effect** combined with an anisotropy effect in the bay region of phenanthrene has already been mentioned. Methyl groups in the bay region enhance the steric interaction. 4,5-Dimethylphenanthrene (**127b**) has a chiral, nonplanar conformation. Its methyl signal shows an up-field shift compared to the monomethyl compound **127a**. The classic example of steric effects on proton shifts is the polycyclic compound **128**. The $\Delta\delta$ value of the two methylene protons is 2.67 ppm. However, when the hydroxy group is in the *exo*-position, the $\Delta\delta$ value is considerably reduced. Examples that are largely free from anisotropy effects are shown by the hydrocarbons **129a** to **129c**. At room temperature, the methyl groups of the

tert-butyl groups are chemically equivalent and show a low-field shift with increasing steric hindrance.

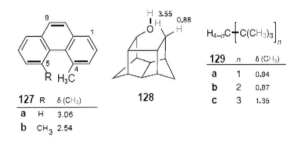

	δ (CH$_3$)
127 R	
a	H 3.06
b	CH$_3$ 2.54

	n	δ (CH$_3$)
a	1	0.94
b	2	0.97
c	3	1.35

The solvent used has a special influence on the ¹H chemical shifts of the compounds, when acid–base interactions, alteration of hydrogen bonds, or strong intermolecular anisotropy effects occur. The latter is the case when C$_6$D$_6$ or perdeuterated pyridine (C$_5$D$_5$N) is applied.

The ¹H NMR measurement of acetic acid ethyl ester (**130**) is taken here as an example for the **solvent effect**. The differences $\Delta\delta$ for different solvents are usually small ($\Delta\delta < 0.1$ ppm) with the exception of the measurement in C$_6$D$_6$, which reveals significant shifts to a higher field ($0.1 < \Delta\delta < 0.4$ ppm).

Solvent	H$_3$C—COO—CH$_2$—CH$_3$ (**130**)		
	δ	δ	δ
CDCl$_3$	2.04	4.11	1.26
CD$_3$OD	2.01	4.09	1.24
CD$_3$SOCD$_3$	1.99	4.03	1.17
CD$_3$CN	1.97	4.06	1.20
CD$_3$COCD$_3$	1.97	4.05	1.20
C$_6$D$_6$	1.65	3.89	0.92

Host–guest effects represent another important feature of the chemical shift. The chemical environment of the protons of a guest molecule is remarkably changed by the complexation to a host molecule. Depending on the formed supramolecular structure, this effect causes $\Delta\delta$ values of variable size ($0.1 \leq \Delta\delta \leq 25$ ppm). Some distinctive examples shall be discussed here.

When acetylene (**131**) and cyclobutadiene (**132**) are generated inside of a hemicarcerand, a $\Delta\delta$ value of $+0.11$ ppm can be measured for **131** in CDCl$_3$. Such a comparison does not exist for **132**, because **132** is unstable in solution. A singlet signal at $\delta = 2.27$ is obtained for the encapsulated cyclobutadiene (**132**). This resonance at high field corresponds to the antiaromatic character.

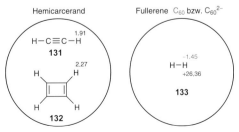

If molecular hydrogen (**133**) is included in fullerene C_{60} or C_{60}^{2-}, a dramatic shift of the NMR signal can be observed. Dissolved in 1,2-dichloromethane, H_2 gives a signal at $\delta = 4.54$ ppm; however, the inclusion in C_{60} leads to $\delta = -1.45$ and in the dianion C_{60}^{2-} to $\delta = 26.36$ ppm. C_{60} has aromatic character to a certain extent. Diamagnetic ring currents can be conceived in the 20 six-membered rings and paramagnetic ring currents in the 12 five-membered rings. The up-field shift of $\Delta\delta \approx 6$ ppm for H_2 included in C_{60} is in accordance with the anisotropy model. The ring current behavior of the five- and six-membered rings is reverted for C_{60}^{2-}, and consequently a very big down-field shift is observed for H_2.

Macrocyclic host molecules, such as crown ethers, cryptands, calixarenes, cyclodextrins, cucurbiturils, and the recently studied pillararenes assume an especial role. **Fig. 3.39** shows the inclusion of 1,6-dibromohexane (**134**) in pillar[5]arene (**135**). The complex **135** ⊃ **134** is characterized by remarkable up-field shifts of the protons e, f, and g of the guest **134** and slight downfield shifts of some protons of the host **135**.

The proton resonance of tetradecan-2-ol (**136**) in solution and encapsulated in **137** is discussed here as a further example. **Fig. 3.40a** shows the normal ^1H NMR solution spectrum (400 MHz) of **137** in $CDCl_3$. Apart from the triplet of H_3C-14 ($\delta = 0.87$), the doublet of H_3C-1 ($\delta = 1.18$), and the multiplet of HC-2 ($\delta = 1.43$), a broad signal ($1.20 \leq \delta \leq 1.35$) for all 22 CH_2 protons is

obtained. If **136** is encapsulated in jackets of **137** from each side, as shown in **Fig. 3.40b**, a far shifted and spread ^1H NMR spectrum can be measured (**Fig. 3.40c**). All signals of **136** are up-field shifted, in particular, those of the terminal groups H_3C-1 and H_3C-14. Moreover, a splitting of the signals of the diastereotopic protons of the methylene groups H_2C-3, -4, and -5 can be seen. The size of the splitting decreases with increasing distance of the center of chirality C-2 and is not discernible beyond H_2C-6.

3.3.3 $^1H,^1H$ Coupling

The magnetic coupling between two nuclei in a molecule is generally transmitted by the intervening bonds. (Scalar coupling through space is, however, also known. This through-space coupling occurs when two nuclei are brought so close together by steric compression that their orbitals overlap.) The quantitative measure for the coupling is the **coupling constant** ^nJ, where n indicates the number of bonds. The most important $^1H,^1H$ couplings are summarized in **Table 3.10**.

The **sign of the coupling constant** cannot be determined from first-order spectra. Relative, but not absolute, signs can sometimes be determined from second-order spectra. Absolute signs are based on the assumption that $^1J(^{13}C,^1H)$ is positive. $^2J(H,H)$ couplings are usually negative, $^3J(H,H)$ couplings positive, and long-range couplings either positive or negative.

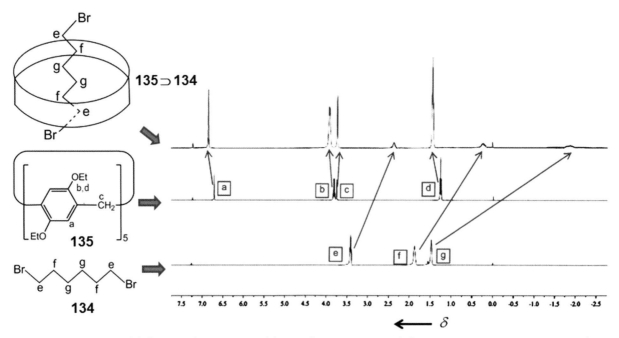

Fig. 3.39 ^1H NMR spectra of the host **135**, the guest **134**, and the complex **135** ⊃ **134** in $CDCl_3$ (from Cao D, Meier H. 2014, Asian J. Org. Chem. **3**, 244).

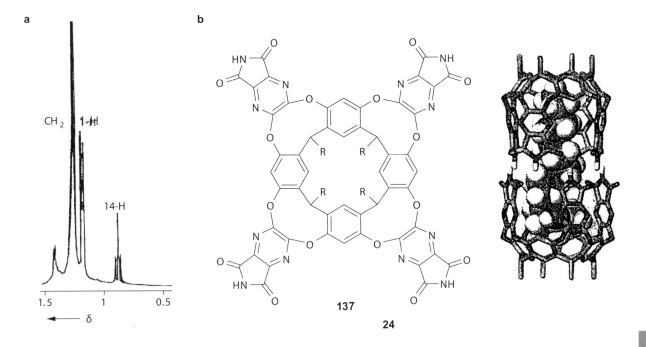

137

24

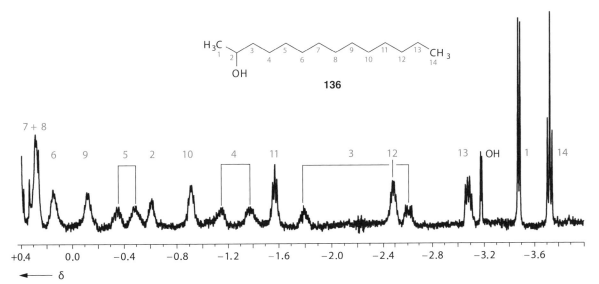

136

Fig. 3.40 (a) ¹H NMR spectrum (CH₂ and CH₃ groups) of tetradecan-2-ol (**136**) in CDCl₃; (**b**) cylindrical encapsulation of **136** by two molecules **137**; (**c**) ¹H NMR spectrum of **136** encapsulated in two molecules **137** (from Siering C, Toraeng J, Kruse H, Grimme S, Waldvogel SR. 2010, Chem. Commun. **46**, 1625).

Nuclear Magnetic Resonance Spectroscopy

Table 3.10 ^{1}H,^{1}H couplings

Type of coupling	Coupling constant ^nJ (order of magnitude)	Structural elements
direct	^1J (276 Hz)	H—H
geminal	^2J (0...30 Hz) usually negative	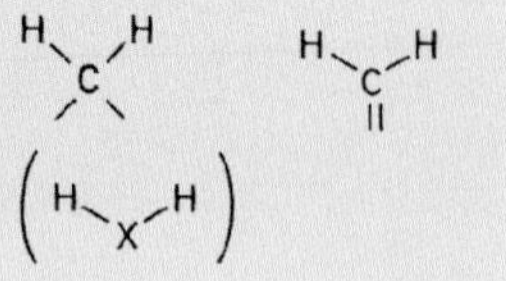
vicinal	^3J (0...20 Hz) positive	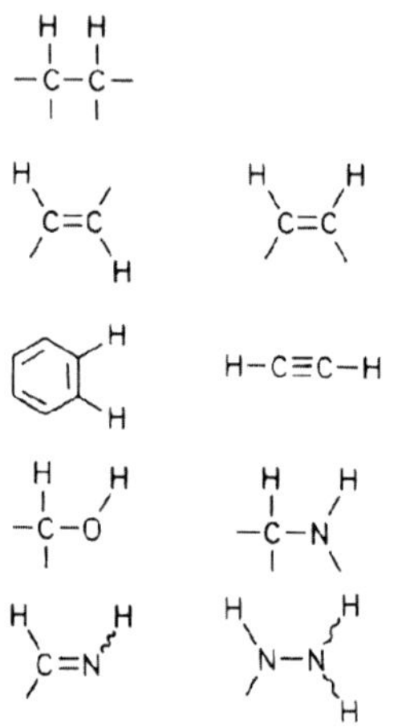
long-range	^4J (0...3 Hz) positive or negative	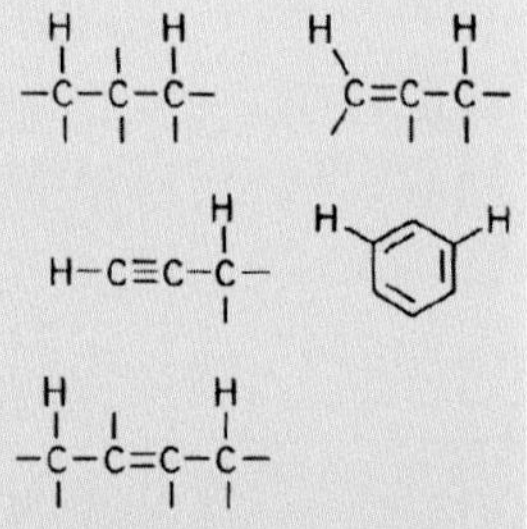
	^5J (0...2 Hz) positive	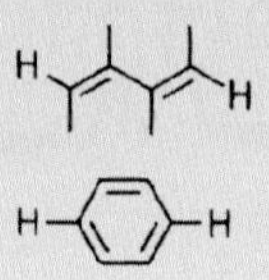

The coupling between magnetically equivalent protons has no effect on ^{1}H NMR spectra. It can be determined by deuteration or from ^{13}C satellites (see p. 152 and p. 155).

As n increases, the coupling generally gets weaker, i.e., $|^nJ|$ decreases.

H—H	H—CH$_2$—H	H—CH$_2$—CH$_2$—H	H—(CH$_2$)$_3$—H
133	**12**	**138**	**90**
276 Hz	12 4.Hz	8.0 Hz	<1 Hz

Geminal Coupling

Geminal coupling increases with the *s*-character of the hybrid orbitals.

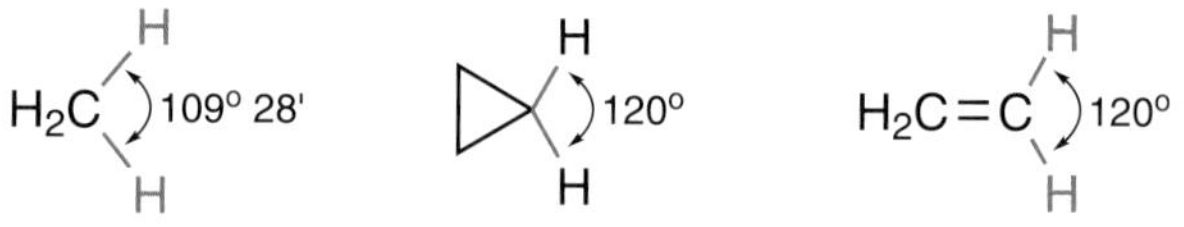

12	**139**	**15**
−12.4 Hz	−4.5 Hz	+2.5 Hz

The following examples show the effect of various substituents.

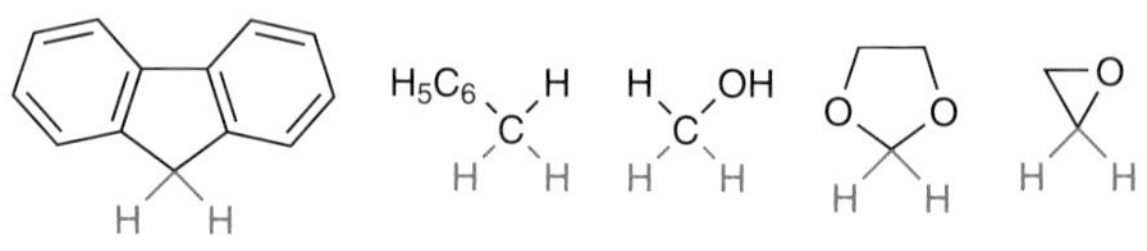

140	**19**	**141**	**142**	**143**
−22.3 Hz	−14.5 Hz	−10.8 Hz	±0 Hz	+5.5 Hz 143

144	**145**	**146**	**147**
−3.2 Hz	−1.3 Hz	+7.1 Hz	+42.2 Hz

More complex effects are shown by the examples **148** and **149**. The effect of a neighboring oxygen atom (more electronegative) on the *geminal* coupling of a methylene group is an increase in the coupling constant (less negative) compared to methane (**12**). A neighboring π-bond, here CO, reduces the coupling constant (more negative).

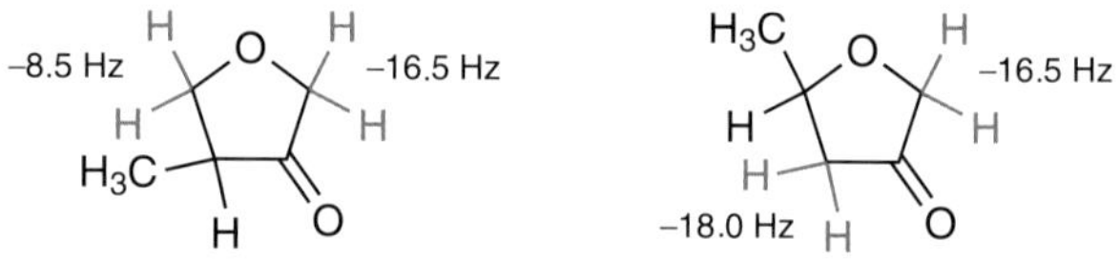

148   **149**

Vicinal Coupling

The generally positive ^3J(H,H) coupling depends, apart from substituent effects, to a large degree on the molecular structure. The bond lengths l, the bond angles α, and the dihedral angles ϕ are all important.

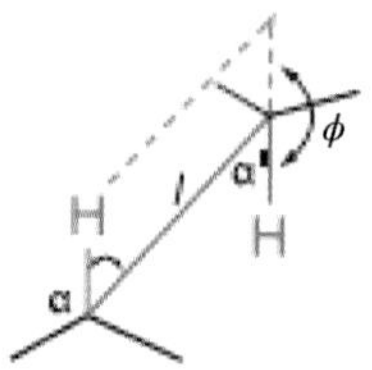

If, as in the cases of olefins or aromatics, there is no possibility of the dihedral angle varying (rotation about the C–C bond), then ^3J decreases with increasing bond length l and increasing bond angle α.

In small rings or rigid bicyclic systems like norbornane, the ^3J coupling of *cis*-protons can however be larger than that of *trans*-protons. For cyclopropane, $^3J_{cis}$ is about 7 to 10 Hz and $^3J_{trans}$ is about 4 to 7 Hz.

15 ^3J: 11.6 Hz
150 8.3 Hz
105 7.54 Hz
150 6.9 Hz

→ increasing bond length l →
(decreasing π bond order)

10 ^3J: 1.3 Hz
151 2.8 Hz
152 5.1 Hz
153 8.8 Hz

← increasing angle α ←

^3J (E) ($\phi = 180°$) is always larger than ^3J (Z) ($\phi = 0°$).

H, R C=C		^3J (Z)	^3J (E)	^2J (gem)
15	R = H	11,6	19,1	2,5
19	R = C₆H₅	11,5	18,6	1,1
154	R = OCH₃	6,7	14,0	−2,0
144	R = F	4,7	12,8	−3,2

155
$^3J(Z) = 12.3$ Hz

8
$^3J(E) = 15.8$ Hz

For ^3J-couplings across "freely" rotating C–C bonds, the size of the coupling constant varies with the dihedral angle ϕ. A quantitative estimate is given by the **Karplus curve** and related equations (**Fig. 3.41**). The measured couplings are particularly for $\phi = 0°$ and 180° somewhat larger than predicted (shaded area).

When there is rapid rotation, an averaged value is obtained for ^3J. For a first approximation, it can be assumed that the three staggered conformations are equally populated. The average value is then given by

$$^3J = \frac{^3J(60°) + ^3J(180°) + ^3J(300°)}{3}$$

$$\approx \frac{3.5 + 14 + 3.5}{3} \approx 7 \text{ Hz}$$

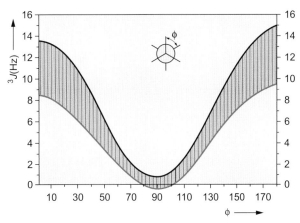

Fig. 3.41 Dependence of the *vicinal* coupling constant ^3J on the dihedral angle ϕ. Karplus curve:

$$^3J = \begin{cases} 8.5 \cos^2\phi - 0.28 & \text{for} \quad 0° \lesseqgtr \phi \lesseqgtr 90° \\ 9.5 \cos^2\phi - 0.28 & \text{for} \quad 90° \lesseqgtr \phi \lesseqgtr 180° \end{cases}$$

In practice, the coupling constants ^3J in ethyl groups vary about this value; the effect of the electronegativity of R being noticeable.

H₂C–CH–R

156	R = Li	$^3J = 8.4$ Hz
138	R = H	8.0 Hz
157	R = C₆H₅	7.6 Hz
90	R = CH₃	7.3 Hz
158	R = OC₂H₅	7.0 Hz

Vicinal couplings from aldehyde protons to neighboring protons on saturated carbon atoms are conspicuously small:

$^3J \approx 1, ..., 3$ Hz

$^3J \approx 5, ..., 8$ Hz

In the chair conformation of cyclohexane (**159**), three *vicinal* couplings can be distinguished:

159

$^3J_{aa} \approx 7 ... 12$ Hz ($\phi = 180°$)
$^3J_{ee} \approx 2 ... 5$ Hz ($\phi = 60°$)
$^3J_{ae} \approx 2 ... 5$ Hz ($\phi = 60°$)

D-Glucose (**160 ⇌ 161**) may be considered as a specific example. An aqueous solution shows two doublets for the proton on C-1 at $\delta = 5.22$ and 4.63, which at lower field is assigned to α-D-glucose (**160**), because it has the smaller coupling to the proton at C-2 ($^3J_{ae} < {}^3J_{aa}$).

α-D-Glucose **160**

($^3J_{1e.2a} = 3.5$ Hz)

β-D-Glucose **161**

($^3J_{1a.2a} = 7.7$ Hz)

If the relevant C−C bond carries electronegative substituents, the coupling constant is reduced. For organometallic compounds, it is correspondingly increased. This **substituent effect** is observed for saturated, unsaturated, and heteroaromatic compounds.

In conformational analysis, the ^3J couplings of a proton H_X with the two nonequivalent protons H_A and H_M of an α-CH$_2$ group are often of interest. In an ideal case, the two dihedral angles φ_1 and φ_2 differ by 120°. This leads to the coupling constants and splitting patterns shown in **Fig. 3.42.**

Vicinal Couplings to Exchanging Protons

The coupling of OH protons with *vicinal* CH protons can only be observed for very pure water- and acid-free alcohols. DMSO has established itself as the favored solvent, since **proton exchange** in DMSO at room temperature is sufficiently slow (cf. p. 117).

Exactly similar circumstances apply to NH protons. The vicinal coupling ^3J(CH−NH) is only observable if the (base catalyzed) proton transfer is slow. This is often the case for the structural element (aromatic amines, enamines, and amides).

In trifluoroacetic acid, the proton exchange of ammonium cations is so slow that the coupling ^3J(CH−NH$^+$) is observable as a splitting of the (C−H) signal.

$$^3J(-NH-\underset{|}{C}H-) \quad \approx \quad {}^3J(-\overset{+}{N}H-\underset{|}{C}H-)$$

The conformational dependence of the ^3J coupling constant in alcohols, amines, and amides is similar to that of the *vicinal* CH−CH coupling. For free rotation:

$$^3J(CH-OH) \quad \approx \quad 4\text{–}5 \text{ Hz}$$

$$^3J(CH-NH) \quad \approx \quad 5\text{–}6 \text{ Hz}$$

$$^3J(CH-NH-\underset{\overset{||}{O}}{C}) \approx \quad 7 \text{ Hz}$$

It is worth mentioning that in amides, for example, the splitting of the CH signal can appear even when the NH proton shows a

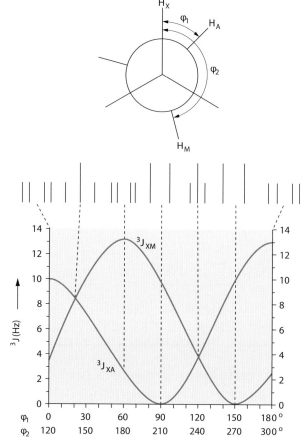

Fig. 3.42 *Vicinal* coupling constants and splitting patterns for the H_X signals in an AMX system >CH$_X$–CH$_A$H$_M$–.

broad signal (cf. the ^{1}H NMR spectrum of *N*-methylacetamide **162**; **Fig. 3.43**).

To conclude this section, a few comments about ^{14}N,^{1}H coupling are appropriate. In the NH$_4^+$ ion it is 52.8 Hz. ^{14}N with a nuclear spin of 1 has an electric quadrupole moment (see Section 3.1.1, p. 86). Apart from a few exceptions (NR$_4^+$ ions, methyl isonitrile with $|^2J| = 2.2$ Hz), the contribution of the quadrupole to the relaxation is so large that ^{14}N,^{1}H coupling is not observed. Nonetheless, this is a further cause, apart from proton exchange, for broad NH signals. It can however be removed by irradiation at the resonance frequency of the ^{14}N nucleus (see **Fig. 3.59**, p. 159).

Long-Range Coupling

Long-range couplings extend over four or more bonds. In open chain, saturated compounds, they are usually smaller than 1 Hz and are not observed. In bicyclic and polycyclic systems, the situation changes when a fixed W-orientation of the four

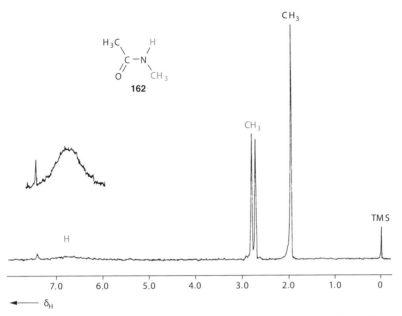

Fig. 3.43 ¹H NMR spectrum of *N*-methylacetamide (**162**) in CDCl₃ (the compound exists exclusively in the conformation with s-*trans*-oriented methyl groups).

bonds in a ^4J coupling is caused by the molecular structure; see for example, bicyclo[2.1.1]hexane (**163**).

$$^4J_{1,4} = +7 \text{ Hz}$$
$$^4J_{1,3} = {}^4J_{2,4} \approx {}^4J_{2,3} \approx 0 \text{ Hz}$$

163

An extreme example is bicyclo[1.1.1]pentane (**164**), in which the bridgehead protons have a ^4J coupling of 18 Hz.

$$^4J = 18 \text{ Hz}$$

164

In unsaturated compounds, **allylic** ^4J **and homoallylic** ^5J **couplings** are observed:

$$^4J (Z) \approx -3 \cdots +2 \text{ Hz}$$
$$^4J (E) \approx -3.5 \cdots +2.5 \text{ Hz}$$

$$^5J (Z) \approx {}^5J (E) \approx 0 \cdots -2.5 \text{ Hz}$$

Larger magnitudes of ^4J and ^5J couplings can occur in alkynes, allenes, and cumulenes, and also in certain carbocycles and heterocycles.

$$H-C\equiv C-\overset{|}{\underset{|}{C}}-H \qquad {}^4J \approx 2-3 \text{ Hz}$$

$$C=C=C \qquad {}^4J \approx -6 \cdots -7 \text{ Hz}$$

$$H-\overset{|}{\underset{|}{C}}-C\equiv C-\overset{|}{\underset{|}{C}}-H \qquad |{}^5J| \approx 1-3 \text{ Hz}$$

$$C=C=C=C \qquad {}^5J \approx 6-8 \text{ Hz}$$
$$(cis \text{ and } trans)$$

Complete sets of coupling constants

A complete evaluation of the ¹H NMR spectrum of 1,3-butadiene (**165**) gives the following data for the three

Nuclear Magnetic Resonance Spectroscopy

chemically nonequivalent kinds of protons and the nine different couplings:

165 (AA'BB'CC')

$$\delta_A = \delta_{A'} = 5.02 \qquad \delta_B = \delta_{B'} = 5.13 \qquad \delta_C = \delta_{C'} = 6.23$$

$$^2J_{AB} = {}^2J_{A'B'} = 1.8\ Hz \qquad\qquad {}^4J_{BC} = {}^4J_{B'C} = -0.8\ Hz$$
$$^3J_{AC} = {}^3J_{A'C'} = 10.2\ Hz \qquad\qquad {}^5J_{AA'} = 1.3\ Hz$$
$$^3J_{BC} = {}^3J_{B'C'} = 17.1\ Hz \qquad\qquad {}^5J_{AB'} = {}^5J_{A'B} = 0.6\ Hz$$
$$^3J_{CC'} = 10.4\ Hz \qquad\qquad\qquad {}^5J_{BB'} = 0.6\ Hz$$
$$^4J_{AC'} = {}^4J_{A'C} = -0.9\ Hz$$

Long-range couplings are of particular importance in the spectra of aromatic and heteroaromatic rings. A comparison of 1,3-cyclohexadiene (**104**) and benzene (**105**) shows the influence of the different bond orders.

104   **105**

In substituted benzenes:

$$^3J_{ortho} = 6.0 \dots 9.0\ Hz$$
$$^4J_{meta} = 0.9 \dots 3.0\ Hz$$
$$^5J_{para} = 0 \ \ \dots 1.0\ Hz$$

Toluene (**19**) is discussed here as an explicit example with nine different coupling constants:

19

$$^3J_{BC} = {}^3J_{B'C'} = 7.7\ Hz \qquad {}^4J_{CC'} = 1.5\ Hz$$
$$^3J_{CD} = {}^3J_{C'D'} = 7.5\ Hz \qquad {}^5J_{AC} = {}^5J_{AC'} = 0.3\ Hz$$
$$^4J_{AB} = {}^4J_{AB'} = -0.7\ Hz \qquad {}^5J_{BC} = {}^5J_{B'C} = 0.6\ Hz$$
$$^4J_{BD} = {}^4J_{B'D'} = 1.3\ Hz \qquad {}^6J_{AD} = -0.6\ Hz$$
$$^4J_{BB'} = 2.0\ Hz$$

For benzene derivatives, the substitution is often recognizable from the splitting patterns. This even applies to identical

spin systems with different relationships of the coupling constants; thus, o-substituted benzenes with two identical substituents have the same AA'BB' or AA'XX' spin system as p-substituted benzenes with two different substituents (cf. **Table 3.5**, p. 103). Nevertheless, as **Fig. 3.44** shows, a distinction between the two cases is easily made from the patterns of lines observed.

A frequent mistake in the interpretation of p-substituted benzenes is the assumption that the separation of the two most intense lines corresponds to the coupling constant ^3J. **Fig. 3.45** shows that, as the m- and p-couplings get smaller, the AA' part of an AA'XX' spin system (just like the XX' part) appears as a doublet. Poor resolution increases the impression. In fact, the separation of the two most intense lines is always equal to the sum of o- and p-couplings: $^3J + {}^5J$.

Condensed benzenoid arenes, e.g., naphthalene (**150**), have similar ^3J, ^4J, and ^5J couplings to benzene.

150

$$\delta(H_A) = 7.66 \qquad\qquad {}^5J_{AA'} = 0.7\ Hz$$
$$\delta(H_B) = 7.30 \qquad\qquad {}^5J_{AA''} = 0.9\ Hz$$
$$^3J_{AB} = 8.3\ Hz \qquad\qquad {}^5J_{AB'''} = 0.2\ Hz$$
$$^3J_{BB'} = 6.9\ Hz \qquad\qquad {}^6J_{AB''} = -0.1\ Hz$$
$$^4J_{AB'} = 1.3\ Hz \qquad\qquad {}^6J_{BB'''} = 0.1\ Hz$$
$$^4J_{AA'''} = -0.5\ Hz \qquad\qquad {}^7J_{BB''} = 0.3\ Hz$$

In heterocyclic systems the deviations are often greater, in particular, certain ^3J coupling constants can become quite small:

89 Pyridine   **166 Pyridinium-Ion**

$\delta(H_A)$	8.60	9.23
$\delta(H_B)$	7.25	8.50
$\delta(H_C)$	7.64	9.04
$^3J_{AB} = {}^3J_{A'B'}$	5.5 Hz	6.0 Hz
$^3J_{BC} = {}^3J_{B'C}$	7.6 Hz	8.0 Hz
$^4J_{AC} = {}^4J_{A'C}$	1.9 Hz	1.5 Hz
$^4J_{AA'}$	0.4 Hz	1.0 Hz
$^4J_{BB'}$	1.6 Hz	1.4 Hz
$^5J_{AB'} = {}^5J_{A'B}$	0.9 Hz	0.8 Hz

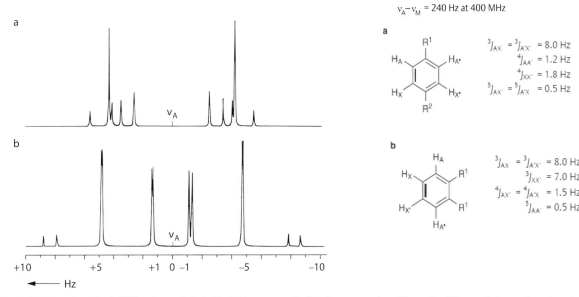

Fig. 3.44 AA′ parts of the AA′XX′ systems of disubstituted benzenes with the given parameters. The major difference between the spin patterns results from $^3J_{XX'}$.

	17 Furan	167 Pyrrole	14 Thiophene
$\delta(H_A) = \delta(H_{A'})$	7.38	6.62	7.20
$\delta(H_B) = \delta(H_{B'})$	6.30	6.05	6.96
$^3J_{AB} = ^3J_{A'B'}$	1.8	2.6	4.8
$^3J_{BB'}$	3.4	3.5	3.5
$^4J_{AB'} = ^4J_{A'B}$	0.9	1.3	1.0
$^4J_{AA'}$	1.5	2.1	2.8

A frequent case of structure determination based on a four-spin system occurs in the cyclodimerization of asymmetric alkenes RCH=CHR′ (head R, tail R′) to cyclobutanes. As well as the regioselectivity (head-to-head or head-to-tail), the stereoselectivity also has to be determined. Even if the substituents R and R′ are in the *trans* orientation in both educt and adduct, the addition of the two head groups can occur either syn or anti; the same applies to the head–tail orientation. There are four possibilities, with symmetries C_s, C_2, C_{2v}, and C_i and the corresponding spin systems AA′BB′ or A_2B_2. The easiest to recognize is the product with C_{2v} symmetry. Its coupling pattern, corresponding to the A_2B_2 system, has the basic appearance of two triplets with roof effect; this is more apparent in the spectrum with a larger linewidth (**Fig. 3.46**, right-hand side). For simplification in the calculated spectra,

all $^3J_{trans}$-couplings have been assumed to be 10.8 Hz, all $^3J_{cis}$-couplings to be 6.9 Hz, and all ^4J-couplings to be −0.9 Hz. (It should be pointed out that the $^3J_{cis}$-coupling in four-membered rings can be larger than the $^3J_{trans}$-coupling.) The variation of the coupling pattern of the AA′BB′ spin system with $\delta(H_A) =$ 5.10 and $\delta(H_B) = 4.90$ results from the different orientations of the cyclobutane protons in **168 a-d**. **Fig. 3.46** also demonstrates the strong influence of the linewidth/resolution on the appearance of the spectrum.

3.3.4 Coupling to Other Nuclei

For coupling between ¹H and other nuclei, the natural abundance of these nuclei must be considered. The isotope ¹³C ($I = 1/2$) makes up 1.1% of natural carbon, alongside 98.9% of the spin-free ¹²C. The ¹H,¹³C coupling therefore leads to signals with only about 1% of the intensity of the corresponding ¹H signals without ¹³C coupling. The measurement of these so-called **¹³C satellites** is described on p. 155. In routine ¹H spectra, they can usually be ignored.

From **Table 3.1** it can be seen that the only nuclei of importance in organic chemistry with appreciable abundances and with $I = 1/2$ are ¹⁹F and ³¹P. They both have a natural abundance of 100%; therefore, proton spectra of fluorine or phosphorus compounds show splittings due to the ¹H,¹⁹F or ¹H,³¹P couplings. Summaries of the sizes of the coupling constants observed are given in **Tables 3.11 and 3.12** (for ¹H,D and ¹H,¹³C couplings, see p. 152 and Section 3.4.5, p. 191).

Nuclear Magnetic Resonance Spectroscopy

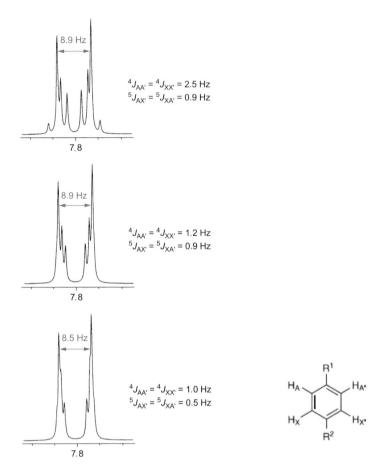

$^4J_{AA'} = {}^4J_{XX'} = 2.5$ Hz
$^5J_{AX'} = {}^5J_{XA'} = 0.9$ Hz

$^4J_{AA'} = {}^4J_{XX'} = 1.2$ Hz
$^5J_{AX'} = {}^5J_{XA'} = 0.9$ Hz

$^4J_{AA'} = {}^4J_{XX'} = 1.0$ Hz
$^5J_{AX'} = {}^5J_{XA'} = 0.5$ Hz

Fig. 3.45 Effect on the spectral pattern (AA′ part of AA′XX′) of a benzene derivative with different substituents in 1,4-positions of changing the *m*-coupling ^4J and the *p*-coupling ^5J [300 MHz, δ (H$_A$) = 7.90, δ (H$_x$) = 7.40, $^3J_{AX} = {}^3J_{A'X'} = 8.0$ Hz].

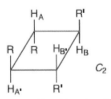

$^3J_{AA'} = {}^3J_{BB'} = 6.9$ Hz
$^3J_{AB} = {}^3J_{A'B'} = 10.8$ Hz
$^4J_{AB'} = {}^4J_{A'B} = -0.9$ Hz

168a Head-head-syn adduct

$^3J_{AA'} = {}^3J_{BB'} = 10.8$ Hz
$^3J_{AB} = {}^3J_{A'B'} = 10.8$ Hz
$^4J_{AB'} = {}^4J_{A'B} = -0.9$ Hz

168b Head-head-anti adduct

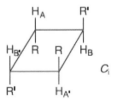

$^3J_{AB} = 10.8$ Hz
$^4J_{AA}, {}^4J_{BB}$
do not appear in the spectrum

168c Head-tail-anti adduct

$^3J_{AB} = {}^3J_{A'B'} = 10.8$ Hz
$^3J_{AB'} = {}^3J_{A'B} = 6.9$ Hz
$^4J_{AA'} = {}^4J_{BB'} = -0.9$ Hz

168d Head-tail-syn adduct

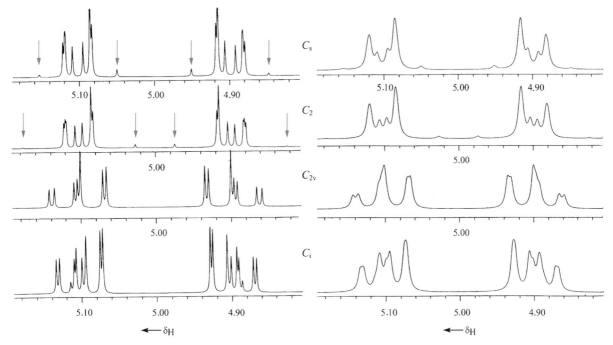

Fig. 3.46 Simulated ¹H spectra of cyclobutanes (**168a–d**) with symmetries C_s, C_2, C_i (all AA'BB'), and $C_{2v}(A_2B_2)$. The spectra on the left are shown with a linewidth of 0.5 Hz, and on the right with 2.0 Hz (the *arrows* mark weaker lines).

3.3.5   Correlation of ¹H Shifts with Structural Features

Methyl Protons

A summary of the chemical shifts of methyl groups in different chemical environments is given in **Table 3.13**. (Here and in the following tables, typical values only are given; extreme chemical shift values are ignored).

Methylene Protons

As explained in Section 3.2.2 (p. 100), the two protons of a methylene group are only chemically equivalent when they are related by a symmetry or pseudosymmetry element in the molecule. This includes the effect of fast internal motion in the molecule. Even when the two protons are chemically equivalent, this does not imply that they are necessarily magnetically equivalent. In **Table 3.14** the shift ranges of methylene protons are given as a function of the substitution.

Methine Protons

The shift ranges for methine protons are considerably wider than those for methyl or methylene groups as a consequence of the larger number of substitution possibilities. The most common combinations of substituents are summarized in **Table 3.15**.

Protons on CC Double and Triple Bonds

Table 3.16 gives information on shifts of olefinic protons. Due to different anisotropy effects, the shift range for acetylenic protons is at higher field than that for olefinic protons (**Table 3.17**).

Protons on Aromatic and Heteroaromatic Rings

Protons bound to C atoms in aromatic or heteroaromatic rings have shifts between $\delta = 6.0$ and 11.0, the majority being between 7 and 8 ppm. The variety of possibilities in this class of compounds is so great that producing tables for this introduction is inappropriate. Selected examples may be found in Section 3.5.3 (p. 232).

Protons on Carbonyl and Imino Groups

In **Table 3.18**, chemical shifts of aldehyde and aldimine protons are summarized.

OH-, SH-, and NH-Protons

The chemical shifts of protons bound to heteroatoms such as O, S, or N depend strongly on the measurement conditions (concentration, temperature, and solvent). The signals are often

Table 3.11 ^{1}H,^{19}F coupling constants of selected fluorine compounds

Compound	Type of coupling	$|J|$ in Hz
H—CH$_2$—F Fluoromethane	^2J(H, F)	46
H—CF$_2$—F Trifluoromethane	^2J(H, F)	80
H—CH$_2$—CH— F (H below) Fluoroethane	^2J(H, F) ^3J(H, F)	47 25
H—CH$_2$—CF$_2$—F 1,1,1-Trifluoroethane	^3J(H, F)	13
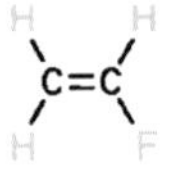 C=C (Fluoroethene)	^2J(H, F) ^3J(Z-H, F) ^3J(E-H, F)	85 20 52
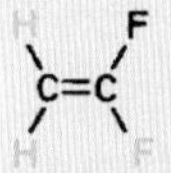 C=C (1,1-Difluoroethene)	^3J(Z-H, F) ^3J(E-H, F)	≈1 34
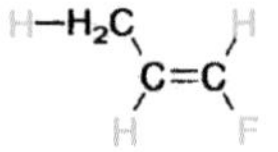 C=C (E)-1-Fluoropropene	^2J(H, F) ^3J(H, F) ^4J(H, F)	85 20 3
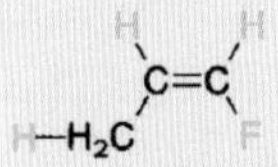 C=C (Z)-1-Fluoropropene	^2J(H, F) ^3J(H, F) ^4J(H, F)	85 42 2
H—CH$_2$—C—F ‖ O Acetylfluoride	^3J(H, F)	7
H—C≡C—F Fluoroacetylene	^3J(H, F)	21
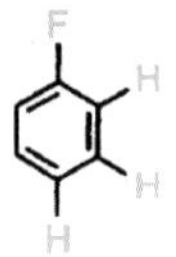 Fluorobenzene	^3J(H, F) ^4J(H, F) ^5J(H, F)	9.0 5.7 0.2

broad and/or show no couplings. The reason for this is the rapid intermolecular **proton exchange** (see also p. 132). If a rapid exchange occurs between two chemically nonequivalent groups within a molecule, only a single, averaged signal is observed.

To identify the signals of XH-protons, **exchange with D$_2$O or CF$_3$COOH** is often useful.

Table 3.12 ^{1}H,^{31}P coupling constants of selected phosphorus compounds

Compound	Type of coupling	$|J|$ in Hz
H—PH—C$_6$H$_5$ Phenylphosphine	^1J(H, P)	201
H—P(CH$_3$)$_2$ Dimethylphosphine	^1J(H, P)	192
H—$\overset{+}{P}$(CH$_3$)$_3$ Cl$^-$ Trimethylphosphonium	^1J(H, P)	506
(H—CH$_2$—CH—)$_3$P Triethylphosphine	^2J(H, P) ^3J(H, P)	0.5 14
(H—CH$_2$—CH—)$_4$P$^+$ Cl$^-$ Tetraethylphosphonium	^2J(H, P) ^3J(H, P)	13 18
(H—CH$_2$—CH—O—)$_3$P Triethyl phosphite	^3J(H, P) ^4J(H, P)	8 1
(H—CH$_2$—CH)$_3$P=O Triethylphosphine oxide	^2J(H, P) ^3J(H, P)	16 12
H—P(OCH$_3$)$_2$ ‖ O Dimethyl phosphite	^1J(H, P)	710
H—P(OC$_2$H$_5$)$_2$ ‖ O Diethyl phosphite	^1J(H, P)	688
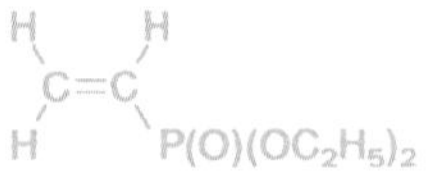 C=C P(O)(OC$_2$H$_5$)$_2$ Diethyl ethenylphosphonate	^3J(Z—H, P) ^3J(E—H, P)	14 30
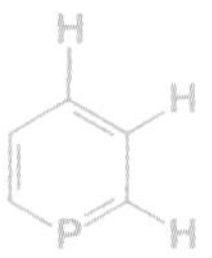 Phosphabenzene (phosphorine)	^2J(H, P) ^3J(H, P) ^4J(H, P)	38 8 4

Table 3.13 ¹H Chemical shifts of methyl groups (δ values measured in $CDCl_3$)

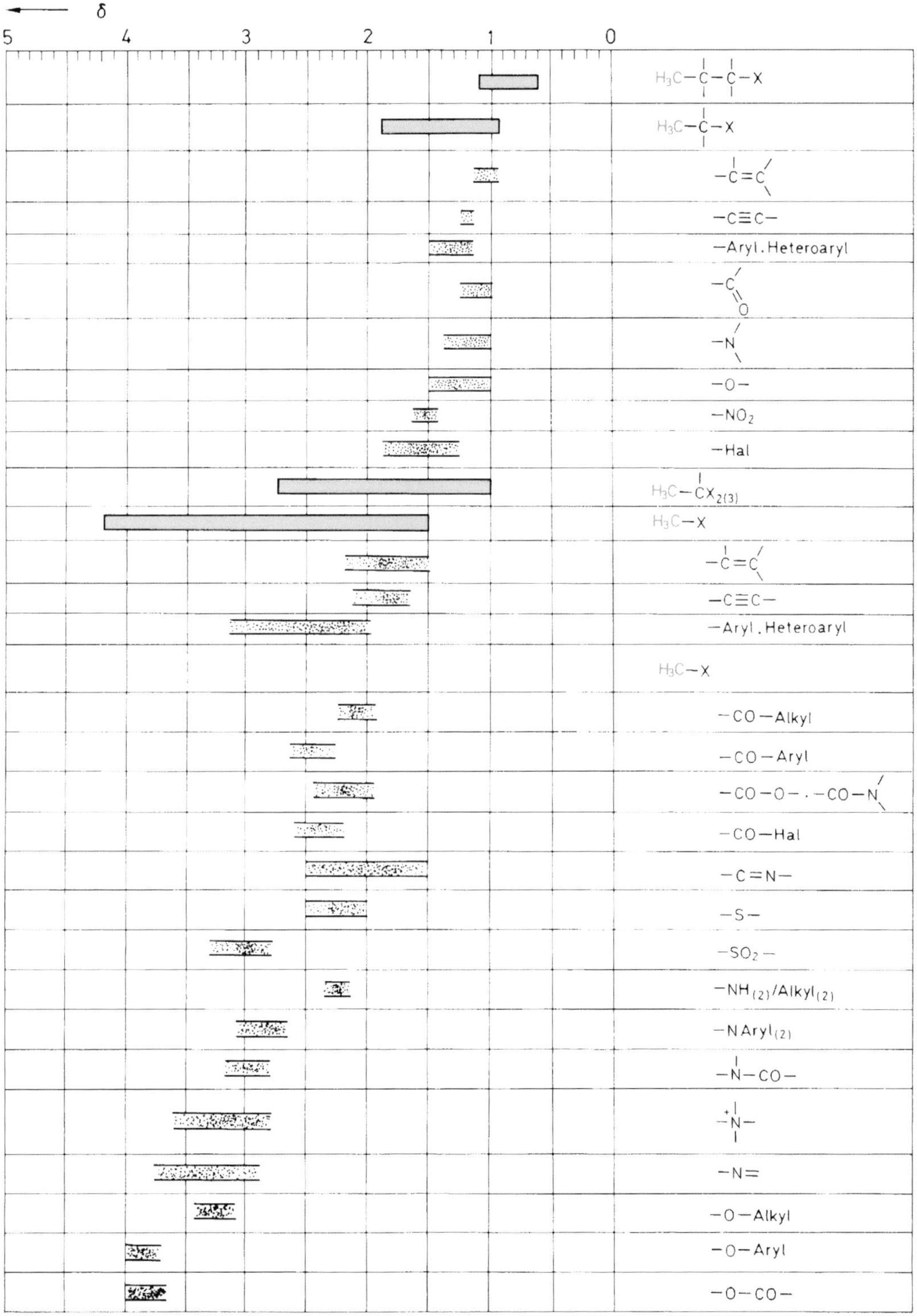

δ scale	
$H_3C-\overset{\shortmid}{C}-\overset{\shortmid}{C}-X$	
$H_3C-\overset{\shortmid}{C}-X$	
$-\overset{\shortmid}{C}=C\overset{/}{}$	
$-C\equiv C-$	
$-$Aryl. Heteroaryl	
$-C\overset{/}{\underset{O}{}}$	
$-N\overset{/}{\underset{\backslash}{}}$	
$-O-$	
$-NO_2$	
$-$Hal	
$H_3C-\overset{\shortmid}{C}X_{2(3)}$	
H_3C-X	
$-\overset{\shortmid}{C}=C\overset{/}{}$	
$-C\equiv C-$	
$-$Aryl. Heteroaryl	
H_3C-X	
$-CO-$Alkyl	
$-CO-$Aryl	
$-CO-O-$. $-CO-N\overset{/}{\backslash}$	
$-CO-$Hal	
$-C=N-$	
$-S-$	
$-SO_2-$	
$-NH_{(2)}/$Alkyl$_{(2)}$	
$-$N Aryl$_{(2)}$	
$-\overset{\shortmid}{N}-CO-$	
$-\overset{+}{\underset{\shortmid}{N}}-$	
$-N=$	
$-O-$Alkyl	
$-O-$Aryl	
$-O-CO-$

Nuclear Magnetic Resonance Spectroscopy

Table 3.14 ^1H Chemical shifts of methylene groups (δ values measured in $CDCl_3$)

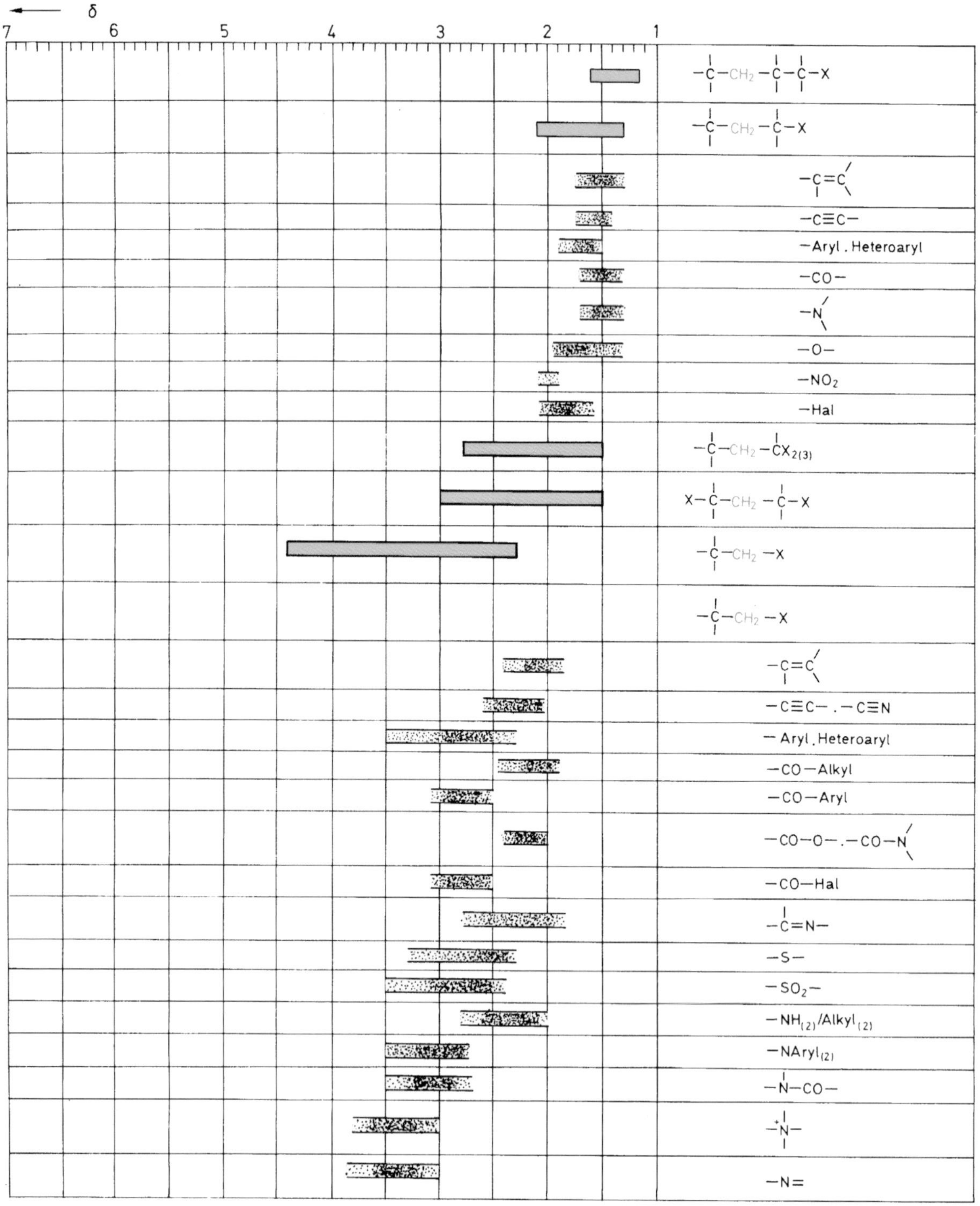

Structure
$-\overset{\mid}{\underset{\mid}{C}}-CH_2-\overset{\mid}{\underset{\mid}{C}}-\overset{\mid}{\underset{\mid}{C}}-X$
$-\overset{\mid}{\underset{\mid}{C}}-CH_2-\overset{\mid}{\underset{\mid}{C}}-X$
$-C{=}C\diagdown$
$-C{\equiv}C-$
$-$Aryl . Heteroaryl
$-CO-$
$-N\diagdown$
$-O-$
$-NO_2$
$-Hal$
$-\overset{\mid}{\underset{\mid}{C}}-CH_2-CX_{2(3)}$
$X-\overset{\mid}{\underset{\mid}{C}}-CH_2-\overset{\mid}{\underset{\mid}{C}}-X$
$-\overset{\mid}{\underset{\mid}{C}}-CH_2-X$
$-\overset{\mid}{\underset{\mid}{C}}-CH_2-X$
$-C{=}C\diagdown$
$-C{\equiv}C-\,.-C{\equiv}N$
$-$Aryl . Heteroaryl
$-CO-Alkyl$
$-CO-Aryl$
$-CO-O-\,.-CO-N\diagdown$
$-CO-Hal$
$-\overset{\mid}{C}{=}N-$
$-S-$
$-SO_2-$
$-NH_{(2)}/Alkyl_{(2)}$
$-NAryl_{(2)}$
$-N-CO-$
$-\overset{+}{\underset{\mid}{N}}-$
$-N{=}$

Table 3.14 (Continued)

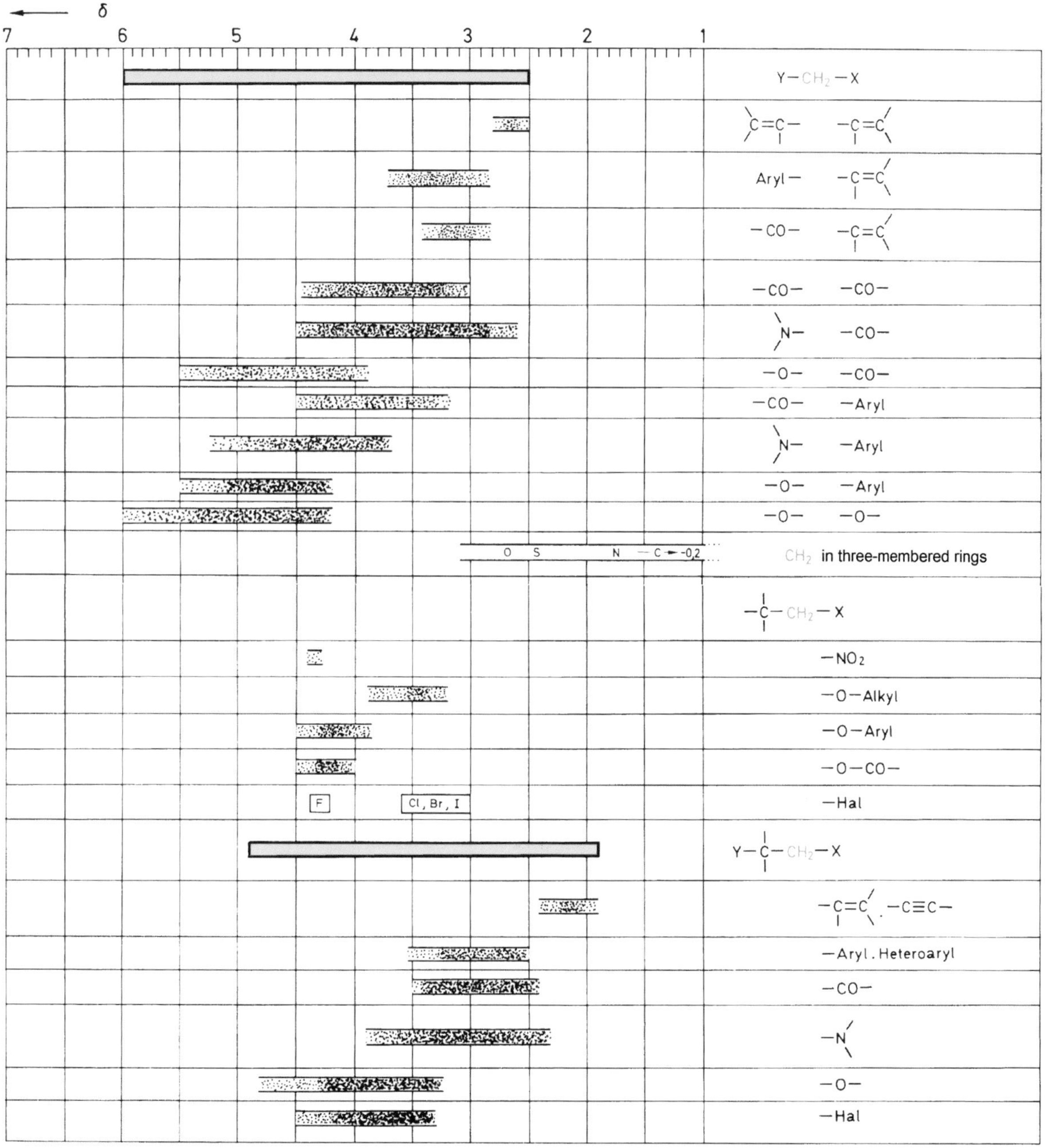

δ

| 7 | 6 | 5 | 4 | 3 | 2 | 1 |

$Y-CH_2-X$

$\overset{|}{C}=C-\quad -C=C\overset{/}{\underset{\backslash}{}}$

$Aryl-\quad -C=C\overset{/}{\underset{\backslash}{}}$

$-CO-\quad -C=C\overset{/}{\underset{\backslash}{}}$

$-CO-\quad -CO-$

$\overset{\backslash}{N}-\quad -CO-$

$-O-\quad -CO-$

$-CO-\quad -Aryl$

$\overset{\backslash}{N}-\quad -Aryl$

$-O-\quad -Aryl$

$-O-\quad -O-$

O   S   N   — C ←-0,2 ... CH_2 in three-membered rings

$-\overset{|}{\underset{|}{C}}-CH_2-X$

$-NO_2$

$-O-Alkyl$

$-O-Aryl$

$-O-CO-$

F   Cl, Br, I $-Hal$

$Y-\overset{|}{\underset{|}{C}}-CH_2-X$

$-C=C\overset{/}{\underset{|}{}}, -C\equiv C-$

$-Aryl.Heteroaryl$

$-CO-$

$-N\overset{/}{\underset{\backslash}{}}$

$-O-$

$-Hal$

Nuclear Magnetic Resonance Spectroscopy

Table 3.15 ^{1}H Chemical shifts of methine groups (δ values measured in $CDCl_3$)

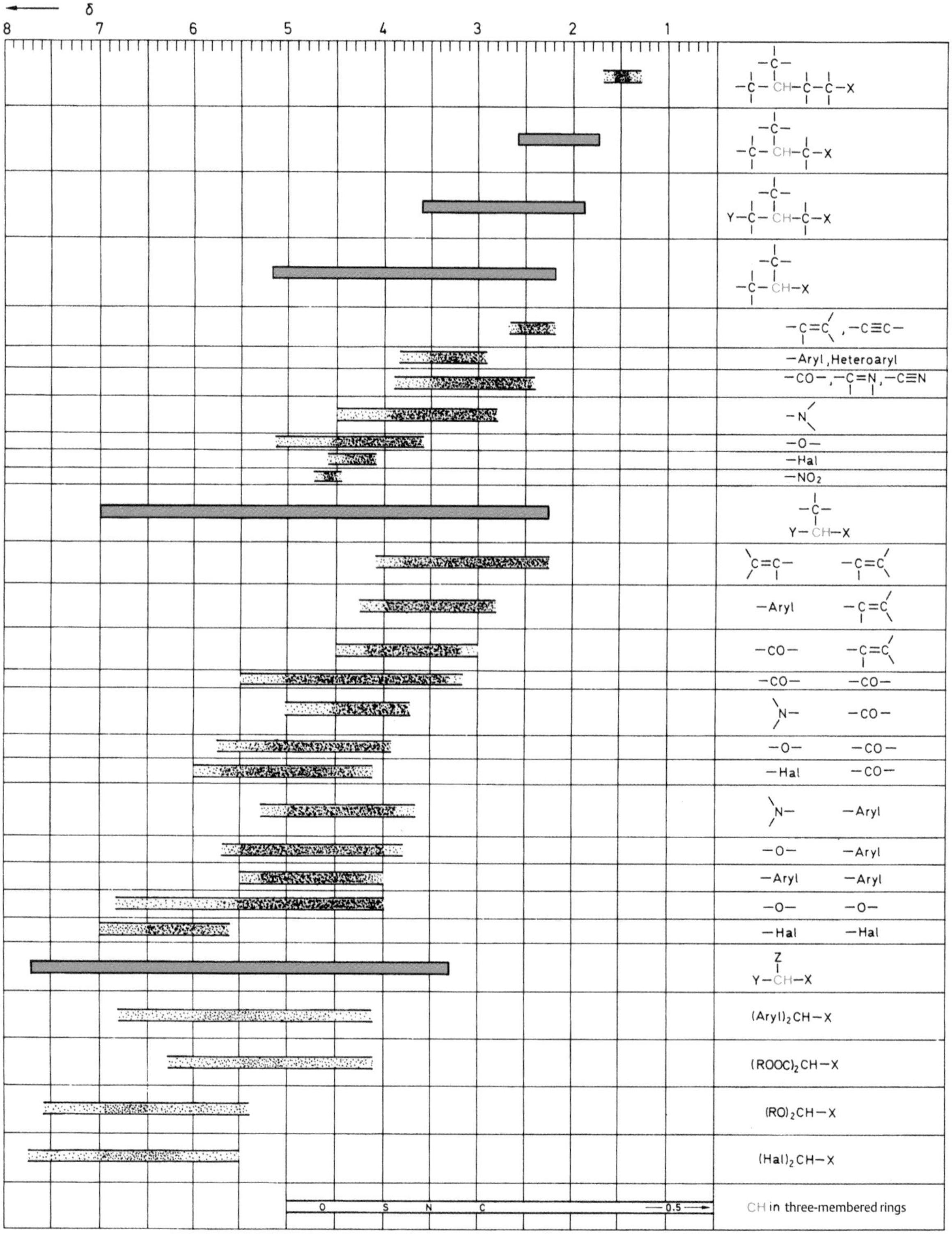

Table 3.16 ¹H Chemical shifts of olefinic protons (δ values measured in $CDCl_3$)

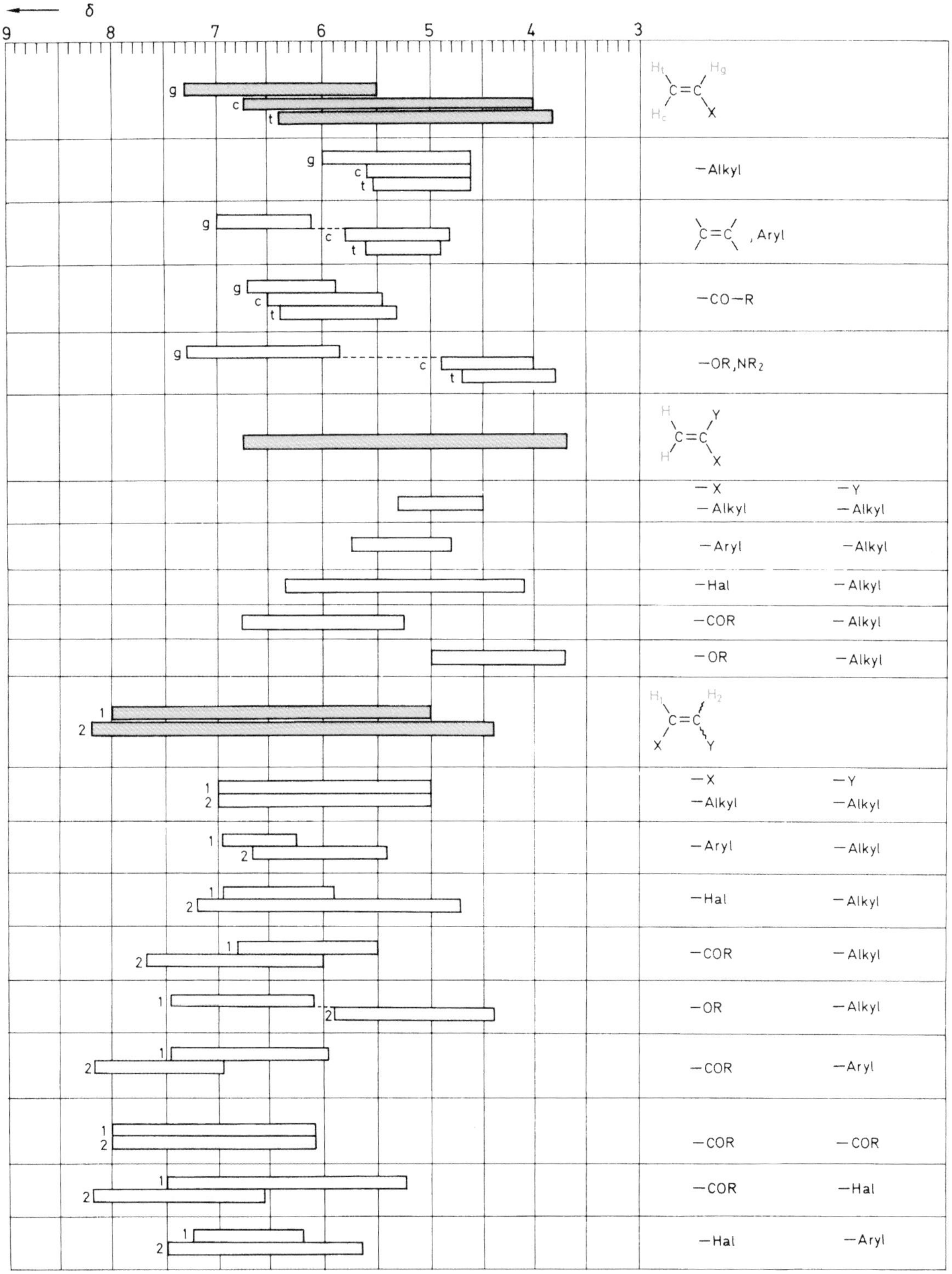

Table 3.16 (Continued)

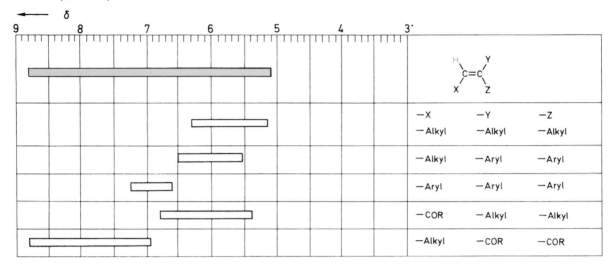

Table 3.17 ¹H Chemical shifts of acetylenic protons (δ values measured in CDCl₃)

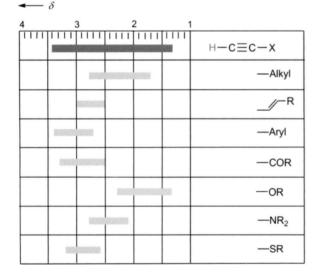

In **Table 3.19**, ranges for chemical shifts of XH-protons are summarized. The δ values given are for carbon tetrachloride or deuteriochloroform as a solvent. The OH-signals of alcohols generally shift to lower field with increasing concentration or decreasing temperature; both effects increase the association of the alcohol molecules and therefore the deshielding effect of the hydrogen bonds. The OH-signal of a very dilute solution of ethanol in CCl₄ lies at δ ≈ 0.9. With increasing concentration, it shifts toward δ ≈ 4.6, the value of pure ethanol. **When assigning X–H signals, the strong dependence of shift on concentration, temperature, and solvent must be taken into consideration.**

3.3.6   Increment Systems for Estimating ¹H Chemical Shifts

The large amount of known ¹H NMR data allows empirical rules for the dependence of the shifts of protons on their chemical environment to be established. These are based on the assumption of additivity of substituent effects, though this will not apply perfectly. Except in cases where particularly strong steric or electronic interactions occur, reasonable estimates of δ values can however be obtained.

A simple increment system for methylene and methyl protons is given in **Table 3.20**. A few examples (**Table 3.21**, p. 147) serve to illustrate the comparison of calculated and observed values.

On a similar basis and with similar "quality," the shifts of olefinic protons can be estimated (**Table 3.22**). Some examples are shown in **Table 3.23**.

The effects of substituents (particularly) have been carefully studied in the case of benzene derivatives. The limits of this approach are met where there are steric interactions of substituents (1,2-disubstituted, 1,2,3-trisubstituted benzenes, etc.) or unusual electronic interactions (**Tables 3.24 and 3.25**, p. 149).

The increment systems allow a rough estimation of ¹H chemical shifts. However, data-bank predictions (p. 230) are normally better.

3.3.7   ¹H NMR Data of Representatives of Common Classes of Compounds

A collection of **¹H NMR data** in tabular form is given together with **¹³C NMR data** of selected structural examples in Section 3.5.3 (p. 232).

Table 3.18 ¹H Chemical shifts of aldehyde and aldimine protons (δ values measured in $CDCl_3$)

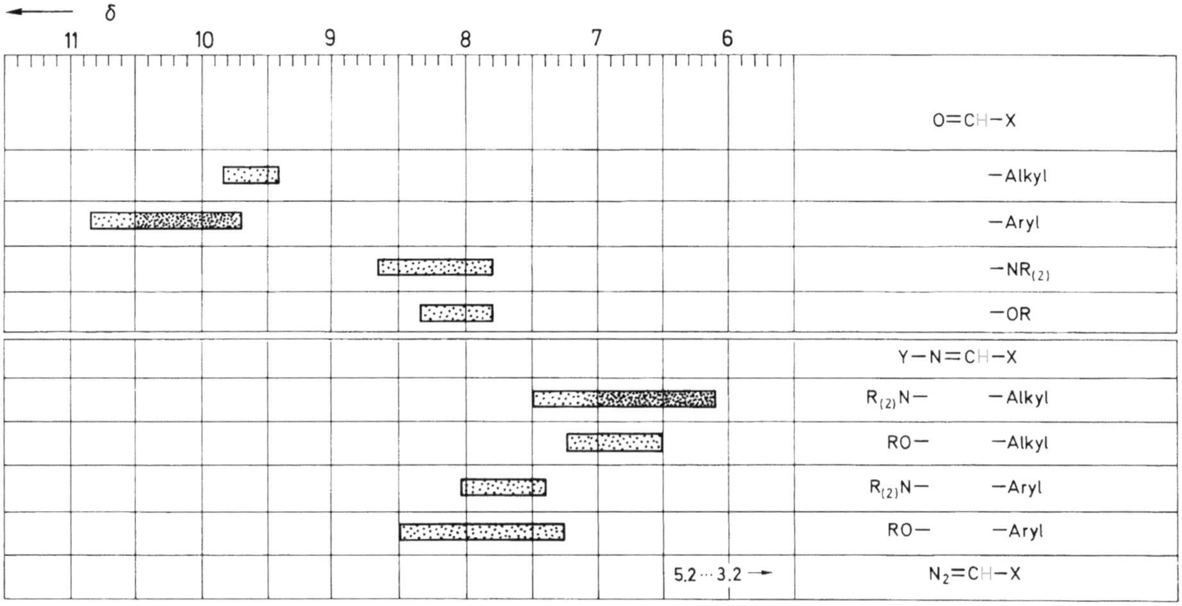

δ							
11	10	9	8	7	6		

$O{=}CH{-}X$
— Alkyl
— Aryl
— $NR_{(2)}$
— OR

$Y{-}N{=}CH{-}X$
$R_{(2)}N{-}$ — Alkyl
$RO{-}$ — Alkyl
$R_{(2)}N{-}$ — Aryl
$RO{-}$ — Aryl
$5.2 \cdots 3.2 \rightarrow$ $N_2{=}CH{-}X$

3.3.8 Specialized Techniques

The following chart summarizes important (one-dimensional) techniques, which are helpful for ¹H NMR studies.

Method	Application
Enhancement of the measurement frequency/field strength	Improvement of sensitivity and/or resolution
Use of microcells and/or special probeheads (p. 147)	Limited amounts of sample, hardly soluble compounds
Variation of solvents, additives (D_2O) (p. 147)	Better separation of signals, identification of OH signals
Addition of shift reagents (p. 153)	Better separation of signals, differentiation of enantiomers
Selective deuteration (p. 152)	Assignment of signals, isotope perturbation method
Measurement of ¹³C satellites (p. 155)	Determination of coupling constants, which are not visible in the normal spectrum, e.g., distinction between symmetrical cis- and trans-configurations
Spin decoupling, double and multiple resonance (p. 155)	Simplification of spectra, determination of the connectivity of protons through bonds
NOE difference spectroscopy (p. 159)	Determination of the connectivity through space
INDOR method (p. 161)	Determination of coupling protons

Increasing the Measurement Frequency/Field Strength

As described in Section 3.1.1 (p. 86), the chemical shift measured in Hz increases linearly with the magnetic field B_0. The coupling constants J however are independent of B_0. If, for example, a 60-MHz spectrum shows groups of signals that overlap, then for an exact analysis a higher field spectrum, e.g., at 400 MHz, is recommended (cf. **Fig. 3.19**, p. 146). Another example is shown in **Fig. 3.47** (p. 150). The complex spectrum of strychnine (**169**) gets dramatically better resolved on going from 60 to 250 MHz.

Of decisive importance is the change in the ratio $\Delta v/J$. In Section 3.1.3 (p. 89), the ratio $\Delta v/J$ was used as a criterion for the appearance of first- or second-order spectra. The increase of this ratio on increasing the measurement frequency (field strength) can change the order and thus the appearance of the spectrum drastically. For example, an ABC spectrum can change into an AMX spectrum. In general, this is a decisive advantage, but information can also be lost as a result: the signs of the coupling constants cannot be determined from an AMX spectrum.

In measurements on stereoisomers, often diastereomers or, in chiral environments, enantiomers are encountered which have some signals which only show very small differences. Moreover, the minor components (in cases of high **diastereomeric excess**, **de**, or high **enantiomeric excess**, **ee**) are often present in very small amounts. In such cases, the use of high-field

Nuclear Magnetic Resonance Spectroscopy

Table 3.19 ^{1}H Chemical shifts of OH-, SH-, and NH-protons (δ values)

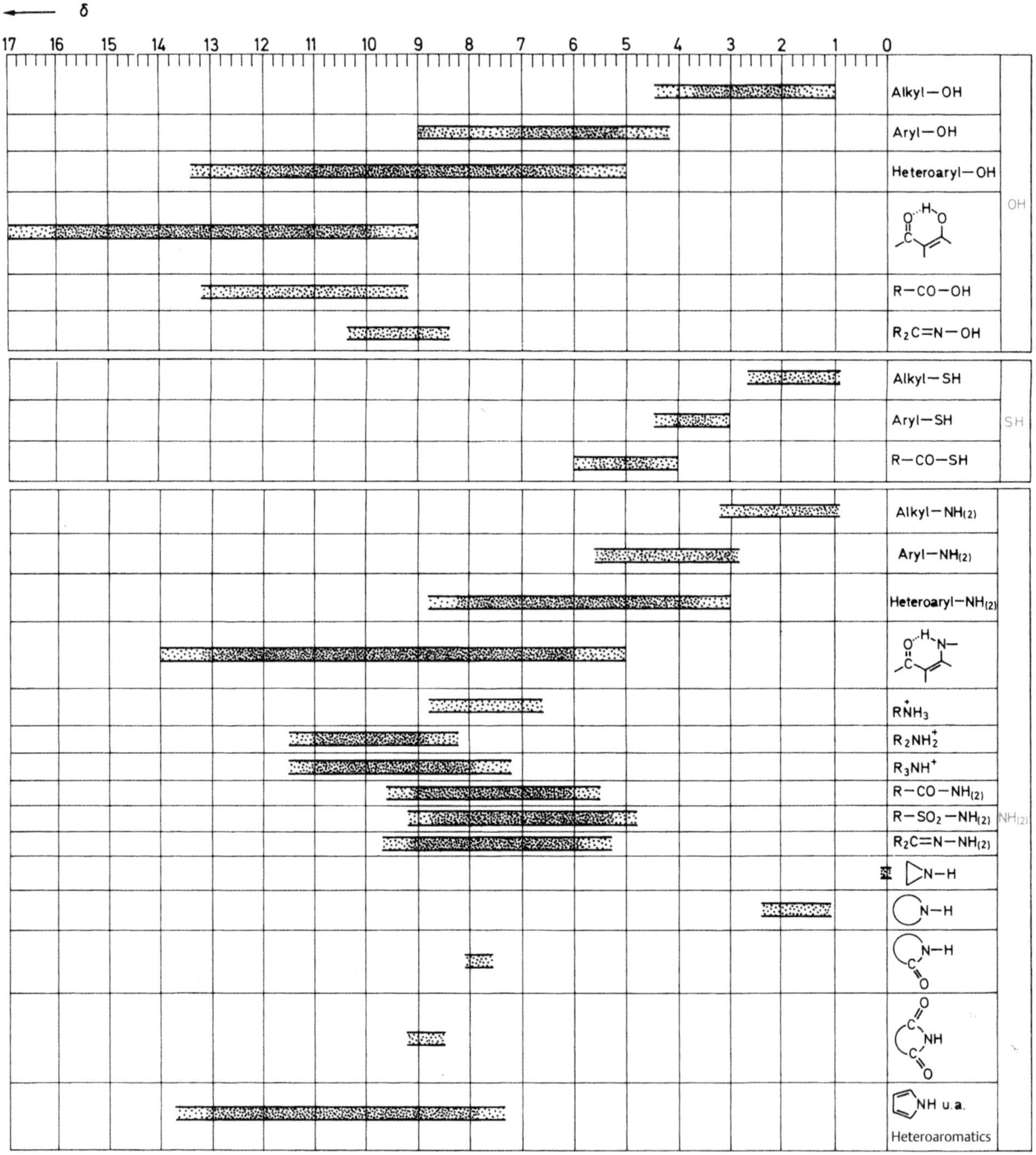

$\longleftarrow$ δ

δ scale	Group
17 16 15 14 13 12 11 10 9 8 7 6 5 4 3 2 1 0	
	Alkyl—OH
	Aryl—OH
	Heteroaryl—OH
	(enol O···H—O chelate)
	R—CO—OH
	$R_2C{=}N{-}OH$
	Alkyl—SH
	Aryl—SH
	R—CO—SH
	$Alkyl{-}NH_{(2)}$
	$Aryl{-}NH_{(2)}$
	$Heteroaryl{-}NH_{(2)}$
	(O···H—N chelate)
	$R\overset{+}{N}H_3$
	$R_2NH_2^+$
	R_3NH^+
	$R{-}CO{-}NH_{(2)}$
	$R{-}SO_2{-}NH_{(2)}$
	$R_2C{=}N{-}NH_{(2)}$
	aziridine N—H
	N—H
	lactam N—H (C=O)
	imide NH
	pyrrole NH u.a. Heteroaromatics

OH

SH

NH$_{(2)}$

Table 3.20 Increment system for the estimation of ¹H chemical shifts of methylene and methine protons (modified Shoolery's rules)

$R^1-CH_2-R^2$	$R^1-\underset{\underset{R^3}{\mid}}{CH}-R^2$
$\delta = 1.25 + l_1 + l_2$	$\delta = 1.50 + l_1 + l_2 + l_3$
Substituent	Increment l
—Alkyl	0.0
—C=C—	0.8
—C≡C—	0.9
—C_6H_5	1.3
—CO—H, —CO-Alkyl	1.2
—CO—C_6H_5	1.6
—COOH	0.8
—CO—O-Alkyl	0.7
—C≡N	1.2
—NH_2, NH-Alkyl, N(Alkyl)$_2$	1.0
—NO_2	3.0
—SH, —S—Alkyl	1.3
—OH	2.0
—O-Alkyl	1.7
—O—C_6H_5	2.3
—O—CO-Alkyl	2.7
—O—CO—C_6H_5	2.9
—F	2.9
—Cl	2.0
—Br	1.9
—I	1.8

Table 3.21 Calculated and measured δ (¹H) values of methylene and methine protons

Compound	$\delta_{calculated}$	$\delta_{measured}$
CH_2Br_2 Dibromomethane	$1.25 + 2 \cdot 1.9 = 5.05$	4.94
$C_6H_5-OCH_2-C_2H_5$ Phenyl propyl ether	$1.25 + 2.3 + 0 = 3.55$	3.86
$Cl-CH_2\underset{a}{}-\overset{\overset{O}{\mid\mid}}{C}-O-CH_2\underset{b}{}-CH_3$ Ethyl chloroacetate	a: $1.25 + 2.0 + 0.7$ $= 3.95$ b: $1.25 + 2.7 + 0$ $= 3.95$	4.05 4.25
$H_3C-CHCl_2$ 1,1-Dichloroethane	$1.50 + 2 \cdot 2.0 + 0$ $= 5.50$	5.75
$C_2H_5-\underset{\underset{NO_2}{\mid}}{CH}-Cl$ 1-Chloro-1-nitropropane	$1.50 + 2.0 + 3.0 + 0$ $= 6.50$	5.80
$(C_6H_5)_3CH$ Triphenylmethane	$1.50 + 3 \cdot 1.3 = 5.40$	5.56
$H_3C-\underset{\underset{b}{}}{\overset{\overset{OH}{\mid}}{CH}}-CH_2-CH_2\underset{a}{}-OH$ 1,3-Butanediol	a: $1.25 + 1.7 + 0$ $= 2.95$ b: $1.50 + 1.7 + 0 + 0$ $= 3.20$	3.80 4.03

spectrometers is recommended, since the signal separation increases linearly with B_0 and the sensitivity with ca. $B_0^{3/2}$. Thus, in going from 200 to 800 MHz, the signal separations (measured in Hz) are quadrupled, and sensitivity increases by a factor of 8.

Measurement of Small Amounts of Sample

For the measurement of ¹H NMR spectra of samples that are only available in limited amounts, microcells with a volume of 100 µL can be used instead of the conventional 5-mm diameter tubes. For samples that are less than 0.1 mg, special probeheads can be employed, in conjunction with as high a field B_0 as possible. A low noise level and therefore an advantageous S/N ratio can be realized in cryo-probeheads. Microgram quantities are then measurable on a routine basis and sub-microgram quantities are possible with overnight accumulations.

Variation of Solvent, Special Additives

Solvent effects can be deliberately used in NMR for solving specific problems. On p. 138, mention has already been made of the recognition of acidic protons by exchange with D_2O or trifluoroacetic acid and the slowing down of proton exchange in dimethylsulfoxide as a solvent. Here the use of aromatic solvents such as hexadeuteriobenzene or pentadeuteriopyridine will be discussed. The formation of preferred encounter complexes in the solvation of the compounds studied combines with the strong intermolecular anisotropy effects of these solvents to produce large changes in shifts compared with those measured in $CDCl_3$ solutions. An illustrative example is O-methyl phenylthioacetate (**170**; **Fig. 3.48**, p. 151).

In $CDCl_3$ the signals of the methyl and methylene protons are coincident. Such coincidental isochronosities can easily lead to errors in structural interpretation. The spectrum measured in C_6D_6 however shows the expected two separate signals, with an intensity ratio of 3:2.

Table 3.22 Increment system for the estimation of the chemical shifts of olefinic protons

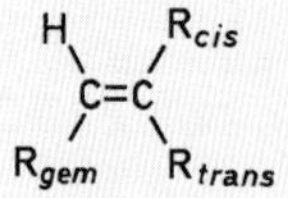

$$\delta = 5.25 + I_{gem} + I_{cis} + I_{trans}$$

Substituent	Increments		
	I_{gem}	I_{cis}	I_{trans}
—H	0	0	0
—Alkyl	0.45	− 0.22	− 0.28
—Alkyl ring*	0.69	− 0.25	− 0.28
—CH$_2$—Aryl	1.05	− 0.29	− 0.32
—CH$_2$OR	0.64	− 0.01	− 0.02
—CH$_2$NR$_2$	0.58	− 0.10	− 0.08
—CH$_2$—Hal	0.70	0.11	− 0.04
—CH$_2$—CO—R	0.69	− 0.08	− 0.06
—C(R)=CR$_2$ (Diene)	1.00	− 0.09	− 0.23
extended conjugation	1.24	0.02	− 0.05
—C≡C—	0.47	0.38	0.12
—Aryl	1.38	0.36	− 0.07
—CHO	1.02	0.95	1.17
—CO—R (Enone)	1.10	1.12	0.87
extended conjugation	1.06	0.91	0.74
—CO—OH (α,β-unsaturated carboxylic acid)	0.97	1.41	0.71
extended conjugation	0.80	0.98	0.32
—CO—OR (α,β-unsaturated carboxylic ester)	0.80	1.18	0.55
extended conjugation	0.78	1.01	0.46
—CO—NR$_2$	1.37	0.98	0.46
—CO—Cl	1.11	1.46	1.01
—C≡N	0.27	0.75	0.55
—OR (saturated)	1.22	− 1.07	− 1.21
—OR (other)	1.21	− 0.60	− 1.00
—O—CO—R	2.11	− 0.35	− 0.64
—S—R	1.11	− 0.29	− 0.13
—SO$_2$—R	1.55	1.16	0.93
—NR$_2$ (saturated)	0.80	− 1.26	− 1.21
—NR$_2$ (other)	1.17	− 0.53	− 0.99
—N—CO—R	2.08	− 0.57	− 0.72
—NO$_2$	1.87	1.32	0.62
—F	1.54	− 0.40	− 1.02
—Cl	1.08	0.18	0.13
—Br	1.07	0.45	0.55
—I	1.14	0.81	0.88

*applies to double bonds in five- or six-membered rings

Table 3.23 Calculated and measured δ (^{1}H) values of olefinic protons

Compound	$\delta_{calculated}$*		$\delta_{measured}$*	
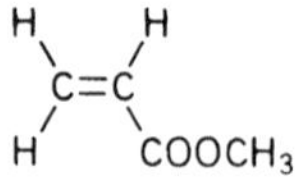 Methyl acrylate	5.80 6.43	6.05 –	5.82 6.38	6.20 –
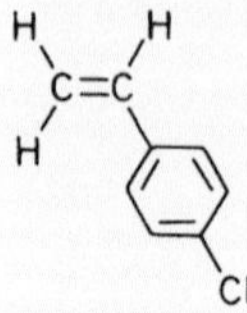 4-Chlorostyrene	5.18 5.61	6.63 –	5.28 5.73	6.69 –
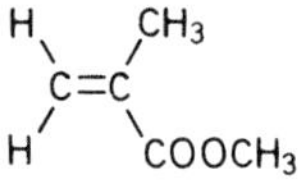 Methyl 2-methacrylate	5.58 6.15	– –	5.57 6.10	– –
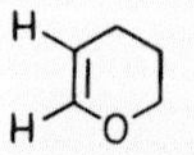 3,4-Dihydro-2H-pyran	4.73 6.19	– –	4.65 6.37	– –
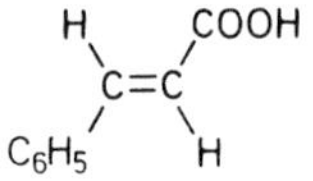 (E)-Cinnamic acid	7.61 –	– 6.41	7.82 –	– 6.47
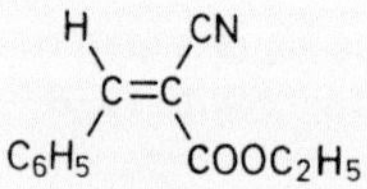 Ethyl 2-cyano-(Z)-cinnamate	7.84 –	– –	8.22 –	– –

*The order of the shift values corresponds to the position of the hydrogen atoms in the structural formula.

To conclude this section, brief mention will be made of the use of **chiral solvents**, such as (R)- or (S)-2,2,2-trifluoro-1-phenylethanol or (R)- or (S)-2,2,2-trifluoro-1-(9-anthryl)ethanol (**Pirkle's alcohol**). They can be used to determine optical purity and in suitable cases can give indications of absolute configurations. The NMR spectra of **enantiomers** (+)A and (−)A are identical in optically inactive solvents. In rare cases, self-discrimination occurs by the formation of associates/aggregates. An RS pair is then diastereomeric to the enantiomeric pairs RR and SS. In an optically active medium S, **diastereomeric solvation complexes** can be formed from the two enantiomers (+)A and (−)A, which will have different spectra.

$$(+)A \ldots (+)S \neq (-)A \ldots (+S)$$

Table 3.24 Increment system for the estimation of chemical shifts of benzene protons

$\delta = 7.26 = \Sigma I$

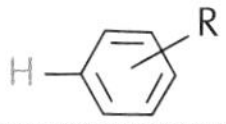

Substituent	I_{ortho}	I_{meta}	I_{para}
—H	0	0	0
—CH$_3$	− 0.18	− 0.10	− 0.20
—CH$_2$CH$_3$	− 0.15	− 0.06	− 0.18
—CH(CH$_3$)$_2$	− 0.13	− 0.08	− 0.18
—C(CH$_3$)$_3$	0.02	− 0.09	− 0.22
—CH$_2$Cl	0.00	0.01	0.00
—CH$_2$OH	− 0.07	− 0.07	− 0.07
—CH$_2$NH$_2$	0.01	0.01	0.01
—CH=CH$_2$	0.06	− 0.03	− 0.10
—C≡CH	0.15	− 0.02	− 0.01
—C$_6$H$_5$	0.30	0.12	0.10
—CHO	0.56	0.22	0.29
—CO—CH$_3$	0.62	0.14	0.21
—CO—CH$_2$—CH$_3$	0.63	0.13	0.20
—CO—C$_6$H$_5$	0.47	0.13	0.22
—COOH	0.85	0.18	0.25
—COOCH$_3$	0.71	0.11	0.21
—CO—O—C$_6$H$_5$	0.90	0.17	0.27
—CO—NH$_2$	0.61	0.10	0.17
—COCl	0.84	0.20	0.36
—CN	0.36	0.18	0.28
—NH$_2$	− 0.75	− 0.25	− 0.65
—NH—CH$_3$	− 0.80	− 0.22	− 0.68
—N(CH$_3$)$_2$	− 0.66	− 0.18	− 0.67
—N(CH$_3$)$_3$I$^-$	0.69	0.36	0.31
—NH—COCH$_3$	0.12	− 0.07	− 0.28
—NO	0.58	0.31	0.37
—NO$_2$	0.95	0.26	0.38
—SH	− 0.08	− 0.16	− 0.22
—SCH$_3$	− 0.08	− 0.10	− 0.24
—S—C$_6$H$_5$	0.06	− 0.09	− 0.15
—SO$_2$—OH	0.64	0.26	0.36
—SO$_2$—NH$_2$	0.66	0.26	0.36
—OH	− 0.56	− 0.12	− 0.45
—OCH$_3$	− 0.48	− 0.09	− 0.44
—OCH$_2$—CH$_3$	− 0.46	− 0.10	− 0.43
—O—C$_6$H$_5$	− 0.29	− 0.05	− 0.23
—O—CO—CH$_3$	− 0.25	0.03	− 0.13
—O—CO—C$_6$H$_5$	− 0.09	0.09	− 0.08
—F	− 0.26	0.00	− 0.20
—Cl	0.03	− 0.02	− 0.09
—Br	0.18	− 0.08	− 0.04
—I	0.39	− 0.21	− 0.03
—Si(CH$_3$)$_3$	0.19	0.00	0.00
—PO(OR)$_2$	0.46	0.14	0.22
—Hg–C$_6$H$_5$	0.14	0.14	− 0.02

Table 3.25 Calculated and measured δ (¹H) values of benzene ring protons

Compound	$\delta_{calculated}$		$\delta_{measured}$	

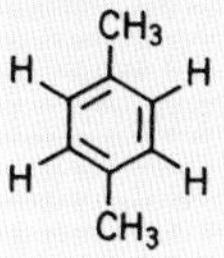

p-Xylene

| | 6.98 | | 6.98 6.97 | |
| | 6.98 | | 6.98 6.97 | |

calculated: 6.98, 6.98, 6.98 ; measured: 6.97, 6.97, 6.97

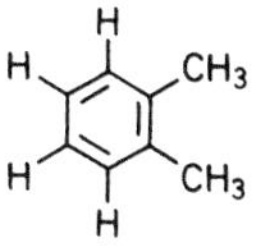

o-Xylene

	6.98	7.05
	6.96	– 7.05 –
	6.96	– 7.05 –
	6.98	7.05

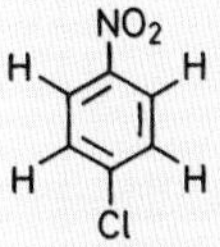

1-Chloro-4-nitrobenzene

| | 8.19 | 8.19 8.17 | 8.17 |
| | 7.55 | 7.55 7.52 | 7.52 |

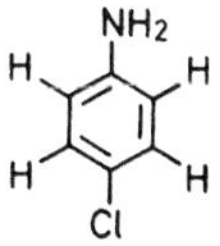

4-Chloroaniline

| | 6.49 | 6.49 6.57 | 6.57 |
| | 7.04 | 7.04 7.05 | 7.05 |

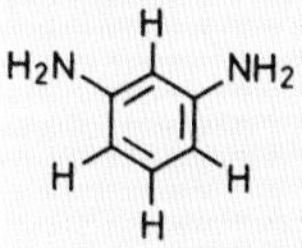

1,3-Diaminobenzene (m-phenylenediamine)

	5.76	6.03
	5.86	5.86 6.11	6.11
	6.76	6.93

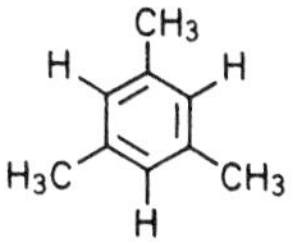

1,3,5-Trimethylbenzene (mesitylene)

| | 6.78 | 6.78 6.78 | 6.78 |
| | 6.78 | 6.78 |

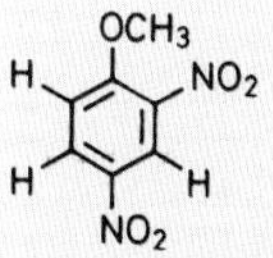

2,4-Dinitro-1-methoxybenzene

| | 7.30 | – 7.28 | – |
| | 8.38 | 9.07 8.47 | 8.72 |

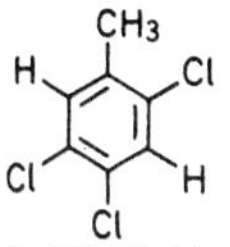

2,4,5-Trichlorotoluene

| | 7.07 | 7.19 | – |
| | 7.20 | 7.31 |

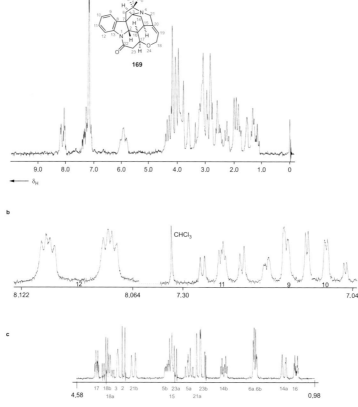

Fig. 3.47 ¹H NMR spectra of strychnine (**169**) in CDCl₃: (**a**) 60-MHz spectrum; (**b** and **c**) sections of a 250-MHz spectrum (Carter JC, Luther GW, Long TC. 1974, J. Magn. Reson. **15**, 122).

¹H NMR data of strychnine (**169**)

Position	δ (±0.004)	J (Hz) (±0.10)	
2	3.846	J_{2-16}	= 10.47
3	3.924		
5a	2.861	J_{5a-5b}	= 9.88
5b	3.185		
6a	1.869	J_{6a-6b}	= 0.02
		J_{6a-5a}	= 10.06
		J_{6a-5b}	= 4.85
6b	1.870	J_{6b-5a}	= 8.54
		J_{6b-5b}	= 3.33
9	7.145	J_{9-10}	= 7.41
		J_{9-11}	= 1.18
		J_{9-12}	= 0.45
10	7.076	J_{10-11}	= 7.46
		J_{10-12}	= 1.12
11	7.230	J_{11-12}	= 8.10
12	8.085		
14a	1.430	$J_{14a-14b}$	= 14.37
		J_{14a-3}	= 1.82
14b	2.338	J_{14b-3}	= 4.11
15	3.126	J_{15-14a}	= 1.98
		J_{15-14b}	= 4.58
		J_{15-18a}	= 2.5
16	1.252	J_{16-15}	= 3.10
17	4.266	J_{17-16}	= 3.12
18a	4.047	$J_{18a-18b}$	= 14.19
18b	4.127		
19	5.881	J_{19-18a}	= 5.74
		J_{19-18b}	= 6.88
21a	2.712	$J_{21a-21b}$	= 14.83
21b	3.691	J_{21b-19}	= 1.2
23a	3.105	$J_{23a-23b}$	= 17.40
		J_{23a-17}	= 8.44
23b	2.657	J_{23b-17}	= 3.28

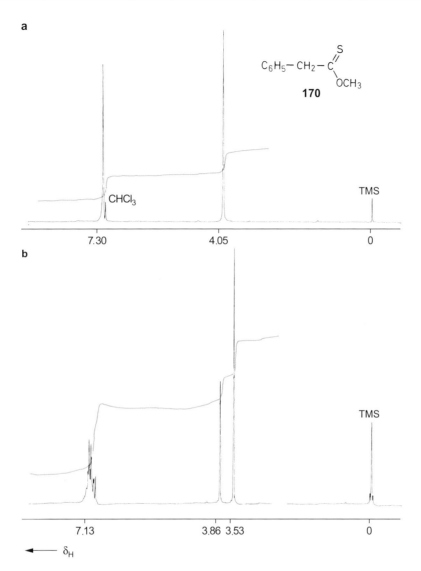

Fig. 3.48 ¹H NMR spectra of *O*-methyl phenylthioacetate (**170**) with integration (**a**) in CDCl₃ and (**b**) in C₆D₆.

In principle, enantiotopic protons in an achiral molecule can become anisochronous when complexed to a chiral (solvent) partner. Alternatively, chemically nonequivalent protons may become coincidentally isochronous. Only homotopic groups remain chemically equivalent in chiral media (see Section 3.2.2, p. 100).

Sometimes it is unavoidable to suppress the signals of nondeuterated solvents. For example, water cannot be replaced as a solvent in many biochemical studies. The addition of 10% D₂O suffices to provide the lock, but there remains a very intense water signal, massive compared to the signals of the substrate. Either presaturation of the water signal or specific pulse sequences for solvent suppression can be employed. Frequently employed is the appropriately named **WATERGATE** sequence (**water** suppression by g**r**adient-**t**ailored **e**xcitation). By combination of this field-gradient method with appropriate FID-manipulation, a complete suppression of the water signal can be achieved while leaving the signals of the substrate unaffected, even when their δ values are very similar to that of water.

Selective Deuteration

The simplification of NMR spectra can occasionally be usefully achieved by the replacement of selected H atoms with deuterium. **Acidic protons** undergo such exchange simply on shaking with D_2O (see p. 138). In other cases, a **specific synthesis of the relevant deuterated compound** must be undertaken. In the ¹H NMR spectrum of the deuterated compound, the signal of the substituted proton will then be absent. The other shifts remain almost unchanged by the deuteration; ²H is slightly more shielding than ¹H, so a small high-field shift is observed. However, the different coupling behaviors of ¹H and D must be taken into account. Deuterium has a nuclear spin $I = 1$, which changes the number of lines and the intensities of the splitting pattern (see **Table 3.2**, p. 92).

Furthermore, the ¹H,D couplings are considerably smaller than the corresponding ¹H,¹H couplings and are therefore often not observed in the spectrum.

$$J(H,H) \approx 6.5\,J(H,D)$$

Apart from the simplification of the spectrum, deuteration can also be utilized for the determination of couplings between magnetically equivalent nuclei. A proton is replaced by a deuteron, the $J(H,D)$ coupling is measured, and the $J(H,H)$ coupling calculated.

For a better understanding of the changes in ¹H NMR spectra caused by H/D exchange, the spectra of cyclopropene (**10**) and its two deuterated derivatives **171** and **172** will be discussed (**Fig. 3.49**).

From $J_{MX} = {}^2J(H,D)$ in **171**, the *geminal* coupling between the chemically and magnetically equivalent H_M protons in cyclopropene (**10**) can be calculated.

If a H_A proton is replaced by deuterium, an AM_2X spin system (**172**) is obtained, corresponding to the right-hand part of **Fig. 3.49** with A and M exchanged. ${}^3J_{AA}$ can then be determined in an analogous fashion to ${}^2J_{MM}$.

(The numbers below the lines indicate the relative intensities.)

A further, very interesting application of deuterium incorporation is seen in the **isotopic perturbation method** developed by Saunders. Because of the identical lineshapes, even at low temperature, it is impossible to distinguish between a degenerate exchange process with sufficiently low activation barrier and a single symmetric structure. The two indistinguishable alternatives are the dynamic model with a double potential energy minimum and the static model with a single minimum. The carbenium ion (**173**) can be considered as an example. The 12 methyl protons give just one doublet at $\delta = 2.93$ when measured in SbF_5/SO_2ClF at -100°C. This can be explained either by a rapid hydride shift or by a bridged ionic structure. The first hypothesis was shown to be correct.

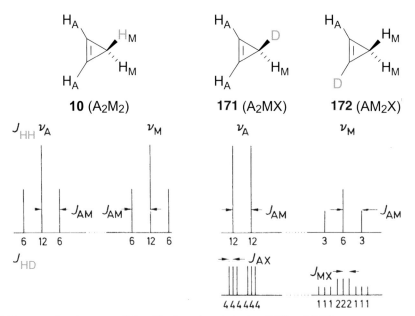

Fig. 3.49 ¹H NMR "stick" spectra of cyclopropene (**10**) and its deuterium derivatives (**171**) and (**172**).

If one of the methyl protons is replaced by deuterium, the degeneracy is lifted, and two doublets, separated by v, are observed.

$$
\begin{array}{ccc}
\text{D-CH}_2 \quad \text{CH}_3 & & \text{D-CH}_2 \quad \text{CH}_3 \\
\text{H} \blacktriangleright \!\!<\!\!+ & \rightleftharpoons & +\!\!>\!\!\blacktriangleleft \text{H} \\
\text{H}_3\text{C} \quad \text{CH}_3 & & \text{H}_3\text{C} \quad \text{CH}_3 \\
\mathbf{A} & \mathbf{174} & \mathbf{B}
\end{array}
$$

The lower field signal is due to the six protons of **174A** shown in blue, which are in fast exchange with the protons of **174B** also shown in blue. At slightly higher field is the signal of the five remaining protons, also a doublet, and also undergoing fast exchange.

The fact that the methyl signal of the six blue protons lies at lower field proves that **174A** is preferred over **174B**. The separation of the signals and the equilibrium constant K are related by

$$
v = \frac{[\delta_1 c(\mathbf{174A}) + \delta_2 c(\mathbf{174B})] - [\delta_2 c(\mathbf{174A}) + \delta_1 c(\mathbf{174B})]}{c(\mathbf{174A}) + c(\mathbf{174B})}
$$

$$
K = \frac{c(\mathbf{174A})}{c(\mathbf{174B})} = \frac{\omega + v}{\omega - v}
$$

v: shift difference between the two signals (from isotopic perturbation).

$\omega = \delta_2 - \delta_1$: hypothetical shift difference if equilibrium were frozen out (must be estimated!).

In this particular case, at −56°C, $K = 1.132 = 53{:}47$. A splitting of the doublet signal would also be expected in the case of a single-bridged carbenium ion; the separation would however be very small, since the effect of deuterium substitution on the ¹H shift is very small. Furthermore, the temperature dependence of K and v is proof of the equilibrium model.

Use of Shift Reagents

The presence of paramagnetic ions in an NMR sample often causes drastic shifts of the signals of nucleophilic species. **Shift reagents** use this effect in a systematic fashion. The most useful ions are Eu(III) and Yb(III), both of which generally cause down-field shifts, and Pr(III), which causes an up-field shift (lanthanide-induced shift). Chelate complexes of these ions with β-diketones are relatively well soluble in organic solvents. Commonly used are as follows:

Eu(dpm)₃

dpm: 2,2,6,6-tetramethyl-3,5-heptanedione (dipivaloyl-methane)

$$
\text{(H}_3\text{C)}_3\text{C}-\overset{\overset{\text{O}}{\|}}{\text{C}}-\text{CH}-\overset{\overset{\text{O}}{\|}}{\text{C}}-\text{C(CH}_3)_3
$$

175

Eu(fod)₃ and Eu(fod)₃-d₂₇

fod: 6,6,7,7,8,8,8-heptafluoro-2,2-dimethyl-3,5-octanedione

$$
\text{F}_3\text{C}-\text{CF}_2-\text{CF}_2-\overset{\overset{\text{O}}{\|}}{\text{C}}-\text{CH}-\overset{\overset{\text{O}}{\|}}{\text{C}}-\text{C(CH}_3)_3
$$

176

and the chiral shift reagents

Eu(facam)₃

facam: 3-trifluoroacetyl-ᴅ-camphor

177

Eu(hfbc)₃

hfbc: 3-heptafluorobutyryl-ᴅ-camphor

178

Compounds with nucleophilic groups complex reversibly with the central lanthanide atom, e.g.:

$$
\text{R}-\text{O}-\text{H} \quad + \quad \text{Eu(fod)}_3 \quad \rightleftharpoons \quad \begin{array}{c} \text{R}-\overset{..}{\text{O}}-\text{H} \\ | \\ \text{Eu(fod)}_3 \end{array}
$$

The resultant shifts of the NMR signals increase with the stability of the complexes and with increased concentration of the shift reagent. In **Fig. 3.50**, the ¹H NMR spectrum of dibutyl ether (**179**) is reproduced. With increasing dosage of Eu(fod)₃, the signals become more and more separated.

A quantitative relationship between the change in chemical shift Δv_i for a nucleus i and the location of this nucleus in a so-called **pseudocontact complex** is given by the **McConnell–Robertson equation**:

$$
\frac{\Delta v_i}{v_i} = K \cdot \frac{3\cos^2 \Theta_i - 1}{r_i^3}
$$

Nuclear Magnetic Resonance Spectroscopy

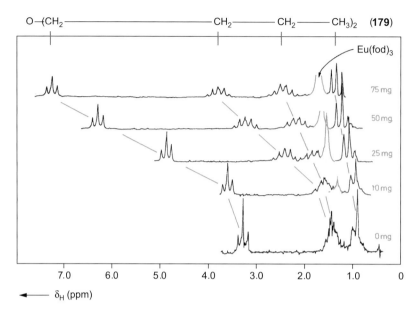

Fig. 3.50 ¹H NMR spectra of a 10⁻⁴ molar solution of dibutyl ether (**179**) in 0.5 mL CCl₄ with increasing addition of Eu(fod)₃. The blue signal arises from the *t*-butyl group in the shift reagent (from Rondeau RE, Rievers RE. 1971, J. Am. Chem. Soc. **93**, 1522).

According to this equation, Δv_i is inversely proportional to the third power of the distance r_i between the nucleus *i* and the central lanthanide atom. *K* is a proportionality constant and θ the angle between the principal magnetic axis of the complex and the line joining the nucleus *i* to the lanthanide ion. It is usually assumed that the principal magnetic axis is coincident with the bond between the central atom and the nucleophilic center.

The angular term in the McConnell–Robertson equation—a kind of a second-order Legendre polynomial—can in no way be ignored: in the region 0 and 55° it is positive, 55 and 125° it is negative.

The statement that europium causes a low-field shift should not be true for the sector shown dark in **Fig. 3.51**. (In the case of the pseudocontact complex with di-*n*-butyl ether, there is however no proton in this region.)

The potential usefulness of shift reagents in the analysis of **NMR** spectra is shown by the example of 2β-androstanol (**180**, **Fig. 3.52**). In the normal spectrum, only the two methyl groups and the H₂α proton can be distinguished. The remaining signals overlap. On addition of Eu(dpm)₃, all the protons in the vicinity of the complexation site can be identified.

On p. 148, it has already been noted that the determination of enantiomer ratios (optical purity) can be carried out by measuring NMR spectra in an optically active solvent. The differences in the chemical shifts of corresponding protons in the two enantiomers are however often very small. This disadvantage

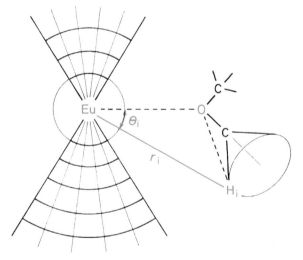

Fig. 3.51 Visualization of the McConnell–Robertson relation in the pseudocontact complex (C₄H₉)₂O . . . Eu(fod)₃.

can often be overcome by measurement in CCl₄ or CDCl₃ with the addition of a chiral shift reagent. **Fig. 3.53** shows the ¹H NMR spectrum of racemic 2-phenyl-2-butanol (**181**) in the presence of an achiral and a chiral shift reagent. The 1:1 ratio of the two enantiomers is clearly shown by the splitting of the methyl signal in the lower spectrum.

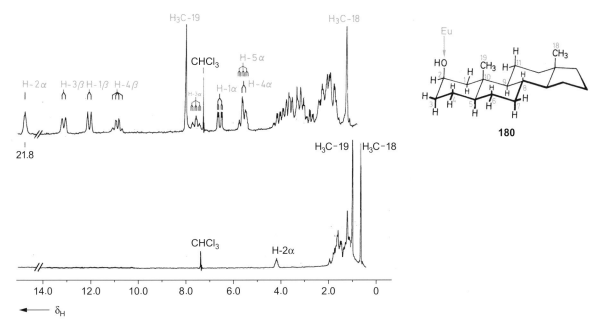

Fig. 3.52 ¹H NMR spectra of 2β-androstanol (**180**; 0.73·10⁻⁴ molar solution in 0.4 mL CDCl₃). The lower part shows the normal spectrum; for the upper spectrum, 40 mg Eu(dpm)₃ were added (from Demarco PV et al. 1970, J. Am. Chem. Soc. **92**, 737).

Nuclear Magnetic Resonance Spectroscopy

Measurement of ¹³C Satellites

Coupling between magnetically equivalent nuclei has no visible effect on the spectrum. The ¹H NMR spectrum of (E)-dichloroethylene (**182**) is an A₂ singlet.

182

182a

If the natural ¹³C content of 1.1% is considered, however, about 2% of (E)-1,2-dichloroethylene is **182a**, the remaining 98% being **182**. Isotopomer **182a** corresponds to an AA′X system. (The isotope effect on the ¹H shift is negligibly small.) **Fig. 3.54** shows the A-part (¹H signals) of the **satellite spectrum**, which lies symmetrically about the singlet signal of **182** with the corresponding intensity. Since $^2J_{A'X}$ is very small, the inner lines lie under the strong main signal of the A₂ system.

From the satellite spectrum the coupling constant $^1J_{XA}$ for the direct ¹³C,¹H coupling can be extracted. The same value can naturally be obtained by measuring the X part, i.e., the coupled ¹³C spectrum.

Furthermore, the ¹H,¹H coupling can be extracted from the satellite spectrum. In contrast to **182**, the two protons in **182a** are no longer magnetically equivalent, so the coupling constant between them can be determined from the spectrum. For the molecule **182**, one can assume—again neglecting any isotope effect—the same value. In this case it is found to be 12.5 Hz, corresponding to the (E)-configuration. In this way, the configuration of symmetrically 1,2-disubstituted ethylenes can be determined. Another solution of this problem exists by applying **gradient-selected** heteronuclear single quantum correlation-NOE spectroscopy (**HSQC-NOESY**) techniques (cf. p. 220).

The coupling between geminal, magnetically equivalent protons can neither be determined from the normal ¹H spectrum nor from the ¹³C satellites.

A₂-System   A₂X-System

The $^2J(H,H)$ coupling in methyl or methylene groups can however be determined by deuteration (see p. 152).

Spin Decoupling (Multiple Resonance)

The spin–spin coupling of magnetically nonequivalent nuclei causes, as described in Section 3.1.3 (p. 89), splitting of the

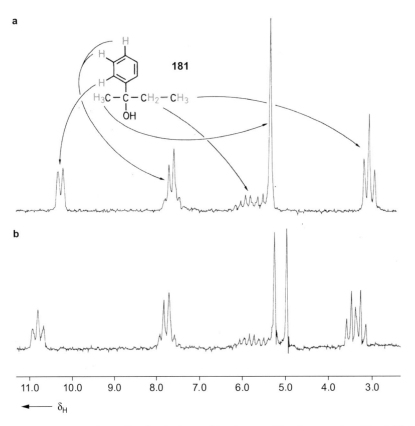

Fig. 3.53 ^{1}H NMR spectra of a 0.54 molar solution of 2-phenyl-2-butanol (**181**) in CCl$_4$: (**a**) with 0.13 molar added Eu(dpm)$_3$; (**b**) with 0.42 molar added Eu(facam)$_3$ (Goering HL et al. 1971, J. Am. Chem. Soc. **93**, 5913).

signals. This often leads to complex multiplets. To simplify such spectra and to determine coupling partners, **spin decoupling** can be performed by **double resonance**.

Let us consider the simple example of an AMX system ($I = 1/2$). The corresponding spectrum consists of four lines of equal intensity for each nucleus.

If the sample is subjected to an additional irradiation at the resonance frequency v_M, then the nuclei A and X do not "see" two separate spin states for M, because the spin orientation of M changes too rapidly. The average value is zero, i.e., M no longer couples with A and X. At the position of the signal of M, an oscillation occurs (in CW spectra); otherwise the spectrum is simplified to two doublets with the coupling constant J_{AX}. The AMX spectrum is thus reduced to an AX spectrum (**Fig. 3.55**). Similarly, the irradiation can be set to v_X or v_A. In **triple resonance**, irradiation is carried out at **two** additional frequencies. The AMX spectrum then reduces to a singlet.

If the signals lie close together, it is not possible to irradiate at the frequency of one nucleus without perturbing the transitions

of other nuclei. Consider an ABX case. If the amplitude of the decoupling field is reduced so that not the whole X part, but only two of the X lines are affected, then only a part of the AB system is affected. This is called **selective decoupling**.

Finally, if the amplitude of the decoupling field is reduced so much that only a single line (e.g., in the X part) is affected, then all the other lines in the spectrum which have common energy levels with the irradiated line are split. This effect is known as **spin tickling**. Further information on these methods is available from the literature quoted in the bibliography. Only the normal double-resonance experiment will be illustrated with a few specific examples.

Compound **183**, mannose triacetate, has the ^{1}H NMR spectrum shown in **Fig. 3.56**. Apart from the acetate methyl groups, there are seven nonequivalent types of protons, H$_A$ to H$_G$. By irradiating at the frequency of H$_C$, all couplings to H$_C$ are eliminated.

Because of the dihedral angle, the largest *vicinal* coupling is J_{CB}. The effect of the decoupling is therefore greatest for H$_B$. But

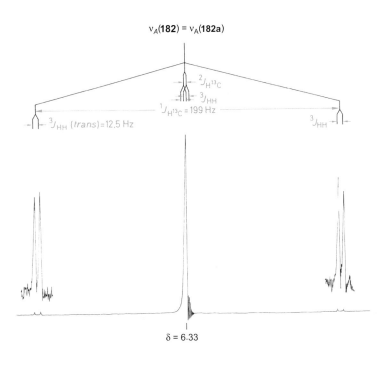

$\nu_A(\mathbf{182}) = \nu_A(\mathbf{182a})$

$^2J_{H^{13}C}$

$^3J_{HH}$

$^1J_{H^{13}C} = 199$ Hz

$^3J_{HH}$ (trans) = 12.5 Hz

$^3J_{HH}$

$\delta = 6.33$

Fig. 3.54 ^{1}H NMR spectrum of (E)-1,2-dichloroethylene (**182**) with the ^{13}C satellites (**182a**).

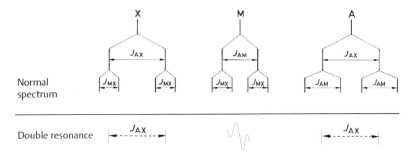

Normal spectrum

Double resonance

Fig. 3.55 Stick spectra of an AMX system and its simplification on additional irradiation at ν_M.

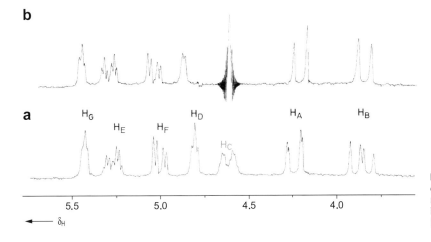

b

a

H$_G$ H$_E$ H$_F$ H$_D$ H$_C$ H$_A$ H$_B$

5.5   5.0   4.5   4.0

δ_H

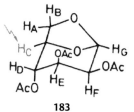

H$_B$ H$_A$ H$_C$ OAc   H$_G$ H$_D$ OAc   AcO   H$_E$ H$_F$

183

Fig. 3.56 (a) 100-MHz ^{1}H NMR spectrum of mannose triacetate (**183**). (**b**) Double-resonance spectrum in frequency sweep on irradiating at ν_{H_C} (Johnson LF. 1965, Varian Inform. Bull. **5**).

the signals of H_A, H_D, and H_E are also recognizably simplified. The ABCDEFG system is transformed to an AB system and a DEFG system. At the position of $\nu(H_C)$, an oscillation is seen which results from the interference between the observation and irradiation frequencies.

In **Fig. 3.56** it is also apparent that the irradiation causes a slight shift of the observed signals (**Bloch–Siegert effect**).

A further example is the polyolefin **184**, which has a complicated ^1H NMR spectrum. In the low-field region, a 400-MHz spectrum shows six signals for the olefinic protons H_A, H_B, H_E, H_F, H_G, and H_H (**Fig. 3.57**). On irradiation at the frequencies of

the saturated protons H_C and H_D, all the multiplets remain unchanged except for H_A and H_B, the signals of which (AB part of an AA'BB'CC'DD' system) are simplified to an AB system with $^3J_{AB} = 11.0$ Hz. The assignment of the signals of the protons in the side chains attached to the four-membered ring is possible from the double-resonance experiments in **Fig. 3.58**. Irradiating H_H leads to the disappearance of the couplings $^2J_{G,H} = 1.8$ Hz, $^3J_{F,H} = 10.1$ Hz, and $^4J_{E,H} = -0.8$ Hz; irradiating H_G eliminates $^3J_{F,G} = 16.6$ Hz and $^4J_{E,G} = -0.9$ Hz as well as $^2J_{G,H}$; irradiating H_E removes $^3J_{E,F} = 11.3$ Hz as well as $^4J_{E,G}$ and $^4J_{E,H}$. Irradiating H_F achieves the greatest simplification of the spectrum and serves as a control experiment.

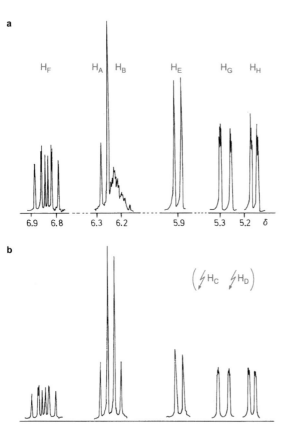

Fig. 3.57 (a) 400-MHz ^1H NMR spectrum (olefinic part) of **184** in CDCl$_3$. **(b)** Triple resonance (irradiation at H_C and H_D).

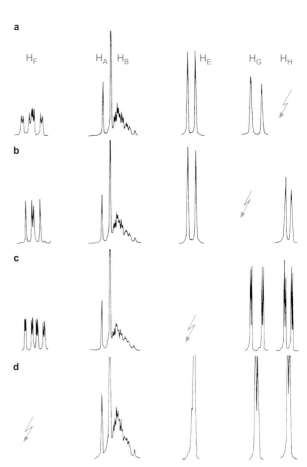

Fig. 3.58 Double-resonance measurements on **184** (400 MHz, CDCl$_3$). Irradiations: **(a)** at H_H; **(b)** at H_G; **(c)** at H_E; **(d)** at H_F.

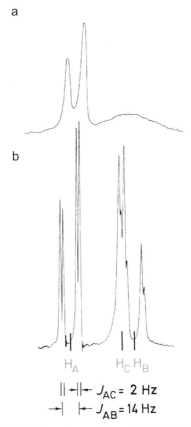

H_A $\delta =$	6,27	
H_B	6,23	
H_C	$\approx 2,3$	
H_D	$\approx 2,3$	
H_E	5,90	
H_F	6,84	
H_G	5,28	
H_H	5,17	

184

All the cases discussed here involve ¹H NMR. Apart from these homonuclear double-resonance experiments, heteronuclear double resonance is also possible. In ¹H NMR, this is occasionally used for compounds with NH groups. Irradiation at the ¹⁴N frequency can remove the line broadening usually observed for such systems. An excellent example is the ¹H NMR spectrum of formamide (**185**; **Fig. 3.59**).

185

Heteronuclear double resonance is especially important in ¹³C NMR spectroscopy (see Section 3.4.1, p. 173).

NOE Difference Spectroscopy

As mentioned in Section 3.1.5 (p. 98), the Nuclear Overhauser effect (**NOE**) is apparent in double-resonance ¹H NMR experiments. The intensity of an observed ¹H signal v_A can be increased by additional irradiation at v_B. The precondition for this effect is that the distance r between the nuclei A and B is small, since the dipole–dipole interaction which is responsible for longitudinal relaxation in such cases is proportional to $1/r^6$.

Small molecules, which undergo rapid motion in solutions of low viscosity, show a positive NOE. The maximum increase in intensity in a homonuclear two-spin system is 50%. Molecules with masses of 1,000 Dalton and more move very slowly, and show negative NOEs (decrease in intensity). A positive **steady-state NOE** (equilibrium NOE, saturation NOE), built up by a sufficiently long irradiation at v_B produces percentage increases in the intensities of the signals of neighbor protons H_A and H_C, which gives information about the relative distances r_{AB} and r_{BC}. Absolute internuclear distances cannot be determined. The observed NOE is often not "symmetrical"; i.e. irradiation at v_B generates a different percentage enhancement of the intensity of the signal of H_A from the enhancement of the signal of H_B caused by irradiation at v_A. Special care is required in cases of **saturation transfer** between exchanging protons,

Fig. 3.59 ¹H NMR spectrum of formamide (**185**). (**a**) Normal spectrum. (**b**) Double-resonance experiment (irradiation at the ¹⁴N resonance frequency).

As well as the **direct NOE** there are also **indirect NOEs**. A positive NOE on H_A caused by irradiation at v_B can generate a negative indirect NOE on H_C, which is superimposed on the positive direct NOE from H_C. In unfavorable geometrical arrangements (angle H_B–H_A–$H_C \approx 90°$), the positive and negative effects can cancel each other out, although the internuclear distance r_{BC} is small. More critical is an overlying negative direct effect. The rapid extension of the originally generated NOE between neighboring nuclei into other regions of the molecule is known as **spin diffusion**. In such situations, under steady-state conditions the information about the molecular structure based on internuclear distances is lost. The **transient NOE** experiment measures the build-up of the NOE, the initial rate of which depends on r^{-6}, and thus allows even the measurement of absolute internuclear distances.

NOE measurements are highly suited for the determination of cis- and trans-positions in (unsymmetrical) double-bond systems or ring compounds. The rigid 1-chloro-2,2-dimethyl-cyclopropane (**186**) exhibits a considerable NOE between the ring proton and the *pro-S* methyl group drawn in red. The blue *pro-R* methyl group is too far away. The same argument is valid for the six-membered ring system 1-chloro-2,2-dimethylcyclohexane (**187**), even when the ring inversion leads to a small population of the conformer with an axial chloro substituent. The ring inversion does not alternate the *cis* arrangement of the drawn proton and the red methyl group.

The use of NOE experiments to determine constitution and/or configuration is extremely important in structural analysis, since NOEs rely on through-space interactions which often nicely complement interactions through bonds (spin–spin coupling).

To decide, for example, whether tricyclic compound **188** exists in the *exo-* or *endo*-configuration, the intensities of the ¹H signals on irradiating the signal of the bridgehead proton H$_e$ can

be compared with the normal spectrum where no irradiation is carried out. For the *exo*-compound **188a**, the intensity of the signal of H$_c$ (as well as H$_d$) should be increased on irradiation; for the *endo*-compound **188b**, on the other hand, the signal of H$_b$ should increase in intensity. Since the effects are often small, it is recommended to use **difference spectroscopy**, i.e., the normal spectrum is subtracted from the double-resonance spectrum. **Fig. 3.60** shows an increase of intensity for H$_c$ and H$_d$, i.e., **188** has the *exo*-configuration.

Apart from configurational determinations, NOE measurements can be used with advantages in conformational analysis. The ester **189** (R' ≠ H, COOR; for example, R' = NH$_2$, OH, Cl) has

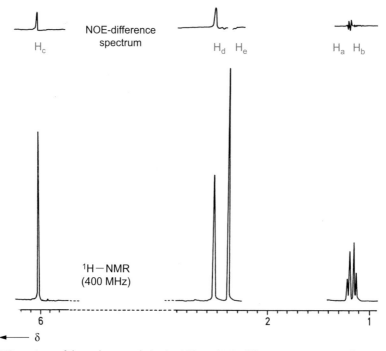

Fig. 3.60 400-MHz ¹H NMR spectrum of the norbornene derivative **188a** and NOE difference spectrum in a deoxygenated CDCl$_3$ solution.

diastereotopic methyl groups. Three rotamers of **189** have to be considered: **189a** to **189c**. Provided that the populations of the rotamers **189b** and **189c** are different, the red (*pro-S*) and the blue (*pro-R*) methyl group can give NOEs of different intensities.

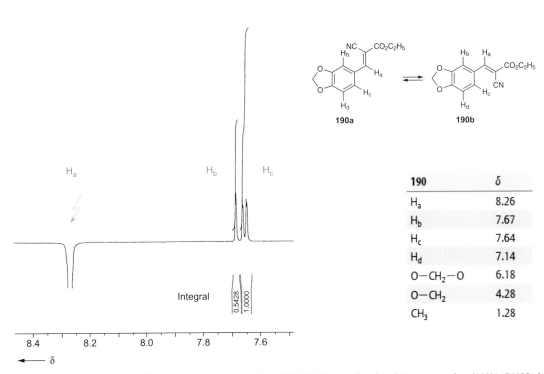

Where two or more conformers are present, however, serious problems can arise. If, for example, one assumes that the nucleus H_A under investigation has a distance r from nucleus H_B (irradiation at v_B) in conformer I, and a distance $2r$ in conformer II, and exchange between conformers I and II (with equal populations) is rapid on the NMR time scale, then the NOE measurement will give an internuclear distance of $1.12r$ due to the averaging of the r^{-6} values. In fact, the average distance is 0.5 $(r + 2r) = 1.5r$, i.e., considerably larger! Nevertheless, useful, if qualitative, results can often be obtained from NOE

measurements in conformational analysis. As an example, the compound **190** will be discussed, in which the olefinic side chain can be found in one of two different configurations. In conformer **190a**, the distance between H_a and H_c is shorter than the distance between H_a and H_b; in **190b**, it is opposite. Since the signals of H_b and H_c are easy to distinguish, the NOE measurement (irradiation at H_a, $\delta = 8.26$) will give simple and fast information about the equilibrium of the two conformers. **Fig. 3.61** shows the result. The integral of the doublet signal is almost twice as large as the integral of the singlet signal. This result proves that conformer **190a** is strongly preferred compared to **190b**.

INDOR Technique

A further useful method is the **INDOR technique** (**IN**ternuclear **DO**uble **R**esonance). This involves monitoring the intensity of one individual line (monitor line) while an additional irradiating field is swept across the remaining spectral region. Whenever this irradiating field reaches the frequency of a line in the spectrum which shares an energy eigenvalue with the monitor line, the intensity of the monitor line is altered as a consequence of the **general Overhauser effect**. If the lower level (starting

190	δ
H_a	8.26
H_b	7.67
H_c	7.64
H_d	7.14
$O-CH_2-O$	6.18
$O-CH_2$	4.28
CH_3	1.28

Fig. 3.61 NOE difference spectra of an oxygen-free solution of ethyl (*E*)-3-(1,3-benzodioxol-5-yl)-2-cyanoacrylate (**190**) in DMSO-d₆ (from Soliman A, Meier H, unpublished).

Nuclear Magnetic Resonance Spectroscopy

level) of the monitor line is more strongly populated, an intensity increase is observed; if the upper level is more strongly populated, then the intensity decreases. The "**spin pumping**" leads in the case of **progressive transitions** to a positive line in the difference spectrum, in the case of **regressive transitions** to a negative line. If a transition has no energy level in common with the monitor line, then its line is not observed in the **INDOR difference spectrum**. To illustrate the effect, consider the simple example of an AX spin system. The four possible spin orientations and the corresponding energy levels are shown in **Fig. 3.3**. If the case where $J < 0$ is considered, then the A_2 line as the monitor line will be increased in intensity when the X_1 transition is irradiated; conversely, it will be decreased in intensity when the X_2 transition is irradiated. If X_2 is chosen as the monitor line, then at A_1 a positive signal is obtained, and at A_2 a negative signal (**Fig. 3.62**).

To determine the coupling partners of a nucleus A with a decoupling experiment, the exact frequency of A must be found. When the signals are complex and/or unsymmetrical, this is far more difficult than selecting an individual line for an INDOR experiment. In cases where signals overlap badly, the INDOR technique is also often preferable to a decoupling experiment. INDOR spectroscopy as described earlier is exclusively a CW experiment; there are however pseudo-INDOR experiments using the PFT method, so that INDOR difference spectroscopy became a routine experiment even with an FT spectrometer. However, the rapid progress in two-dimensional (2D)-NMR decreased the significance of the INDOR technique. Nevertheless, an impressive example of an INDOR measurement shall be discussed here. The methylene protons of indene (**191**) give a complex, symmetrical multiplet at $\delta = 3.48$. If its maximum is used for an INDOR experiment, the spectrum shown in **Fig. 3.63** is obtained. The principal coupling partners of the CH_2 protons are the olefinic protons ($^3J(H_a,H_b) = +2.02$ Hz,

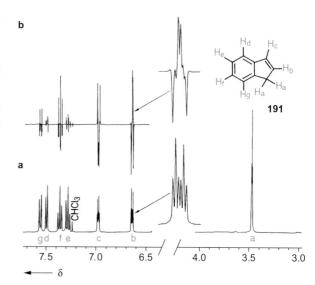

Fig. 3.63 400-MHz ^{1}H NMR spectrum of indene (**191**) in $CDCl_3$: (**a**) normal spectrum; (**b**) INDOR difference spectrum.

$^4J(H_a,H_c) = -1.98$ Hz); couplings to the aromatic protons, particularly H_f, are however also apparent. Even the small ^6J coupling to H_e can be seen.

3.3.9 Two-Dimensional ^{1}H NMR Spectroscopy

Two-dimensional techniques play an outstanding part in modern structure elucidations. The following 2D-^{1}H NMR methods will be discussed in this section: **J-resolved**, correlated spectroscopy (**COSY**), double-quantum filtered-COSY (**DQF-COSY**), **long-range COSY**, exclusive-**COSY** (**E-COSY**), **H-relayed COSY**, total correlation spectroscopy (**TOCSY**), 2D exchange spectroscopy (**EXSY**), Nuclear Overhauser Effect Spectroscopy (**NOESY**), and rotating-frame Overhauser effect spectroscopy (**ROESY**); diffusion-ordered spectroscopy (**DOSY**) measurement will also be briefly mentioned. Most important are COSY and NOESY measurements. While **COSY** spectra are based on the connectivity of protons through the bonds, which are lying between them, **NOESY** spectra reveal the connectivity through the space between the protons. Both methods (including some variants of them) are nowadays routine measurements.

As explained in Section 3.3.1 (p. 117), the PFT method involves the measurement of the decay of the transverse magnetization (FID). The detected signal is a function of the **detection time** t_2. If the **evolution time** t_1, i.e., the time between the first pulse of the chosen pulse sequence and the start of data acquisition, is systematically varied, then the detected signal is a function of t_1 and t_2.

The Fourier transformation now has two frequency variables F_1 and F_2, the requirement for a 2D spectrum. There is a basic distinction between J-resolved and shift-correlated 2D spectra.

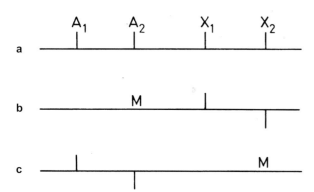

Fig. 3.62 Illustration of INDOR difference spectroscopy on an AX spin system with $J < 0$. (**a**) Four-line spectrum; (**b**) INDOR difference spectrum using A_2 as the monitor line; (**c**) INDOR difference spectrum using X_2 as the monitor line.

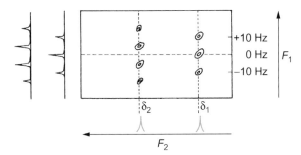

Fig. 3.64 Schematic contour plot for a *J*-resolved 2D ¹H NMR spectrum with the projections onto the *F* axes.

For **homonuclear *J*-resolved 2D experiments**, the initial 90° pulse is followed by an evolution time t_1, in the middle of which a 180° pulse is inserted. In successive measurements, t_1 is increased by a constant amount. During the evolution time t_1, only the scalar coupling develops, i.e., the F_1 information only contains the couplings. The F_2 information contains the whole spectrum as usual. After suitable mathematical processing, the result is presented graphically as a **contour plot**, as shown in **Fig. 3.64** for a triplet and a quartet. It can be considered as a view from above onto the pattern of peaks, which appear as contour lines.

Projection onto the F_2-axis produces a **fully decoupled ¹H NMR spectrum**. Each signal appears as a singlet. It is obvious that this method is interesting for large molecules with many overlapping multiplets. The projection of individual signals onto the F_1-axis (cross-sections) allows the multiplets to be drawn out separately. Weakly coupled spectra with much overlap of multiplets can be resolved with this *J*, δ-spectroscopy. Disadvantages of the method are a rather long measurement time and the appearance of artifacts for strongly coupled spin systems.

As an explicit example of 2D ¹H NMR, the spectra of sucrose (**192**) will be considered. **Fig 3.65a** shows the normal (one-dimensional, 1D) spectrum. Underneath it is shown the projection on the F_2-axis of the *J*-resolved 2D spectrum. For the α-D-glucose part G and the β-D-fructose part F, a total of 11 signals are observed (g_1-g_6, f_1, and f_3-f_6). The diastereotopic protons of the methylene groups are effectively isochronous.

¹H NMR spectra, which exclusively consist of singlet signals, can also be obtained by frequency- and space-selective pulses. A field gradient changes thereby the effective magnetic field in the direction of the tube which contains the sample (**Zangger–Stark method**).

In **shift-correlated 2D NMR spectra**, both frequency axes represent chemical shifts. Without going into the detail of the individual experiments, many different methods can be used (COSY, TOCSY, NOESY, ROESY, EXSY, etc.), each with their own pulse sequence, and with different graphical forms. The **homonuclear**

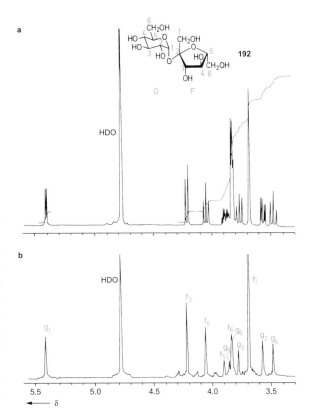

Fig. 3.65 400-MHz ¹H NMR spectrum of sucrose (**192**) in D₂O. (**a**) Normal (1D) spectrum. (**b**) 2D *J*-resolved spectrum (from Bruker brochure: Two-Dimensional NMR).

¹H shift correlations obtained from these experiments show the connectivities in the molecule. A single such experiment can replace a whole series of double-resonance experiments and can also be superior at recognizing long-range or weak correlations.

Consider for example the COSY spectrum of a three-spin system (**Fig. 3.66**). In the contour plot, the normal spectrum lies on the diagonal; these contours are surrounded by black circles.

Further contours, such as those in the solid blue circles, result from coupling between nuclei; in this case between H_3 and H_1 and H_2. The two empty crossing points (empty blue circles) show that there is no detectable coupling between H_1 and H_2.

The ¹H shift correlated 2D spectrum **COSY** (correlated spectroscopy) of sucrose (**192**) in **Fig. 3.67** provides the key to the total coupling network. For example, proton g_1 couples with $g_2(^3J)$, $g_3(^4J)$, $g_5(^4J)$, and $f_1(^5J)$. In the normal 400-MHz spectrum (**Fig. 3.65a**), g_1 only appears as a doublet, i.e., only ^3J is resolved. The doublet for f_3 however is "genuine"; the 2D spectrum shows no other correlations than that to f_4, the ^3J coupling.

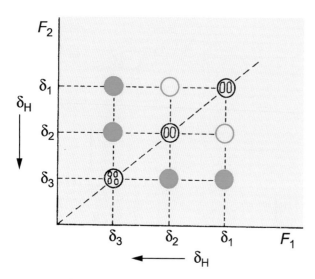

Fig. 3.66 Schematic contour plot for a ^{1}H shift correlated 2D NMR spectrum of a three-spin system with two couplings.

(^{1}H,^{1}H)COSY spectra are nowadays a routine procedure for the structural determination of organic compounds. The measurement often takes no longer than a few minutes and produces valuable information about ^{1}H,^{1}H coupling, the larger (in magnitude) geminal and vicinal couplings being easiest to observe.

Values of coupling constants cannot be obtained from normal COSY spectra. **Phase-sensitive COSY spectra (Ph-COSY, E-COSY;** see p. 166) are required for these data.

A variety of pulse sequences are available for the (^{1}H,^{1}H)COSY experiment; depending on the second pulse angle, there are the COSY-90, COSY-60, and COSY-45 sequences. The COSY-45 spectrum is recommended where crosspeaks lie close to the diagonal, since in this variation the diagonal peaks are weaker. Furthermore, it is frequently possible to distinguish between geminal and vicinal couplings by using the COSY-45 pulse sequence. Crosspeaks arising from geminal couplings are generally extended in a direction parallel to the diagonal (positive slope), whereas vicinal couplings are oriented at 90° (negative slope). **Fig. 3.68** shows this effect clearly using a schematic contour diagram; the difference in the orientation of the crosspeaks results from the different signs of the geminal and vicinal ^{1}H,^{1}H couplings.

Fig. 3.69 shows the distinction of geminal and vicinal couplings in the sulfoxide **193**. In this experiment the crosspeaks for the geminal coupling have a "negative slope" and the crosspeaks

for the vicinal coupling a positive slope. Contingent upon the sulfoxide group with S as the center of chirality, the geminal protons on C-2, -3, -4, -7, and -8 are diastereotopic and show geminal and vicinal couplings.

rac-**193**

COSY contour plots contain in many cases strongly superimposed signals close to the diagonal. This results from the so-called autocorrelation peaks that are crosspeaks, which belong to the lines of a multiplet. An important improvement, which facilitates the interpretation of such contour diagrams, is achieved by use of a **D**ouble-**Q**uantum **F**ilter (DQF). All singlet signals are eliminated by this technique (**DQF-COSY**), because singlets do not have a double-quantum coherence. The superposition of signals is reduced by this method, which can be extended to the triple-quantum filter (TQF) and generally to the multiple-quantum filter technique. A **TQF-COSY** spectrum contains neither singlet signals nor signals of AB or AX spin systems.

An experimental modification, **long-range COSY,** allows observation of correlations due to the smaller long-range couplings, often even when these cannot be directly observed in the 1D spectrum. **Fig. 3.70** shows the long-range COSY spectrum of 6-hexyloxy-10-methylphenanthrene-2-carboxaldehyde (**194**). Apart from the crosspeaks for the two vicinal couplings, it contains crosspeaks for the ^4J couplings of 1-H with 3-H, 5-H with 7-H, 8-H with 9-H, 9-H with CH$_3$, and for the ^5J couplings of 1-H with 4-H, 4-H with 5-H, 4-H and CHO, and 5-H with 9-H, and weak indications of the couplings ^5J(1-H,9-H) and ^5J(5-H,8-H). With this information an unambiguous assignment of the signals is relatively easy.

It is not always possible to dispense with the normal COSY spectrum and only measure a long-range COSY spectrum. Under the measurement conditions required to observe long-range couplings, the crosspeaks for larger coupling constants can disappear.

As a further example, L-serine (**195**) is a simple molecule, but its behavior in 2D experiments is more complicated. This is always

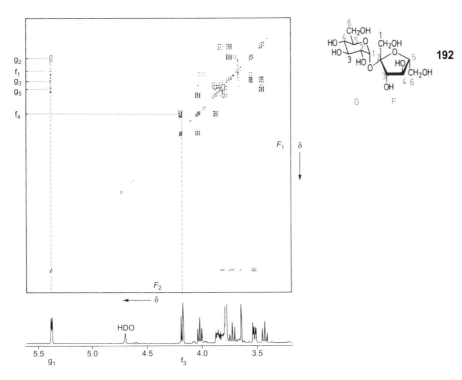

Fig. 3.67 500-MHz ¹H shift correlated 2D NMR spectrum of sucrose (**192**) in D₂O (COSY-45 experiment; from Bruker brochure: Two-Dimensional NMR).

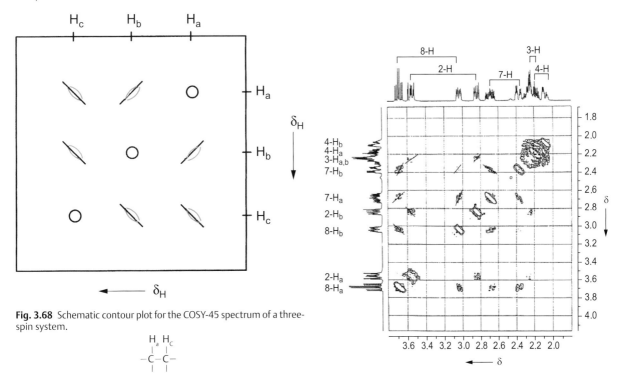

Fig. 3.68 Schematic contour plot for the COSY-45 spectrum of a three-spin system.

Fig. 3.69 COSY-45 NMR spectrum of *rac*-**193** in CDCl₃.

Nuclear Magnetic Resonance Spectroscopy

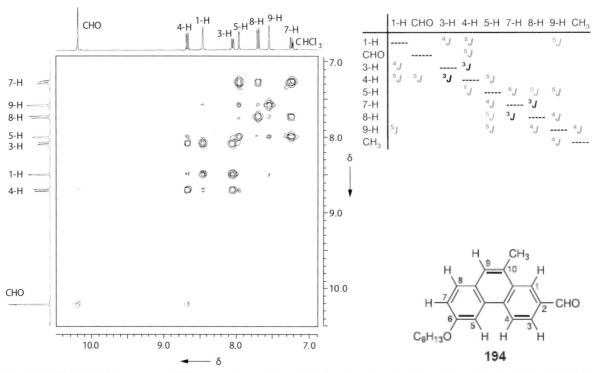

Fig. 3.70 400-MHz ^{1}H,^{1}H long-range COSY spectrum (aromatic part) of 6-hexyloxy-10-methylphenanthrene-2-carboxaldehyde (**194**). The coupling matrix shows the detectable long-range couplings ^4J and ^5J in *blue* .

the case when strong couplings are present, as is the case for all the three relevant spin–spin interactions in L-serine.

195

At 60 MHz, the signals of H_A, H_B, and H_C lie practically together (ABC system). At 360 MHz, the spectrum is that of an ABX system (**Fig. 3.71a**). The 2D J-resolved spectrum is shown in **Fig. 3.71c** as a stack plot (panoramic view). As a result of the strong coupling, more than three multiplets can be observed, and correspondingly there are more than three singlets in the projection onto the δ-axis (**Fig. 3.71b**). The slices through the individual multiplets (cross-sections) in the J direction can be seen in **Fig. 3.71d**. Despite the extra signals caused by the strong coupling, the chemical shifts and couplings can be unambiguously determined.

A procedure for the exact measurement of small coupling constants (down to $|J| = 0.2$Hz) or couplings that are difficult to

get because of multiplet superpositions is provided by **E-COSY** (**E**xclusive **C**orrelation **S**pectroscopy). Homo- or heteronuclear three-spin systems AMX are particularly appropriate for this method, which obtained large significance for polypeptides.

The small ^3J(H,H) coupling can be determined by means of the larger ^1J(C,H) coupling. Parts of the crosspeaks disappear by the superposition of in-phase and out-of-phase subspectra. Compared to DQF-COSY measurements, the number of peaks is divided in half. A 2D pattern remains out of which the coupling constants can be determined as distances of the signals (maxima). When the straight line between the peaks has a positive slope, as shown in **Fig. 3.72**, the two coupling constants have the same sign (^1J(C,H), ^3J(H,H) > 0). Different signs of the coupling constants lead to a negative slope. The method is more complicated for spin systems higher than AMX.

In the **relayed technique**, the magnetization is not directly transferred from one nucleus to the coupling nuclei, as in the

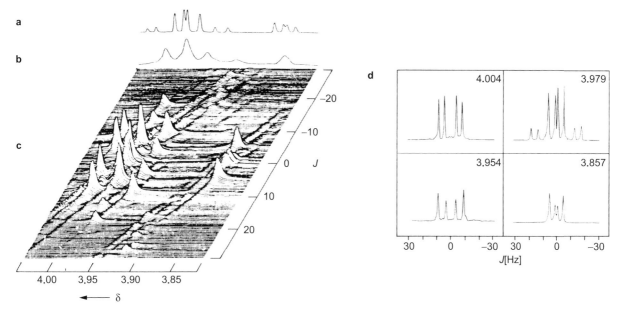

Fig. 3.71 360-MHz ¹H NMR spectrum of L-serine (**195**) in D₂O: (**a**) normal (1D) spectrum; (**b**) 2D *J*-resolved projection spectrum; (**c**) panorama plot of the 2D *J*-resolved spectrum; (**d**) cross-sections (slices down the *J*-axis) at the given δ values (e.g., δ$_X$ = 3.857) (Wider G, Baumann R, Nagayama K, Ernst RR, Wüthrich K. 1981, J. Magn. Res. 42, 73).

normal (H,H)COSY; instead an intervening nucleus serves as a "relay." The polarization can be transferred stepwise across several nuclei. As an example a ¹H→¹H→¹H case will be discussed. **Fig. 3.73a** shows a section of the ¹H shift-correlated NMR spectrum of sucrose (**192**). The same section from a measurement using the relayed COSY method is shown in **Fig. 3.73b**. It contains correlations which are not present in (**a**). They correspond to the ⁴*J* couplings f₃/f₅, f₄/f₆, g₂/g₄, and g₃,g₅. The correlation g₁/g₃ is also observed, but lies outside the area shown. In contrast, no correlation is observed for f₁/f₃, since C-2 of fructose has no "relay proton."

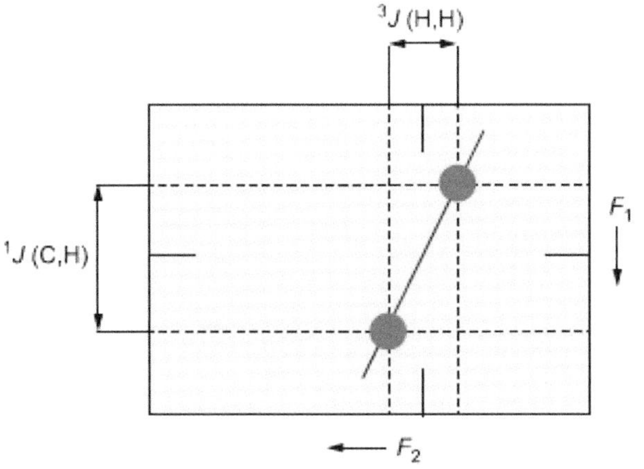

Even more information can be obtained from the **TOCSY** (**T**otal **C**orrelation **S**pectroscopy) experiment and the closely related **HOHAHA** (**Ho**monuclear **Ha**rtmann–**Ha**hn spectroscopy).

In the TOCSY experiment, magnetization transfer is possible along a spin sequence A–B–C–D–. . ., even when proton A, for example, does not couple with proton D. This method is particularly suitable for molecules which consist of discrete, possibly similar building blocks. Oligosaccharides and peptides are typical examples. Starting from the α-protons in an oligopeptide for example, it is possible to assign the resonances in the side chains R¹, R², R³ etc., since a magnetization transfer across a peptide bond does not occur.

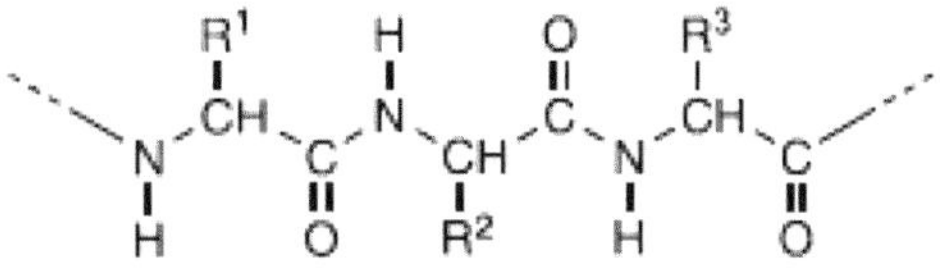

In di- and oligosaccharides, TOCSY spectra permit the separation of the resonances of a single ring, because the magnetization transfer does not take place across the oxygen bridges.

The TOCSY technique is also very useful for the analysis of mixtures. Excellent examples have been established for mixtures of epimers such as anomeric saccharides.

Fig. 3.74a contains a schematic diagram of two independent transfer routes A–B–C–D and A'–B'–C' for the magnetization.

Fig. 3.72 E-COSY crosspeaks of a heteronuclear AMX system.

Nuclear Magnetic Resonance Spectroscopy

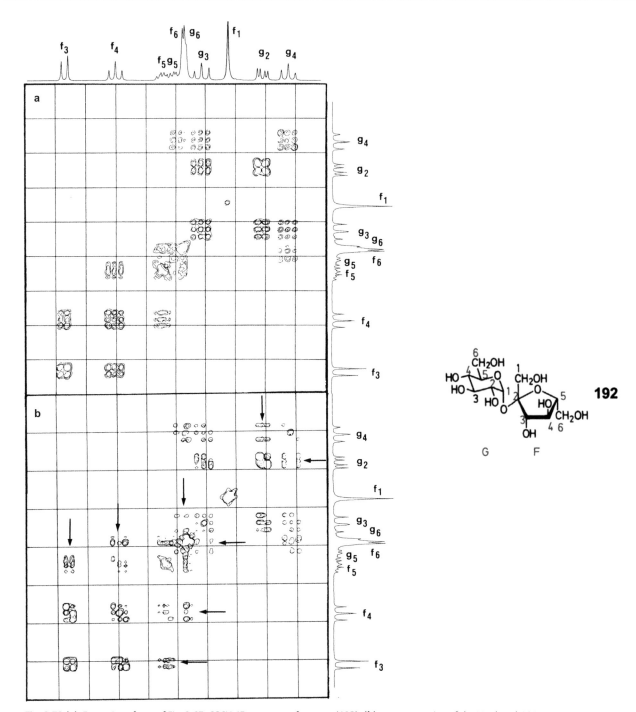

Fig. 3.73 (a) Expansion of part of Fig. 3.67: COSY-45 spectrum of sucrose (**192**); **(b)** same expansion of the H-relayed COSY spectrum.

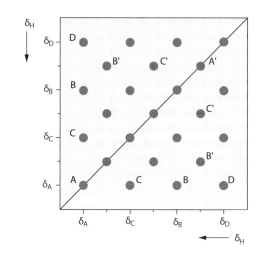

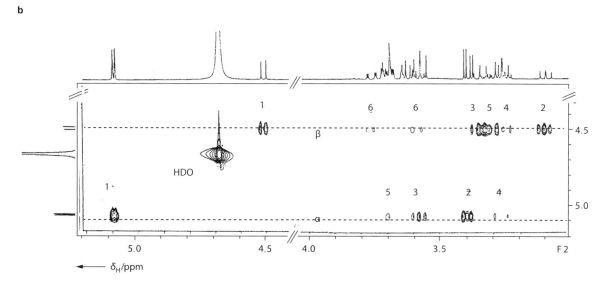

Fig. 3.74 (a) Schematic diagram of a TOCSY spectrum which contains two traces of magnetization transfer: A–B–C–D and A′–B′–C′. These traces can belong to two molecules or to noncoupling parts of one molecule; **(b)** TOCSY spectrum of a mixture of α- and β-D-glucose (**160** and **161**) in D_2O (mixing time: 80 ms).

The TOCSY spectrum does not tell anything about the sequence of the couplings (A with B, B with C, etc.). This information can be obtained from the COSY spectrum. The efficiency of the magnetization transfer depends on the size of the coupling constants. Small coupling constants ($|J| < 4$ Hz)

weaken the transfer or stop it completely ($|J| < 0.5$ Hz). The length of the transfer routes shown in the figure depends on the mixing time, in which a "spin-lock" is reached by a fast sequence of 180° pulses. A mixing time of 20 ms is normally sufficient for a single step. In order to pick up the whole spin systems A–B–C–D and A′–B′–C′, the mixing time must be long enough.

Fig. 3.74b contains parts of the TOCSY spectrum of a nonequilibrated mixture of α- and β-D-glucose. The β-isomer **161** has large coupling constants along its ring, solely the couplings to the exocyclic methylene protons are small. All crosspeaks between 1-H and 2-, 3-, 4-, 5-H, and even 6-H are visible in the depicted spectrum. The α-anomer **160** has a relatively small coupling constant between the equatorial proton 1-H and the axial proton 2-H (see p. 132). Crosspeaks to 2-H and 3-H are clearly visible and those to 4-H and 5-H are already weak. Anomers of glucose as building blocks in di- or oligosaccharides can be identified by this mode.

While (¹H,¹H)COSY and related methods rely on correlation of signals by through-bond couplings, **NOESY** measurements make use of the NOE. The crosspeaks in the 2D spectrum then show the spatial proximity of nuclei. The through-space connectivities are not only useful in difficult structural problems, but also particularly useful for the determination of conformations present in solution. Since the NOE enhancements can be either positive or negative, particular attention must be paid to the possibility of zero values occurring. This difficulty can

be avoided by the **ROESY** experiment (rotating frame NOESY) where the enhancements are always positive. The condensed [18]annulene (**196**) with three phenanthrene systems serves as an example. As in unsubstituted [18]annulenes, the problem is the identification of the inner and outer protons of the 18-membered ring. The crosspeaks in **Fig. 3.75** show, on the one hand, the spatial proximity of 8-H and 28-H and, on the other hand, the proximity of 9-H to 7-H and 1-H. There is no exchange between internal and external protons.

Chemically equivalent or accidentally isochronous protons give only one signal. The NOE between them cannot be measured in NOESY or ROESY spectra. However, such measurements are possible via gs-NOESY or gs-ROESY techniques by means of an isotopomer (cf. p. 220).

The **EXSY** method (**t**wo-**d**imensional e**x**change **s**pectroscopy) uses the magnetization transfer as NOESY does. For this, intramolecular exchange processes are considered, which are based on rotations, inversions, rearrangements, etc., but also intermolecular exchange processes to the point of equilibria of free and complexed or free and aggregated molecules. However, the intra- or intermolecular exchange has to be so slow in the sense of the NMR time scale that separate signals for the exchanging

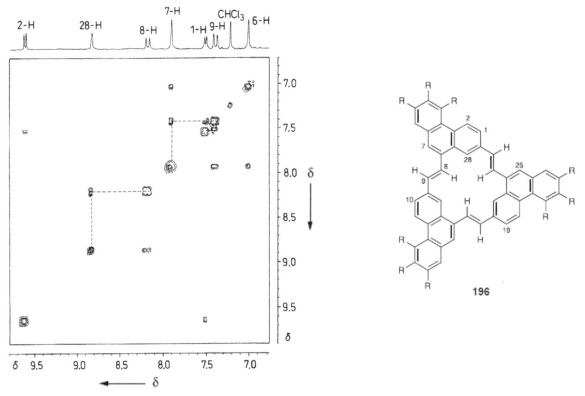

Fig. 3.75 Section of the 2D ROESY spectrum of **196**, measured at 400 MHz in CDCl₃ (Kretzschmann H, Müller K, Kolshorn H, Schollmeyer D, Meier H. 1994, Chem. Ber. **127**, 1735).

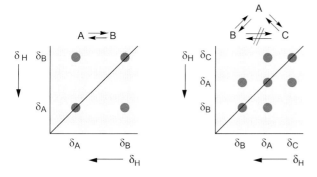

Fig. 3.76 Schematic contour plots for the investigation of exchange processes by applying the EXSY method. The *blue* crosspeaks indicate exchanging signals.

protons exist. The time frame for the rate constant k_{ex} ranges from 10^{-2} to 10^2 s.

The application of the EXSY method can be merely qualitative to demonstrate exchange processes. Quantitative applications enable the evaluation of rate constants k_{ex}.

Fig 3.76 shows a schematic contour plot for an exchange A $\rightleftharpoons$ B (δ_A/δ_B) and for a more complicated system B $\rightleftharpoons$ A $\rightleftharpoons$ C, B $\not\rightleftharpoons$ C $(\delta_A/\delta_B/\delta_C)$. The first case is characterized by two and the second case by four crosspeaks.

N,N-Dimethylacetamide (**197**) represents a simple example for the type A $\rightleftharpoons$ B. The partial double bond character of the CN bond gives rise to two CH₃ signals in the ¹H NMR as well as the ¹³C NMR spectrum at ambient temperature. The exchange process of the two methyl groups can be studied in both spectra.

197

DOSY (**D**iffusion **O**rdered **S**pectroscopy) shall be mentioned here as a further, relatively new 2D technique to study mixtures. The NMR signals of different components are separated according to their diffusion coefficient in solution.

Two-dimensional spectroscopy provides such a wealth of structural information that it has now established itself as one of the foremost experimental methods. Several experiments are now routine NMR techniques, and three-dimensional (3D) spectroscopy is also starting to prove its usefulness as an analytical method (see also p. 221).

3.3.10 Simulation of ¹H NMR Spectra

The interpretation of the spectra of complicated spin systems often requires the use of **simulated spectra** calculated by a computer. Starting with estimated values for the chemical shifts and coupling constants of the nuclei involved, iteration is then used to improve the comparison of the experimental and simulated spectra. Many programs have been developed for such purposes (LAOCOON, etc.). When the observed and simulated spectra agree in both position and intensity, the values of the parameters used for the final simulation can be taken as valid.

¹H NMR spectra of higher order are obtained for compounds which contain chemically equivalent but magnetically nonequivalent protons, and for compounds with $\Delta v / J < 10$ (see p. 93). Both cases shall be discussed here for a four-spin system, each.

The aliphatic protons of the hydrocarbon **48** (p. 172) form an AA′BB′ spin system. The nonequivalence of the four protons shows that the system is rigid and there is no rapid inversion of the ring at room temperature. The spin system can be caused by either a chiral C_2-conformation or an achiral C_s-conformation. The Karplus equation or a modified form thereof (see p. 131) yields very different values of the *vicinal* couplings. From the parameters obtained from a spectral simulation, it is clear that the compound must have the C_2-conformation (**Fig. 3.77**). Spin simulation is also very useful in distinguishing between A_2B_2/A_2M_2 and AA′BB′/AA′MM′ spin systems (cf. p. 136, 137).

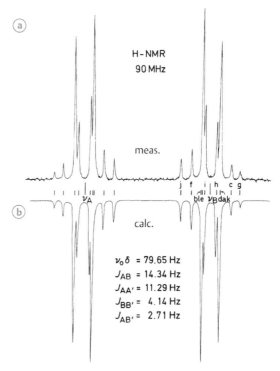

$v_0 \delta$ = 79.65 Hz
J_{AB} = 14.34 Hz
$J_{AA'}$ = 11.29 Hz
$J_{BB'}$ = 4.14 Hz
$J_{AB'}$ = 2.71 Hz

Fig. 3.77 ¹H NMR spectrum of **48** (aliphatic part): (**a**) measured; (**b**) simulated AA′BB′ system (from Meier H, Gugel H, Kolshorn H. 1976. Z. Naturforsch., Part B, **31**, 1270).

Nuclear Magnetic Resonance Spectroscopy

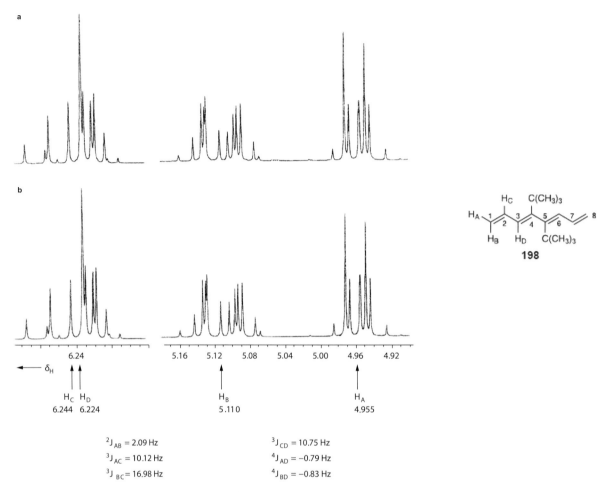

Fig 3.78 ^{1}H NMR resonances of **198** in the region 4 < δ < 7 ppm (400 MHz, CDCl$_3$): (**a**) measured; (**b**) simulated (from Betz M, Hopf H, Ernst L, Jones P.G., Okamoto Y. 2011, Chem. Eur. J. **17**, 231).

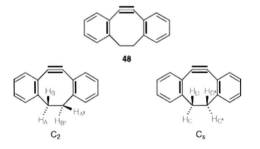

The hydrocarbon **198** [(3*E*,5*E*)-4,5-di-*t*-butylocta-1,3,5,7-tet-raene] gives an ABCD spin pattern in the olefinic part of the 400-MHz ^{1}H NMR spectrum (**Fig. 3.78**). The shift difference between H$_C$ and H$_D$ is small (8 Hz). The torsion at the central bond between C-4 and C-5 leads to two buta-1,3-diene

substructures, between which a coupling cannot be detected. **Fig. 3.78** shows the measured and the calculated spectrum. The exact accordance of both proves the listed chemical shifts and coupling constants.

3.3.11 NMR Spectra of Oriented Phases and Solids

Certain compounds form liquid crystalline states (nematic, smectic, or cholesteric) between the melting point and the clearing point; in these phases the molecules have a preferred orientation. Apart from the **thermotropic liquid crystalline phases**, there are **lyotropic liquid crystalline phases**, which can be formed by amphiphilic compounds, e.g., tensides, and water or other solvents.

Nematic phases are often composed of linear molecules, which are orientated in a specific direction. Guest molecules introduced into the phase then have their Brownian motion restricted and are forced into a similar preferred orientation. In the NMR spectra this preferred orientation is apparent from the direct dipolar couplings between the nuclear spins of the guest molecule; the result is additional splittings of the lines. Partially oriented molecules consequently give multiline spectra, several kHz broad, since the dipolar couplings are much larger than the coupling constants which result from scalar interactions. The analysis of such spectra can yield valuable information about structural parameters such as bond angles and bond lengths, which are related to the geometry of the molecules in the liquid phase.

In going from partially oriented phases to the solid phase, the number of spin–spin interactions increase even more, because now **intermolecular** interactions come into play which for guest molecules in oriented liquid crystalline phases are eliminated by the translational and rotational motions. The signals therefore become very broad in solids. In Section 3.4.10 (see p. 222 ff.), the use of **magic-angle spinning** (**MAS**) to obtain high-resolution spectra from solids is described. Rotation speeds of ca. 30 KHz can produce spectral resolution sufficient for high-resolution ¹H spectroscopy. Particularly useful is the **¹H double quantum MAS technique**.

Finally, it should be mentioned that NMR can be used not only for observations on individual molecules or aggregates of molecules, but that methods such as NMR imaging and nuclear spin tomography have been developed which can provide pictures of macroscopic objects. For biology, and above all for medicinal purposes, these methods have become immensely important, allowing images (cross-sections) of internal organs to be acquired without physical intervention or the use of dangerous radiation.

3.3.12 Combination of Separation Methods with NMR Measurements

The on-line technique combining the separation of complex mixtures with detection and characterization by NMR is an extremely promising recent development, which is especially recommended for small amounts and/or sensitive compounds. Suitable separation methods are high-performance liquid chromatography (HPLC), capillary electrophoresis, permeation chromatography, supercritical liquid chromatography, etc.

As an example, the separation and identification of the *E/Z* isomers of vitamin-A acetate (**199a-e**) will be discussed. **Fig. 3.79** (p. 174) shows the contour diagram of the **on-line HPLC-NMR separation**; the ¹H NMR shifts in the region $4 < \delta < 7$ ppm are plotted against the retention time. The separation was carried out on a modified silica gel column with un-deuterated

n-heptane. The saturated proton region is therefore obscured. Five components are clearly visible in **Fig. 3.79**, the signals of which can be assigned to those of the isomers **199a-e** obtained from the on-flow NMR spectra.

In capillary chromatography, deuterated solvents can be used; in other cases, the unwanted solvent signals must either be suppressed or avoided by appropriate choice of the observation region.

The direct coupling of separation methods and spectroscopic structural analysis is a development that in future will gain in significance.

Fully integrated HPLC-UV/Vis-NMR-MS systems are already available. The separated LC-fractions can either be passed on in parallel to NMR- and MS-measurement or primary mass measurements can be used to decide which fractions to pass on to the NMR measurement.

3.4   ¹³C NMR Spectroscopy

3.4.1   Sample Preparation and Measurement of Spectra

For the measurement of a ¹³C NMR spectrum, the sample needs to be fairly concentrated, but not viscous. As a rule of thumb, a fast routine measurement requires 1 to 2 mg of sample (dissolved in 0.6 mL of solution) for each carbon atom in the molecule. However, when cryo-probeheads or special micro-probeheads are used, 0.1 mg = 100 μg of sample are enough for the measurement of a ¹³C PFT-NMR spectrum (see also Section 3.3.1).

The usual **solvent** is deuteriochloroform. A summary of alternative solvents is given in **Table 3.26**. Deuterated solvents are used to facilitate the measurement by employment of the deuterium resonance of the solvent as a **lock signal** to stabilize the field–frequency relationship in the spectrometer. A further advantage of deuterated solvents is that their ¹³C signals are weaker than those of protonated solvents, as a combined result of the splitting of the ¹³C signal by deuterium (see **Table 3.26**), the lack of the NOE enhancement, and the longer relaxation times T_1. The relatively strong ¹³C signals of protonated solvents or of deuterated solvents in dilute solutions can lead to problems of dynamic range (ability to observe weak signals in the presence of strong ones) in the computer or can obscure weak signals of the sample. The first problem is largely avoided in modern spectrometers. (Additionally, there are instrumental techniques available to reduce the intensity of solvent signals: presaturation, Redfield technique, etc.)

Nuclear Magnetic Resonance Spectroscopy

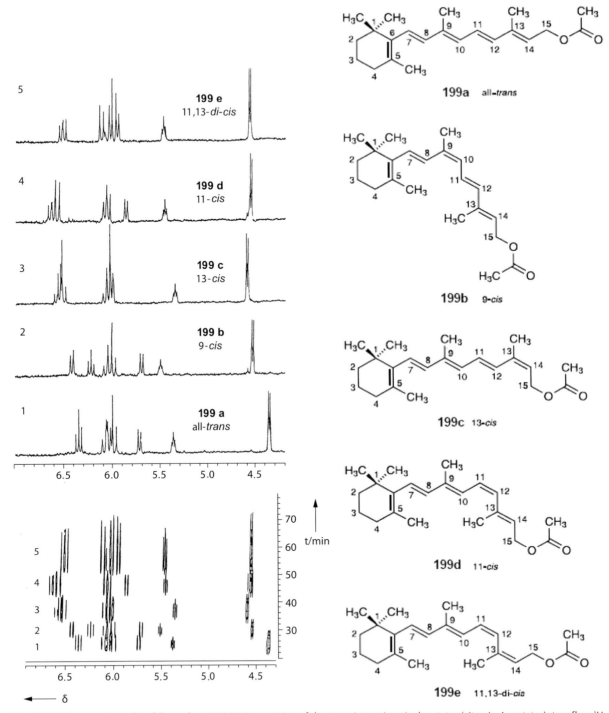

Fig. 3.79 Bottom: contour plot of the on-line HPLC-NMR separation of the stereoisomeric retinol acetates (vitamin-A acetates); top: flow-^{1}H-NMR spectra (olefinic region) of the separated components **199a–e** (from Albert K, Schlotterbeck G, Braumann U, Händel H, Spraul M, Krack G. 1995, Angew. Chem. **107**, 1102).

Table 3.26 Solvents for ¹³C NMR spectroscopy

Solvent	δ (¹³C)	Multiplicity	J(¹³C,D) Hz	
[D$_1$] Chloroform (CDCl$_3$)	77.0	triplet	32	
[D$_4$] Methanol (CD$_3$OD)	49.3	septet	21	T
[D$_6$] Acetone (CD$_3$COCD$_3$)	29.3	septet	20	T
	206.3	multiplet	<1	
[D$_6$] Benzene (C$_6$D$_6$)	128.0	triplet	24	
[D$_2$] Dichloromethane (CD$_2$Cl$_2$)	53.5	quintet	27	T
[D$_3$] Acetonitrile (CD$_3$CN)	1.3	septet	21	
	117.7	multiplet	<1	
[D$_1$] Bromoform (CDBr$_3$)	10.2	triplet	31.5	H
[D$_2$] 1,1,2,2-Tetrachloroethane (C$_2$D$_2$Cl$_4$)	74.0	triplet	28.1	H
[D$_8$] Tetrahydrofuran (C$_4$D$_8$O)	25.5	quintet	21	
	67.7	quintet	22	
[D$_8$] Dioxane (C$_4$D$_8$O$_2$)	66.5	quintet	22	
[D$_6$] Dimethylsulfoxide (CD$_3$SOCD$_3$)	39.7	septet	21	
[D$_5$] Pyridine (C$_5$D$_5$N)	123.5	triplet	25	
	135.5	triplet	24	
	149.5	triplet	27	
[D$_2$] Water (D$_2$O)	–	–	–	
[D$_4$] Acetic acid (CD$_3$COOD)	20.0	septet	20	
	178.4	multiplet	<1	
[D$_{18}$] Hexamethylphosphoric triamide (HMPT, [(CD$_3$)$_2$N]$_3$PO)	35.8	septet	21	H,C
Carbon tetrachloride Tetrachloromethane (CCl$_4$)	96.0	singlet	–	T
Carbon disulfide (CS$_2$)	192.8	singlet	–	T
Trichlorofluoromethane (CCl$_3$F)	117.6	doublet	(^{1}J(C,F) = 337)	
[D$_1$]Trifluoroacetic acid (CF$_3$COOD)	116.5	quartet	(^{1}J(C,F) = 283)	
	164.4	quartet	(^{2}J(C,F) = 44)	

T suitable for low temperature measurements (see Table 3.6)
H suitable for high temperature measurements (see Table 3.6)
C highly carcinogenic

As in ¹H NMR, TMS is used as a **reference** to establish the zero-point of the δ scale (internal or external standard). Frequently but less exactly, the ¹³C signals of the solvent with their known shifts suffice for the determination of the δ values of the sample.

¹³C chemical shifts are presented normally with one decimal digit. A second one does not make much sense because spectral dispersion, digital resolution, and moreover solvent and temperature affect the δ values. Even differences (Δδ) of several ppm are no rarity. Acetic acid ethyl ester (**130**) serves here again as an example how δ values vary in different solvents.

	H$_3$C—COO—CH$_2$—CH$_3$ (130)			
Solvent	δ values			
CDCl$_3$	21.0	171.4	60.5	14.2
CD$_3$OD	20.9	172.9	61.5	14.5
CD$_3$SOCD$_3$	20.7	170.3	59.7	14.4
CD$_3$CN	21.2	171.7	61.0	14.5
CD$_3$COCD$_3$	20.8	171.0	60.6	14.5
C$_6$D$_6$	20.6	170.4	60.2	14.2

Nuclear Magnetic Resonance Spectroscopy

In measurements of ^{13}C spectra of organic compounds there will normally be hydrogen nuclei present in the molecule, which give rise to spin–spin couplings; this means that for a set of isochronous ^{13}C nuclei a multiplet will be expected, caused by *direct, geminal, vicinal,* and *long-range* couplings. The signal intensity will thereby be distributed over several lines. In ^1H NMR, these $^{13}C,^1H$ couplings are normally not apparent because of the low natural abundance of ^{13}C (see however Section 3.3.8, p. 145, on ^{13}C satellites). This problem is generally avoided in ^{13}C NMR by using ^1H **broadband decoupling.** In contrast to **homonuclear spin decoupling** (Section 3.3.8, p. 145), this is a **heteronuclear spin decoupling** technique.

In order to simultaneously remove all the $^{13}C,^1H$ couplings present, a powerful irradiation is used which covers the whole proton chemical shift range. The decoupler frequency is modulated with low-frequency noise. This has led to the use of the term ^1H **noise decoupling** as an alternative term to ^1H **broadband decoupling.** Even more effective methods of decoupling using phase modulation are now replacing noise modulation (e.g., GARP, globally optimized alternating-phase rectangular pulses).

The decoupling effect, as described on p. 155 ff, depends on the coupling ^1H nuclei changing their precession direction (spin orientation) so rapidly as a result of the irradiation at their resonance frequency that all their coupling partners (here the coupling ^{13}C nuclei) only experience an average value of zero. All the lines of the multiplet of a ^{13}C signal form a singlet as a result. Its intensity can amount to 300% of the intensity of the individual lines of the multiplet. Part of this increased intensity is a consequence of the heteronuclear Overhauser effect (**HNOE;** see Section 3.1.5, p. 98).

The ^1H broadband decoupling simplifies the ^{13}C NMR spectra considerably, and the signals gain in intensity. These advantages outweigh the loss of information from $^{13}C,^1H$ coupling so much that routine ^{13}C spectra are always acquired with broadband decoupling. The $^{13}C,D$ couplings remain unaffected. **Fig. 3.80** shows as an example a comparison of coupled and ^1H broadband decoupled ^{13}C spectra for mesitylacetylene (**200**).

In the decoupled spectrum **3.80b**, the signals d, f, and c remain in the same positions as in **3.80a**. They are singlets (s). The corresponding C atoms do not bear hydrogen atoms. In the coupled spectrum **3.80a**, their signals are only broadened by couplings to more distant protons. The signals *e, a,* and *b* each appear in the coupled spectrum as a doublet (d), i.e., the corresponding C nuclei

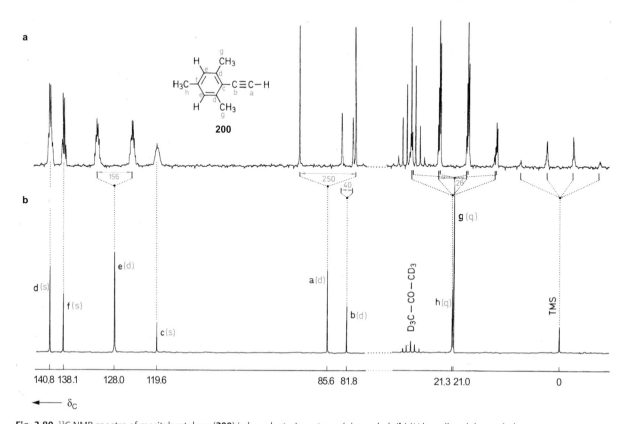

Fig. 3.80 ^{13}C NMR spectra of mesitylacetylene (**200**) in hexadeuterioacetone: (**a**) coupled; (**b**) ^1H broadband decoupled.

couple, apart from long-range couplings, to only one proton. The three coupling constants, 156, 250, and 40 Hz, differ widely. For *e* and *a*, the coupling is to a directly bonded proton [$^1J(C,H)$] for an aromatic carbon and an acetylenic carbon respectively. For *b*, an unusually large $^2J(C,H)$ is observed (see Section 3.4.5, p. 191). The methyl C nuclei *g* and *h* each show a quartet (*q*) from coupling to the three methyl protons. Only in the decoupled spectrum is it easily apparent that there are two signals, with such different intensities that the higher peak can be assigned to the two carbon atoms g. (A CH$_2$ group would give a triplet (*t*), but there is none in this molecule.)

For the coupled spectrum, some 10 times as many scans were accumulated as for the broadband decoupled spectrum, although the concentration of **200** remained unchanged. This is apparent from the increased intensity of the solvent signal (septet for hexadeuterioacetone at δ = 29.3; the CO signal at δ = 206.3 is not reproduced). The measurement of coupled spectra therefore requires a longer measurement time or a higher sample concentration. For a ¹H broadband decoupled spectrum of a reasonably concentrated solution, a few hundred or thousand scans are needed. To determine the time required, it must be remembered that the S/N ratio is proportional to $\sqrt{n}$, where n is the number of scans. The practical effect of this is that if the concentration is halved, then not twice as many, but four times as many scans are necessary to attain the same S/N ratio.

When these considerations are applied to the measurement of fully coupled spectra, it turns out that at a certain concentration about the 10-fold numbers of scans, i.e. the 10-fold time, is required in comparison to the measurement of broadband decoupled spectra.

An increased signal intensity by virtue of the NOE can be attained in coupled ¹³C NMR spectra by the **gated decoupling** technique (cf. p. 208).

3.4.2   ¹³C Chemical Shifts

In the introductory Section 3.1.2 (p. 88), the relationship between the **chemical shift** δ of a nucleus X and the **shielding constant** σ was described.

$$10^{-6} \cdot \delta(X) = \sigma(TMS) - \sigma(X)$$

$$\sigma = \sigma_d + \sigma_p + \sigma'$$

In contrast to ¹H NMR, the σ_{para} term is particularly important in ¹³C NMR. Since this term involves **electronic excitation**, the necessary energy ΔE is involved in σ_{para}. With decreasing ΔE, a down-field shift is observed.

The normal δ (¹³C) scale is about 220 ppm broad. When extreme δ values are taken into account, the shift range is widened to about 400 ppm. In comparison to the δ(¹H) scale, this is a big advantage because chemically nonequivalent ¹³C nuclei are less likely isochronous than ¹H nuclei. How important the ¹³C NMR spectroscopy for structure determinations is becomes particularly apparent for complex molecules in natural product chemistry. **Fig. 3.81** shows the broadband decoupled ¹³C NMR spectrum of quinine ($C_{20}H_{24}N_2O_2$, **201**). The 20 ¹³C signals for the 20 different C atoms are easily discernible.

The dependence of ¹³C chemical shifts on the hybridization becomes evident in the series of the following C_8 hydrocarbons **202** to **205**:

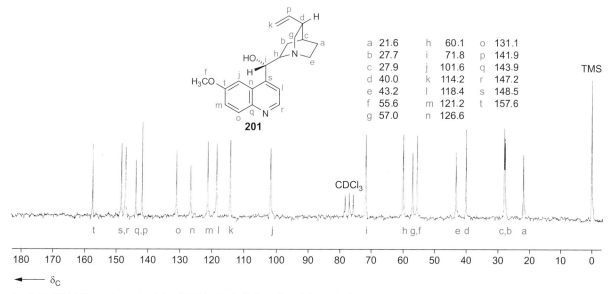

a	21.6	h	60.1	o	131.1
b	27.7	i	71.8	p	141.9
c	27.9	j	101.6	q	143.9
d	40.0	k	114.2	r	147.2
e	43.2	l	118.4	s	148.5
f	55.6	m	121.2	t	157.6
g	57.0	n	126.6		

201

CDCl₃

TMS

t   s,r q,p   o n m l k   j   i   h g,f   e d   c,b a

180 170 160 150 140 130 120 110 100 90 80 70 60 50 40 30 20 10 0

$\longleftarrow$ δ_C

Fig. 3.81 ¹³C NMR spectrum of quinine (**201**) in CDCl₃ (¹H broadband decoupled).

Nuclear Magnetic Resonance Spectroscopy

$H_{13}C_6-CH_2-CH_3$ Octane (**202**)
   22.8   14.1

$H_{13}C_6-CH=CH_2$ Oct-1-ene (**203**)
   139.2   114.1

$H_{13}C_6-C\equiv CH$ Oct-1-yne (**204**)
   84.0   67.8

$H_{11}C_5-CH=C=CH_2$ Octa-1,2-diene (**205**)
   89.5   208.5   74.1

The sequence δ (sp³-C) < δ (sp-C) < δ (sp²-C) in alkanes, alkynes, and alkenes corresponds to the δ (¹H) values of saturated, acetylenic, and olefinic protons. In contrast to the δ (sp-C) values of alkynes, the δ (sp-C) values of allenes and cumulenes are very large.

The chemical shifts of ¹³C nuclei and the protons attached to them often show parallel behavior. However, this comparison should not be taken too far. The ¹³C and ¹H shifts of cyclobutane (**206**) and cyclopropane (**139**), for example, do show parallel behavior.

δ_H: 1.96
δ_C: 23.1

206

δ_H: 0.22
δ_C: −2.8

139

Benzene (**105**) and cyclooctatetraene (**123**) however show opposite effects.

δ_H: 7.26
δ_C: 128.5

105

δ_H: 5.80
δ_C: 131.5

123

The **ring current**, a special case of an anisotropy effect, causes a marked down-field shift in the ¹H NMR of aromatics as compared to olefinic systems. In ¹³C NMR, this effect is apparently negligible for the ring carbon atoms. Olefinic and aromatic carbons give signals in the same region. In the earlier example, the ¹³C shift of cyclooctatetraene is even 3 ppm to low field of that of benzene.

Increasing alkyl substitution generally leads to a down-field shift:

$$\delta\,(CH_4) < \delta\,(C_{prim.}) < \delta\,(C_{sek.}) < \delta\,(C_{tert.}) < \delta\,(C_{quart.})$$

The series of C_8H_{18} alkanes **202**, **207**, and **208** acts here as an example.

$H_{15}C_7-CH_3$ Octane (**202**)
   14.1

$H_{13}C_6-CH_2-CH_3$ Octane (**202**)
   22.8

$H_{11}C_5-CH(CH_3)_2$ 2-Methylheptane (**207**)
   28.0

$H_9C_4-C(CH_3)_3$ 2,2-Dimethylhexane (**208**)
   30.3

An analogous dependence of the ¹³C chemical shift on the number of alkyl substituents can be observed for sp²-C and sp-C atoms:

$H_{13}C_6-CH=CH_2$ Oct-1-ene   (**209**)
   139.2   114.1

$H_{11}C_5-CH=CH-CH_3$
   E   131.0   124.1   (E)-Oct-2-ene (**210**)
   Z   130.5   123.0   (Z)-Oct-2-ene (**211**)

$H_9C_4-CH=C\begin{smallmatrix}CH_3\;131.9\\CH_3\end{smallmatrix}$ 2-Methylhept-2-ene (**212**)
   124.0

$H_{13}C_6-C\equiv CH$ Oct-1-yne (**213**)
   83.1   67.3

$H_{11}C_5-C\equiv C-CH_3$ Oct-2-yne (**214**)
   78.5   74.8

Functional groups normally deshield the ¹³C nucleus they are directly bound to. **Table 3.27** shows the effect of various substituents on a C_8 chain. The electronegativity of the key atom causes a down-field shift for the α-C-atom. An exception is given for iodine by the **heavy-atom effect**. The anisotropy effect of triple bonds in ethynyl and cyano groups induces relatively low δ values for the α-C-atoms.

The comparison with the δ values of n-octane reveals that the resonance of the β-C-atoms is much less down-field shifted by substituents in the 1-position.

The up-field shift of γ-C-atoms, called **γ-effect**, results from steric compression of a gauche interaction and from hyperconjugation in the anti-conformation. The influence of the substituents on the chemical shifts of δ-C atoms and higher positions is very small (**Table 3.27**).

Two or three substituents X on the same carbon atom induce effects which are not always strictly additive. **Table 3.28**

Table 3.27 Influence of terminal substituents on the δ (^{13}C) values of saturated C_8 chains

Substituent	α $-CH_2$	β CH_2	γ CH_2	δ CH_2	ε CH_2	ς CH_2	η CH_2	θ CH_3
H-	14.1	22.8	32.1	29.5	29.5	32.1	22.8	14.1
F-	84.2	30.6	25.3	29.3	29.3	31.9	22.7	14.1
Cl-	45.2	32.8	27.0	29.0	29.2	31.9	22.8	14.1
Br-	33.8	33.0	28.3	28.8	29.2	31.8	22.7	14.1
I-	6.9	33.7	30.6	28.6	29.1	31.8	22.7	14.1
HO-	63.1	32.9	25.9	29.5	29.4	31.9	22.8	14.1
H$_3$C-O-	73.1	31.9	26.3	29.7	29.7	31.9	22.7	14.2
H$_3$C-CO-O-	64.3	29.0	25.7	29.0	28.4	31.6	22.5	14.0
H$_2$N-	42.4	34.1	27.0	29.6	29.4	31.9	22.7	14.1
(H$_3$C)$_2$N-	60.0	29.6	27.7	27.7	29.3	31.9	22.7	14.1
H$_3$C-CO-NH-	39.3	29.1	26.6	29.1	28.9	31.7	22.2	14.0
O$_2$N-	75.8	26.2	27.9	29.6	29.6	31.4	22.7	14.0
HS-	24.7	34.2	28.5	29.2	29.2	31.9	22.8	14.1
H$_2$C=CH-	34.5	29.5	29.5	29.6	29.6	32.2	23.0	13.9
HC≡C-	18.3	28.5	28.5	28.9	29.0	31.7	22.5	13.9
H$_5$C$_6$-	36.2	31.7	29.6	29.7	29.6	31.1	22.8	14.1
OHC-	44.0	22.2	29.3	29.4	29.3	31.9	22.7	14.1
HOOC-	34.2	24.8	29.2	29.2	29.2	31.9	22.7	14.1
NC-	16.9	25.1	28.6	29.4	28.8	31.7	22.5	13.9

Table 3.28 ^{13}C Signals of halogenated methanes

CH$_{4-n}$X$_n$	X = F	Cl	Br	I
CH$_3$X	75.0	24.9	9.8	− 20.8
CH$_2$X$_2$	109.0	54.0	21.4	− 54.0
CHX$_3$	116.4	77.0	12.1	− 139.9
CX$_4$	118.6	96.5	− 29.0	− 292.5

demonstrates this for halogenated methanes. An increasing number of F or Cl substituents causes an increasing down-field shift. An opposite effect can be observed for increasing numbers of I substituents. Tetraiodomethane shows its resonance at an extremely high field. The trend is not uniform for brominated methanes. Electronic effects and heavy-atom effects are antagonists.

The sequential substitution of the hydrogen atoms of CH$_4$ by NO$_2$ (**215a–d**) has a similar effect as the substitution by fluorine. An increasing number of ethoxy groups on a carbon atom leads to an increasing down-field shift for this C atom. The δ (^{13}C) values of the ethoxy groups themselves exhibit small up-field shifts in the series methyl ethyl ether (**216a**), formaldehyde

diethyl acetal (**216b**), orthoformic acid triethyl ester (**216c**), and orthocarbonic acid tetraethyl ester (**216d**). Due to the anisotropy effect, the cyano group leads to a relatively high δ value for the neighboring α-C atom. Nevertheless, an increasing number of CN groups on a carbon atom causes a slight deshielding of this C atom. The δ values of the cyano groups themselves are up-field shifted in the series acetonitrile (**217a**), malonic acid dinitrile (**217b**), tricyanomethane (**217c**, unstable at room temperature), and tetracyanomethane (**217d**).

		215	216			217	
		H$_{4-n}$C(NO$_2$)$_n$	H$_{4-n}$C(O − CH$_2$ − CH$_3$)$_n$			H$_{4-n}$ C(CN)$_n$	
a	$n = 1$	61.2	59.1	69.2	16.1	1.8	116.5
b	$n = 2$	98.4	94.9	63.1	15.3	8.6	110.5
c	$n = 3$	114.2	112.9	59.5	15.2	16.9	106.1
d	$n = 4$	118.4	119.7	58.3	14.8	22.7	103.7

A detailed assessment of the substituent effects on the δ (^{13}C) values of primary, secondary, tertiary, and quaternary C atoms can be made by means of the empirical increment system (Section 3.4.4, p. 189). In principle, all effects discussed earlier are also valid for saturated carbocycles. **Table 3.29** contains the

Table 3.29 ^{13}C Chemical shifts of monosubstituted cyclohexanes

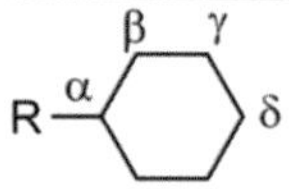

45 (R = H, δ = 26.9)

R—	α	β	γ	δ
F—	90.5	33.1	23.5	26.0
Cl —	60.0	37.3	25.2	25.6
Br—	52.6	37.9	26.1	25.6
I—	31.7	39.7	27.3	25.5
HO —	70.1	36.0	25.0	26.4
H$_3$CO—	78.8	32.3	24.4	26.7
H$_3$C—CO—O—	72.3	32.2	24.6	26.1
H$_2$N—	51.1	37.6	25.8	26.2
(H$_3$C)$_2$N—	64.3	29.2	26.5	26.9
O$_2$N—	84.6	31.4	24.7	25.5
HS —	38.4	38.1	26.6	25.3
H$_2$C=CH—	42.0	32.4	26.0	26.5
HC≡C—	28.7	32.4	25.8	24.8
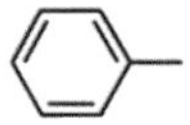 (phenyl)—	44.7	34.6	27.0	26.2
O=HC—	50.1	25.8	25.3	26.3
HOOC—	43.1	29.0	25.4	25.8
N≡C—	27.7	29.2	24.6	24.4

δ (^{13}C) values of cyclohexane and monosubstituted cyclohexanes. It is assumed that the substituents listed in the table preferentially have an equatorial position.

The equilibrium of conformers shall be discussed here for chlorocyclohexane (**218**). A **218a**:**218b** ratio of 80:20 exists at −85°C in CDCl$_3$/CFCl$_3$ (1:1).

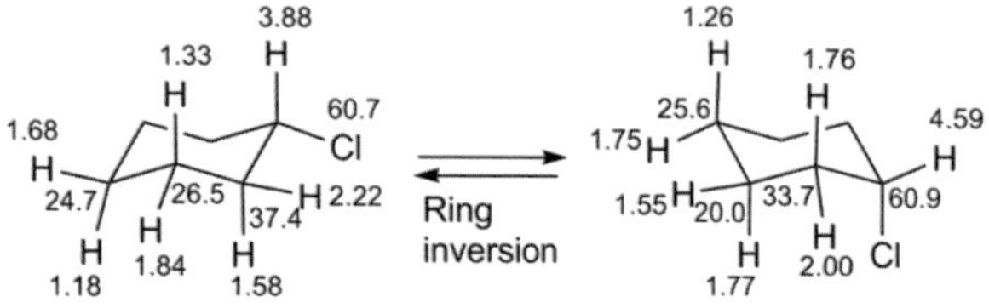

218a

Ring inversion

218b

80 : 20

The fast equilibration at room temperature leads to weighted averages of the δ values of both conformers. The diastereotopicity of the geminal protons is maintained when all equatorial positions become axial and vice versa (ring inversion). The largest difference Δδ (^{13}C) between the two diastereomers is observed for the γ-position. The fast equilibration leads to the average δ value shown in **Table 3.29** (26.5 × 80/100 + 20.0 × 20/100 = 25.2).

Steric factors play an important role as demonstrated here by *cis*- and *trans*-decahydronaphthalene (**219** and **220**). Particularly large Δδ values are found for *exo*- and *endo*-tricyclo[3.2.1.0$^{2.4}$]octane (**221** and **222**) and *exo*- and *endo*-3-oxa-tricyclo[3.2.1.0$^{2.4}$]octane (**223** and **224**).

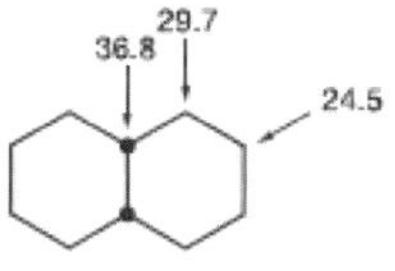

219
cis

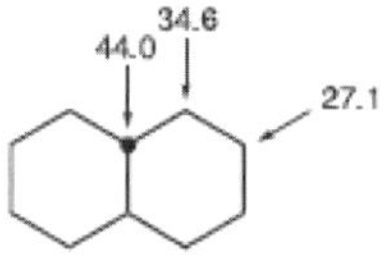

220
trans

221
exo

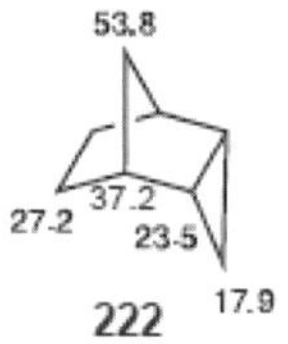

222
endo

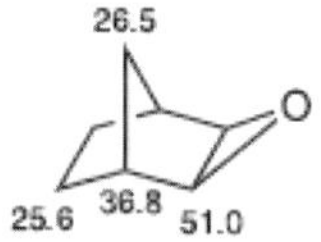

223
exo

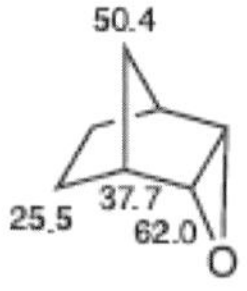

224
endo

Diastereomers in the open chain series often have smaller Δδ (^{13}C) values. Nevertheless, these differences are important for the identification of the stereoisomers. 2,3-Dibromosuccinic acid (**225**) exists as a pair of enantiomers [(*R,R*)- and (*S,S*)-**225**] and the diastereomeric *meso*-isomer (*meso*-**225**).

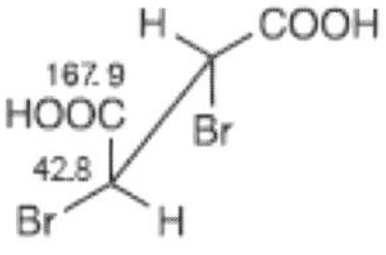

(*R,R*)-**225**
C_2

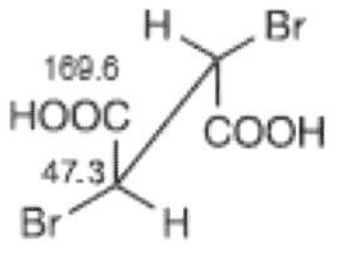

meso-**225**
C_i

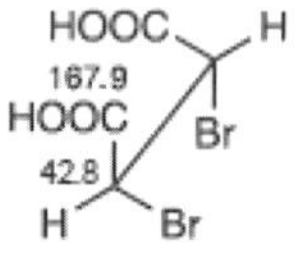

(*SS*)-**225**
C_2

The influence of electron densities on the $\delta\,(^{13}C)$ values of olefinic and acetylenic carbon atoms is particularly remarkable when strong electron-donor groups D and/or electron-acceptor groups A are present. A corresponding discussion has already been made in the ^{1}H NMR section (p. 123f). If for example ethylene is compared with its formyl or methoxy derivatives, effects are observed which are typical for unsaturated carbonyl compounds (e.g., acrolein, **78**) and enol ethers (e.g., methyl vinyl ether, **226**)

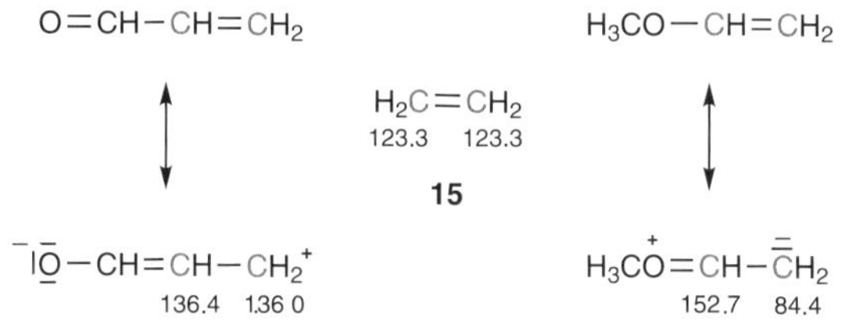

$$O=CH-CH=CH_2 \qquad H_3CO-CH=CH_2$$

$$\begin{array}{c} H_2C=CH_2 \\ 123.3 \quad 123.3 \end{array}$$

15

$$\bar{|\underline{O}}-CH=CH-CH_2^+ \qquad H_3\overset{+}{C}O=CH-\overset{=}{C}H_2$$
$$\begin{array}{cc} 136.4 & 136\ 0 \end{array} \qquad \begin{array}{cc} 152.7 & 84.4 \end{array}$$

78 **226**

An analogous comparison can be made for (E)-but-2-ene (**227**), (E)-pent-3-en-2-one (**228**), and (E)-1-methoxypropene (methyl propenyl ether, **229**).

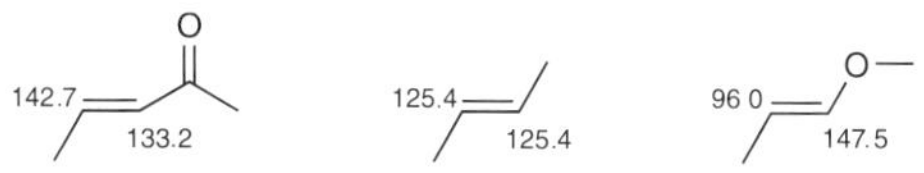

228 **227** **229**

While the inductive effect is largely effective in the α-position, the mesomeric effect causes in the β-position a decrease of the electron density in enones and an increase in enol ethers. The stronger deshielding is apparent as a down-field shift, and the stronger shielding as an up-field shift. Extreme cases are observed for keteneacetals and related compounds:

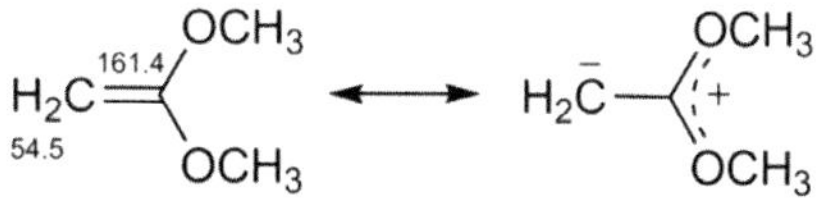

1,1-Dimethoxyethene (**82**)

6,6-Bis(dimethylamino)pentafulvene (**230**)

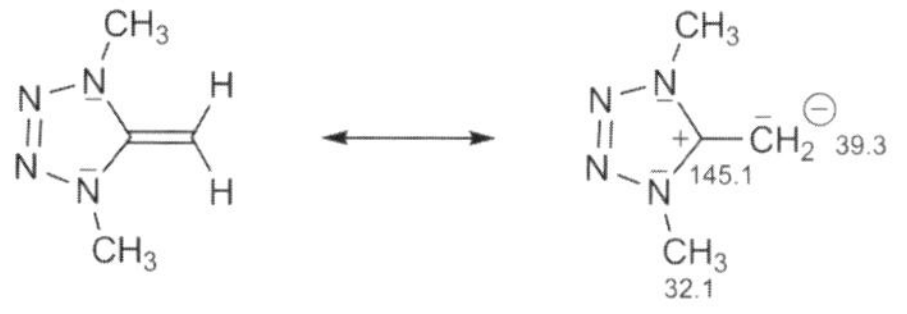

1,4-Dimethyl-5-methylenetetrazole (**231**)

The aromatic character of the five-membered rings in **230** and **231** plays a minor role. Two acceptor groups are present in the diester **80**.

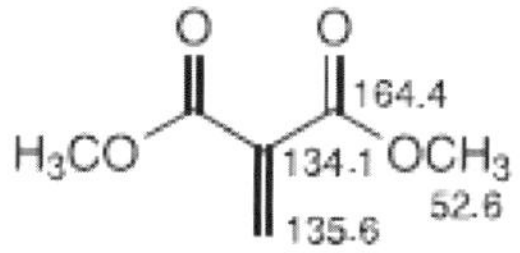

2-Methylenemalonic acid dimethyl ester (80)

Electron-releasing and electron-withdrawing effects are superimposed in push–pull systems, such as **54**, **232**, and **233**, and in captodative compounds, such as **234**.

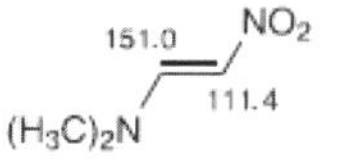

N,N-Dimethyl-2-nitro-1-ethenylamine (**54**)

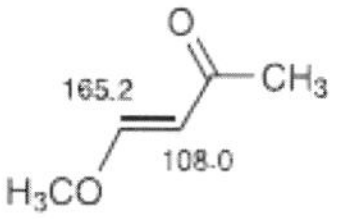

1-Methoxybut-1-en-3-one (**232**)

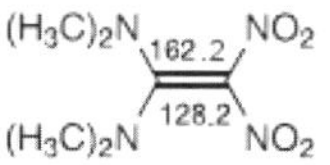

1,1-Bis(dimethylamino)-2,2-dinitroethene (**233**)

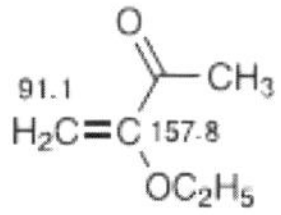

2-Ethoxybut-1-en-3-one (**234**)

When cyano groups represent the electron-withdrawing groups of push–pull compounds, one olefinic signal at very low and one signal at unusually high field can be observed:

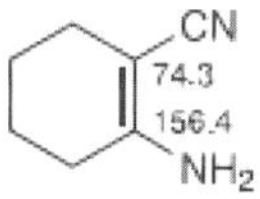

(**235**)

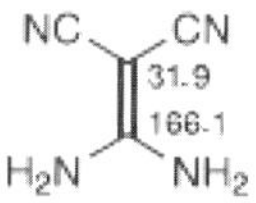

(**236**)

A reliable estimation of the substituent effect on olefinic carbon atoms is granted by the empirical increment system in **Table 3.41** (p. 193). The influence of substituents not directly bound to CC double bonds is much smaller. **Table 3.30** shows the δ values of allyl compounds.

The ring size of cycloalkenes has a dramatic influence on the δ values of the olefinic C atoms:

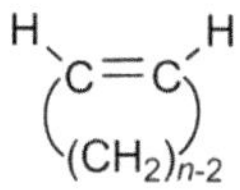

237	**a**	**b**	**c**	**d**
n	3	4	5	6
δ	108.7	137.2	130.6	127.4

Table 3.30 ¹³C chemical shifts of 3-substituted propenes

	H₂C=CH–CH₂–R			
R	H₂C=	=CH	CH₂	R
H	115.9	133.4	19.4	
Br	118.9	134.4	32.7	
OH	115.1	137.3	63.7	
NH₂	113.6	139.9	44.8	
CN	119.6	125.8	21.4	117.1
COCH₃	118.9	130.5	48.6	206.8; 29.5

The ring strain in *trans*-cycloalkenes has a minor influence on the ¹³C chemical shifts:

cis-Cyclooctene (238)   *trans*-Cyclooctene (239)

Normal steric effects and strain are superimposed in the ¹³C chemical shifts of the *anti*-Bredt systems **240** and **241**:

Bicyclo[3.3.1]non-1-ene   9-Oxabicyclo[4.2.1]non-6-ene
(240)   (241)

The influence of electron-withdrawing and electron-releasing substituents on CC triple bonds is similar to their influence on CC double bonds:

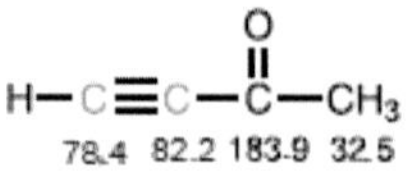

H–C≡C–C(=O)–CH₃   HC≡CH   H–C≡C–O–CH₂–CH₃
78.4 82.2 183.9 32.5   71.9   26.4 90.6   72.6   10.0

83 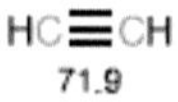   84 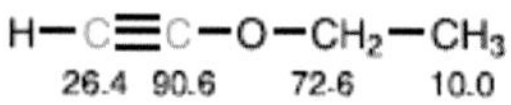   85

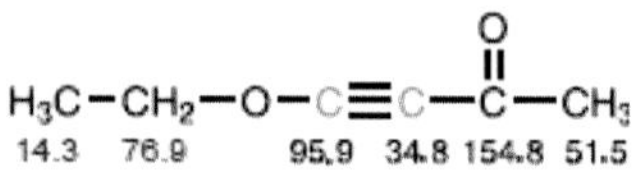

H₃C–CH₂–O–C≡C–C(=O)–CH₃
14.3   76.9   95.9 34.8 154.8 51.5

242

(H₃C)₂N–C≡C–C≡C–C(=O)–CH₃
42.5   100.0 53.5 83.0 85.1 183.3 32.0

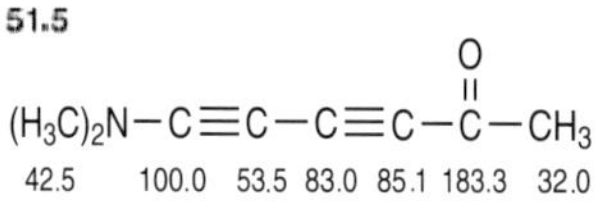

243

However, CC triple bonds cause an up-field shift for sp³-, sp²-, and sp-C atoms in the α-position:

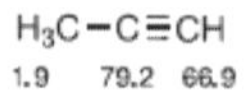

H₃C–C≡CH   H₂C=CH–C≡CH
1.9   79.2 66.9   129.3 117.3 82.9 80.1
        100

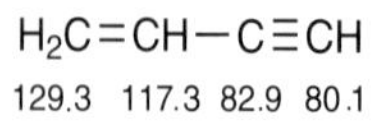

244

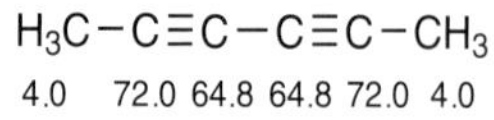

H₃C–C≡C–C≡C–CH₃
4.0   72.0 64.8 64.8 72.0 4.0

245

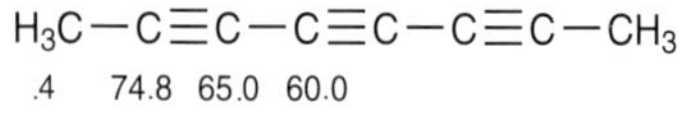

H₃C–C≡C–C≡C–C≡C–CH₃
.4   74.8 65.0 60.0

246

247 (30.2, 77.8, 70.1)

248 (80.1, 89.8, 119.1)

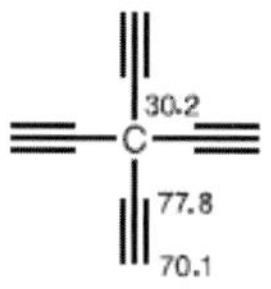

An analogous statement is valid for CN triple bonds (see also compounds **235** and **236**):

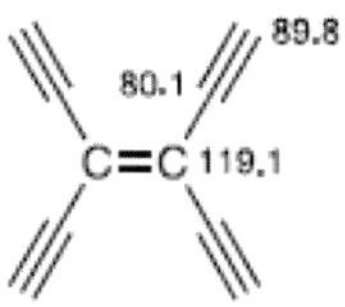

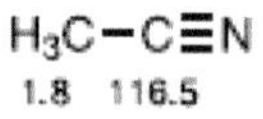

H₃C–C≡N   H₂C=CH–C≡N   HC≡C–C≡N
1.8 116.5   137.5 107.8 117.2   73.8 57.2 104.6

216a 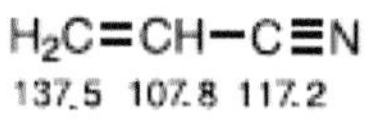   249 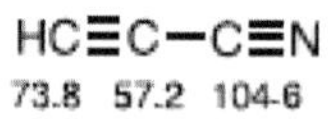   250

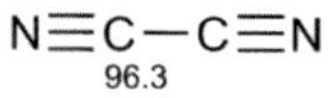

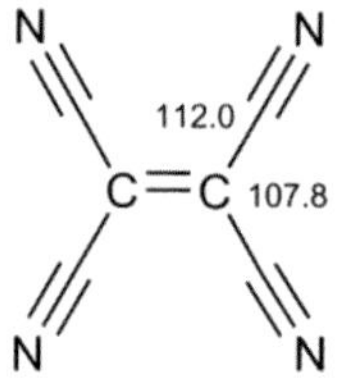

N≡C–C≡N
96.3

251

252 (112.0, 107.8)

The ring strain in cycloalkynes leads to a down-field shift for the sp-C atoms:

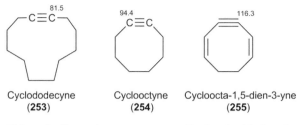

Cyclododecyne (**253**) Cyclooctyne (**254**) Cycloocta-1,5-dien-3-yne (**255**)

Although allenes have two perpendicular double bonds, the effect of polar substituents affects both double bonds (**Table 3.31**).

A low ring strain, as in cyclonona-1,2-diene (**257**), has a minor effect on the δ (¹³C) values:

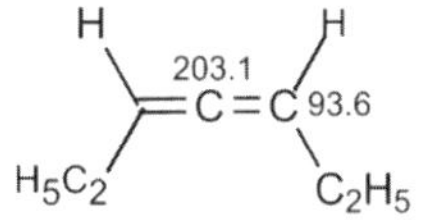

Hepta-3,4-diene (**256**) Cyclonona-1,2-diene (**257**)

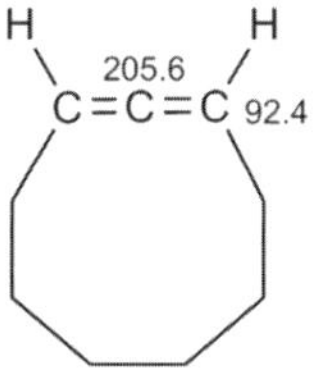

An extremely high ring strain together with a substitution by heteroatoms is present in **258**:

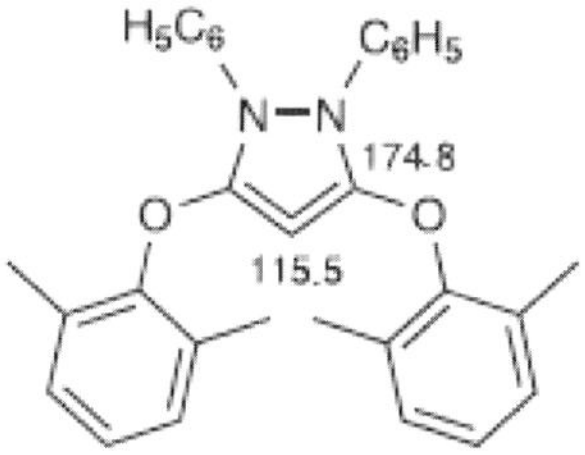

258

Mesomeric and inductive effects change the charge densities and thus the ¹³C shifts in a characteristic fashion. The examples of nitrobenzene (**259**) and the benzenediazonium salt **260**, however, show that only the δ values in the p-position are reliable indicators of changes in charge densities. In the o-position, other factors need to be considered.

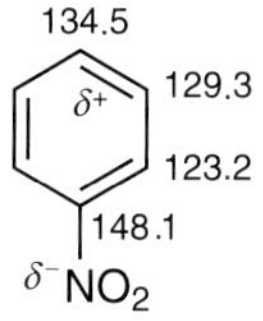

259 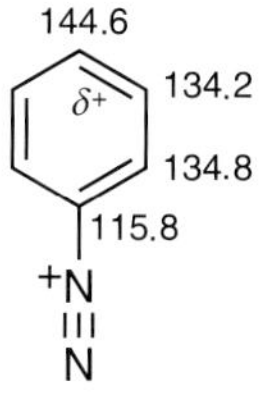 **260**

Aniline (**261**) as an electron-rich ring system and its electron-poor trimethylammonium salt **262** or pyridine (**262**) and its

Table 3.31 δ (¹³C) values of allenes

Allene (Propa-1,2-diene)	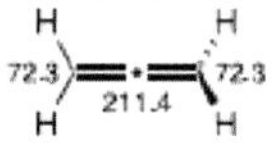 H, H, 72.3, 211.4, 72.3, H, H
3-Methylbuta-1,2-diene	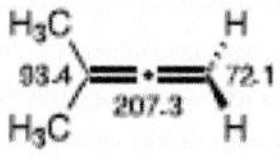 H_3C, H, 93.4, 207.3, 72.1, H_3C, H
1,1-Difluoroallene	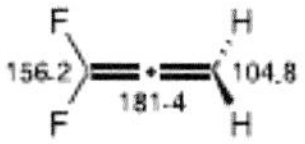 F, H, 156.2, 181.4, 104.8, F, H
Tetrafluoroallene	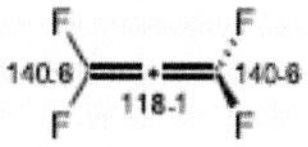 F, F, 140.6, 118.1, 140.6, F, F
Tetramethoxyallene	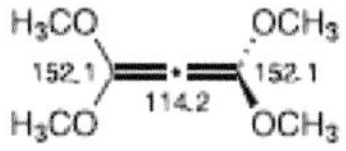 H_3CO, OCH_3, 152.1, 114.2, 152.1, H_3CO, OCH_3
1,1-Diethoxy-4,4,4-trifluoro-3-trifluoromethylbuta-1,2-diene	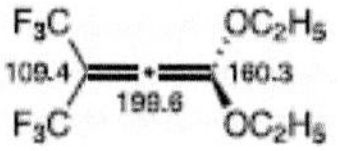 F_3C, OC_2H_5, 109.4, 198.6, 160.3, F_3C, OC_2H_5

N-oxide **263** are further examples how substituents change the electron density and accordingly the ¹³C chemical shifts:

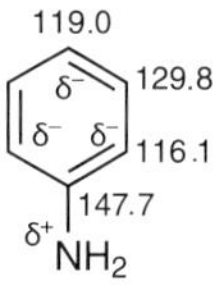

261 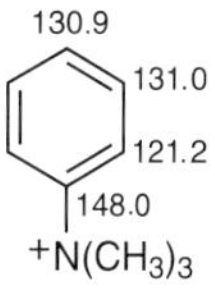 **262**

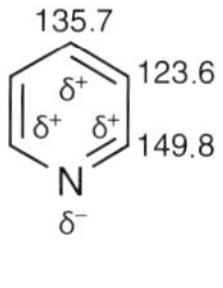

118 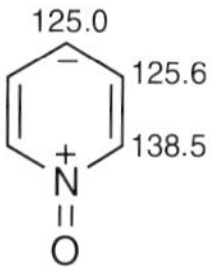 **263**

The push–pull substitution of benzene derivatives is demonstrated here by 4-nitrophenol (**96**) and its isomers **264** and **265**. The mesomeric interaction of donor and acceptor groups is present in **96** and **265**; in the $ortho$-isomer it is superimposed by the effect of the intramolecular hydrogen bond.

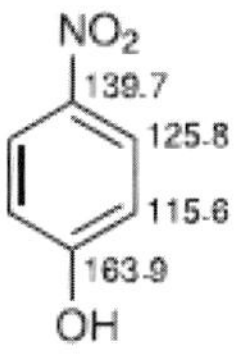

96 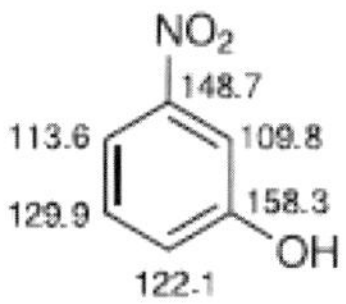 **264** 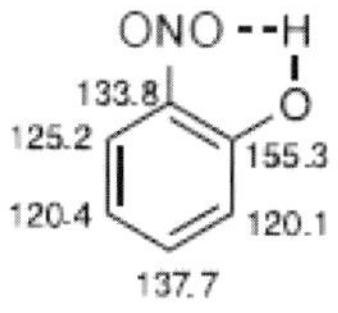 **265**

The terminal push–pull substitution of conjugated oligomers has a far-reaching effect on the chemical shifts. A comparison of the all-(*E*)-configured oligo(1,4-phenylenevinylene)s **267** to **269** with (*E*)-stilbene (**266**) reveals an alternating polarization, which is in accordance with quantum-chemically calculated partial charges.

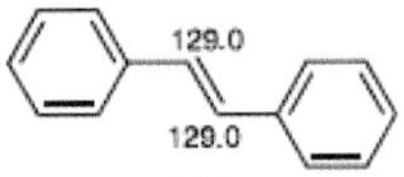

266

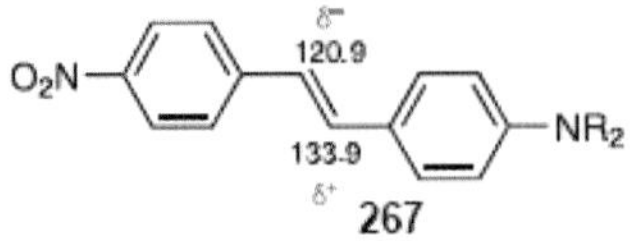

267

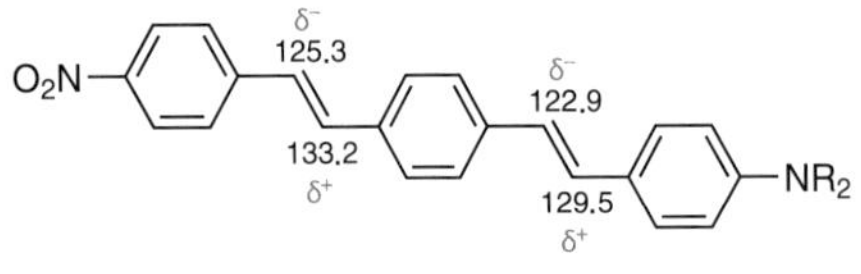

268

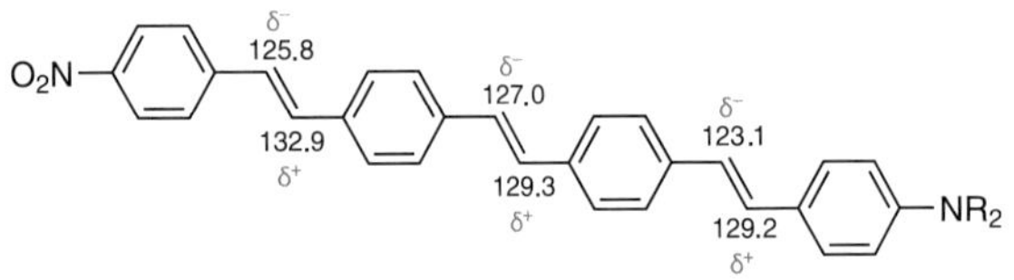

269

A reliable estimation of the effect of substituents, which are directly bound to a benzene ring, is possible by application of the empirical increment system (**Table 3.43**, p. 194). The influence of substituents in the α-position of toluene is shown in **Table 3.32**.

Anisotropy effects play an important role in the ^{1}H NMR spectroscopy of aromatics. The ring current effect is however negligible in the ^{13}C NMR spectroscopy. Bicyclo[10.2.2]-1(15),12(16),13-triene (**109**) is an instructive example. While the δ(^{1}H) values of the methylene groups decrease from the α- to the ε-position, the corresponding δ(^{13}C) values do not manifest a uniform trend.

δ(^{1}H)
δ(^{13}C)

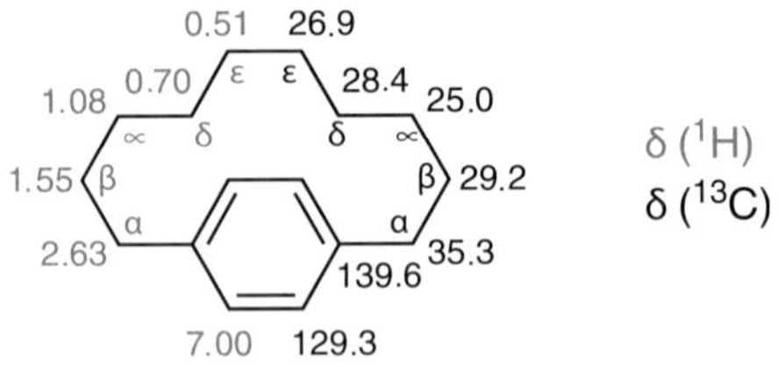

109

How the strain of aromatic compounds affects the ^{13}C chemical shifts is not always easy to discriminate. [2.2](1,4) Cyclophane (**271**) and superphane (**272**) are here compared to dibenzyl (**270**):

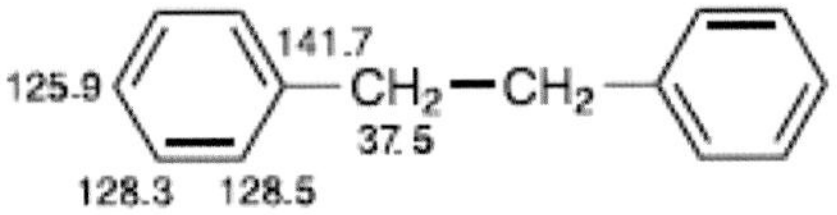

270

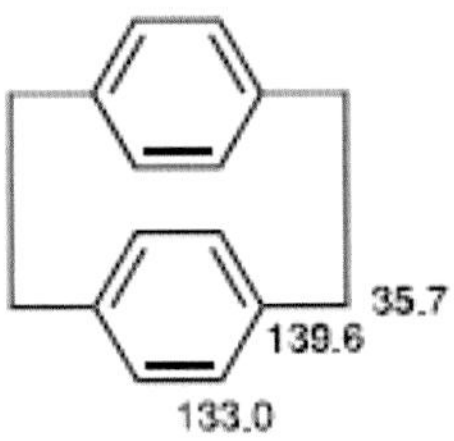

271

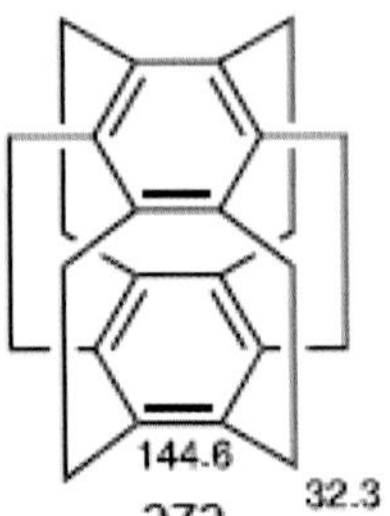

272

The *ansa* compounds **273**, **274**, and **109** represent further examples:

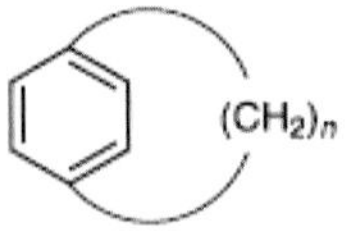

	n	δ(C_i)	δ(CH)
273	6	143.3	131.8
274	8	140.5	129.9
109	10	139.6	129.3

Depending on the class of compounds, the δ(^{13}C) values of carbonyl C-atoms vary between 140 and 220 ppm (**Table 3.33**).

Heteroatoms bound to the CO group cause smaller δ values. The same effect can be observed for unsaturated or aromatic groups. The conjugation has in 1,3-dienes a much smaller influence than in enones:

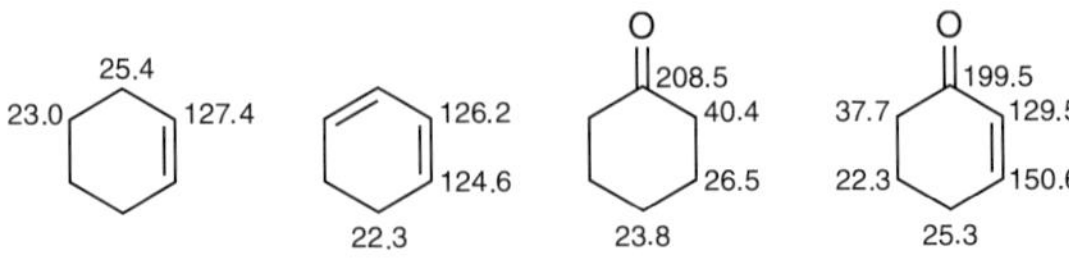

153

104

275

276

Table 3.32 ^{13}C Chemical shifts of α-substituted toluenes

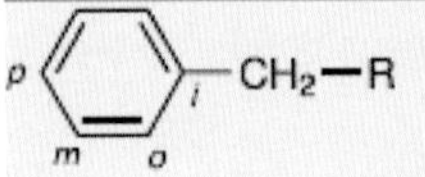

R	*i*-C	*o*-C	*m*-C	*p*-C	CH$_2$	R
H	137.7	129.3	128.5	125.6	21.3	
I	138.5	128.5	128.5	127.6	5.9	
OH	140.8	126.8	128.2	127.2	64.5	
NH$_2$	143.4	126.9	128.3	126.5	46.3	
CN	130.0	129.1	127.9	128.0	23.5	117.8
COOH	133.4	129.4	128.6	127.3	41.0	178.4

Table 3.33 ¹³C Chemical shifts of carbonyl C-atoms in various types of compounds

Class of compound	R—CO—X	R: CH₃	CH=CH₂	C₆H₅
Ketones	R—CO—CH₃	206.0	197.5	195.7
Aldehydes	R—CO—H	199.7	193.3	197.6
Thiocarboxylic acid S-ethyl esters	R—CO—SC₂H₅	195.0	190.2	191.2
Carboxylic acid S-anhydrides	R—CO—SCOR	191.7	190.0	190.1
Carboxylates	R—CO—O⁻	181.7	177.8	175.5
Carboxylic acids	R—CO—OH	178.1	171.9	172.6
Carboxylic acid amides	R—CO—NH₂	172.7	171.5	169.7
Carboxylic acid methyl esters	R—CO—OCH₃	170.7	166.4	166.8
Carboxylic acid chlorides	R—CO—Cl	170.5	166.3	168.0
Carboxylic acid anhydrides	R—CO—OCOR	166.9	162.2	162.9

Class of compound	X—CO—X/Y			
Dithiocarbonic acid esters	H₃CS—CO—SCH₃	189.8		
Thiocarbonic acid esters	H₅C₂S—CO—OC₂H₅	170.7		
Ureas	H₂N—CO—NH₂ (H₃C)₂N—CO—N(CH₃)₂	161.2 165.4		
Urethanes	H₃CNH—CO—OC₂H₅	157.8		
Carbonates	H₃CO—CO—OCH₃	156.5		
Chloroformates	Cl—CO—OC₂H₅	149.9		
Phosgene	Cl—CO—Cl	142.1		

CF₃, CCl₃, and CO groups in the α-position to the carbonyl carbon atom induce a strong up-field shift for δ(CO):

H₃C—CO—CH₃
206,6 30,6

277

H₃C—CO—CF₃
23,1 187,4 115,6

278

Cl₃C—CO—CCl₃
175,5 90,2

279

H₃C—CO—CO—CH₃
197,7 23,2

280

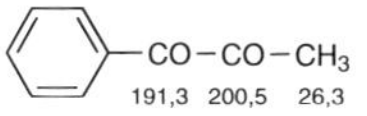

[phenyl]—CO—CO—CH₃
191,3 200,5 26,3

281

[phenyl]—CO—CO—[phenyl]
194,3

282

Branched alkyl groups, on the other hand, lead to a down-field shift:

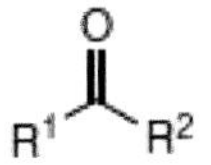

		R¹	R²	δ(CO)
	277	CH₃	CH₃	206.6
	283	CH₃	C(CH₃)₃	212.8
	284	C(CH₃)₃	C(CH₃)₃	218.0

The situation for the cycloaliphatic ketones is more complex, because the ring size has a strong influence (**Fig. 3.82**).

In the series of cycloalkane-1,2-diones (**286**), the ring size determines the enolization tendency. Cyclooctane-1,2-dione exists as diketone, whereas the monoenol tautomers predominate

for the corresponding five- and six-membered ring systems. Cycloheptane-1,2-dione and 2-hydroxycyclohept-2-en-1-one equilibrate in CDCl₃ at 20°C to the ratio 60:40.

286	n	C-1	C-2	C-3	C-4	C-5	C-6	C-7	C-8
a	8 A	209.4	209.4	39.5	20.9	26.0	26.0	20.9	39.5
b	7 A	204.6	204.6	40.0	24.5	29.2	24.5	40.0	–
b	7 B	198.4	149.1	118.2	25.2	25.1	20.6	40.1	–
c	6 B	195.5	147.0	118.3	23.7	23.1	36.4	–	–
d	5 B	204.6	153.6	130.6	21.9	32.4	–	–	–

Cyclic and acyclic 1,3-diketones exist in tautomeric equilibria (see, for example, **Fig. 3.25**, p. 111); 1,4- and higher diketones have an enol proportion which resembles the corresponding monoketones.

When the O atom of carbonyl groups is replaced by NR or S, significant changes of the δ values occur. The sequence δ(C=N) < δ(C=O) < δ(C=S) can be observed.

Nuclear Magnetic Resonance Spectroscopy

The compounds **287** to **298** demonstrate the wide range of δ (C=N) between 130 and 170 ppm.

H$_5$C$_6$ C$_6$H$_5$
C=N
168.5
H$_5$C$_6$

N-(Diphenylmethylidene)aniline (**287**)

H$_3$C   CH$_3$
C=N
168.0
H$_3$C

N-(Isopropylidene)methylamine (**288**)

H$_3$C
C=NH
165.4
H$_2$N

Acetamidine (**289**)

H$_3$C   CH$_3$
CH=N
162.7
H$_3$CO

(*E*)-*N*-Methylacetimino acid methyl ester (**290**)

H$_2$N
C=NH
159.6
H$_2$N

Guanine (**291**)

H$_5$C$_6$
C=N   C$_6$H$_5$
H$_5$C$_6$ N=C
159.1   C$_6$H$_5$

Benzophenone azine (**292**)

H$_2$C=N
155.4   C$_6$H$_5$

N-Methyleneaniline (**293**)

H$_3$C   OH
C=N
155.4
H$_3$C

Acetone oxime (**294**)

H$_3$C   NH$_2$
C=N
152.4
H$_3$C

Acetone hydrazone (**295**)

H$_3$C   OH
C=N
C=N
153.3
H$_3$C   OH

Butane-2,3-dione dioxime (**296**)

H   NH$_2$
C=N
146.1
H$_3$C

(*E*)-Acetaldehyde hydrazone (**297**)

H   N(CH$_3$)$_2$
C=N
134.9
H$_3$C

(*E*)-Acetaldehyde *N,N*-dimethylhydrazone (**298**)

The differences of the chemical shifts for (*E*)- and (*Z*)-configurations are demonstrated for acetaldehyde oxime (**299**):

H   OH
15.0   C=N
147.9
H$_3$C

(*E*)-**299**

H$_3$C   OH
11.2   C=N
147.6
H

(*Z*)-**299**

The δ (CS) values of the thiocarbonyl compounds **300** to **309** vary between 180 and 260 ppm.

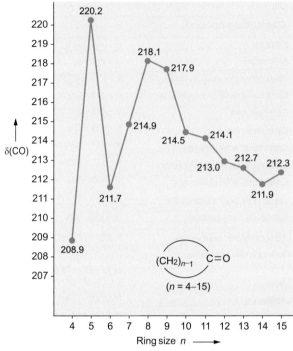

Fig. 3.82 ^{13}C Chemical shifts of carbonyl groups in cycloalkanones **285** (*n* = 4–15)

H$_3$C
C=S   255.0 (–30 °C)
H$_3$C

Thioacetone (**300**)

H$_5$C$_6$
C=S   238.3
H$_5$C$_6$

Thiobenzophenone (**301**)

S
‖
H$_3$C–C–SC$_2$H$_5$
233.1

Dithioacetic acid ethyl ester (**302**)

S
‖
H$_3$CS–C–SCH$_3$
224.1

Trithiocarbonic acid dimethyl ester (**303**)

S
‖
H$_3$C–C–OCH$_2$–C$_6$H$_5$
219.6

Thioacetic acid-O-benzyl ester (**304**)

S
‖
H$_3$C–C–NH$_2$
205.6

Thiourea (**305**)

S
‖
H$_3$CO–C–OCH$_3$
197.2

Thioacetic acid O,O'-dimethyl ester (**306**)

S
‖
H–C–N(CH$_3$)$_2$
187.9

N,N-Dimethylthioformamide (**307**)

S
‖
H$_3$C–HN–C–NH–CH$_3$
182.5

N,N'-Dimethylthiourea (**308**)

S
‖
H$_2$N–C–NH–NH$_2$
181.6

Thiosemicarbazide (**309**)

When the S atom in thioketones is replaced by higher elements of the sixth group, further down-field shifts can be observed:

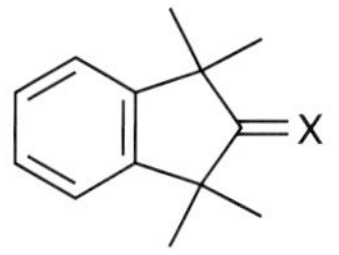

	X	δ
310	O	226
311	S	282
312	Se	294
313	Te	301

Table 3.34 gives a survey over the δ (¹³C) values of ketenes and **Table 3.35** a survey over some other heterocumulenes.

The ¹³C chemical shifts of aliphatic diazo compounds are summarized in **Table 3.36**:

C≡N triple bonds are present in different classes of compounds, such as nitriles, cyanates, and cyanamides. However, their δ (¹³C) values are in a relatively narrow range between 110 and 120 ppm (**Table 3.37**). Dicyan and propiolonitrile are exceptions due to the anisotropy of the adjacent triple bond.

Carbo- and heterocyclic three-membered rings give very low δ (¹³C) values; even negative δ values can be measured—as the examples **139**, **315**, and **316** show. The corresponding four- and five-membered ring systems are listed here for comparison.

Table 3.34 ¹³C Chemical shifts of ketenes

Ketene	$H_2C=C=O$ 2,5 194,0
Methylketene	$H_3C-CH=C=O$ 10,9 200,0
Dimethylketene	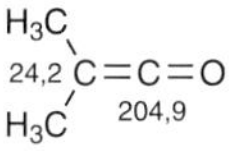 H_3C 24,2 C=C=O H_3C 204,9
Methyl-phenylketene	H_3C 33,8 C=C=O H_5C_6 205,6
Diphenylketene	H_5C_6 57,9 C=C=O H_5C_6 196,9
Dipivaloylketene (2,2,6,6-Tetramethyl-4-oxo-methylene-heptane-3,5-dione) (unstable)	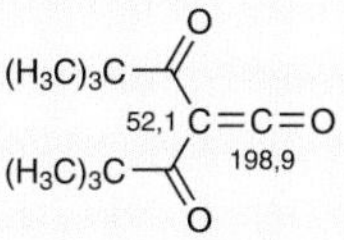
Bis(ethoxycarbonyl)ketene (2-Oxomethylenmalonic acid diethylester)	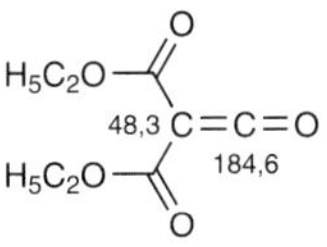

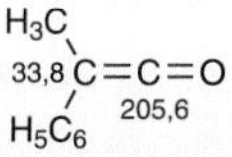

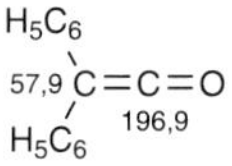

Table 3.35 ¹³C Chemical shifts of heterocumulenes

Carbon dioxide	$O=C=O$ 123.9
Carbon disulfide	$S=C=S$ 192.3
Methylisocyanate	$H_3C-N=C=O$ 121.5
Methylisothiocyanate	$H_3C-N=C=S$ 128.7
Dicyclohexylcarbodiimide	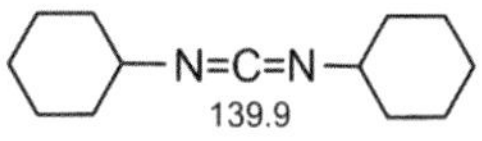 139.9

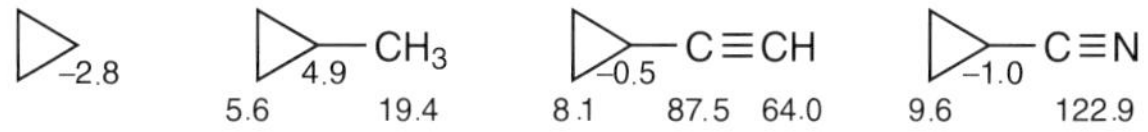

-2.8	CH₃ 5.6 19.4 4.9	C≡CH 8.1 87.5 64.0 -0.5	C≡N 9.6 122.9 -1.0
139	**314**	**315**	**316**

 NH	S	O
28.7	18.0	40.5
317	**318**	**143**

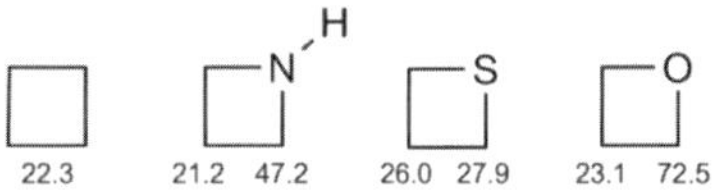

	N–H	S	O
22.3	21.2 47.2	26.0 27.9	23.1 72.5
206	**319**	**320**	**321**

	NH	S	O
25.8	25.7 47.1	31.4 31.2	26.5 68.4
322	**323**	**324**	**325**

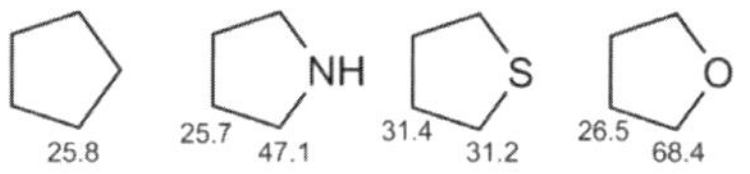

Finally, the ¹³C chemical shifts of ions and carbenes shall be discussed. The aromatic ions cyclopropenylium (**121**), cyclopentadienide (**123**), and cycloheptatrienylium (tropylium; **124**) exhibit together with benzene (**105**) a linear relationship between their δ values and the π electron density (**Fig. 3.83**).

In studies of carbocations, the measurement of their ¹³C spectra in "magic acid" played a decisive role. As shown by the examples **326** to **332**, the extent of delocalization of the charge is easily recognized from the ¹³C shift of the central C atom.

Table 3.36 ^{13}C Chemical shifts of diazoalkanes, diazoketones, and diazoesters

Diazomethane	$H_2C = N_2$ 23.1	Phenyldiazomethane	H_5C_6 47.2 $\overset{\diagdown}{\underset{\diagup}{C}} = N_2$ H
Diphenyldiazommethane	H_5C_6 62.5 $\overset{\diagdown}{\underset{\diagup}{C}} = N_2$ H_5C_6	2-Diazo-1,2-diphenyl-ethanone	$H_5C_6 \diagdown \diagup N_2$ 72.4 188.1 $O \diagdown C_6H_5$
2-Diazoacetic acid methyl ester	$H \diagdown \diagup N_2$ 46.3 167.6 $H_3CO \diagup O$	2-Diazo-1,3-diphenyl-propane-1,3-dione	$\overset{O}{\diagdown} N_2$ H_5C_6 84.5 186.7 $O \diagdown C_6H_5$

Table 3.37 δ ($^{13}C \equiv N$) Values of different classes of compounds

Acetonitrile	$H_3C - C \equiv N$ 116.5
Benzonitrile	$H_5C_6 - C \equiv N$ 112.6
2-Oxopropionitrile	173.2 $H_3C - \overset{\overset{O}{\|\|}}{C} - C \equiv N$ 113.0
Phenylcyanate	$H_5C_6 - O - C \equiv N$ 115.2
Phenylthiocyanate	$H_5C_6 - S - C \equiv N$ 110.5
N,N-Dimethylcyanamide	$(H_3C)_2N - C \equiv N$ 119.2

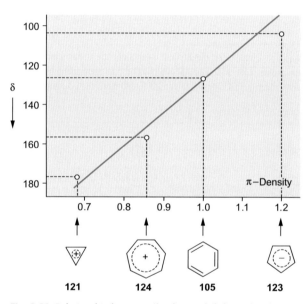

Fig. 3.83 Relationship between ^{13}C chemical shifts and π electron density.

compounds will be strongly pH-dependent. Simplified assumptions of electron densities can however easily lead to errors. The protonation of linear alkyl amines generally causes an up-field shift of C_α, C_β, and C_γ; conversely a down-field shift is observed on deprotonation of carboxylic acids!

$$-\overset{|}{C_\gamma}-\overset{|}{C_\beta}-\overset{|}{C_\alpha}-NH_2 \, / \, \overset{+}{N}H_3$$

$$\varDelta\delta \approx -0.5 \quad -3.8 \quad -2.3$$

$$-\overset{|}{C_\gamma}-\overset{|}{C_\beta}-\overset{|}{C_\alpha}-COOH \, / \, COO^-$$

$$\varDelta\delta \approx +0.6 \quad +1.6 \quad +3.5 \qquad +4.7$$

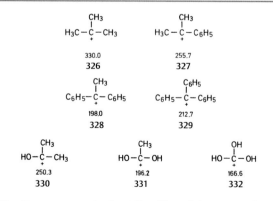

After these comments about the effect of charges on chemical shift, it is understandable that the signals of acidic or basic

In the amphoteric amino acids, the δ values generally increase when the pH increases, as shown by the example of alanine (**23**):

23

pH	0.43	4.96	12.52
C–1	174.0	177.0	185.7
C–2	50.1	51.9	52.7
C–3	16.5	17.5	21.7

It was already referred to the problems in measuring NMR spectra of doublet and triplet ground states (radicals, carbenes). However, stable singlet carbenes including mesoionic species can be facilely measured:

−R	δ(C−2)	δ(C−2)	δ(C−5)
−CH (CH₃)₂	236.8	210.5	
H₃C ... CH₃ (aryl)	243.8	219.7	
(di-tert-butylphenyl)	244.0		201.9

3.4.3 Correlation of ¹³C Chemical Shifts with Structural Features

The ranges of the ¹³C chemical shifts for the most important structural types of organic compounds are collected in **Table 3.38**. Extreme values are ignored. For the interpretation of ¹³C spectra, it is recommended that this table is used in conjunction with the data in Section 5.3 (see p. 440), arranged by type of compound, and the increment systems shown in Section 3.4.4.

If more than one functional group is attached to a single saturated C atom, then for a first approximation the additivity of the shift effects can be assumed. Exceptions, present for example for push–pull substitution, are discussed in Section 3.4.2. As a matter of principle, the use of modern databases can be recommended (see Section 3.5.2, p. 229).

3.4.4 Increment Systems for the Estimation of ¹³C Chemical Shifts

The ¹³C chemical shifts of aliphatic compounds and benzene derivatives can be estimated with **empirical increment systems.** If there is more than one substituent, an additive behavior is assumed. The following increment systems are a useful aid for the assignment of ¹³C signals. The tabulated examples however show the possible deviations between calculated and observed shifts.

For the estimation of the ¹³C shifts of saturated C atoms, the simplest procedure is to base the calculation on a corresponding hydrocarbon and add the increments for the various functional groups.

If the ¹³C shifts δ_i of the hydrocarbon itself are not known, they can be calculated from the **Grant–Paul rules** as follows:

$$\delta_i = -2.3 + \sum_k A_k n_k + S_{i\alpha} \quad \text{for all } C_i.$$

The increments $A_k n_k$ are added to the shift value for methane $\delta = -2.3$. Increments are added for all the positions $k = \alpha, \beta, \gamma, \delta, \varepsilon$, relative to the relevant carbon. nk is the number of C atoms at position k. The increments A_k have the following values:

$$A_\alpha = +9.1, A_\beta = +9.4, A_\gamma = -2.5, A_\delta = +0.3, A_\varepsilon = +0.2.$$

For tertiary and quaternary C atoms and their immediate neighbors, a steric correction factor $S_{i\alpha}$, must be added. This is derived from the most substituted carbon C_α next to the carbon C_i for which the shift is being calculated. The correction values $S_{i\alpha}$ are then as follows.

C_i (carbon atom under consideration)	C_α (highest substituted neighbouring C atom)			
	—CH₃	—CH₂—	—CH—	—C—
primary —CH₃	0	0	−1.1	−3.4
secondary —CH₂—	0	0	−2.5	−7.5
tertiary —CH—	0	−3.7	−9.5	(−15.0)
quaternary —C—	−1.5	−8.4	(−15.0)	(−25.0)

The system will become clearer from a consideration of 2-methylbutane's example.

Table 3.38 ¹³C chemical shift regions of important structural elements; δ values (ppm)

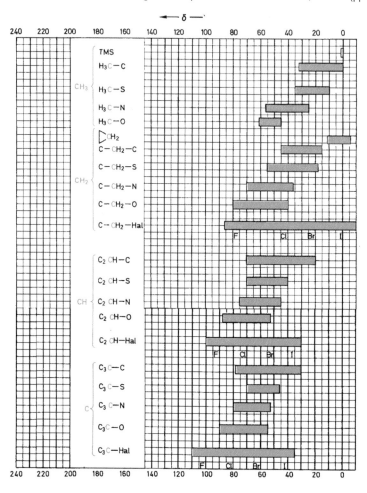

C_i	−2.3	$+\sum_{k}A_k n_k$	$+ S_{i\alpha}$	$= \delta_{calculated}$	$\delta_{observed}$
C–1	−2.3	+9.1			
		+9.4 · 2			
		−2.5	−1.1	= 22.0	21.9
C–2	−2.3	+9.1 · 3			
		+9.4	−3.7	= 30.7	29.7
C–3	−2.3	+9.1 · 2			
		+9.4 · 2	−2.5	= 32.2	31.7
C–4	−2.3	+9.1			
		+9.4			
		−2.5 · 2		= 11.2	11.4

In sterically hindered hydrocarbons, the agreement between $\delta_{calculated}$ and $\delta_{observed}$ is poorer. If rotation around a C–C bond is restricted or prevented, then additional **conformational corrections** must be made.

Using the measured or calculated δ_i values of an alkane C_nH_{2n+2} as a basis, the ¹³C shifts of the substituted compounds $C_nH_{2n+1}X$, $C_nH_{2n}XY$, etc. can be calculated. **Table 3.39** gives the increments I for a selection of substituents X depending on the position of substitution relative to the carbon C_i for which the shift is being calculated.

In **Table 3.40**, the observed shifts of some representative examples are given, together with the values calculated with the increment system. (For the substituents –OR, –NR₂, and –SR, it is recommended to apply the steric corrections $S_{i\alpha}$ as for hydrocarbons.)

Table 3.38 (Continued)

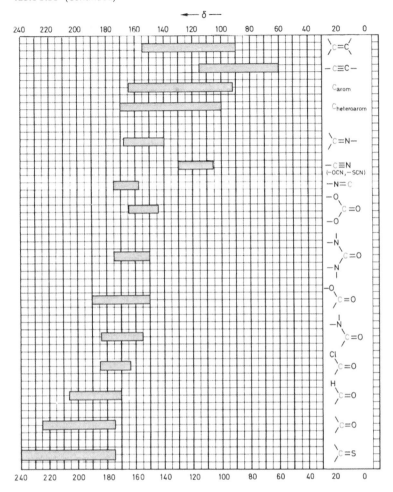

The ¹³C shifts of olefinic carbon atoms can be calculated using the values given in **Table 3.41**.

Table 3.42 gives an impression of the "quality" of the estimations possible with this increment system. (The application of the system to trisubstituted olefins, not to mention tetrasubstituted olefins, is not recommended.)

The increment system for benzene derivatives in **Table 3.43** (p. 194) works on similar principles. The increments I_1, I_2, etc. for each substituent X_1, X_2, etc. are added to the basis value for benzene ($\delta = 128.5$). The value of I depends on the nature of the substituent and its position relative to the C atom for which the shift is being calculated.

Several illustrative examples are given in **Table 3.44** (p. 194). Larger deviations between calculated and observed values can be expected when the additivity of the substituent effects is disturbed by steric or electronic interactions.

The control and prediction of chemical shifts by applying increment systems has lost a lot of significance by the use of modern databases (see Section 3.5.2).

3.4.5 ¹³C,¹H Couplings

The coupling $^1J(C,H)$ is a measure of the s-character of the hybrid orbitals of the relevant C–H bonds. The following empirical relation holds approximately:

$$^1J(C, H) = 500p \quad \text{where} \quad p = \begin{cases} 0.25 \text{ for } C-sp^3 \\ 0.33 \text{ for } C-sp^2 \\ 0.50 \text{ for } C-sp \end{cases}$$

Table 3.45 shows measured coupling constants and values calculated from this formula for ethane, ethylene, and acetylene.

Table 3.39 Increment system for the estimation of ^{13}C chemical shifts of aliphatic compounds $\delta_i(RX) = \delta_i(RH) + I_{xk} + S_{i\alpha}$ ($k = \alpha, \beta, \gamma, \delta$) for all C_i

Substituent X	$k = \alpha$	β	γ	δ
$-C{=}C-$	20.0	6.9	-2.1	0.4
$-C{\equiv}C-$	4.4	5.6	-3.4	-0.6
$-C_6H_5$	22.1	9.3	-2.6	0.3
$-CH{=}O$	29.9	-0.6	-2.7	0
$-C{=}O$ R	22.5	3.0	-3.0	0
$-COOH$	20.1	2.0	-2.8	0
$-COOR$	22.6	2.0	-2.8	0
$-CO{-}NR_2$	22.0	2.6	-3.2	-0.4
$-COCl$	33.1	2.3	-3.6	0
$-C{\equiv}N$	3.1	2.4	-3.3	-0.5
$-OH$	49.0	10.1	-6.2	0
$-OR$	58.0	7.2	-5.8	0
$-O{-}CO{-}R$	54.0	6.5	-6.0	0
$-NR_2$	28.3	11.3	-5.1	0
$-\overset{+}{N}R_3$	30.7	5.4	-7.2	-1.4
$-NO_2$	61.6	3.1	-4.6	-1.0
$-SH$	10.6	11.4	-3.6	-0.4
$-SCH_3$	20.4	6.2	-2.7	0
$-F$	70.1	7.8	-6.8	0
$-Cl$	31.0	10.0	-5.1	-0.5
$-Br$	18.9	11.0	-3.8	-0.7
$-I$	-7.2	10.9	-1.5	-0.9

If the $^1J(C,H)$ coupling constants of cyclohexane (125 Hz) and cyclopropane (160 Hz) are compared, the similarity of cyclopropane to olefinic systems, as suggested by the Walsh orbital model, for example, is apparent. These estimates of s-character of hybrid orbitals should however be restricted to hydrocarbons.

The effect of substituents on the $^1J(C,H)$ couplings constants is shown in **Table 3.46** (p. 195) for a series of methane derivatives.

Since $^1J(C,H)$ is of some importance for spectral interpretation, a few further examples are given in **Table 3.47** (p. 196).

While the $^1J(C,H)$ coupling constants lie between ca. $+320$ and $+100$ Hz, values of $^2J(C,H)$ between ca. $+70$ and -20 are known. The *vicinal* couplings $^3J(C,H)$ are always positive and less than 15 Hz. Their values depend on the dihedral angle according to the Karplus curve (see p. 131). Some characteristic data for $^2J(C,H)$, $^3J(C,H)$, and $^nJ(C,H)$ ($n \geq 4$) are given in **Table 3.48** (p. 197).

3.4.6 Coupling of ^{13}C to Other Nuclei (D, F, N, and P)

In **Table 3.26** a series of $^1J(C,D)$ **coupling constants** have already been given for deuterated solvents. Approximately,

$$J(C, H): J(C, D) \approx \gamma_H : \gamma_D \approx 6.5 : 1.$$

The ^{13}C,D couplings are therefore considerably smaller than the corresponding ^{13}C,^{1}H couplings. **Fig. 3.84** (see p. 198) shows the ^{13}C NMR spectrum of trifluoroacetic acid (**333**) in deuteriochloroform.

Table 3.40 Calculated and observed δ values for selected aliphatic compounds

Compound	$\delta_{calculated}$[a]	$\delta_{observed}$
1-Hexyne		
$\overset{1}{H}C{\equiv}\overset{2}{C}-\overset{3}{C}H_2-\overset{4}{C}H_2-\overset{5}{C}H_2-\overset{6}{C}H_3$	C–1 –	67.4
	C–2 –	82.8
	C–3 17.4 (18.1)	17.4
	C–4 30.4 (30.9)	29.9
	C–5 21.4 (21.9)	21.2
	C–6 12.4 (13.1)	12.9
Basis:		
$\overset{1}{H}_3C-\overset{2}{C}H_2-CH_2-CH_3$	C–1 13.7	13.0
	C–2 25.3	24.8
2-Butanol[b]		
$\overset{1}{H}_3C-\overset{2}{C}H-\overset{3}{C}H_2-\overset{4}{C}H_3$ OH	C–1 22.0 (22.7)	22.6
	C–2 70.1 (70.6)	68.7
	C–3 32.4 (32.9)	32.0
	C–4 6.8 (7.5)	9.9
Basis:		
$\overset{1}{H}_3C-\overset{2}{C}H_2-CH_2-CH_3$	C–1 13.7	13.0
	C–2 25.3	24.8
2-Chloro-2-methylbutane		
$\overset{1}{H}_3C-\overset{2}{\underset{CH_3}{\overset{Cl}{C}}}-\overset{3}{C}H_2-\overset{4}{C}H_3$	C–1 31.9 (32.0)	32.0
	C–2 60.7 (61.7)	71.1
	C–3 41.7 (42.2)	38.8
	C–4 6.3 (6.1)	9.4
Basis:		
$\overset{1}{H}_3C-\overset{2}{\underset{CH_3}{C}H}-\overset{3}{C}H_2-\overset{4}{C}H_3$	C–1 22.0	21.9
	C–2 30.7	29.7
	C–3 32.2	31.7
	C–4 11.2	11.4
Leucine[b]		
$HOOC-\overset{2}{C}H-\overset{3}{C}H_2-\overset{4}{C}H-\overset{5}{C}H_3$ NH₂ CH₃(6)	C–1 –	176.6
	C–2 56.1 (55.9)	54.8
	C–3 42.5 (43.0)	41.0
	C–4 21.8 (22.8)	25.4
	C–5,	
	C–6 21.9 (22.0)	23.2/22.1
Basis:		
$\overset{1}{H}_3C-\overset{2}{\underset{CH_3}{C}H}-\overset{3}{C}H_2-\overset{4}{C}H_3$	C–1 22.0	21.9
	C–2 30.7	29.7
	C–3 32.2	31.7
	C–4 11.2	11.4

[a] The basis values used are the **observed** ^{13}C shifts of the hydrocarbon; the δ_i values in brackets are related to the **calculated** ^{13}C shifts of the hydrocarbon.

[b] For C–1, C–2, and C–3 in 2-butanol and for C–2 and C–3 in leucine steric correction factors $S_{i\alpha}$ have been included

Table 3.41 Increment system for the estimation of ^{13}C shifts of olefinic carbons

$\overset{1}{X}-\overset{2}{CH}=\overset{}{CH_2}$	$\delta_1 = 123.3 + I_1, \delta_2 = 123.3 + I_2,$
$\overset{1}{X}-\overset{2}{CH}=CH-Y$	$\delta_1 = 123.3 + I_{X1} + I_{Y2},$
	$\delta_2 = 123.3 + I_{Y1} + I_{X2}$

Substituent	Increments	
	I_1	I_2
—H	0	0
—CH_3	10.6	– 8.0
—C_2H_5	15.5	– 9.7
—CH_2—CH_2—CH_3	14.0	– 8.2
—$CH(CH_3)_2$	20.3	– 11.5
—$(CH_2)_3$—CH_3	14.7	– 9.0
—$C(CH_3)_3$	25.3	– 13.3
—$CH=CH_2$	13.6	– 7.0
—$C\equiv C$—R	– 7.5	8.9
—C_6H_5	12.5	– 11.0
—CH_2Cl	10.2	– 6.0
—CH_2Br	10.9	– 4.5
—CH_2OR	13.0	– 8.6
—CH_2SR	10.0	– 6.0
—$CH=O$	13.1	12.7
—CO—CH_3	15.0	5.9
—$COOH$	4.2	8.9
—$COOR$	6.0	7.0
—$COCl$	8.1	14.0
—CN	– 15.1	14.2
—OR	28.8	–39.5
—O—CO—R	18.0	– 27.0
—NR_2	16.0	– 29.0
—$\overset{+}{N}(CH_3)_3$	19.8	– 10.6
—NO_2	22.3	– 0.9
—SR	19.0	– 16.0
—SO_2R	14.2	7.7
—$Si(CH_3)_3$	16.9	6.7
—F	24.9	– 34.3
—Cl	2.6	– 6.1
—Br	– 7.9	– 1.4
—I	– 38.1	7.0

The $^1J(^{13}C,^{19}F)$ **couplings** have a larger magnitude than the comparable $^1J(^{13}C,^1H)$ couplings, but have a negative sign. The $^1J(C,F)$ values lie between –150 and –400 Hz. A selection of $^nJ(C,F)$ coupling constants is given in **Table 3.49** (p. 199).

A few comments on ^{13}**C,^{15}N and ^{13}C,^{31}P couplings** will complete this section. While ^{13}C,^{19}F and ^{13}C,^{31}P couplings can be directly measured from ^{1}H broadband decoupled ^{13}C spectra of fluorine and phosphorus compounds, ^{13}C,^{15}N couplings are only accessible via ^{15}N enrichment. The natural abundance of both ^{19}F and ^{31}P is 100%, but only 0.37% for ^{15}N (see **Table 3.1**, p. 85). Some idea of

Table 3.42 Calculated and observed δ values for the ^{13}C shifts of olefinic C atoms

Compound	δ_i calculated		δ_i observed	
	C–1	C–2	C–1	C–2
2-Butene				
	125.9	125.9	124.8	124.8
	125.9	125.9	123.4	123.4
Methyl methacrylate				
	122.3	139.9	124.7	136.9
(E)-Crotonaldehyde				
	146.6	128.4	153.7	134.9
Fumaric acid				
	136.4	136.4	134.5	134.5
Maleic acid				
	136.4	136.4	130.8	130.8
2-Phenyl-1-butene				
	102.6	151.3	109.7	148.9

typical values of ^{13}C,^{15}N and ^{13}C,^{31}P coupling constants can be gained from the examples given in **Tables 3.50 and 3.51** (p. 199, 200).

$J(^{13}C,^{14}N)$ couplings are seldom observable since the ^{14}N nucleus has a quadrupole moment, which causes such rapid relaxation that the coupling is lost. Exceptions occur when the electric field gradients are very small. **Fig. 3.85** (p. 199) shows methylisonitrile (**334**) as an example; both signals are split into three lines. The couplings $J(^{13}C,^{14}N)$ are 6.2 Hz for the isocyanide carbon and 7.5 Hz for the methyl carbon. Further examples of systems that show observable $J(^{13}C,^{14}N)$ couplings are diazo compounds and tetraalkylammonium salts.

3.4.7 ^{13}C,^{13}C Couplings

The coupling of a ^{13}C nucleus with a neighboring ^{13}C nucleus is generally not observable in routine ^{13}C NMR spectra. The

Nuclear Magnetic Resonance Spectroscopy

Table 3.43 Increment system for the estimation of ^{13}C shifts of substituted benzenes

$$\delta_i = 128.5 + I_{1i} + I_{2i} + \dots$$

Substituent	ipso	ortho	meta	para
—H	0.0	0.0	0.0	0.0
—CH$_3$	9.3	0.6	0.0	− 3.1
—C$_2$H$_5$	15.7	− 0.6	− 0.1	− 2.8
—CH(CH$_3$)$_2$	20.1	− 2.0	0.0	− 2.5
—C(CH$_3$)$_3$	22.1	− 3.4	− 0.4	− 3.1
—CH=CH$_2$	7.6	− 1.8	− 1.8	− 3.5
—C≡CH	− 6.1	3.8	0.4	− 0.2
—C$_6$H$_5$	13.0	− 1.1	0.5	− 1.0
—CF$_3$	2.6	− 2.6	− 0.3	− 3.2
—CH$_2$Cl	9.1	0.0	0.2	− 0.2
—CH$_2$Br	9.2	0.1	0.4	− 0.3
—CH$_2$OR	13.0	− 1.5	0.0	− 1.0
—CH$_2$—NR$_2$	15.0	− 1.5	− 0.2	− 2.0
—CH$_2$—PO(OR)$_2$	1.6	3.6	− 0.2	3.4
—CH$_2$—SR	9.8	0.4	− 0.1	− 1.6
—CF$_3$	2.6	− 3.1	0.4	3.4
—CH=O	7.5	0.7	− 0.5	5.4
—CO—CH$_3$	9.3	0.2	0.2	4.2
—CO—C$_6$H$_5$	9.1	1.5	− 0.2	3.8
—CH=NC$_6$H$_5$	8.7	1.4	0.9	3.5
—COOH	2.4	1.6	− 0.1	4.8
—COOR	2.0	1.0	0.0	4.5
—CO—NR$_2$	5.5	− 0.5	− 1.0	5.0
—COCl	4.6	2.9	0.6	7.0
—C≡N	− 16.0	3.5	0.7	4.3
—OH	26.9	− 12.6	1.6	− 7.6
—OCH$_3$	31.3	− 15.0	0.9	− 8.1
—OC$_6$H$_5$	29.1	− 9.5	0.3	− 5.3
—O—CO—R	23.0	− 6.0	1.0	− 2.0
—NH$_2$	19.2	− 12.4	1.3	− 9.5
—NR$_2$	21.0	− 16.0	0.7	− 12.0
—NH—CO—CH$_3$	11.1	− 9.9	0.2	− 5.6
—N=CH—C$_6$H$_5$	24.5	− 6.9	1.0	− 1.9
—N=N—C$_6$H$_5$	24.0	− 5.8	0.3	2.2
—N=C=O	5.7	− 3.6	1.2	− 2.8
—NO$_2$	19.6	− 5.3	0.8	6.0
—SH	2.2	0.7	0.4	− 3.1
—SCH$_3$	10.1	− 1.6	0.2	− 3.5
—SC$_6$H$_5$	6.8	0.5	2.2	− 1.6
—SO$_3$H	15.0	− 2.2	1.3	3.8
—F	35.1	− 14.3	0.9	− 4.4
—Cl	6.4	0.2	1.0	− 2.0
—Br	− 5.4	3.3	2.2	− 1.0
—I	− 32.3	9.9	2.6	− 0.4
—Si(CH$_3$)$_3$	13.4	4.4	− 1.1	− 1.1

Table 3.44 Calculated and observed δ values for the ^{13}C shifts of substituted benzenes

Compound		C atom	$\delta_{calculated}$	$\delta_{observed}$
p-Xylene		C–1	134.7	134.5
		C–2	129.1	129.1
Mesitylene		C–1	137.8	137.6
		C–2	126.6	127.4
Hexamethylbenzene		C	135.9	132.3
p-Cresol		C–1	152.3	152.6
		C–2	115.9	115.3
		C–3	130.7	130.2
		C–4	130.2	130.5
3,5-Dimethoxybenzaldehyde		C–1	137.8	138.4
		C–2	106.1	107.0
		C–3	160.2	161.2
		C–4	103.9	107.0
1-Chloro-2,4-dimethoxy-5-nitrobenzene		C–1	112.5	113.8
		C–2	166.9	159.9
		C–3	100.3	97.2
		C–4	153.4	154.5
		C–5	126.0	131.9
		C–6	125.2	127.7

Table 3.45 Relationship between ^1J(C,H) and s-character of the hybrid orbitals of carbon

	Hybridisation			^1J(C,H)	
	sp^{λ^2}	λ^2	$p = \dfrac{1}{1+\lambda^2}$	calculated	measured
Ethane	sp^3	3	0.25	125	124.9
Ethylene	sp^2	2	0.33	167	156.4
Acetylene	sp	1	0.50	250	248

low natural abundance of ^{13}C of 1.1% means that the nuclear combination ^{13}C...^{13}C is only ca. 1/100 times as likely as ^{13}C...^{12}C and only 10^{-4} times as likely as ^{12}C...^{12}C. Apart from weak satellites in long accumulations of intense signals of pure liquids, the measurement of ^{13}C,^{13}C couplings requires the use of ^{13}C-enriched compounds or the use of the INADEQUATE method (see p. 219).

^{13}C,^{13}C couplings show a similar dependence on hybridization and electronic effects as ^{1}H,^{13}C couplings. The compounds in **Table 3.52** (p. 201) demonstrate that ^1J(C,C) increases strongly with the s-character of the orbitals involved.

Table 3.46 ¹J(C,H) coupling constants in substituted methanes

Compound	¹J(C,H) (Hz)
CH_4	125
CH_3F	149
CH_3Cl	150
CH_3Br	152
CH_3I	151
CH_3NH_2	133
$CH_3N^+H_3$	145
CH_3NO_2	147
CH_3OH	141
CH_3O^-	131
CH_3OCH_3	140
CH_3SCH_3	138
$CH_3Si(CH_3)_3$	118
CH_3Li	98
CH_2Cl_2	177
$CHCl_3$	209
CHF_3	239

The effect of substituents on ¹J(¹³C,¹³C) is generally small for saturated C atoms; larger effects are observed for olefinic, aromatic, or carbonyl carbon atoms.

¹J(C,C) Hz	X = H	CH₃	Cl	OC₂H₅
H_3C-CH_2-X	34.6	34.6	36.1	38.9
$H_2C{=}CH-X$	67.6	70.0	77.6	78.1
$HC{\equiv}CH-X$	171.5	175.0		216.5
H_6C_5-X	56.0	57.1	65.2	67.0
$H_3C-CO-X$	39.4	40.1	56.1	58.8

In polycyclic compounds unexpected effects can occur. While the ²J coupling between the bridgehead carbon atoms in bicyclo[1.1.1]pentane (**164**) is 25.1 Hz, the ¹J coupling between C-1 and C-3 in the corresponding [1.1.1]propellane (**335**) is extremely small, viz. 0.6 Hz.

164   **335**

Geminal couplings ²J(C,C) can be positive or negative; their magnitude is generally less than 5Hz. Exceptions occur particularly for carbonyl compounds, alkynes, nitriles, and organometallic compounds:

$H_3C-CO-CH_3$ $H_3C-C{\equiv}CH$ H_3C-CH_2-CN
16 Hz   12 Hz   33 Hz
277   **100**   **336**

$H_3C-C{\equiv}C-\overset{\overset{O}{\|}}{C}-OCH_3$ $H_3C-Hg-CH_3$
20 Hz   22 Hz
337   **338**

³J(C,C) coupling constants are positive and generally smaller than 5 Hz. Exceptions occur for conjugated systems; thus, the ³J coupling between C-1 and C-4 in butadiene for example is 9.1 Hz.

A few further ⁿJ (¹³C,¹³C) couplings (absolute values in Hz) are listed below and continued on p. 201.

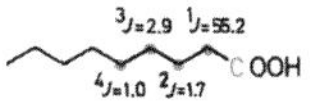

³J=2.9 ¹J=55.2
⁴J=1.0 ²J=1.7   COOH   **339**

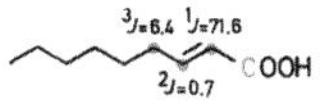

³J=6.4 ¹J=71.6
²J=0.7   COOH   **340**

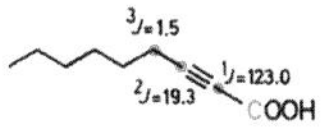

³J=1.5
²J=19.3   ¹J=123.0
COOH   **341**

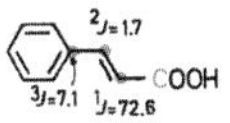

²J=1.7
³J=7.1 ¹J=72.6   COOH
8

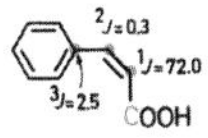

²J=0.3
¹J=72.0
³J=2.5   COOH
155

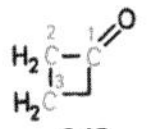

$H_2C{-}C{\overset{O}{\|}}$ ¹J(C-1, C-2) = 29.5   ²J(C-1, C-3) = 9.7
H_2C ¹J(C-2, C-3) = 28.5
342

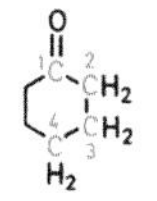

¹J(C-1, C-2) = 37.7   ²J(C-2, C-4) = 2.2
¹J(C-2, C-3) = 30.4
¹J(C-3, C-4) = 33.1
343

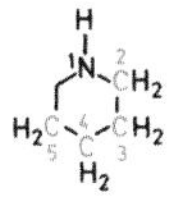

¹J(C-2, C-3) = 35.2   ²J(C-2, C-4) = 2.6
¹J(C-3, C-4) = 33.0   ³J(C-2, C-5) = 1.7
344

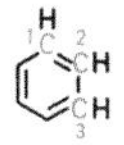

¹J(C-1, C-2) = 56.0   ²J(C-1, C-3) = 2.5
105

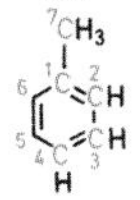

¹J(C-1, C-2) = 57.3   ²J(C-1, C-3) = 0.8
¹J(C-1, C-7) = 44.2   ²J(C-2, C-4) = 2.6
²J(C-2, C-7) = 3.1
³J(C-1, C-4) = 9.5
³J(C-3, C-7) = 3.8
⁴J(C-4, C-7) = 0.9
19

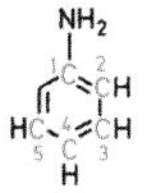

¹J(C-1, C-2) = 61.3   ³J(C-2, C-5) = 7.9
¹J(C-2, C-3) = 58.1
¹J(C-3, C-4) = 56.2
261

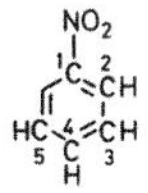

¹J(C-1, C-2) = 55.4   ³J(C-2, C-5) = 7.6
¹J(C-2, C-3) = 56.3
¹J(C-3, C-4) = 55.8
259

Table 3.47 $^1J(C,H)$ couplings of selected compounds

Compound	$^1J(C,H)$ (Hz)
$H—CH_2—CH_3$	125 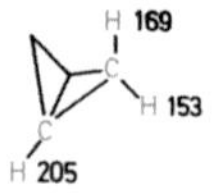 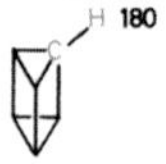 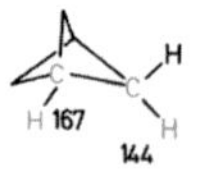
$H—CH(CH_3)_2$	119
$H—CH(CH_3)_3$	114
$H—CH_2—CH{=}CH_2$	122
$H—CH_2—C_6H_5$	129
$H—CH_2—C{\equiv}CH$	132
$H—CH_2—C{\equiv}N$	136
$H—CH_2—COOH$	130
$H—CH(OH)—C_6H_5$	140

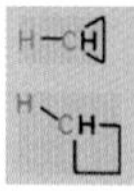

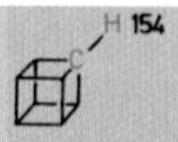

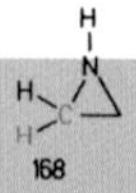

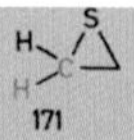

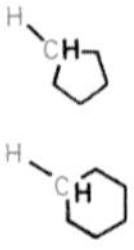

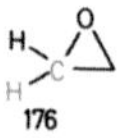

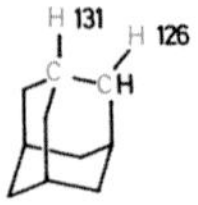

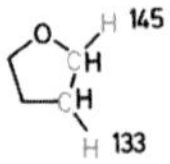

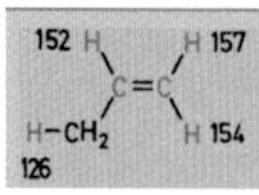

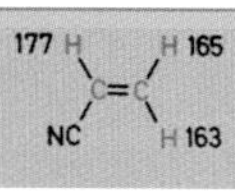

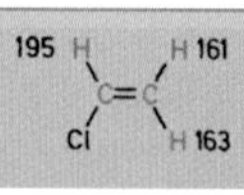

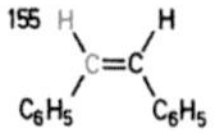

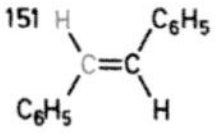

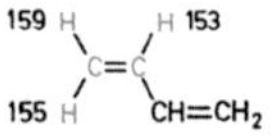

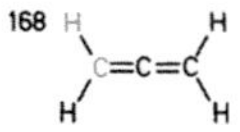

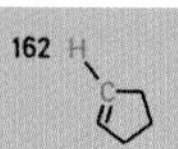

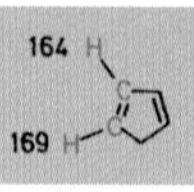

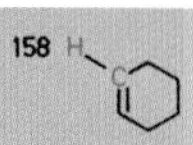

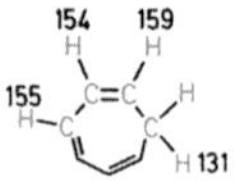

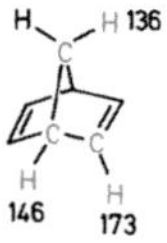

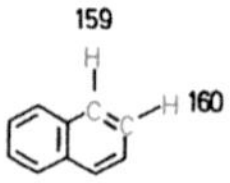

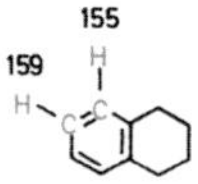

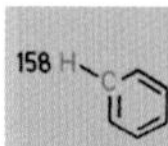

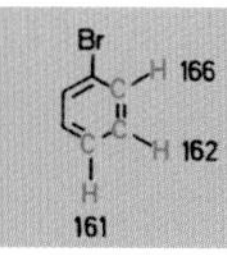

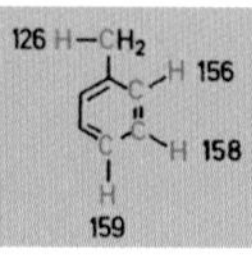

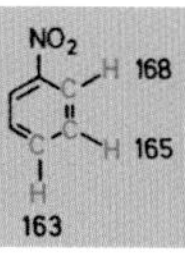

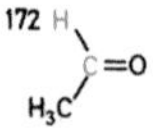

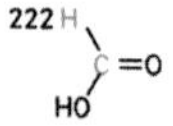

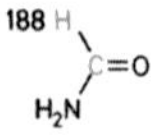

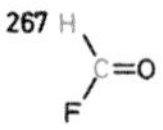

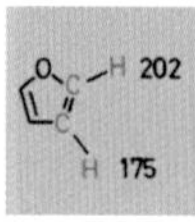

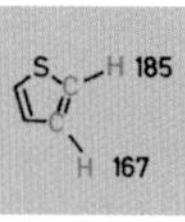

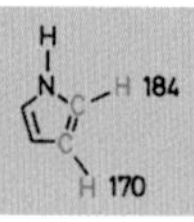

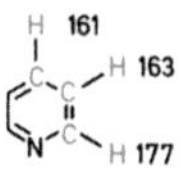

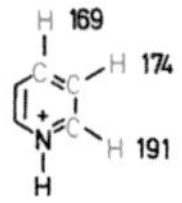

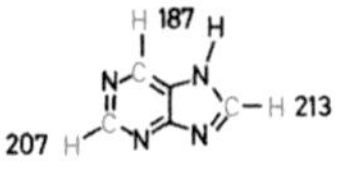

Table 3.48 $^2J(C, H)$, $^3J(C, H)$, and higher C, H couplings of selected compounds

Compound	^2J(C, H) (Hz)	^3J(C, H) (Hz)	^nJ(C, H) (Hz)
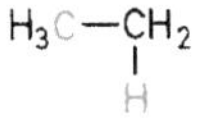 H$_3$C—CH$_2$ H	– 4.5		
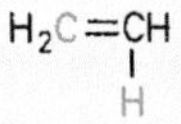 H$_2$C=CH H	– 2.4		
 HC≡C—H	+ 49.6		
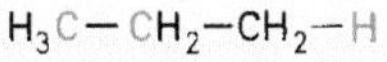 H$_3$C—CH$_2$—CH$_2$—H	– 4.4	+ 5.8	
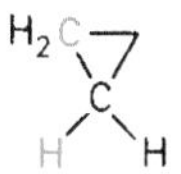 H$_2$C⟍C⟋ (cyclopropane) H H	– 2.5		
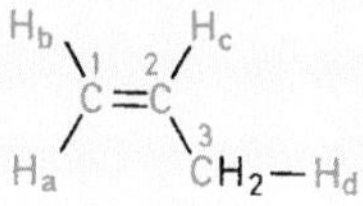 (allyl structure H$_b$, H$_c$, H$_a$, CH$_2$—H$_d$)	C$_1$H$_c$: + 0.4 C$_2$H$_a$: – 2.6 C$_2$H$_b$: – 1.2 C$_2$H$_d$: – 6.8 C$_3$H$_c$: + 5.0	C$_1$H$_d$: + 6.7 C$_3$H$_a$: + 7.6 C$_3$H$_b$: + 12.7	
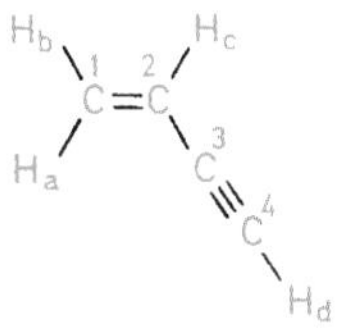 (enyne structure H$_b$, H$_c$, H$_a$, C≡C, H$_d$)	C$_1$H$_c$: + 8.8 C$_2$H$_a$: – 3.7 C$_2$H$_b$: – 0.3 C$_3$H$_c$: + 2.0	C$_2$H$_d$: + 4.8 C$_3$H$_a$: + 9.5 C$_3$H$_b$: + 16.3	C$_1$H$_d$: + 2.8 (n = 4) C$_4$H$_a$: <1 (n = 4) C$_4$H$_b$: <1 (n = 4)
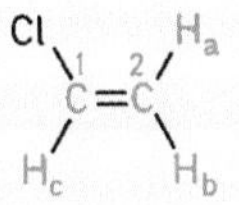 (Cl, H$_a$ / C=C / H$_c$, H$_b$)	C$_1$H$_a$: – 8.3 C$_1$H$_b$: + 7.1 C$_2$H$_c$: + 6.8		
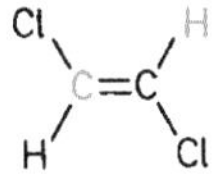 (Cl, H / C=C / H, Cl)	+ 0.8		
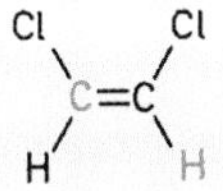 (Cl, Cl / C=C / H, H)	+ 16.0		
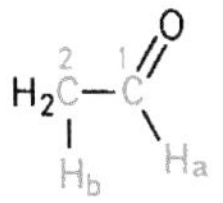 (H$_2$C—C(=O) / H$_b$, H$_a$)	C$_1$H$_b$: – 6.6 C$_2$H$_a$: + 26.7		
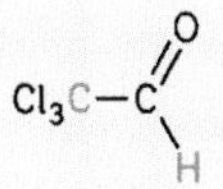 Cl$_3$C—C(=O)—H	+ 46.3		

Nuclear Magnetic Resonance Spectroscopy

Table 3.48 (Continued)

Compound	$^2J(C,H)$ (Hz)	$^3J(C,H)$ (Hz)	$^nJ(C,H)$ (Hz)
	CH_o: +1.1	CH_m: +7.6	CH_p: −1.2 (n = 4)
	C_1H_a: −3.4 C_2H_b: +1.4 C_3H_a: +0.3 C_3H_c: +1.6 C_4H_b: +0.9	C_1H_b: + 11.1 C_2H_c: + 7.8 $C_2H'_a$: + 5.1 $C_3H'_b$: + 8.2 C_4H_a: + 7.4	C_1H_c: − 2.0 (n = 4) $C_2H'_b$: − 1.2 (n = 4) $C_3H'_a$: − 0.9 (n = 4)
		CH_o: +4.1	CH_m: + 1.1 (n = 4) CH_p: + 0.5 (n = 5)
	C_2H_b: +7.4 C_3H_a: +4.7 $C_3H'_b$: +5.9	$C_2H'_b$: + 10.0 $C_2H'_a$: + 5.0 $C_3H'_a$: + 9.8	
	C_2H_b: +3.1 C_3H_a: +8.5 C_3H_c: +0.9 C_4H_b: +0.7	$C_2H'_a$: + 11.1 C_2H_c: + 6.8 $C_3H'_b$: + 6.6 C_4H_a: + 6.4	$C_2H'_b$: − 0.9 (n = 4) $C_3H'_a$: − 1.7 (n = 4)

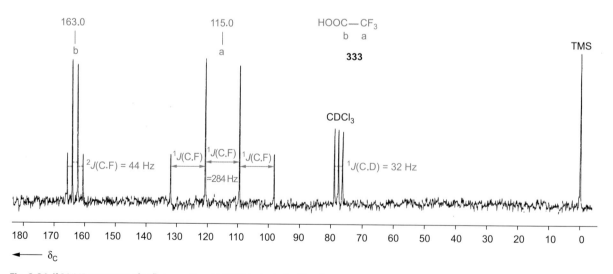

Fig. 3.84 ^{13}C NMR spectrum of trifluoroacetic acid (**333**) in deuteriochloroform.

Table 3.49 $^n\!J(^{13}C, {}^{19}F)$ coupling constants of selected compounds (in Hz)

Compound	$|{}^1\!J(C,F)|$	$|{}^2\!J(C,F)|$	$|{}^n\!J(C,F)|$
F—CH₃	162		
F—CFH₂	235		
F—CF₂H	274		
F—CF₃	259		
F—CF₂—C(OH)(OH)—CF₃	286	34	
F—CH₂—CH₂—CH₂—C₃H₇	167	20	³J(C, F): 5 ⁴J(C, F): < 2 ⁵J(C, F): ≈ 0
F₂C=CH₂	287		
F₂C=O (H)	369		
cyclohexane-F	171	19	³J(C, F): 5 ⁴J(C, F): ≈ 0
fluorobenzene	245	21	³J(C, F): 8 ⁴J(C, F): 3
CHF₂-benzene	272	32	³J(C, F): 4 ⁴J(C, F): 1 ⁵J(C, F): 0

Table 3.50 $^{13}C,{}^{15}N$ coupling constants of selected compounds (in Hz)

Compound	$|{}^1\!J(C,N)|$	$|{}^2\!J(C,N)|$	$|{}^n\!J(C,N)|$
H₃C—NH₂	5		
H₃C—⁺N(CH₃)₃	6		
H₃C—C(NH₂)=O	14	10	
H₃C—C≡N	18	3	
pyridine (NH₂)	11.4	2,7	³J(C, N): 1.3 ⁴J(C, N): < 1
pyrrole	0.5	2.4	³J(C, N): 3.9
pyridinium	12.0	2.1	³J(C, N): 5.3

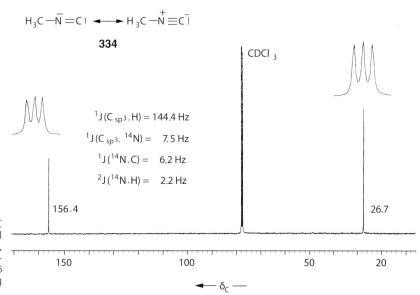

$$H_3C-\bar{N}=C\,I \longleftrightarrow H_3C-\overset{+}{N}\equiv C\,\bar{I}$$

334

CDCl₃

¹J(C$_{sp3}$.H) = 144.4 Hz

¹J(C$_{sp3}$.¹⁴N) = 7.5 Hz

¹J(¹⁴N.C) = 6.2 Hz

²J(¹⁴N.H) = 2.2 Hz

156.4

26.7

150   100   50   20

$\longleftarrow \delta_C \longrightarrow$

Fig. 3.85 ¹H-Broadband decoupled ¹³C NMR spectrum of methylisonitrile (methyl isocyanide, **334**) in deuteriochloroform, showing splitting (1:1:1) of the methyl signal (δ = 26.7) and of the isocyanide signal (δ = 156.4) as a result of the ¹³C,¹⁴N-coupling (Reggelin M, unpublished).

Nuclear Magnetic Resonance Spectroscopy

Table 3.51 $^{13}C,^{31}P$ coupling constants of selected compounds (in Hz)

Compound	$|^{1}J(C, P)|$	$|^{2}J(C, P)|$	$|^{n}J(C, P)|$
$P(CH_2-CH_2-CH_2-CH_3)_3$	11	12	$^{3}J(C, P)$: 13 $^{4}J(C, P)$: ≈0
$^{+}P(CH_2-CH_2-CH_2-CH_3)_4\ Br^{-}$	48	4	$^{3}J(C, P)$: 15 $^{4}J(C, P)$: ≈0
$Cl_2P-CH_2-CH_2-CH_2-CH_3$	44	14	$^{3}J(C, P)$: 11 $^{4}J(C, P)$: ≈0
$(H_5C_6)_3P{=}C(CH_3)_2$	123	14	
$P{\equiv}C-C(CH_3)_3$	38		
$O{=}P(CH_2-CH_2-CH_2-CH_3)_3$	66	5	$^{3}J(C, P)$: 13 $^{4}J(C, P)$: ≈0
$O{=}P(OC_2H_5)_2-CH_2-CH_2-CH_2-CH_2-CH_2-CH_3$	Hexyl: 141 Ethoxy:	5 6	$^{3}J(C, P)$: 16 $^{4}J(C, P)$: 1 $^{5,6}J(C, P)$: ≈0 $^{3}J(C, P)$: 5
$O{=}P(OC_2H_5)_2-C{\equiv}C-CH_3$	300	53	$^{3}J(C, P)$: 5
$(H_5C_2O)_2P-O-CH_2-CH_3$		11	$^{3}J(C, P)$: 5
$O{=}P(OC_2H_5)_2-O-CH_2-CH_3$		6	$^{3}J(C, P)$: 7
$(C_6H_5)_2P-$ phenyl ring	13	20	$^{3}J(C, P)$: 7 $^{4}J(C, P)$: 0.3
$O{=}P(C_6H_5)_2-$ phenyl ring	104	10	$^{3}J(C, P)$: 12 $^{4}J(C, P)$: 2
$(C_6H_5O)_2P-O-$ phenyl ring		3	$^{3}J(C, P)$: 7 $^{4}J(C, P)$: ≈0 $^{5}J(C, P)$: 1
$O{=}P(OC_6H_5)_2-O-$ phenyl ring		7	$^{3}J(C, P)$: 5 $^{4}J(C, P)$: 1 $^{5}J(C, P)$: 2

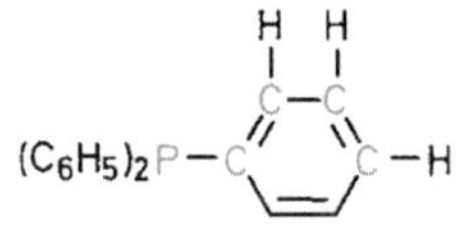

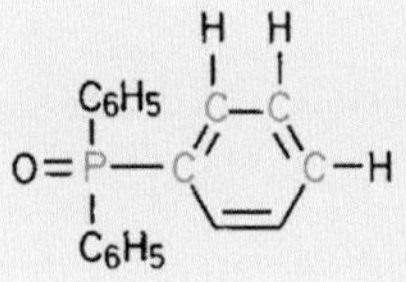

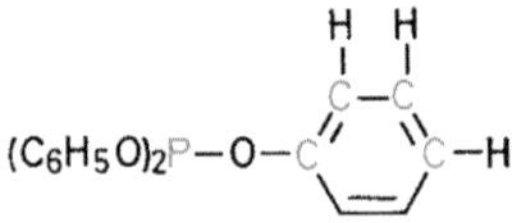

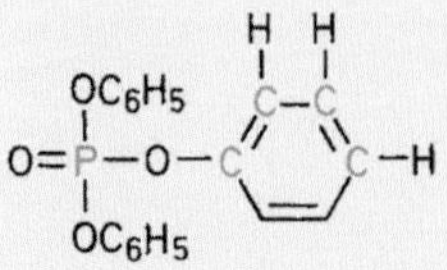

Table 3.52 $^1J(^{13}C,^{13}C)$ coupling constants for carbon atoms with different hybridization

sp-sp	$HC\equiv C-C\equiv CH$	+ 190.3 Hz
	$HC\equiv CH$	+ 171.5 Hz
	$HC\equiv C-C\equiv CH$	+ 153.4 Hz
sp^2-sp	$H_2C\!=\!C\!=\!CH_2$	+ 98.7 Hz
	$HC\equiv C-CH\!=\!CH_2$	+ 86.7 Hz
sp^2-sp^2	$H_2C\!=\!CH-CH\!=\!CH_2$	+ 68.6 Hz
	$H_2C\!=\!CH_2$	+ 67.6 Hz
sp^3-sp	$H_3C-C\equiv CH$	+ 67.4 Hz
sp^2-sp^2	(benzene ring structure)	+ 56.0 Hz
	$H_2C\!=\!CH-CH\!=\!CH_2$	+ 53.7 Hz
sp^3-sp^2	H_3C-C (cyclopentadiene)	+ 44.2 Hz
	$H_3C-CH\!=\!CH_2$	+ 41.9 Hz
sp^3-sp^3	H_3C-CH_3	+ 34.6 Hz
	H_2C-CH_2 (cyclopropane)	+ 12.4 Hz

(structure of a naphthalene/chromene derivative with numbering 9, 10, OCH$_3$, 8, 8a, 10a, 7, 4b, 4a, 1, 2, 345, 6, 5, 4, 3)

$^1J(C-1, C-2)$ = 70.0
$^1J(C-1, C-10a)$ = 67.2
$^2J(C-1, C-3)$ = 0.6
$^2J(C-1, C-4a)$ = 2.1
$^2J(C-1, C-10)$ = 0.7
$^3J(C-1, C-4)$ = 7.7
$^3J(C-1, C-4b)$ = 3.8

$^3J(C-1, C-9)$ = 4.4
$^4J(C-1, C-5)$ = 0.4
$^4J(C-1, C-8a)$ = 0.9
$^5J(C-1, C-6)$ = 0.7
$^5J(C-1, C-8)$ = 0.3
$^6J(C-1, C-7)$ = 0.4

3.4.8 Special Techniques

Since the very beginning, in the middle of the 20th century, NMR spectroscopy has developed terrifically. Many methods qualified originally as special measurements are nowadays routine. Sections 3.4.8 and 3.4.9 contain a survey of 1D and 2D techniques. Applications for structure determinations are in the foreground, whereas pulse sequences and other measurement details are only mentioned briefly. The following methods are discussed here:

- Enhancement of the measurement frequency/field strength.
- Application of special probeheads for the measurement of small quantities of samples or hardly soluble compounds.
- Spin decoupling: heteronuclear double resonance.
- J-Modulated spin-echo (attached proton test, APT).
- Spectrum integration.
- Use of lanthanide shift reagents.
- Specific isotope labeling.
- NOE measurements.
- Polarization transfer.

The majority of these methods have already been described for ^{1}H NMR spectroscopic applications (section 3.3.8, p. 145). Two-dimensional techniques are discussed in Section 3.4.9.

Enhancement of the Measurement Frequency/Field Strength

The **sensitivity** of the measurement scales with $B_0^{3/2}$. When the frequency for ^{13}C is enhanced from 100 to 150 MHz (by going from a 400- to a 600-MHz spectrometer), the sensitivity increases by a factor of $1.5^{1.5} \approx 1.8$. A questionable signal, which just disappears in the electronic noise, can be unambiguously identified by this procedure. Moreover, the enhancement of the strength B_0 of the magnetic field leads to a linear increase of $\Delta\nu$ of neighboring signals and thus to an improvement of the **resolution**. (Apart from the **spectral dispersion**, the **digital resolution** has to be considered, i.e. a sufficient density of the data points in the Hz scale must be assured).

Measurement of Small Quantities of Compounds

When only small amounts of a compound are accessible or when the solubility of a compound is very low, the use of a cryo-probehead at high field strength is advisable. Low temperatures guarantee in this case a low noise. Thus, viable ^{13}C NMR spectra of 0.1 mg = 100 µg of an organic compound (of not too high molecular mass) can be obtained.

Spin Decoupling: Heteronuclear Double Resonance

In ^{1}H NMR spectroscopy, coupled spectra are normally measured. Decoupling is only employed as a special experiment to add signal assignment in difficult structural problems (see p. 155 ff.). Routine ^{13}C NMR spectra are in contrast **proton-broadband-decoupled** (see Section 3.4.1, p. 173). This method of

Nuclear Magnetic Resonance Spectroscopy

measurement is frequently referred to as $^{13}C-\{^1H\}$ NMR (**proton noise decoupling**).

Because of the low natural abundance of ^{13}C of 1.1%, $^{13}C,^{13}C$ couplings are normally not observed. Couplings to ^1H, D, ^{19}F, ^{31}P, etc. however are observed. The proton broadband decoupling causes all couplings to protons to be removed, so that all the multiplets caused by $^{13}C,^1H$ coupling collapse to singlets. This simplifies the ^{13}C spectra enormously. There is also a considerable gain in signal intensity, caused by the removal of the splittings and additionally by the **NOE** (see Section 3.4.1, p. 173 ff.). To achieve proton broadband decoupling, a high intensity of radiation must be applied to the sample to cover the whole of the proton frequency range. This is in fact a case of heteronuclear multiple resonance. If the decoupling power is too low, only the quaternary C atoms give sharp, intense signals, all other C atoms give relatively broad signals. This phenomenon can be used systematically to identify the quaternary C atoms (**low power noise decoupling**).

A serious disadvantage of proton broadband decoupling is the total loss of information about the multiplicity of the signals derived from direct $^{13}C,^1H$ couplings, i.e., the differentiation between primary (CH_3), secondary (CH_2), tertiary (CH), and quaternary (C_q) carbon atoms. This disadvantage can be overcome by **proton off-resonance decoupling**, the J-modulated spin-echo experiment described on p. 204, or the distortionless enhancement by polarization transfer (DEPT) experiment described on p. 212. For off-resonance decoupling, a decoupling frequency outside the shift range of the protons is chosen and no noise modulation is applied. This leads to a reduction of the coupling constants, so that in general only the direct $^1J^R(C,H)$ couplings are visible. The magnitude of the reduced coupling constants $^1J^R$ increases with the magnitude of ^1J, with the difference between the relevant ^1H resonance frequency and the irradiation frequency, and with decreasing irradiation power; typical values of $^1J^R$ are in the range of ca. 30 to 50 Hz. It is useful to vary the conditions as appropriate to produce the minimal overlap of the **quartet, triplet, doublet,** and **singlet signals** of the CH_3, CH_2, CH groups, and quaternary C atoms. This assumes the equivalence of *geminal* protons. A H_A-C-H_B group has the spectrum of the X part of an ABX system, which is not always a 1:2:1 triplet.

Since the NOE is still effective in the case of off-resonance decoupling, the measurement time required is less than for

a fully coupled ^{13}C NMR spectrum (but still longer than for a normal, broadband decoupled spectrum).

In **Fig. 3.86**, the proton broadband decoupled, off-resonance, and coupled spectra of ethylbenzene (**346**) are shown for comparison. In the off-resonance spectrum **3.86b**, the methyl signal a is split into a quartet, the methylene signal b into a triplet, and the signals of the benzene CH groups c, d, and e into doublets. The quaternary C atom remains as a singlet. All the coupling constants are reduced. (Since the decoupler frequency was placed on the low-field side, the reduction of the coupling constants is stronger on the low-field side of the spectrum than on the high-field side.)

In the coupled spectrum **3.86c**, the multiplets overlap. As a result, the aromatic part is very confusing. **Fig. 3.86d, e** shows expansions of the aliphatic and aromatic parts of the spectrum, respectively. More splittings than those due to $^1J(C,H)$ couplings are apparent. The exact interpretation of coupled ^{13}C spectra is often made difficult by two factors: firstly by the already mentioned overlap of multiplets (a particular problem in the case illustrated for the signals d and e) and secondly by the appearance of coupling patterns which do not obey the **rules for first-order spectra**. This is clearest for C-atom c. The left- and right-hand parts of this doublet are different. The asymmetry of such spin multiplets is a result of the mixing of nuclear spin states of similar energies. This phenomenon can even occur when the ^1H NMR spectrum of the ^{12}C isotopomer is first order.

A heteronuclear double-resonance experiment can naturally also be carried out in such a way that the frequency of a single proton signal is irradiated. In the ^{13}C spectrum only the $^{13}C,^1H$ couplings arising from this proton signal are removed. For the remaining ^{13}C signals, the conditions are those of off-resonance decoupling, i.e., multiplets with reduced coupling constants are observed. This **selective decoupling (single frequency decoupling)** relies on the proton signals being reasonably well separated. (If necessary, this may be achieved by employing a higher magnetic field strength or by the use of lanthanide shift reagents.)

As an aid to understanding the various possibilities, a schematic representation of the different decoupling methods is given in **Fig. 3.87**.

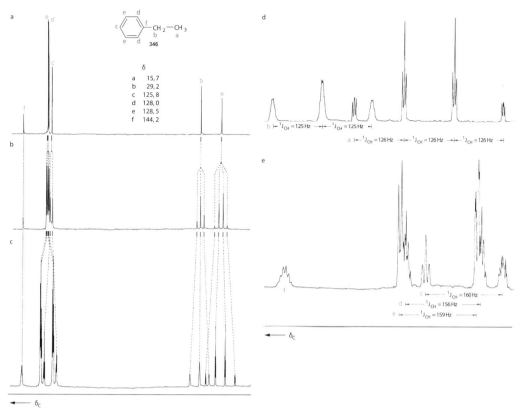

	δ
a	15,7
b	29,2
c	125,8
d	128,0
e	128,5
f	144,2

Fig. 3.86 ¹³C NMR spectra of ethylbenzene (**346**). (**a**) ¹H broadband decoupled; (**b**) off-resonance decoupled; (**c**) coupled spectrum (single resonance); (**d**) expansion of the aliphatic part of the coupled spectrum **c**; (**e**) expansion of the aromatic part of the coupled spectrum **c**.

Nuclear Magnetic Resonance Spectroscopy

Decoupling technique (irradiation in the ¹H spectral region)	¹³C spectra

a Single resonance

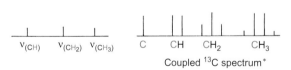

b Proton broadband decoupling

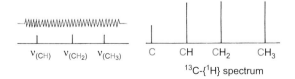

c Selective decoupling

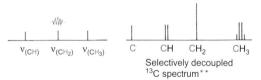

d Off-resonance decoupling

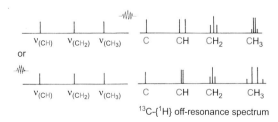

Fig. 3.87 Schematic diagram showing the various decoupling techniques in the measurement of ¹³C NMR spectra. *Only direct ¹³C,¹H couplings are represented in the diagrams. **Similar decoupling at the frequency ν(CH) leads to the collapse of the doublet.

A further example of the practical application of selective heteronuclear double resonance is shown for the heterocyclic compound **347**.

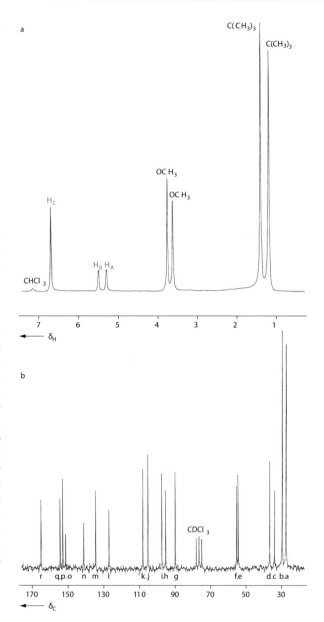

347

The ^{1}H NMR spectrum is depicted in **Fig. 3.88a**. While clearly separated signals are observed for the *t*-butyl groups, the methoxy groups, and the protons H_A and H_B of the oxepin ring, the signals of the benzene protons H_C are coincidentally isochronous.

In the proton broadband decoupled ^{13}C spectrum (**Fig. 3.88b**), 18 singlets are observed, as expected for the 18 types of chemically nonequivalent carbon atoms in the molecule (a–r). The interesting question is which ^{13}C signals correlate with the signals of H_A, H_B, and H_C. This was determined by successively irradiating at the frequency *v* of the protons; the ^{13}C spectra obtained from these heteronuclear double-resonance experiments are shown in **Fig. 3.88c–e**. On irradiating at the frequency of H_A (**Fig. 3.88c**), C_j appears as a singlet and C_g, C_i, and C_k as doublets. This allows the first unambiguous assignment: H_A is bonded to C_j. From the magnitudes of the reduced couplings for C_k, C_i, and C_g, it can additionally be concluded that H_B belongs to C_g. The final proof is provided by irradiating at v_B (**Fig. 3.88d**). For the protons H_C, the C atoms C_i and C_k are the only remaining possibilities. As a control experiment, an irradiation was carried out at v_C (**Fig. 3.88e**). As expected, both the doublets of C_i and C_k collapse to singlets. The quaternary C-atoms can be assigned with the aid of increment systems and from comparisons with related systems. Overall, this gives the following assignments:

C-2: r: δ = 166.7
C-3: j: δ = 106.0
C-4/7: p, q: δ = 154.5/156.0
C-5: g: δ = 90.6
C-5a: h: δ = 96.1
C-5b: l: δ = 128.0
C-6: i: δ = 98.3
C-7/4: p, q: δ = 154.5/156.0
C-8: k: δ = 108.9
C-9: m: δ = 135.6
C-9a: n: δ = 142.4
C-10a: o: δ = 152.9

OCH$_3$: e, f: δ = 54.8/55.6
C(CH$_3$)$_3$: a, b: δ = 27.5/29.6
c, d: δ = 34.1/36.9

Fig. 3.88 NMR spectra of **347** in CDCl$_3$: (**a**) ^{1}H spectrum; (**b**) ^{13}C–[^{1}H] spectrum (proton broadband decoupling).

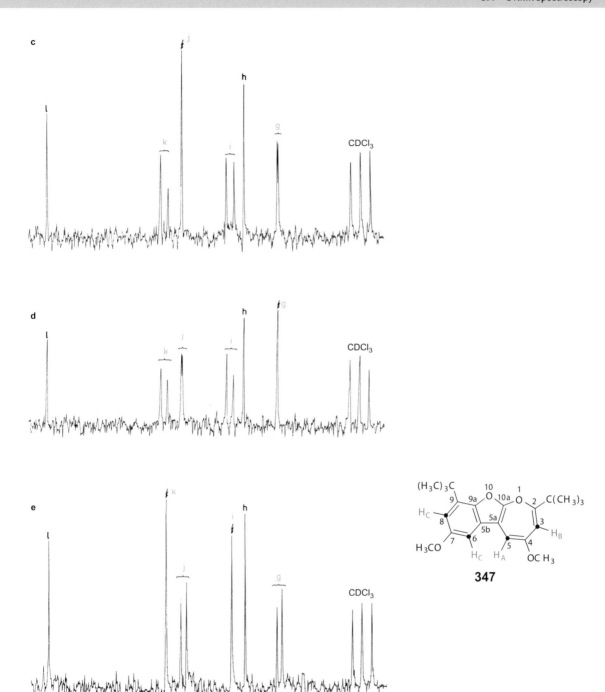

Fig. 3.88 (continued) (**c**) expansion of ¹³C spectrum from double-resonance experiment v_A (**d**) expansion of ¹³C spectrum from double-resonance experiment v_B; (**e**) expansion of ¹³C spectrum from double-resonance experiment v_C (Meier H, Schneider H-P, Rieker A, Hitchcock PB. 1978, Angew. Chem. **90**, 128).

Nuclear Magnetic Resonance Spectroscopy

Proton broadband decoupled spectra are the most often used ^{13}C NMR spectra. **Off-resonance spectra** and **spectra with selective proton decoupling** have lost a lot of their former significance because of the development of **APT** (below), **DEPT** (p. 211), **INEPT**, (p. 211), and **HSQC** (p. 212) techniques.

J-modulated Spin-Echo, APT

The *J*-modulated spin-echo, called **APT** (**A**ttached **p**roton **T**est), is an excellent method for the differentiation of the ^{13}C signals of CH_3, CH_2, CH groups, and quaternary carbon atoms C_q. The so-called spin-echo pulse sequence represents the spectrometer basis: a 90° excitation pulse is followed after a time τ by a 180° pulse, which inverts the transverse magnetization. After a further time interval τ the FID is observed. The ^1H decoupler is switched off during the second τ-period, which causes a modulation of the signal intensity by the ^{13}C,^1H coupling. The decoupler is switched back on immediately before the FID is accumulated, so the observed spectrum is decoupled, but the intensities of the signals of the CH_n groups depend on τ, or more exactly on expressions containing the functions $\cos(n\pi\tau J)$. If τ = 1/*J* is chosen (τ = 8 ms for $^1J(^{13}C,^1H)$ = 125 Hz), then positive signals are obtained for C_q and CH_2- groups ($\cos 0$ and $\cos 2\pi > 0$) and negative signals for CH- and CH_3- groups ($\cos \pi$ and $\cos 3\pi < 0$). (An analogous experiment with τ = 1/2*J* leads to a spectrum only showing the signals of quaternary carbons.) As an example, the broadband decoupled ^{13}C NMR spectrum of isobornyl chloride [exo-2-chloro-1,7,7-trimethyl-bicyclo[2.2.1]heptane (**348**)] and the corresponding APT spectrum are depicted in **Fig. 3.89**.

The corresponding *endo*-isomer [bornyl chloride (**349**)] gives an APT spectrum, which contains again five positive and five negative signals; however, the δ values of both isomers differ. The weaker shielding of C-6, caused by the *exo*-position of Cl in **348** in comparison to the *endo*-standing Cl in **349** is the most obvious difference.

exo-2-Chloro-1,7,7-tri-methyl-bicyclo [2.2.1]-heptane [isobornyl chloride (**348**)]

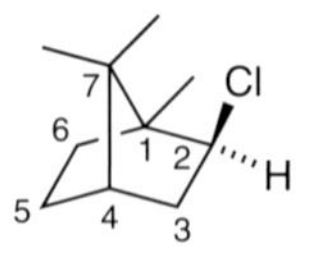

endo-2-Chloro-1,7,7-trimethyl-bicyclo[2.2.1]-heptane [bornyl chloride (**349**)]

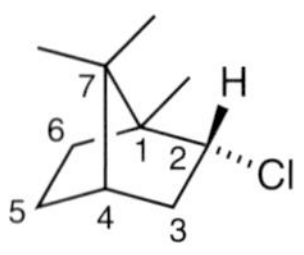

Position	δ	δ
C–1	49.7	50.7
C–2	68.2	67.1
C–3	42.4	40.3
C–4	46.0	45.3
C–5	26.9	28.3
C–6	36.2	28.3
C–7	47.3	47.8
1-CH_3	13.4	13.3
7-CH_3	20.4/20.1	20.5/18.4

Spectrum Integration

In contrast to ^1H NMR spectra, ^{13}C NMR spectra are not normally integrated (see Section 3.1.5, p. 98). This is because the signal intensities depend on the **relaxation times** of the various ^{13}C nuclei and in decoupled spectra on the different **NOEs**.

Fig. 3.90 shows the decisive **influence of the measurement conditions**. All four spectra were measured on the same solution of 4-ethyl-5-methyl-1,2,3-thiadiazole (**350**) using different values of the pulse length (PW). The other instrumental conditions remained unchanged (800 scans, 8 K, etc.). It can be seen that the relative intensity of the quaternary C atoms increases as the pulse length decreases. But decreasing the pulse length also worsens the S/N ratio considerably.

One possibility of obtaining spectra which yield useful integrations is the addition of paramagnetic **relaxation reagents**.

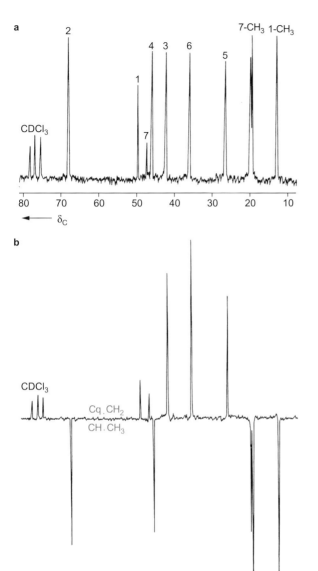

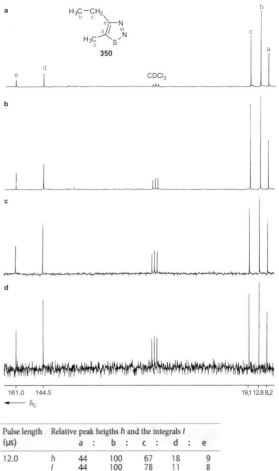

Fig. 3 .89 ^{13}C NMR spectrum of isobornyl chloride (**348**) in CDCl$_3$. (**a**) ^{1}H broadband decoupling. (**b**) *J*-modulated spin-echo spectrum (τ = 1/*J* = 8 ms).

Pulse length (µs)	Relative peak heigths *h* and the integrals *I*		a	:	b	:	c	:	d	:	e
12.0	*h*		44		100		67		18		9
	I		44		100		78		11		8
3.5	*h*		53		100		92		28		18
	I		53		100		97		21		18
1.0	*h*		66		100		83		69		39
	I		62		100		101		43		43
0.5	*h*		65		100		86		81		44
	I		57		100		120		52		48

Fig. 3.90 ^{1}H broadband decoupled ^{13}C spectrum of 4-ethyl-5-methyl-1,2,3-thiadiazole (**350**) in CDCl$_3$ under identical measurement conditions apart from pulse length; PW: 12 µs (**a**), 3.5 µs (**b**), 1.0 µs (**c**), 0.5 µs (**d**); *h* peak height, *I* peak integral.

Chromium and iron acetylacetonate have been found to be particularly useful (ca. 0.05 molar solution). Nondegassed samples contain dissolved oxygen, which also increases relaxation rates, particularly for the most slowly relaxing ^{13}C nuclei. The magnetic moment of the unpaired electrons makes a new relaxation process effective, which dominates over other mechanisms. The different relaxation times of the different nuclei all become the same, and the NOE also becomes ineffective. The relaxation reagent must not react with the sample or even form weak complexes with it; otherwise, the signals would be shifted when paramagnetic shift reagents are used. **Fig. 3.91** shows the integrated spectrum of 4-ethyl-5-methyl-1,2,3-thiadiazole (**350**) with added Cr(acac)$_3$.

The second method of obtaining useful integrations from ^{13}C spectra is the **inverse gated decoupling method**. (Normal **gated decoupling** is a method for measuring coupled spectra. The decoupler is only switched on during the period between the ending accumulation of one FID and the next pulse [pulse delay or relaxation delay]. This allows the NOE to enhance the intensity of signals of ^{13}C nuclei with attached protons without decoupling the signals. Fully coupled ^{13}C spectra are therefore best measured using this method of **pulsed proton decoupling**.) For inverse gated decoupling, the decoupler is switched on during the accumulation of the FID (so that a decoupled spectrum is obtained) but switched off during the pulse delay. The latter should be long enough to allow complete relaxation of the nuclei and decay of the NOE, which means that it must be longer than the longest ^{13}C T_1 in the sample (occasionally $T_1 \geq 100$ s!). Spin lattice relaxation and NOE then have no effect on the populations of the nuclear energy levels and therefore no effect on the signal intensities. The disadvantage of this method compared to the use of relaxation reagents is the long measurement times caused by the long pulse delays needed to ensure full relaxation.

Use of Lanthanide Shift Reagents

An introduction to the use of lanthanide shifts reagents in ^1H NMR was given on p. 153. Essentially the same considerations apply to their use in ^{13}C spectroscopy. The total shift is made up of a **pseudocontact** and a **contact term**. For chelate complexes of the lanthanides, the (Fermi) contact term is generally small, and the **McConnell–Robertson equation** applies approximately

(see p. 154). It therefore follows that ^1H and ^{13}C nuclei in comparable positions (relative to the lanthanide central atom) should have the same shift. The literature values for isoborneol (**351**) have been chosen as an example. The upper shift value refers to $Eu(fod)_3$, the middle value to $Eu(dpm)_3$, and the lower value to $Pr(fod)_3$. The blue figures give the lanthanide-induced shifts for ^{13}C, and the black figures the ^1H values. ($Eu(fod)_3$ and $Eu(dpm)_3$ shift to low field, and $Pr(fod)_3$ to high field.

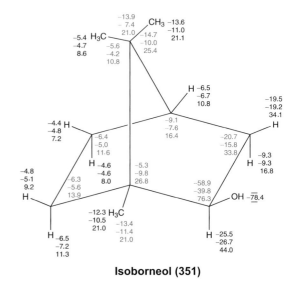

Isoborneol (351)

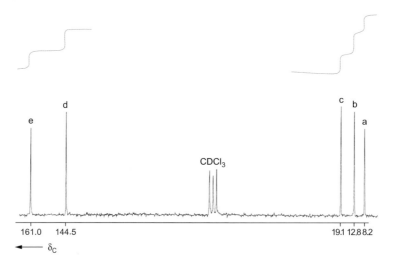

Fig. 3.91 ^{13}C NMR spectrum (proton broadband decoupled) of 4-ethyl-5-methyl-1,2,3-thiadiazole (**350**) in $CDCl_3$ containing $Cr(acac)_3$ with an integration curve.

Specific Isotopic Labeling

If one or more H atoms in a compound are substituted by **deuterium**, the ^{13}C spectrum changes in a very characteristic fashion. While CH, CH$_2$, and CH$_3$ groups only show singlets in the ^{1}H broadband decoupled ^{13}C spectrum, a 1:1:1 triplet is observed for CD–, CHD–, and CH$_2$D– groups, a 1:2:3:2:1 quintet for CD$_2$– and CHD$_2$– groups, and a 1:3:6:7:6:3:1 septet for CD$_3$– groups (see p. 92). The ^{13}C,D couplings are smaller than the corresponding ^{13}C,^{1}H couplings by a factor $\gamma_H/\gamma_D \approx 6.5$. ^{13}C,D long-range couplings are therefore usually not observed. The H/D exchange is also noticeable by a small isotope effect on the ^{13}C shift: for ^{13}C−D a high-field shift of ca. 0.2 to 0.7 ppm is observed, for ^{13}C−C−D the isotope shift is even less (0.11–0.15 ppm).

More important is the fact that the NOE of fully deuterated ^{13}C nuclei (i.e., with no C–H bond) is reduced and the relaxation time is increased. These two factors combine to cause a considerable loss of signal intensity. Together with the splitting caused by ^{13}C,D coupling described earlier, this often leads to the signals of fully deuterated ^{13}C nuclei disappearing partially or wholly into the noise. The signals of undeuterated ^{13}C nuclei are scarcely affected; at most they are split or broadened by the small ^{13}C,D coupling over two or three bonds.

H/D exchange is therefore a very valuable method for making unambiguous **signal assignments**. This goal can be approached even more directly with ^{13}C **labeling**. The incorporation of ^{13}C enriched carbon into specific positions in a molecule leads to intense ^{13}C signals. With a small number of accumulations, these may often be the only signals observed. ^{13}C labeling also affords a convenient method of measuring ^{13}C,^{13}C **coupling constants**.

In **Fig. 3.92**, this is demonstrated using cyclopentane carboxylic acid (**352**) as an example. **Fig. 3.92a** shows the ^{13}C NMR spectrum of the normal, unlabeled material. In **Fig. 3.92b**, where C atom a has been enriched with ^{13}C to a level of 50%, the signal of this carboxy C atom has greatly increased in intensity. The tertiary C atom b shows three peaks in the spectrum. The central line is the same as in **Fig. 3.92a**, being due to a ^{13}C nucleus at b with only ^{12}C neighbors. The two remaining lines are a doublet, caused by ^{13}C nuclei at b with neighboring ^{13}C nuclei at a. From the separation of the lines of this doublet, the coupling constant $^1J(C_b, C_a) = 56.8$ Hz can be obtained. (Because of the natural abundance of ^{13}C of 1.1% at nucleus b and the enrichment of nucleus a to a level of 50% every second ^{13}C nucleus at b has a ^{13}C neighbor at a, but only every 90th ^{13}C nucleus, C_a has a ^{13}C neighbor b).

The converse procedure, ^{13}C **depletion** below the natural abundance of 1.1%, is also of interest. The incorporation of ^{13}C depleted carbon leads to a disappearance of the relevant signals in the ^{13}C spectrum.

Since a wide variety of D-, ^{13}C-, and ^{12}C-labeled compounds are commercially available, the synthesis of suitable labeled compounds is becoming an increasingly popular method. Such labeling techniques are of special importance for following reaction pathways of organic and biochemical processes by NMR.

The **isotopic perturbation** method already described on p. 152 ff. for ^{1}H NMR spectroscopy can naturally also be applied to ^{13}C NMR. The central problem of **nonclassical ions**, the norbornyl cation (**353**), will be discussed as an example. There are two alternatives (**Fig. 3.93**):

(a) The double energy minimum model, with two "classical" ions in equilibrium.

(b) The single energy minimum model of a symmetrical, nonclassical ion.

After the incorporation of deuterium (X = D, Y = H/X = H, Y = D), only a very small splitting of the ^{13}C signals of C-1 and C-2 is observed. This is only compatible with the static model b. If there were a "perturbed equilibrium" of a fast Wagner–Meerwein rearrangement (Model a), the difference between the chemical shifts would be an order of magnitude larger.

Here borderline cases are naturally also feasible, involving a fast equilibrium between two species which only deviate very slightly from the symmetrical form.

HNOE Measurements

In order to measure the ^{13}C,^{1}H **HNOE**, the signal intensities of ^{1}H decoupled and ^{1}H coupled spectra would normally need to be measured. The intensity determination is particularly inaccurate for coupled spectra. Therefore, it is preferable in practice to compare the intensities of two decoupled spectra, one with and one without NOE. To produce the decoupled spectrum without the Overhauser effect, **gated decoupling** is used, with the decoupler only switched on during the pulse and the measurement of the FID. In the (long) pulse delay following accumulation of the FID, the decoupler is switched off. This method relies on the fact that the NOE builds up relatively slowly, whereas the decoupling occurs almost spontaneously.

If the ^{13}C relaxation only takes place through the dipole–dipole mechanism (i.e., as a result of the effect of the protons), the NOE η_c is given by the factor

$$\eta_c = \frac{\gamma_H}{2\gamma_c} = 1.988.$$

Much smaller values than this maximum are observed if other relaxation mechanisms are important.

An explicit example of the use of a heteronuclear NOE experiment is dimethylformamide (**49**; **Fig. 3.24**, p. 109). At 25°C the following NOE factors are observed:

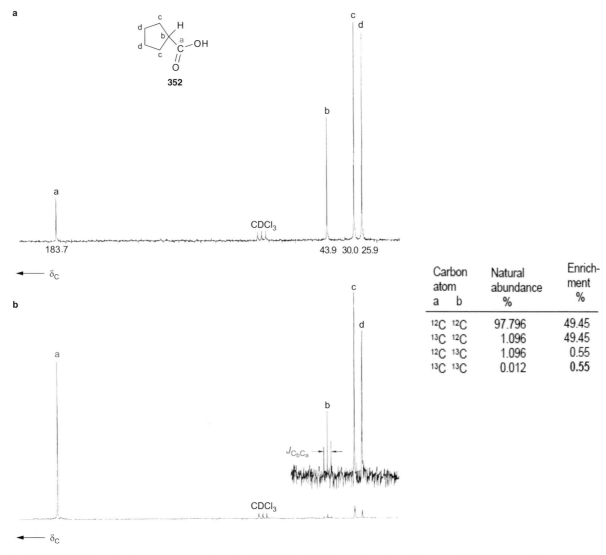

Fig. 3.92 ^{13}C NMR spectra of cyclopentane carboxylic acid (**352**) in CDCl$_3$ (proton broadband decoupled) (**a**) with natural ^{13}C content and (**b**) with 50% enrichment of the carboxy C atom a (Timm U, Zeller K-P, Meier H. 1977, Tetrahedron **33**, 453).

	CO	CH$_3$ (b)	CH$_3$ (a)
η	1.4	1.8	1.4

49

(a) is the CH$_3$- group *anti* to the formyl hydrogen. Its signal lies at higher field and has a lower intensity than that of (b) in the broadband decoupled spectrum. With increasing temperature, the exchange rate of the two methyl groups increases, and the difference between the two NOE factors decreases.

NOE measurements are of great importance for structural assignments, since **connectivities through space** usefully complement the **connectivities through bonds** (coupling; cf. p. 89 and 191). **Fig. 3.94** for example shows a heteronuclear NOE measurement of compound **354**. Signal assignments are difficult because of the many heteroatoms in 2-bromo-5-methyl-7H-1,3,4-thiadiazolo[3,2-a]pyrimidin-7-one. This is especially so for the low-field signals of C-7 and C-8a. A heteronuclear difference NOE-spectrum with irradiation of the singlet proton signal of H-6 (δ = 6.09) shows strong enhancements of the intensities of the signals of C-7 and C-5 (the latter is broadened by coupling

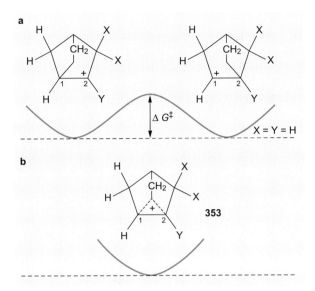

Fig. 3.93 Norbornyl cation: (**a**) classical; (**b**) nonclassical.

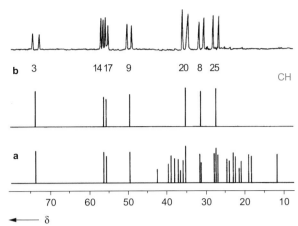

Fig. 3.95 ¹³C spectra of cholesteryl acetate (**355**) in CDCl³. (**a**) Normal broadband decoupled spectrum of the saturated carbon atoms. (**b**) Subspectrum of the methine groups (CH) coupled and decoupled (DEPT technique).

C	δ
C-2	132.5
C-5	147.5
C-6	109.2
C-7	166.1
C-8a	163.7
CH₃	17.4

354

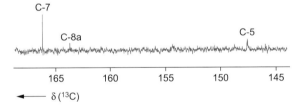

355

CH₃	δ	CH₂	δ	CH	δ	Cq	δ
C–18	12.0	C–1	37.3	C–3	73.7	C–5	139.9
C–19	19.3	C–2	28.2	C–6	122.6	C–10	36.7
C–21	18.9	C–4	38.4	C–8	32.2	C–13	42.5
C–26	22.7	C–7	32.2	C–9	50.4	C–28	169.6
C–27	22.9	C–11	21.3	C–14	57.0		
C–29	20.9	C–12	32.5	C–17	56.6		
		C–15	24.6	C–20	36.1		
		C–16	40.1	C–25	28.2		
		C–22	36.7				
		C–23	24.3				
		C–24	39.8				

Fig. 3.94 Heteronuclear NOE-difference spectrum of **354** in CD₃SOCD₃: increased signal intensity for C-7 and C-5 on irradiation of 6-H [δ (¹H) = 6.09]. (Safarov S, Kukaniev MA, Karpuk E, Meier H. 2007, J. Heterocycl. Chem. **44**, 269.)

to the methyl protons), but only a weak enhancement of the signal of C-8a, which is considerably further away from H-6.

Polarization Transfer, DEPT, and INEPT

From **Table 3.1** (see p. 87) it can be seen that the natural abundance and the sensitivity of certain NMR measurable nuclei like ¹³C and ¹⁵N are very low. One reason for the low sensitivity is the low population difference between the two spin states (see Section 3.1.1, p. 86). Various methods are available **SPI** (**s**elective **p**opulation **i**nversion), **INEPT** (**i**nsensitive **n**uclei **e**nhanced by **p**olarization **t**ransfer), and **DEPT** (**d**istortionless **e**nhancement by **p**olarization **t**ransfer) **pulse sequences**) which allow the transfer of the larger population difference for protons to a less sensitive nucleus (¹³C, ¹⁵N, etc.) present in the same molecule. This **polarization transfer** causes the transitions (absorption and emission!) to become stronger. The effect exceeds that caused by the NOE.

Polarization transfer can therefore be used to increase the signal intensity of ^{1}H decoupled or ^{1}H coupled ^{13}C spectra. A second application of **INEPT** (**i**nsensitive **n**uclei **e**nhanced by **p**olarization **t**ransfer) or **DEPT** (**d**istortionless **e**nhancement by **p**olarization **t**ransfer) is the measurement of spectra in which peaks can be selected by their multiplicity (spectral editing). For example, a ^{1}H decoupled or ^{1}H coupled ^{13}C spectrum of cholesteryl acetate (**355**) can be obtained in which only the CH- groups are present (**Fig. 3.95**), or alternatively only the CH$_2$- or CH$_3$- groups; **355** has a total of 29 chemically nonequivalent C atoms. The congestion of signals in the region between δ = 20 and 40 is particularly high. In the normal decoupled or off-resonance decoupled spectra, the signals are hopelessly overlapped.

The **DEPT** technique is normally preferred over the **INEPT** method, because the measurement is somewhat easier and coupling patterns have the genuine intensity distribution. The D in DEPT stands for distortionless; the triplet of a CH$_2$ group has in a coupled DEPT spectrum the expected distribution 1:2:1 and the quadruplet of a CH$_3$ group the distribution 1:3:3:1. **^{1}H-decoupled DEPT** spectra have become important for the determination of the multiplicities of ^{13}C signals. According to the pulse angle used in the proton channel, three variants have to be distinguished: DEPT 45, DEPT 90, and DEPT 135. **Fig. 3.96** demonstrates that the ^{13}C signals emerge positively, negatively, or not at all in the corresponding DEPT subspectra.

A selective CH subspectrum is obtained by DEPT 90, a CH$_2$ subspectrum by the subtraction [DEPT 45 – DEPT 135], and a CH$_3$ subspectrum by [DEPT 45 + DEPT 135 – x DEPT 90].

Fig. 3.97 shows as an example the DEPT 135 spectrum of 4-propoxybenzaldehyde (**356**). It contains positive signals for CHO, HC-2, and HC-3, negative signals for α-CH$_2$ and β-CH$_2$, and no signal for the quaternary carbon atoms C-1 and C-4.

DEPTQ, a modified DEPT sequence, additionally indicates signals for the quaternary C-atoms. A variation of the INEPT sequence, which shows the same effect, is known as **PENDANT** (polarization enhancement nurtured during attached nucleus testing). The PENDANT spectrum resembles the DEPT 135 measurement, but contains negative, mostly weak additional signals for quaternary C atoms.

3.4.9　Multidimensional ^{13}C NMR Spectra

There are two major categories of 2D ^{13}C NMR techniques, namely ^{13}C,^{1}H shift correlation and J-resolved spectroscopy. The ^{13}C,^{1}H shift correlation represents a very powerful method for structure determinations. First, correlations of directly bound nuclei ^{13}C–^{1}H shall be discussed here. Three techniques have to

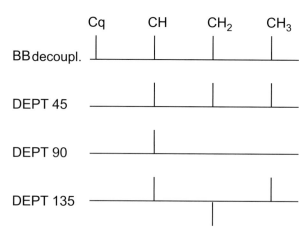

Fig. 3.96 Schematic representation of the signal phases in DEPT measurements.

be mentioned, which differ in their pulse sequences in the ^{13}C as well as in the ^{1}H channel:

- **HETCOR** (**het**eronuclear shift **cor**relation)
- **HMQC** (**h**eteronuclear **m**ultiple **q**uantum **c**orrelation)
- **HSQC** (**h**eteronuclear **s**ingle **q**uantum **c**orrelation)

The long-known **HETCOR** experiment uses the ^{13}C detection (acquisition of the FID), the polarization transfer ^{1}H → ^{13}C, and broadband decoupling which eliminates ^1J (^{13}C,^{1}H).

HMQC and **HSQC** are inverse methods, which use the ^{1}H detection (acquisition of the FID), polarization transfer, and normally work with ^1J(C,H) decoupling. The decoupling can be omitted in all three methods. ^{1}H and ^{13}C nuclei have large differences in natural abundance and sensitivity (p. 87). Since the S/N ratio is proportional to $\gamma^{5/2}$, the sensitivity increase of inverse techniques amounts to $[\gamma(^1\mathrm{H})/\gamma(^{13}\mathrm{C})]^{5/2} \approx 30$. Although the real factor is lower, HMQC and HSQC guarantee a large saving of time in comparison to HETCOR. An overnight HETCOR measurement corresponds to about 1 hour measurement of HMQC or HSQC.

HSQC has a more complex pulse sequence than **HMQC**, but leads to smaller crosspeaks. This effect has an important advantage for closely lying ^{13}C signals. Originally HSQC was used preferentially for biological probes, but the better resolution also has advantages for small organic molecules. Nowadays, HSQC measurements are performed more than **h**eteronuclear **m**ultiple-**b**ond **c**orrelation (HMBC) and much more than HETCOR.

As in (^{1}H,^{1}H)COSY spectra (p. 164), two **frequency domains** F$_1$ and F$_2$ exist. Since they belong here to different nuclei, the 2D plot does not have a diagonal with the corresponding symmetry. The F$_2$ domain is used normally for the nucleus whose FID acquisition is made. Illustration of crosspeaks:

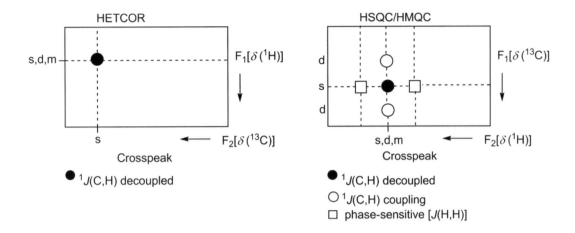

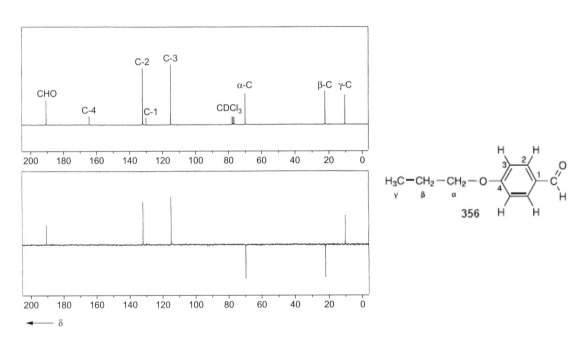

Fig. 3.97 Proton broadband decoupled ¹³C spectrum (top) and DEPT-135 spectrum of 4-propoxybenzaldehyde (**356**).

The combination **HSQC-TOCSY** (**Fig. 3.98**) or the **relayed techniques** (see also p. 166) reveal besides the directly bound protons the correlation to other, in general, vicinal protons of the spin system.

HMBC (**h**eteronuclear **m**ultiple-**b**ond **c**orrelation) spectra show the correlation of ¹³C and ¹H through two, three and, seldom, four bonds [²J(C,H), ³J(C,H), ⁴J(C,H)]; in general, ³J(C,H) couplings are preferred.

$$^1H-C-C-^{13}C$$

Fig. 3.98 HSQC-TOCSY scheme for a C_3H_3 segment.

The parameters of these measurements can be selected in a way that the ^1J(C,H) couplings are visible as crosspeaks, which have a doublet structure. Long-range HETCOR measurements are analogous to HMBC spectra but afford a much longer time than the inverse HMBC technique.

Fig. 3.99 shows the HMQC and the HMBC spectrum of the dithienylethyne **357**. The protons 3′-H and 4′-H as well as the protons 3″-H and 4″-H form AX spin systems, each. The differentiation between 4′-H and 4″-H is possible by the electronic effects of the push–pull compound. The crosspeaks in **Fig. 3.99a** lead to the assignment of 3′-H, 4′-H, 3″-H, and 4″-H; the crosspeaks in **Fig. 3.99b** to the assignment of C-2′, C-5′, C-2″ and C-5″ and of the acetylenic carbon atoms C-1 and C-2. The vertical and horizontal lines in **Fig. 3.99a** mark the ^1J(C,H) couplings and in **3.99b** the ^3J(C,H) couplings. The HMBC spectrum contains additionally six ^2J(C,H) couplings.

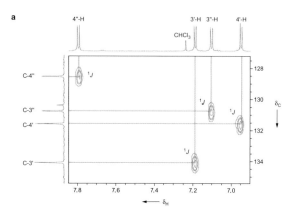

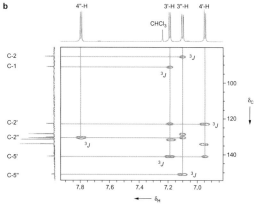

Fig 3.99 (a) HMQC spectrum of **357** in CDCl$_3$ (heteroaromatic part); **(b)** HMBC spectrum of **357** in CDCl$_3$ (heteroaromatic and alkyne part) (Meier H, Mühling B, Theisinger S, unpublished).

	$\delta(^1H)$	^3J(H,H)	$\delta(^{13}C)$
1			91.2
2			85.4
2′			122.8
3′	7.19	3.9 Hz	134.1
4′	6.95	3.9 Hz	131.6
5′			140.9
2″			130.4
3″	7.10	4.4 Hz	130.8
4″	7.80	4.4 Hz	128.6
5″			150.9

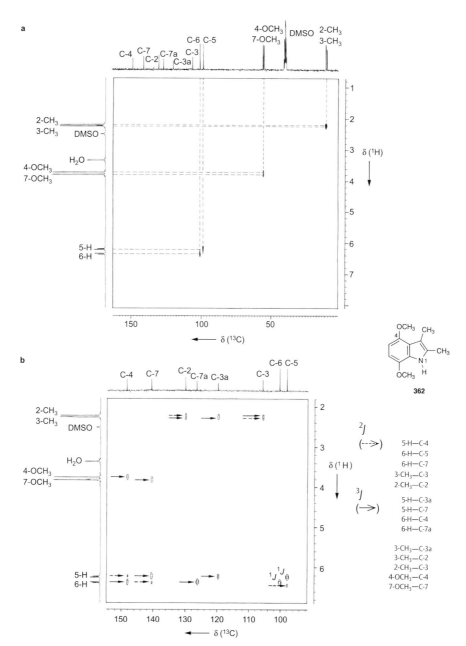

Fig. 3.100 ¹H,¹³C-Heteronuclear shift correlation of 4,7-dimethoxy-2,3-dimethylindole (**358**) in DMSO-d₆ (Pindur UF, unpublished). (**a**) HETCOR; (**b**) long-range HETCOR.

Fig. 3.100a shows the HETCOR and **Fig. 3.100b** the long-range HETCOR spectrum of the indole derivative **358**. The obtained information corresponds principally to the results obtained by HSQC or HMQC and HMBC; however, the inverse measurement of **357** needs much less time than the measurement of **358**. The best-known long-range HETCOR technique is called **COLOC** (correlation via **lo**ng-range **c**ouplings). Nowadays, it is completely replaced by HMBC.

Fig. 3.101 illustrates the ¹³C,¹H-heteronuclear shift correlation of the ester **190**. In this case it is better to dispense with the decoupling, so that all ¹J(¹³C,¹H) couplings become clearly visible. Moreover, a number of higher couplings ⁿJ(¹³C,¹H) for n = 2, 3, and 4 are also visible. The relatively large ³J(CN,α-H) is even resolved as a doublet.

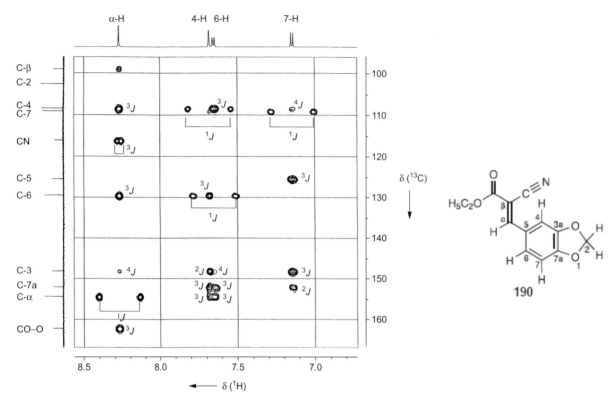

Fig. 3.101 Section of the $^{13}C,^{1}H$-heteronuclesr shift correlation of (*E*)-3-(1,3-benzodioxol-5-yl)-2-cyanoacrylic acid ethyl ester (**190**) with $^{13}C,^{1}H$ couplings (Soliman A, Meier H, unpublished).

In an analogy to the H-relayed ($^{1}H,^{1}H$)COSY experiment, the magnetization can be carried over from one proton, via another proton, to a ^{13}C nucleus. A part of an H-relayed ($^{1}H,^{13}C$)COSY spectrum of sucrose (**192**) is shown in **Fig. 3.102**. By comparison with a normal $^{13}C,^{1}H$ shift correlated 2D NMR spectrum, the contour plot contains additional peaks, arising from the **relayed technique** employed, which correspond to ^{13}C nuclei and protons on neighboring carbon atoms ($^{2}J_{C,H}$):

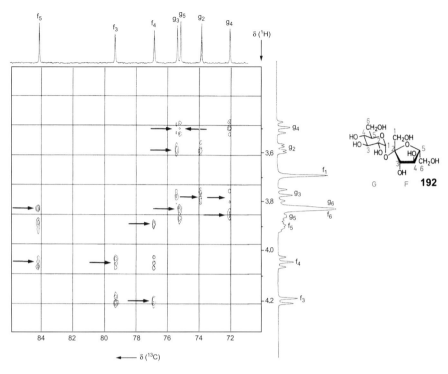

Fig 3.102 Expansion of part of an H-relayed H,C-COSY spectrum of sucrose (**192**). The *arrows* indicate correlation peaks which correspond to ²*J*(C,H) couplings and therefore do not appear in the normal HETCOR spectrum.

Fructose ring	C-3/4-H (f₃ → f₄)
	C-4/3-H and 5-H
	C-5/4-H and 6-H
	(C-1, 2, 6 not shown)
Glucose ring	C-2/3-H
	C-3/2-H and 4-H
	C-4/3-H and 5-H
	C-5/4-H and 6-H
	(C-1, 6 and 1-H not shown)

Fig. 3.103 depicts a phase-sensitive HSQC spectrum of Lepistol (**359**), a natural product. The diterpene structure contains 20 different C atoms, whereof four represent stereogenic centers (asymmetric C atoms). Moreover **359** contains 28 different H atoms. Many NMR signals lie very close together—in particular in the ranges 20 < δ (¹³C) < 50 and 0.9 < δ (¹H) < 2.8. Exactly this section of the HSQC spectrum is reproduced in **Fig. 3.103**. The red crosspeaks belong to CH₂ groups. H₂C-1, H₂C-6, H₂C-8, and H₂C-9 contain diastereotopic protons. Therefore, C-1, C-6, C-8, and C-9 have two crosspeaks each. The crosspeaks of CH and CH₃ groups are drawn in blue. The phase-sensitive

measurement preserves the ¹H,¹H couplings and makes a DEPT 135 spectrum unnecessary. The complete signal assignment of the identified relative configuration of **359** is based on the depicted HSQC spectrum and on additional HMBC and NOESY measurements.

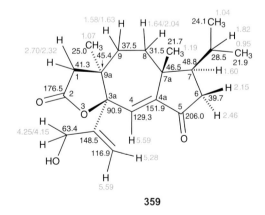

359

Heteronuclear *J*-resolved spectroscopy is represented by contour plots, which show the chemical shifts of ¹³C nuclei in the F₂

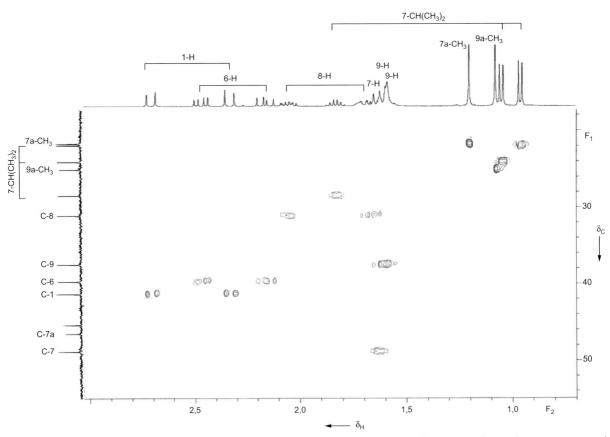

Fig. 3.103 Phase-sensitive HSQC spectrum of Lepistol (**359**) in CDCl$_3$. (The numbering corresponds to 3-oxaazulene as the parent compound and not to the terpene nomenclature). (Kolshorn H, Opatz T, unpublished .)

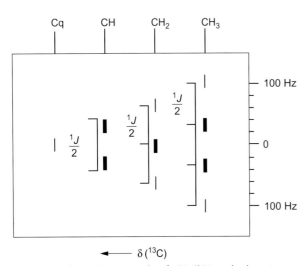

Fig. 3.104 Schematic contour plot of a ^{1}H–^{13}C J-resolved spectrum.

domain and the ^{13}C^{1}H spin patterns separated in the F_1 domain. **Fig. 3.104** contains a schematic diagram of this 2D technique. The advantage of this method—compared to a normal coupled ^{13}C NMR spectrum—is given by the separation of otherwise superimposed spin patterns. Owing to the **gated decoupling** (p. 208), the line distances of the doublets, triplets, or quadruplets correspond to 0.5 $^1J(^{13}$C,^{1}H).

As outlined in Section 3.4.7 (see p. 194), ^{13}C,^{13}C couplings give valuable information about the C–C bonds present in a molecule. Because of the low natural abundance of ^{13}C (1.1%), only about 0.01% of the molecules contain **two** anisochronous ^{13}C nuclei and therefore fulfill the condition for the appearance of ^{13}C,^{13}C coupling in the spectrum. The main signal arising from molecules with **one** ^{13}C nucleus is accompanied by low-intensity satellites, which are however difficult to measure. This is particularly so when the coupling constants are small and the satellites lie under the wings of the main signal.

The **INADEQUATE** technique (**i**ncredible **n**atural **a**bundance **d**ouble **qua**ntum **t**ransfer **e**xperiment) suppresses the main signal. A special pulse sequence excites the **double quantum**

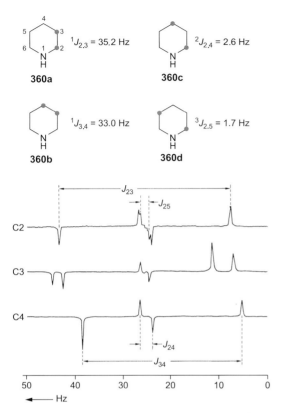

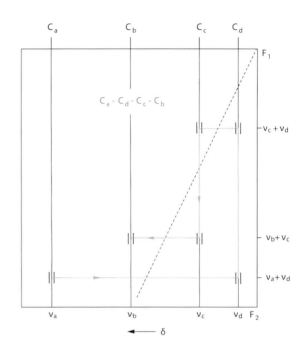

Fig. 3.106 Schematic diagram of a 2D INADEQUATE spectrum of a C_4 chain with unknown connectivity. Each pair of ¹³C nuclei connected by a ¹J coupling gives an AX spin pattern at the F_1 position which corresponds to the sum of the chemical shifts. (The center points of the connected contours lie on the straight line $F_1 = 2F_2$.)

Fig. 3.105 Sections of the ¹³C NMR spectrum of piperidine (**360**): INADEQUATE experiment for the determination of the ¹³C,¹³C couplings (Bax A, Freeman R, Kempsell ST. 1980, J. Am. Chem. Soc. **102**, 4850).

transitions. First, a practical, 1D example, piperidine (**360**), is shown in **Fig. 3.105**. Four couplings are observed, arising from the isotopomers (**360a–d**) which are present in low concentration in normal piperidine.

Each AX (AB) spectrum is made up of four lines. In **Fig. 3.105** the two components of each doublet appear in antiphase. The experimental conditions can be adjusted so that they appear as in phase doublets.

If the 1D INADEQUATE spectrum becomes too complicated, a 2D INADEQUATE spectrum is recommended. This separates the AX spin systems of neighboring ¹³C nuclei (see **Fig. 3.106**). The observed δ values of the ¹³C nuclei lie on the F_2-axis, and the double quantum frequencies on the F_1-axis. The **connectivities** over the C–C bonds follow the blue arrow and prove in this case that the carbon atoms are arranged in the chain in the sequence C_a–C_d–C_c–C_b. The line distances within the A and X parts correspond to the respective ¹J(¹³C,¹³C) couplings.

Fig. 3.107 shows the 2D INADEQUATE spectrum of the 1,6-naphthyridine derivative **361**, which was measured with

a low-noise cryo-probehead (150 MHz, 28 h). The vertical axis gives the sum of δ values of two coupling nuclei, for example 137.6 + 160.3 = 297.9 ppm for the pair C-1′–C-2. Starting at C-4′ (δ = 130.4 ppm), the ¹J(¹³C,¹³C) couplings lead to the trace C-4′–C-3′/5′–C-2′/6′–C-1′–C-2–C-3–C-4–C-4a –C-5 and C-8a. Hence the carbon scaffold of **361** is largely established. The δ values of C-7 and C-8 can be obtained by making use of a 2D ¹⁵N measurement. (The signal of C-8 at 67.9 ppm cannot be seen in **Fig. 3.107**. This extremely low δ value of an sp² carbon atom results from the effect of the enamine and the cyano group. The ¹H chemical shifts of **361** are drawn in blue, the ¹³C chemical shifts in black, and the ¹J(¹³C,¹³C) couplings in red).

Nuclear Magnetic Resonance Spectroscopy

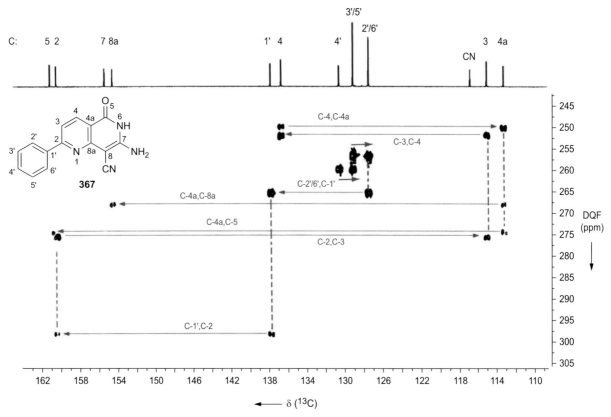

Fig. 3.107 2D-INADEQUATE spectrum of **361** in CD$_3$SOCD$_3$. The lines mark the trace of the $^1J(^{13}C,^{13}C)$ couplings (from Liermann JC, Elnagdi MH, Meier H. 2013, Magn. Res. Chem. **51**, 16).

Two neighboring ^{13}C nuclei have a low natural abundance of 0.01%, which is a big handicap. When in addition the solubility of the studied compound is low, INADEQUATE measurements cannot be performed.

While INADEQUATE and the other 2D methods discussed so far give information about **connectivity through bonds**, NOE experiments answer questions of **connectivity through space**.

Two-dimensional heteronuclear Overhauser spectroscopy (HOESY) for spin pairs $^{13}C,^1H$ is well established. It certainly has interesting applications whenever suitable $^1H,^1H$ pairs do not exist for the proof of a through-space connectivity. However, the sensitivity of HOESY is relatively low.

Chemically equivalent or randomly isochronous nuclei give only one signal in the NMR spectrum. Therefore, NOEs between such nuclei cannot be measured directly. However, the NOE of chemically equivalent protons can be measured, when the isotopomer is used, which contains a ^{13}C nucleus (natural abundance) on one side. The corresponding 2D techniques are called **gs-NOESY-HSQC** and **gs-HSQC-NOESY**, respectively; gs is the abbreviation of gradient-selected. In the first variant, a

magnetization transfer occurs from ^{12}CH to ^{13}CH groups; and in the second variant vice versa. An important example was published for the light-sensitive component **362** of the red pigment of human hair. The yellow (Z)-configuration is converted by light to the (E)-configuration.

The gs-HSQC-NOESY measurement of a mixture of maleic acid diethyl ester (Z)-**363** and fumaric acid diethyl ester (E)-**363** is presented here as an explicit example.

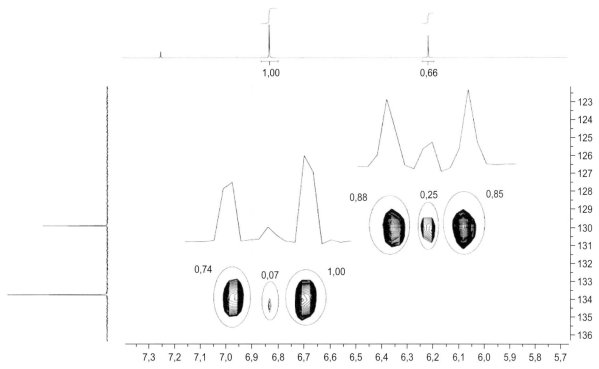

Fig. 3.108 gs-HSQC-NOESY measurement of the olefinic signals of a mixture of maleic acid diethyl ester (Z)-**363** and fumaric acid diethyl ester (E)-**363**. Integration of the singlet crosspeaks reveals that the minor component [δ (¹H) = 6.2 ppm, δ (¹³C) = 130 ppm] has the larger NOE and is therefore (Z)-**363** (Meier H, Liermann JC, unpublished).

The olefinic protons of (Z)-**363** have a much shorter distance than those of (E)-**363**. Hence (Z)-**363** should exhibit a much stronger NOE than (E)-**363**. **Fig. 3.108** illustrates the olefinic ¹H and ¹³C signals of the mixture of (Z)- and (E)-**363**. The ¹²C¹H singlets lie in the center of the ¹³C¹H doublets. Integration of the singlet crosspeaks yields the ratio 7:25, and integration of the singlets in the ¹H NMR spectrum yields the ratio 100:66. The minor component has obviously the stronger NOE and is therefore the (Z)-isomer. Integration of the signals is an essential part for such an investigation, when the distances of the studied nuclei do not differ drastically. If only one component is present, the signal ratio of ¹³C¹H and ¹²C¹H must be evaluated.

Exchange processes can also be studied in the ¹³C NMR by applying a 2D **EXSY** method. Bullvalene (**66**) shows a fast degenerate valence isomerization between 10!/3 identical valence isomers (p. 112). While a single sharp signal for the 10 carbon atoms can be observed for the fluctuating structures at 128°C, four signals (a, b, c, and d) appear at −60°C for the four different types of carbon atoms (**Fig. 3.109**). The arrows in formula **66** indicate the process (Cope rearrangement), in which the six conversions

(d → a, a → d, etc.) occur. Therefore, six crosspeaks can be expected in the ¹³C-EXSY spectrum. **Fig. 3.109** shows four crosspeaks (d → a, a → d, 2a → 2c, and 2c → 2a). The crosspeaks for c → b and b → c are difficult to see, because the signals b and c of the olefinic C atoms are lying very close together. The size of the crosspeaks indicates that two C atoms of type a rearrange to type c and vice versa, but only one d → a and one a → d take place. This NMR study performed by Nobel Prize winner R. Ernst is an excellent proof for the degenerate Cope rearrangement as a mechanism of the fluctuating bullvalene structure.

For very complex structures, a variety of 3D techniques have been developed. For example, it is of interest to combine the connectivity through bonds (2D) and the connectivity through space (2D) in a **3D measurement**. A further application is the correlation of three different nuclear types. **Fig. 3.110** shows a **3D ¹H−¹³C−¹⁵N correlation spectrum** of ¹³C and ¹⁵N labeled ribonuclease T1, measured at 750 MHz. Each crosspeak indicates the ¹H and ¹⁵N shift (cf. Section 3.6.3) of an N–H group in this higher protein. In the third dimension, the connectivity to the α-C atom of the next amino acid is indicated. The method allows a signal assignment related to the amino acid sequence.

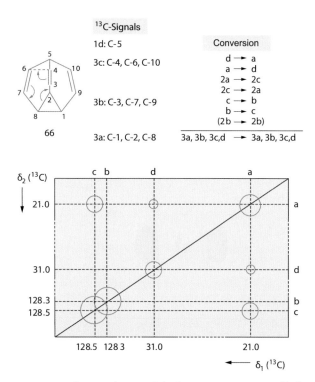

Fig. 3.109 Schematic diagram of the ^{13}C-EXSY measurement of bullvalene (**66**) at −60°C in [D$_8$]THF/CS$_2$.

3.4.10 Solid-State Spectra

For various reasons, high-resolution spectra cannot be obtained from solids using the normal PFT technique. Dipole–dipole- and quadrupole-field gradient interactions, which average to zero in solution, lead to extremely broad lines: the anisotropy of the chemical shift leads to broad, complex line shapes, and the spin–lattice relaxation times of nuclei such as ^{13}C are very long in solids.

The chemical shift is described by a second-order **shift tensor**, the matrix of which can be diagonalized. In solution the average value σ_{iso} is measured, which is a third of the trace of the matrix.

$$\begin{pmatrix} \sigma_{11} & 0 & 0 \\ 0 & \sigma_{22} & 0 \\ 0 & 0 & \sigma_{23} \end{pmatrix} \quad \sigma_{iso} = 1/3\,(\sigma_{11} + \sigma_{22} + \sigma_{23}).$$

In a single crystal the individual components σ_{ii} can be determined. The completely symmetrical case $\sigma_{11} = \sigma_{22} = \sigma_{33}$ occurs in methane; cylindrical symmetry, e.g., in acetylene, leads to $\sigma_\perp$ and $\sigma_\parallel$; otherwise, the tensor is asymmetric with the three main components σ_{ii} (i = 1, 2, 3). For polycrystalline or amorphous solids broad signals are observed, which cover the whole of the range of the main components σ_{ii} (**Fig. 3.111**).

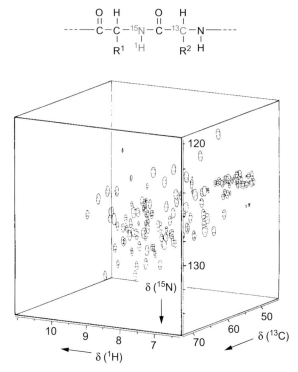

Fig. 3.110 750-MHz 3D ^{1}H–^{13}C–^{15}N–NH(CO)CA correlation spectrum of a 2-mM solution of ^{13}C and ^{15}N enriched ribonuclease T1 in H$_2$O/D$_2$O (Bruker, Analytische Messtechnik).

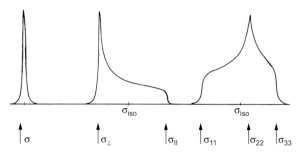

Fig. 3.111 ^{13}C NMR signals in powder measurements for the cases: $\sigma_{11} = \sigma_{22} = \sigma_{33} = \sigma$, $\sigma_{11} = \sigma_{22} = \sigma_\perp$, $\sigma_{33} = \sigma_\parallel$, $\sigma_{11},\sigma_{22},\sigma_{33}$ different (ignoring dipolar couplings).

Particularly large **chemical shift anisotropies** are observed for sp^2 and sp carbon atoms. In the following diagram the main components σ_{11}, σ_{22}, and σ_{33} of the ^{13}C shift tensor of naphthalene (**150**) are given for the individual carbon atoms, one below the other. The ppm values are referred to TMS. Their average in each case gives values close to the σ values of the solution spectrum (given in blue). It is apparent that in the crystalline state of naphthalene the symmetry elements, except the center of symmetry, are absent. C-1 and C-8 for example have quite different values of the component σ_{22}.

150

The **dipolar coupling**, a through-space effect, can also be described by a tensor. Since its trace is zero, this coupling has no effect in solution. The magnitude of the dipolar coupling depends on the magnetogyric ratios γ of the nuclei involved, the internuclear distance r, and the angle θ between the line joining the nuclei and the magnetic field $\boldsymbol{B}_0$. For powder measurements with a statistical angular distribution, linewidths are typically up to 30 kHz. Normal ^{13}C,^{1}H couplings in CH-, CH$_2$-, or CH$_3^-$ groups are around 10 kHz. Because of the low natural abundance, ^{13}C,^{13}C coupling can be ignored. Double ^{13}C labeling can afford an accurate method for determining the C–C distance from this coupling.

The angular dependence of the chemical shift σ and the dipolar coupling D is expressed by the term $(3\cos^2\theta - 1)$, a kind of second-order Legendre polynomial:

$$\sigma = \sigma_{iso} + 1/3 \sum_{i=1}^{3} (3\cos^2\Theta - 1)\,\sigma_{ii}$$

$$D = \pm h/2 \cdot \gamma_1 \cdot \gamma_2 \cdot \frac{1}{r^3}\,(3\cos^2\Theta - 1)$$

High-resolution solid state spectra can therefore be obtained, if $3\cos^2\theta - 1 = 0$. This can be achieved by spinning the sample at high speed about an axis which is at the "magic" angle θ_m to the field $\boldsymbol{B}_0$. From $\cos^2\theta_m = 1/3$ it follows that $\theta_m = 54.73°$. This magic angle spinning (**MAS**) makes $\sigma = \sigma_{iso}$ and $D = 0$; this means that the powder spectrum is essentially the same as the solution spectrum. (The symmetry properties in the crystal, which may be different from the free molecule, are naturally unaffected.) Spinning side bands (and their envelope) can be suppressed by increasing the spinning frequency.

Scalar couplings and the remnants of the dipolar coupling between ^{13}C and ^{1}H nuclei can be eliminated by high power proton broadband decoupling. The homonuclear coupling between the protons in the sample is a far more serious problem because of the magnitude of γ and the number of protons in typical organic molecules. Consequently, **^{13}C solid-state NMR spectroscopy** is much preferred to **^{1}H solid-state NMR spectroscopy**. However, higher spinning frequencies ($v_{rot} \geq 30$ kHz) and improved pulse sequences have made proton measurements, particularly ^{1}H–^{13}C-HETCOR spectra, more attractive.

The final problem, of long relaxation times, can be solved by **cross polarization (CP)**. This utilizes polarization transfer from surrounding protons to the ^{13}C nuclei. The effective relaxation times are thus drastically reduced, and the intensity of the ^{13}C signals is increased. The S/N ratio increases by a maximum factor of $\gamma_H/\gamma_C = 3.98$.

Fig. 3.112 shows a comparison of the ^{13}C CP-MAS and solution spectra of the enol (**364**), which can exist in the chelated (Z)-form as well as the (E)-configuration.

The spectrum observed for the solid corresponds to the pure (Z)-form with an intramolecular hydrogen bond. Carbons 1 and 3 are coincidentally isochronous within the limits of resolution. The restricted rotation of the mesityl groups in the solid state causes the two o- and m-carbons and the two o-methyl groups of each mesityl group to become chemically nonequivalent. This increases the number of aromatic signals from 8 to 12 and the number of methyl signals from 4 to 6. In total, 10 and 4 signals respectively are resolved, some showing double intensity because of overlap.

Temperature-dependent dynamic processes can also take place in crystals. The valence tautomerism of bullvalene in the solid state has already been discussed on p. 112 and 221. There are additional flips around the C$_3$-axis, which have an activation energy of ca. 88 kJ·mol^{-1}. Many benzene derivatives show processes involving flipping by 180°.

High-resolution solid-state NMR spectroscopy is useful for structural determination particularly for insoluble or almost insoluble compounds, e.g.. for polymers or biopolymers, for catalysts, and also for molecular structure problems where solvation has an important effect on the configuration or conformation. Low-temperature measurements on solid-state matrices can also provide useful information about short-lived intermediates.

An example of an accurate structural assignment for an insoluble material is shown for the polyesters **365** and **366** in **Fig. 3.113** on p. 225. The introduction of a second methoxy group into the repeat unit leads to $\Delta\delta$ values which are accurately reproduced by the increment system for benzene derivatives. Special measurement conditions even allow the distinction between amorphous and polycrystalline states of polymers.

Nuclear Magnetic Resonance Spectroscopy

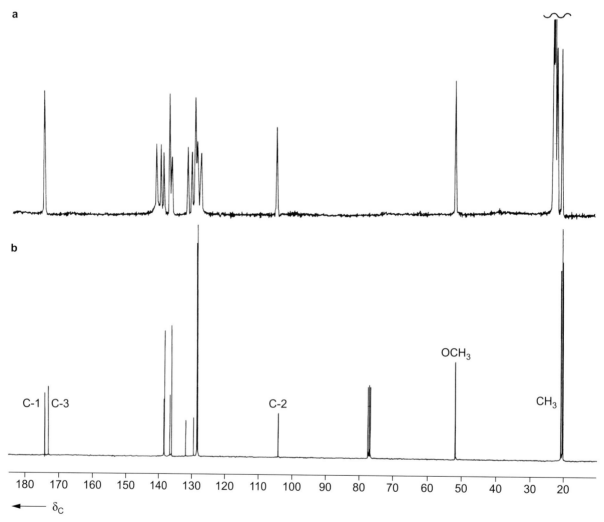

Fig. 3.112 ^{13}C NMR spectra of methyl (Z)-3-hydroxy-2,3-dimesityl-2-propenoate (**364**). **(a)** CP-MAS solid-state spectrum (without spinning sidebands); **(b)** normal PFT spectrum of a solution in CDCl$_3$.

The methodology of combinatorial chemistry has led to increased interest in solid-state synthesis. For these studies analytical techniques are required which allow investigation of the reaction products while still attached to the carrier. The resins used are insoluble in organic solvents, at best they swell up to form gels. A good example is the case of the stereoisomeric norbornane-2-carboxylic acids **367**, which are bonded via a "linker" to a polystyrene resin R.

exo　　　　endo

367

Fig. 3.114 shows a comparison of the ^{13}C spectrum obtained in the gel phase with the **high-resolution MAS (HRMAS)** spectrum. The advantages of the latter technique are immediately apparent; especially in the carbonyl region, signals are visible which clearly show the bonding of the norbornane-carboxylic acids to the linker of the resin. From these signals the occupancy of the resin-bonding sites can be determined. Signals of the resin in the regions around 40 and 130 ppm (and those of the benzene used as a swelling agent) can be suppressed by using special techniques (e.g., spin echo). ^{13}C,^{1}H-shift correlations and other 2D techniques based on MAS enlarge the scope for the investigation of heterogeneous samples, where the analytical target has restricted mobility because it is attached to a carrier.

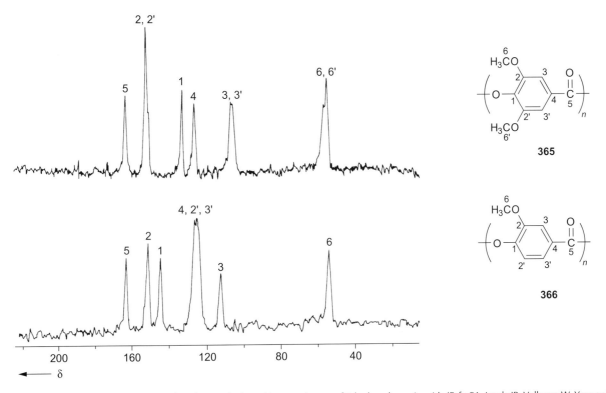

Fig. 3.113 ¹³C CP-MAS NMR spectra of methoxy-substituted homopolymers of 4-hydroxybenzoic acids (Fyfe CA, Lyerla JR, Volk-sen W, Yannoni CS. 1979, Macromol. **12**, 757).

A novel method for the measurement of crystalline or amorphous probes (in particular for biological samples or materials) was established by **d**ynamic **n**uclear **p**olarization (**DNP-NMR**). The probe is doped with a stable paramagnetic agent such as TEMPO (cf p. 122), bi- or polyradicals, and irradiated with microwaves (150–650 GHz). The high polarization of the electron spin of the radical is thereby transferred to the nuclei of the probe. MAS measurements at low temperature (80–140 K) in a high magnetic field gain by this method a factor of 200 to 300 in sensitivity.

3.5 Combination of ¹H and ¹³C NMR Spectroscopy

3.5.1 Complete Assignment of ¹H and ¹³C NMR Signals

NMR data can be used for the **identification of compounds** with known NMR data. Compared to identification by comparison of IR data, there is the advantage that impurities can be more readily identified. The detection limit in routine measurements lies somewhere between 3 and 10%, depending on the compounds involved; however, it can be brought down considerably below 1% by special measurements.

More important is the use of NMR for the **determination of new structures**. Using increment systems for the prediction of chemical shifts, recognition of coupling patterns, and estimates of coupling constants, working hypothetical structures can be deduced—for a final structural determination in the more difficult cases as many analytical methods as possible may need to be used, including a wide range of 1D and 2D NMR experiments. Modern NMR offers the possibility of an unambiguous and essentially complete assignment of the signals of the individual nuclei of molecules of low molecular weight. The most difficult cases are those where very similar signals occur for similar but chemically nonequivalent groups. The pyrazole **368** serves as an example; it has two benzene rings I and II and four different propoxy groups. The complete assignment, shown in the following figure, depends on the use of a combination of 1D and 2D measurement techniques.

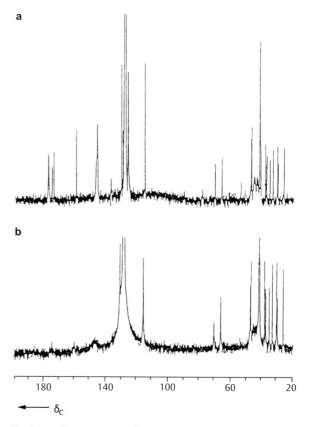

a

b

← δ_C

Fig. 3.114 ^{13}C NMR spectra of *exo/endo*-norbornane-2-carboxylic acid (**367a**), which is bonded *via* a linker to a polystyrene resin. (**a**) HRMAS spectrum with benzene as the swelling agent. (**b**) Gel-phase spectrum (Anderson RC, Jarema JA, Shapiro MJ, Stokes JP, Ziliox M. 1995, J. Org. Chem. **60**, 2650).

The routine 400 MHz ^{1}H NMR spectrum of **368** shows two partially overlapping patterns for the ABM spin systems of the total of six protons of the benzene rings, four overlapping patterns for the propoxy chains, and two singlet signals (**Fig. 3.115**). The ^{1}H-broadband decoupled ^{13}C spectrum of **368** (**Fig. 3.116**) shows, in addition to the strongly overlapping signals of the saturated C-atoms, seven signals for the aromatic and heteroaromatic methine groups (CH) and eight signals for the quaternary C-atoms (C$_q$).

The singlet signals can be immediately assigned to 4-H and N–CH$_3$. This provides two "anchor points." Then, for example, the following strategy can be employed: an HMBC spectrum allows the identification of the signal of C-5 by means of the crosspeak arising from the ^3J(C,H) coupling with the N-methyl group (**Fig. 3.117**). The same spectrum (in a part not illustrated) leads from C-5 via another ^3J(C,H) coupling to 6″-H [and via ^4J(C,H) to

3″-H]. Thus, the ABM spin system of ring II in the 1D ^{1}H NMR spectrum can be identified. Confirmation can be obtained from the (^{1}H,^{1}H)COSY spectrum (**Fig. 3.118**).

An HMQC spectrum then yields the assignment of all the H-bonded C-atoms. The aromatic and heteroaromatic part of this spectrum is shown in **Fig. 3.119** on p. 229.

The blue correlation lines indicate the correlated ^{1}H- and ^{13}C-signals of the relevant methine groups (CH). Still to be resolved are the assignments of the quaternary C-atoms. This can be achieved from the crosspeaks from ^3J(C,H) coupling to the benzene protons (and to 4-H) in the HMBC spectrum. In this way, not only the assignments of C-3, C-1′, and C-1″ but also the assignments of the four oxygen-bonded C$_q$ are achieved; C-2″ for example has ^3J(C,H) coupling to 6″-H and 3″-H. From the C$_q$O assignments, the triplet signals of the OCH$_2$ groups can be assigned via the ^3J(C,H) couplings (**Fig. 3.117**). The COSY spectrum of the saturated region then gives the assignments of the

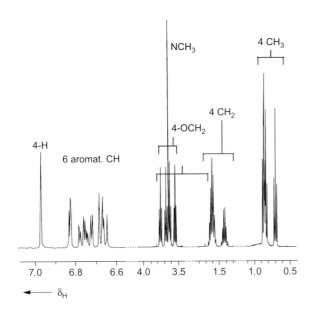

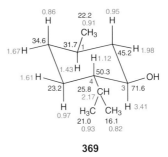

The complete assignment of NMR signals is particularly helpful for complex structures, which are often found in natural products. (–)Menthol (**369**) is discussed here as an example. It has (1*R*, 3*R*, 4*S*)-configuration. (The numbering in formula **369** corresponds to the terpene nomenclature.)

369

Fig. 3.115 ¹H NMR spectrum of the pyrazole **368** (measurement at 400 MHz in C_6D_6/CDCl₃, 1:1).

CH_2 groups attached to the OCH_2 groups, which in turn allow the assignments of the neighboring CH_3 groups to be made. This completes the assignment problem.

It is sometimes difficult to assign the crosspeaks of signals which lie very close together. This applies for example to C-2′ and C-2″ in **Fig. 3.117**. Using only the coupling through bonds, there is no way to decide which OC_3H_7 group is bonded to which aromatic C-atom. Here the connectivity through space can help. A ROESY-measurement (**Fig. 3.120**, p. 230) shows clearly the proximity of 2″-OCH_2 to 3″-H (and similarly the proximity of 2′-OCH_2 to 3′-H).

Starting at 3-H as an anchor point, the ¹H NMR data can be established on the basis of COSY and NOESY spectra. An HSQC spectrum serves then for the assignment of the ¹³C signals of CH, CH₂, and CH₃ groups. Quaternary carbon atoms C_q are not present in **369**. The diastereotopic geminal protons of H_2C-2, H_2C-5, and H_2C-6 and even the diastereotopic methyl groups of the isopropyl substituent can be definitely assigned by NOEs. A great advantage thereby is that the inversion of the cyclohexane ring would lead to a hardly populated conformer, which has all three substituents in unfavorable axial positions.

As demonstrated earlier, **the majority of NMR structure determinations can be solved by measuring and interpreting the following series of spectra:**

- ¹H NMR
- (¹H,¹H)COSY, eventually NOESY or ROESY

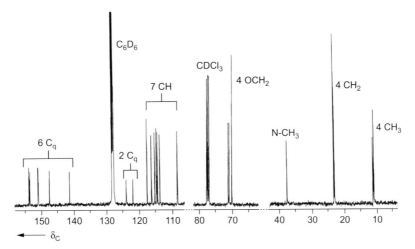

Fig. 3.116 ¹H coupled ¹³C NMR spectrum of the pyrazole **368** (measurement at 100 MHz in C_6D_6/CDCl₃, 1:1).

Nuclear Magnetic Resonance Spectroscopy

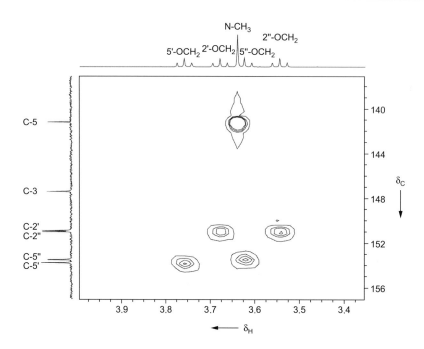

Fig. 3.117 Part of the HMBC spectrum of **368** in C$_6$D$_6$/CDCl$_3$ (1:1).

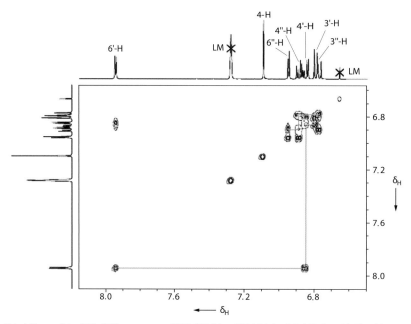

Fig. 3.118 Part of the ^{1}H-shift correlated 2D-NMR spectrum of **368** (COSY at 600 MHz in C$_6$D$_6$/CDCl$_3$,1:1). The blue correlation lines allow the two ABM spectra to be recognized.

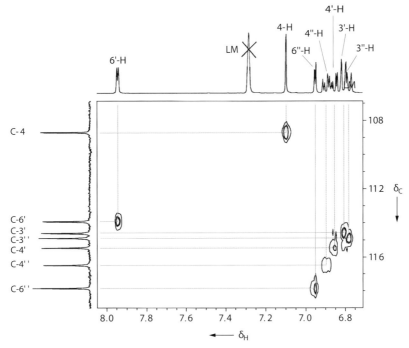

Fig. 3.119 Part of the HMQC spectrum of **368** in $C_6D_6/CDCl_3$ (1:1).

	R
370	C_2H_5 ⟶ References
371	$CH_2-C_6H_5$

References

- ¹³C NMR (broadband decoupled)
- DEPT 135
- HSQC or HMQC (a phase-sensitive HSQC can substitute the DEPT)
- HMBC

Unfortunately, many publications of new structures contain insufficient NMR characterizations, for example, just a list of ¹³C signals. Reliable assignments of all ¹H and ¹³C signals can be very helpful to avoid wrong structure propositions.

3.5.2 Use of Databases

The use of information technology is becoming increasingly important in the area of instrumental analysis.

Several free or commercially accessible databases exist, which can be used for an NMR inquiry. Either NMR databases (NMR Shift DB, ChemBioDraw, NMR Predict Specsurf, ACD, and SDBS) can be directly applied or a search of NMR data can be performed via structure databases (ChemSpider, PubChem, SciFinder, STN [CAS], and Reaxys).

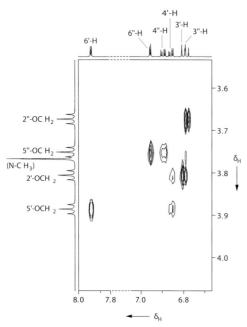

Fig. 3.120 Part of the ROESY spectrum (600 MHz) of **368** in C_6D_6/ $CDCl_3$ (1:1).

Application of Structure Databases

The use of structure databases (Reaxys: more than 125×10^6, ChemSpider: more than 67×10^6, SciFinder(CAS): more than 154×10^6 compounds) reveals whether the inquired compound is known or not. Given links should then be checked. The options "spectroscopic data" and more precise "NMR" help to narrow the search. When the compound is new or when the NMR data of a known compound are unknown, substructures may help. Two simple examples explain the proceeding:

The search concerns the two esters **370** and **371**. Reaxys indicates for **370** the references for ^1H and ^{13}C but not ^{15}N data. On the contrary ester **371** is new. The inquired structure **371** can be split into two substructures, which bear open positions. The common parts of the substructures should be as large as possible. Several examples of compounds having the drawn substructures can be found in Reaxys and their ^1H and ^{13}C chemical shifts can then be composed to predict the chemical shifts of **371**. More than two substructures may be taken into account for large, complex molecules.

Application of NMR Databases

As an example, the freely accessible database NMR Shift DB shall be discussed briefly. It contains more than 40,000 structures with (predominating) ^{13}C NMR spectra. The option "Search" permits to find out whether the ^{13}C NMR shifts (or eventually ^1H, ^{15}N, ^{19}F, ^{31}P, ... chemical shifts) are stored in the database or not. If data of the inquired compound are not

obtainable, the option "Predict" should be selected. The drawn structure delivers then predicted NMR data (^{13}C, ^1H, ^{19}F, ^{31}P, ^{15}N, ...) 3,5-Divinylpyridine (**372**) serves here as an example.

372
(predicted, measured)

The predicted δ (^1H) and δ (^{13}C) values correspond well to the measured values. However, the database does not permit to distinguish between *cis* and *trans* standing protons.

The prediction for the compound **373** is more difficult.

373
(predicted, measured)

The comparison of measured and predicted ^1H and ^{13}C chemical shifts is for the most positions satisfying. Major divergences exist for HC-1 and C-3 of the indolizine. These are the positions in which the electron density is influenced by the resonance participation of the free electron pair on the nitrogen atom. Indolizine has a heteroaromatic character. Such electronic effects as well as steric effects are disregarded in the database.

The tool "Quick Check" in NMR Shift DB permits a fast and easy control of the ^1H and ^{13}C chemical shifts of an assumed structure. When the assigned δ values are inserted besides the drawn structure, the program predicts the expected δ (^1H) and δ (^{13}C) values.

The final quality report distinguishes between: accept/revision/reject. The tool works very well for the vast majority of compounds.

Another possibility given by the database NMR Shift DB consists in entering of the found δ(¹³C) values and searching which complete spectrum or which subspectrum fits best to the data (similarity factor).

The next database, NMR Predict/Specsurf, shall be discussed here. It contains the ¹³C shifts and, where known, also coupling constants and relaxation times of many organic compounds. A fundamental problem is that the spectral data are dependent on sample-specific (solvent, concentration, purity, and temperature) and instrument-specific measurement conditions. This fact can be allowed for by defining a tolerance range ("window method") for the chemical shifts.

SPECINFO permits a variety of searches:

(a) Search for reference spectra on the basis of measured chemical shifts.
(b) Search for reference spectra on the basis of names of compounds or compound types.
(c) Search for reference spectra on the basis of molecular formula.
(d) Search for similar spectra.
(e) Search for compounds and spectra for defined structures or partial structures (substructures).
(f) The estimation of chemical shifts (and also coupling constants where appropriate) for assumed structures.

As an illustration, a simple example will be chosen. The chemical shifts of a compound are known, with the signal multiplicities derived from ¹J(C,H) couplings:

Observed δ values 154.7 (singlet)
139.7 (singlet)
129.1 (doublet)
121.5 (doublet)
115.9 (doublet)
112.3 (doublet)
20.8 (quartet)

Which compound can it be? To answer this question the choice (a), search for reference spectra, is chosen. The specific question is: is there a set of data among the millions of stored data which agrees with the observed shifts of the seven signals within a chosen error limit (here 0.5 ppm for example)? As each signal is input, the number of possible compounds must reduce. Finally, in this case only one compound remains, namely *m*-cresol (**22**) with the following stored data (δ values in CDCl₃):

C	δ
1	139.8
2	116.2
3	155.0
4	112.5
5	129.4
6	121.8
7	21.1

22

An even more important application is offered by choice (f). For a newly synthesized compound, a specific structure is assumed. What ¹³C chemical shifts are predicted? The example of 3,8-dithiabicyclo[8.3.1]tetradeca-1(14),10,12-triene-5-yne (**374**) and its fluoro derivative **375** show the possibilities but also the limits of such an inquiry. The blue δ values give the predictions with their standard deviations, and the δ values printed in black give the ¹³C shifts observed in CDCl₃.

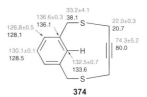

374

For the aromatic C atoms, the agreement is good. The calculated δ values in the bridge of the *m*-cyclophane are interpolated; the criteria for similarity, on which the incorporation of reference data is based, are apparently dubious. The program gives unsatisfactory values for the bridge carbon atoms when fluorine is included in the structure. Nonetheless, the large differences between observed

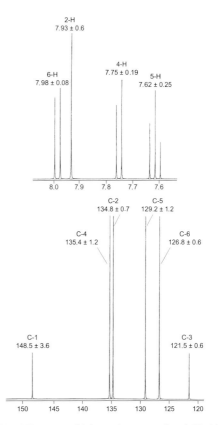

Fig. 3.121 NMR spectra of 3-bromobenzenesulfonyl chloride (**376**) calculated by the ACD database. Upper: ¹H NMR with ³J couplings at 400 MHz; lower: ¹³C NMR (broadband decoupled).

Nuclear Magnetic Resonance Spectroscopy

and calculated δ values are usually associated with larger standard deviations. The same considerations apply here as in the application of increment systems: a critical use can still be of great assistance in structure determination.

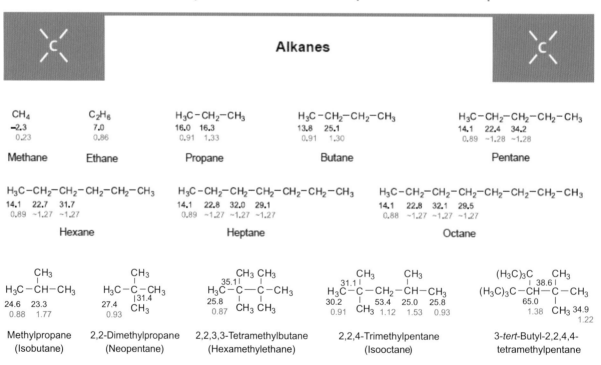

375

Of particular interest is the combination of ^{1}H and ^{13}C NMR data for structural predictions. As an example, the application of the ACD in-house database to 3-bromobenzenesulfonyl chloride (**376**) will be discussed. **Fig. 3.121** shows the predicted spectra. Since compound **376** is not included in the database, the data must be derived from related compounds, with the error limits shown.

376

DFT-based calculations of the chemical shifts can be a valuable support for the assignment of the observed signals. Even large molecules, such as chlorophyll a can be calculated with some time expanditure. Many of its 55 ^{13}C signals were calculated with deviations of less than 5 ppm from the measured values. However, for some C atoms the deviations are so large, that an assignment, which is exclusively based an the *ab initio* calculation, would fail.

Every structure determination using analytical instrumentation, including computer-aided methods, should be based on as many methods as possible. The idealized objective is to use computers to derive a structure directly from the observed UV, IR, NMR, and MS data. This requires programs that not only rely on databases but that can adopt the logical and empirical thought processes of the analytical chemist. The automatic operation of such programs can at best be expected only for relatively simple structural problems; for more complex problems a dialogue with the analytical chemist—a semiautomatic solution—is desirable. This offers new, promising perspectives for instrumental analysis.

3.5.3 ^{1}H and ^{13}C NMR Data of Representatives of the Most Important Classes of Compounds

Alkanes

CH$_4$
−2.3
0.23

Methane

C$_2$H$_6$
7.0
0.86

Ethane

H$_3$C−CH$_2$−CH$_3$
16.0 16.3
0.91 1.33

Propane

H$_3$C−CH$_2$−CH$_2$−CH$_3$
13.8 25.1
0.91 1.30

Butane

H$_3$C−CH$_2$−CH$_2$−CH$_2$−CH$_3$
14.1 22.4 34.2
0.89 ~1.28 ~1.28

Pentane

H$_3$C−CH$_2$−CH$_2$−CH$_2$−CH$_2$−CH$_3$
14.1 22.7 31.7
0.89 ~1.27 ~1.27

Hexane

H$_3$C−CH$_2$−CH$_2$−CH$_2$−CH$_2$−CH$_2$−CH$_3$
14.1 22.8 32.0 29.1
0.89 ~1.27 ~1.27 ~1.27

Heptane

H$_3$C−CH$_2$−CH$_2$−CH$_2$−CH$_2$−CH$_2$−CH$_2$−CH$_3$
14.1 22.8 32.1 29.5
0.88 ~1.27 ~1.27 ~1.27

Octane

CH$_3$
|
H$_3$C−CH−CH$_3$
24.6 23.3
0.88 1.77

Methylpropane
(Isobutane)

CH$_3$
|
H$_3$C−C−CH$_3$
|31.4
CH$_3$
27.4
0.93

2,2-Dimethylpropane
(Neopentane)

CH$_3$ CH$_3$
35.1| |
H$_3$C−C−C−CH$_3$
25.8 | |
0.87 CH$_3$ CH$_3$

2,2,3,3-Tetramethylbutane
(Hexamethylethane)

CH$_3$ CH$_3$
31.1 | |
H$_3$C−C−CH$_2$−CH−CH$_3$
30.2 | 53.4 25.0 25.8
0.91 CH$_3$ 1.12 1.53 0.93

2,2,4-Trimethylpentane
(Isooctane)

(H$_3$C)$_3$C   CH$_3$
| 38.6 |
(H$_3$C)$_3$C−CH−C−CH$_3$
65.0 |
1.38 CH$_3$ 34.9
1.22

3-*tert*-Butyl-2,2,4,4-
tetramethylpentane

The numbers printed in black give the δ values of the ^{13}C signals, the blue figures the δ values of the ^{1}H signals. The most compounds are measured in CDCl$_3$ with TMS as internal standard.

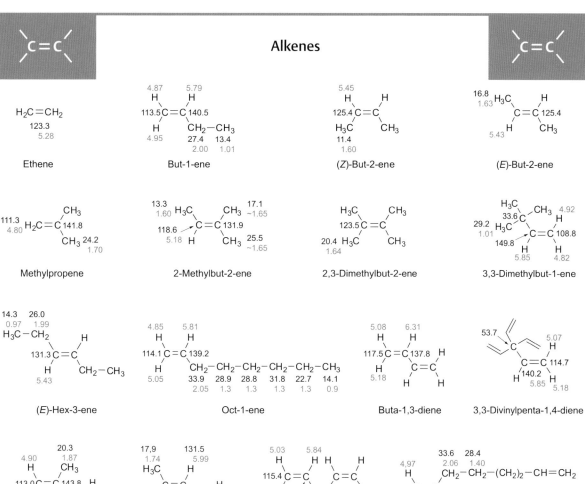

Alkenes

$H_2C=CH_2$
123.3
5.28

Ethene

4.87   5.79
H   H
113.5C=C140.5
H   CH₂—CH₃
4.95   27.4 13.4
       2.00 1.01

But-1-ene

5.45
H   H
125.4C=C
H₃C   CH₃
11.4
1.60

(Z)-But-2-ene

16.8 H₃C   H
1.63   C=C 125.4
       H   CH₃
5.43

(E)-But-2-ene

111.3 H₂C=C141.8
4.80   CH₃ 24.2
       CH₃   1.70

Methylpropene

13.3 H₃C   CH₃ 17.1
1.60   C=C131.9   ~1.65
118.6 H   CH₃ 25.5
5.18   ~1.65

2-Methylbut-2-ene

H₃C   CH₃
123.5C=C
20.4 H₃C   CH₃
1.64

2,3-Dimethylbut-2-ene

H₃C   CH₃ 4.92
33.6   H
29.2 H₃C
1.01   C=C 108.8
149.8 H   H
       5.85 4.82

3,3-Dimethylbut-1-ene

14.3 26.0
0.97 1.99
H₃C—CH₂   H
131.3C=C
H   CH₂—CH₃
5.43

(E)-Hex-3-ene

4.85   5.81
H   H
114.1C=C139.2
H   CH₂—CH₂—CH₂—CH₂—CH₃
5.05   33.9 28.9 28.8 31.8 22.7 14.1
       2.05 1.3   1.3   1.3   1.3   0.9

Oct-1-ene

5.08   6.31
H   H
117.5C=C137.8 H
H   C=C
5.18   H   H

Buta-1,3-diene

53.7
        C
              5.07
         C=C 114.7
H 140.2 H
   5.85   5.18

3,3-Divinylpenta-1,4-diene

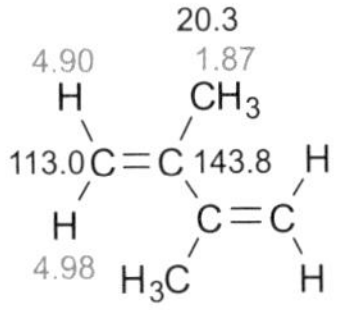

20.3
4.90 1.87
H   CH₃
113.0C=C 143.8 H
H   C=C
4.98 H₃C   H

2,3-Dimethylbuta-1,3-diene

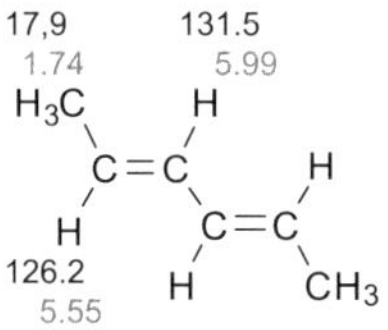

17.9   131.5
1.74   5.99
H₃C   H
C=C
H   H
126.2 H   CH₃
5.55

(E,E)-Hexa-2,4-diene

5.03   5.84
H   H H   H
115.4C=C   C=C
H   CH₂ H
5.05 136.3
      37.9
      2.80

Penta-1,4-diene

33.6 28.4
4,97 2.06 1.40
H   CH₂—CH₂—(CH₂)₂—CH=CH₂
114.3C=C
H   CH₂ 138.9
4.94   5.80

Octa-1,7-diene

Alkynes

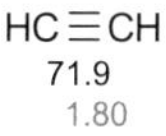

HC≡CH
71.9
1.80

Acetylene (Ethyne)

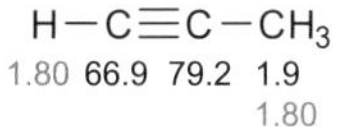

H—C≡C—CH₃
1.80 66.9 79.2 1.9
          1.80

Propyne

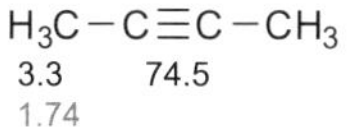

H₃C—C≡C—CH₃
3.3   74.5
1.74

But-2-yne

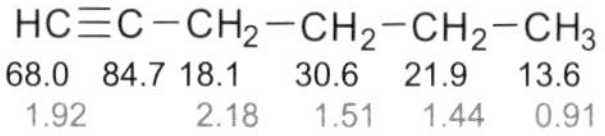

HC≡C—CH₂—CH₂—CH₂—CH₃
68.0 84.7 18.1 30.6 21.9 13.6
1.92   2.18 1.51 1.44 0.91

Hex-1-yne

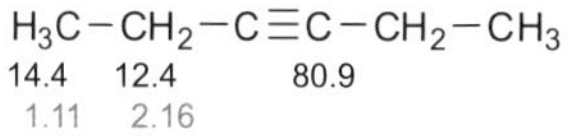

H₃C—CH₂—C≡C—CH₂—CH₃
14.4 12.4   80.9
1.11 2.16

Hex-3-yne

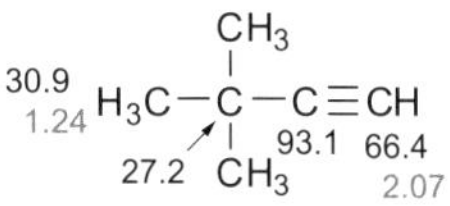

30.9 H₃C—C—C≡CH
1.24 | 93.1 66.4
27.2 CH₃   2.07
       CH₃

3,3-Dimethylbut-1-yne

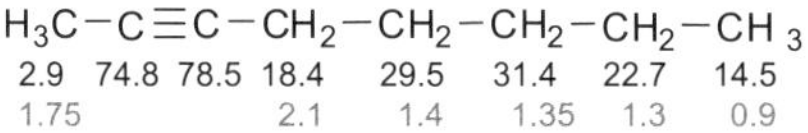

H₃C—C≡C—CH₂—CH₂—CH₂—CH₂—CH₃
2.9 74.8 78.5 18.4 29.5 31.4 22.7 14.5
1.75   2.1   1.4   1.35 1.3   0.9

Oct-2-yne

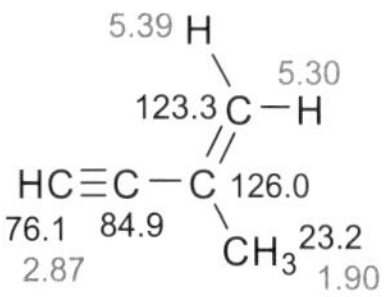

5.39 H
         5.30
123.3 C—H
HC≡C—C 126.0
76.1 84.9   CH₃ 23.2
2.87   1.90

2-Methylbut-1-en-3-yne

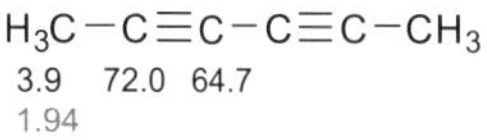

H₃C—C≡C—C≡C—CH₃
3.9 72.0 64.7
1.94

Hexa-2,4-diyne

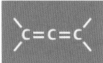

Allenes, Cumulenes

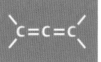

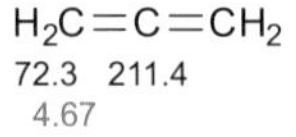

$H_2C=C=CH_2$
72.3 211.4
4.67

Allene

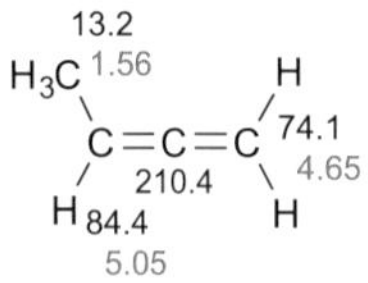

13.2
H_3C 1.56
$C=C=C$ 74.1
210.4 4.65
H 84.4 H
5.05

Buta-1,2-diene

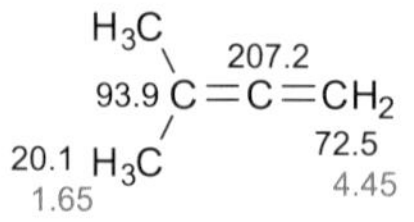

H_3C 207.2
93.9 $C=C=CH_2$
20.1 H_3C 72.5
1.65 4.45

3-Methylbuta-1,2-diene

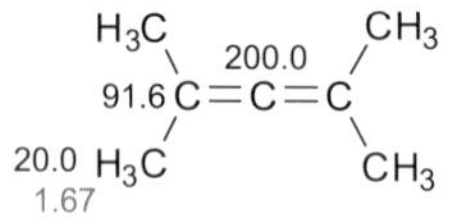

H_3C 200.0 CH_3
91.6 $C=C=C$
20.0 H_3C CH_3
1.67

2,4-Dimethylpenta-2,3-diene

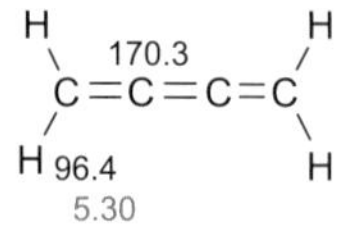

H 170.3 H
$C=C=C=C$
H 96.4 H
5.30

Butatriene

Cycloalkanes, Cycloalkenes, Cycloalkynes

Cyclopropane	C_3H_6	–2.8	0.22	Cyclohexane	C_6H_{12}	26.9	1.44
Cyclobutane	C_4H_8	22.3	1.96	Cyclooctane	C_8H_{16}	27.5	1.54
Cyclopentane	C_5H_{10}	25.8	1.51	Cyclododecane	$C_{12}H_{24}$	23.8	1.34

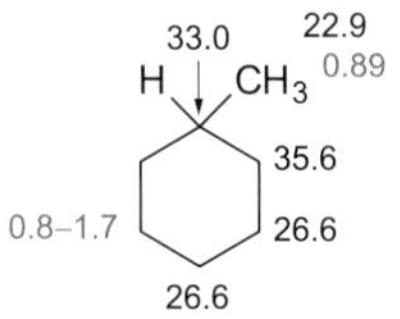

33.0 22.9
H CH_3 0.89
35.6
0.8–1.7 26.6
26.6

Methylcyclohexane

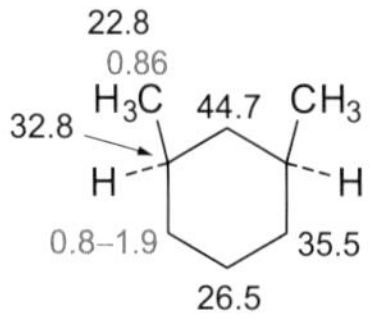

22.8
0.86
32.8 H_3C 44.7 CH_3
H H
0.8–1.9 35.5
26.5

***cis*-1,3-Dimethyl-cyclohexane**

20.5
0.90
27.1 H_3C 41.5 H
H CH_3
0.9–1.9 33.9
20.8

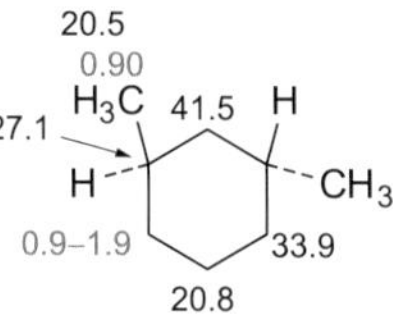

***trans*-1,3-Dimethyl-cyclohexane**

28.7
0.88 H_3C CH_3
H 30.2
39.6 H
1.22
22.6 H
~1.4 H
H H 26.6
~1.4

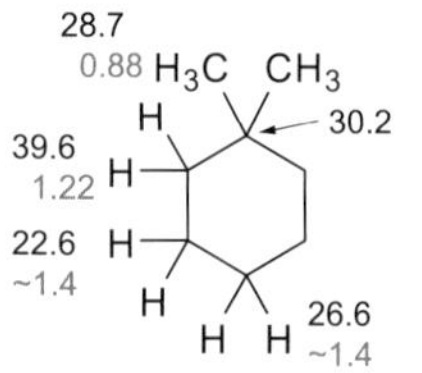

1,1-Dimethyl-cyclohexane

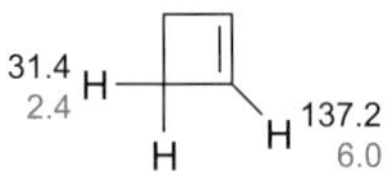

31.4 H
2.4
H 104.8 H
H 137.2
6.0

Cyclobutene

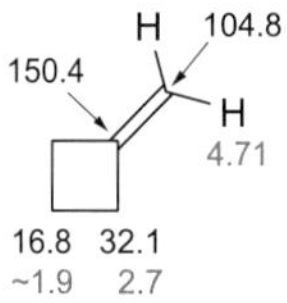

150.4
H 104.8
H
4.71
16.8 32.1
~1.9 2.7

Methylencyclobutane

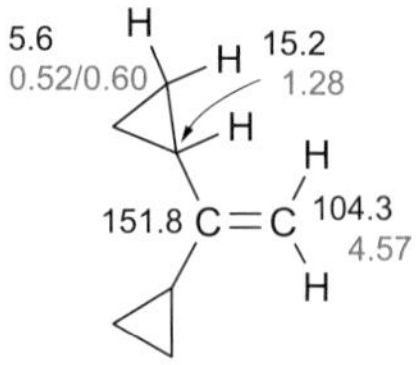

5.6 H
0.52/0.60 H 15.2
H 1.28
H
151.8 $C=C$ 104.3
4.57
H

1,1-Dicyclopropylethene

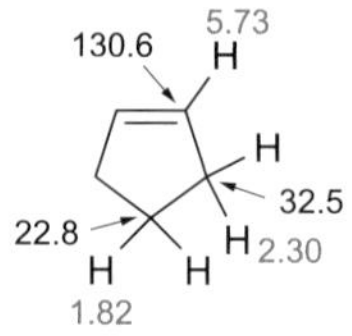

130.6 5.73
H
H
22.8 32.5
H H 2.30
1.82

Cyclopentene

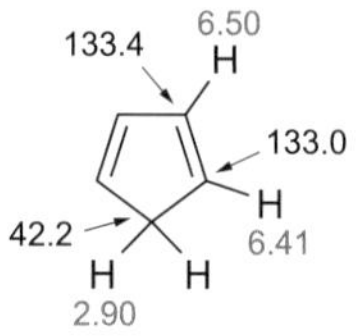

133.4 6.50
H
133.0
H
42.2 H
H 6.41
2.90

Cyclopentadiene

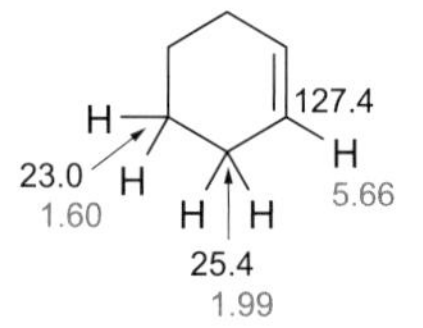

H 127.4
23.0 H H
1.60 H H 5.66
25.4
1.99

Cyclohexene

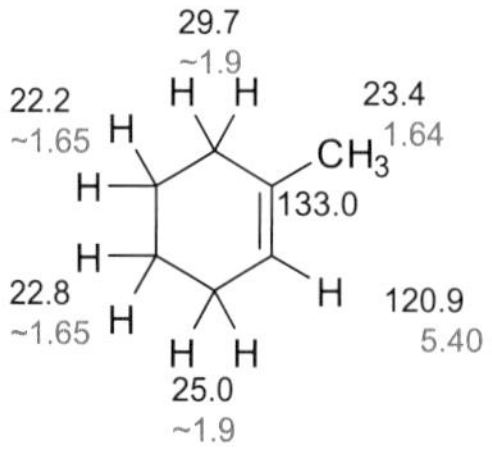

29.7
~1.9
22.2 H H 23.4
~1.65 H CH_3 1.64
H 133.0
22.8 H 120.9
~1.65 H H 5.40
H H
25.0
~1.9

1-Methylcyclohexene

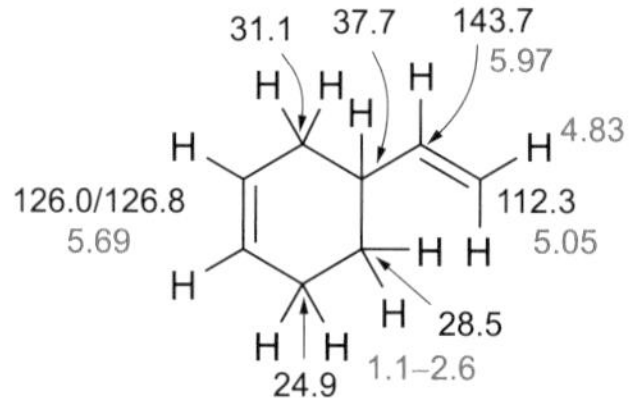

31.1 37.7 143.7
H H H 5.97
H H 4.83
126.0/126.8 112.3
5.69 H 5.05
H H 28.5
H H 1.1–2.6
24.9

4-Vinylcyclohexene

28.9 74.3
2.1 2.75 H
21.4 H H
1.55 H 85.7
H 119.9
22.1 H 136.3
1.55 H H 25.5 6.15
2.1

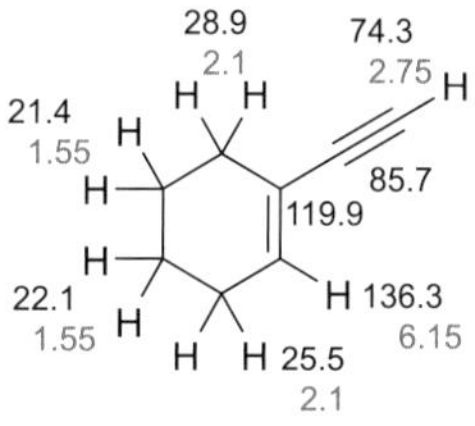

1-Ethynylcyclohexene

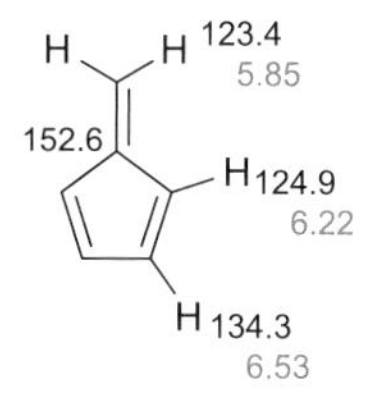

H 123.4
5.85

152.6

H 124.9
6.22

H 134.3
6.53

Pentafulvene

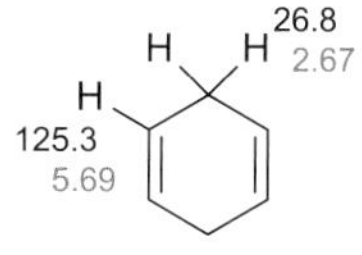

26.8
2.67

125.3
5.69

H

Cyclohexa-1,4-diene

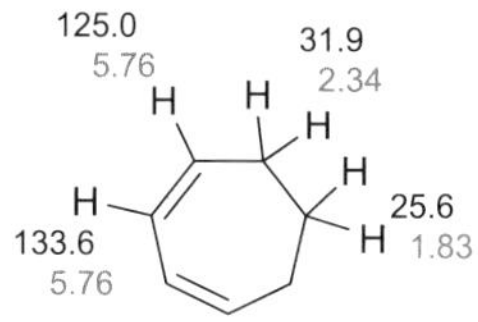

125.0
5.76

31.9
2.34

133.6
5.76

25.6
1.83

Cyclohepta-1,3-diene

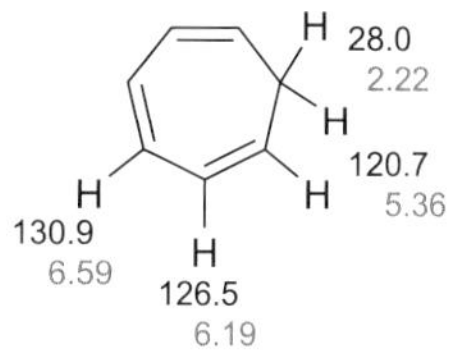

28.0
2.22

120.7
5.36

130.9
6.59

126.5
6.19

Cycloheptatriene

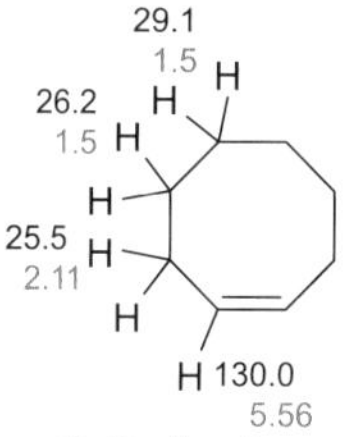

29.1
1.5

26.2
1.5

1.5

25.5
2.11

H 130.0
5.56

cis-Cyclooctene

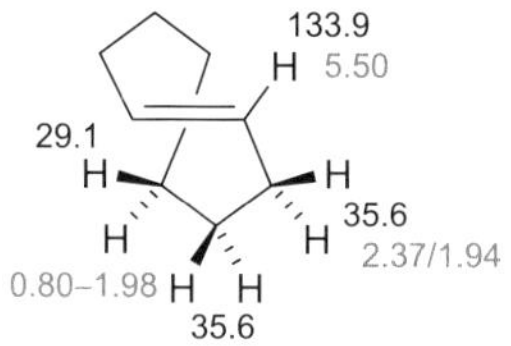

133.9
5.50

29.1

35.6
2.37/1.94

0.80–1.98

35.6

trans-Cyclooctene

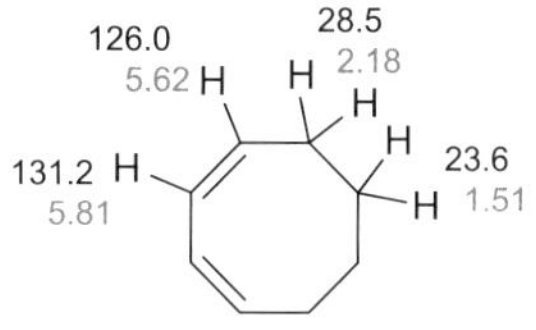

126.0
5.62

28.5
2.18

23.6
1.51

131.2
5.81

Cycloocta-1,3-diene

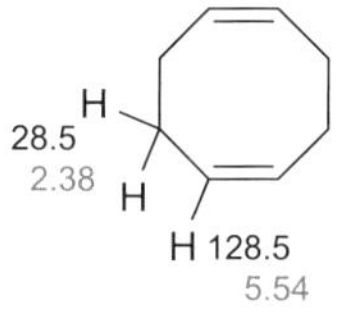

28.5
2.38

H 128.5
5.54

Cycloocta-1,5-diene

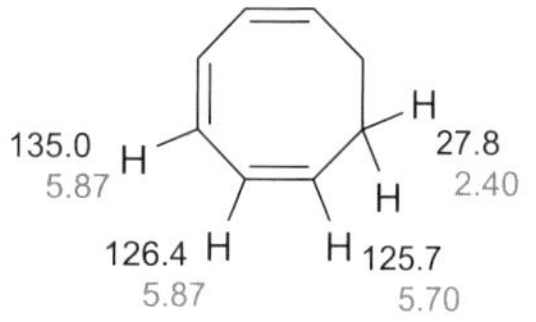

135.0
5.87

27.8
2.40

126.4
5.87

125.7
5.70

Cycloocta-1,3,5-triene

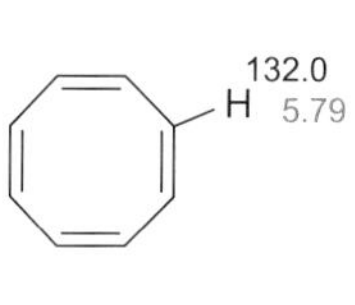

132.0
5.79

Cyclooctatetraene

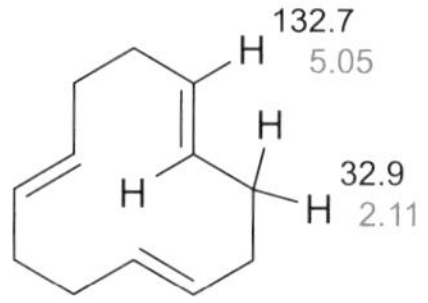

132.7
5.05

32.9
2.11

(E,E,E)-Cyclo-
dodeca-1,5,9-triene

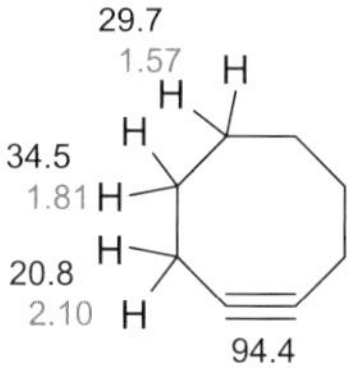

29.7
1.57

34.5
1.81

20.8
2.10

94.4

Cyclooctyne

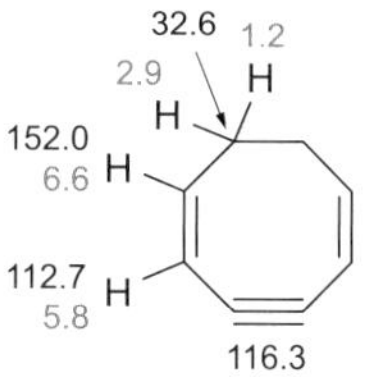

32.6
1.2

2.9

152.0
6.6

112.7
5.8

116.3

Cycloocta-1,5-dien-3-yne

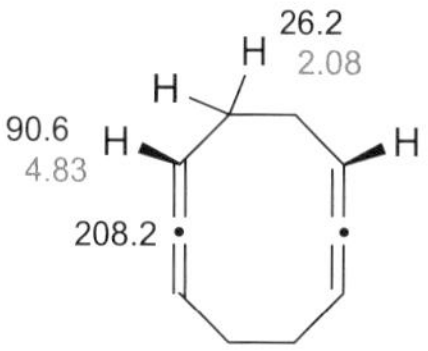

26.2
2.08

90.6
4.83

208.2

meso-Cyclodeca-1,2,6,7-tetraene
[(1R*,6S*)-Cyclodeca-1,2,6,7-tetraene]

Bi- and Polycyclic Hydrocarbons

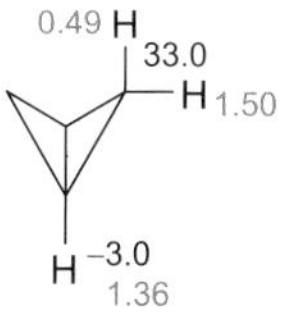

0.49 H
33.0
H 1.50

H –3.0
1.36

Bicyclo[1.1.0]butane

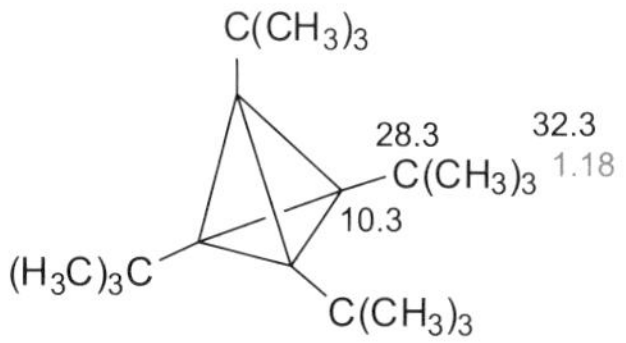

C(CH₃)₃

28.3   32.3
C(CH₃)₃ 1.18

(H₃C)₃C   10.3

C(CH₃)₃

Tetra-tert-butyltetrahedrane

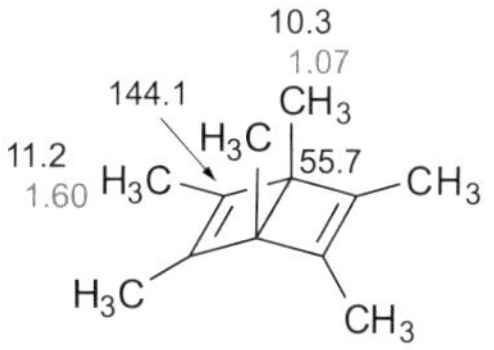

10.3
1.07
144.1   CH₃
11.2   H₃C   55.7
1.60 H₃C   CH₃

H₃C   CH₃

Hexamethyldewarbenzene

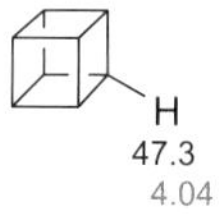

H
47.3
4.04

Cuban
(Pentacyclo[4.2.0.
0²,⁵.0³,⁸.0⁴,⁷]octane)

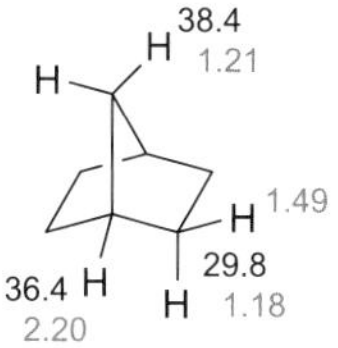

38.4
H 1.21

H 1.49
36.4 H   29.8
2.20   H 1.18

Norbornane
(Bicyclo[2.2.1]heptane)

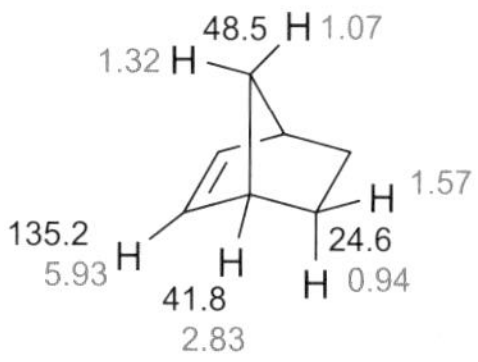

48.5 H 1.07
1.32 H

H 1.57
135.2 H   24.6
5.93   H 0.94
41.8
2.83

Norbornene
(Bicyclo[2.2.1]hept-2-ene)

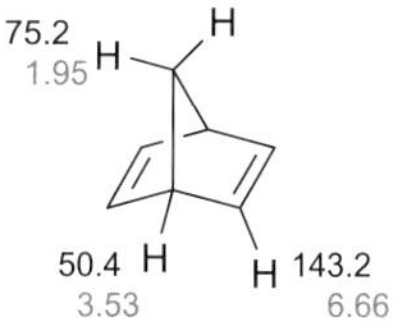

75.2   H
1.95

50.4 H   H 143.2
3.53   6.66

Norbornadiene
(Bicyclo[2.2.1]hepta-2,5-diene)

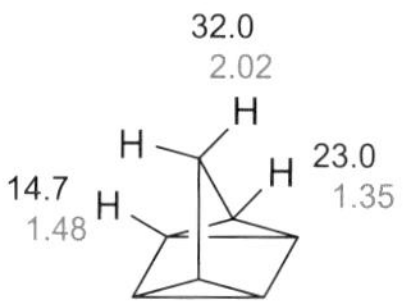

32.0
2.02

H   23.0
14.7   1.35
1.48

Quadricyclane

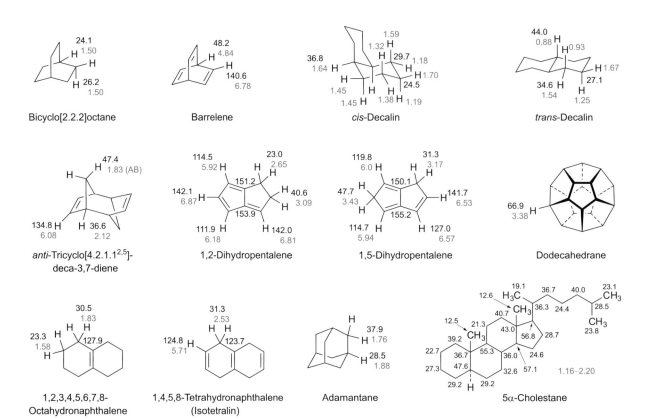

Bicyclo[2.2.2]octane

Barrelene

cis-Decalin

trans-Decalin

anti-Tricyclo[4.2.1.1²,⁵]-
deca-3,7-diene

1,2-Dihydropentalene

1,5-Dihydropentalene

Dodecahedrane

1,2,3,4,5,6,7,8-
Octahydronaphthalene

1,4,5,8-Tetrahydronaphthalene
(Isotetralin)

Adamantane

5α-Cholestane

Spirohydrocarbons

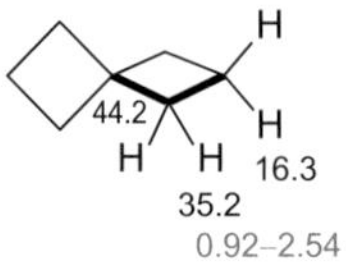

Spiro[3.3]heptane

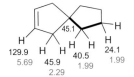

Spiro[4.4]non-2-ene

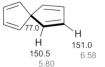

Spiro[4.4]nonatetraene

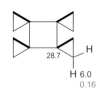

Tetraspiro[2.0.2.0.2.0.2.0]
dodecane

Aromatic Hydrocarbons

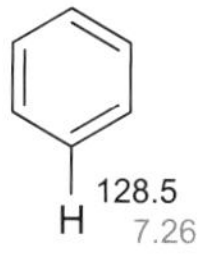

Benzene
128.5 H 7.26

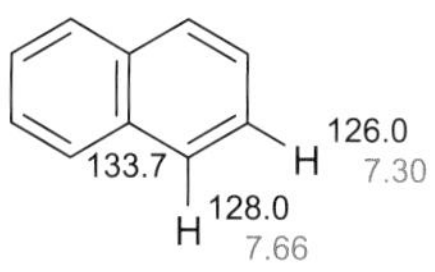

Naphthalene
133.7 126.0 7.30
128.0 7.66 H

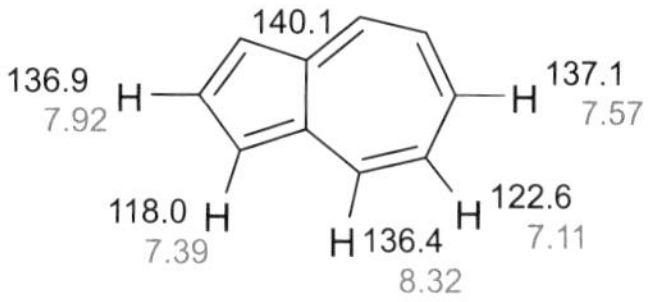

Azulene
140.1
136.9 H 7.92
137.1 H 7.57
118.0 H 7.39
H 136.4 8.32
122.6 7.11

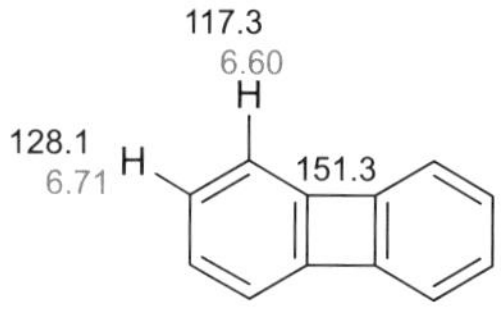

Biphenylene
117.3 6.60 H
128.1 H 6.71
151.3

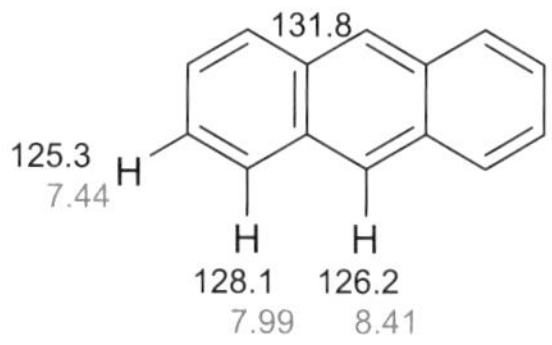

Anthracene
131.8
125.3 H 7.44
128.1 7.99
126.2 8.41

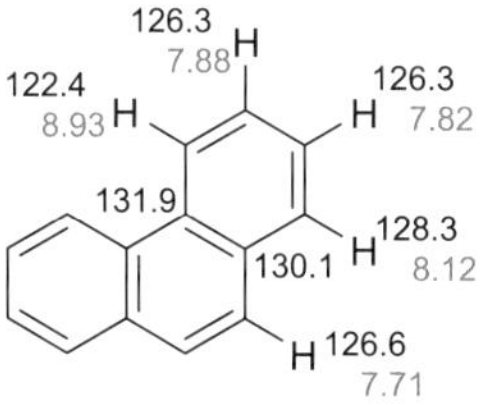

Phenanthrene
126.3 H 7.88
122.4 8.93 H
126.3 H 7.82
131.9
130.1
128.3 8.12
126.6 7.71

Triphenylene
123.2 8.54
129.7
127.1 7.56

Pyrene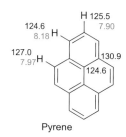
125.5 H 7.90
124.6 8.18 H
130.9
127.0 7.97 H
124.6

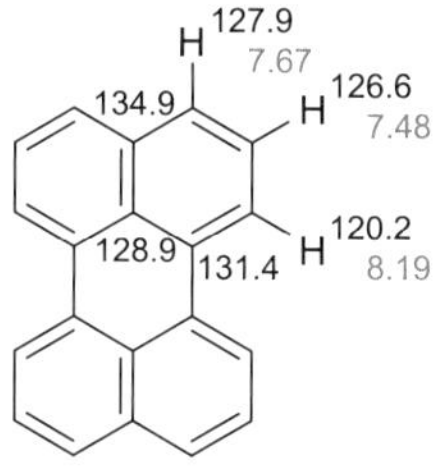

Perylene
127.9 H 7.67
126.6 H 7.48
134.9
120.2 8.19
128.9 131.4

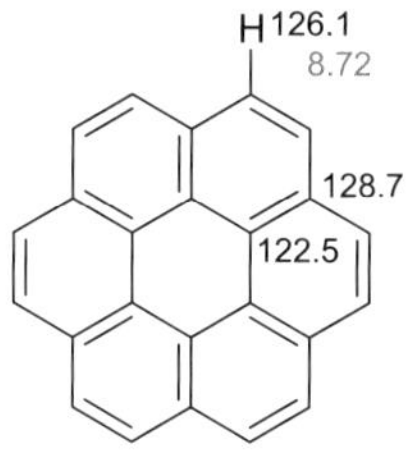

Coronene
126.1 H 8.72
128.7
122.5

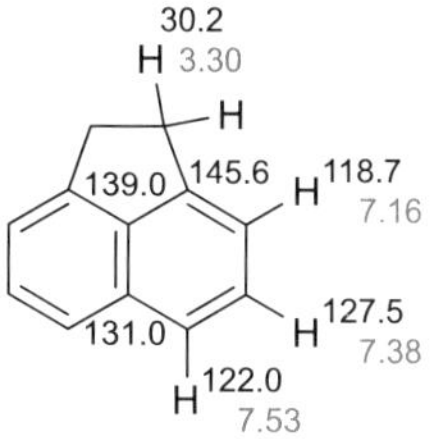

Acenaphthene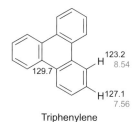
30.2 H 3.30
H
139.0 145.6
118.7 H 7.16
131.0
127.5 H 7.38
122.0 7.53

Fluorene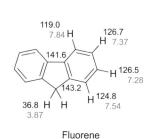
119.0 H 7.84
126.7 H 7.37
141.6
143.2
126.5 H 7.28
36.8 H H 3.87
124.8 H 7.54

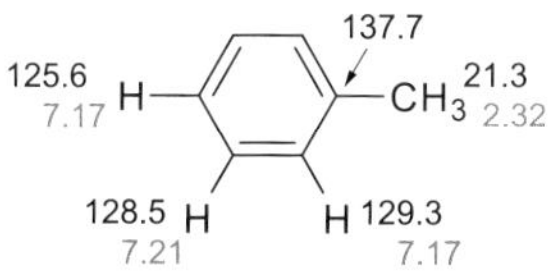

Toluene
125.6 H 7.17
137.7
CH₃ 21.3 2.32
128.5 H 7.21
H 129.3 7.17

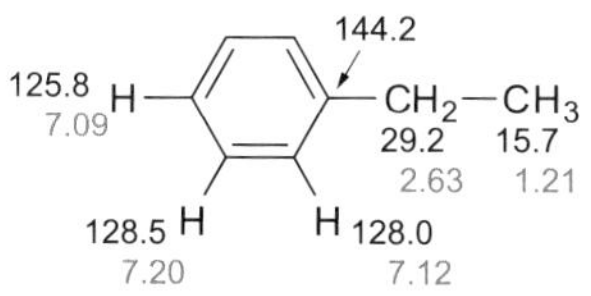

Ethylbenzene
125.8 H 7.09
144.2
CH₂—CH₃
29.2 15.7
2.63 1.21
128.5 H 7.20
H 128.0 7.12

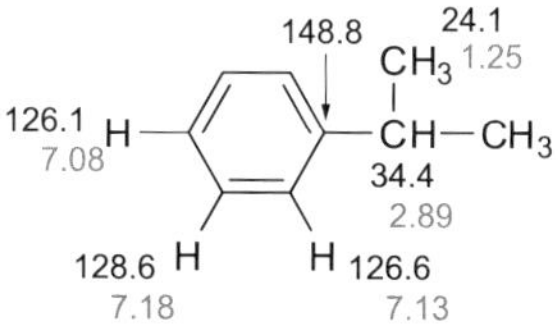

Isopropylbenzene
126.1 H 7.08
148.8 24.1
CH₃ 1.25
CH—CH₃
34.4 2.89
128.6 H 7.18
H 126.6 7.13

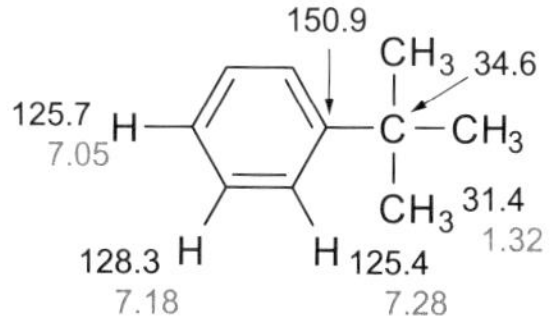

***tert*-Butylbenzene**
125.7 H 7.05
150.9
CH₃ 34.6
C—CH₃
CH₃ 31.4 1.32
128.3 H 7.18
H 125.4 7.28

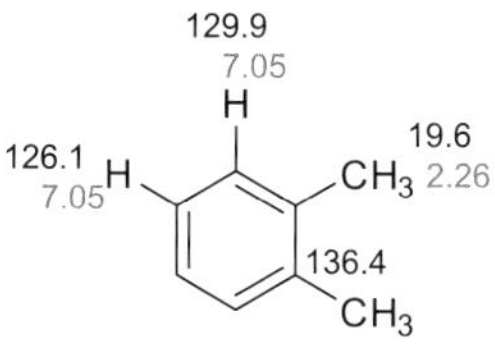

1,2-Dimethylbenzene
(*o*-Xylene)
129.9 7.05
H
126.1 H 7.05
CH₃ 19.6 2.26
136.4
CH₃

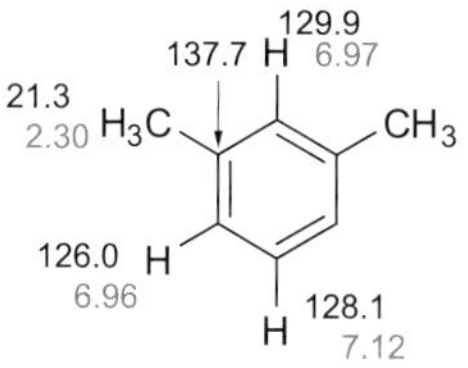

1,3-Dimethylbenzene
(*m*-Xylene)
129.9 H 6.97
137.7
21.3 2.30 H₃C
CH₃
126.0 H 6.96
128.1 H 7.12

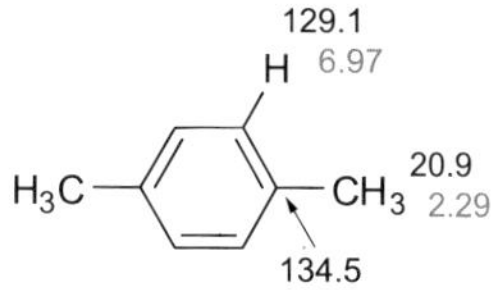

1,4-Dimethylbenzene
(*p*-Xylene)
129.1 H 6.97
H₃C
CH₃ 20.9 2.29
134.5

1,3,5-Trimethylbenzene
(Mesitylene)
137.6
H₃C
CH₃ 21.2 2.23
127.4 H 6.78
CH₃

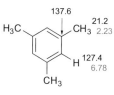

Nuclear Magnetic Resonance Spectroscopy

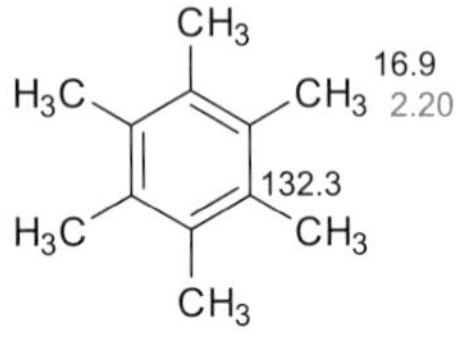

Hexamethylbenzene

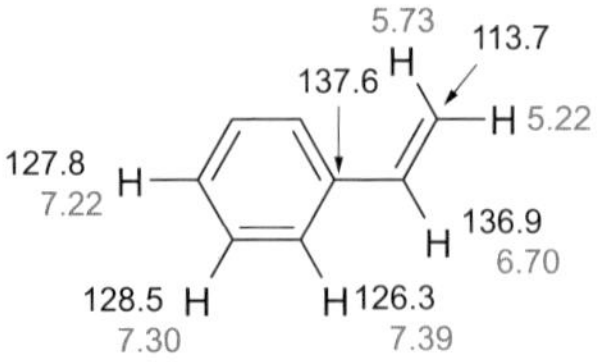

CH₃ 16.9 2.20
132.3

Styrene

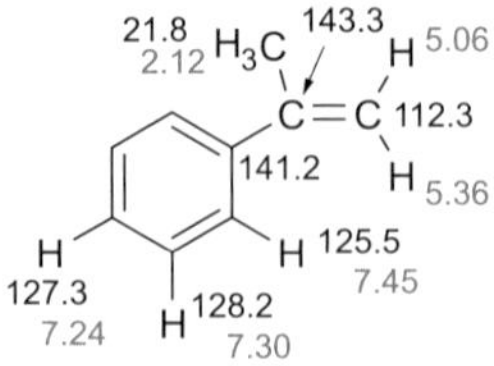

5.73 113.7
137.6
127.8 7.22
128.5 H 7.30
H 126.3 7.39
136.9 6.70
H 5.22

α-Methylstyrene

21.8 H₃C 143.3 H 5.06
2.12
141.2
C=C 112.3
H 5.36
127.3 7.24
128.2 7.30
125.5 7.45

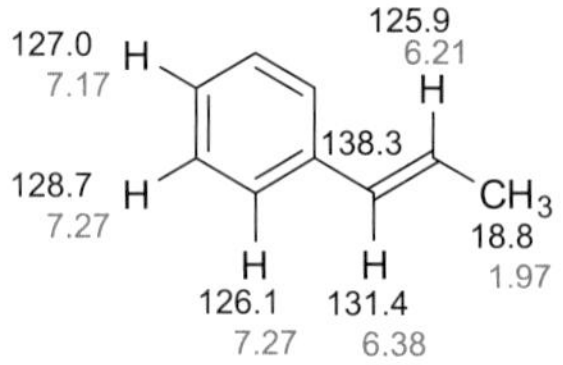

(E)-β-Methylstyrene
[(1E)-1-Propen-1-ylbenzene]

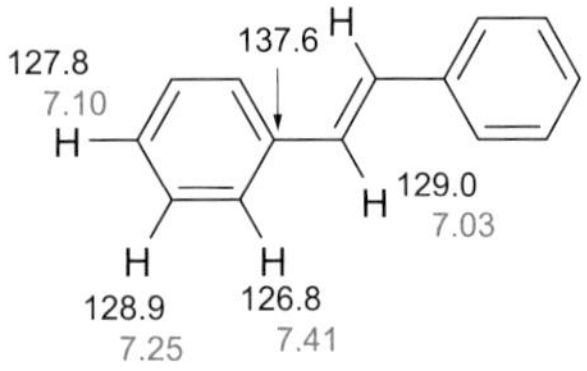

127.0 H 7.17
125.9 6.21
138.3
128.7 H 7.27
126.1 7.27
131.4 6.38
CH₃ 18.8 1.97

(E)-1,2-Diphenylethene
(E-Stilbene)

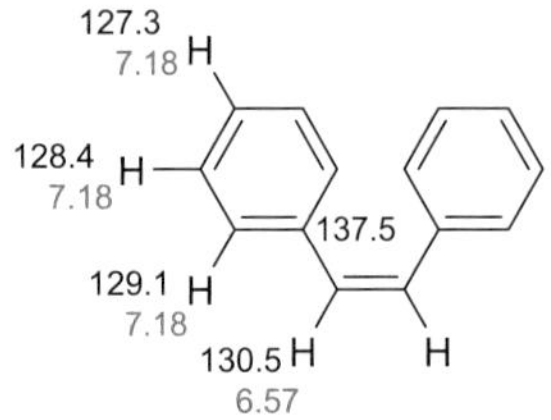

127.8 7.10
137.6 H
129.0 7.03
128.9 7.25
126.8 7.41

(Z)-1,2-Diphenylethene
(Z-Stilbene)

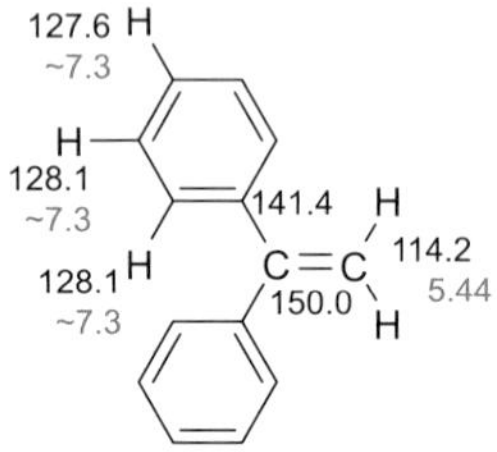

127.3 7.18
128.4 7.18
129.1 7.18
137.5
130.5 H 6.57

1,1-Diphenylethene

127.6 H ~7.3
128.1 ~7.3
128.1 ~7.3
141.4 H
114.2
150.0
C=C
H 5.44

Allylbenzene

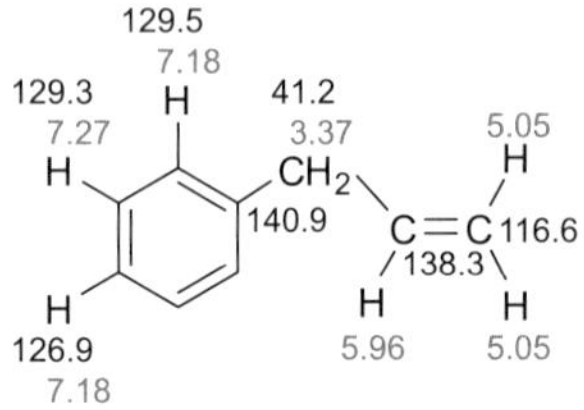

129.5 7.18
129.3 7.27
41.2 3.37
CH₂
140.9
138.3
5.05
C=C 116.6
126.9 7.18
5.96 5.05

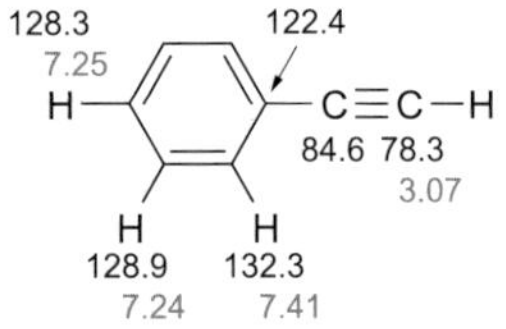

Phenylacetylene
(Ethynylbenzene)

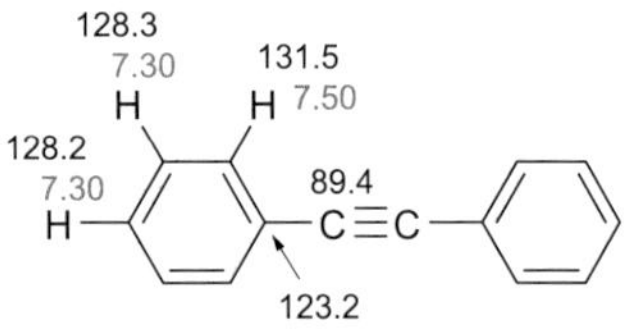

128.3 7.25
122.4
C≡C—H
84.6 78.3 3.07
128.9 7.24
132.3 7.41

Diphenylacetylene
(Tolane)

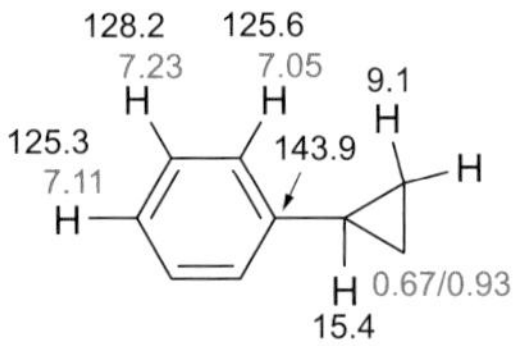

128.3 7.30
131.5 7.50
128.2 7.30
89.4
C≡C
123.2

Cyclopropylbenzene

128.2 7.23
125.6 7.05
9.1 H
125.3 7.11
143.9
H
0.67/0.93
15.4

Cyclohexylbenzene

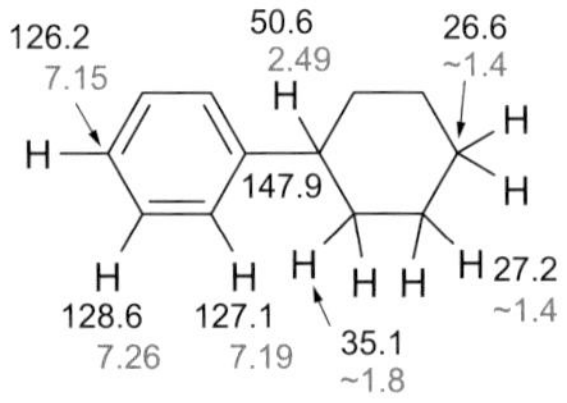

126.2 7.15
50.6 2.49
26.6 ~1.4
147.9
128.6 7.26
127.1 7.19
35.1 ~1.8
27.2 ~1.4

Biphenyl

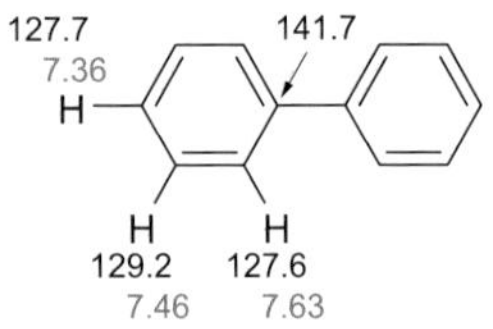

127.7 7.36
141.7
129.2 7.46
127.6 7.63

Diphenylmethane

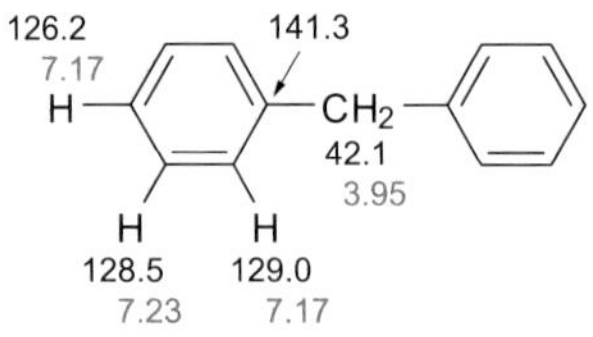

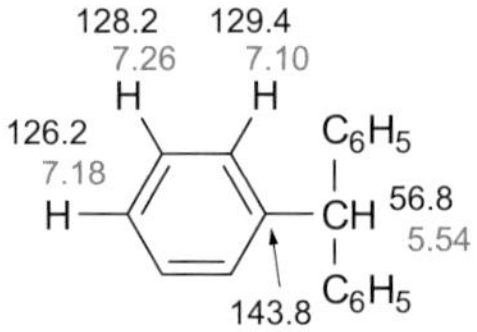

126.2 7.17
141.3
CH₂ 42.1 3.95
128.5 7.23
129.0 7.17

Triphenylmethane

128.2 7.26
129.4 7.10
C₆H₅
126.2 7.18
CH 56.8 5.54
143.8 C₆H₅

1,2-Diphenylethane

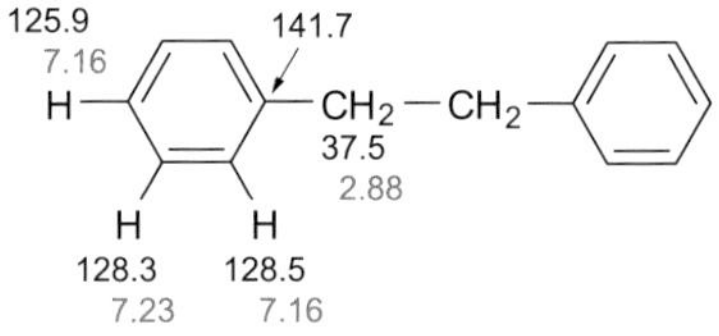

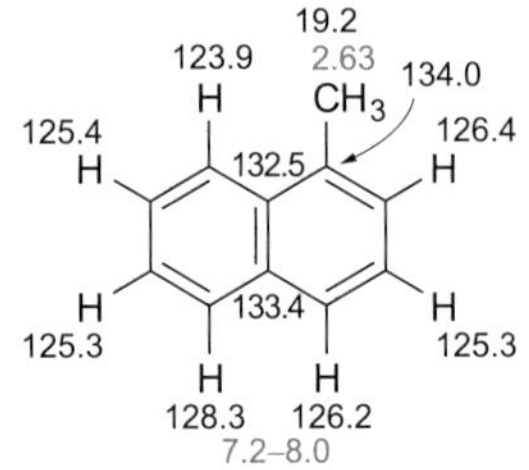

125.9 7.16
141.7
CH₂—CH₂
37.5 2.88
128.3 7.23
128.5 7.16

1-Methylnaphthalene

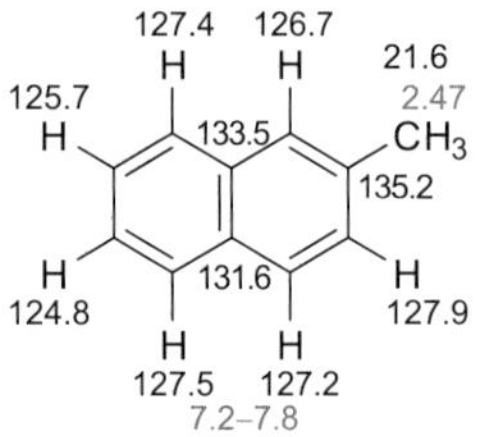

123.9 H
19.2 2.63 CH₃
134.0
125.4 H
132.5
126.4 H
125.3 H
133.4
125.3 H
128.3 H 7.2–8.0
126.2 H

2-Methylnaphthalene

127.4 H
126.7 H
21.6 2.47
125.7 H
133.5
CH₃
135.2
124.8
131.6
127.9
127.5 H 7.2–7.8
127.2 H

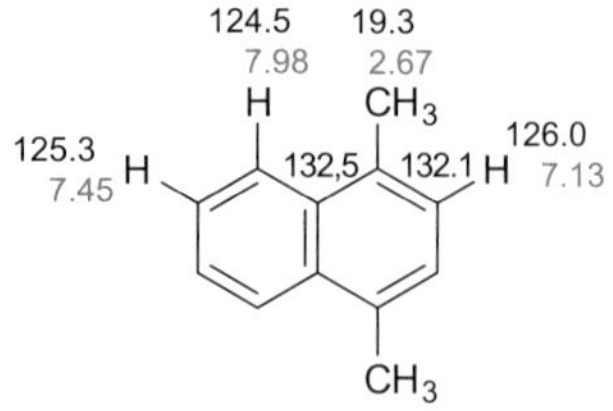

1,4-Dimethylnaphthalene

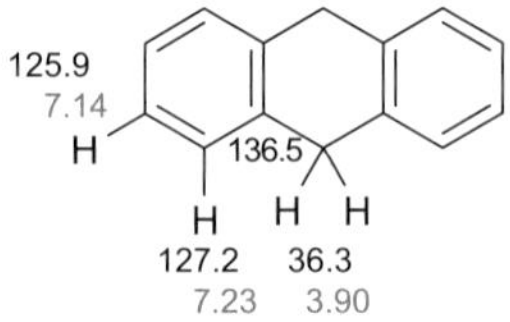

124.5 7.98
19.3 2.67
CH₃
126.0
125.3 7.45 H
132.5
132.1 H 7.13
CH₃

9,10-Dihydroanthracene

125.9 7.14
136.5
127.2 7.23
36.3 3.90

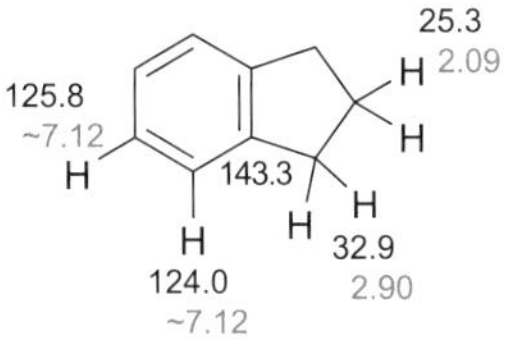

Indane

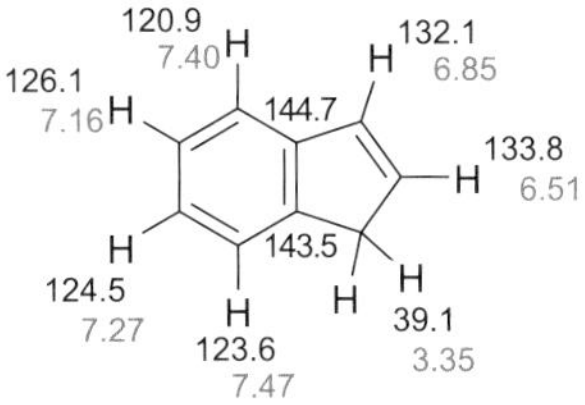

125.8 ~7.12 | 143.3 | 124.0 ~7.12 | 32.9 2.90 | 25.3 2.09

Indene

126.1 7.16 | 120.9 7.40 | 144.7 | 132.1 6.85 | 133.8 6.51 | 124.5 7.27 | 143.5 | 123.6 7.47 | 39.1 3.35

Tetralin

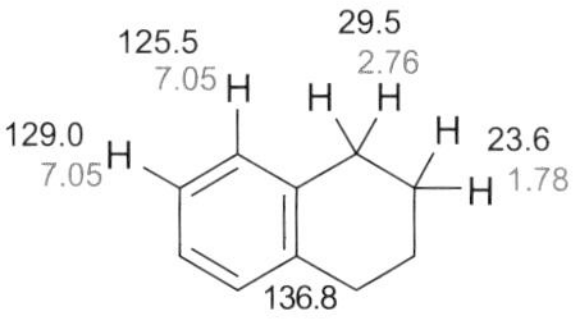

125.5 7.05 | 29.5 2.76 | 129.0 7.05 | 23.6 1.78 | 136.8

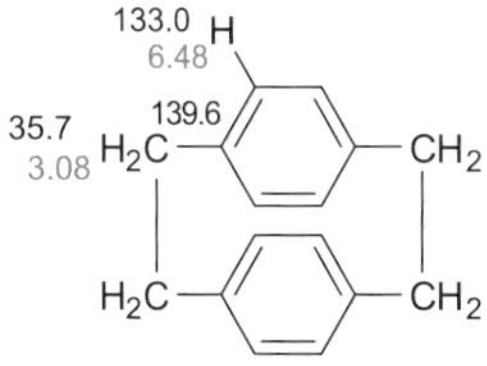

[2,2](1,4)Benzocyclophane (Paracyclophane)

133.0 6.48 | 139.6 | 35.7 3.08 | H₂C | CH₂ | H₂C | CH₂

Heterocycles (O, S, N, P)

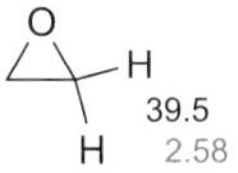

Oxirane (Ethylenoxide)

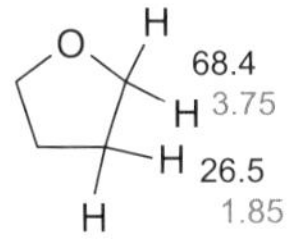

39.5 2.58

Tetrahydrofuran

68.4 3.75 | 26.5 1.85

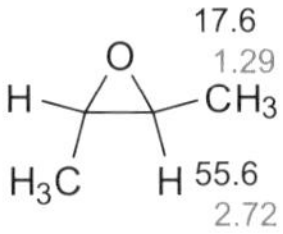

trans-2,3-Dimethyloxirane

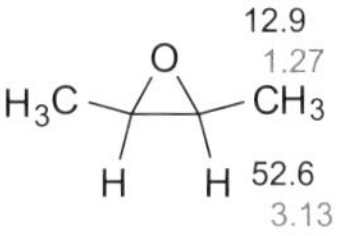

17.6 1.29 | CH₃ | H₃C | 55.6 2.72

cis-2,3-Dimethyloxirane

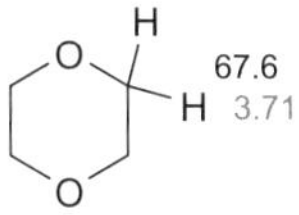

12.9 1.27 | H₃C | CH₃ | 52.6 3.13

1,4-Dioxane

67.6 3.71

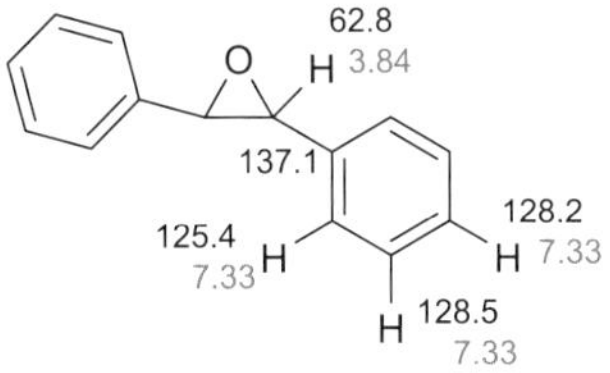

trans-2,3-Diphenyloxirane (trans-Stilbene oxide)

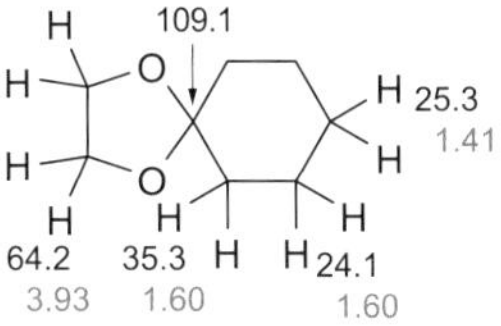

62.8 3.84 | 137.1 | 125.4 7.33 | 128.2 7.33 | 128.5 7.33

Cyclohexanonethan-1,2-diolketal (1,4-Dioxaspiro[4.5]decane)

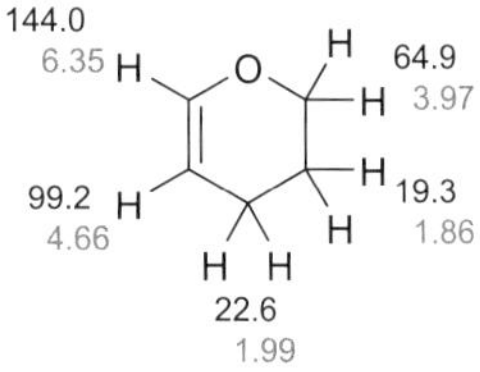

109.1 | 25.3 1.41 | 64.2 3.93 | 35.3 1.60 | 24.1 1.60

3,4-Dihydro-2H-pyran

144.0 6.35 | 64.9 3.97 | 99.2 4.66 | 19.3 1.86 | 22.6 1.99

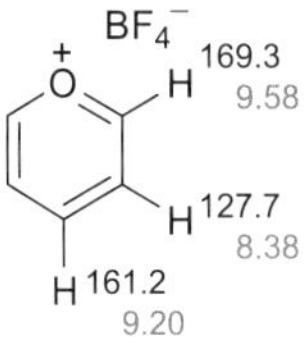

Pyryliumtetrafluoroborate

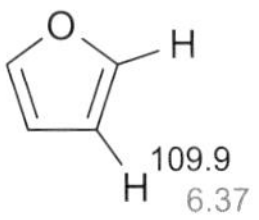

BF₄⁻ | 169.3 9.58 | 127.7 8.38 | 161.2 9.20

Furan

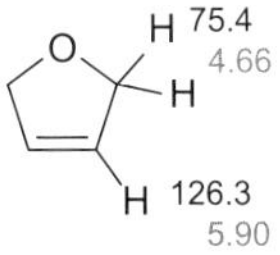

109.9 6.37

2,5-Dihydrofuran

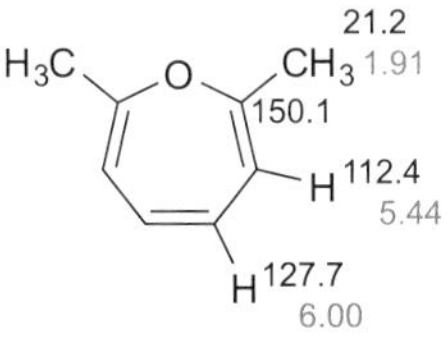

75.4 4.66 | 126.3 5.90

2,7-Dimethyloxepin

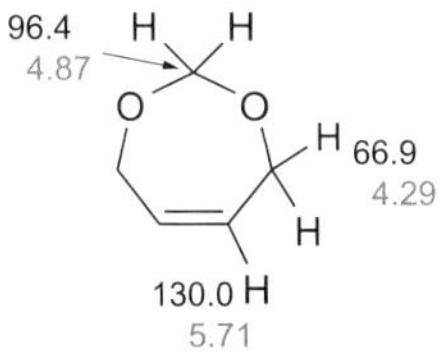

21.2 1.91 | H₃C | CH₃ | 150.1 | 112.4 5.44 | 127.7 6.00

4,7-Dihydro-1,3-dioxepin

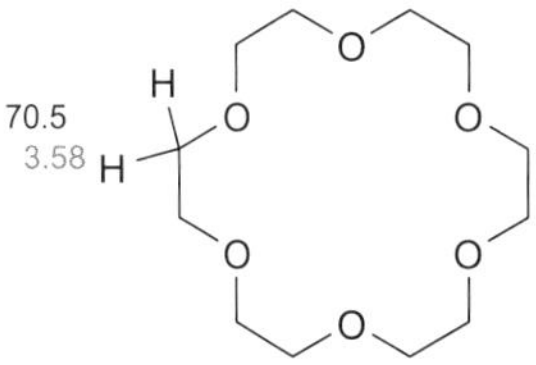

96.4 4.87 | 66.9 4.29 | 130.0 5.71

18-Crone-6

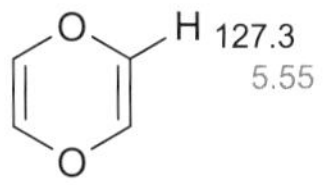

70.5 3.58

1,4-Dioxin

127.3 5.55

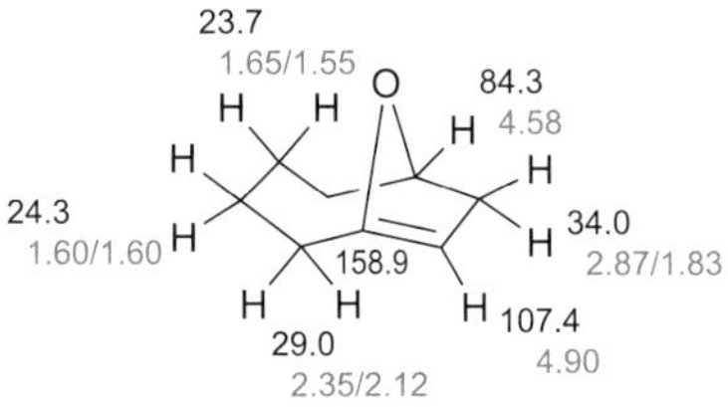

9-Oxabicyclo[4.2.1]non-6-ene

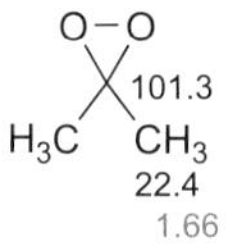

23.7 1.65/1.55 | 84.3 4.58 | 24.3 1.60/1.60 | 34.0 2.87/1.83 | 158.9 | 29.0 2.35/2.12 | 107.4 4.90

Dimethyldioxirane (explosive)

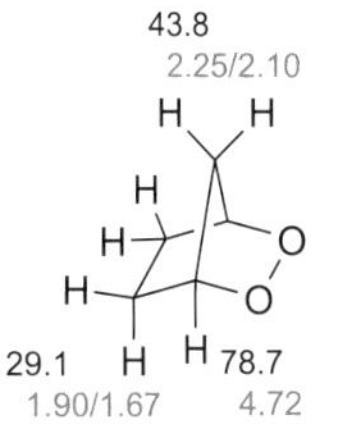

O–O | 101.3 | H₃C | CH₃ | 22.4 1.66

2,3-Dioxa-bicyclo[2.2.1]heptane

43.8 2.25/2.10 | 29.1 1.90/1.67 | 78.7 4.72

Nuclear Magnetic Resonance Spectroscopy

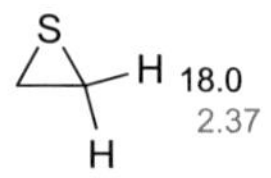

Thiirane
(Ethylene sulfide)

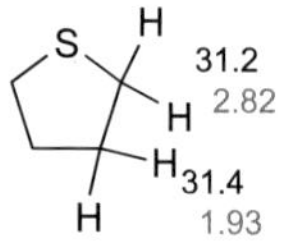

Tetrahydrothiophene

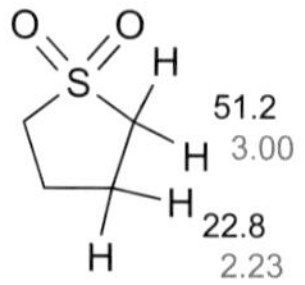

Sulfolane

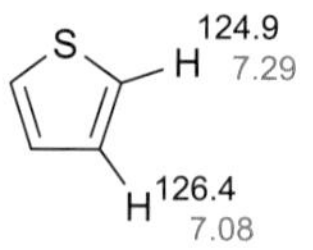

Thiophene

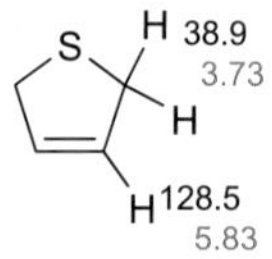

2,5-Dihydrothiophene

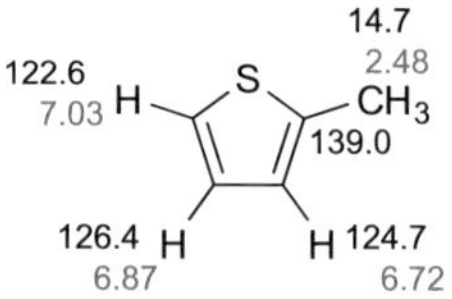

2-Methylthiophene

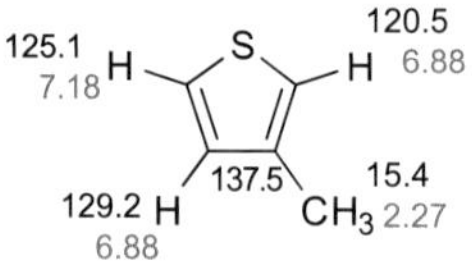

3-Methylthiophene

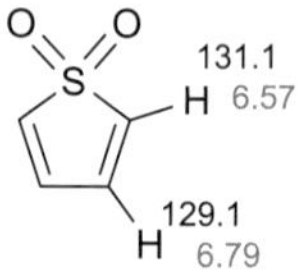

Thiophene-1,1-dioxide
(not isolable)

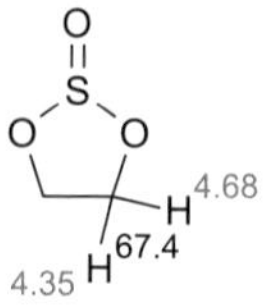

Glycol sulfite
(1,3,2-Dioxathiolane-
2-oxide)

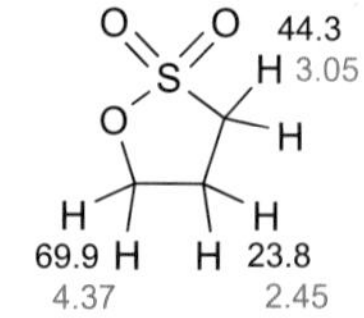

1,3-Propane sultone
(1,2-Oxathiolane-
2,2-dioxide)

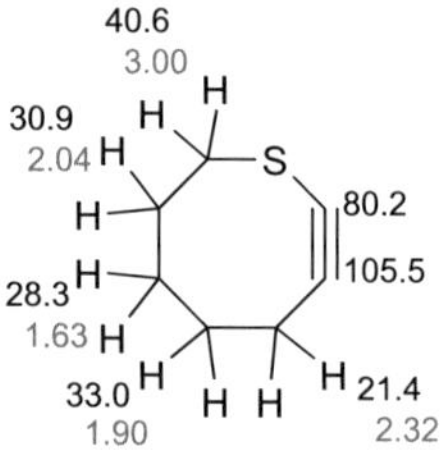

1-Thiacyclooct-2-yne

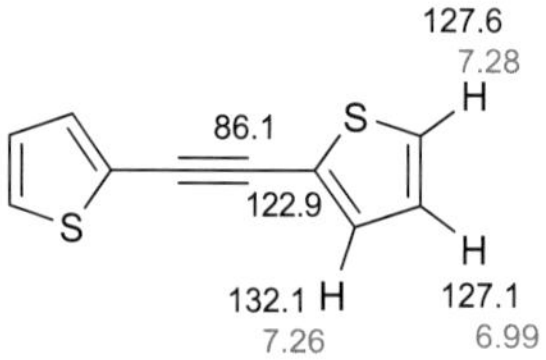

Bis(2-thienyl)ethyne

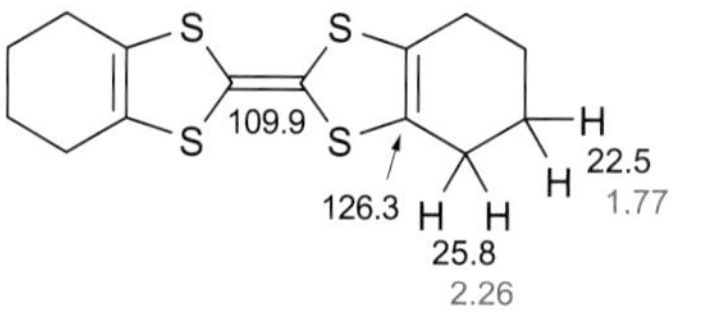

4,4',5,5',6,6',7,7'-Octahydro-
dibenzotetrathiafulvalene

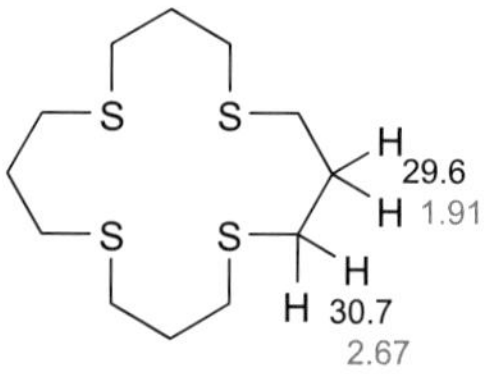

1,5,9,13-Tetrathiacyclo-
hexadecane

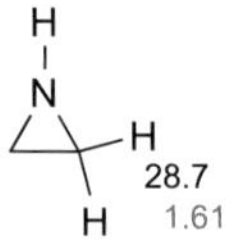

Aziridine (Ethylenimine)

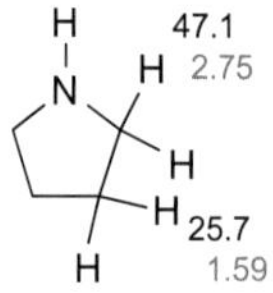

Pyrrolidine

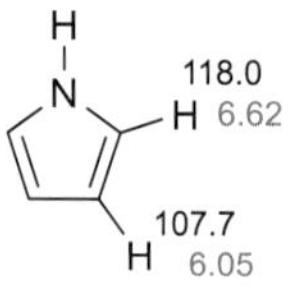

Pyrrole

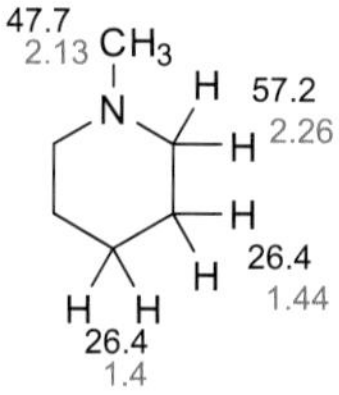

N-Methylpiperidine

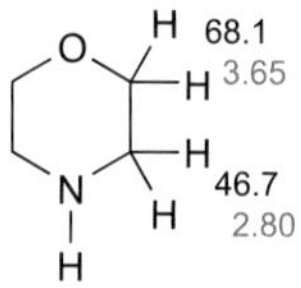

Morpholine

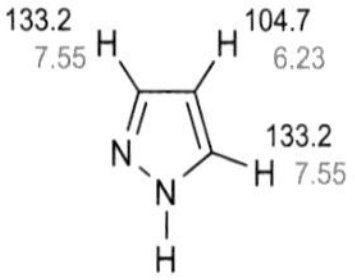

Pyrazole[a]

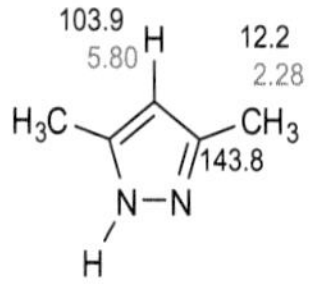

3,5-Dimethylpyrazole[a]

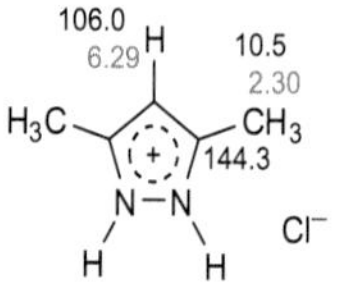

3,5-Dimethylpyrazol
hydrochloride(DMSO)

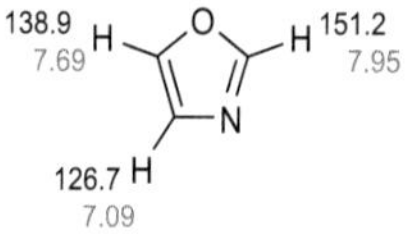

Oxazole

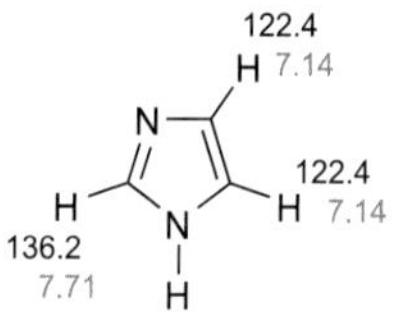

Imidazole[a]

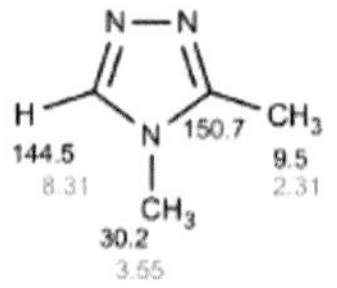

3,4-Dimethyl-4*H*-1,2,4-
triazole

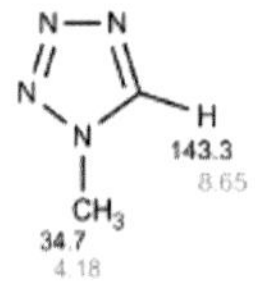

1-Methyltetrazole

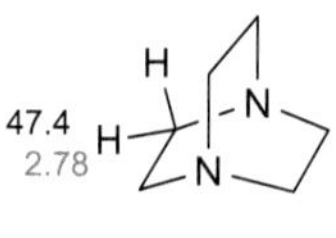

1,4-Diazabicyclo[2.2.2]-
octane

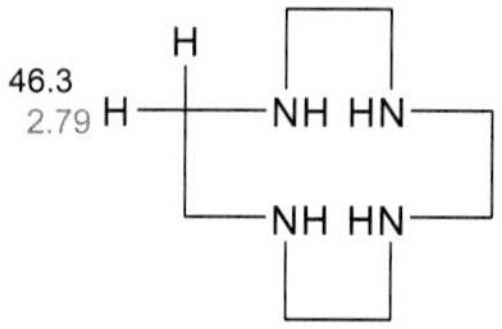

1,4,7,10-Tetraazacyclododecane
(Cyclene)

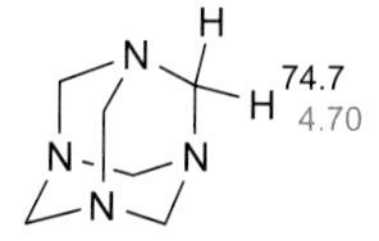

Hexamethylentetramine
(Urotropine)

Thiazole
118.6 H 7.40 · S · 152.6 H 8.86 · 143.3 H 7.96

Pyridine
149.8 H 8.59 · N · 123.6 H 7.38 · 135.7 H 7.75

Pyridin-N-oxide
O · 138.5 H 8.36 · 125.6 H 7.46 · 125.0 H 7.46

2-Ethylpyridine
31.4 / 13.8 · 2.86 / 1.26 · CH₂—CH₃ · 149.1 H 8.62 · N · 163.4 · 120.7 H 7.23 · 121.8 H 7.29 · 136.1 H 7.76

Phosphole
P · 131.2 H 7.10 · 140.7 H 7.33

Phosphorin
154.1 H 8.61 · P · 136.6 H 7.72 · 128.2 H 7.38

1,2-Dihydropyridine
42.2 H 4.06 · N · 136.2 H 6.42 · 110.5 H 5.08 · 95.5 H 4.81 · 125.2 H 5.98

1,4-Dihydropyridine
N · 128.6 H 6.13 · 96.1 H 4.35 · 22.3 H 3.10

2,5-Dihydropyridine
49.3 H 4.13 · N · 160.1 H 7.85 · 28.3 H 2.49 · 120.8 H 5.60 · 126.1 H 5.60

4-Vinylpyridine
150.1 H 8.55 · N · 120.7 H 7.24 · 144.6 · 134.7 H 6.64 · C=C 118.6 · H 5.94 · H 5.46

1,3-Benzodioxole
121.6 H 6.80 · O · 100.5 H 5.90 · 147.4 · O · 108.6 H 6.80

Pyrimidine
N · 159.0 H 9.26 · 121.4 H 7.36 · N · 156.4 H 8.78

2H-Benzo[b]thiete
122.7 H 6.86 · 123.9 H 7.12 · 139.5 · 36.5 H 4.29 · 128.4 H 7.26 · 142.5 S · 120.8 H 7.06

Indole
120.5 H 7.64 · 121.7 H 7.10 · 102.1 H 6.50 · 127.6 · 124.1 H 7.03 · 119.6 H 7.17 · 135.5 · N · 111.0 H 7.25

Benzo[b]furan
121.6 H 7.49 · 123.2 H 7.13 · 106.9 H 6.66 · 127.9 · 145.0 H 7.52 · 124.6 H 7.19 · 155.5 O · 111.8 H 7.42

Benzo[b]thiophene
123.8 H 7.78 · 124.3 H 7.33 · 124.0 H 7.29 · 139.8 · 126.4 H 7.40 · 124.4 H 7.31 · 139.9 S · 122.6 H 7.86

Benzimidazole[a]
115.4 H 7.73 · 122.9 H 7.23 · 137.9 N · 141.5 H 8.40 · N · H

Benzotriazole[a]
114.8 H 7.98 · 125.3 H 7.48 · 138.7 N · N · 138.7 N · H · 125.3 H 7.48 · 114.8 H 7.98

Purine
144.8 H 9.19 · 128.4 · N · 147.9 H 8.68 · N · 152.0 H 8.99 · N · 154.9 N · H

Quinoline
129.2 H 8.05 · 129.2 H 7.61 · 148.1 · 150.0 H 8.81 · N · H · 126.3 H 7.43 · 128.0 · 120.8 H 7.26 · 127.6 H 7.68 · 135.7 H 8.00

Isoquinoline
127.3 H 7.87 · 152.2 H 9.15 · 127.0 H 7.50 · 128.5 · N · 130.1 H 7.57 · 135.5 H · 142.7 H 8.45 · 126.2 H 7.71 · 120.2 H 7.50

Nuclear Magnetic Resonance Spectroscopy

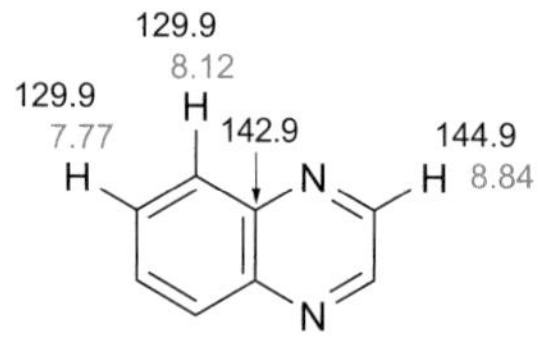

129.9
8.12

129.9
7.77 H 142.9   144.9
H   H 8.84

Quinoxaline

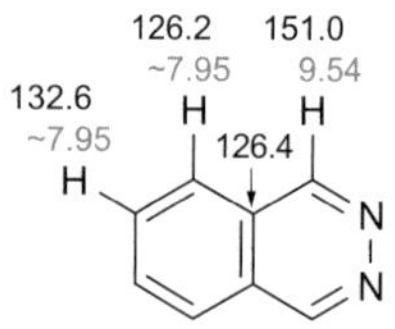

126.2 151.0
~7.95 9.54

132.6
~7.95 H 126.4
H

Phthalazine

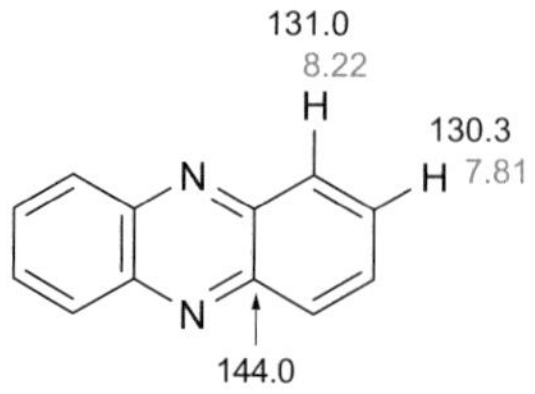

131.0
8.22
H   130.3
H 7.81

Phenazine

144.0

135.8   126.3
8.15 H   H 7.68
            129.0
123.0 H
7.57 H

150.1 H   146.5
9.17

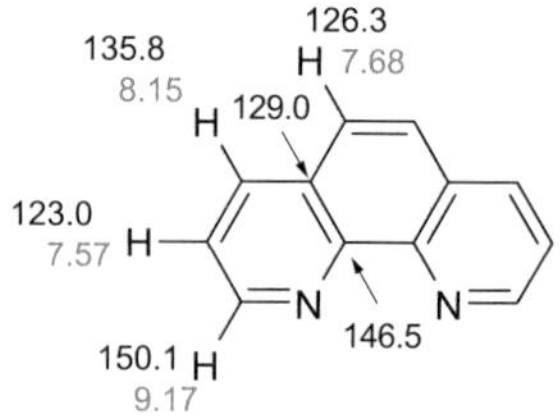

1,10-Phenanthroline

[a]Fast proton transfer between the N-atoms.

C — Hal   Halogenated Hydrocarbons [b]   C — Hal

H_3C-CH_2F
13.3   78.0
1.24   4.36
Fluoroethane

H_3C-CH_2Cl
17.5   38.7
1.33   3.47
Chloroethane

H_3C-CH_2Br
19.3   27.9
1.66   3.37
Bromoethane

H_3C-CH_2I
20.5   −1.0
1.88   3.18
Iodoethane

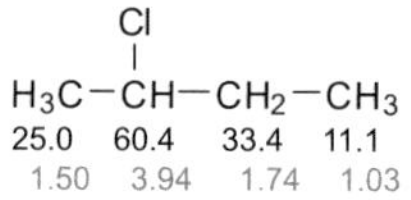

Cl
|
$H_3C-CH-CH_2-CH_3$
25.0   60.4   33.4   11.1
1.50   3.94   1.74   1.03
2-Chlorobutane

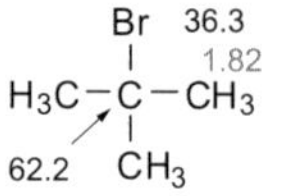

Br   36.3
| 1.82
$H_3C-C-CH_3$
62.2 CH_3
2-Bromo-2-methylpropane

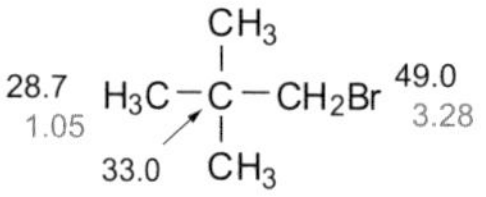

CH_3
|
28.7 $H_3C-C-CH_2Br$ 49.0
1.05   3.28
33.0 CH_3
1-Bromo-2,2-dimethylpropane

$H_3C-CHBr_2$
34.2   39.1
2.47   5.86
1,1-Dibromoethane

$Br-CH_2-CH_2-C_2H_4Br$
33.6   31.8
3.47   2.05
1,4-Dibromobutane

$Cl-CH_2-CH_2-CH_2-C_3H_6Cl$
44.9   32.4   26.2
3.53   1.79   1.48
1,6-Dichlorohexane

$CHCl_2-CH_2Cl$
70.4   50.1
5.77   3.98
1,1,2-Trichloroethane

$F_3C-CF_2-CF_2-C_2F_5$
118.5 109.8   111.1
Dodecafluoropentane
(Perfluoropentane)

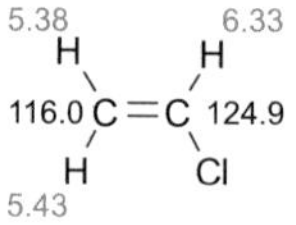

5.38   6.33
H   H
116.0 C = C 124.9
H   Cl
5.43
Vinyl chloride

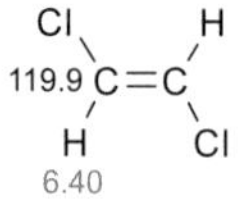

Cl   H
119.9 C=C
H   Cl
6.40
(E)-1,2-Dichloroethene

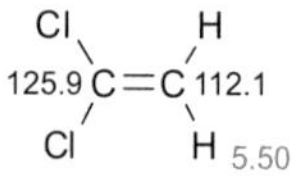

Cl   H
125.9 C=C 112.1
Cl   H 5.50
1,1-Dichloroethene

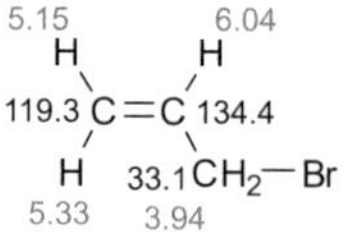

5.15   6.04
H   H
119.3 C=C 134.4
H   33.1 CH_2—Br
5.33   3.94
Allyl bromide

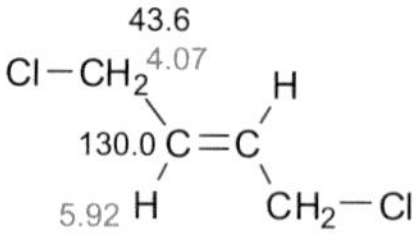

43.6
$Cl-CH_2$ 4.07
          H
130.0 C=C
5.92 H CH_2-Cl
(E)-1,4-Dichlorobut-2-ene

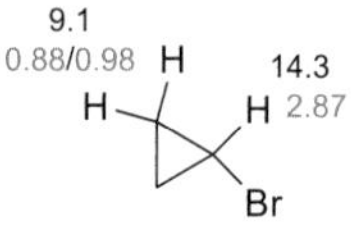

9.1
0.88/0.98 H   14.3
H   H 2.87
Br
Bromocyclopropane

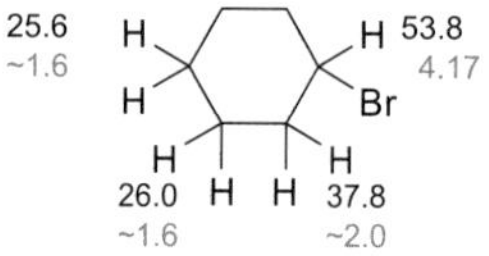

25.6
~1.6 H   H 53.8
      H   4.17
      H   Br
      26.0 H H 37.8
      ~1.6   ~2.0
Bromocyclohexane

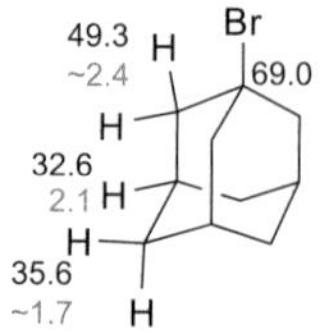

49.3 H Br
~2.4   69.0
32.6 H
2.1 H
H
35.6 H
~1.7 H
1-Bromoadamantane

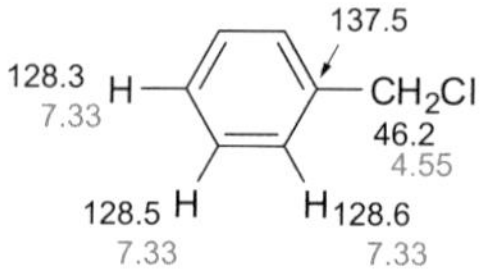

137.5
128.3 H CH_2Cl
7.33   46.2
             4.55
128.5 H   H 128.6
7.33   7.33
Benzyl chloride
(Chloromethylbenzene)

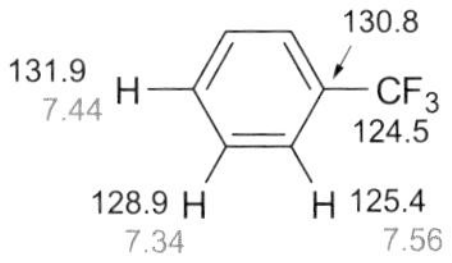

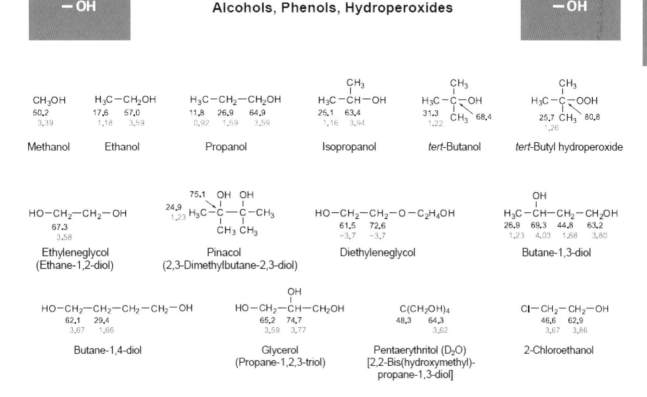

131.9
H 7.44
130.8
CF₃
124.5
128.9 H 7.34
H 125.4 7.56

Trifluoromethylbenzene

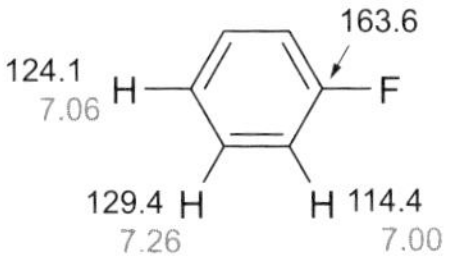

124.1
H 7.06
163.6
F
129.4 H 7.26
H 114.4 7.00

Fluorobenzene

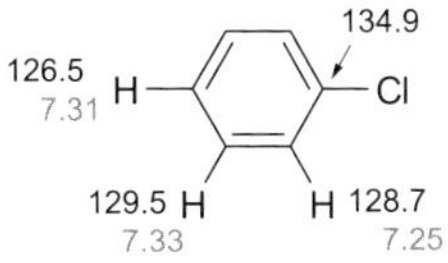

126.5
H 7.31
134.9
Cl
129.5 H 7.33
H 128.7 7.25

Chlorobenzene

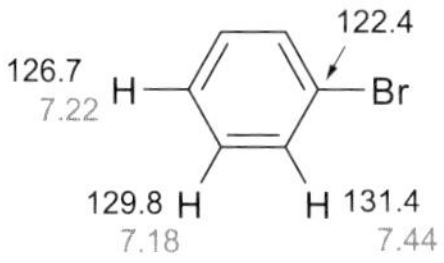

126.7
H 7.22
122.4
Br
129.8 H 7.18
H 131.4 7.44

Bromobenzene

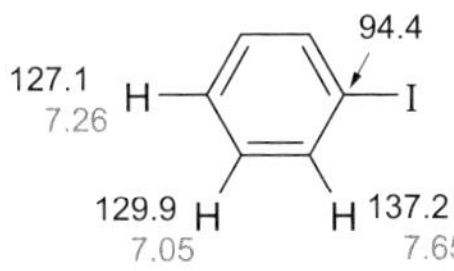

127.1
H 7.26
94.4
I
129.9 H 7.05
H 137.2 7.65

Iodobenzene

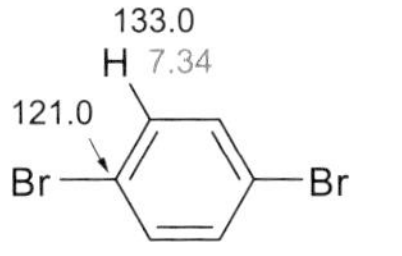

133.0
H 7.34
121.0
Br
Br

1,4-Dibromobenzene

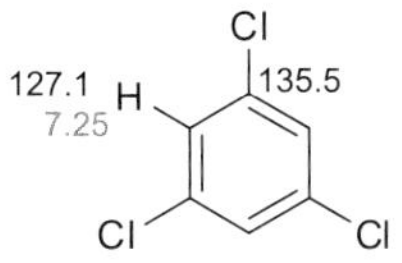

127.1
H 7.25
Cl
135.5
Cl
Cl

1,3,5-Trichlorobenzene

28.6
4.60
133.6
H 7.37
137.6
Br—CH₂
CH₂—Br
Br—CH₂
CH₂—Br

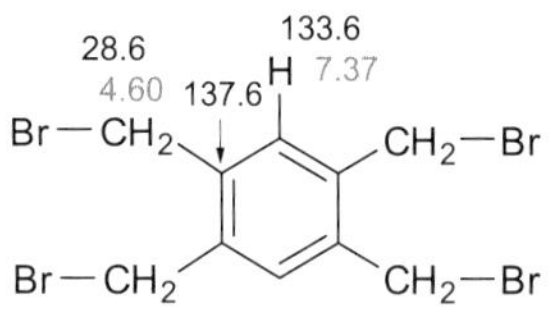

1,2,4,5-Tetrakis(bromomethyl)benzene

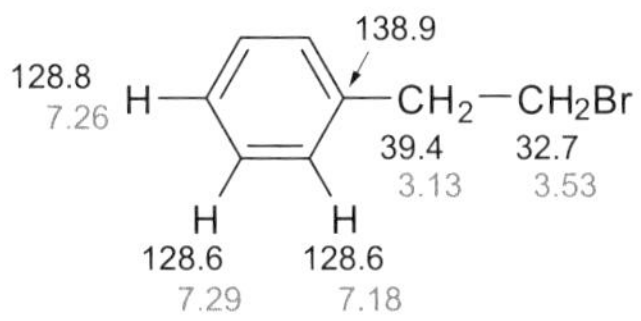

128.8
H 7.26
138.9
CH₂—CH₂Br
39.4   32.7
3.13   3.53
H   H
128.6   128.6
7.29   7.18

(2-Bromoethyl)benzene

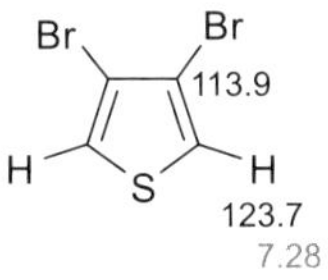

Br   Br
113.9
H   H
S
123.7
7.28

3,4-Dibromothiophene

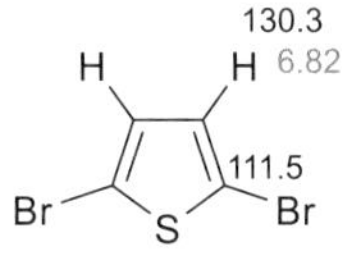

130.3
H   H 6.82
111.5
Br   Br
S

2,5-Dibromothiophene

138.4
7.63 H
125.1
Cl   H 7.28
130.9
148.4 H   Cl
8.36   N
149.4

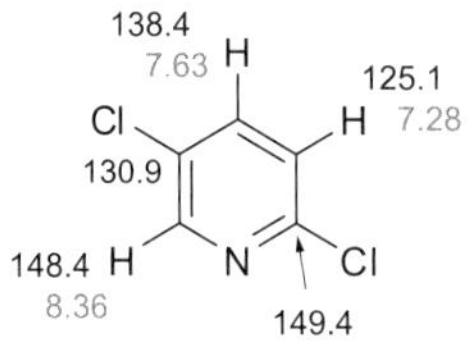

2,5-Dichloropyridine

ᵇ For the halogenated methanes see Tables 3.8 and 3.28

—OH   Alcohols, Phenols, Hydroperoxides   —OH

CH₃OH
50.2
3.39

Methanol

H₃C—CH₂OH
17.6   57.0
1.18   3.59

Ethanol

H₃C—CH₂—CH₂OH
11.8   26.9   64.9
0.92   1.59   3.59

Propanol

CH₃
H₃C—CH—OH
25.1   63.4
1.16   3.94

Isopropanol

CH₃
H₃C—C—OH
31.3   CH₃   68.4
1.22

***tert*-Butanol**

CH₃
H₃C—C—OOH
25.7 CH₃   80.8
1.26

***tert*-Butyl hydroperoxide**

HO—CH₂—CH₂—OH
67.3
3.58

**Ethyleneglycol
(Ethane-1,2-diol)**

75.1   OH   OH
24.9
1.23   H₃C—C—C—CH₃
CH₃ CH₃

**Pinacol
(2,3-Dimethylbutane-2,3-diol)**

HO—CH₂—CH₂—O—C₂H₄OH
61.5   72.6
−3.7   −3.7

Diethyleneglycol

OH
H₃C—CH—CH₂—CH₂OH
26.9   69.3   44.8   63.2
1.23   4.03   1.68   3.80

Butane-1,3-diol

HO—CH₂—CH₂—CH₂—CH₂—OH
62.1   29.4
3.67   1.66

Butane-1,4-diol

OH
HO—CH₂—CH—CH₂OH
65.2   74.7
3.59   3.77

**Glycerol
(Propane-1,2,3-triol)**

C(CH₂OH)₄
48.3   64.3
3.62

**Pentaerythritol (D₂O)
[2,2-Bis(hydroxymethyl)-
propane-1,3-diol]**

Cl—CH₂—CH₂—OH
46.6   62.9
3.67   3.86

2-Chloroethanol

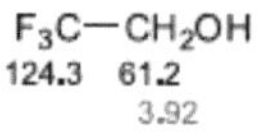

F₃C—CH₂OH
F_3C-CH_2OH
124.3 61.2
3.92

2,2,2-Trifluoroethanol

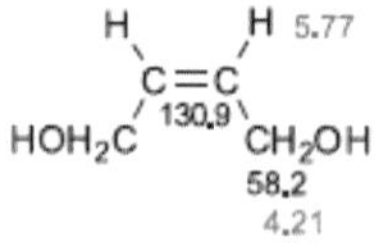

H   H 5.77
C=C 130.9
HOH₂C   CH₂OH
58.2
4.21

(Z)-But-2-ene-1,4-diol

$HC\equiv C-CH_2OH$
73.8 82.0 50.6
2.50   4.28

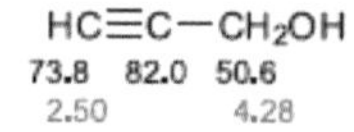

Propargyl alcohol
(Prop-2-yne-1-ol)

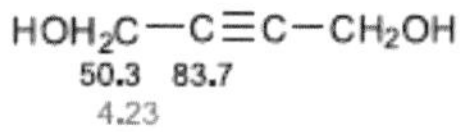

$HOH_2C-C\equiv C-CH_2OH$
50.3 83.7
4.23

But-2-yne-1,4-diol

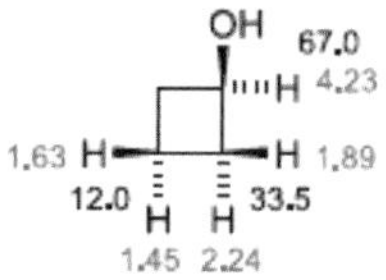

OH 67.0
H 4.23
1.63 H   H 1.89
12.0 H H 33.5
1.45 2.24

Cyclobutanol

2.7
0.52/0.20 H
H   13.5
1.09
CH₂OH
67.5
H 3.42

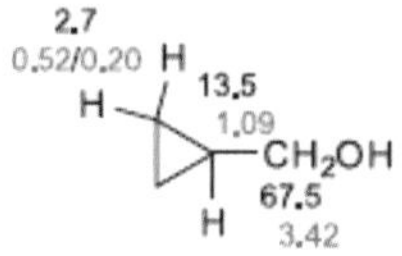

Cyclopropane methanol

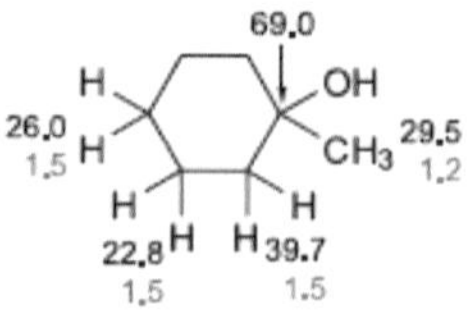

69.0
H   OH
26.0 H   CH₃ 29.5
1.5   1.2
H   H
22.8 H H 39.7
1.5   1.5

1-Methylcyclohexanol

36.1 19.9
1.3–1.7
H H
H   H
HO
69.8
3.56 H
OH
H

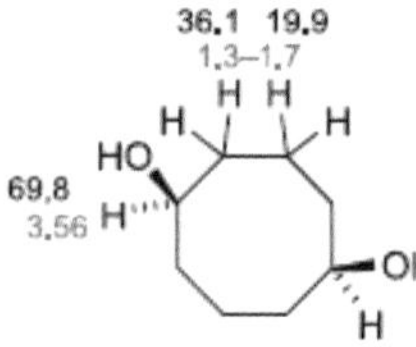

cis-Cyclooctane-1,5-diol

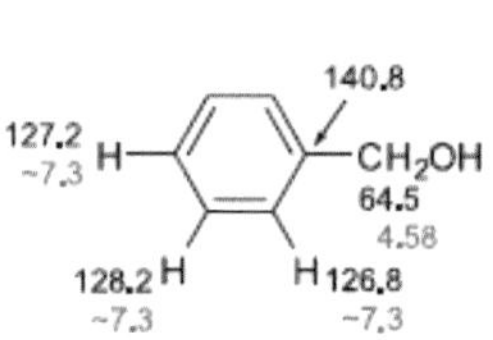

140.8
127.2 H   CH₂OH
~7.3   64.5
4.58
128.2 H H 126.8
~7.3   ~7.3

Benzyl alcohol

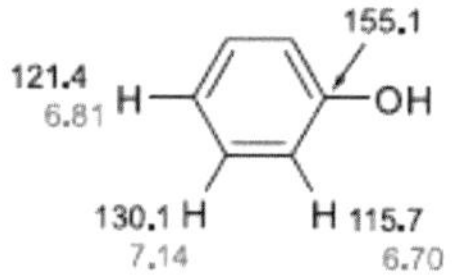

155.1
121.4 H   OH
6.81
130.1 H H 115.7
7.14   6.70

Phenol

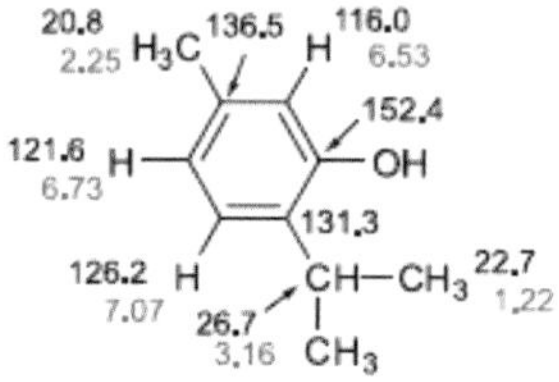

20.8 136.5 116.0
2.25 H₃C   H 6.53
152.4
121.6 H   OH
6.73
131.3
126.2 H CH—CH₃ 22.7
7.07   26.7   1.22
3.16 CH₃

Thymol

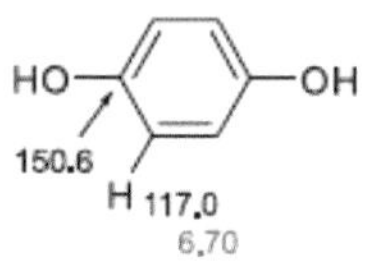

HO   OH
150.6
H 117.0
6.70

Hydroquinone

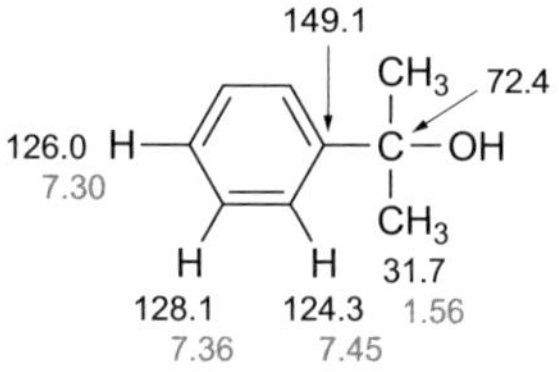

149.1
CH₃ 72.4
126.0 H   C—OH
7.30
CH₃
H H 31.7
128.1 124.3 1.56
7.36 7.45

2-Phenylisopropanol

145.0
CH₃ 84.3
127.8 H   C—OOH
7.30
CH₃
H H 26.5
128.9 125.8 1.58
7.35 7.46

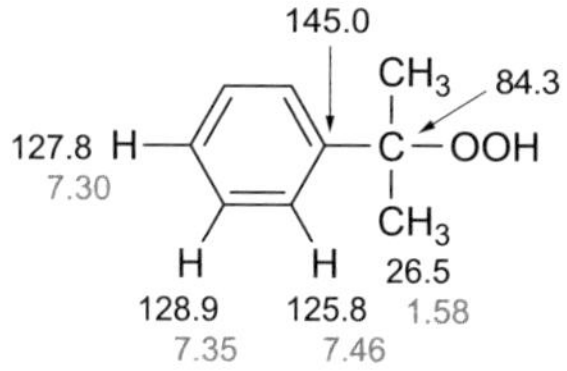

2-Phenyl-isopropyl-hydroperoxide
(Cumene hydroperoxid)

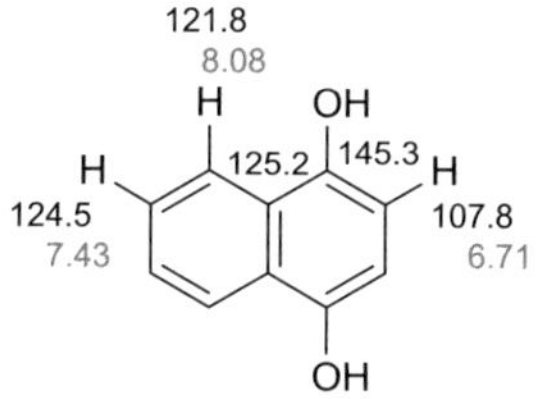

121.8
8.08
H OH
H 125.2 145.3 H
124.5   107.8
7.43   6.71
OH

1,4-Dihydroxy-
naphthalene

OH
126.0 152.8 H
108.3
6.84
H
124.5
OH H 7.18
112.8
7.64

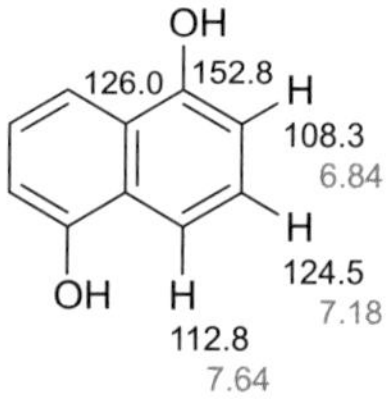

1,5-Dihydroxy-
naphthalene

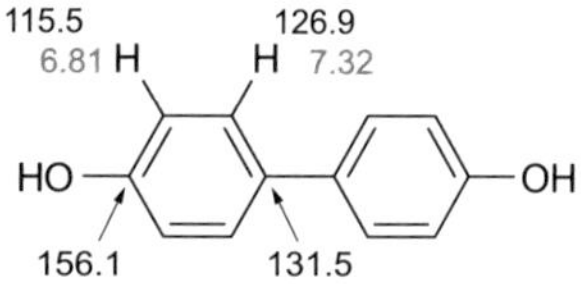

115.5   126.9
6.81 H   H 7.32
HO   OH
156.1   131.5

4,4'-Biphenol

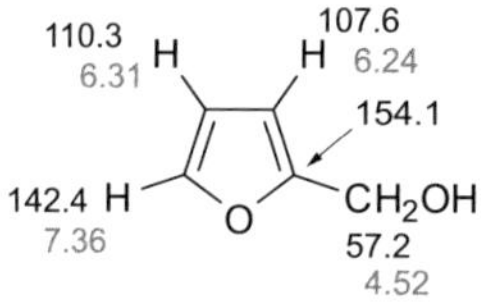

110.3   107.6
6.31 H H 6.24
154.1
142.4 H   CH₂OH
7.36   O 57.2
4.52

Furfuryl alcohol

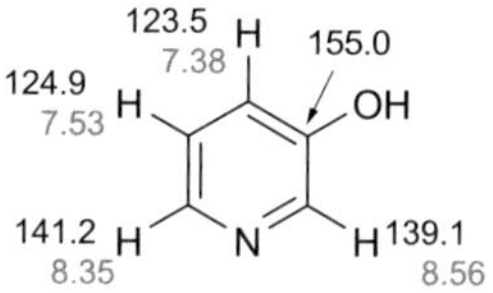

123.5 H 155.0
7.38
124.9 H   OH
7.53
141.2 H H 139.1
8.35   N   8.56

3-Hydroxypyridine

OH
H
130.2 H   N
6.97   N
OH 156.1

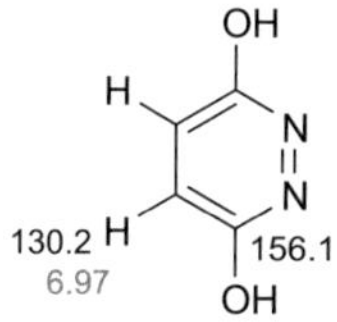

3,6-Dihydroxypyridazine

— OR

Ethers, Acetals, Orthoesters

— OR

H₃C—CH₂—OC₂H₅
$H_3C-CH_2-OC_2H_5$
14.6 65.2
1.16 3.36

Diethyl ether

H₃C—CH₂—CH₂—OC₃H₇
$H_3C-CH_2-CH_2-OC_3H_7$
11.1 24.0 73.2
0.94 1.57 3.40

Dipropyl ether

CH₃
H₃C—CH—O—CH(CH₃)₂
24.3 69.2
1.14 3.65

Diisopropyl ether

CH₃
H₃C—O—C—CH₃
50.1 73.6 28.2
3.20   CH₃ 1.19

tert-Butyl methyl ether

Cl₂CH—O—CH₃
$Cl_2CH-O-CH_3$
98.9 52.6
7.33 3.68

α,α-(Dichlormethyl)-

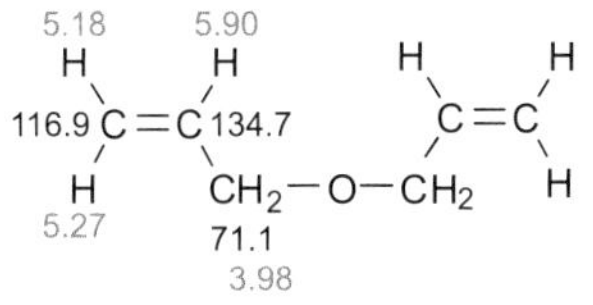

Diallyl ether

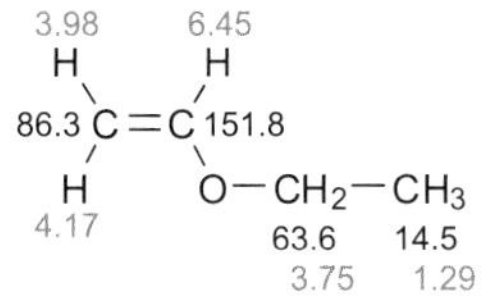

Ethyl vinyl ether

$H-C\equiv C-O-CH_2-CH_3$
26.4 90.9 74.6 14.2
1.54 4.13 1.39

Ethynyl ethyl ether

$H_3C-O-CH_2-CH_2OCH_3$
58.4 71.3
3.39 3.54

Ethylenglycol-dimethyl ether

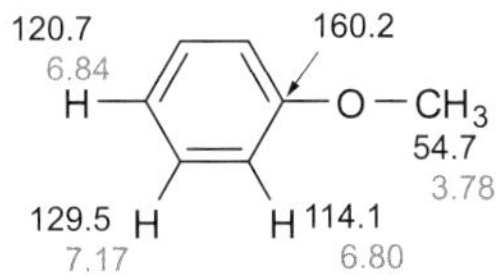

Anisol
(Methoxybenzene)

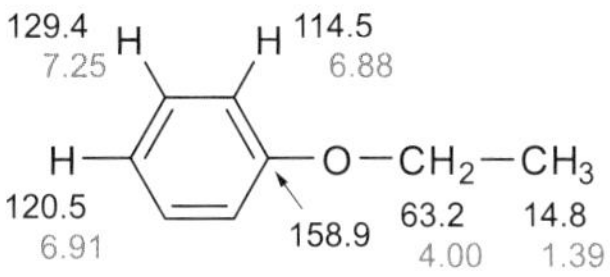

Phenetol
(Methoxybenzene)

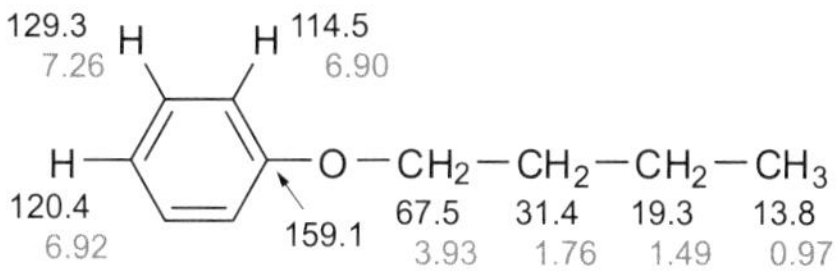

Butyl phenyl ether

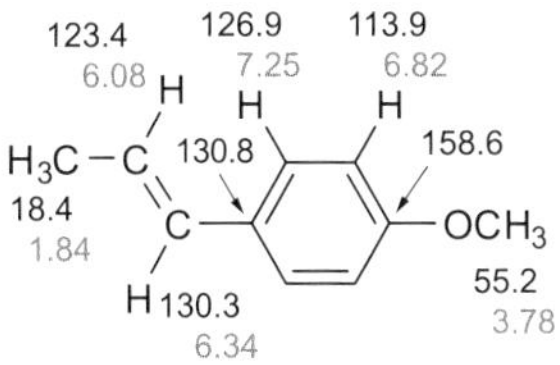

Anethol
[1-Methox-4-(prop-en-1-yl)-
benzene

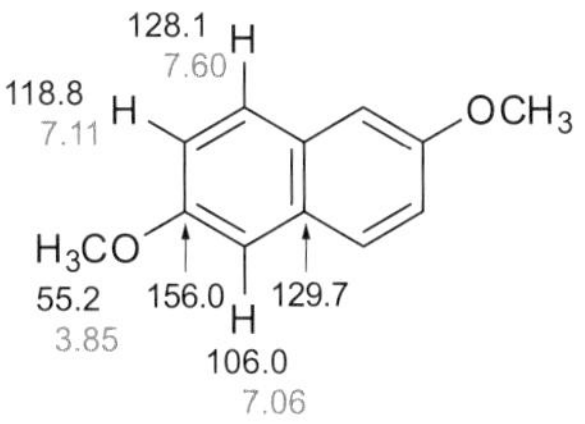

2,6-Dimethoxynaphthalen

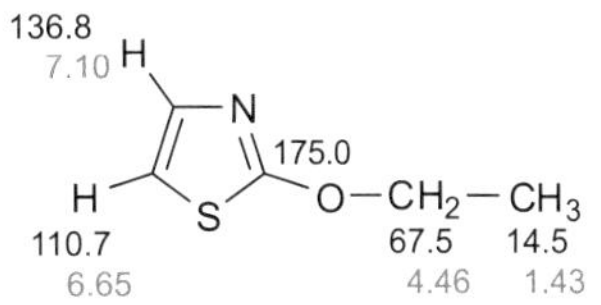

2-Ethoxythiazole

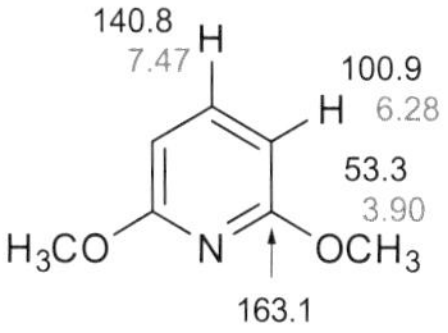

2,6-Dimethoxypyridin

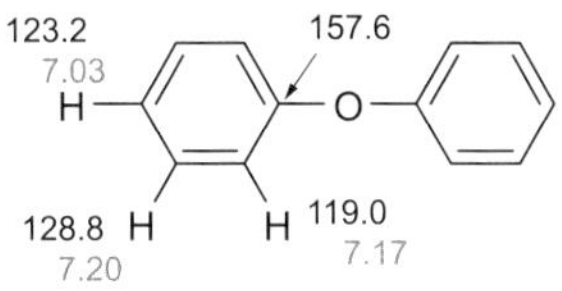

Diphenyl ether

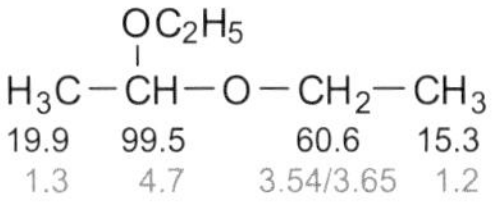

Acetaldehyde diethyl acetal

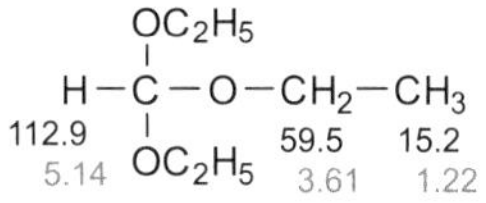

Triethyl orthoformate

$C(OCH_3)_4$
121.0 50.4
3.31

Tetramethyl
orthocarbonate

$-NR_2$

Amines, Ammonium salts, Amine oxides, Hydrazines, Hydroxylamines, Aminales Orthoamides, Nitramines, Nitrosamines

$-NR_2$

H_3C-NH_2
28.3
2.47

Methylamine

$H_3C-CH_2-NH_2$
19.0 36.8
1.10 2.74

Ethylamine

$H_3C-CH_2-NH-C_2H_5$
15.4 44.1
1.10 2.64

Diethylamine

$H_3C-CH_2-N(C_2H_5)_2$
12.4 47.0
1.00 2.54

Triethylamine

$[H_3C-\overset{+}{N}H_3]\ Cl^-$
24.4
2.63 (D_2O)

Methylamine
hydrochloride

$[H_3C-\overset{+}{N}(CH_3)_3]\ Cl^-$
54.3
3.19 (D_2O)

Tetramethyl-
ammonium chloride

$H_3C-CH_2-CH_2-CH_2-NH_2$
13.9 20.0 36.1 42.0
0.91 1.36 1.40 2.69

Butylamine

$Cl-CH_2-CH_2-CH_2-NH_2$
41.5 29.7 37.9
3.61 2.07 3.08

3-Chloropropylamine

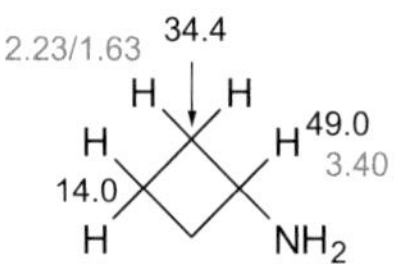

Cyclobutylamine

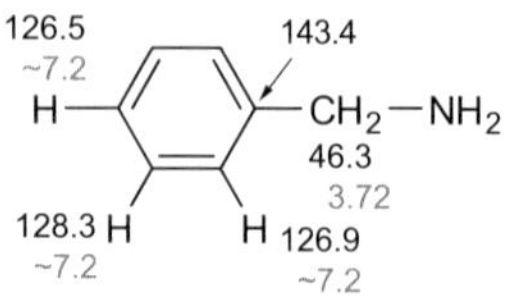

2.23/1.63 34.4
49.0 3.40
14.0
14.0
NH₂

Benzylamine

126.5
~7.2
143.4
CH₂—NH₂
46.3
3.72
128.3 H   H 126.9
~7.2   ~7.2

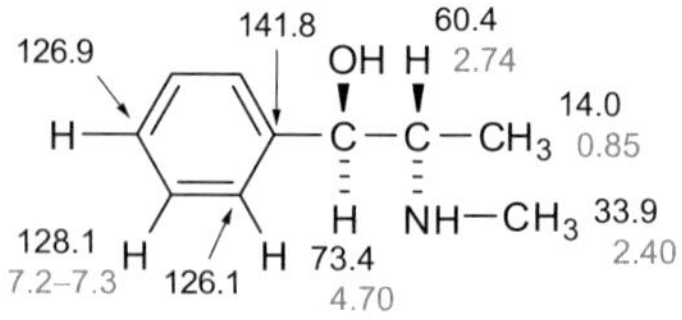

Aniline

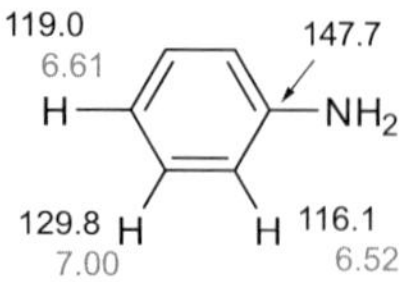

119.0
6.61
147.7
NH₂
129.8 H   H 116.1
7.00   6.52

(1R,2S)-(−)-Ephedrine

126.9   141.8   60.4
OH H 2.74
14.0
CH₃ 0.85
NH—CH₃ 33.9
2.40
128.1 H   H 73.4
7.2–7.3   126.1
4.70

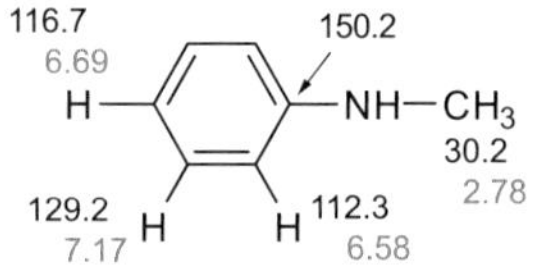

N-Methylaniline

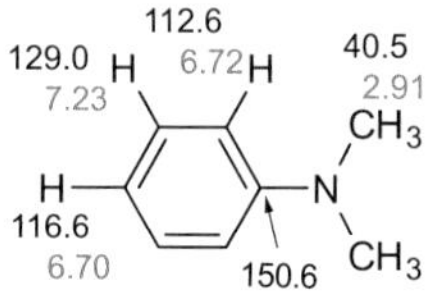

116.7
6.69
150.2
NH—CH₃
30.2
2.78
129.2 H   H 112.3
7.17   6.58

N,N-Dimethylaniline

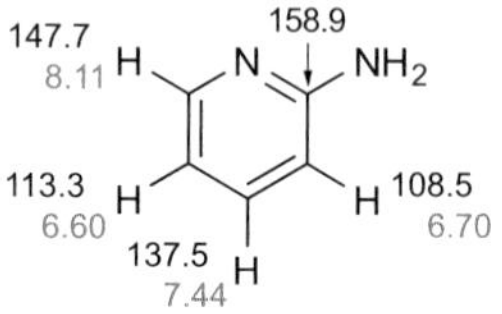

112.6
129.0 H   6.72 H   40.5
7.23   2.91
CH₃
N
116.6 H   150.6
6.70
CH₃

2-Aminopyridine

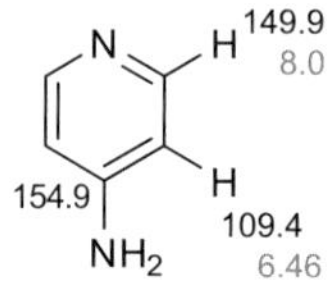

147.7   158.9
8.11 H   N   NH₂
113.3
6.60
108.5
6.70
137.5 H
7.44

4-Aminopyridine

N   149.9
8.0
154.9
H
NH₂   109.4
6.46

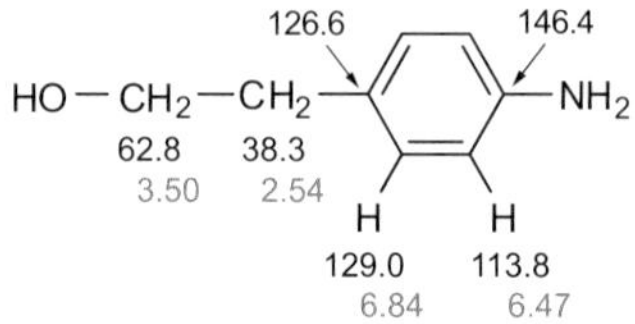

2-(4-Aminophenyl)ethanol

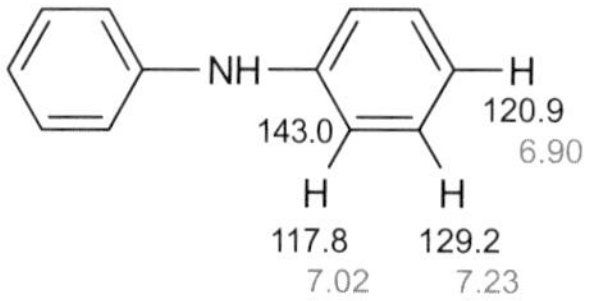

HO—CH₂—CH₂   126.6   146.4
NH₂
62.8   38.3
3.50   2.54
H   H
129.0   113.8
6.84   6.47

Diphenylamine

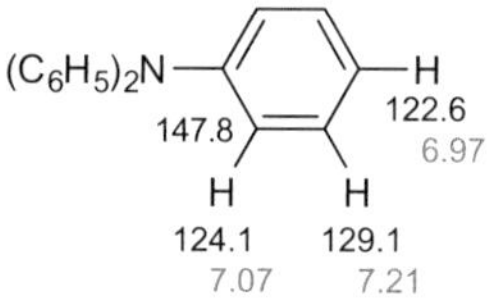

NH
143.0
120.9
6.90
H   H
117.8   129.2
7.02   7.23

Triphenylamine

(C₆H₅)₂N
147.8
H
122.6
6.97
H   H
124.1   129.1
7.07   7.21

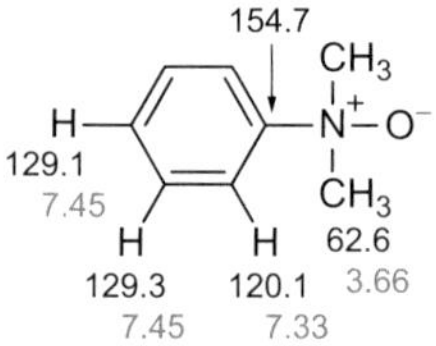

N,N-Dimethylaniline
N-oxide

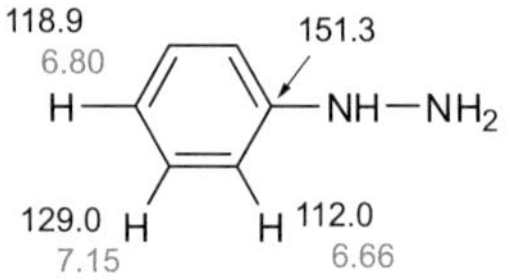

154.7
CH₃
H   N⁺—O⁻
129.1   CH₃
7.45   62.6
H   H   3.66
129.3   120.1
7.45   7.33

Phenylhydrazine

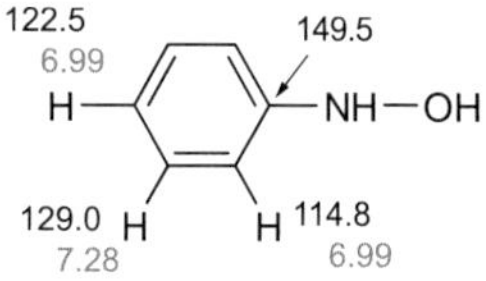

118.9
6.80
151.3
NH—NH₂
129.0 H   H 112.0
7.15   6.66

Phenylhydroxylamine

122.5
6.99
149.5
NH—OH
129.0 H   H 114.8
7.28   6.99

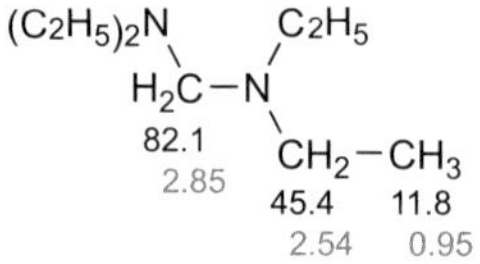

N,N,N',N'-Tetraethyl-
methanediamine

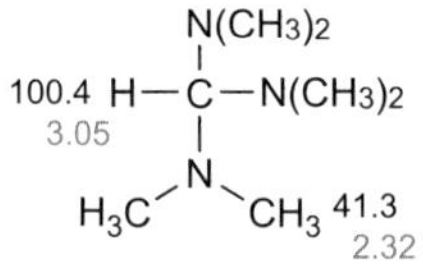

(C₂H₅)₂N   C₂H₅
H₂C—N
82.1   CH₂—CH₃
2.85   45.4   11.8
2.54   0.95

Tris(dimethylamino)-
methane

N(CH₃)₂
100.4 H—C—N(CH₃)₂
3.05
H₃C   CH₃ 41.3
N   2.32

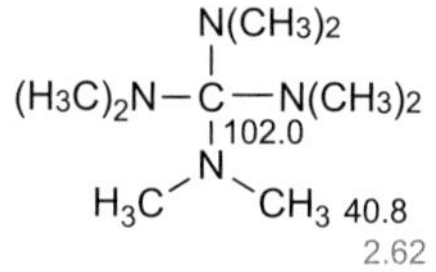

Tetrakis(dimethylamino)-
methane

N(CH₃)₂
(H₃C)₂N—C—N(CH₃)₂
102.0
H₃C   CH₃ 40.8
N
2.62

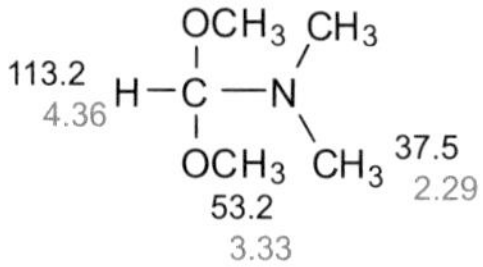

N,N-Dimethylformamide
dimethylacetal

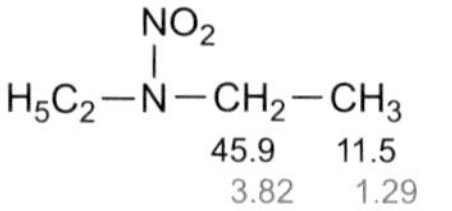

OCH₃ CH₃
113.2 H—C—N
4.36
OCH₃ CH₃ 37.5
53.2   2.29
3.33

N-Nitrodiethylamine

NO₂
H₅C₂—N—CH₂—CH₃
45.9   11.5
3.82   1.29

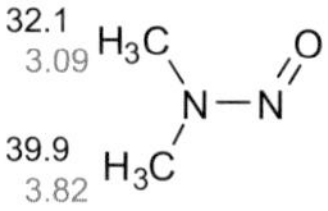

Nitrosodimethylamine

32.1 H₃C   O
3.09   N—N
39.9 H₃C
3.82

— N₃   **Azides** a) b)   — N₃

Hexylazide
(1-Azidohexane)

H₃C—CH₂—CH₂—CH₂—CH₂—CH₂—N₃
13.9   22.5   28.8   26.4   31.3   51.5
0.90   1.26–1.39   1.60   3.26

3-Azidopropanol

HO—CH₂—CH₂—CH₂—N₃
59.6   31.4   48.4
3.77   1.85   3.47

5-Azidopentanenitrile

NC—CH₂—CH₂—CH₂—CH₂—N₃
119.3   17.0   22.8   28.0   50.6
2.40   1.75   1.75   3.36

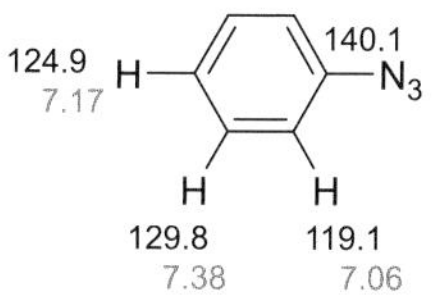

124.9 7.17

140.1 N₃

129.8 7.38

119.1 7.06

Phenylazide

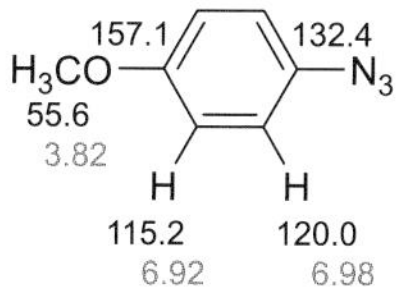

157.1 H₃CO 55.6 3.82

132.4 N₃

115.2 6.92

120.0 6.98

4-Methoxyphenylazide

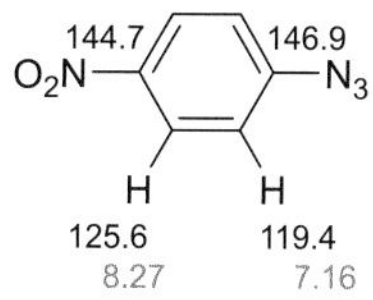

144.7 O₂N

146.9 N₃

125.6 8.27

119.4 7.16

4-Nitrophenylazide

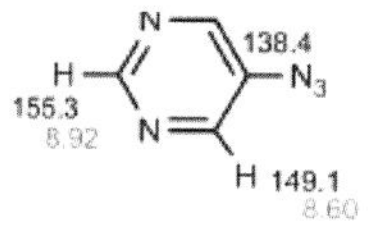

155.3 8.92

138.4 N₃

149.1 8.60

5-Azidopyrimidine

[a]Azides are explosive, in particular strong heating is dangerous
[b]See also carboxylic acid azides

— NO₁,₂,₃

Nitro compounds, Nitroso compounds[a], Nitrous acid esters, Nitric acid esters

— NO₁,₂,₃

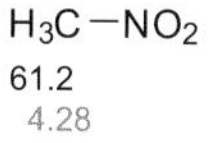

H₃C—NO₂
61.2
4.28

Nitromethane

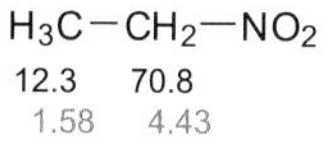

H₃C—CH₂—NO₂
12.3   70.8
1.58   4.43

Nitroethane

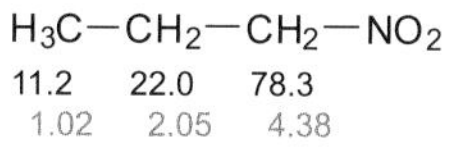

H₃C—CH₂—CH₂—NO₂
11.2   22.0   78.3
1.02   2.05   4.38

1-Nitropropane

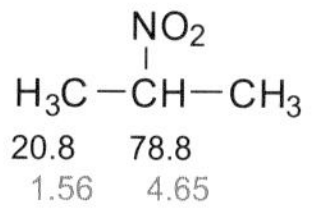

NO₂
|
H₃C—CH—CH₃
20.8   78.8
1.56   4.65

2-Nitropropane

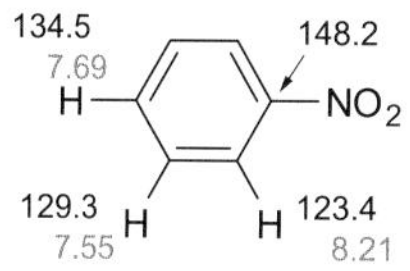

134.5 7.69 H

148.2 NO₂

129.3 H 7.55

123.4 H 8.21

Nitrobenzene

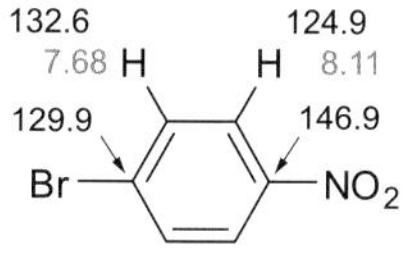

132.6 7.68 H

124.9 H 8.11

129.9

146.9

Br

NO₂

1-Bromo-4-nitrobenzene

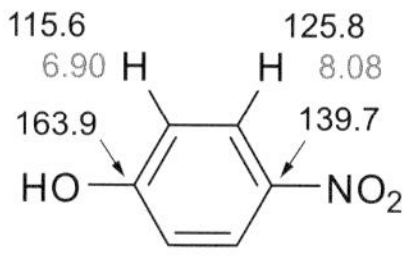

115.6 6.90 H

125.8 H 8.08

163.9

139.7

HO

NO₂

4-Nitrophenol

113.4 6.69 H

111.5 H 7.34

152.6

145.0 H 7.58

O   NO₂

2-Nitrofuran

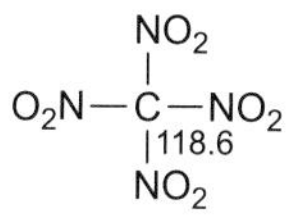

NO₂
|
O₂N—C—NO₂
|118.6
NO₂

Tetranitromethane

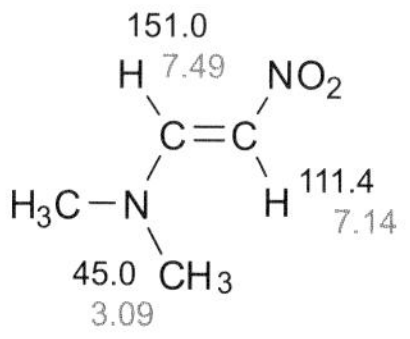

151.0 H 7.49   NO₂
C=C
H₃C—N   H 111.4
45.0 CH₃   7.14
3.09

(E)-1-Dimethylamino-2-nitroethene

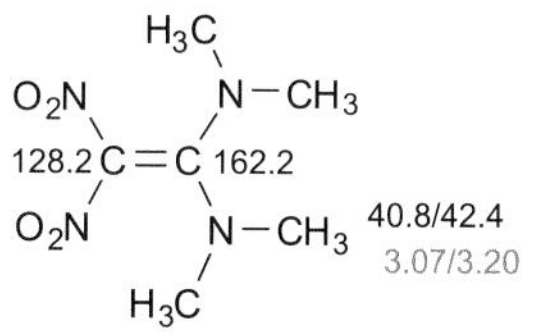

H₃C   N—CH₃
O₂N
128.2 C=C 162.2
O₂N
N—CH₃   40.8/42.4
H₃C   3.07/3.20

1,1-Bis(dimethylamino)-2,2-dinitroethene

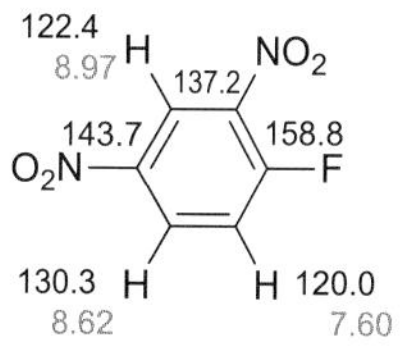

122.4 H 8.97   NO₂
137.2
143.7   158.8
O₂N   F
130.3 H   H 120.0
8.62   7.60

2,4-Dinitrofluorbenzene

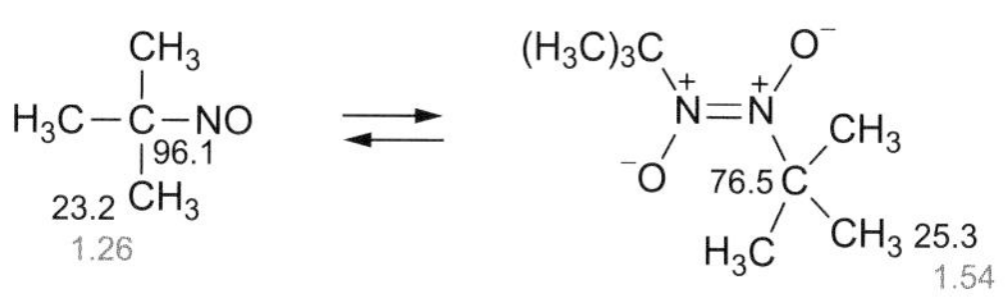

CH₃
|
H₃C—C—NO
|96.1
23.2 CH₃
1.26

⇌

(H₃C)₃C   O⁻
N⁺=N⁺
O⁻   76.5 C   CH₃
H₃C   CH₃ 25.3
1.54

2-Methyl-2-nitrosopropane

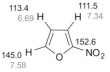 —NO

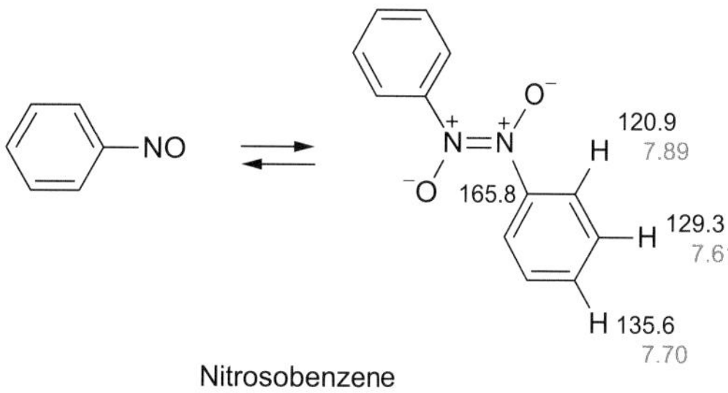

⇌

N⁺=N⁺   120.9
O⁻   165.8   H 7.89
H 129.3
7.61
H 135.6
7.70

Nitrosobenzene

Nuclear Magnetic Resonance Spectroscopy

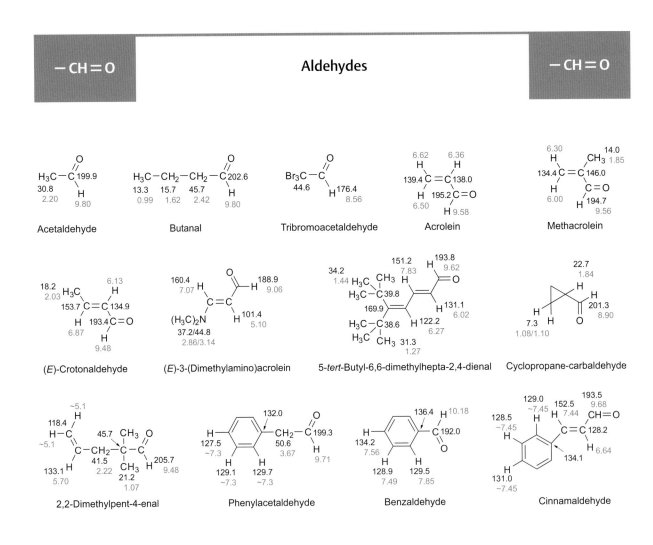

Isoamyl nitrite
(Nitrous acid 3-methylbutyl ester)

1-Propyl nitrate
(Nitric acid propyl ester)

Benzyl nitrate
(Nitric acid benzyl ester)

---

[a] The majority of C-nitroso compounds exist in solution in an equilibrium with their dimers, the azo compound N,N-dioxides.

−CH=O

Aldehydes

−CH=O

Acetaldehyde

Butanal

Tribromoacetaldehyde

Acrolein

Methacrolein

(E)-Crotonaldehyde

(E)-3-(Dimethylamino)acrolein

5-tert-Butyl-6,6-dimethylhepta-2,4-dienal

Cyclopropane-carbaldehyde

2,2-Dimethylpent-4-enal

Phenylacetaldehyde

Benzaldehyde

Cinnamaldehyde

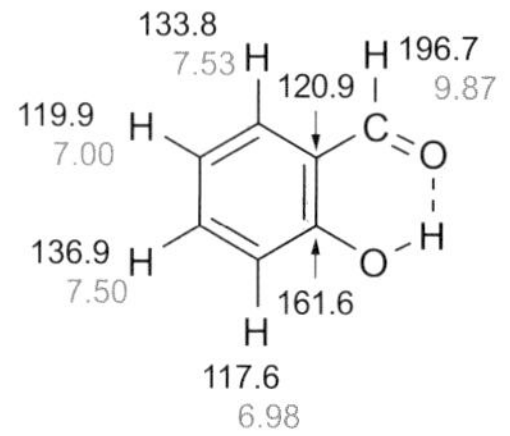

Salicylaldehyde

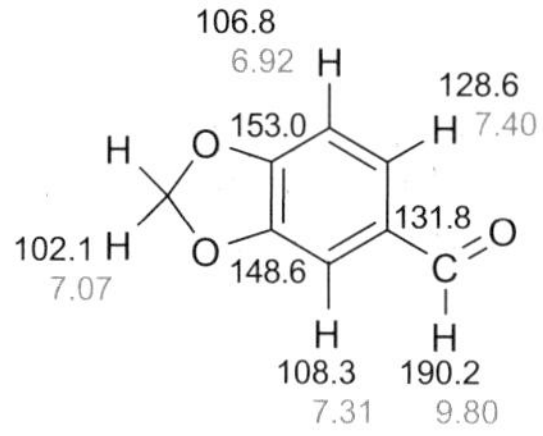

133.8 7.53 H, H 196.7 9.87, 120.9, 119.9 7.00 H, 136.9 7.50 H, 117.6 6.98, 161.6, O—H

Piperonal

106.8 6.92 H, 153.0, 128.6 7.40 H, 102.1 7.07 H, 148.6, 131.8, 108.3 7.31 H, 190.2 9.80

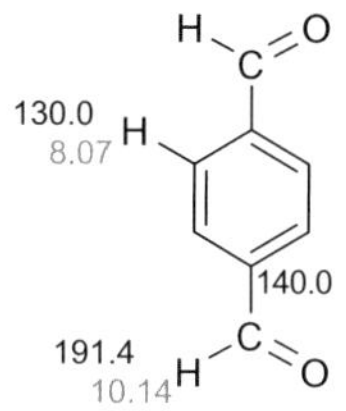

Terephthaldicarbaldehyde

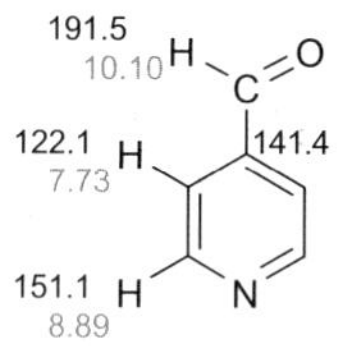

130.0 8.07 H, 140.0, 191.4 10.14

Pyridine-4-carbaldehyde

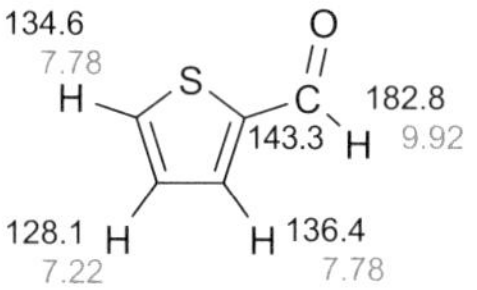

191.5 10.10, 122.1 7.73 H, 141.4, 151.1 8.89 H, N

Thiophene-2-carbaldehyde

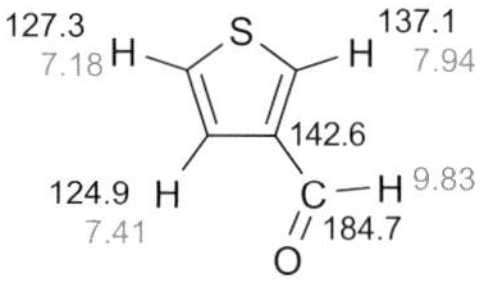

134.6 7.78 H, 143.3, 182.8 9.92, 128.1 7.22 H, 136.4 7.78 H

Thiophene-3-carbaldehyde

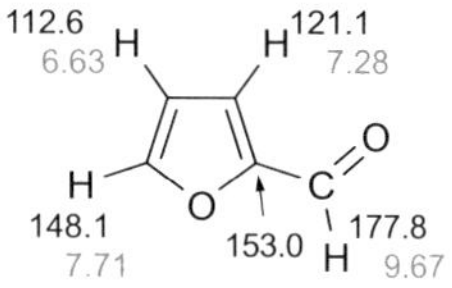

127.3 7.18 H, 137.1 7.94 H, 142.6, 124.9 7.41 H, 9.83 H, 184.7

Furfural (Furan-2-carbaldehyde)

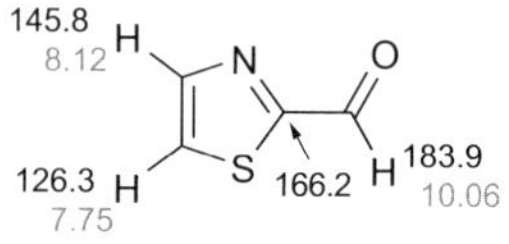

112.6 6.63 H, 121.1 7.28 H, 148.1 7.71 H, 153.0, 177.8 9.67

Thiazole-2-carbaldehyde

145.8 8.12 H, N, 126.3 7.75 H, 166.2, 183.9 10.06

Ketones

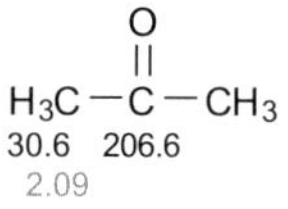

Acetone: H₃C–C–CH₃, 30.6 206.6, 2.09

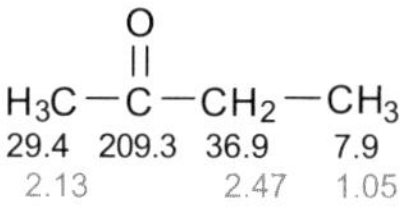

Butan-2-one: H₃C–C–CH₂–CH₃, 29.4 209.3 36.9 7.9, 2.13 2.47 1.05

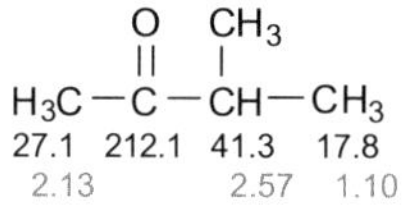

3-Methylbutan-2-one: H₃C–C–CH–CH₃, 27.1 212.1 41.3 17.8, 2.13 2.57 1.10

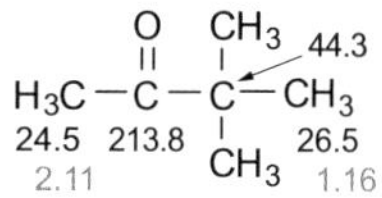

3,3-Dimethylbutan-2-one (Pinacolone): H₃C–C–C–CH₃, 24.5 213.8 44.3 26.5, 2.11 1.16

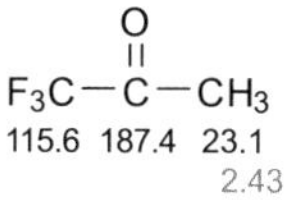

1,1,1-Trifluoracetone: F₃C–C–CH₃, 115.6 187.4 23.1, 2.43

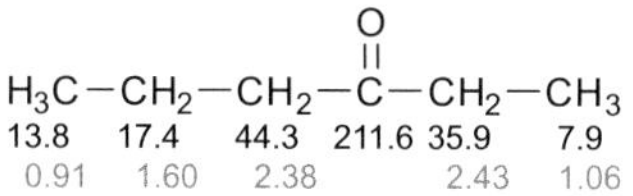

Hexan-3-one: H₃C–CH₂–CH₂–C–CH₂–CH₃, 13.8 17.4 44.3 211.6 35.9 7.9, 0.91 1.60 2.38 2.43 1.06

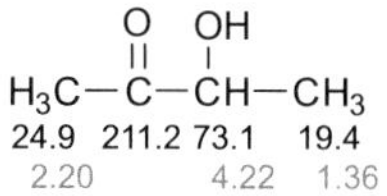

3-Hydroxybutan-2-one: H₃C–C–CH–CH₃, 24.9 211.2 73.1 19.4, 2.20 4.22 1.36

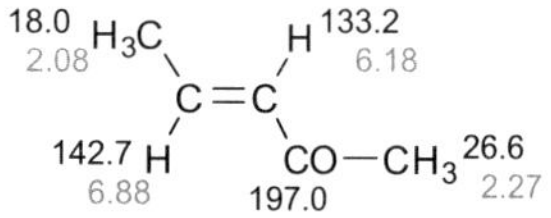

(E)-Pent-3-en-2-one: 18.0 2.08, 133.2 6.18, 142.7 6.88, 197.0, 26.6 2.27

Cyclopentanone: 213.6, 22.0 2.02, 36.7 2.06

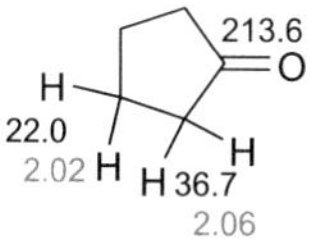

Cyclohexanone: 23.8 1.74, 208.5, 26.5 1.88, 40.4 2.22

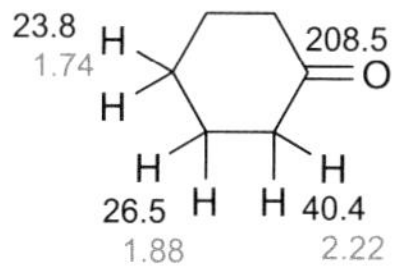

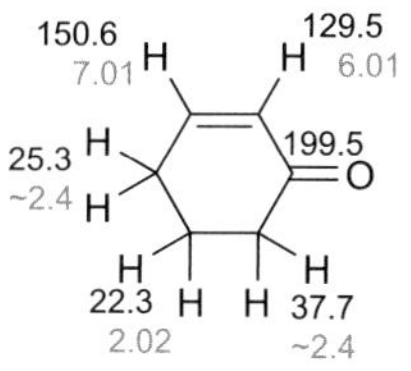

Cyclohex-2-enone: 150.6 7.01, 129.5 6.01, 25.3 ~2.4, 199.5, 22.3 2.02, 37.7 ~2.4

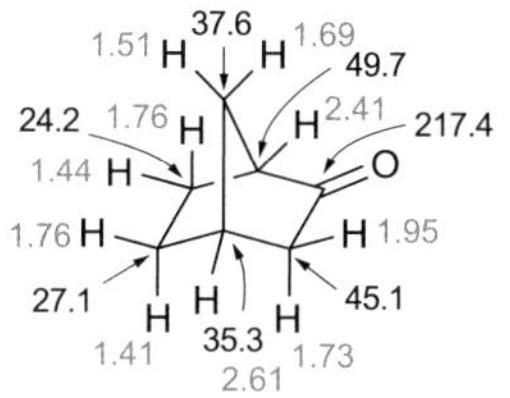

Bicyclo[2.2.1]heptan-2-one (Norcampher): 37.6, 49.7, 1.51, 1.69, 24.2 1.76, 2.41 217.4, 1.44, 1.76, 1.95, 27.1, 45.1, 1.41, 35.3 2.61, 1.73

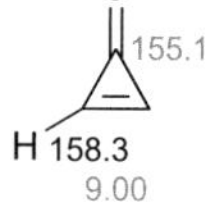

Cyclopropenone: 155.1, 158.3 9.00

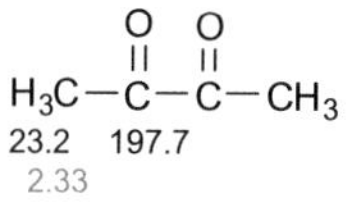

Butane-2,3-dione (Diacetyl): H₃C–C–C–CH₃, 23.2 197.7, 2.33

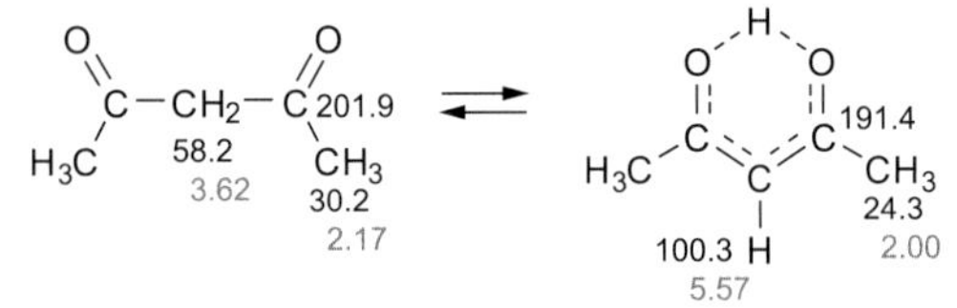

Acetylacetone
(Pentane-2,4-dione)

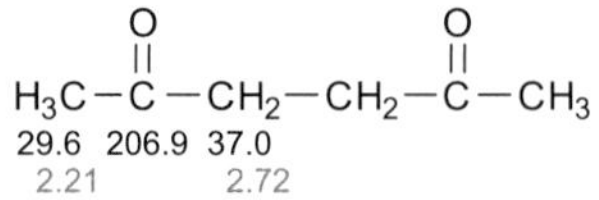

Hexane-2,5-dione
(Acetonylacetone)

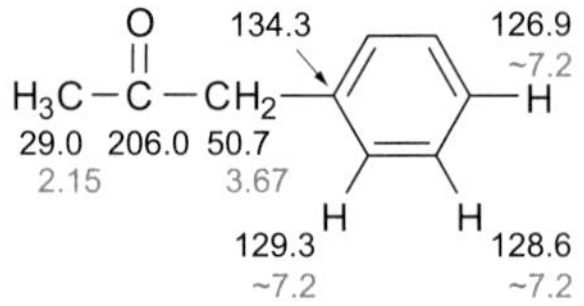

1-Phenylpropan-2-one
(Benzyl methyl ketone)

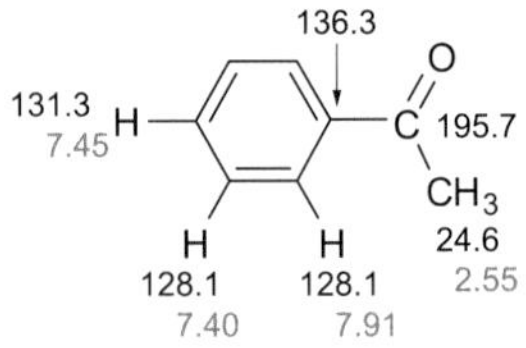

Acetophenone

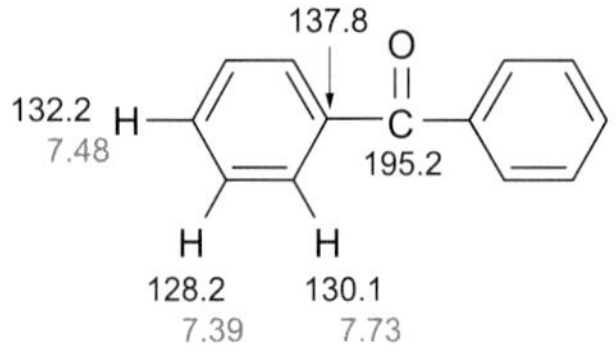

Benzophenone

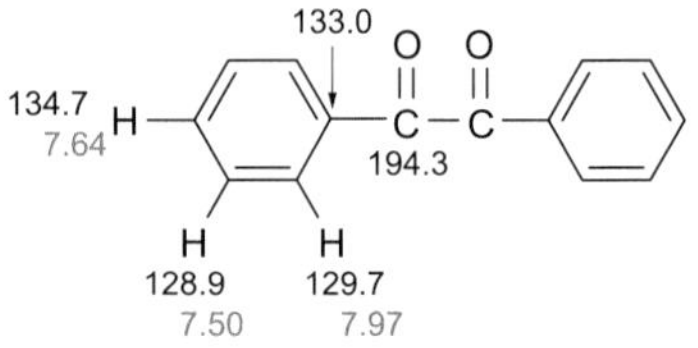

Benzil
(Dibenzoyl)

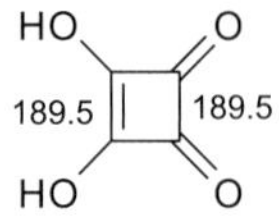

Squaric acid

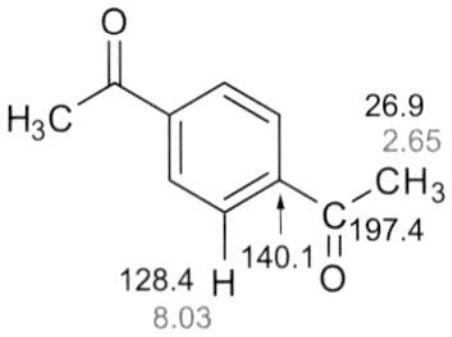

1,4-Diacetylbenzene

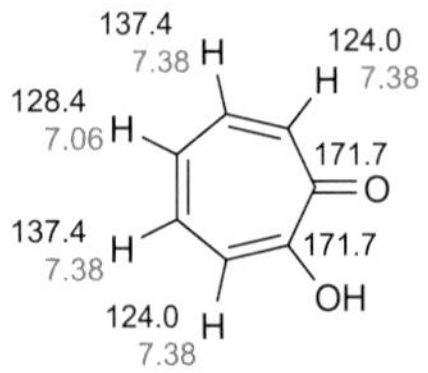

Tropolone[a]

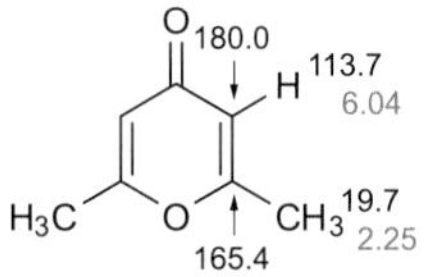

2,6-Dimethyl-γ-pyrone

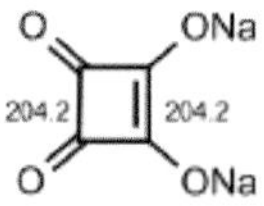

Sodium squarate

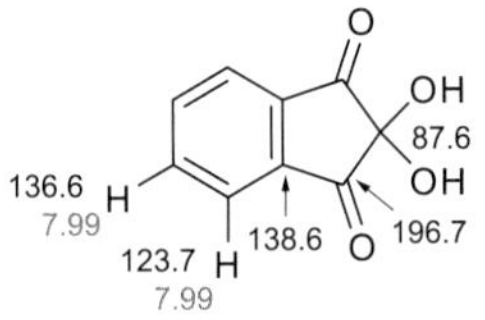

Ninhydrine

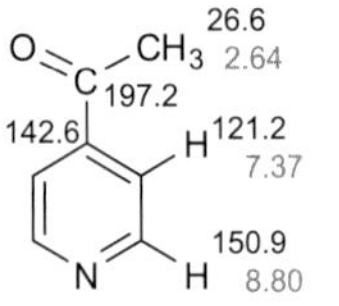

4-Acetylpyridine

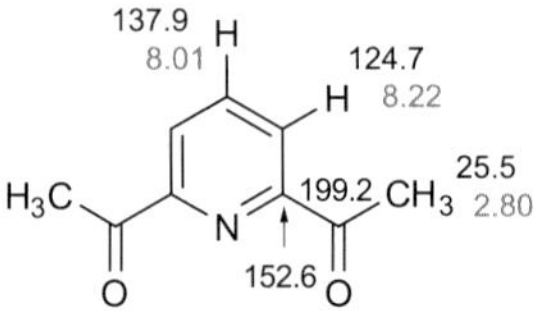

2,6-Diacetylpyridine

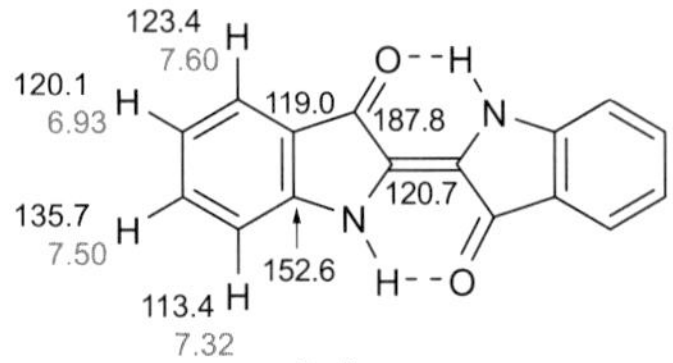

Indigo

_____

[a] Tropolone shows in solvents such as CDCl₃ a fast tautomerism, which increases the symmetry.

Quinones

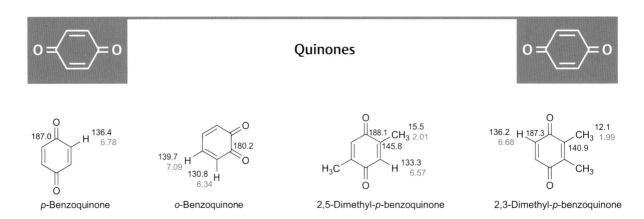

p-Benzoquinone

o-Benzoquinone

2,5-Dimethyl-p-benzoquinone

2,3-Dimethyl-p-benzoquinone

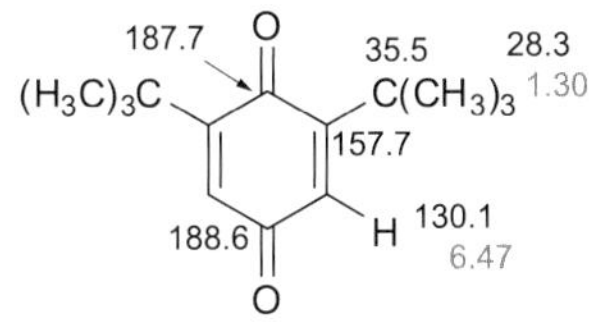

187.7 35.5 28.3
(H₃C)₃C C(CH₃)₃ 1.30
157.7
188.6 H 130.1
6.47

2,6-Di-*tert*-butyl-*p*-benzoquinone

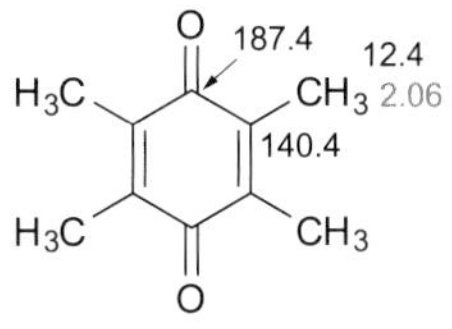

187.4 12.4
H₃C CH₃ 2.06
140.4
H₃C CH₃

Tetramethyl-*p*-benzoquinone
(Duroquinone)

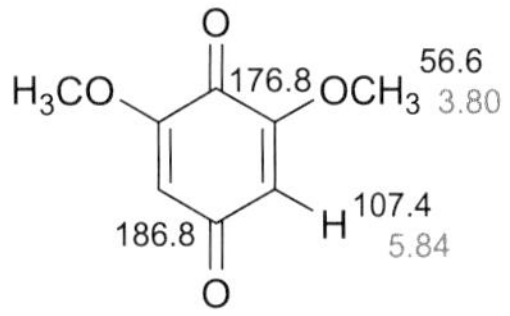

56.6
H₃CO 176.8 OCH₃ 3.80
186.8 H 107.4
5.84

2,6-Dimethoxy-*p*-benzoquinone

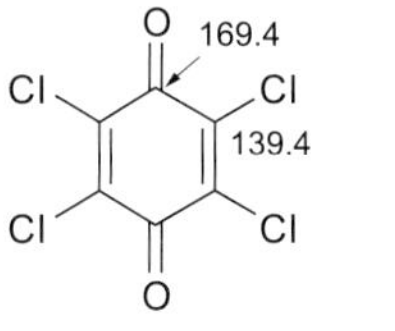

169.4
Cl Cl
139.4
Cl Cl

Tetrachloro-*p*-benzoquinone
(Chloranil)

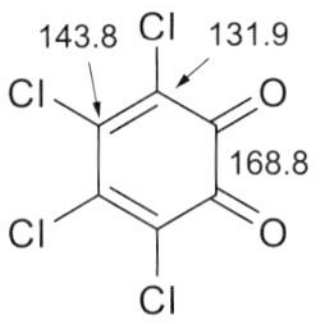

143.8 Cl 131.9
Cl O
168.8
Cl O
Cl

Tetrachloro-*o*-benzoquinone

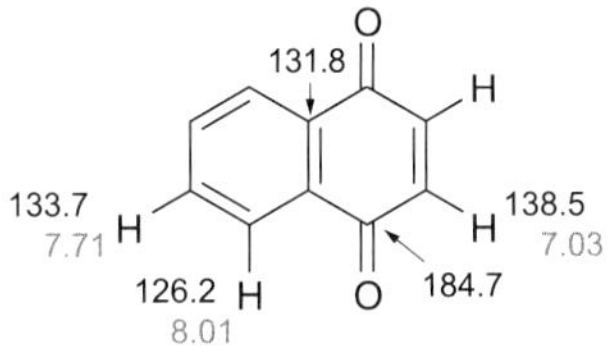

131.8
H
133.7 H
7.71 138.5
7.03
126.2 H H
8.01 O 184.7

1,4-Naphthoquinone

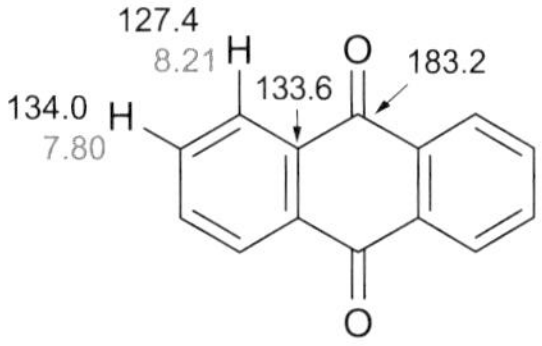

127.4
8.21 H 133.6 O 183.2
134.0 H
7.80

9,10-Anthraquinone

135.9
H 7.44
124.0 H H 130.4
7.97 7.60
135.7
130.9 H 129.6
8.24
180.3

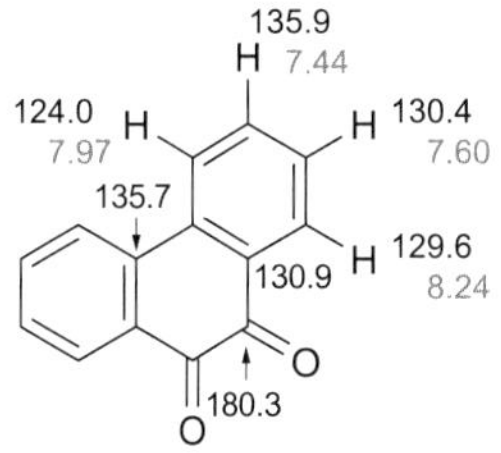

9,10-Phenanthrenequinone

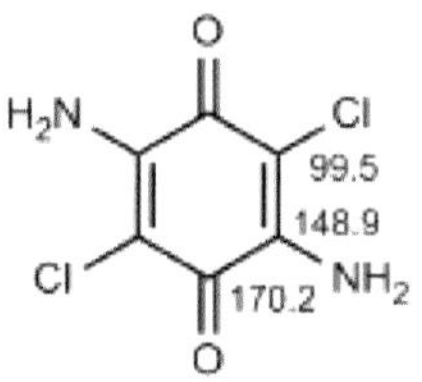

H₂N O
Cl
99.5
148.9
Cl NH₂
170.2 O

2,5-Diamino-3,6-dichloro-
p-benzoquinone

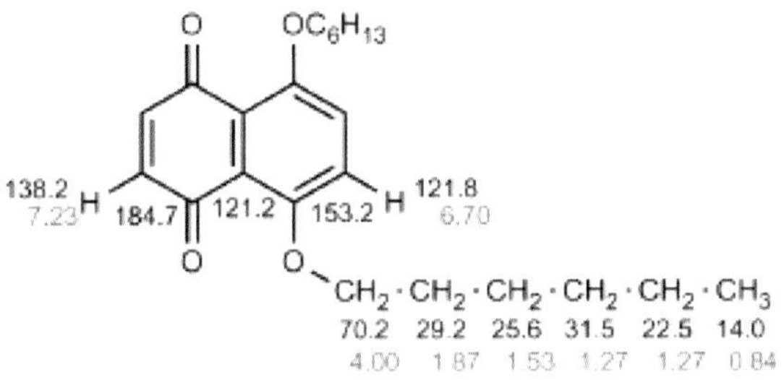

O OC₆H₁₃
138.2 H 121.8
7.23 H 184.7 121.2 153.2 H 6.70
O
CH₂·CH₂·CH₂·CH₂·CH₂·CH₃
70.2 29.2 25.6 31.5 22.5 14.0
4.00 1.87 1.53 1.27 1.27 0.84

5,8-Dihexyloxy-1,4-naphthoquinone

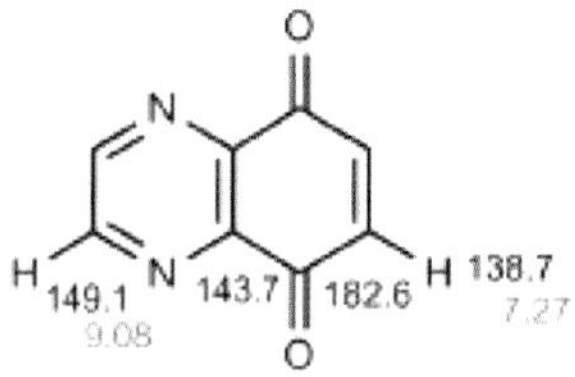

N O
H H 138.7
149.1 143.7 182.6 7.27
9.08 N O

Quinoxaline-5,8-dione

C=C=O

Ketenes

C=C=O

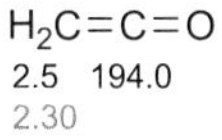

H₂C=C=O
2.5 194.0
2.30

Ketene

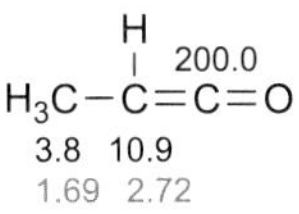

H 200.0
H₃C—C=C=O
3.8 10.9
1.69 2.72

Methylketene

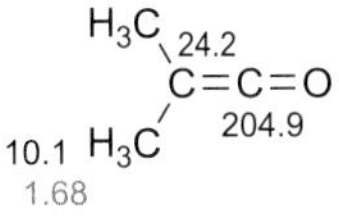

H₃C 24.2
C=C=O
204.9
10.1 H₃C
1.68

Dimethylketene

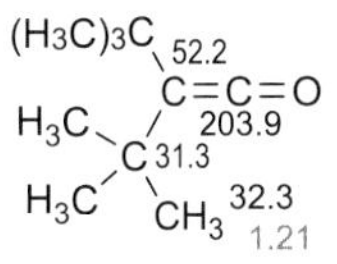

(H₃C)₃C 52.2
C=C=O
H₃C 203.9
C 31.3
H₃C CH₃ 32.3
1.21

Di-*tert*-butylketene

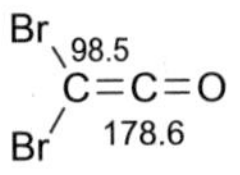

Br
  \98.5
   C=C=O
Br/   178.6

Dibromoketene

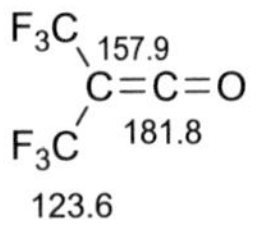

F₃C 157.9
   \
   C=C=O
   /
F₃C   181.8

123.6

Bis(trifluoromethyl)ketene

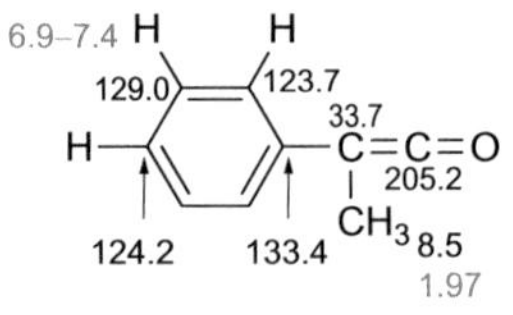

6.9–7.4 H   H
            |
129.0   123.7
              33.7
H—   C=C=O
          | 205.2
124.2 133.4 CH₃ 8.5
                1.97

Methylphenylketene

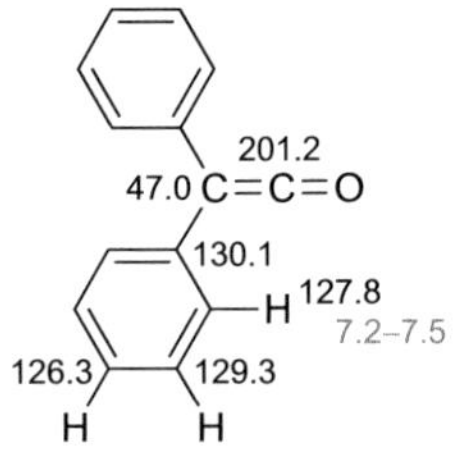

          201.2
47.0 C=C=O

130.1
      H 127.8
        7.2–7.5
126.3   129.3
H   H

Diphenylketene

— C ⟨=O, —OH

Carboxylic acids

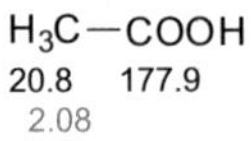

— C ⟨=O, —OH

H₃C—COOH
20.8   177.9
2.08

Acetic acid

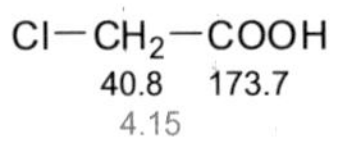

Cl—CH₂—COOH
40.8   173.7
4.15

Chloroacetic acid

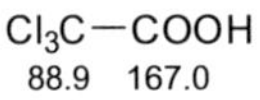

Cl₃C—COOH
88.9   167.0

Trichloroacetic acid

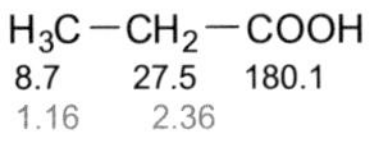

H₃C—CH₂—COOH
8.7   27.5   180.1
1.16   2.36

Propionic acid

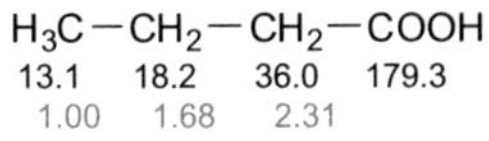

H₃C—CH₂—CH₂—COOH
13.1   18.2   36.0   179.3
1.00   1.68   2.31

Butanoic acid

18.8 CH₃
1.19 |
H₃C—CH—COOH
34.1   184.1
2.57

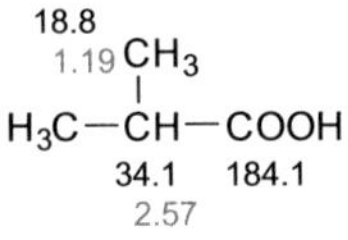

2-Methylpropanoic acid

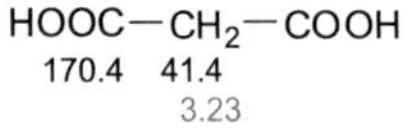

HOOC—CH₂—COOH
170.4   41.4
        3.23

Malonic acid

HOOC—CH₂—CH₂—CH₂—CH₂—COOH
174.2   33.9 24.0
        2.20 1.50

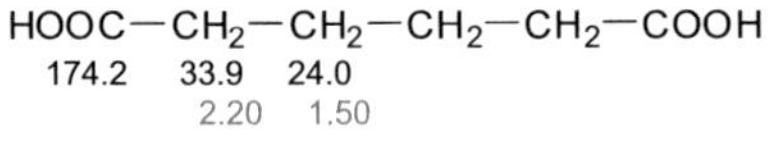

Adipic acid
(Hexanedioic acid)

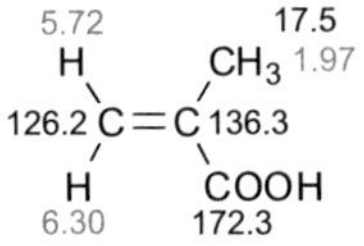

5.72   17.5
H   CH₃ 1.97
126.2 C=C 136.3
H   COOH
6.30   172.3

Methacrylic acid

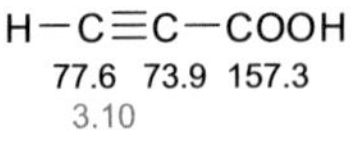

H—C≡C—COOH
77.6 73.9 157.3
3.10

Propiolic acid

          OH
          |
H₃C—CH—COOH
19.9 66.0 176.8
1.56 4.40

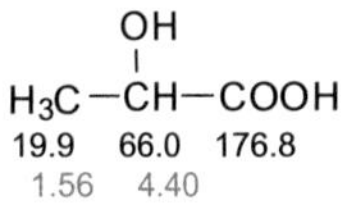

2-Hydroxypropionic acid
(Lactic acid)

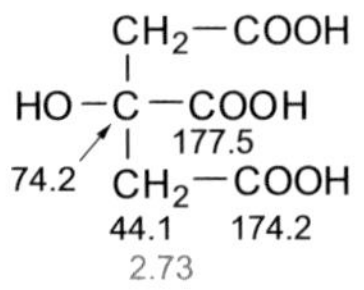

CH₂—COOH
HO—C—COOH
       177.5
74.2 CH₂—COOH
     44.1   174.2
     2.73

Citric acid (D₂O)

O
‖
H₃C—C—CH₂—CH₂—COOH
29.8 206.9 37.7   27.8   178.5
2.20   2.76   2.62

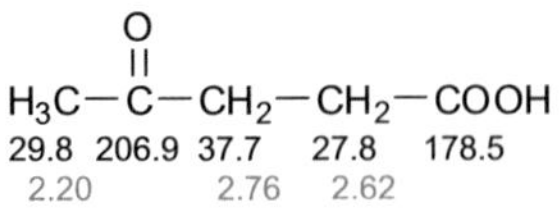

Levulinic acid
(4-Oxopentanoic acid)

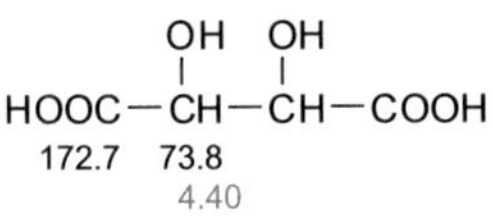

OH OH
| |
HOOC—CH—CH—COOH
172.7 73.8
       4.40

***meso*-Tartaric acid**

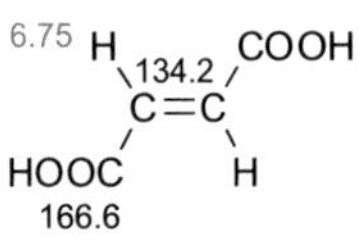

6.75 H   COOH
      \134.2/
       C=C
HOOC/   \H
166.6

Fumaric acid

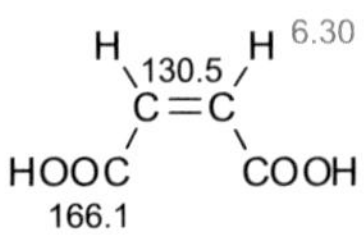

H   H 6.30
 \130.5/
  C=C
HOOC   COOH
166.1

Maleic acid

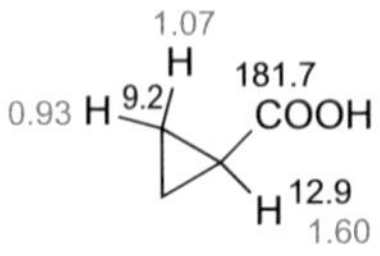

1.07
H 181.7
0.93 H 9.2/ COOH
       H 12.9
         1.60

Cyclopropanecarboxylic acid

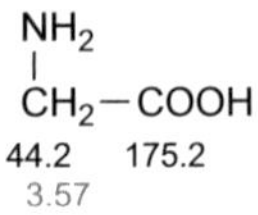

NH₂
|
CH₂—COOH
44.2   175.2
3.57

Glycine (D₂O)

CH₃
|
H₃C—N—CH₂—COOH
46.3   62.6   173.0
2.92   3.72

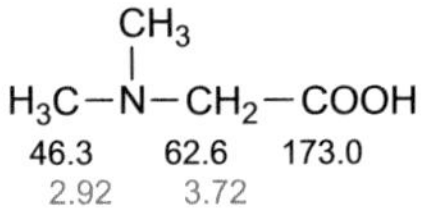

N,N-Dimethylglycine (D₂O)

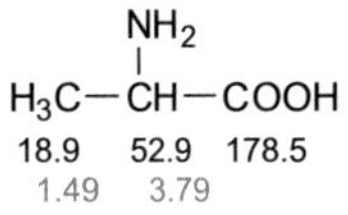

NH₂
|
H₃C—CH—COOH
18.9 52.9 178.5
1.49 3.79

Alanine (D₂O)

15.9
1.2 CH₃ NH₂
        | |
H₃C—CH₂—CH—CH—COOH
12.5 25.7 39.7 60.9 175.2
1.1 1.6/1.7 2.3 3.8

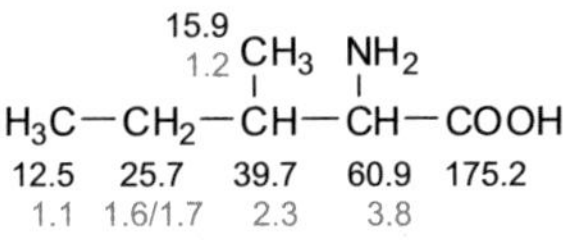

Isoleucine (D₂O)

OH NH₂
| |
H₃C—CH—CH—COOH
22.3 68.7 63.2 175.6
1.33 4.27 3.58

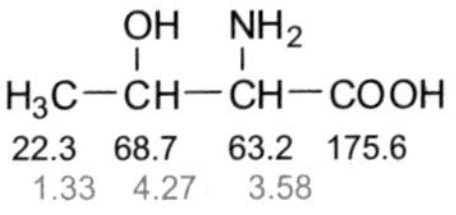

Threonine (D₂O)

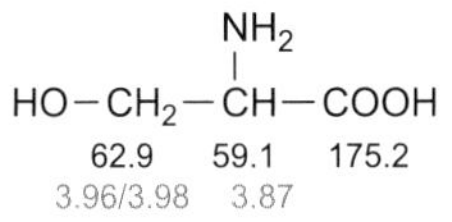

Serine (D₂O)

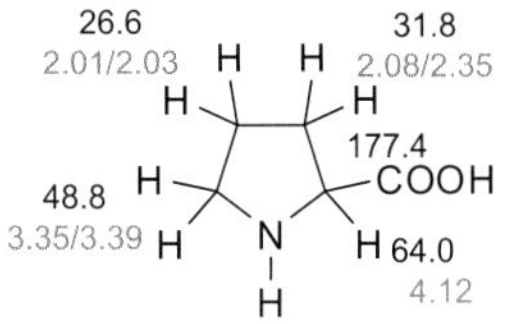

NH₂
HO–CH₂–CH–COOH
62.9 59.1 175.2
3.96/3.98 3.87

Proline (D₂O)

26.6 31.8
2.01/2.03 H H 2.08/2.35
H H
177.4
48.8 H COOH
3.35/3.39 H N H 64.0
H
4.12

Histidine (D₂O)

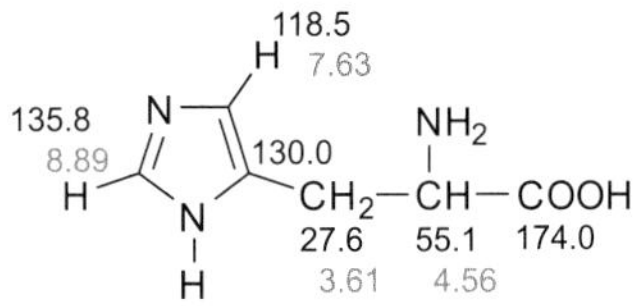

118.5
H 7.63
135.8 N
8.89
130.0 NH₂
H CH₂–CH–COOH
27.6 55.1 174.0
N 3.61 4.56
H

Phenylalanine (D₂O)

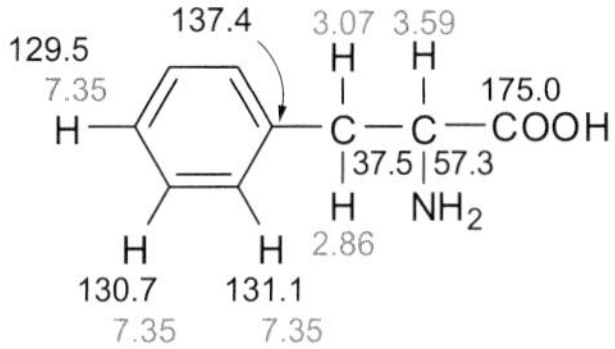

137.4 3.07 3.59
129.5 H H 175.0
7.35 C–C–COOH
H |37.5|57.3
H H NH₂
130.7 131.1 2.86
7.35 7.35

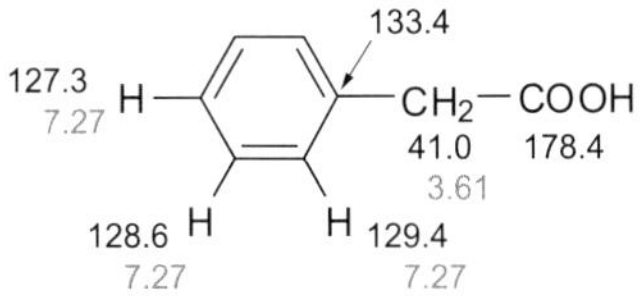

Phenylacetic acid

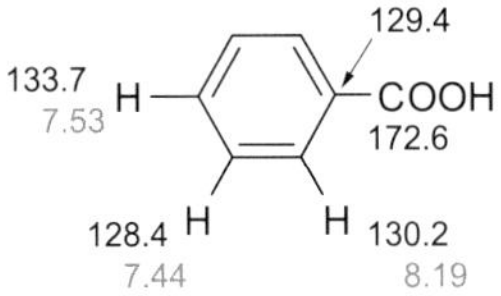

133.4
127.3 H CH₂–COOH
7.27 41.0 178.4
3.61
128.6 H H 129.4
7.27 7.27

Benzoic acid

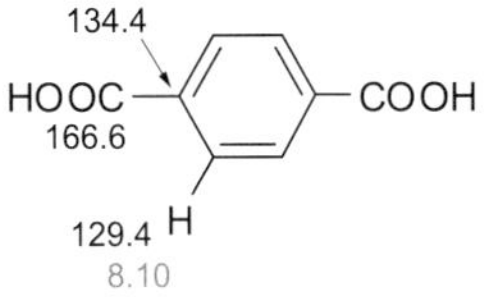

129.4
133.7 H COOH
7.53 172.6
128.4 H H 130.2
7.44 8.19

Terephthalic acid

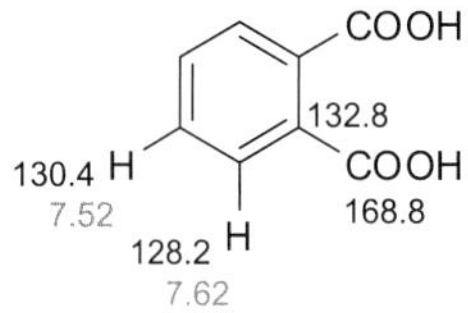

134.4
HOOC COOH
166.6
129.4 H
8.10

Phthalic acid

COOH
132.8
130.4 H COOH
7.52 168.8
128.2 H
7.62

Carboxylic acid esters, Lactones

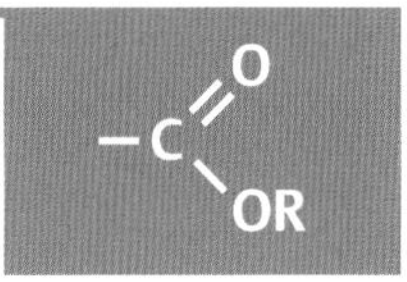

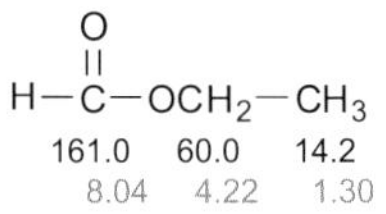

Formic acid ethyl ester

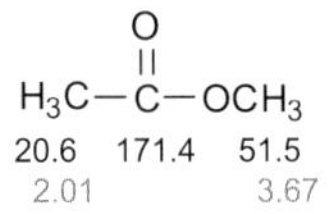

O
H–C–OCH₂–CH₃
161.0 60.0 14.2
8.04 4.22 1.30

Acetic acid methyl ester

O
H₃C–C–OCH₃
20.6 171.4 51.5
2.01 3.67

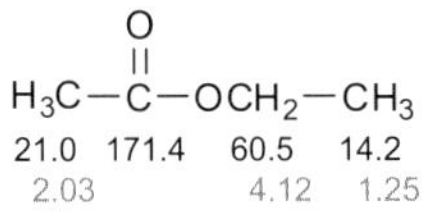

Acetic acid ethyl ester

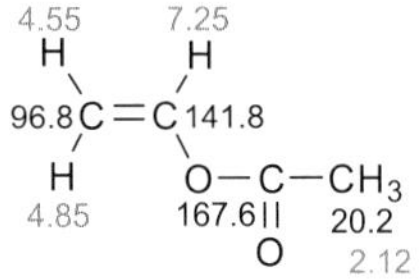

O
H₃C–C–OCH₂–CH₃
21.0 171.4 60.5 14.2
2.03 4.12 1.25

Acetic acid vinyl ester

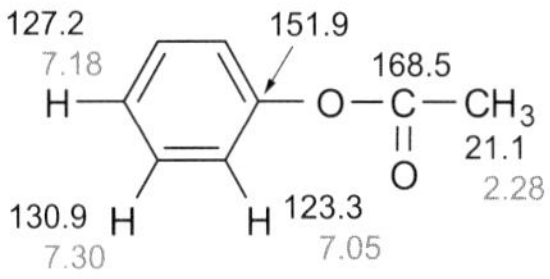

4.55 7.25
H H
96.8 C=C 141.8
H O–C–CH₃
4.85 167.6|| 20.2
O 2.12

Acetic acid phenyl ester

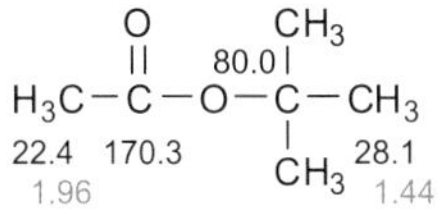

127.2 151.9
7.18 168.5
H O–C–CH₃
21.1
130.9 H H 123.3 2.28
7.30 7.05

Acetic acid *tert*-butyl ester

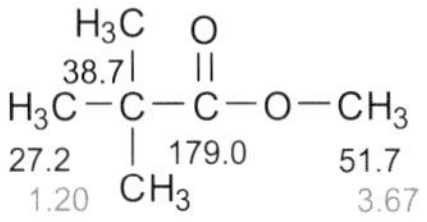

O CH₃
80.0|
H₃C–C–O–C–CH₃
22.4 170.3 CH₃ 28.1
1.96 1.44

2,2-Dimethylpropanoic acid methyl ester

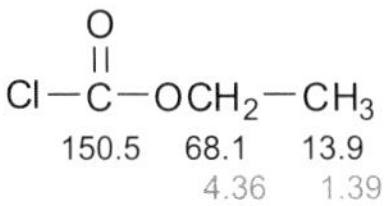

H₃C O
38.7| ||
H₃C–C–C–O–CH₃
27.2 179.0 51.7
1.20 CH₃ 3.67

Carbonochloridic acid ethyl ester

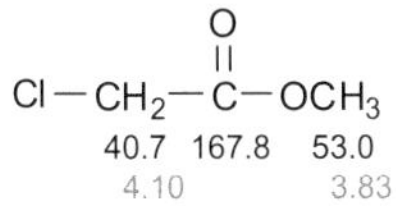

O
Cl–C–OCH₂–CH₃
150.5 68.1 13.9
4.36 1.39

Chloroacetic acid methyl ester

O
Cl–CH₂–C–OCH₃
40.7 167.8 53.0
4.10 3.83

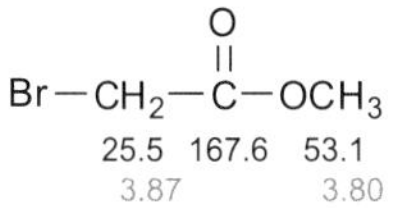

Bromoacetic acid methyl ester

O
Br–CH₂–C–OCH₃
25.5 167.6 53.1
3.87 3.80

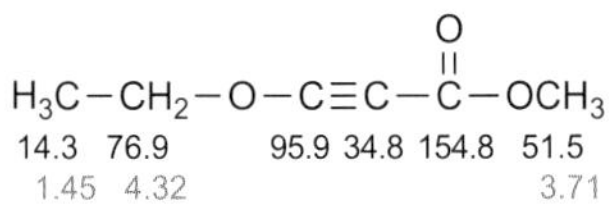

3-Ethoxy-propiolic acid methyl ester

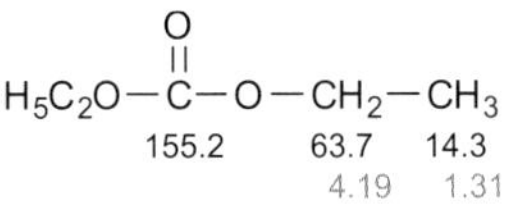

O
H₃C–CH₂–O–C≡C–C–OCH₃
14.3 76.9 95.9 34.8 154.8 51.5
1.45 4.32 3.71

Carbonic acid diethyl ester

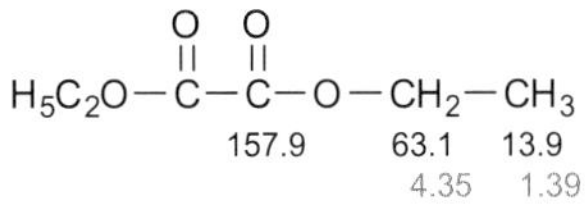

O
H₅C₂O–C–O–CH₂–CH₃
155.2 63.7 14.3
4.19 1.31

Oxalic acid diethyl ester

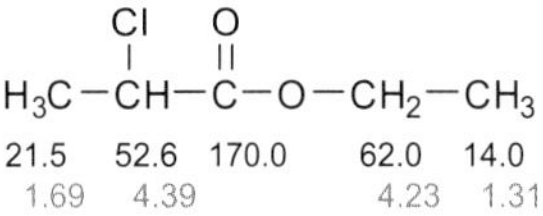

O O
H₅C₂O–C–C–O–CH₂–CH₃
157.9 63.1 13.9
4.35 1.39

2-Chloropropanoic acid ethyl ester

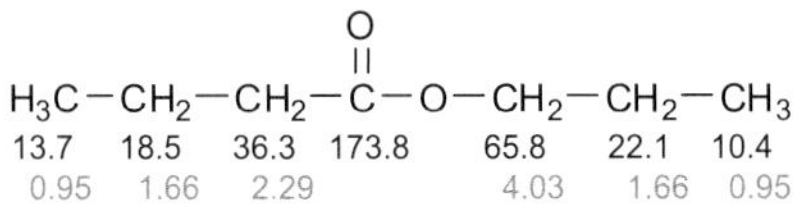

Cl O
H₃C–CH–C–O–CH₂–CH₃
21.5 52.6 170.0 62.0 14.0
1.69 4.39 4.23 1.31

Butanoic acid propyl ester

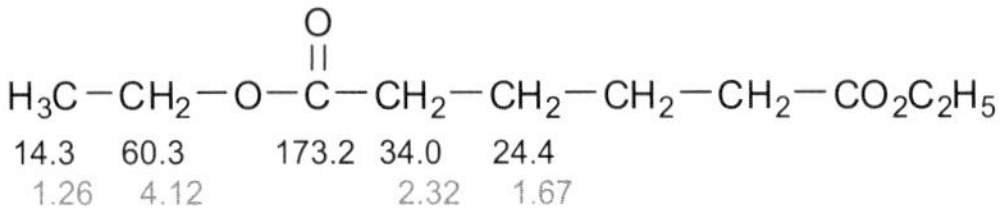

O
H₃C–CH₂–CH₂–C–O–CH₂–CH₂–CH₃
13.7 18.5 36.3 173.8 65.8 22.1 10.4
0.95 1.66 2.29 4.03 1.66 0.95

Adipic acid diethyl ester

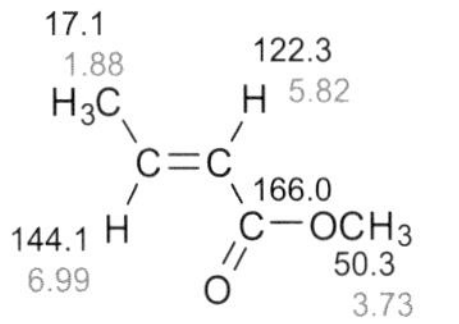

O
H₃C–CH₂–O–C–CH₂–CH₂–CH₂–CH₂–CO₂C₂H₅
14.3 60.3 173.2 34.0 24.4
1.26 4.12 2.32 1.67

(E)-Crotonic acid methyl ester

17.1
1.88 122.3
H₃C H 5.82
C=C
166.0
144.1 H C–OCH₃
6.99 O 50.3
3.73

$H_3C-CH_2-O-\overset{\overset{O}{\|}}{C}-CH_2-COOC_2H_5$

14.1   61.4   166.5 41.7
1.38   4.20   3.36

Malonic acid diethyl ester

Acrylic acid 2-hydroxyethyl ester

6.44
131.3 $C=C$ 128.0
166.5 $C-O-CH_2-CH_2-OH$
66.1   60.8
4.29   3.86
5.87   6.16

Butanoic acid allyl aster

5.30
$H_3C-CH_2-CH_2-\overset{\overset{O}{\|}}{C}-OCH_2-C=C$ 117.9
13.7 18.5 36.2 173.2 64.9
0.96 1.68 2.32   4.58
132.4 5.22
5.91

Fumaric acid diethyl ester

14.2   61.3
1.33   4.27
H_3C-CH_2O
164.9 $C-C$
$C-CO_2C_2H_5$
133.8
6.87

Acetylene dicarboxylic acid
dimethyl ester

$H_3CO-\overset{\overset{O}{\|}}{C}-C\equiv C-COOCH_3$
53.6   152.3 74.6
3.88

Propiolic acid ethyl
ester

$HC\equiv C-\overset{\overset{O}{\|}}{C}-OCH_2-CH_3$
74.5 74.8 152.7 62.3   14.0
2.92   4.26   1.33

Cyclopropane carboxylic acid
methyl ester

1.67   51.7
3.67
OCH_3
175.3
8.3 H
0.87/1.00

Acetoacetic acid methyl ester (Keto form, (Z)- and (E)-Enol)

$H_3C-\overset{\overset{O}{\|}}{C}-CH_2-\overset{\overset{O}{\|}}{C}-OCH_3$ ⇌
29.7 200.3 49.7 167.9 51.9
1.93   3.16   3.46

176.3 C C 173.2
H_3C OCH_3
20.9   55.3
89.7
1.69   4.91 3.49

18.8
2.28 CH_3
173.5 C C 168.1
HO OCH_3
91.0   50.5
4.99   3.26

Levulinic acid ethyl ester

$H_3C-\overset{\overset{O}{\|}}{C}-CH_2-CH_2-\overset{\overset{O}{\|}}{C}-O-CH_2-CH_3$
29.9 206.6 37.9   28.0 172.7   60.6   14.2
2.20   2.76   2.57   4.13   1.26

Maleic acid diallyl ester

5.35 H
118.8 $C=C-CH_2-O-C$ 164.7
5.27 H   H   65.9
4.78
131.7   129.7 H
5.95   6.28
$C-O-CH_2-CH=CH_2$
$C=C$
H

Dihydroxycyclohexa-1,4-diene-
1,4-dicarboxylic acid dimethyl ester

H_3CO O
OH
168.4
H
28.4
HO
93.1   3.17
C CH_3 51.8
O   171.5 O   3.80

Benzoic acid methly ester

132.8
7.47
130.3
H
166.8
$C-OCH_3$
51.8
3.88
128.3 H   H 129.5
7.37   7.97

Terephthalic acid diethyl ester

H_5C_2O
134.2
165.8
$O-CH_2-CH_3$
129.5 H   61.4   14.3
8.10   4.40   1.41

Furan-2-carboxylic acid methyl ester

111.9 H   117.9 H
6.46   7.14
159.0
H $C-OCH_3$
146.4   51.7
7.53   144.8 O   3.90

Nicotinic acid ethyl ester

136.9
8.29
H   O
123.2   126.3 $C-O-CH_2-CH_3$
7.39   61.4   14.3
165.2   4.42   1.42
153.9 H   N   H 150.8
8.78   9.23

β-Propiolactone

59.1   H H   39.2
4.28   3.55
H   H
169.4
O   O

γ-Butyrolactone

22.2   27.8
2.28 H   H 2.50
H   H
177.8
68.6 H   O   O
4.36 H

Cumarin

128.5   143.5
~7.5 H   H 7.72
124.4   116.7
~7.3 H   118.9   H 6.41
131.8   160.6
~7.5 H
154.1 O   O
116.8 H
~7.3

L(+)-Ascorbic acid
(Vitamin C)

77.4 HO   OH
157.2
4.72   118.8
H   174.7
O
HO-CH 70.1
3.73
HO-CH_2 63.4
3.43

Carboxylic acid Amides, Imides, Hydrazides, Azides, Hydroxamic acids, Ureas, Urethanes, Lactames

$H_3C-\overset{O}{\underset{}{C}}-NH_2$
22.3 172.7
1.92

Acetamide

$H_3C-NH-\overset{O}{\underset{}{C}}-CH_3$
25.9   170.9 22.4
2.77   2.0

N-Methylacetamide

$H_3C-\overset{O}{\underset{}{C}}-N\overset{CH_3}{\underset{CH_3}{}}$ 35.0 2.93
21.5 170.5   38.0 3.03
2.09

N,N-Dimethylacetamide

$H_3C-\overset{O}{\underset{}{C}}-NH$ 138.2   124.1 7.0
24.1 169.5
2.1   120.4 H   128.7
             7.4   7.2

Acetanilide

$H_3C-\overset{CH_3}{\underset{}{CH}}-\overset{O}{\underset{}{C}}-NH_2$
19.5 34.0 179.1
1.06   2.39

Isobutyramide
(2-Methylpropanamide)

Acrylamide
5.73 H 167.5 $\overset{O}{C}-NH_2$
128.7 C=C
6.30 H   H 130.1
                6.19

4-Aminobenzamide
151.5   120.8
H₂N $\overset{O}{C}-NH_2$
             168.0
112.3 H   H 129.0
6.54   7.62

Nicotinamide
        135.1
        8.25 H   O
123.3   166.4
7.53 H   NH₂
          129.6
148.6 H   H 151.8
8.74 N   9.10

N-Acetylacetamide
(Diacetamide)
$H_3C-\overset{O}{\underset{}{C}}-N-\overset{O}{\underset{}{C}}-CH_3$
         H
175.2 30.3
      2.05

Acetic acid hydrazide
$H_3C-\overset{O}{\underset{}{C}}-NH-NH_2$
20.6 171.0
1.92

Benzoic acid hydrazide
131.9 H   132.6 $\overset{O}{C}-NH-NH_2$
7.75   108.7
    H   H
  128.7 126.8
  7.46   7.79

Pyridine-4-carboxylic acid hydrazide
       O H
       ‖ |
169.5 C N-NH₂
143.1   H 124.2
              7.63
        H 152.3
    N   8.63

N,N-Dimethylurea
$H_3C-HN-\overset{O}{\underset{}{C}}-NH-CH_3$
163.4   27.8
            2.71

Phenylacetylazide
127.8   141.3
7.23 $CH_2-\overset{O}{\underset{}{C}}-N_3$
        38.2 181.5
129.3 129.9 3.98
7.28 7.34

Carbonazidic acid ethyl ester
$H_3C-CH_2-O-\overset{O}{\underset{}{C}}-N_3$
13.9 64.6   157.3
1.32 4.27

4-Fluorobenzoylazide
167.3   O
F   ‖
          C-N₃
          171.6
116.2 H   H 127.2
7.11   8.03

1,4-Benzenedicarbonyl diazide
$N_3-\overset{O}{\underset{}{C}}$ $\overset{O}{\underset{}{C}}-N_3$
                171.9
          H 129.8
            8.08

(Z)- and (E)-Propionohydroxamic acid
(Z)- and (E)-N-Hydroxypropanamide]
       O   OH
       ‖ |
       C-N
      /170.0 \
H_3C-CH_2 H
9.9   25.5
1.00 1.96

       O   H
       ‖ |
       C-N
      /176.6 \
H_3C-CH_2 OH
8.3   24.2
1.00 2.28

N,N,N'N'-Tetraethylurea
H_5C_2 O C_2H_5
       \ ‖ /
        N-C-N
        164.8
H_5C_2 CH_2-CH_3
          42.2 13.1
          3.10 1.05

N,N-Diphenylurea
            O
            ‖
       NH-C-NH   H 122.3
         153.0 140.2   6.97
          H   H
        118.6   129.2
        7.45   7.28

1,1'-Carbonyldiimidazole
        138.3
        H 8.30
   N   O
   ‖   ‖   N
   N-C-N
      144.3
130.8 H   119.0 H
7.70   7.17

N-Ethylcarbamic acid ethyl ester
       O
       ‖
$H_3C-CH_2-O-C-NH-CH_2-CH_3$
14.7 60.5 156.6 35.8 15.3
1.23 4.10   3.20 1.13

Succinimide
       H
       N
O= ‖ =O
       183.6
  H
30.3 H
2.73

N-Methyl-maleimide
23.2 CH₃
2.88
O= N =O
    171.0
H 134.5
  7.00

Phthalimide
134.2   O
7.85   ‖
           NH
      132.6 169.1
122.8 H   O
7.85

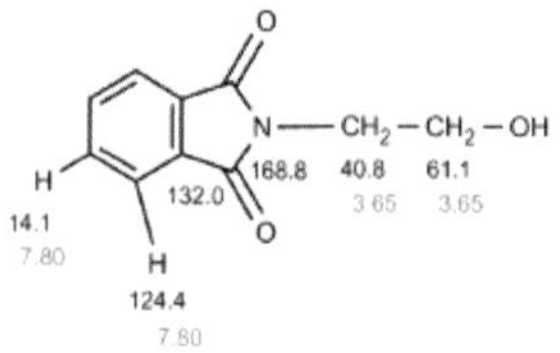

N-(2-Hydroxyethyl)phthalimide

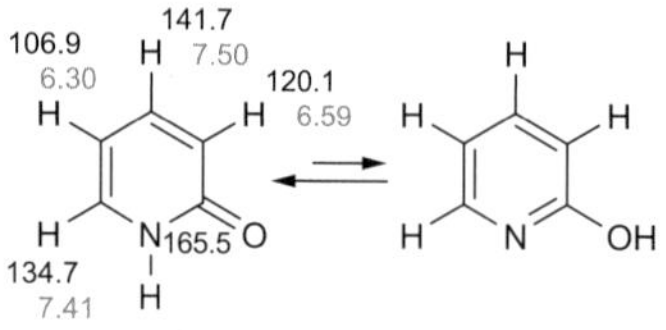

168.8 40.8 61.1
3.65 3.65
132.0
14.1
7.80
124.4
7.80

α-Pyridinone

141.7
106.9 H 7.50
6.30
120.1
H 6.59
134.7
7.41 H
165.5 O

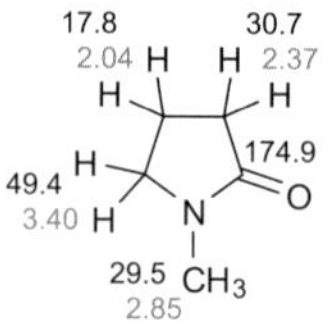

N-Methylpyrrolidin-2-one

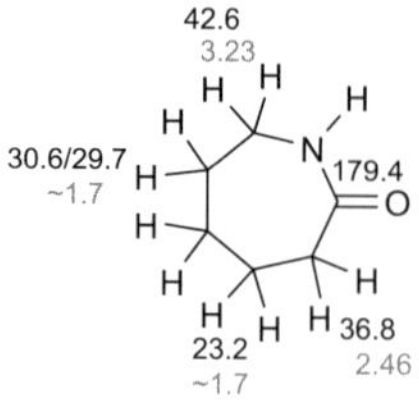

17.8 30.7
2.04 H H 2.37
174.9
49.4 H
3.40 H
29.5 CH₃
2.85

ε-Caprolactam

42.6
3.23
30.6/29.7 H
~1.7
179.4
O
23.2 36.8
~1.7 2.46

Carboxylic acid chlorides, Carboxylic acid anhydrides

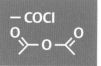

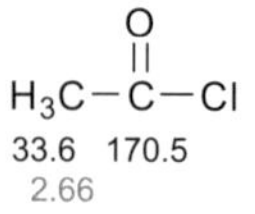

$H_3C-\overset{O}{\overset{\|}{C}}-Cl$
33.6 170.5
2.66

Acetyl chloride

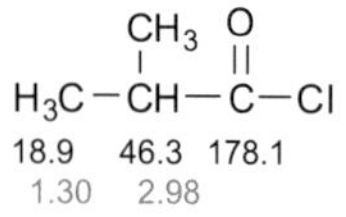

$H_3C-\overset{CH_3}{\overset{|}{CH}}-\overset{O}{\overset{\|}{C}}-Cl$
18.9 46.3 178.1
1.30 2.98

Isobutyryl chloride

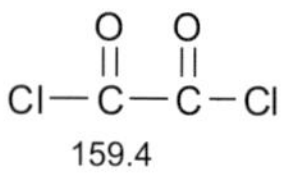

$\overset{O}{\overset{\|}{Cl-C}}-\overset{O}{\overset{\|}{C}}-Cl$
159.4

Oxalyl chloride

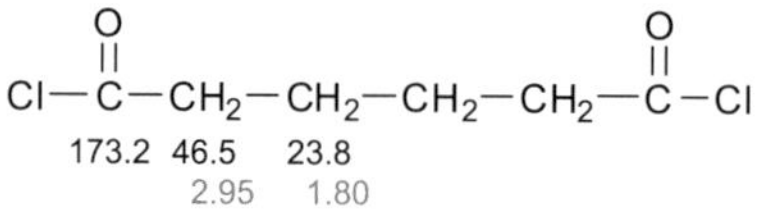

$Cl-\overset{O}{\overset{\|}{C}}-CH_2-CH_2-CH_2-CH_2-\overset{O}{\overset{\|}{C}}-Cl$
173.2 46.5 23.8
2.95 1.80

Adipic acid dichloride

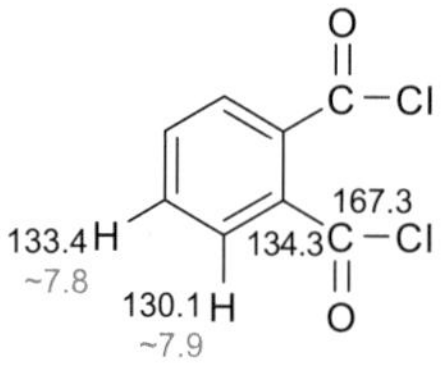

167.3
133.4 H 134.3 C-Cl
~7.8
130.1 H
~7.9

Phthalic acid dichloride

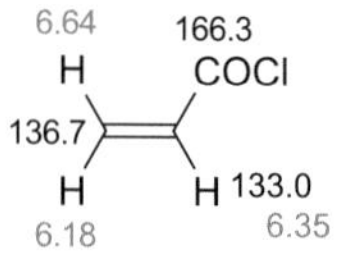

6.64 166.3
H COCl
136.7
H H 133.0
6.18 6.35

Acryloyl chloride

165.4 125.4
H₃CO COCl
55.8 167.1
3.90
134.0 H H 114.3
6.94 8.06

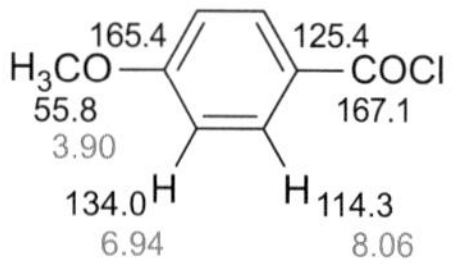

4-Methoxybenzoyl chloride

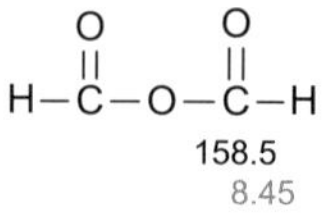

$H-\overset{O}{\overset{\|}{C}}-O-\overset{O}{\overset{\|}{C}}-H$
158.5
8.45

Formic acid anhydride (unstable)

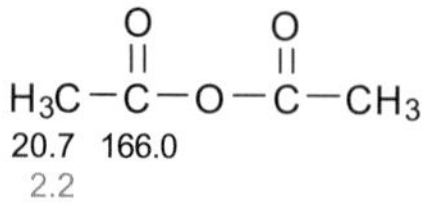

$H_3C-\overset{O}{\overset{\|}{C}}-O-\overset{O}{\overset{\|}{C}}-CH_3$
20.7 166.0
2.2

Acetic acid anhydride

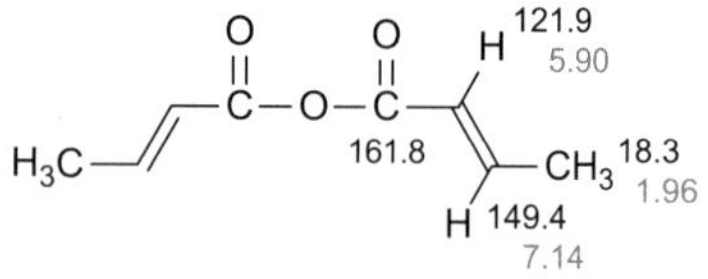

121.9
H 5.90
H₃C
161.8 CH₃ 18.3
H 149.4 1.96
7.14

Crotonic acid anhydride

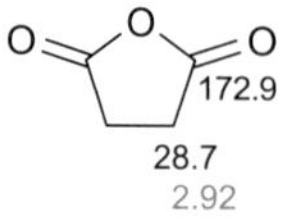

172.9
28.7
2.92

Malonic acid anhydride

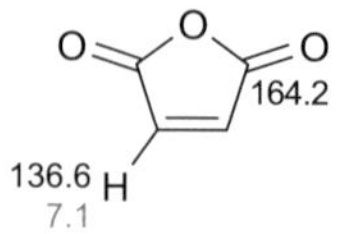

164.2
136.6 H
7.1

Maleic acid anhydride

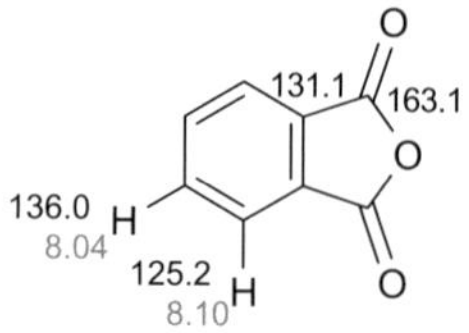

131.1 163.1
136.0 H
8.04
125.2 H
8.10

Phthalic acid anhydride

Imines (Azomethines), Nitrones, Oximes, Hydrazones, Azines, Semicarbazones, Isonitriles (Isocyanides)

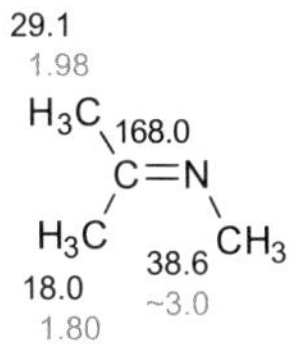

N-Isopropylidene-
methylamine

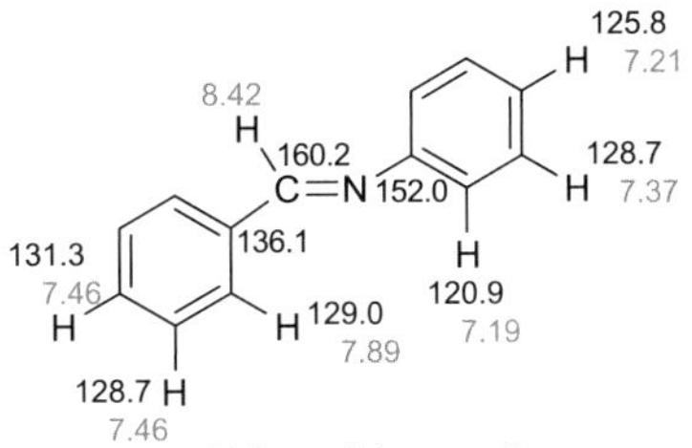

29.1
1.98
H₃C 168.0
C=N
H₃C 38.6 CH₃
18.0 ~3.0
1.80

N-Benzylidene-aniline

125.8
H 7.21
8.42
H 160.2
C=N 152.0 128.7
H 7.37
131.3 136.1
7.46 H
H 120.9
129.0 7.19
128.7 H 7.89
7.46

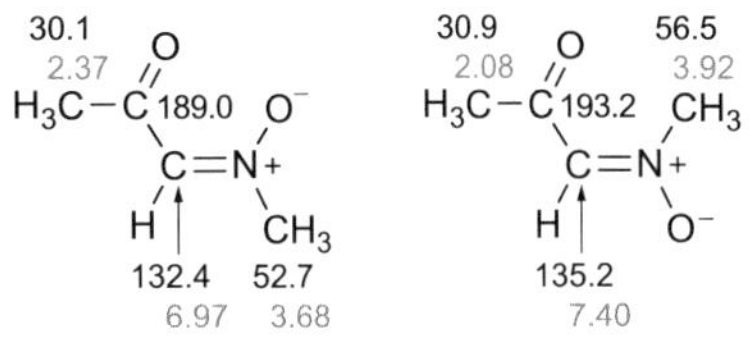

(Z)- and (E)-N-Methyl-(2-oxo-propylidene)amine-
N-oxide

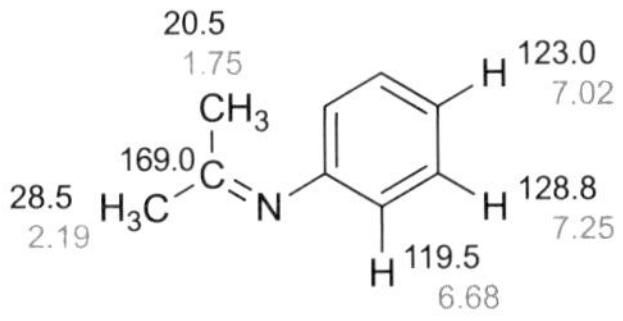

30.1 O
2.37
H₃C—C 189.0 O⁻
C=N⁺
H | CH₃
132.4 52.7
6.97 3.68

30.9 O 56.5
2.08 3.92
H₃C—C 193.2 CH₃
C=N⁺
H | O⁻
135.2
7.40

N-Isopropylidenaniline

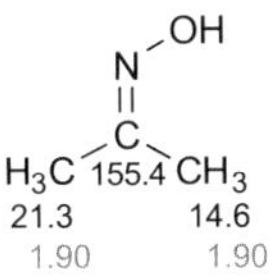

20.5
1.75
CH₃ 123.0
H 7.02
169.0
28.5 H₃C C=N 128.8
2.19 H 7.25
H 119.5
6.68

Acetone oxime

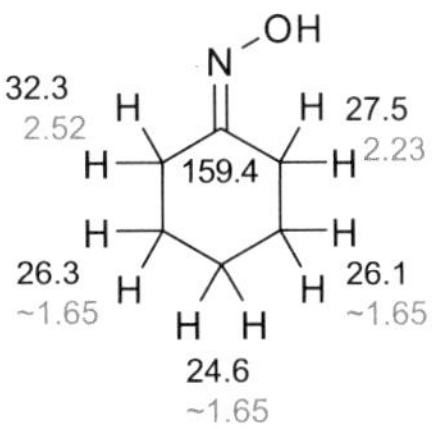

N—OH
H₃C—C—CH₃
21.3 155.4 14.6
1.90 1.90

Cyclohexanone oxime

N—OH
32.3 H 27.5
2.52
H 159.4 H 2.23
26.3 H H 26.1
~1.65 ~1.65
H H
24.6
~1.65

Butan-2,3-dione dioxime

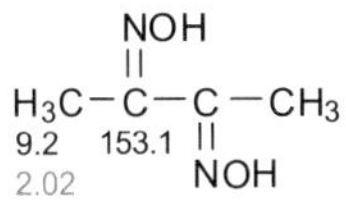

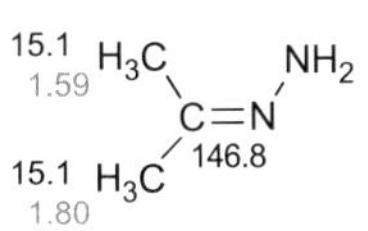

NOH
H₃C—C—C—CH₃
9.2 153.1 ||
2.02 NOH

Acetone hydrazone

15.1 H₃C N—NH₂
1.59
C=N
15.1 H₃C 146.8
1.80

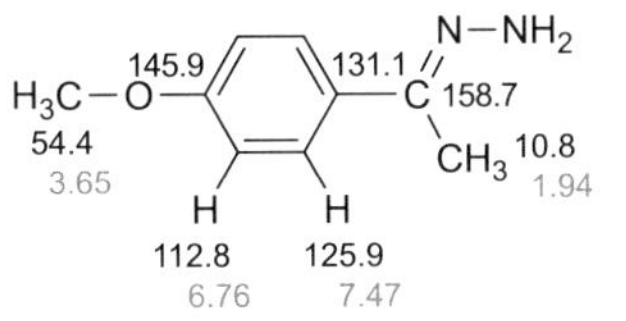

4-Methoxyacetophenone hydrazone

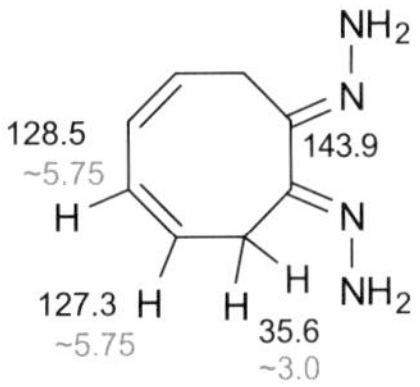

145.9 131.1 N—NH₂
H₃C—O C 158.7
54.4 CH₃ 10.8
3.65 1.94
H H
112.8 125.9
6.76 7.47

Cycloocta-4,6-diene-1,2-dione
(E,E)-dihydrazone

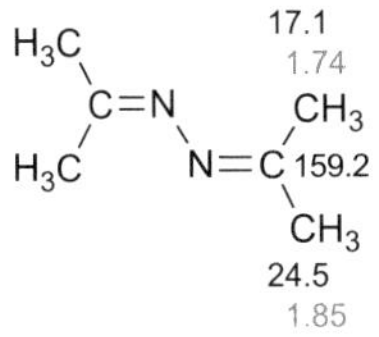

NH₂
N
128.5 143.9
~5.75
H
N
127.3 H H NH₂
~5.75 35.6
~3.0

Acetone azine

17.1
H₃C 1.74
C=N CH₃
H₃C N=C 159.2
CH₃
24.5
1.85

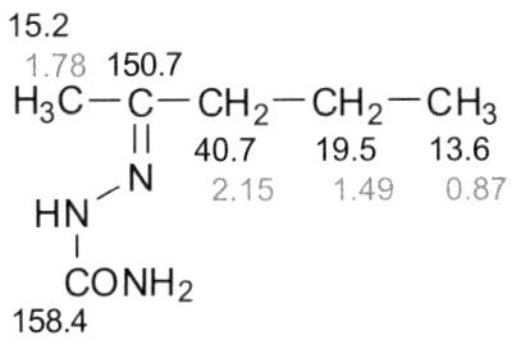

(E)-Pentan-2-one semicarbazone

15.2 150.7
1.78
H₃C—C—CH₂—CH₂—CH₃
|| 40.7 19.5 13.6
N 2.15 1.49 0.87
HN
|
CONH₂
158.4

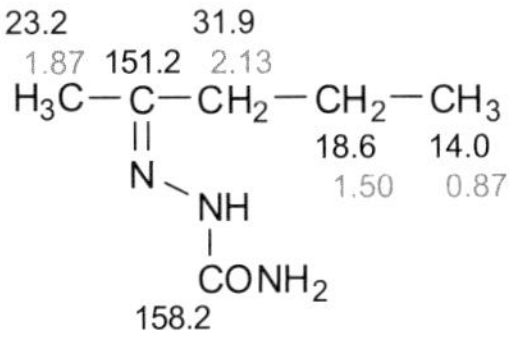

(Z)-Pentan-2-one semicarbazone

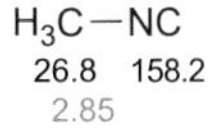

23.2 31.9
1.87 151.2 2.13
H₃C—C—CH₂—CH₂—CH₃
|| 18.6 14.0
N 1.50 0.87
NH
|
CONH₂
158.2

Methylisocyanide

H₃C—NC
26.8 158.2
2.85

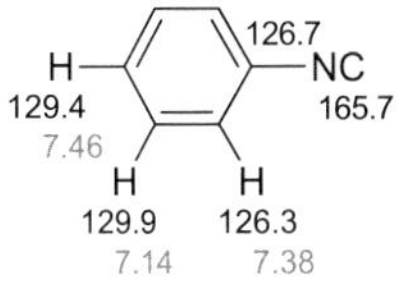

Phenylisocyanide

126.7
H NC
129.4 165.7
7.46
H H
129.9 126.3
7.14 7.38

Diazo Compounds, Azo Compounds

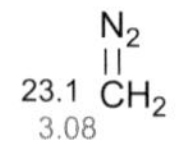

N₂
23.1 CH₂
3.08

Diazomethane

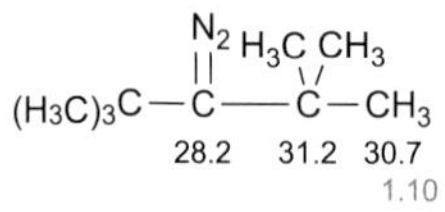

N₂ H₃C CH₃
(H₃C)₃C—C—C—CH₃
28.2   31.2 30.7
             1.10

3-Diazo-2,2,4-tetramethylpentane

N₂ O   O N₂
H—C—C—CH₂—CH₂—CH₂—CH₂—C—C—H
          24.2   40.2 194.4 54.2
          1.60   2.30   5.25

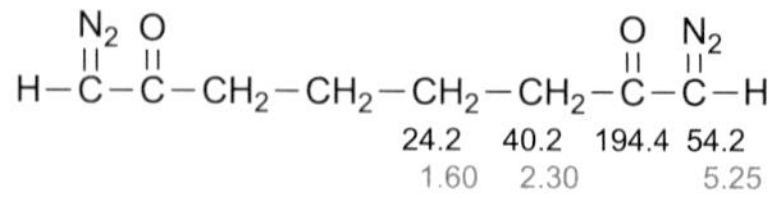

1,8-Bis(diazo)-2,7-octandione

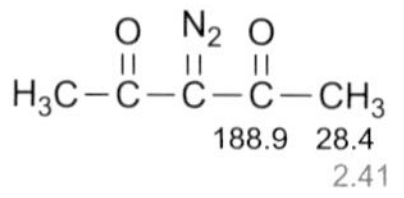

O N₂ O
H₃C—C—C—C—CH₃
188.9 28.4
        2.41

3-Diazopentane -2,4-dione[a]

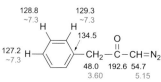

127.2 H   CH₂—C—CH=N₂
~7.3   48.0 192.6 54.7
                   3.60   5.15

1-Diazo-3-phenyl-propan-2-one

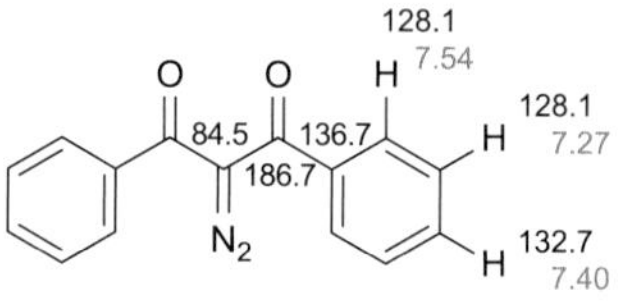

2-Diazo-1,3-diphenyl-1,3-propanedione

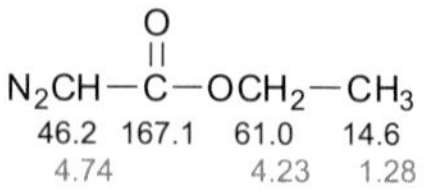

O
‖
N₂CH—C—OCH₂—CH₃
46.2 167.1 61.0   14.6
4.74   4.23   1.28

Diazoacetic acid ethyl ester

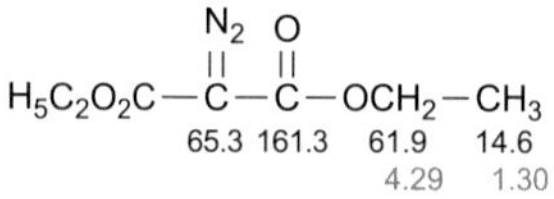

N₂ O
H₅C₂O₂C—C—C—OCH₂—CH₃
65.3 161.3 61.9   14.6
            4.29   1.30

Diazomalonic acid diethyl ester

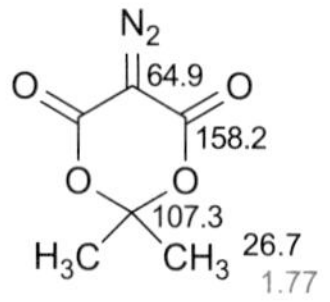

5-Diazo-2,2-dimethyl-1,3-dioxane-4,6-dione
(5-Diazo-Meldrum's acid)

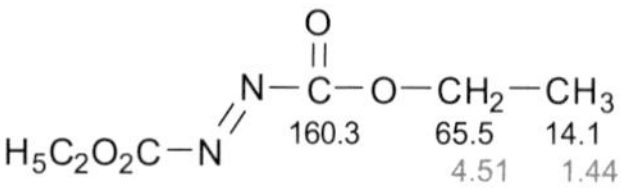

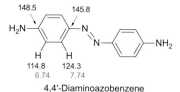

Azodicarboxylic acid diethyl ester

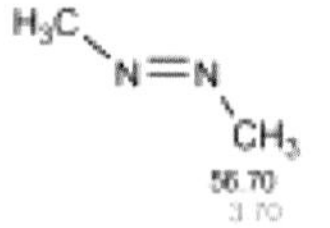

H₃C
  ╲
   N=N
      ╲
       CH₃
56.70
3.70

trans-Dimethyldlazene
(trans-Azomethane)

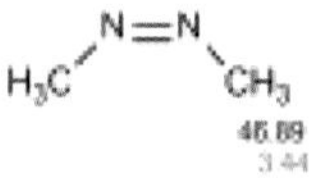

H₃C   N=N
              ╲
               CH₃
46.89
3.44

cis-Dimethyldlazene (D₂O)
(cis-Azomethane)

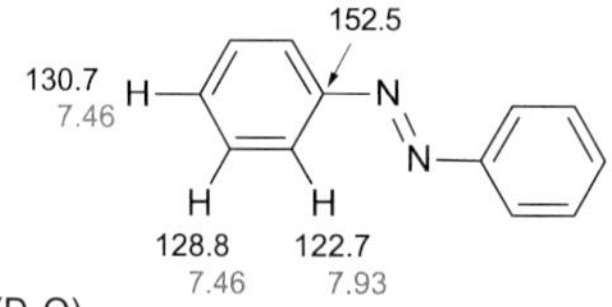

152.5
130.7 H
7.46

128.8   122.7
7.46   7.93

Azobenzene

148.5   145.8
H₂N   N=N—NH₂

114.8   124.3
6.74   7.74

4,4'-Diaminoazobenzene

---

[a] The ¹³C NMR Signals of signals of diazo groups are strongly broadend and difficult to detect.

Nitriles, Nitrile oxides, Cyanates, Thiocyanates, Cyanamides

$-C\equiv N$

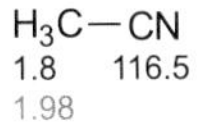

$-C\equiv N$

H_3C-CN
1.8   116.5
1.98

Acetonitrile

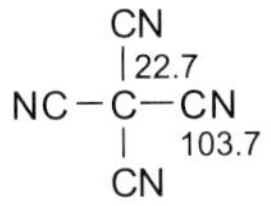

CN
| 22.7
$NC-C-CN$
| 103.7
CN

2,2-Dcyano-malononitrile

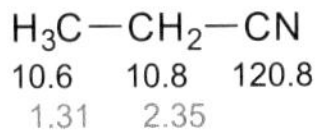

H_3C-CH_2-CN
10.6   10.8   120.8
1.31   2.35

Propanenitrile
(Propionitrile)

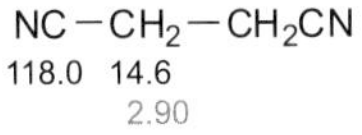

$NC-CH_2-CH_2CN$
118.0 14.6
2.90

Butane dinitrile
(Succinodinitrile)

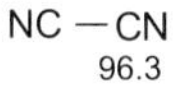

$NC-CN$
96.3

Dicyane

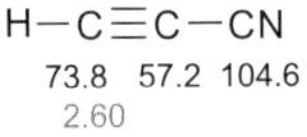

$H-C\equiv C-CN$
73.8   57.2 104.6
2.60

Propiolonitrile

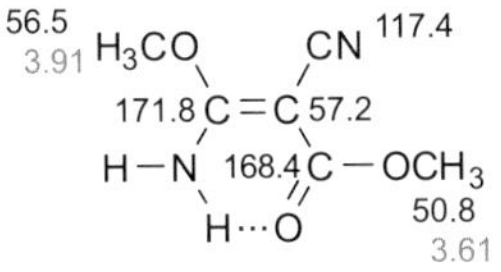

56.5 H_3CO CN 117.4
3.91
171.8 C=C 57.2
$H-N$ 168.4 C$-OCH_3$
H···O   50.8
3.61

3-Amino-2-cyano-3-methoxy-acrylic methyl ester

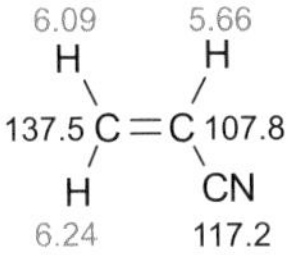

6.09   5.66
H   H
137.5 C=C 107.8
H   CN
6.24   117.2

Acrylonitrile

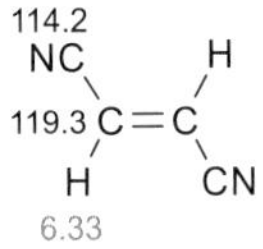

114.2
NC   H
119.3 C=C
H   CN
6.33

Fumaronitrile

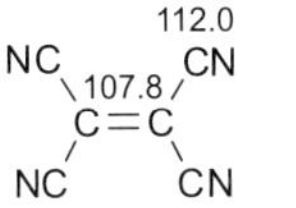

112.0
NC   CN
107.8
C=C
NC   CN

Tetracyanoethene

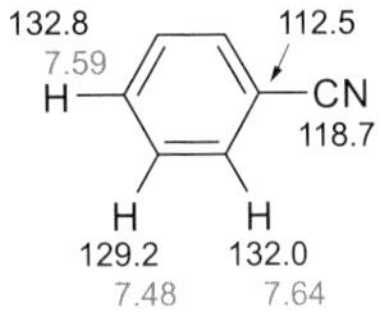

132.8   112.5
7.59
H   CN
118.7
H   H
129.2 132.0
7.48   7.64

Benzonitrile

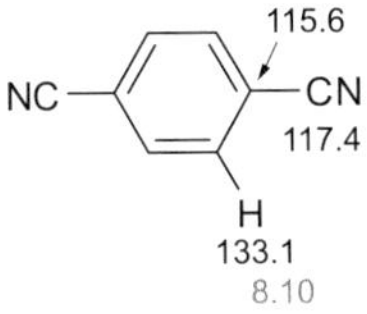

115.6
NC   CN
117.4
H
133.1
8.10

1,4-Benzodinitrile

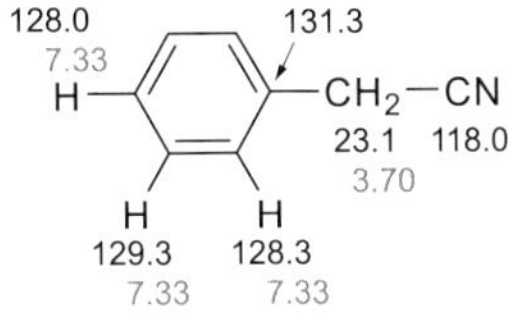

128.0   131.3
7.33
H CH_2-CN
23.1 118.0
3.70
H   H
129.3 128.3
7.33   7.33

Phenylacetonitrile
(Benzylcyanide)

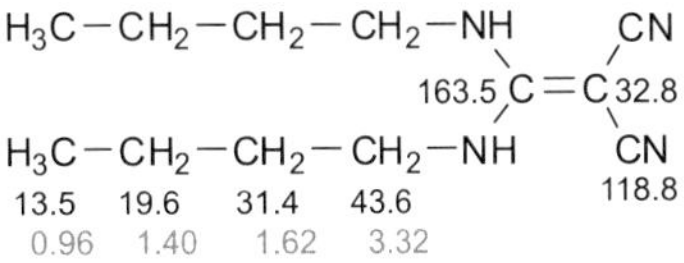

$H_3C-CH_2-CH_2-CH_2-NH$ CN
163.5 C=C 32.8
$H_3C-CH_2-CH_2-CH_2-NH$ CN
13.5 19.6   31.4 43.6   118.8
0.96 1.40   1.62 3.32

2-[Bis(dibutylamino)-methylene]-malononitrile

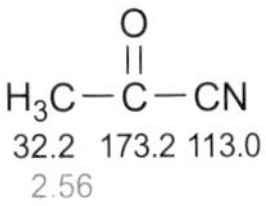

O
||
$H_3C-C-CN$
32.2 173.2 113.0
2.56

2-Oxopropionitrile

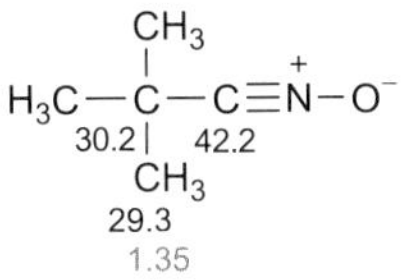

CH_3
| +
$H_3C-C-C\equiv N-O^-$
30.2| 42.2
CH_3
29.3
1.35

Dimethylpropionitrile N-oxide

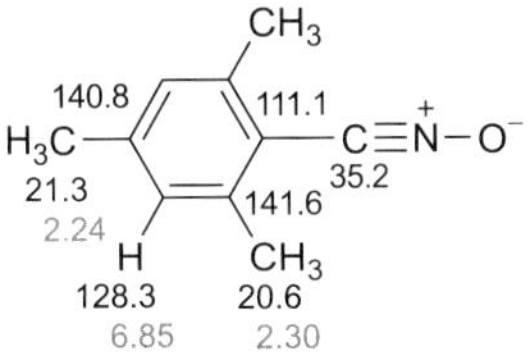

CH_3
140.8   111.1
H_3C $C\equiv N-O^-$
21.3   35.2
141.6
2.24 H CH_3
128.3 20.6
6.85   2.30

2,4,6-Trimethylbenzonitrile N-oxide

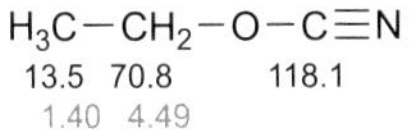

$H_3C-CH_2-O-C\equiv N$
13.5 70.8   118.1
1.40 4.49

Ethyl cyanate

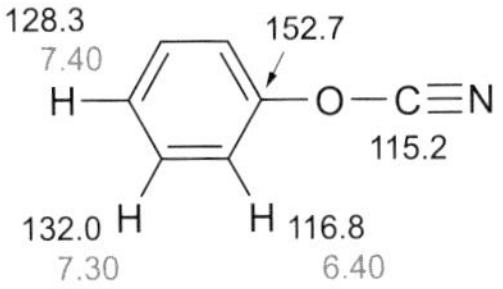

128.3   152.7
7.40
H $O-C\equiv N$
115.2
132.0 H   H 116.8
7.30   6.40

Phenyl cyanate

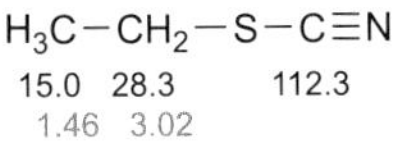

$H_3C-CH_2-S-C\equiv N$
15.0   28.3   112.3
1.46   3.02

Ethyl thiocyanate

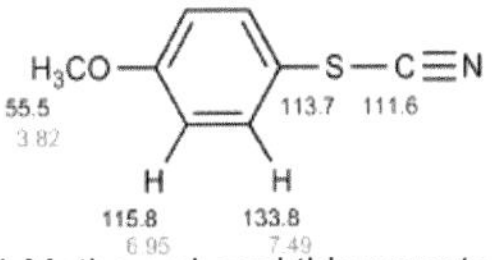

H_3CO $S-C\equiv N$
55.5   113.7 111.6
3.82
H   H
115.8 133.8
6.95   7.49

4-Methoxyphenyl thiocyanate

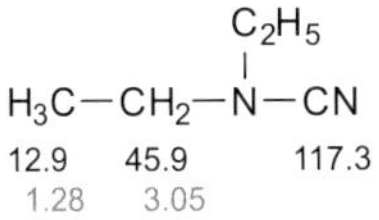

C_2H_5
|
H_3C-CH_2-N-CN
12.9   45.9   117.3
1.28   3.05

Diethyl cyanamide

Isocyanates, Isothiocyanates,Carbodiimides (X = O, S, NR)

N=C=X

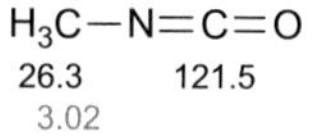

N=C=X

H₃C—N=C=O
26.3   121.5
3.02

Methyl isocyanate

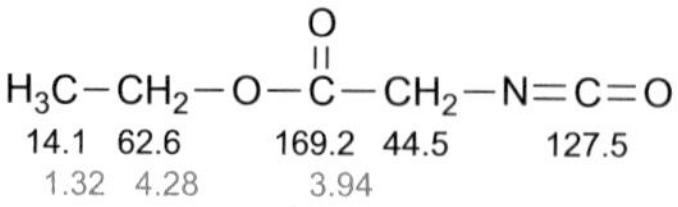

H₃C—CH₂—O—C(=O)—CH₂—N=C=O
14.1 62.6 169.2 44.5   127.5
1.32 4.28   3.94

Isocyanatoacetic acid ethyl ester

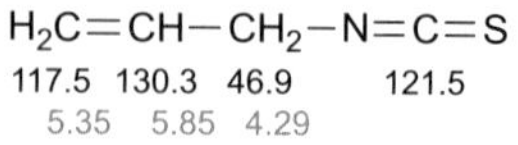

H₂C=CH—CH₂—N=C=S
117.5 130.3 46.9   121.5
5.35 5.85 4.29

Allyl isothiocyanate

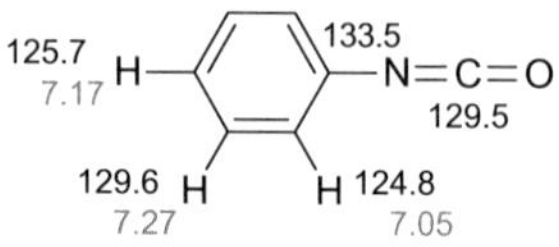

125.7 H ... 133.5 —N=C=O
7.17   129.5
129.6 H   H 124.8
7.27   7.05

Phenyl isocyanate

127.5 H ... 131.2 —N=C=S
7.31   135.3
129.4 H   H 126.9
7.30   7.19

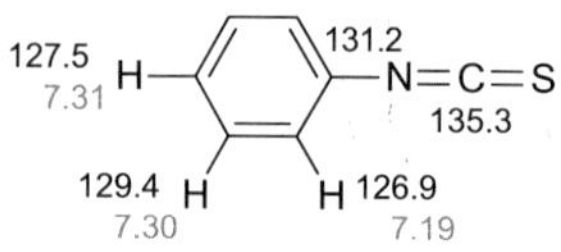

Phenyl isothiocyanate

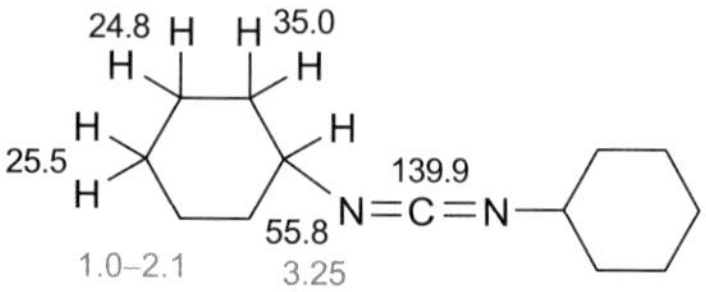

24.8 H   H 35.0
25.5 H
H
55.8 139.9 N=C=N
1.0–2.1 3.25

Dicyclohexylcarbodiimide

Sulfur Compounds [a)]

—S—
C=S
—SOₙ—

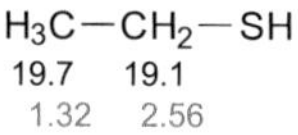

—S—
C=S
—SOₙ—

H₃C—CH₂—SH
19.7   19.1
1.32   2.56

Ethanethiol

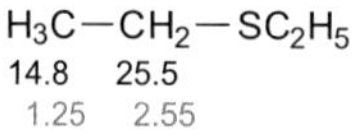

H₃C—CH₂—SC₂H₅
14.8   25.5
1.25   2.55

Diethyl sulfide

H₃C—SCH₃
18.2
2.14

Dimethyl sulfide

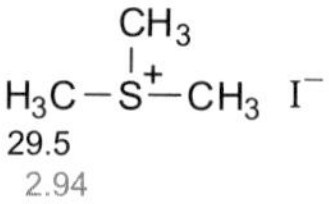

CH₃
H₃C—S⁺—CH₃ I⁻
29.5
2.94

Trimethylsulfonium iodide

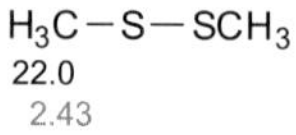

H₃C—S—SCH₃
22.0
2.43

Dimethyl disulfide

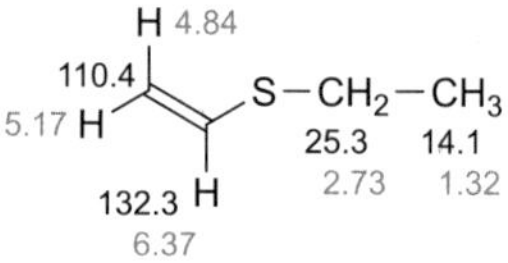

H 4.84
110.4 S—CH₂—CH₃
5.17 H 25.3   14.1
       2.73   1.32
132.3 H
6.37

Ethyl vinyl sulfide

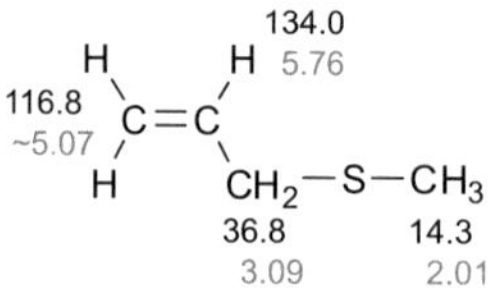

134.0
H   H 5.76
116.8 C=C
~5.07
H   CH₂—S—CH₃
    36.8   14.3
    3.09   2.01

Allyl methyl sulfid

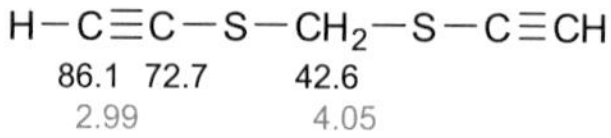

H—C≡C—S—CH₂—S—C≡CH
86.1 72.7   42.6
2.99   4.05

Bis(ethinylthio)methane

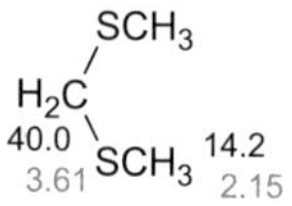

SCH₃
H₂C
40.0 SCH₃ 14.2
3.61   2.15

Bis(methylthio)methane

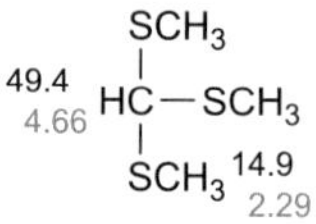

SCH₃
49.4 HC—SCH₃
4.66
SCH₃ 14.9
     2.29

Tris(methylthio)methane

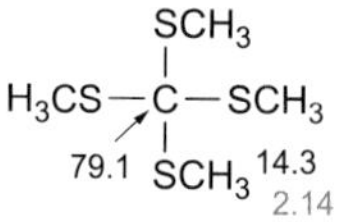

SCH₃
H₃CS—C—SCH₃
79.1 SCH₃ 14.3
          2.14

Tetrakis(methylthio)methane

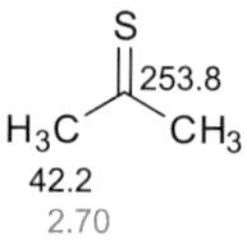

S
253.8
H₃C   CH₃
42.2
2.70

Thioacetone (Propane-2-thione)

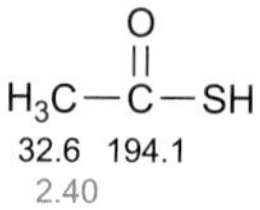

O
H₃C—C—SH
32.6 194.1
2.40

Ethanethloic acid
(Thioacetic acid)

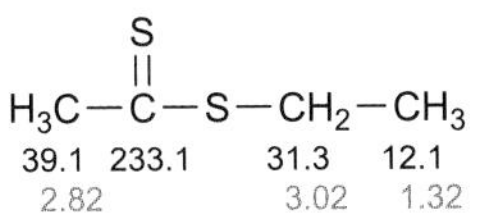

H₃C—C(=S)—S—CH₂—CH₃
39.1 233.1 31.3 12.1
2.82   3.02 1.32

Dithioacetic acid ethyl ester

H₅C₂—S—C(=S)—S—CH₂—CH₃
      224.2 31.5 13.5
             3.31 1.29

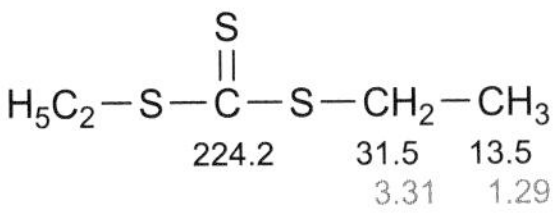

Trithlocarbonic acid diethyl ester

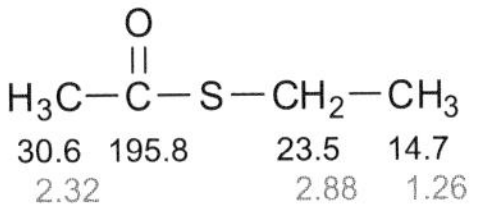

H₃C—C(=O)—S—CH₂—CH₃
30.6 195.8 23.5 14.7
2.32   2.88 1.26

Thioacetic acid S-ethyl ester

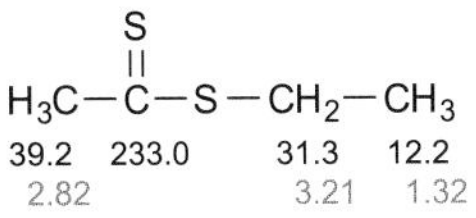

H₃C—C(=S)—S—CH₂—CH₃
39.2 233.0 31.3 12.2
2.82   3.21 1.32

Dithioacetic acid ethyl ester

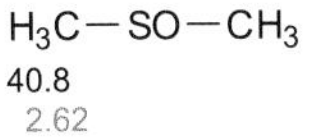

H₃C—SO—CH₃
40.8
2.62

Dimethyl sulfoxide

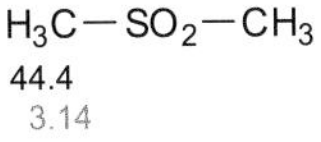

H₃C—SO₂—CH₃
44.4
3.14

Dimethyl sulfone

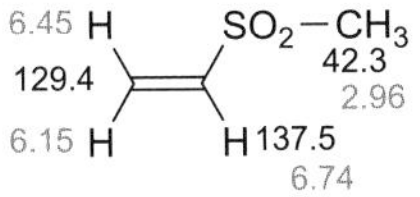

6.45 H ⟩C=C⟨ SO₂—CH₃
129.4   42.3
                   2.96
6.15 H   H 137.5
          6.74

Methyl vinyl sulfone

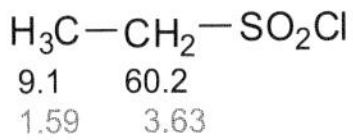

H₃C—CH₂—SO₂Cl
9.1   60.2
1.59 3.63

Ethansulfonic acid chloride

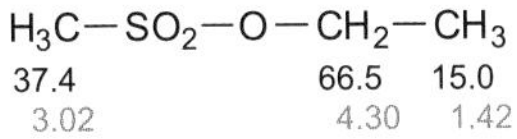

H₃C—SO₂—O—CH₂—CH₃
37.4   66.5 15.0
3.02   4.30 1.42

Methansulfonic acid ethyl
ester

F₃C—SO₂—O—CH₂—CH₃
118.7   74.2 15.3
             4.63 1.52

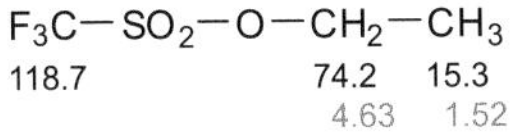

Trifluoromethansulfonic acid
ethyl ester

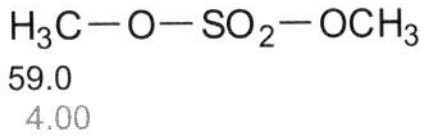

H₃C—O—SO₂—OCH₃
59.0
4.00

Dimethyl sulfate

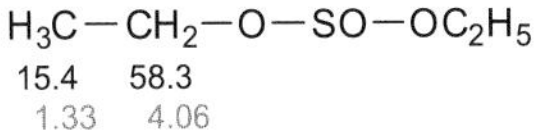

H₃C—CH₂—O—SO—OC₂H₅
15.4 58.3
1.33 4.06

Diethyl sulfite

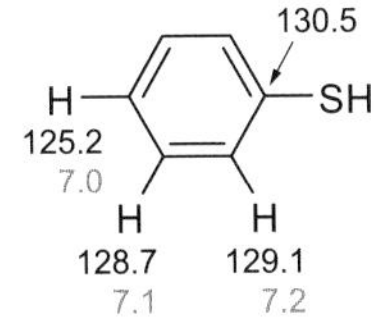

Thiophenol
130.5
—SH
125.2
7.0
128.7 129.1
7.1   7.2

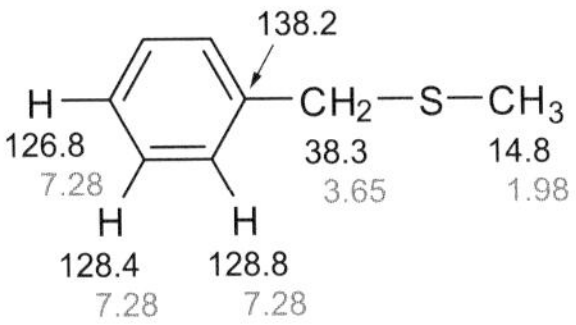

Benzyl-methyl-sulfide
138.2
—CH₂—S—CH₃
126.8   38.3 14.8
7.28   3.65 1.98
128.4 128.8
7.28   7.28

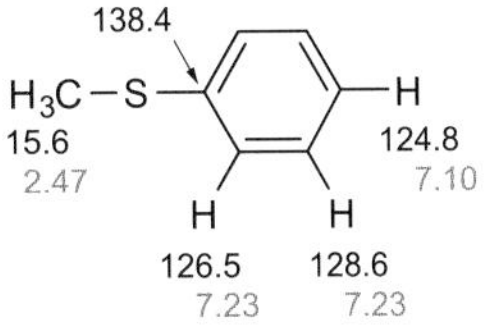

Methyl phenyl sulfide
138.4
H₃C—S—
15.6   124.8
2.47   7.10
126.5 128.6
7.23   7.23

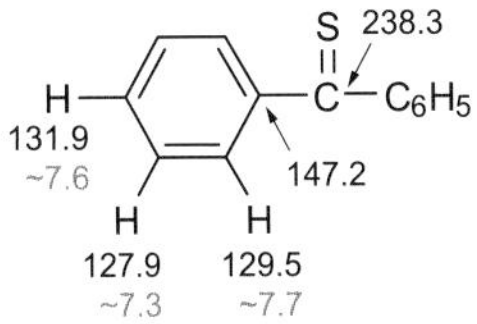

Thiobenzophenone
S 238.3
C—C₆H₅
131.9   147.2
~7.6
127.9 129.5
~7.3   ~7.7

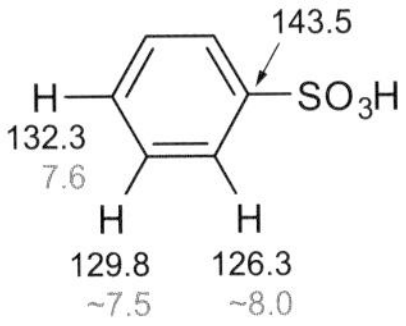

Benzenesulfonic acid
143.5
—SO₃H
132.3
7.6
129.8 126.3
~7.5   ~8.0

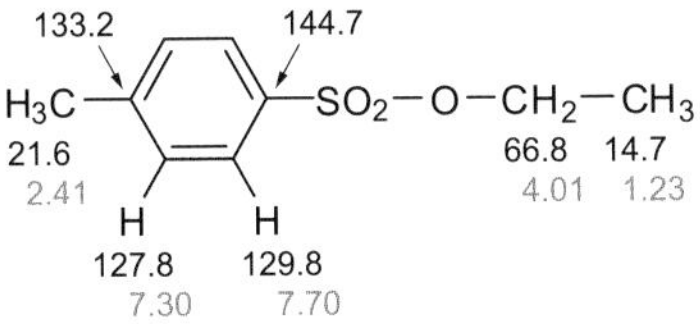

p-Toluensulfonic acid ethyl ester
133.2   144.7
H₃C—   —SO₂—O—CH₂—CH₃
21.6   66.8 14.7
2.41   4.01 1.23
127.8 129.8
7.30   7.70

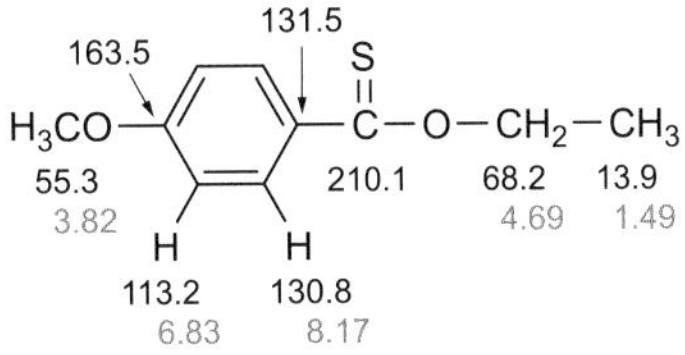

4-Methoxybenzenecarbothioic
acid O-ethyl ester
163.5   131.5
H₃CO—   —C(=S)—O—CH₂—CH₃
55.3   210.1 68.2 13.9
3.82   4.69 1.49
113.2 130.8
6.83   8.17

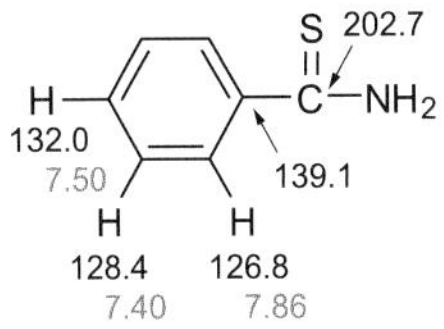

Thiobenzamide
S 202.7
C—NH₂
132.0   139.1
7.50
128.4 126.8
7.40   7.86

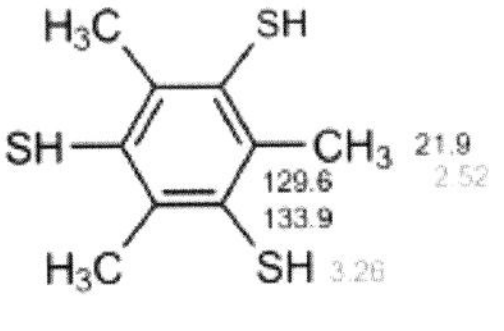

2,4,6-Trimethylbenzene-1,3,5-
trithiol
H₃C   SH
SH—   —CH₃ 21.9
129.6   2.52
133.9
H₃C   SH 3.26

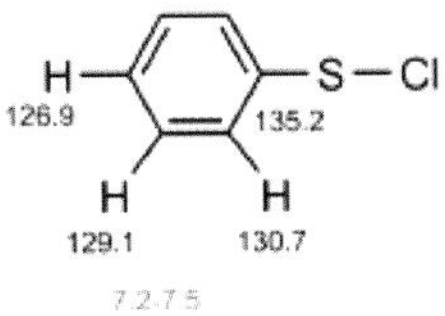

Benzene sulfenyl chloride
H—   —S—Cl
126.9   135.2
129.1 130.7
7.2 7.5

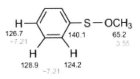
Benzenesulfenic acid methyl
ester
H—   —S—OCH₃
126.7   140.1 65.2
~7.21   3.55
128.9 124.2
~7.21

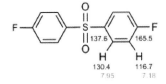
Bis(4-fluoropheny)sulfone
F—   —S(=O)(=O)—   —F
        137.6   165.5
130.4 116.7
7.95   7.18

ᵃ See also thiocyanates, isothiocyanates and heterocycles

Phosphorus Compounds

H$_3$C—CH$_2$—P(C$_2$H$_5$)$_2$
10.3   19.5
0.99   1.27

Triethylphosphine

H$_3$C—CH$_2$—P(C$_2$H$_5$)Cl
8.9   27.6
1.18   1.85

Chloro(diethyl)phosphine

H$_3$C—CH$_2$—PCl$_2$
7.2   36.2
1.30   2.31

Dichloro(ethyl)phosphine

H$_3$C—CH$_2$—$\overset{+}{P}$(C$_2$H$_5$)$_3$ Br$^-$
6.1   12.0
1.34   2.56

Tetraethylphosphonium-bromide

H$_3$C$\diagdown$
     $\overset{\delta^+}{C}\overset{\delta^-}{}$
H$_3$C—P=CH$_2$
     $\diagup$ −1.5
19.7 H$_3$C   −0.6
1.36

Trimethyl-methylene-
phosphorane

H$_3$C—CH$_2$—C≡P
14.8   19.4   177.0
1.17   2.34

Propylidyne phosphine

O
‖
H$_3$C—P—O—CH$_2$—CH$_3$
11.2   61.4   16.4
1.47 OC$_2$H$_5$ 4.09   1.33

Methylphosphonic acid diethyl
ester

O   OCH$_2$—CH$_3$
‖ ╱
H$_3$C—CH$_2$—P
6.8 19.1   ╲OCH$_2$—CH$_3$
0.93 1.16   61.3   16.8
             3.90   1.13

Ethylphosphonic acid diethyl
ester

H$_3$C$\diagdown$ O   OCH$_2$—CH$_3$
      CH—CH$_2$—P‖╱
H$_3$C$\diagup$24.2 34.9 ╲OCH$_2$—CH$_3$
24.1 2.03 1.63 61.0 16.7
1.01   4.05 1.28

2-Methylpropylphosphonic acid
diethyl ester

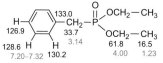

H—⟨⟩—133.0 O OCH$_2$—CH$_3$
126.9   CH$_2$—P‖╱
          33.7   ╲OCH$_2$—CH$_3$
128.6 H H 3.14   61.8 16.5
7.20–7.32 130.2   4.00 1.23

Benzylphosphonic acid
diethyl ester

H$_3$C—CH$_2$—O—P(OC$_2$H$_5$)$_2$
17.4   58.0
1.28   3.91

Triethyl phosphite
(Phosphorous acid ethyl ester)

O
‖
H$_3$CO—P—OCH$_3$
52.1 |
3.79   H
       6.80

Phosphorous acid
dimethyl ester

O
‖
H$_3$C—CH$_2$—O—P(OC$_2$H$_5$)$_2$
15.9   63.4
1.35   4.12

Triethyl phosphate
(Phosphoric acid triethyl ester)

S
‖
H$_3$C—CH$_2$—O—P—OC$_2$H$_5$
15.8   64.3 |
1.40   4.24 SH

Dithiophosphoric acid
O,O-diethyl ester

CH$_3$
|
H$_3$C—CH—O—P$\left(\begin{array}{c}O\diagdown\\O\diagup\end{array}\right)_4$
24.1   68.5
1.25   4.35

Pentaisopropoxy-λ^5-phosphine

128.5   137.2
7.30
H—⟨⟩—P(C$_6$H$_5$)$_2$
128.4 H   H 133.6
7.30   7.30

Triphenylphosphine

132.0   132.7 O
7.51   ‖
H—⟨⟩—P(C$_6$H$_5$)$_2$
128.6 H   H 132.2
7.43   7.64

Triphenylphosphine oxide

135.8   117.4
7.95   +
H—⟨⟩—P(C$_6$H$_5$)$_3$ Br$^-$
130.9 H   H 134.3
7.83   7.64

Tetraphenylphosphonium bromide

135.1   133.3
7.64
H—⟨⟩—PO$_3$H$_2$
131.5 H   H 133.3
7.55   7.81

Phenylphosphonic acid

131.9 127.2 C$_6$H$_5$ O
7.53 $\overset{\delta^+}{|}\overset{\delta^-}{}$ ‖190.7
H—⟨⟩—P=CH—C—CH$_3$
        51.5 28.5
        C$_6$H$_5$ 3.70 2.10
128.7 H H 132.9
7.43   7.64

1-(Triphenylphosphoranylidene)propan-2-one

55.0   161.6 126.1
3.77   O
H$_3$CO—⟨⟩—P—C$_6$H$_4$—OCH$_3$
                 |
113.6 H H 132.6 OH
6.93   7.66

P,P-Bis(4-methoxyphenyl)phosphinic acid

125.5   150.4 O
~7.20   ‖
H—⟨⟩—O—P(OC$_6$H$_5$)$_2$
129.7 H H 120.1
~7.31   ~7.22

Triphenyl phosphate
(Phosphoric acid triphenyl ester)

OH   NH$_2$
| |
H$_3$C—P—CH$_2$—CH$_2$—CH—COOH
14.4 ‖ 26.0 23.5 53.6 172.0
1.42 O   2.10 1.80 4.06

2-Amino-4-(hydroxy-methyl-phosphinoyl)-
butanoic acid (D$_2$O)

H$_3$C$\diagdown$ CH$_3$
      P—P
H$_3$C$\diagup$ $\diagdown$CH$_3$
              9.53
              0.88

Tetramethyldiphosphine

O   OC$_2$H$_5$
H$_3$C—CH$_2$—CH$_2$—CH$_2$—CH$_2$—CH$_2$—P‖╱
13.8 21.7 22.1 23.8 24.7 25.2 ╲OCH$_2$—CH$_3$
0.81 ~1.2 ~1.2 ~1.3 1.53 1.67 61.5 16.6
                                    4.02 1.24

Hexylphosphonic acid diethyl ester

Silicon Compounds

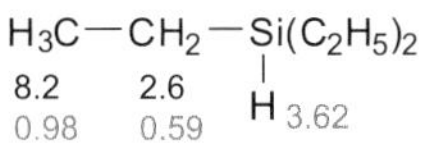

H₃C—CH₂—Si(C₂H₅)₂
8.2   2.6   H 3.62
0.98 0.59

Triethylsilane

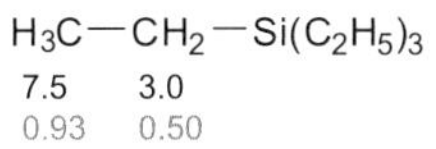

H₃C—CH₂—Si(C₂H₅)₃
7.5   3.0
0.93 0.50

Tetraethylsilane

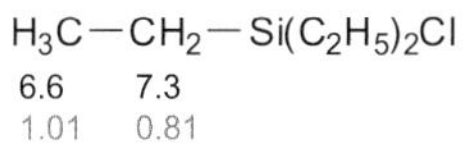

H₃C—CH₂—Si(C₂H₅)₂Cl
6.6   7.3
1.01 0.81

Chlortriethylsilane

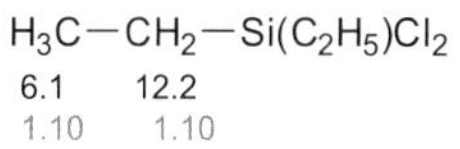

H₃C—CH₂—Si(C₂H₅)Cl₂
6.1   12.2
1.10 1.10

Dichlordiethylsilane

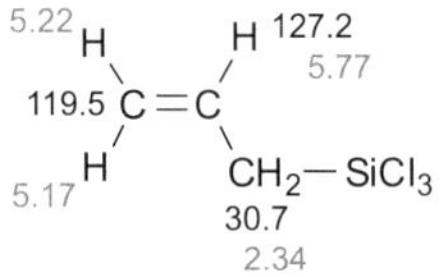

5.22 H   H 127.2
                   5.77
119.5 C═C
   H   CH₂—SiCl₃
5.17   30.7
               2.34

Allyltrichlorsilane

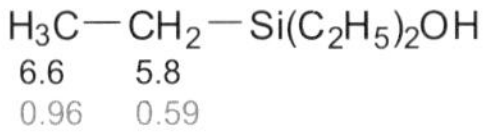

H₃C—CH₂—Si(C₂H₅)₂OH
6.6   5.8
0.96 0.59

Triethylsilanol

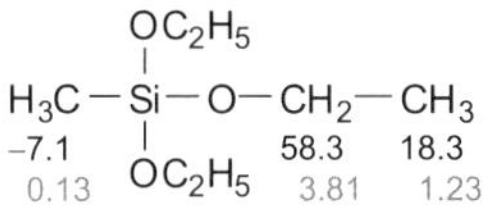

OC₂H₅
H₃C—Si—O—CH₂—CH₃
-7.1 | 58.3 18.3
0.13 OC₂H₅   3.81 1.23

Methyltriethoxysilane

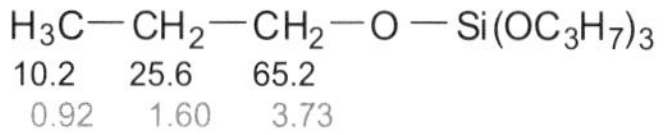

H₃C—CH₂—CH₂—O—Si(OC₃H₇)₃
10.2 25.6 65.2
0.92 1.60 3.73

Tetrapropylorthosilicate

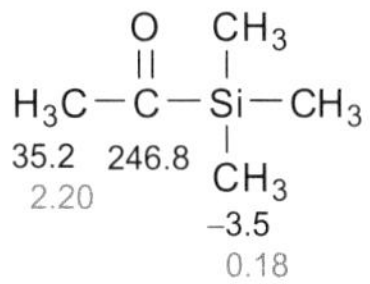

        O   CH₃
        ‖ |
H₃C—C—Si—CH₃
35.2 246.8 CH₃
2.20   -3.5
             0.18

1-(Trimethylsilanyl)ethanone

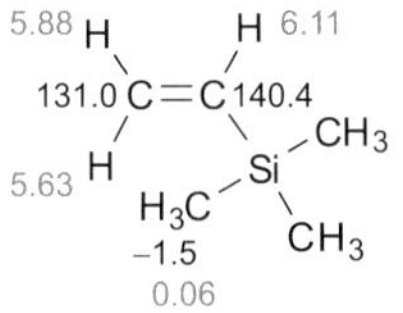

5.88 H   H 6.11
131.0 C═C 140.4
     H   CH₃
5.63   H₃C Si
          -1.5 CH₃
          0.06

Trimethyl-vinyl-silane

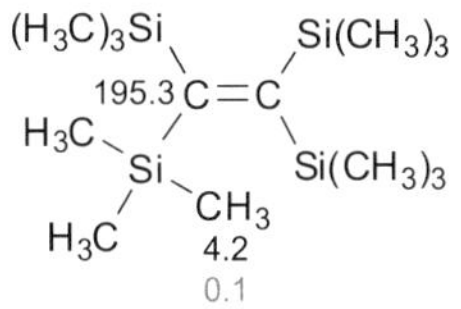

(H₃C)₃Si   Si(CH₃)₃
        195.3 C═C
H₃C   Si(CH₃)₃
     Si   CH₃
H₃C   4.2
        0.1

1,1,2,2-Tetrakis(trimethylsilanyl)ethene

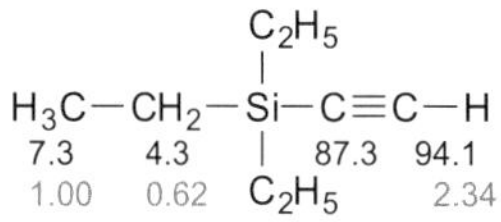

              C₂H₅
              |
H₃C—CH₂—Si—C≡C—H
7.3   4.3 | 87.3 94.1
1.00 0.62 C₂H₅   2.34

Triethylethynylsilane

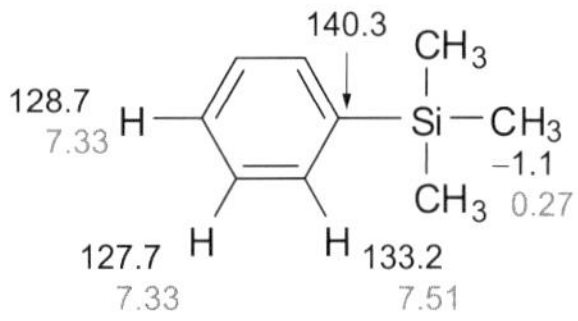

         140.3 CH₃
128.7 |
7.33 H   Si—CH₃
     [benzene ring] -1.1
127.7 H   H 133.2   CH₃ 0.27
7.33   7.51

Trimethylphenylsilane

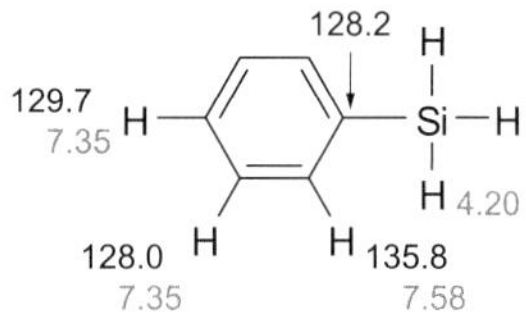

           128.2 H
129.7 |
7.35 H   Si—H
     [benzene ring]
               H 4.20
128.0 H   H 135.8
7.35   7.58

Phenylsilane

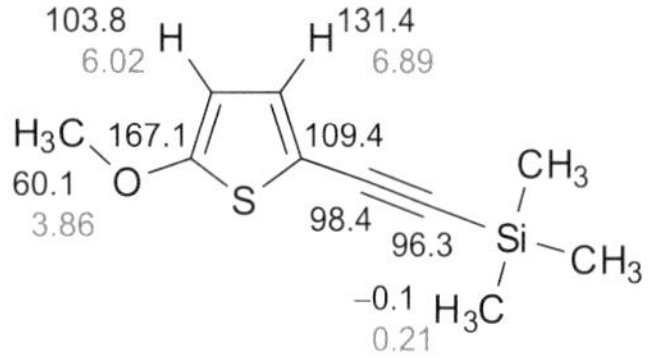

103.8   H 131.4
   6.02   6.89
H₃C 167.1   109.4
60.1 O   S   CH₃
3.86   98.4   Si—CH₃
              96.3
           -0.1 H₃C
           0.21

2-Methoxy-5-(trimethylsilylethynyl)thiophene

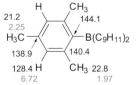

Boron Compounds [a]

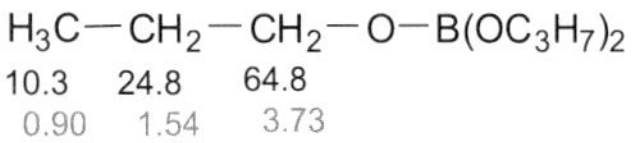

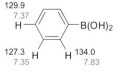

H₃C—CH₂—CH₂—O—B(OC₃H₇)₂
10.3 24.8 64.8
0.90 1.54 3.73

Boric acid tripropyl ester

        H   CH₃
21.2   144.1
2.25
H₃C   B(C₉H₁₁)₂
138.9   140.4
128.4 H   CH₃ 22.8
6.72   1.97

Trimesitylborane

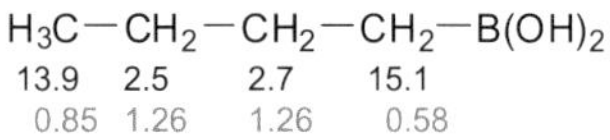

H₃C—CH₂—CH₂—CH₂—B(OH)₂
13.9 2.5 2.7 15.1
0.85 1.26 1.26 0.58

Butylboronic acid

129.9
7.37
H   B(OH)₂
[benzene ring]
127.3 H   H 134.0
7.35   7.83

Phenylboronic acid

Nuclear Magnetic Resonance Spectroscopy

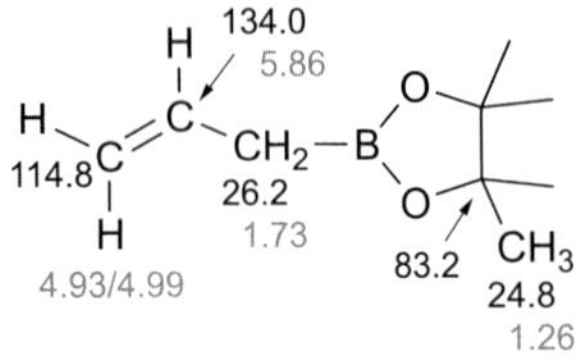

134.0
5.86

114.8
26.2
1.73
4.93/4.99
83.2
24.8
1.26

2-Allyl-4,4,5,5- tetramethyl-1,3,2-dioxaborolane

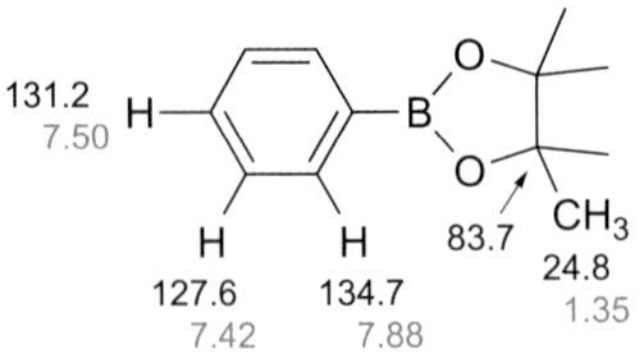

131.2
7.50
127.6
7.42
134.7
7.88
83.7
24.8
1.35

4,4,5,5-Tetramethyl-2-phenyl-1,3,2-dioxaborolane

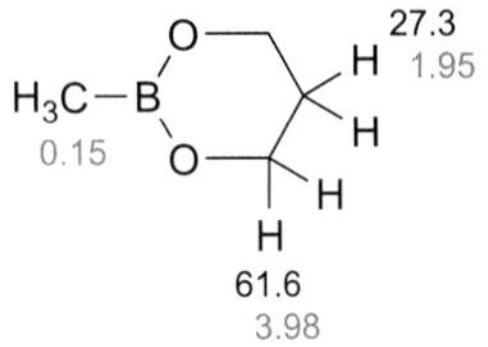

27.3
1.95
H_3C—B
0.15
61.6
3.98

2-Methyl-1,3,2-dioxaborinane

a) Due to quadrupolar B-induced relaxation the signals of $^{13}C_q$-B are broad and often not easy to detect.

C—Met	**Organometallic Compounds, Complexes**	C—Met

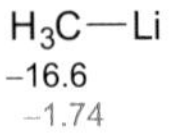

H_3C—Li
−16.6
−1.74

Methyllithium

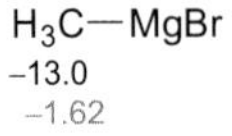

H_3C—MgBr
−13.0
−1.62

Methylmagnesium bromide

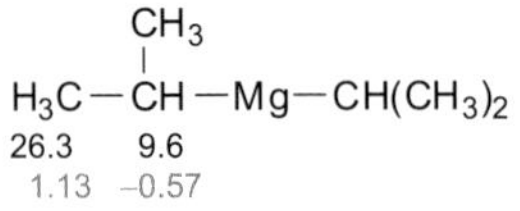

CH_3
|
H_3C—CH—Mg—CH(CH_3)$_2$
26.3   9.6
1.13 −0.57

Diisopropylmagnesium

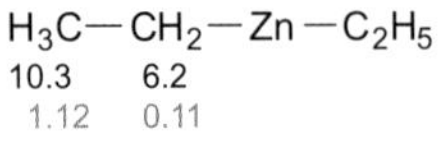

H_3C—CH_2—Zn—C_2H_5
10.3   6.2
1.12   0.11

Diethylzinc

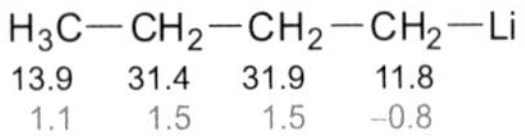

H_3C—CH_2—CH_2—CH_2—Li
13.9   31.4   31.9   11.8
1.1   1.5   1.5   −0.8

Butyllithium

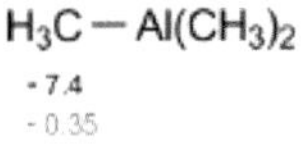

H_3C—Al(CH_3)$_2$
−7.4
−0.35

Trimethylaluminium

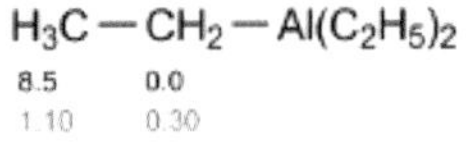

H_3C—CH_2—Al(C_2H_5)$_2$
8.5   0.0
1.10   0.30

Triethylaluminium

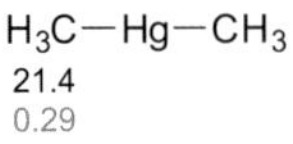

H_3C—Hg—CH_3
21.4
0.29

Dimethylmercury

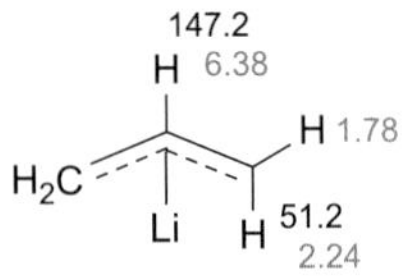

147.2
H 6.38
H 1.78
H_2C
Li   H 51.2
2.24

Allyllithium

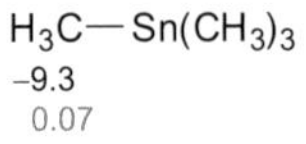

H_3C—Sn(CH_3)$_3$
−9.3
0.07

Tetramethylstannan

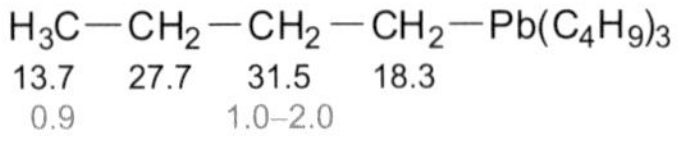

H_3C—CH_2—CH_2—CH_2—Pb(C_4H_9)$_3$
13.7   27.7   31.5   18.3
0.9   1.0–2.0

Tetrabutyllead

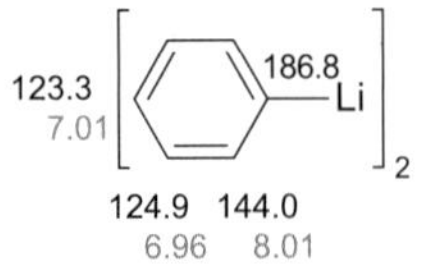

123.3
7.01
186.8
Li
124.9 144.0
6.96   8.01

Phenyllithium

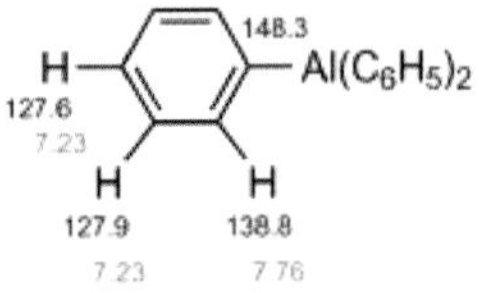

148.3
H—
127.6
7.23
—Al(C_6H_5)$_2$
H   H
127.9   138.8
7.23   7.76

Triphenylaluminium

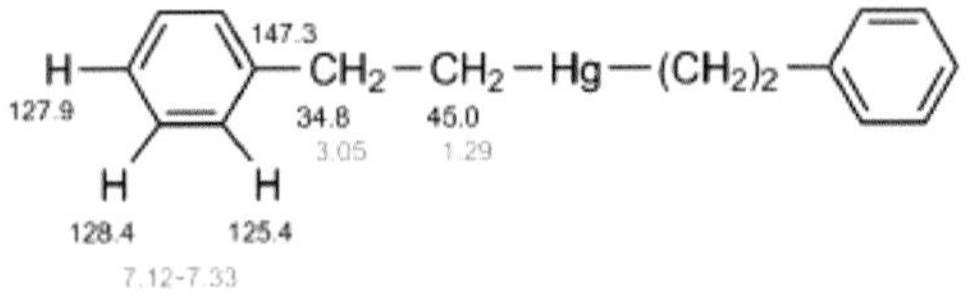

147.3
H—
127.9
—CH_2—CH_2—Hg—(CH_2)$_2$—
H   H
128.4   125.4
34.8   45.0
3.05   1.29
7.12-7.33

Bis(2-phenylethyl)mercury

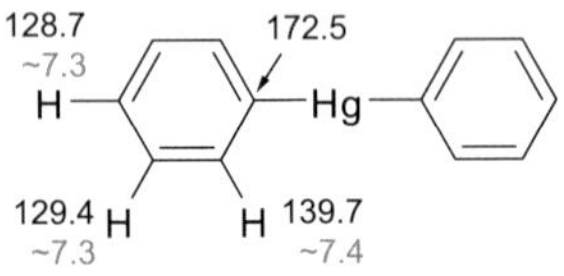

128.7
~7.3
H—
172.5
—Hg—
129.4 H
~7.3
H 139.7
~7.4

Diphenylmercury

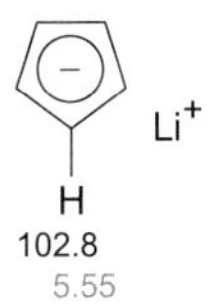

**Lithium cyclo-
pentadienide**

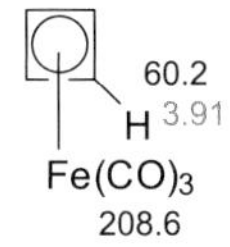

102.8
5.55
H

Li⁺

60.2
H 3.91
Fe(CO)₃
208.6

**(Cyclobutadiene)Iron
tricarbonyl**

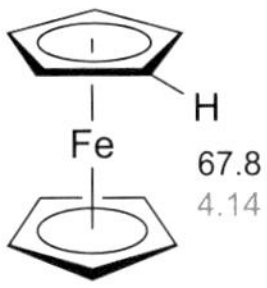

Fe
H
67.8
4.14

Ferrocene

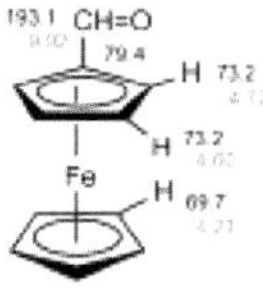

193.1 CH=O
79.4
H 73.2
H 73.2
Fe
H 69.7

**Ferrocenecarb-
oxaldehyde**

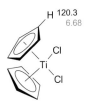

H 120.3
6.68
Cl
Ti
Cl

Titanocene dichloride

93.4 H
5.6
(CO)₃Fe
67.4 H
3.2
Fe(CO)₃

**(Cyclooctatetraene)diiron
hexacarbonyl**

CO
OC CO 212.2
Fe
99.9
H
5.23

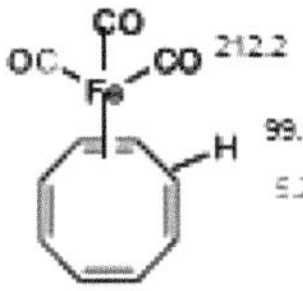

**(Cyclooctatetraene)Iron
tricarbonyl**

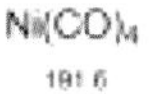

Ni(CO)₄
191.6

**Nickeltetra-
carbonyl**

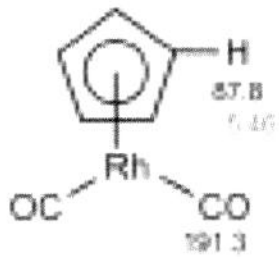

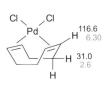

H
87.8
5.46
Rh
OC CO
191.3

**(η⁵-2,4-Cyclopenta
dien-1-yl)rhodium
dicarbonyl**

232.7
Cr(CO)₃
H 1.08
24.2
H 2.17
100.2 H
5.30
H 58.2
102.3 H 2.67
4.18

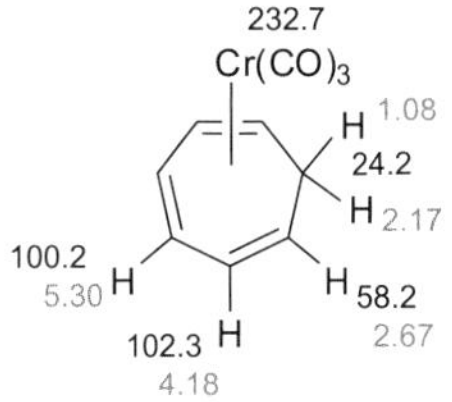

**(Cycloheptatriene)chromium
tricarbonyl**

78.6 H
H H 24.4
H
PdCl₂
118.9 H
31.5 H
H H
34.1

**(Cycloocta-1,4-diene)
palladium dichloride**

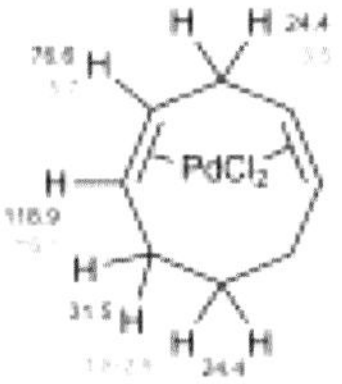

Cl Cl
Pd
116.6
H 6.30
H 31.0
2.6
H

**(Cycloocta-1,5-diene)
palladium dichloride**

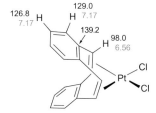

126.8 H
7.17
129.0
H 7.17
139.2
H 98.0
6.56
Pt
Cl
Cl

**(Dibenzo[a,e]cyclooctatetraene)
platinum dichloride**

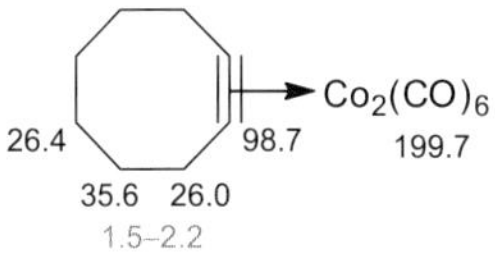

Co₂(CO)₆
26.4 98.7 199.7
35.6 26.0
1.5–2.2

**(μ-Cyclooctyne)dicobalthexa
carbonyl**

H₃C
89.9 109.8
5.12
H CH₃
20.5
H₃C 2.44
OC W CO 212.0
CO

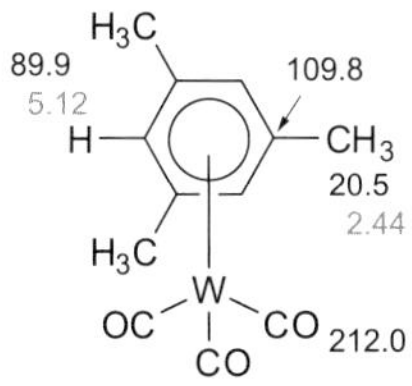

**(Mesitylene)tungsten-
tricarbonyl**

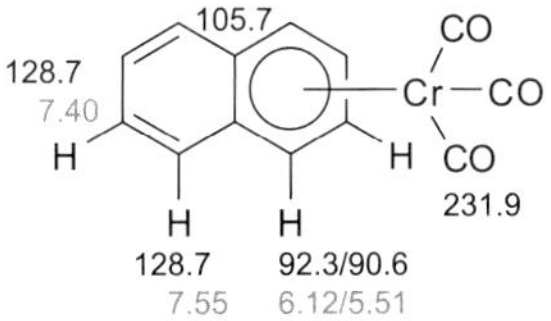

105.7
128.7 CO
7.40 Cr CO
H CO
231.9
H H
128.7 92.3/90.6
7.55 6.12/5.51

**(Naphthalene)chromium
tricarbonyl**

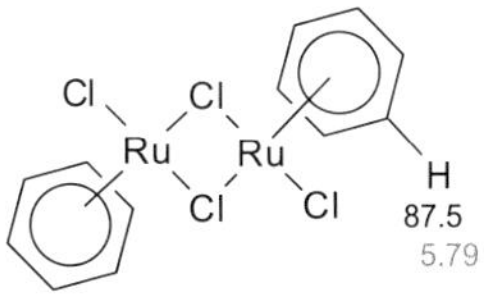

Cl Cl
Ru Ru
Cl Cl H
87.5
5.79

**Bis(η⁶-benzene)dichloro-
dichlorodiruthenium**

CO 218.0
214.1 OC Mo(CO)₂
49.0 H H 78.4
3.78 4.93
H 64.9
1.35

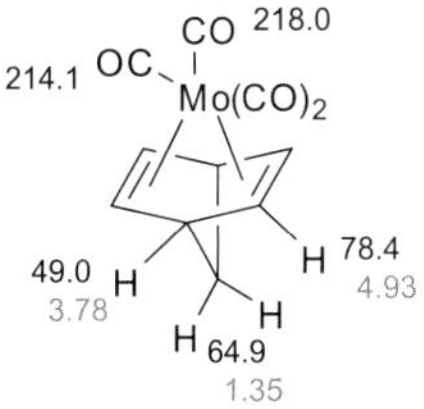

**(Norbornadiene)molybdenum
tetracarbonyl**

*Nuclear Magnetic Resonance
Spectroscopy*

3.6   NMR of other Nuclei

3.6.1 ^{19}F NMR Spectroscopy

^{19}F, the only naturally occurring isotope of fluorine, has a **nuclear spin quantum number** of $1/2$. The **relative sensitivity** is somewhat less than that of ^1H (see **Table 3.1**, p. 87); the **shift range** however is considerably wider. For carbon–fluorine compounds, it extends over 300 to 400 ppm; if inorganic fluorine compounds are included, it is a factor of 10 greater. The **shielding** is dominated by the σ_{para} term. The most common **reference compound** for establishing the zero-point of the δ scale is trichlorofluoromethane ($CFCl_3$, Freon 11). Signals at higher field have a negative value, and signals at lower field have a positive δ value. Measurement parameters have a big influence on the $\delta\,(^{19}F)$ values. Variations $\Delta\delta$ of more than 1 ppm can often be observed. Therefore, **Table 3.53** contains only integer δ values. Exact chemical shifts can be obtained, when for example in $CDCl_3$ the following internal reference compounds are used, whose δ values were determined in relation to external, pure $FCCl_3$:

$FCCl_3$	$F\text{-}C_6H_5$	$F_3C\text{-}C_6H_5$	C_6F_6	$F_3C\text{-}COOH$
+0.65	−112.96	−62.96	−161.61	−75.39

As a result of $^{19}F,^1H$ and/or $^{19}F,^{19}F$ coupling, the ^{19}F NMR spectra of organofluorine compounds often show many lines. Since the quotients $\Delta\nu/J$ however are usually large, the spectra are mostly first order. Typical $^{19}F,^{19}F$ **couplings** are summarized in **Tables 3.53 and 3.54**. The signs of the coupling constants are not always known. Therefore, **Table 3.54** contains the absolute values. An example that illustrates the signs of $^nJ(F,F)$ is presented here by 1,2,4-trifluorobenzene (**377**).

$^5J = +15.2$ Hz

$^3J = -20.7$ Hz

377

$^4J = +3.1$ Hz

$^{19}F,^1H$ and $^{19}F,^{13}C$ couplings have already been considered in Sections 3.3.4 and 3.4.6 (see p. 136 and p. 192).

The spectrum of 1,1,2-trichloro-1,2-difluoro-2-iodoethane (**378**) will be discussed as an example of a ^{19}F spectrum (**Fig. 3.122**). At room temperature an AB spectrum is obtained, with the lines of the B part being particularly broad. These signals arise from averaging of the signals of the three rotamers **378a–c**. At −90°C the rotation around the C−C bond is frozen out, and three separate AB systems are seen. The intensities correspond to the **population** of the **rotamers**.

Many special ^1H NMR techniques also can be applied in the ^{19}F spectroscopy. However, ^{19}F nuclei often exhibit a significant coupling through space. Accordingly (**$^{19}F,^{19}F$)COSY** spectra contain couplings through the bonds and through the space between chemically nonequivalent ^{19}F nuclei. Thus, they resemble (**$^{19}F,^{19}F$)NOESY** measurements.

3.6.2 ^{31}P NMR Spectroscopy

^{31}P with the **nuclear spin quantum number** $1/2$ is the only naturally occurring isotope of phosphorus. Its **relative sensitivity** is only 6.6% that of ^1H. The **shift range** is ca. 1,000 ppm wide. This does however include extreme shift values, for example for P_4 and diphosphenes, which have shifts of −488 and up to +600 ppm, respectively.

Eighty-five percent phosphoric acid is the most usual (external) standard for referencing the δ scale. Signals at lower field have a positive signal, and at higher field a negative δ value.

Despite the wide total range of **^{31}P chemical shifts**, many classes of compounds have relatively narrow shift ranges:

Primary phosphines	PH_2R	$-170 < \delta < -110$
Secondary phosphines	PHR_2	$-100 < \delta < -10$
Tertiary phosphines	PR_3	$-70 < \delta < +70$
Phosphonium salts	$PR_4^{\oplus}$	$-20 < \delta < +40$
Phosphates	$OP(OR)_3$	$-20 < \delta < 0$
Phosphonates	$RP(OR)_2$	$-30 < \delta < +60$
	$\overset{\|}{O}$	
Phosphinates	$R_2P(OR)$	$0 < \delta < +70$
	$\overset{\|}{O}$	
Phosphine oxides	OPR_3	$+10 < \delta < +70$
Phosphites	$P(OR)_3$	$+125 < \delta < +145$

Some typical ^{31}P chemical shifts are summarized in **Table 3.55**.

In the spectra of organic phosphorus compounds, **spin–spin coupling** is most commonly seen to ^1H nuclei. (For sizes of $J(P,H)$ and $J(P,C)$ couplings, see Sections 3.3.4 and 3.4.6, p. 136 and p. 192.) Some $^{31}P,^{31}P$ coupling constants are given in **Table 3.56**.

As an exercise a coupled spectrum will now be discussed. Compound **379** forms an ABX system.

379

378

		Temper- ature	Chemical shifit (δ values rel. to CFCl₃)		$^3J(F,F)$- Coupling constants
		(°C)	F_A	F_B	(Hz)
378	CFCl₂CIFCl	+ 28	– 65.21	– 63.20	– 22.3
a		– 90	– 64.4	– 59.6	– 19.5
b		– 90	– 68.8	– 63.5	– 27.1
c		– 90	– 75.4	– 67.6	– 22.0

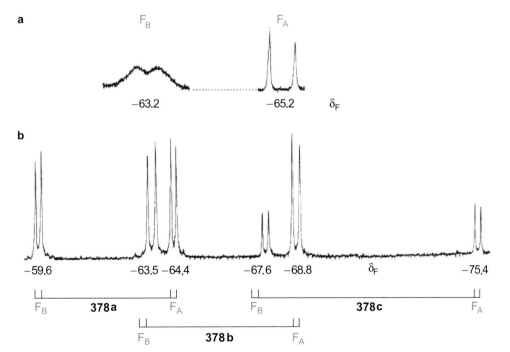

Fig. 3.122 ¹⁹F NMR spectrum of CFCl₂–CFICl (**378**; 94.1 MHz): (**a**) at room temperature; (**b**) at –90 °C (Cavalli, L. 1972, J. Magn. Res. **6**, 298).

Table 3.53: ^{19}F chemical shifts of selected organofluorine compounds (δ values referred to $CFCl_3$)[a]

Group	Compound	δ	Group	Compound	δ
—C—F	H—CH$_2$—F	− 268	=C—F	H$_2$C=CHF	− 114
	CH$_3$—CH$_2$—F	− 212		(H, H / H$_3$C, F cis)	− 132
	C$_3$H$_7$—CH$_2$—F	− 219		(H$_3$C, H / H, F)	− 130
	(CH$_3$)$_2$CH—F	− 165		(H, CH$_3$ / H, F)	− 89
	(CH$_3$)$_3$C—F	− 131		(H$_5$C$_6$, C$_6$H$_5$ / F, F)	− 133
	CH$_2$Cl—CH$_2$—F	− 220		(H$_5$C$_6$, F / F, C$_6$H$_5$)	− 158
	CCl$_3$—CH$_2$—F	− 198		(H, H / F, F)	− 165
	CH$_3$—CHCl—F	− 123		(F, H / H, F)	− 186
	CH$_3$—CCl$_2$—F	− 46	=C(F)$_2$	(H, F / H, F)	− 81
	CFCl$_2$—CCl$_2$—F	− 68		(F$_3$C, F / F$_3$C, F)	− 66
	cyclopropyl–F	− 218		(F, F / F, F)	− 135
	cyclohexyl–F	− 174	≡C—F	HC≡C—F	− 273
—CF$_2$—	H—CF$_2$—H	− 144	—C(=O)—F	H—CO—F	+ 41
	CH$_3$—CF$_2$—H	− 110		CH$_3$—CO—F	+ 49
	CH$_3$—CF$_2$—CH$_3$	− 85		C$_6$H$_5$—CO—F	+ 17
	C$_6$H$_5$—CF$_2$—C$_6$H$_5$	− 89	Aryl-F	fluorobenzene	− 113
	H—CF$_2$—CN	− 120		H$_2$N—C$_6$H$_4$—F	− 129
	H—CF$_2$—OCH$_3$	− 88			
	CH$_3$—CF$_2$—Cl	− 47			
	Br—CF$_2$—Br	+ 7			
	difluorocyclohexane	− 96			
	bicyclic (F)	− 87			
	bicyclic (F)	− 110			
	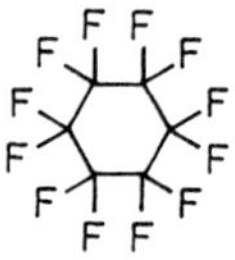 perfluorocyclohexane	− 133			
—CF$_3$	H—CF$_3$	− 79			
	CH$_3$—CF$_3$	− 64			
	C$_6$H$_5$—CF$_3$	− 64			
	CF$_3$CO—CF$_3$	− 85			
	HOOC—CF$_3$	− 79			
	HO—CF$_3$	− 55			
	H$_2$N—CF$_3$	− 49			
	Cl—CF$_3$	− 29			
	F—CF$_3$	− 67			
	CF$_3$—CF$_3$	− 89			
	F$_3$C—C≡C—CF$_3$	− 57			
CF$_4$	CF$_4$	− 63			

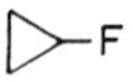

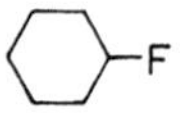

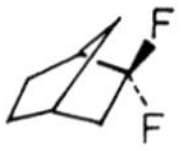

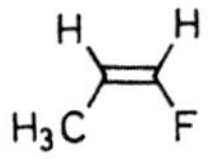

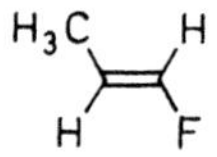

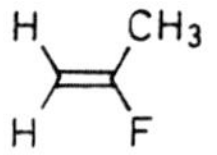

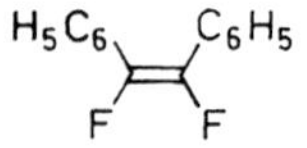

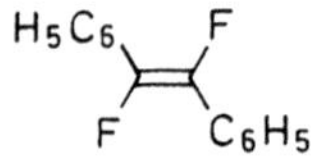

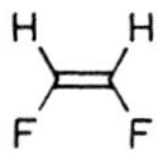

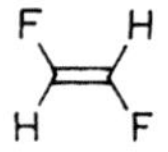

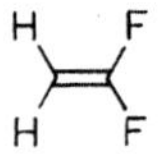

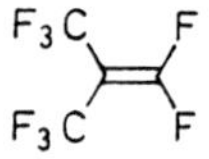

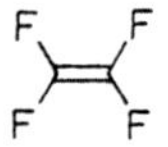

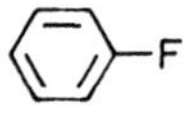

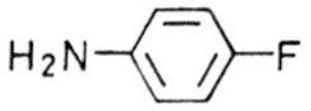

Table 3.53 (Continued)

Group	Compound	δ
	O_2N–C$_6$H$_4$–F	− 102
	F–C$_6$H$_4$–F	− 119
	(difluorobenzene, meta)	− 110
	(difluorobenzene, ortho)	− 139
	(hexafluorobenzene)	− 162
	(1-fluoronaphthalene)	− 124
	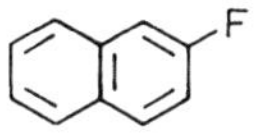 (2-fluoronaphthalene)	− 115
Hetaryl-F	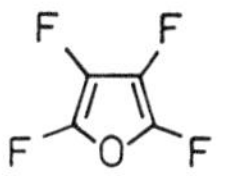 (tetrafluorofuran)	− 196
		− 137
	(tetrafluorothiophene)	− 156
		− 165
	(2-fluoropyridine)	− 61
	(3-fluoropyridine)	− 132
	(4-fluoropyridine)	− 106
	(pentafluoropyridine)	− 134
		− 162
		− 88

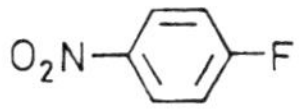

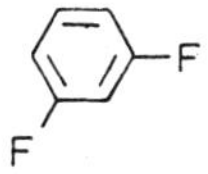

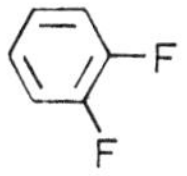

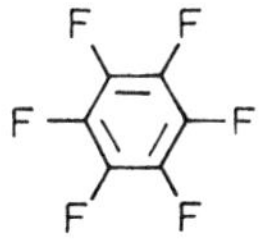

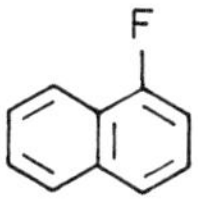

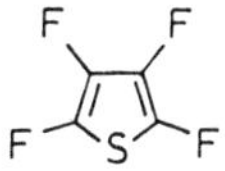

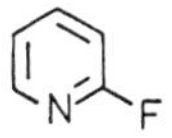

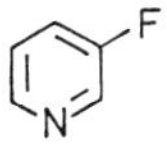

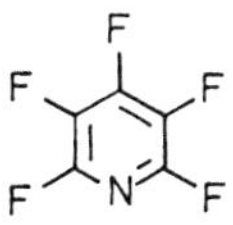

Heteroatom-F

Compound	δ
$H_3C{-}O{-}F$	+ 120
$C_6H_5{-}SF_5$	+ 85
	+ 63
$F_3C{-}NF_2$	− 19
$(F_3C)_2C{=}NF$	− 48
$H_3C{-}SiF_3$	− 134
$(H_5C_2)_2O^{+}{-}^{-}BF_3$	− 153
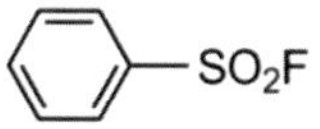 $C_6H_5{-}SO_2F$	− 200

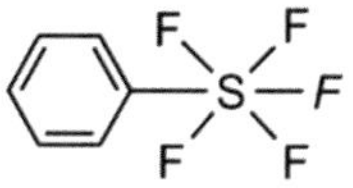

[a] Solvent effects can cause large deviations of the δ values; therefore only integer numbers of δ are given.

Of the 15 theoretical transitions, one (a combination line) has intensity 0. A maximum of 14 lines is therefore expected. In the ^{31}P spectrum the AB part has eight lines, in the ^{1}H spectrum the X part has four lines (**Fig. 3.123**). Two further lines in the X part are so weak that they are lost in the noise. The spin of X (ignoring the spins of A and B) is equally distributed between the two possible orientations. The AB part can therefore be divided into two equally intense subspectra and, using the rules for AB spectra, the effective Larmor frequencies v_a^* and v_b^* can be determined (see p. 272 ff.). In the X part the two intense lines 9 and 12 are separated by $J_{AX} + J_{BX}$. The separation of the centers of the two subspectra is half this amount. The parameters of the ABX system can be obtained from the following formulae:

$$2v_A = v_{1a}^* + v_{2a}^* \quad \text{or} \quad 2v_A = v_{1a}^* + v_{2b}^*$$
$$2v_B = v_{1b}^* + v_{2b}^* \qquad\qquad 2v_B = v_{1b}^* + v_{2a}^*$$
$$\pm J_{AX} = 2(v_a^* - v_A)$$
$$\pm J_{BX} = 2(v_b^* - v_B)$$

One of the two solutions must be eliminated either from a consideration of the line intensities or from criteria of plausibility of the size of the coupling constants.

The analysis of the spectrum of **379** gives the parameters in **Fig. 3.123**, p. 272.

Table 3.54 ^{19}F,^{19}F coupling constants of selected compounds (in Hz)

a) through bonds

Compound	$\lvert{}^{2}J(F,F)\rvert$	$\lvert{}^{3}J(F,F)\rvert$
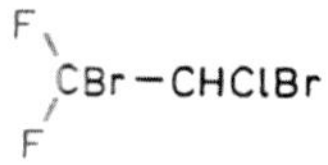 CBr—CHClBr	154	–
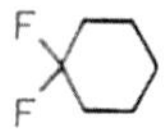	244	–
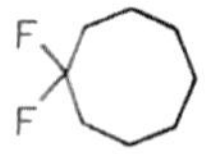	297	–
CF$_3$—CFH$_2$	–	15
CHF$_2$—CHF—CHF$_2$	–	13
CHF$_2$—CF$_2$—CHF$_2$	–	4
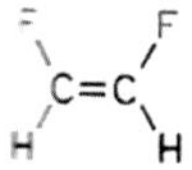	–	19
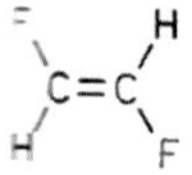	–	133
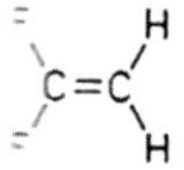	33	–
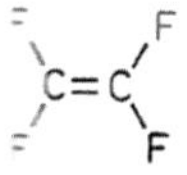	124	73 (cis) 111 (trans)

Compound	$\lvert{}^{3}J(F,F)\rvert$	$\lvert{}^{4}J(F,F)\rvert$	$\lvert{}^{5}J(F,F)\rvert$
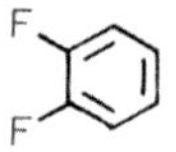	21	–	–
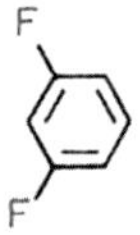	–	7	–
	–	–	18
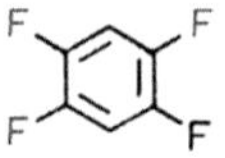	21	0	13
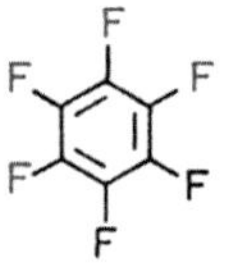	20	3	4
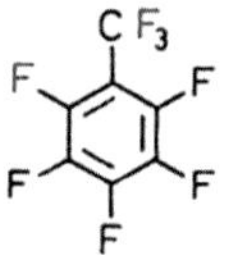		23	

b) through space

Compound	$\lvert J(F,F)\rvert$
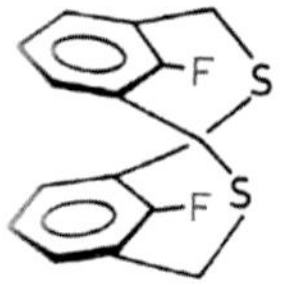	42
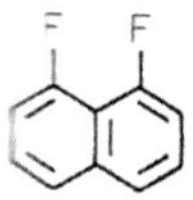	59

Compound	$\lvert J(F,F)\rvert$
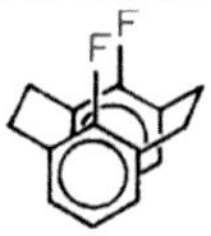	99
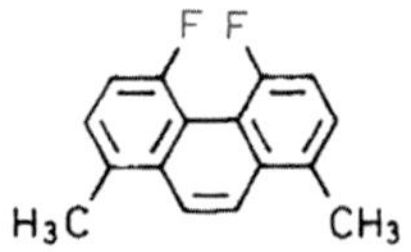	170

Table 3.55 ^{31}P shifts of selected organophosphorus compounds (δ values referred to 85% H_3PO_4 as the external standard)[a]

Group	Compound	δ	Group	Compound	δ
P—	$PH_2(CH_3)$	− 164	P≡	$P\equiv C-C(CH_3)_3$	− 69
	$PH_2(C_2H_5)$	− 127			
	$PH_2(C_6H_5)$	− 122	P	$PH_3(CH_3)^+Cl^-$	− 62
	$PH(CH_3)_2$	− 99		$PH(CH_3)_3^+Cl^-$	− 3
	$PH(C_2H_5)_2$	− 55		$P(CH_3)_4^+I^-$	+ 25
	$PH(C_6H_5)_2$	− 41		$P(C_6H_5)_4^+I^-$	+ 23
	$P(CH_3)_3$	− 62			
	$P(C_2H_5)_3$	− 20	P	$(C_6H_5)_3P(OC_2H_5)_2$	− 55
	$P(C_6H_5)_3$	− 6		$P(OCH_3)_5$	− 67
	$P(CH_2-C_6H_5)_3$	+ 23		$P(OC_6H_5)_5$	− 85
	$P[C(CH_3)_3]_3$	+ 62			
	$P(CH_3)F_2$	+ 245	P	$(C_6H_5)_3P=CH_2$	+ 20
	$P(CH_3)_2F$	+ 186		$(C_6H_5)_3P=C(C_2H_5)_2$	− 11
	$P(C_6H_5)F_2$	+ 208			
	$P(C_2H_5)Cl_2$	+ 196		$(CH_3)_3P=O$	+ 36
	$P(C_6H_5)Cl_2$	+ 161		$(C_2H_5)_3P=O$	+ 48
	$P(C_2H_5)_2Cl$	+ 81		$(C_6H_5)_3P=O$	+ 27
	$C_6H_5P(OCH_3)_2$	+ 159		$(C_6H_5)_3P=S$	+ 43
	$(C_6H_5)_2P(OCH_3)$	+ 116		$(C_6H_5)_2PO(OCH_3)$	+ 32
	$P(OCH_3)Cl_2$	+ 181		$C_6H_5PO(OCH_3)_2$	+ 20
	$P(OCH_3)_3$	+ 141		$(C_2H_5)_2POCl$	+ 76
	$P(OC_6H_5)_3$	+ 127		$C_6H_5POCl_2$	+ 24
	$P[N(CH_3)_2]_3$	+ 123		$PO(OC_2H_5)Cl_2$	+ 3
	$(C_6H_5)_2P-CH=CH_2$	− 12		$PO(OC_2H_5)_3$	− 1
	$Cl_2P-CH=CH_2$	+ 159		$PO(OC_6H_5)_3$	− 18
	$P(C=C-C_6H_5)_3$	− 91		$PO[N(CH_3)_2]_3$	+ 27
				$PS(C_6H_5)Cl_2$	+ 80

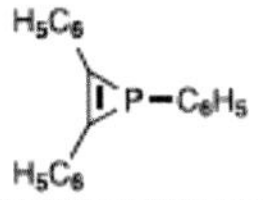

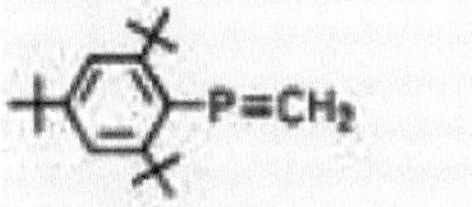

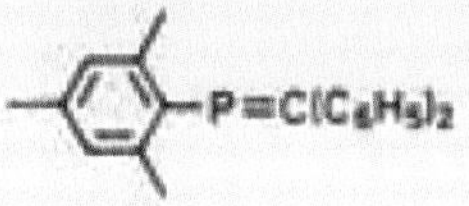

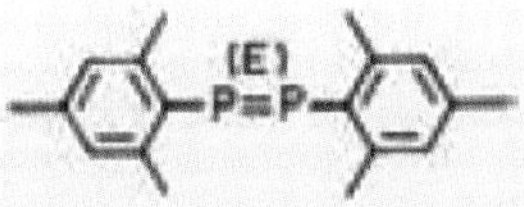

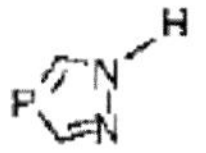

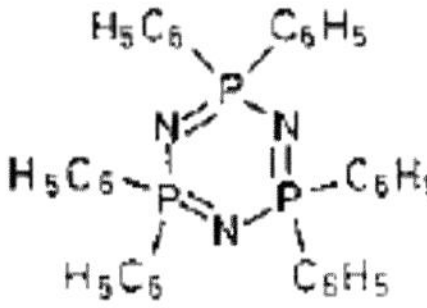

Group	Compound	δ
P	(triangular diphenylcyclopropene phosphine) H_5C_6 / H_5C_6 —P—C$_6$H$_5$	− 330

		δ
(mesityl)—P=CH$_2$		+ 290
(mesityl)—P=C(C$_6$H$_5$)$_2$		+ 233
(mesityl)—(E)—P=P—(mesityl)		+ 492
(phospholyl anion)		+ 211
(diazaphospholide)		+ 83

(cyclotriphosphazene/triphenyl structure): + 9

P (hexaphenyl spirophosphorane anion): − 181

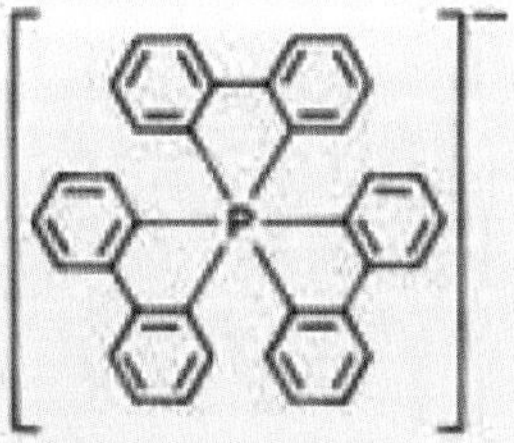

[a] Because of the strong solvent dependency only whole-number values are given

Table 3.56 $^{31}P,^{31}P$ coupling constants of selected organophosphorus compounds (in Hz)

Compound	$\vert^{1}J(P, P)\vert$	$\vert^{n}J(P, P)\vert$
$[(CH_3)_3C]_2P—P[C(CH_3)_3]_2$	451	–
$(H_3C)_2P—P(CH_3)_2$	180	–
$(C_2H_5O)_2\overset{\parallel}{\underset{S}{P}}—\overset{\parallel}{\underset{O}{P}}(OC_2H_5)_2$	583	–
$(H_3C)_2\overset{\parallel}{\underset{S}{P}}—\overset{\parallel}{\underset{S}{P}}(CH_3)_2$	19	–
$(C_2H_5O)_2\overset{\parallel}{\underset{O}{P}}—CH_2—\overset{\parallel}{\underset{O}{P}}(OC_2H_5)_2$	–	$^2J < 1$
$(C_6H_5)_3P{=}CH—P(C_6H_5)_2$	–	$^2J : 150$
$CH_2-\overset{O}{\overset{\parallel}{P}}(OC_2H_5)_2$ ... $CH_2-\overset{\parallel}{\underset{O}{P}}(OC_2H_5)_2$		
ortho	–	$^5J : 9$
meta	–	$^6J : 3$
para	–	$^7J : 8$

As a final comprehensive example of NMR, phosphorine (phosphabenzene, **380**) will be discussed:

380

Symmetry considerations show that the total NMR of the compound (^1H, ^{13}C, and ^{31}P) depends on 7 chemical shifts and 31 coupling constants. In the ^{1}**H NMR spectrum** the AA'BB'C part of an AA'BB'CX system is observed. The protons show six different couplings between themselves; in addition, there are three couplings to phosphorus. Couplings to ^{13}C have no observable effect because of the low natural abundance of ^{13}C.

In the ^{31}**P spectrum**, the X part of the AA'BB'CX system is correspondingly observed (**Fig. 3.124**).

In the coupled ^{13}**C spectrum**, the $^{13}C,^1H$ and $^{13}C,^{31}P$ couplings appear. Of the 13 different $^{13}C,^1H$ couplings, the direct $^1J(C,H)$ couplings are easily distinguished. The determination of the three $^{13}C,^{31}P$ couplings is facilitated by proton broadband

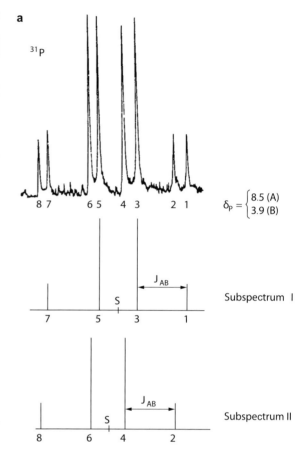

$$\delta_P = \begin{cases} 8.5\ (A) \\ 3.9\ (B) \end{cases}$$

Subspectrum I

Subspectrum II

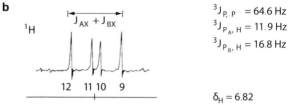

$^3J_{P,\,P} = 64.6\ Hz$

$^3J_{P_A,\,H} = 11.9\ Hz$

$^3J_{P_B,\,H} = 16.8\ Hz$

$\delta_H = 6.82$

Fig. 3.123 NMR spectra of **379** in water: (**a**) ^{31}P spectrum; (**b**) ^1H spectrum (Maier, L. 1973, Phosphorus 2, 229).

decoupling. The chemical shifts of **380** are summarized in **Table 3.57** and the coupling constants in **Table 3.58**.

3.6.3 ^{15}N NMR Spectroscopy

The isotope ^{14}N with a natural abundance of 99.6% has a nuclear spin $I = 1$ and gives broad signals which are of little use for structural determinations. The ^{15}N nucleus with $I = 1/2$ is therefore preferred. However, the low natural abundance of about 0.4% and the extremely low relative sensitivity (cf. **Table 3.1**) make measurements so difficult that ^{15}N NMR spectroscopy was slow

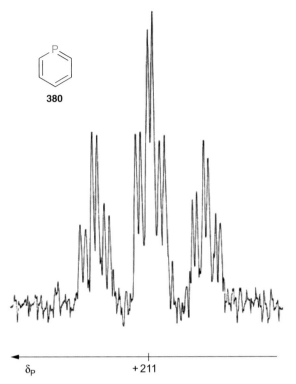

Fig. 3.124 ³¹P spectrum of phosphorine (phosphabenzene, **380**) (Ashe AJ, Sharp RR, Tolan JW. 1976, J. Am. Chem. Soc. **98**, 5451)

to become an accepted analytical tool. A further peculiarity is the negative magnetogyric ratio; in proton decoupled spectra the NOE can strongly reduce the signal intensity. DEPT and INEPT pulse techniques and the inverse detection via coupling protons are therefore particularly important for ¹⁵N NMR spectroscopy.

The range of ¹⁵N chemical shifts is about 600 ppm wide. If extreme values for metal complexes are included, it extends to over 1,400 ppm. The isotope effect between ¹⁴N and ¹⁵N is negligible. Nitromethane is frequently used as a **reference substance**, which can be added in a sealed-off capillary. Values are also frequently quoted with respect to a saturated aqueous solution of ammonium chloride or ammonium nitrate. Liquid ammonia as a reference has the advantage of positive δ (¹⁵N) values for nitrogen-containing organic compounds. Nitromethane and liquid ammonia are the most common reference substances. If not otherwise stated the δ values are related here to CH_3NO_2. The following ¹⁵N shift values can be used to convert the δ values:

CH_3NO_2	$\delta = 0.0$
NH_4Cl	$\delta = -352.9$
NH_4NO_3	$\delta(NH_4^+) = -359.5$
	$\delta(NO_3^-) = -3.9$
NH_3 (liquid)	$\delta = 380.5$

Table 3.57 Chemical shifts for phosphorine (phosphabenzene, **380**)

			δ values		
referred to TMS		referred to TMS		referred to H_3PO_4	
H–2	8.61	C–2	154.1	P–1	+211
H–3	7.72	C–3	133.6		
H–4	7.38	C–4	128.8		
H–5	7.72	C–5	133.6		
H–6	8.61	C–6	154.1		

Table 3.58 Spin–spin couplings in phosphorine (phosphabenzene, **380**), absolute values of coupling constants in Hz

J	H-2	H–3	H–4	H–5	H–6	C–2	C–3	C–4	C–5	C–6	P
H–2	–	10.0	1.2	1.2	1.9	157					38
H–3		–	9.1	1.8	1.2		156				8
H–4			–	9.1	1.2			161			3.5
H–5				–	10.0				156		8
H–6					–					157	38
C–2						–					53
C–3							–				14
C–4								–			22
C–5									–		14
C–6										–	53
P											–

Table 3.59 ¹⁵N Chemical shifts of 1,2,4-triazine (**381**) in different solvents (δ values relative to H_3CNO_2)

Solvent	N–1	N–2	N–4
Cyclohexane	+ 51.2	+ 6.8	– 79.9
Tetrachloromethane	+ 48.4	+ 3.8	– 80.0
Trichloromethane (Chloroform)	+ 42.1	– 2.4	– 80.2
Dimethylsulfoxide	+ 39.2	– 4.9	– 80.3
Ethanol	+ 36.7	– 8.8	– 79.5
Water	+ 18.5	– 21.6	– 85.5

In principle, ¹⁵N NMR spectroscopy has great importance for structural analysis, since N-containing functional groups and N atoms in molecular skeletons are frequently encountered. When quoting specific δ values, it should be remembered that the ¹⁵N NMR signals often depend strongly on the concentration and temperature, and particularly on the solvent. Moreover, intermolecular hydrogen bonds can play an important role.

Table 3.59 illustrates the strong solvent dependence of the ¹⁵N chemical shifts of 1,2,4-triazine (**381**) as an example.

381

The influence of the hybridization on the **¹⁵N chemical shifts** is often similar to the ¹³C NMR.

The series of amines, imines, and nitriles reveals: $\delta\,(^{15}\text{N-sp}^3) < \delta\,(^{15}\text{N-sp}) < \delta\,(^{15}\text{N-sp}^2)$.

The **Tables 3.60 to 3.62** present lists of **^{15}N chemical shifts** of typical examples in different classes of compounds.

Fig. 3.125 shows as a specific example the ^{15}N NMR spectrum of 2-diazo-1,3-diphenyl-1,3-propanedione (**382**). The inner nitrogen atom gives a signal at $\delta = -117.2$ ppm, the outer at $\delta = -9.4$ppm (referred to CH$_3$NO$_2$).

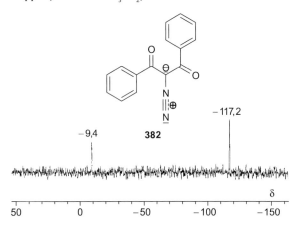

Fig. 3.125 40.5-MHz ^{15}N NMR spectrum of 2-diazo-1,3-diphenyl-1,3-propanedione (**382**) in C$_6$D$_6$.

^{15}N shifts can often show remarkable variations. Thus, $\Delta\delta$ in the azene **383** is almost 600 ppm. The charge distribution would, as in the diazo compound **382**, suggest the reversed signal assignment; however, the large paramagnetic term, which exists for low-energy electronic transitions ($n\pi^*$ transitions), is decisive for the chemical shift. The nitrene nitrogen of the aminonitrene therefore appears at very low field.

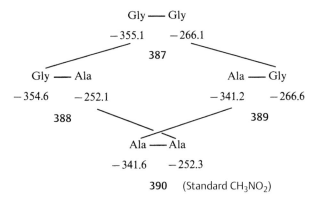

383

(Standard CH$_3$NO$_2$)

On conversion of diethylamine (**384**) into its hydrochloride (**385**), a down-field shift $\Delta\delta$ of less than 4 ppm is observed.

H$_5$C$_2$

N—H $\xrightarrow[- \text{HCl}]{+ \text{HCl}}$

H$_5$C$_2$

−333.7

384   (Standard CH$_3$NO$_2$)

H$_5$C$_2$ H

N$^+$ Cl$^-$

H$_5$C$_2$ H

−330.1

385

A comparison with the ^{1}H and ^{13}C NMR of ammonium salts shows that the positive charge is essentially located not on the central nitrogen atom, but on the ligands.

Amino acids show the expected dependence of the $\delta\,(^{15}\text{N})$ values on pH. Glycine (**386**) serves here as an example:

$$\overset{\oplus}{\text{H}_3}\text{N}-\text{CH}_2-\text{COOH} \underset{+\text{H}^+}{\overset{-\text{H}^+}{\rightleftharpoons}} \overset{\oplus}{\text{H}_3}\text{N}-\text{CH}_2-\text{COO}^{\ominus}$$

−347.3   −345.4

386 a   **386 b**

$$\underset{+\text{H}^+}{\overset{-\text{H}^+}{\rightleftharpoons}} \text{H}_2\text{N}-\text{CH}_2-\text{COO}^{\ominus}$$

−354.7

(Standard CH$_3$NO$_2$)   **386 c**

If an amino acid is built into a peptide, then the δ value at the N-terminus is largely unaffected, whereas the nitrogen in the peptide bond is strongly shifted to low field.

Gly — Gly

−355.1   −266.1

387

Gly — Ala   Ala — Gly

−354.6   −252.1   −341.2   −266.6

388   **389**

Ala — Ala

−341.6   −252.3

390   (Standard CH$_3$NO$_2$)

The rapid tautomerism due to the intermolecular exchange of proton between two N atoms in azoles (**Table 3.79**) can be slowed down by using DMSO as a solvent to such an extent that even at room temperature different ^{15}N signals appear. DMSO functions as a hydrogen bond acceptor. The effect becomes evident by the measurement of pyrazole (**391**) in DMSO and CDCl$_3$:

−173.1
−134.0

−79.8
−134.0

−173.1
−134.0

−79.8
−134.0

$\delta\,(^{15}\text{N})$ in DMSO
$\delta\,(^{15}\text{N})$ in CDCl$_3$

391

Table 3.60 ^{15}N chemical shifts of typical examples, which contain different N-functionalities (δ values referred to CH$_3$NO$_2$)

Class of compound	Example		Solvent	δ
Alkylamines 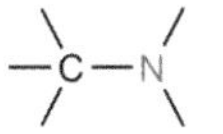	Ethylamine	C$_2$H$_5$—NH$_2$	Methanol	– 355.4
	Isopropylamine	(H$_3$C)$_2$CH—NH$_2$	Methanol	– 338.1
	tert-Butylamine	(CH$_3$)$_3$C—NH$_2$	Methanol	– 324.3
	Diethylamine	(C$_2$H$_5$)$_2$NH	Methanol	– 333.7
	Di-*tert*-butylamine	[(CH$_3$)$_3$C]$_2$NH	neat	– 292.8
	Triethylamine	(C$_2$H$_5$)$_3$N	Methanol	– 332.0
Tetraalkylammonium salts 	Tetramethylammonium chloride	(CH$_3$)$_4$N$^+$Cl$^-$	H$_2$O	– 337.0
	Tetraethylammonium chloride	(C$_2$H$_5$)$_4$N$^+$Cl$^-$	H$_2$O	– 316.5
Enamines 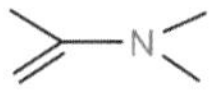	Dimethyl-1-propenylamine	H$_3$C—CH=CH—N(CH$_3$)$_2$	neat	– 349.3
	3-Dimethylamino-acrolein	CHO—CH=CH—N(CH$_3$)$_2$	neat	– 287.5
Arylamines 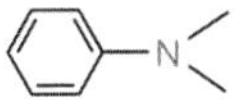	Aniline	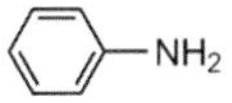	DMSO	– 320.3
	N-Methylaniline	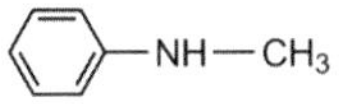	neat	– 324.0
	N,N-Dimethylaniline	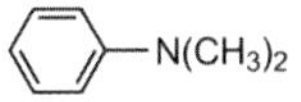	neat	– 332.2
	Diphenylamine	(C$_6$H$_5$)$_2$NH	DMSO	– 288.8
Hydrazines 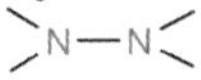	Tetramethylhydrazine	(CH$_3$)$_2$N—N(CH$_3$)$_2$	Acetone	– 284.8
Amides, Lactams 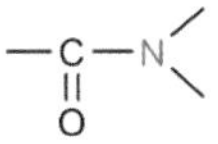	Formamide	HCO—NH$_2$	neat	– 267.6
	N-Methylformamide	HCO—NH—CH$_3$	neat	– 270.1
	N,N-Dimethylformamide	HCO—N(CH$_3$)$_2$	neat	– 275.2
	Benzamide	C$_6$H$_5$—CONH$_2$	DMF	– 279.3
	ε-Caprolactam	(CH$_2$)$_5$—C—NH, $\overset{\|}{O}$	Chloroform	– 262.6
Imides 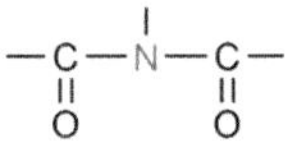	Succinimide	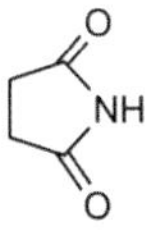	DMSO	– 204.1
Ureas, Urethanes 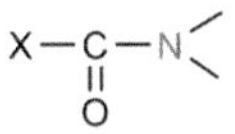	Urea (X=N)	H$_2$N—CO—NH$_2$	H$_2$O	– 305.0
	Trimethylurethane (X=O)	H$_3$CO—CO—N(CH$_3$)$_2$	Chloroform	– 314.2
Sulfonamides 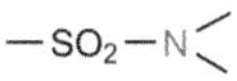	Benzenesulfonamide	C$_6$H$_5$—SO$_2$—NH$_2$	DMSO	– 285.9
Thioamides 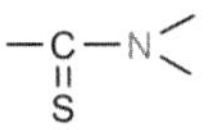	Thioacetamide	H$_3$C—CS—NH$_2$	CDCl$_3$	– 230.0
	Thiourea	H$_2$N—CS—NH$_2$	H$_2$O	– 273.3
	Trimethylthiourethane	H$_3$CO—CS—N(CH$_3$)$_2$	CDCl$_3$	– 269.8

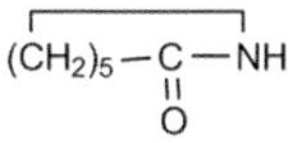

Nuclear Magnetic Resonance Spectroscopy

Table 3.60 (Continued)

Class of compound	Example		Solvent	δ
Aminoxides $\ce{>N+-O^-}$	N-Methylpiperidine N-oxide	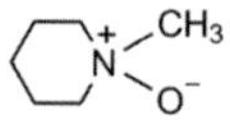 (structure)	CDCl₃	− 272.7
Imines $\ce{>C=N}$	N-Methylbenzaldimine N-Phenylbenzaldimine	$C_6H_5-CH=N-CH_3$ $C_6H_5-CH=N-C_6H_5$	Chloroform Chloroform	− 62.1 − 54.1
Oximes $\ce{>C=N-OH}$	Acetone oxime Benzaldehyde oxime	$(H_3C)_2C=NOH$ $C_6H_5-CH=NOH$	Chloroform Chloroform	− 45.9 − 26.3
Nitrones $\ce{>C=N+-O^-}$	N-(Benzylidene)aniline N-oxide	$C_6H_5-CH=\overset{+}{N}-C_6H_5$ (O^-)	CDCl₃	− 96.8
Keteneimines $\ce{-N=C=C<}$	N-(2-Methyl-1-propenyl= idene)aniline	$C_6H_5-N=C=C(CH_3)_2$	D₂O	− 164.7
Isocyanates $\ce{-N=C=O}$	Phenylisocyanate	$C_6H_5-N=C=O$	neat	− 333.7
Isothiocyanates $\ce{-N=C=S}$	Phenylisothiocyanate	$C_6H_5-N=C=S$	neat	− 273.1
Carbodiimides $\ce{-N=C=N-}$	Dicyclohexylcarbo-diimide	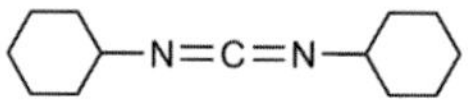 (structure) $-N=C=N-$	neat	− 280.7
Isonitriles $\ce{-N=C}$	Methylisonitrile Phenylisonitrile	$H_3C-N=C$ $C_6H_5-N=C$	neat neat	− 218.0 − 200.0
Nitriles $\ce{-C#N}$	Acetonitrile Benzonitrile	$H_3C-C\equiv N$ $C_6H_5-C\equiv N$	neat DMSO	− 137.1 − 121.5
Cyanates $\ce{-O-C#N}$	Phenylcyanate	$C_6H_5-O-C\equiv N$	neat	− 212.0
Thiocyanates $\ce{-S-C#N}$	Phenylthiocyanate	$C_6H_5-S-C\equiv N$	neat	− 97.0
Nitrile N-oxides $\ce{-C#N+-O^-}$	Benzonitrile N-oxide	$C_6H_5-C\equiv\overset{+}{N}-O^-$	CDCl₃	− 181.9
Azo compounds $\ce{-N=N-}$	(E)-Azobenzene (Z)-Azobenzene	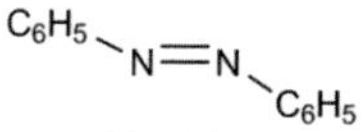 (structures)	Chloroform Chloroform	+ 129.0 + 146.5

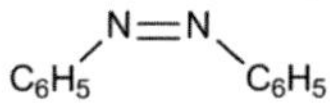

Table 3.60 (Continued)

Class of compound	Example		Solvent	δ
Nitrates $-O-NO_2$	Ethyl nitrate	$C_2H_5-O-NO_2$	$CDCl_3$	$-$ 38.0
Nitro compounds $-NO_2$	Nitroethane Nitrobenzene (Trinitromethyl)benzene	$C_2H_5-NO_2$ $C_6H_5-NO_2$ $C_6H_5-C(NO_2)_3$	$CDCl_3$ DMSO $CDCl_3$	$+$ 10.1 $-$ 9.8 $-$ 28.9
Nitrites $-O-NO$	Butyl nitrite	C_4H_9-ONO	neat	$+$ 189.0
Nitroso compounds $-N=O$	2-Methyl-2- nitrosopropane Nitrosobenzene	$(CH_3)_3C-NO$ C_6H_5-NO	neat neat	$+$ 578.0 $+$ 532.0

Table 3.61 ^{15}N chemical shifts of compounds with functionalities having two or three different nitrogen atoms (δ values referred to CH_3NO_2)

Class of compound	Example		Solvent
Hydrazones	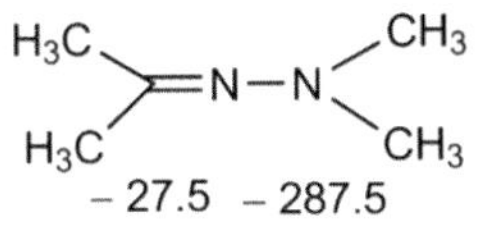 $\begin{array}{c} H_3C \\ \\ H_3C \end{array}\!\!>\!\!=\!N\!-\!N\!\!<\!\!\begin{array}{c} CH_3 \\ \\ CH_3 \end{array}$ $-27.5\ \ -287.5$	Acetone 2,2-dimethylhydrazone	neat
Guanidines	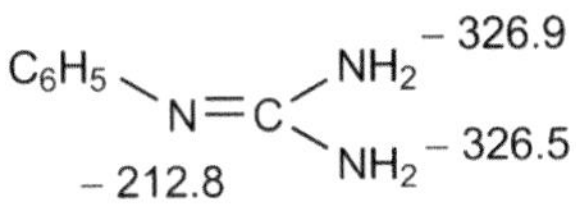 $\begin{array}{c} C_6H_5 \\ \end{array}\!\!>\!N\!=\!C\!\!<\!\!\begin{array}{c} NH_2 \ \ {}^{-326.9} \\ NH_2 \ \ {}^{-326.5} \end{array}$ -212.8	N-Phenylguanidine	$CDCl_3$
Cyanamides	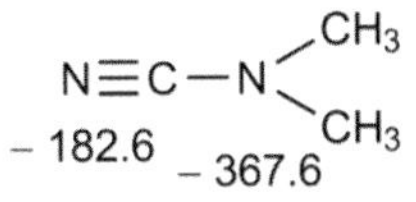 $N\!\equiv\!C\!-\!N\!\!<\!\!\begin{array}{c} CH_3 \\ CH_3 \end{array}$ -182.6 $\ \ -367.6$	N,N-Dimethylcyanamide	DMSO
Nitramines	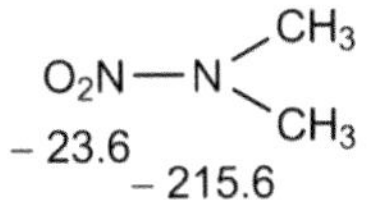 $O_2N\!-\!N\!\!<\!\!\begin{array}{c} CH_3 \\ CH_3 \end{array}$ -23.6 $\ \ -215.6$	N-Nitrodimethylamine	neat
Nitrosamines	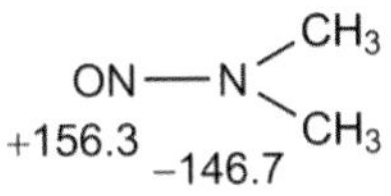 $ON\!-\!N\!\!<\!\!\begin{array}{c} CH_3 \\ CH_3 \end{array}$ $+156.3$ $\ \ -146.7$	Dimethylnitrosamine	neat

Nuclear Magnetic Resonance Spectroscopy

Table 3.61 (Continued)

Class of compound	Example		Solvent
Diazo compounds	$\overset{H}{\underset{H}{>}} \overset{-}{C} - \overset{+}{N} \equiv N$ − 96.0 + 7.8	Diazomethane	Ether
Diazonium salts	(phenyl)$-\overset{+}{N} \equiv N$ BF_4^- − 149.8 − 66.3	Benzenediazonium tetrafluoroborate	Acetonitrile
Azoxy compounds	$H_5C_6 \sim\sim N \overset{+}{=} N \overset{C_6H_5}{\underset{O^-}{<}}$ −19.8 −36.0	(E)-Azoxybenzene	Chloroform
	−46.7 −57.1	(Z)-Azoxybenzene	Chloroform
Azides	$H_3C - \overset{-}{N} - \overset{+}{N} \equiv N$ −321.2 −171.0 −129.7	Methyl azide	Benzene
	H_3C-(phenyl)$-SO_2 - \overset{-}{N} - \overset{+}{N} \equiv N$ Tosyl azide −240.4 −138.3 −146.0		DMSO

Table 3.62 ^{15}N chemical shifts of selected heterocycles (δ values referred to CH_3NO_2; DMSO as the solvent except otherwise specified)

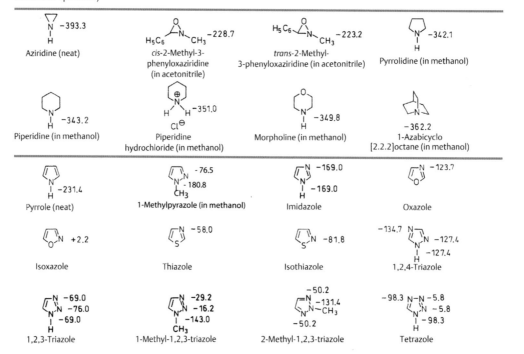

Aziridine (neat) −393.3	cis-2-Methyl-3-phenyloxaziridine (in acetonitrile) −228.7	trans-2-Methyl-3-phenyloxaziridine (in acetonitrile) −223.2	Pyrrolidine (in methanol) −342.1
Piperidine (in methanol) −343.2	Piperidine hydrochloride (in methanol) −351.0	Morpholine (in methanol) −349.8	1-Azabicyclo[2.2.2]octane (in methanol) −362.2
Pyrrole (neat) −231.4	1-Methylpyrazole (in methanol) − 76.5 − 180.8	Imidazole −169.0 −169.0	Oxazole −123.7
Isoxazole +2.2	Thiazole −58.0	Isothiazole −81.8	1,2,4-Triazole −134.7 −127.4 −127.4
1,2,3-Triazole −69.0 −76.0 −69.0	1-Methyl-1,2,3-triazole −29.2 −16.2 −143.0	2-Methyl-1,2,3-triazole −50.2 −131.4 −50.2	Tetrazole −98.3 −5.8 −5.8 −98.3

Table 3.62 (Continued)

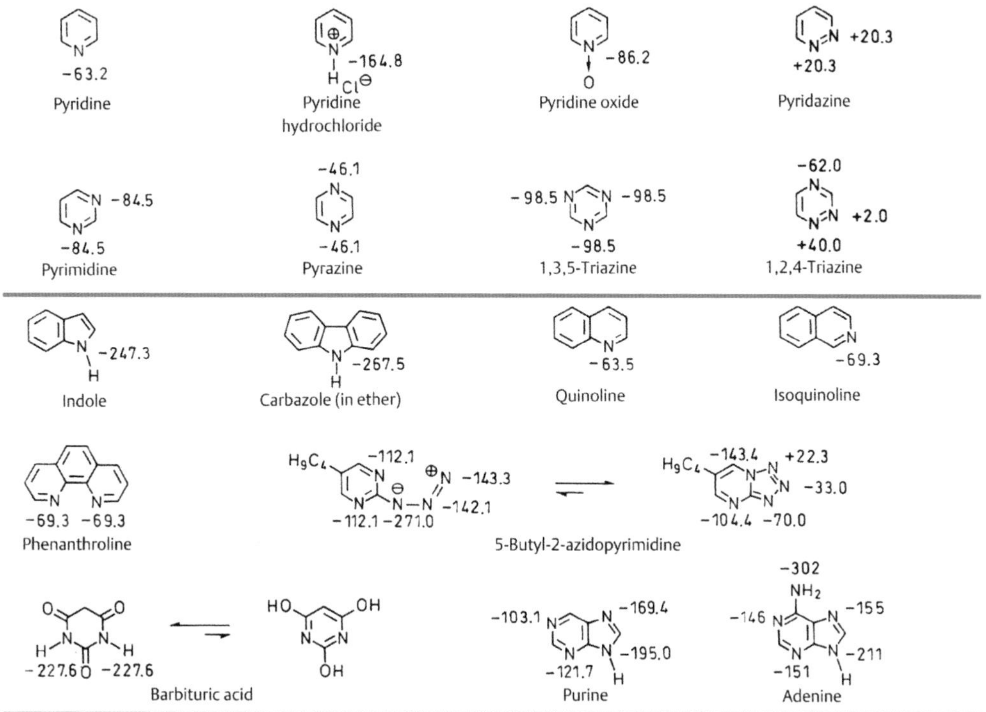

Pyridine −63.2

Pyridine hydrochloride −164.8 H Cl⊖

Pyridine oxide −86.2

Pyridazine N=N +20.3 / +20.3

Pyrimidine −84.5 / −84.5

Pyrazine −46.1 / −46.1

1,3,5-Triazine −98.5 N / N −98.5 / −98.5

1,2,4-Triazine −62.0 / N=N +2.0 / +40.0

Indole −247.3 H

Carbazole (in ether) −267.5 H

Quinoline −63.5

Isoquinoline −69.3

Phenanthroline −69.3 −69.3

5-Butyl-2-azidopyridimidine H_9C_4 −112.1 ⊕N −143.3 N−N −142.1 −112.1 −271.0 ⇌ H_9C_4 −143.4 +22.3 N −33.0 −104.4 −70.0

Barbituric acid H N −227.6 O −227.6 N H ⇌ HO OH N N OH

Purine −103.1 N / N −169.4 / −195.0 / −121.7 H

Adenine −302 NH₂ / −146 N −155 / −151 N −211 H

A fast tautomerism of methylpyrazole (**392/393**) in $CDCl_3$ gives two averaged δ (^{15}N) values, because it does not correspond to an automerization:

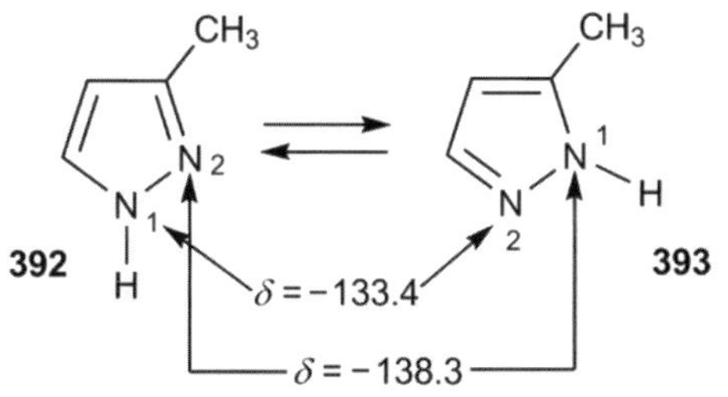

392 ⇌ **393**

$\delta = -133.4$

$\delta = -138.3$

Tautomeric equilibria of compounds, which show a proton exchange between N and O or between NO and O, can be efficiently studied by ^{15}N NMR. The ring-chain tautomerism of the oxazolidine **394** and its open form **395** and the tautomeric equilibria **396/397** and **398/399** represent typical examples that are characterized by large $\Delta\delta$ (^{15}N) values:

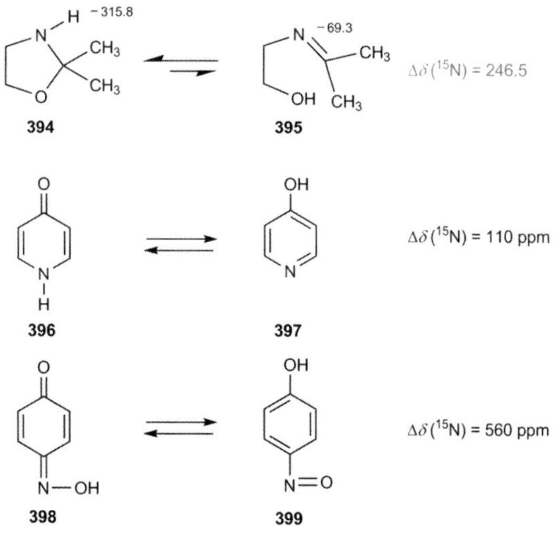

N −315.8 H CH₃ CH₃ **394** ⇌ N −69.3 CH₃ OH CH₃ **395** $\Delta\delta$ (^{15}N) = 246.5

396 ⇌ **397** $\Delta\delta$ (^{15}N) = 110 ppm

398 N—OH ⇌ **399** N=O $\Delta\delta$ (^{15}N) = 560 ppm

Finally, ^{15}N NMR can be used for investigations on molecular dynamics; inversion at nitrogen is a particularly suitable case.

While isotope effects on ^{15}N and ^{14}N chemical shifts are negligible, the spin–spin coupling constants of ^{15}N and ^{14}N differ both in sign and magnitude. The following relation applies:

$$J(^{15}N,X) = -1.4027\,J(^{14}N,X)$$

$$X = {}^1H, {}^{13}C,...$$

Table 3.63 ^nJ (^{15}N,^{1}H) coupling constants (n = 1, 2, 3, and 4) of selected compounds in Hz

Compound	Solvent	^1J	^2J	^3J	^4J
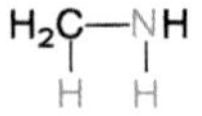 H_2C-NH (H, H)	neat liquid	−64.5	1.0		
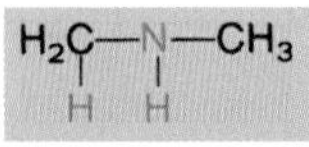 $H_2C-N-CH_3$ (H, H)	neat liquid	−67.0	0.9		
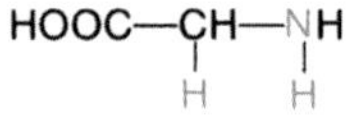 $HOOC-CH-NH$ (H, H)	Water	−74.7	0.5		
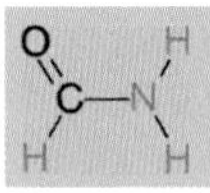 O=C(H)−N(H)(H) (formamide)	neat liquid	−88.3 / −90.7	14.6		
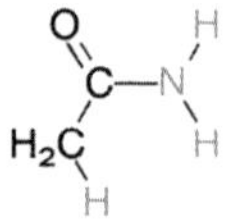 O=C(H)−N(H)−CH$_2$(H) (acetamide)	Water	−88.4 / −90.9		1.3	
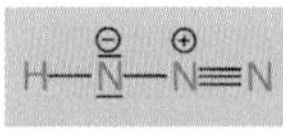 $H-\overset{\ominus}{N}-\overset{\oplus}{N}\equiv N$	Ether	−70.2	2.3	2.2	
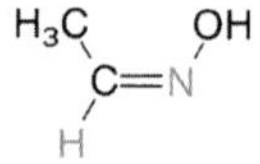 H_3C(H)$C=N-OH$	Water		−15.9		
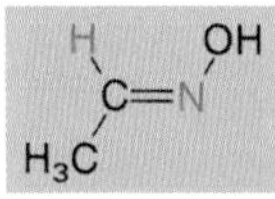 H(H$_3$C)$C=N-OH$	Water			+ 2.9	
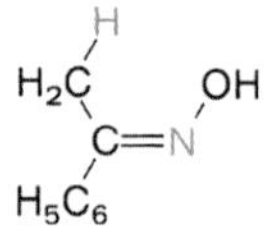 H_2C(H_5C_6)$C=N-O$(H) (H)	Chloroform			−2.0	
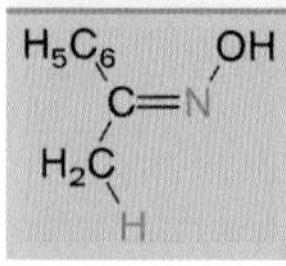 H_5C_6(H_2CH)$C=N-OH$	Chloroform			−4.2	
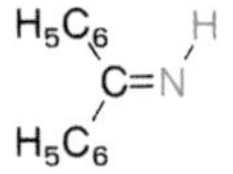 H_5C_6(H_5C_6)$C=N-H$	Pentane	−51.2			
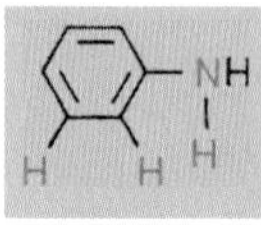 aromatic−NH (H, H)	Chloroform / Benzene	−78.0		−1.9	−0.5
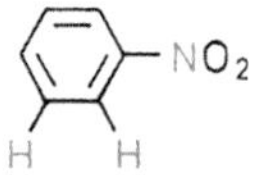 aromatic−NO_2 (H, H)	Acetone			−1.9	−0.8
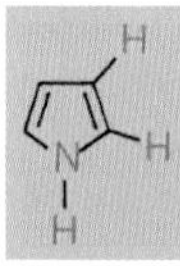 pyrrole	Benzene / neat liquid	−96.5	− 4.5	−5.4	
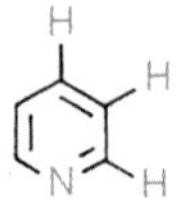 pyridine	neat liquid		−10.8	−1.5	0.2
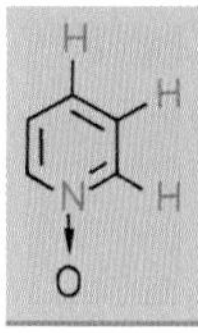 pyridine N-oxide	Chloroform		0.5	−5.3	1.1

Typical ^1J(^{15}N,^{1}H) coupling constants lie in the range of (−80 ± 15) Hz. Some examples, including exceptions, are tabulated in **Table 3.63**. ^2J(^{15}N,^{1}H) coupling constants are generally less than 2 Hz in magnitude; only where there are sp^2-hybridized C and/or N atoms, do the magnitudes reach 3 to 12 Hz. In compounds with C=N double bonds, the coupling constant can be as large as −16 Hz (**Table 3.63**). The sign of the ^2J(^{15}N,^{1}H) coupling can be positive or negative. The same applies to ^3J(^{15}N,^{1}H) couplings and long-range couplings ^nJ(^{15}N,^{1}H). The latter only have significant values when there are multiple coupling pathways.

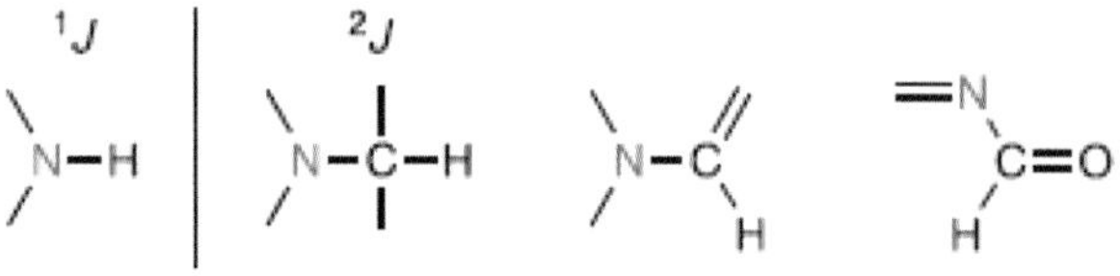

^1J | ^2J

(structural diagrams: N–H ; N–C–H ; N–C with H ; ≡N ... C=O with H)

Couplings between ^{15}N and ^{13}C are difficult to measure without isotopic enrichment. Values of ^1J(^{15}N,^{13}C) are generally less than 20 Hz. The sign can be positive or negative. If the value of ^1J is close to zero, it can be exceeded by ^2J or ^3J couplings. Some examples of known ^nJ(^{15}N,^{13}C) couplings are given in **Table 3.64**.

^{15}N,^{15}N couplings will not be treated here. They can only be measured for singly or doubly labeled compounds and are not important for structural purposes.

Many 2D techniques of ^{13}C,^{1}H NMR find analogies in ^{15}N,^{1}H NMR. **Fig. 3.126** presents a schematic **(^{15}N,^{1}H)HSQC** spectrum of an oligopeptide, which contains a glycine and an alanine building block. The two crosspeaks indicate ^1J (N,H) couplings. The differentiation of the signals is easy, because the ^3J (^{1}H,^{1}H) coupling

Table 3.64 $^n J$ (^{15}N,^{13}C) coupling constants (n = 1, 2, and 3) of selected compounds in Hz. (If the sign is not known, the magnitude is given)

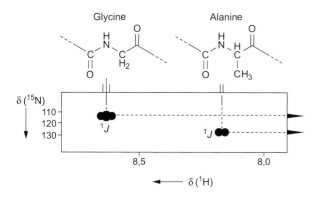

in the ^{1}H NMR spectrum gives a triplet of glycine and a doublet of alanine.

Fig. 3.127 depicts an example of a (**^{15}N,^{1}H)HMBC** spectrum. The heterocyclic compound **400** gives nine crosspeaks. The $^1 J$ couplings for H–N-2 and H–N-4 are easy to identify. N-4 shows additionally $^3 J$ couplings to 2-H, 5-CH$_2$, and 5-CH$_3$. The CH$_2$ group gives the AB part of an ABX$_3$ spin pattern in the ^{1}H NMR spectrum. The two methylene protons are diastereotopic, because C-5 is a chiral center. N-2 shows only a $^3 J$ coupling to 4-H. All other protons are separated from N-2 by too many bonds. Consequently, the third series of crosspeaks must belong to N-1, which possesses a $^2 J$ coupling to 2-H and a $^3 J$ coupling to 6-CH$_3$.

3.6.4 Complete Assignment of the NMR Signals of a Compound Containing ^{1}H, ^{13}C, ^{15}N, …

A complete assignment of all accessible NMR signals can be recommended, whenever a difficult structure determination has to be solved. Apart from ^{1}H and ^{13}C, ^{15}N nuclei are often important for organic scaffolds. Normally such an assignment can be performed on the basis of (^{1}H,^{1}H)COSY, (^{13}C,^{1}H)HSQC, (^{13}C,^{1}H)HMBC, and (^{15}N,^{1}H)HMBC/HSQC spectra. In special cases additional NOESY or ROESY spectra are necessary.

The pyrazolo[1,5-a]pyrimidine **401** shall be discussed here as an example. The anellation of a pyrimidine and a pyrazole ring and the substitution positions of the two phenyl groups can be unambiguously determined by the HSQC and the two HMBC measurements mentioned earlier. The (^{15}N,^{1}H)HMBC spectrum is depicted in **Fig. 3.128**. All four ^{15}N nuclei deliver crosspeaks, whereby the size of the coupling constants becomes obvious: $|^1 J| > |^2 J| ≈ |^3 J| > |^4 J|$.

The protons of the amino group show a $^3 J$ coupling to the nodal atom ^{15}N-7a (in addition to the $^1 J$ coupling). Moreover, ^{15}N-7a exhibits a $^3 J$ coupling to 2-H and a weak $^4 J$ coupling to 5-H. N-4 couples with 5-H ($^3 J$) and with 2-H ($^4 J$). N-1 couples with 2-H ($^2 J$). The ^{15}N chemical shift of the amino group is at high field, whereas N-1, N-4, and N-7a give low-field signals. All three N atoms are included in the heteroaromatic ring system. The full assignment of all ^{1}H, ^{13}C, and ^{15}N signals is indicated in the formula.

Fig. 3.126 Partial view of a schematic (^{15}N,^{1}H)HSQC spectrum of an oligopeptide, which contains a glycine and an alanine building block [(δ (^{1}H) values relative to TMS; δ (^{15}N) values relative to NH$_3$(liquid)].

401

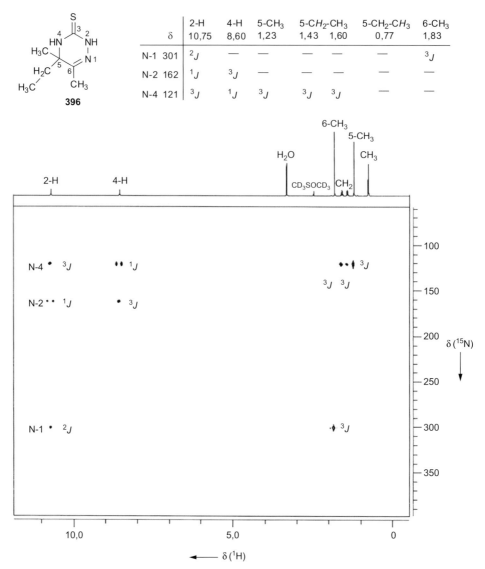

	2-H	4-H	5-CH₃	5-C*H₂*-CH₃	5-CH₂-C*H₃*	6-CH₃
δ	10,75	8,60	1,23	1,43 1,60	0,77	1,83
N-1 301	^2J	—	—	—	—	^3J
N-2 162	^1J	^3J	—	—	—	—
N-4 121	^3J	^1J	^3J	^3J	^3J	—

Fig. 3.127 (^{15}N,^{1}H)HMBC spectrum of the 1,2,4-triazine derivative **400**. [Measurement in CD₃SOCD₃, δ (^{1}H) values relative to TMS, δ (^{15}N) values relative to NH₃ (liquid)], (Safarov S, Pulatov E, Kukaniev MH, Kolshorn H, Meier H. 2009, J. Heterocycl. Chem. **46**, 552).

(−)Nicotine (**402**) represents a further example for a complete assignment of all ^{1}H, ^{13}C, and ^{15}N chemical shifts. The results based on (^{1}H,^{1}H)COSY, NOESY, (^{13}C,^{1}H)HSQC, and (^{15}N,^{1}H)HSQC spectra are shown in the formula.

3.6.5 Other Nuclei

Other nuclei with a spin quantum number 1/2 are ^{29}Si, ^{77}Se, ^{117}Sn, ^{119}Sn, ^{195}Pt, ^{199}Hg, ^{203}Tl, ^{205}Tl, and ^{207}Pb.

A combination of the natural abundance and the magnitude of the magnetic moment leads to the following order for the relative sensitivities:

$$^{205}Tl > {}^{203}Tl > {}^{119}Sn > {}^{195}Pt > {}^{207}Pb > {}^{117}Sn > {}^{199}Hg > {}^{77}Se > {}^{29}Si > {}^{13}C$$

Among nuclei with spin quantum numbers $I > 1/2$, the most important are ^{2}H, ^{7}Li, ^{11}B, ^{14}N, ^{17}O, and ^{33}S. Their nuclear quadrupole moments often cause extreme line broadenings.

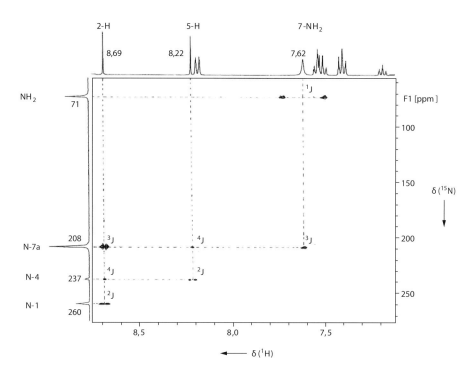

Fig. 3.128 (^{15}N,^{1}H)HMBC spectrum of 7-amino-3,6-diphenylpyrazolo[1,5-*a*]pyrimidine (**401**) in CD$_3$SOCD$_3$ (Anwar HF, Fleita DH, Kolshorn H, Meier H, Elnagdi MH. 2006, Arkivoc 133)

Literature

General Works

Akitt, J. W. (2000), N. M. R. and Chemistry, An Introduction to Nuclear Magnetic Resonance Spectroscopy, CRC Press, Cleveland.

Bakhmutov, V. I. (2005), Practical NMR Relaxation for Chemists, J. Wiley & Sons, New York.

Balci, M. (2005), Basic 1H- and 13C-NMR Spectroscopy, Elsevier, Amsterdam.

Berger, S., Braun, S. (2004), 200 and More NMR Experiments, Wiley-VCH, Weinheim.

Blümich, B. (2005), Essential NMR, Springer, Berlin.

Blümich, B., Haber-Pohlmeier, S., Zia, W. (2014), Compact NMR, Walter de Gruyter, Berlin.

Breitmaier, E. (2002) Structure Elucidation by NMR in Organic Chemistry: A Practical Guide, J. Wiley & Sons, New York.

Brey, W. S. (1996), Magnetic Resonance in Perspective, Highlights of a Quarter Century, Academic Press, San Diego.

Bruch, M. D. (1996), NMR Spectroscopy Techniques, M. Dekker, New York.

Canet, D. (1994), NMR-Konzepte und Methoden, Springer, Heidelberg.

Claridge, T. D. W. (2008), High-Resolution NMR Techniques in Organic Chemistry, Pergamon, Amsterdam.

Contreras, R.H. (2013), High Resolution NMR Spectroscopy: Understanding Molecules and Their Electronic Structures, Elsevier, Amsterdam.

Ernst, R., Bodenhausen, G., Wokaun, A. (1990), Principles of Nuclear Magnetic Resonance in One and Two Dimensions, Clarendon Press, London.

Findeisen, M., Berger, S. (2013), 50 and More Essential NMR Experiments: A Detailed Guide, Wiley, Hoboken.

Freeman, R. (2003), Magnetic Resonance in Chemistry and Medicine, Oxford University Press, Oxford.

Freeman, R. (1998), Spin Choreography: Basic Steps in High Resolution NMR, Oxford University Press, Oxford.

Friebolin, H. (2005), Basic One- and Two-Dimensional NMR Spectroscopy, Wiley-VCH, Weinheim.

Günther, H. (1991), NMR-Spektroskopie, Georg Thieme Verlag, Stuttgart.

Heise, H., Matthews, S. (2013), Modern NMR Methodology, Springer: Top. Curr. Chem. 335, Berlin.

Herzog, W.-D., Messerschmidt, M. (1995), NMR-Spektroskopie für Anwender, VCH, Weinheim.

Hore, P. J., Jones, J. A., Wimperis, S. (2000), NMR, Oxford University Press, Oxford.

Jacobsen, N. E. (2007), NMR Spectroscopy Explained, J.Wiley & Sons, New York.

Kalinowski, H.-O., Berger, S., Braun, S. (1984), 13C-NMR-Spektroskopie, Georg Thieme Verlag, Stuttgart.

Keeler, J. (2010), Understanding NMR Spectroscopy, J. Wiley & Sons, New York.

McDermott, A. E., Polenova, T. (2010), Solid State NMR Studies of Biopolymers, J. Wiley & Sons, Chichester.

Mitchell, T. N., Costisella, B. (2007), NMR – From Spectra to Structures, Springer, Berlin.

Nakanishi, K. (1990), One-dimensional and Two-dimensional NMR Spectra by Modern Pulse Techniques, University Science Books, London.

Nelson, J. H. (2003), Nuclear Magnetic Resonance Spectroscopy, Prentice-Hall, Princton.

Pihlaja, K., Kleinpeter, E. (1994), Carbon – 13 NMR Chemical Shifts in Structural and Stereochemical Analysis, VCH, Weinheim.

Rao, D. K. (2014), Nuclear Magnetic Resonance (NMR): Theory, Applications and Technology, CRC Press, Boca Raton.

Richards, S., Hollerton, J. (2011), Essential Practical NMR for Organic Chemistry, Wiley, Weinheim.

Roberts, J. D. (2000), ABC of FT-NMR, Freeman, Houndmills.

Sanders, J. K. M., Hunter, B. K. (1994), Modern NMR Spectroscopy, Oxford University Press, Oxford.

Slichter, C. P. (1990), Principles of Magnetic Resonance, Springer-Verlag, Berlin.

Trepalin, S., Yarkov, A. (2004), Proton NMR of Organic Compounds 1-3, Wiley, New York.

Van de Ven, F. J. M. (1995), Multidimensional NMR in Liquids, VCH, Weinheim.

Webb, G. A. (2006), Modern Magnetic Resonance, Vol. 1 – 3, Springer Netherlands, Dordrecht.

Wrackmeyer, B. (2002), Guide to Multinuclear Magnetic Resonance, J. Wiley & Sons, Chichester.

Zerbe, O., Jurt, S. (2013), Applied NMR Spectroscopy for Chemists and Life Scientists, Wiley, Hoboken.

Special Methods and Effects

Abraham, R., Mobli, M. (2008), Modelling ^{1}H NMR Spectra of Organic Compounds, J. Wiley, New York.

Albert, K. (2002), On-line LC-NMR and Related Techniques, J. Wiley & Sons, New York.

Bakhmutov, V. I. (2011), Solid-State NMR in Materials Science: Principles and Applications, CRC Press, Boca Raton.

Bakhmutov, V. (2004), Practical Nuclear Magnetic Resonance Relaxation for Chemists, J. Wiley & Sons, Chichester.

Bargon, J., Kuhn, L. T. (2007), In situ NMR Methods in Catalysis, Springer, Berlin.

Bigler, P. (1997), NMR Spectroscopy: Processing Strategies, VCH, Weinheim.

Burnell, E. E., de Lange, C. A. (2003), NMR of Ordered Liquids, Kluwer Academic Publishers, Dordrecht.

Chan, J. C. (2012), Solid State NMR, Springer: Top. Curr. Chem. 306, Berlin.

Delpuech, J. J. (1995), Dynamics of Solutions and Fluid Mixtures by NMR, Wiley, New York.

Duer, M.J. (2004), Introduction to Solid-State NMR Spectroscopy, Blackwell, Oxford.

Duer, M.J. (2002), Solid State NMR Spectroscopy: Principles and Applications, Blackwell, Oxford.

Freeman, R., Hill, H. D. W. (1975), Dynamic Nuclear Magnetic Resonance Spectroscopy, Academic Press, New York, London.

Fyfe, C. A. (1983), Solid State NMR for Chemists, C. F. C. Press, Ontario.

Gadian, D. G. (1981), Nuclear Magnetic Resonance and Its Applications to Living Systems, Oxford University Press, Oxford.

Gonella, N. C. (2013), LC – NMR: Expanding the Limits of Structure Elucidation, CRC Press, Boca Raton.

Hägele, G., Engelhardt, M., Boenigk, W. (1987), Simulation und automatisierte Analyse von Kernresonanzspektren, VCH, Weinheim.

Harris, R. K. (2010), NMR Crystallography, J. Wiley, New York.

Jackman, L. M., Cotton, F. A. (1975), Dynamic Nuclear Magnetic Resonance Spectroscopy, Academic Press, New York, London.

Kaplan, J. I., Fraenkel, G. (1980), NMR of Chemically Exchanging Systems, Academic Press, New York, London.

Lepley, A. R., Closs, G. L. (1973), Chemically Induced Magnetic Polarization, Wiley, New York.

Levitt, M. (2008), Spin Dynamics, Wiley, New York.

Marshall, J. L. (1983), Carbon-Carbon and Carbon-Proton NMR Couplings, Verlag Chemie, Weinheim, Deerfield Beach, Florida, Basel.

Mehring, M. (1983), Principles of Resolution NMR in Solids, Springer-Verlag, Berlin.

Morrill, T. C. (1987), Lanthanide Shift Reagents in Stereochemical Analysis, VCH, Weinheim.

Morris, G. (2010), Multidimensional NMR Methods for the Solution State, Wiley, New York.

Neuhaus, D., Williamson, M. P. (1989), The Nuclear Overhauser Effect in Structural and Conformational Analysis, VCH, Weinheim.

Oki, M. (1985), Applications of Dynamic NMR Spectroscopy to Organic Chemistry, Verlag Chemie, Weinheim, Deerfield Beach, Florida, Basel.

Sandström, J. (1982), Dynamic NMR Spectroscopy, Academic Press, New York.

Schmidt-Rohr, K., Spiess, H. W. (1994), Multidimensional Solid-State NMR and Polymers, Academic Press, New York.

Schorn, C. (2001), NMR-Spectroscopy: Data Acquisition, Wiley-VCH, Weinheim.

Sievers, R. E. (1973), Nuclear Magnetic Shift Reagents, Academic Press, New York, London.

Silva Elipe, M. V. (2011), LC – NMR and Other Hyphenated NMR Techniques: Overview and Applications, Wiley, Hoboken.

Takeuchi, Y., Marchand, A. P. (1986), Applications of NMR Spectroscopy to Problems in Stereochemistry and Conformational Analysis, Verlag Chemie, Weinheim, Deerfield Beach, Florida, Basel.

Special Compound Classes, Applications

Alberts, K. (2002), On-line LC-NMR and Related Techniques, J. Wiley & Sons, New York.

Bargon, J., Kuhn, L. T. (2007), In situ NMR Methods in Catalysis, Springer, Berlin.

Batterham, T. J. (1973), NMR Spectra of Simple Heterocycles, Wiley, New York.

Berger, S., Sicker, D. (2009), Classics in Spectroscopy, Isolation and Structure Elucidation of Natural Products, Wiley-VCH, Weinheim.

Bertini, I., McGreevy, K.S., Parigi, G. (2012), NMR of Biomolecules: Towards Mechanistic Systems Biology, Wiley-Blackwell, Weinheim.

Bertini, I., Molinari, H., Niccolai, N. (1991). NMR and Biomolecular Structure, VCH, Weinheim.

Bovey, F., Mirau, P. (1996), NMR of Polymers, Academic Press, London.

Burnell, E. E., de Lange, C. A. (2003), NMR of Ordered Liquids, Kluwer Academic Publishers, Dordrecht.

Cavanagh, J. (2007), Protein NMR Spectroscopy, Academic Press, New York, London.

Chamberlain, N. F., Reed, J. J. R. (1971), Nuclear-Magnetic-Resonance Data of Sulfur Compounds, Wiley-Interscience, New York.

Cheng, H. N., English, A. D. (2002), NMR Spectroscopy of Polymers in Solution and in the Solid State, Oxford University Press, Oxford.

de Certaines, J. D., Bovée, W. M. M. J., Podo, F. (1992), Magnetic Resonance Spectroscopy in Biology and Medicine, Pergamon Press, Oxford.

de Graaf, R. A. (2007), In Vivo NMR Spectroscopy, J. Wiley & Sons, New York.

Downing, K. A. (2004), Protein NMR Techniques, Springer, Berlin.

Dwek, R. A. (1977), Nuclear Magnetic Resonance in Biochemistry, Clarendon Press, London.

Gielen, M., Willem, R., Wrackmeyer, B. (1996), Advanced Applications of NMR to Organometallic Chemistry, Wiley, New York.

Hatada, K. (2004), NMR Spectroscopy of Polymers, Springer, Berlin.

Holzgrabe, U., Wawer, I., Diehl, B. (2008), NMR Spectroscopy in Pharmaceutical Analysis, Elsevier, Oxford.

Holzgrabe, U., Wawer, I., Diehl, B. (1999), NMR Spectroscopy in Drug Development and Analysis, Wiley-VCH, Weinheim.

Ibett, R. N. (1993), NMR Spectroscopy of Polymers, Chapman and Hall, London.

Iggo, J. A. (2000), NMR Spectroscopy in Inorganic Chemistry, Oxford University Press, Oxford.

Jiminez-Barbero, J. (2002), NMR of Glycoconjugates, J. Wiley & Sons, Chichester.

Jue, T. (2006), NMR in Biomedicine: Basic and Experimental Principles, Humana Totowa, USA.

Komoroski, R. A. (1986), High Resolution NMR Spectroscopy of Synthetic Polymers in Bulk, Verlag Chemie, Weinheim, Deerfield Beach, Florida, Basel.

LaMar, G. N., Horrocks, W. D., Holm, R. H. (1973), NMR of Paramagnetic Molecules, Academic Press, New York, London.

Lian, L.-Y., Roberts, G. (2011), Protein NMR Spectroscopy: Practical Techniques and Applications, J. Wiley & Sons, Chichester.

Mackenzie, K.J.D. (2002), Multinuclear Solid-State Nuclear Magnetic Resonance of Inorganic Materials, Pergamon, New York.

Mann, B. E., Taylor, B. F. (1981), ^{13}C-NMR Data of Organometallic Compounds, Academic Press, New York, London.

Marchand, A. P. (1983), Stereochemical Applications of NMR Studies in Rigid Bicyclic Systems, Verlag Chemie, Weinheim.

Pons, M. (1999), NMR in Supramolecular Chemistry, Springer Netherlands, Dordrecht.

Pregosin, P.S. (2008), Special Issue: applications of NMR to inorganic and organometallic chemistry, Elsevier, Amsterdam.

Pregosin, P.S. (2012), NMR in Organometallic Chemistry, Wiley-VCH, Weinheim.

Rabidean, P. (1989), The Conformational Analysis of Cyclohexenes, Cyclohexadienes and Related Hydroaromatic Compounds, VCH, Weinheim.

Ramamoorthy, A. (2006), NMR Spectroscopy of Biological Solids, CRC Press, Boca Raton.

Reid, D. G. (1996), Protein NMR Techniques, Chapman and Hall, London.

Roberts, G. C. K. (1993), NMR of Macromolecules, IRL Press, Oxford.

Rule, G. S., Hitchens, T. K. (2005), Fundamentals of Protein NMR Spectroscopy, Springer Netherlands, Dordrecht.

Schmidt-Rohr, K., Spiess, H. W. (1994), Multidimensional Solid-State NMR and Polymers, Academic Press, New York.

Tonelli, A. E. (1989), NMR Spectroscopy and Polymer Microstructure, VCH, Weinheim.

Wenzel, T. J. (2007), Discrimination of Chiral Compounds Using NMR Spectroscopy, J. Wiley & Sons, New York.

Whitesell, J. K., Minton, M. A. (1987), Stereochemical Analysis of Alicyclic Compounds by C-13 Nuclear Magnetic Resonance Spectroscopy, Chapman and Hall, London.

Wüthrich, K. (1986), NMR of Proteins and Nucleic Acids, J. Wiley & Sons, New York.

Zerbe, O. (2003), Bio-NMR in Drug Research, Wiley-VCH, Weinheim.

Catalogues

Pham, Q. T., Péraud, R., Waton, H., Llauro-Darricades, M.-F. (2002), Proton and Carbon NMR Spectra of Polymers, Wiley, New York.

Nakanishi, K. (1980), One-dimensional and Two-dimensional NMR Spectra by Modern Pulse Techniques, W. H. Freeman & Co, Oxford.

Bremser, W., Franke, B., Wagner, H. (1982), Chemical Shift Ranges in Carbon-13 NMR Spectroscopy, Verlag Chemie, Weinheim.

Bremser, W., Ernst, L., Franke, B., Gerhards, R., Hardt, A. (1981), Carbon-13 NMR Spectral Data, Verlag Chemie, Weinheim.

Sasaki, S., Handbook of Proton-NMR Spectra and Data, Vol 1–10 and Index, Academic Press, London.

Ault, A., Ault, M. R. (1980), A Handy and Systematic Catalog of NMR Spectra, University Science Books, Mill Valley.

Brügel, W. (1979), Handbook of NMR Spectral Parameters, Heyden, London.

Breitmaier, E., Haas, G., Voelter, W. (1975, 1979), Atlas of Carbon-13 NMR Data, 2 Vol., Heyden, London.

Pouchert, C. J., Campbell, J. R. (1974), The Aldrich Library of NMR Spectra, 3 Bde., Aldrich Chemical Comp., Milwaukee.

Johnson, L. F., Jankowski, W. C. (1972), Carbon-13 NMR Spectra, A Collection of Assigned Coded and Indexed Spectra, Wiley, New York.

Bovey, F. A. (1967), NMR Data Tables for Organic Compounds, Wiley-Interscience, New York.

Simons, W. W. (1967), The Sadtler Handbook of Proton NMR Spectra, Sadtler Research Laboratories, Philadelphia.

Hershenson, H. M. (1965), NMR and ESR Spectra Index, Academic Press, New York.

Howell, M. G., Kende, A. S., Webb, J. S. (1965), Formula Index to NMR Literature Data I, II, Plenum Press, New York.

Bhacca, N. S., Johnson, L. F., Shoolery, J. N. (1962/63), High Resolution NMR Spectra Catalogue, I, II, Varian Associates, Palo Alto.

Serials, Periodicals

Advances in Magnetic Resonance

Advances in Magnetic and Optical Resonance

Analytical Chemistry Annual Reviews: NMR Spectroscopy

Annual Reports on NMR Spectroscopy

Annual Review of NMR Spectroscopy

Applications of NMR Spectroscopy

Applied Magnetic Resonance

Biological Magnetic Resonances

BioMagnetic Research and Technology

Bulletin of Magnetic Resonance

Canadian NMR Research News

Nuclear Magnetic Resonance Spectroscopy

Chemical Abstracts Selects: Carbon and Heteroatom NMR
Chinese Journal of Magnetic Resonance
Concepts in Magnetic Resonance, Part A, B
Encyclopedia of NMR
Encyclopedia of Nuclear Magnetic Resonance
Journal of Biomolecular NMR
Journal of Magnetic Resonance, Series A and B
Magnetic Resonance in Chemistry (früher: Organic Magnetic Resonance
Magnetic Resonance Review

Magnetic Resonance in Solids
Magnetic Resonance Materials in Physics, Biology and Medicine
Nuclear Magnetic Resonance Abstracts and Index
NMR, Basic Principles, Progress
(Grundlagen und Fortschritte)
Nuclear Magnetic Resonance, Specialist Periodical Report
Progress in NMR Spectroscopy
Solid State Nuclear Magnetic Resonance
The Open Magnetic Resonance Journal

Stefan Bienz, Laurent Bigler

4 Mass Spectrometry

4 Mass Spectrometry

4.1 Introduction

Although the method of mass spectrometry (MS) is relatively old—already in 1910, J. J. Thomson was able to separate the ^{20}Ne and ^{22}Ne isotopes—it did not achieve recognition as an important analytical method in organic chemistry until 1960. Two features have helped to bring it to prominence. Firstly, with MS it is possible to determine the relative molecular mass and even the elemental composition of a compound using only the smallest amount of substance. Furthermore, fragmentation patterns, i.e., the decomposition of the material being analyzed under the influence of electron bombardment or other techniques, depicted in mass spectra allow one to make important deductions about the structures of the investigated compounds. Both of these aspects have been—and still are—crucial to the development and the application of this method.

There are limits to the mass spectrometric determination of relative molecular masses. These limits used to be at rather low mass numbers because in the early days of MS it was necessary to transfer the analytes by evaporation into the gas phase. This was possible only for rather volatile and slightly polar compounds. The larger a compound is, the greater, in general, is the number of functional groups and therewith the danger of thermal decomposition upon vaporization. Additionally, highly functionalized compounds—if they reached the mass analyzer still intact—usually showed such strong fragmentation upon electron ionization (EI), the method used at that time, that reliable information about their molecular masses was often inaccessible. To solve this problem, various procedures have been developed that allow also the ionization and transfer of rather large and highly functionalized nonvolatile molecules into the gas phase (e.g., electrospray ionization (ESI) or matrix-assisted laser desorption ionization (MALDI), Section 4.3.2), which makes the mass spectrometric determination of masses of many more analytes possible. Modern ionization techniques allow, e.g., the investigation of biopolymers such as proteins or nucleic acids with masses of 1,000,000 u or above directly from solution. This opens previously undreamt and nowadays routinely applied possibilities for research in emergent areas such as functional genomics, functional proteomics, and metabolomics.

These days, sophisticated new methods also allow us to directly analyze molecules from surfaces or within cells. With the possibility to separate and select particles according to their masses in the spectrometer, components of complex mixtures can often be analyzed independently directly from the matrix without the need to previously separate them physically. These recent developments led to a new revival of MS, which was considered to be largely exhausted in the 1980s.

Also, the second aspect of MS, the utilization of the highly informative data that can be gained from fragmentation reactions, is also gaining increasing attention. Fragmentation reactions were extensively studied in the golden age of EI-MS (Section 4.3.2, and 4.4.3), and many reference spectra were deposited in databases. As a result, we currently know quite a lot about the behavior of radical cations in mass spectrometers, and we can readily identify compounds by their mass spectrometric "fingerprint." With the advent of the modern and mild ionization methods, however, the focus of MS was shifted more to the determination of molecular masses, and the investigation of fragmentation reactions was somewhat neglected. This era now belongs to the past, and the value of fragmentation reactions for the structural elucidation of molecules is again appropriately recognized. Modern MS spectra are, however, much less comparable with each other and, thus, less well searchable in MS databases than standardized EI-MS.

This shortcoming, however, is well compensated for by the technical advances that have been attained with respect to sensitivity, resolution, and mass accuracy of the instruments: Modern mass spectrometers allow the measurements of masses with resolutions of above 40,000 and accuracies of <1 ppm! This, together with elaborate data-processing engines, allows the reliable determination of chemical formulas of molecules and fragments with masses of up to ca. 1,000 u. (Section 4.4.2). With the high sensitivities and reproducibilities of the methods available these days, MS becomes also an important method for accurate quantification of analytes that may even occur in small amounts only and within complex matrices.

We will give you in the following sections an introduction to the most important aspects of MS and will provide you with tools and tables for your everyday use. While the next section (Section 4.2) focuses on fundamentals and basic aspects of MS and, thus, is still rather theoretical in its presentation, the subsequent sections will cover particularly the practical and directly applicable aspects of the method. The diverse ionization methods and mass analyzer systems are in the focus of Section 4.3, as well as the question of which method is best suited to solve which problem. Section 4.4 presents concrete possibilities and options provided by MS to be used for the structural elucidation of molecules, Section 4.5 gives advice regarding appropriate sample preparation, and Section 4.6 draws attention to important problems that often arise in connection with MS and may lead to erroneous interpretations of spectra. Focus is particularly given there to the problem of formation and significance of artifacts. A number of tables that compile important data for mass spectrometric applications are finally provided for your convenience in Section 4.7.

4.2 General Aspects of Mass Spectrometry

4.2.1 The Principle of Mass Spectrometry

Charged particles in the gas phase can be accelerated and deflected, e.g., by a homogeneous magnetic field, to separate them according to their mass. The technical realization of this endeavor, however, is not trivial, but the detailed knowledge of the involved processes and their technical realization is not crucial for a preparative working user. We thus explain the methods and instrumentations just to the extent that is required for a practitioner to understand the method sufficiently well for his practical work.

The principle of MS is exemplarily explained with the schematic representation shown in **Fig. 4.1**. As is seen in the illustration, a mass spectrometer can be subdivided into four different functional units: the sample introduction (inlet), ion generation (ion source), ion separation and detection (mass analyzer), and data processing. An additional unit that may be used to initiate fragmentations or other reactions can be inserted in between the sections of ion formation and ion separation.

Sample introduction and ion generation are often inextricably linked and, thus, not independent of each other. Vaporizable samples can, e.g., be directly submitted to an instrument or introduced via a gas inlet. They may be subsequently—independently of the inlet system—converted to cations or anions by ionization methods such as EI or chemical ionization (CI). On the other hand, nonvaporizable samples are usually fed to the instrument from solution or from a fixed matrix with simultaneous ionization, e.g., by spraying, desorption, or laser activation. Common for all methods, however, is the fact that analytes have to be ionized and transferred into the gas phase prior to their entry into the analyzer section of the instruments,

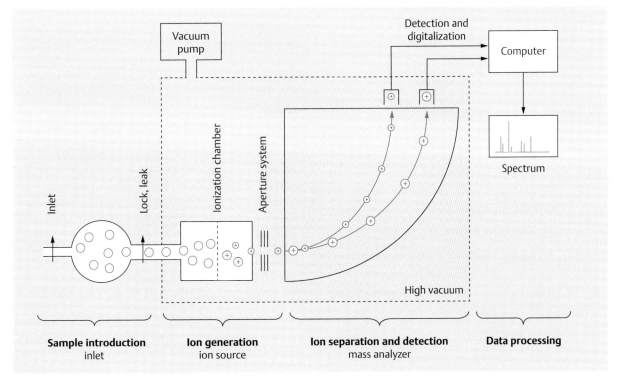

Fig. 4.1 Schematic representation of a mass spectrometer.

where mass separation takes place. You may learn more about the different ionization methods and the respective compatible inlet systems in Section 4.3.

The processes in the analyzer unit of the mass spectrometer (mass separation and ion detection) in all cases take place under high vacuum, typically at ca. 10^{-8} Pa. This prevents undesirable collisions of ions, molecules, and atoms. The transfer of the analytes from ambient pressure to the high vacuum portion of the analyzer represents one of the major problems in MS. Several possibilities are available for this purpose, depending on the nature of the inlet system and the ionization method used. Detailed information on several analyzer systems can also be found in Section 4.3.

More relevant than profound knowledge about the analyzer systems is for chemists to know that the ions are actually not separated and detected according to their mass but rather according to their ratio of mass to charge (m/z, given in mass units of u^a). In the example of the magnetic field spectrometer shown in **Fig. 4.1**, the ions to be analyzed possess, depending on their mass and charge, different but, through the experimental parameters, well-defined velocities. The separation of the ions in the analyzer unit is then realized by directing the ion beam through a strong magnetic field, which effects a stronger deflection of lighter and more highly charged ions as compared to heavier and less highly charged particles. In this case, the squares of the turning radius are directly proportional to the m/z values of the ions.

Hence, a signal of a mass spectrum does not exhibit directly the mass of the detected ions! Multiply charged ions—which are more often observed with the modern MS methods and with larger molecules—first need to be recognized as such, and the respective detected signals (in m/z) must be converted to the "effective" masses by calculation. Multiply charged ions can be easily recognized by their isotopic signals or, if an ensemble of differently charged ions of the same sort is detected (e.g., with peptides, nucleic acids, or other polymers), directly by the pattern of the respective signal distribution.

4.2.2   The Mass Spectrum

Terminology in Mass Spectrometry

The special terminology used in MS is introduced and illustrated with the two mass spectra shown in **Fig. 4.2**.

Spectrum. A *mass spectrum* is a two-dimensional representation in which the abundance of ions detected in the

---

[a] The unit u (*unified atomic mass unit*) is equivalent to the unit Dalton (Da), defined as 1/12 of the mass of a ^{12}C atom in its ground state ($= 1.660538782(83) \times 10^{-27}$ kg). The unit amu (*atomic mass unit*), which is often used as another equivalent to u, is obsolete and rejected by IUPAC.

spectrometer (signal intensity, ordinate) is plotted versus the respective m/z values (abscissa). The spectrum can be recorded as a *profile spectrum* (**Fig. 4.2**, spectrum (a)) or a *centroid spectrum* (**Fig. 4.2**, spectrum (b)). The profile spectrum corresponds to the continuous writing of the detected ion abundance and thus represents the complete set of the original data, while the centroid spectrum is a bar graph representation of the peak centroids, calculated from the original data. The m/z values given in the centroid spectra thus correspond to the m/z values of peak maxima of the related signals in the profile spectra—sometimes also to the rounded nominal masses (see below). The bar lengths represent the areas below these peaks. Profile spectra have the advantage that they show the shapes of the individual signals, which can be highly important in certain cases—in particular in connection with high-resolution MS (HR-MS). Centroid spectra, on the other hand, are more clear and less heavy on data, which is the reason why they are often preferred.

Intensity and Mass Scale. The *signal intensities* in mass spectra are usually given in relative intensities (rel. int. (%), **Fig. 4.2**, spectrum (b)). The scale refers then to the signal of highest intensity in the spectrum, called the *base peak* ($= 100$ rel.%; signals at $m/z = 442$ and 118 in the spectra of **Fig. 4.2**). Alternatively, the signal intensities can also be given in absolute values of the measured ion currents after amplification (**Fig. 4.2**, spectrum (a)). The use of absolute values is particularly required for quantitative measurements, but it is also often helpful when measurements close to the instrumental limits are performed (e.g., with analytes of very high or very low concentrations). The scaling of the abscissa in a mass spectrum corresponds to values of m/z with m being the mass of the particles in u and z the number of the elementary charges located on these units.

Ion Types. Three types of ions are distinguished in MS: (i) molecular ions and other ions for which the molecular composition of the analyte is retained, (ii) fragment ions, which represent just partial structures of the original molecules, and (iii) ions that can be regarded as components of analyte salts. *Molecular ions* are positively or negatively charged ions resulting from the removal or addition of electrons from or to the intact sample molecules. They comprise the complete structural entity of the analyte (e.g., $M^{+\bullet}$ of **2**, registered at $m/z = 136$ in spectrum (b), **Fig. 4.2**) and thus allow direct conclusions regarding the molecular mass of the analyzed compound. Radical cationic molecular ions are formed, e.g., by the exposure of vaporized analyte molecules to a beam of high-energetic electrons (usually 70 eV in EI-MS, Section 4.3.2). Like molecular ions, *adduct ions* also comprise the full composition of the analyte's molecule. *Cationized* and *anionized molecules* are usually formed under mild ionization conditions: cations by the addition of one or several cationic species to the neutral molecules (e.g., $+ H^+$, $+ Na^+$, $+ NH_4^+$, etc.) and anions by deprotonation or by the addition of anionic species (e.g., + Cl⁻, + $MeCO_2^-$, etc.). Such ionized molecules used to be called "*quasi-molecular ions*," which is a term no longer accepted. Signals for cationized molecules **1** of the types $[M + H]^+$ and $[M + Na]^+$ are found, e.g., in spectrum (a) of **Fig. 4.2** at m/z = 420 and 442.

Adduct ions are cations or anions that contain one or more neutral species (e.g., MeOH, H_2O, MeCN, NH_3, NaI, etc.) in addition to the charge-carrying particle. Such ions are, e.g., $[M + H + MeOH]^+$ and $[M + Na + MeOH]^+$ that are registered at m/z = 452 and 474 in spectrum (a) of compound **1** in **Fig. 4.2**.

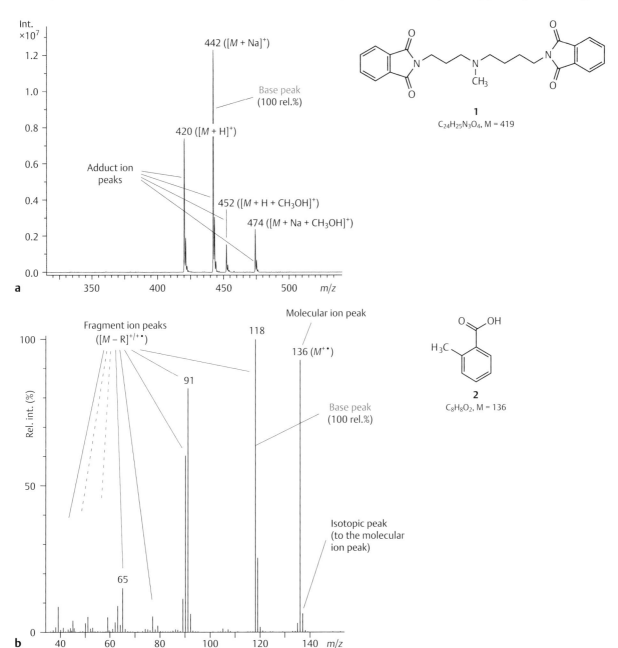

1

$C_{24}H_{25}N_3O_4$, M = 419

2

$C_8H_8O_2$, M = 136

Fig. 4.2 (a) ESI-MS of triamine derivative **1** (dissolved in MeOH) recorded as a profile spectrum with intensities given in absolute units (ion current after amplification), **(b)** EI-MS of 2-methylbenzoic acid (**2**) as a centroid spectrum with intensities given in units relative to the base peak (rel. int. (%) or rel.%).

The neutral species found in the adduct ions most often stem from the solvent, as in the case shown, from buffers, matrices, or reactant gases.

In most cases, the signal at the highest mass number of a spectrum represents the peak of the molecular ion or of an adduct ion. If the ion type is known (charge, way of ion formation, and presence and nature of added neutral or cationizing/anionizing species), the mass of the analyte molecules can be readily determined.

Fragment ions are ions that are formed in a mass spectrometer from *precursor ions* (which might be molecular ions, adduct ions, or fragment ions themselves) directly or via intermediary species by the loss of molecular portions (e.g., the ions derived from **2** that are registered at $m/z = 118, 91$, and 65 in spectrum (b) of **Fig. 4.2** are fragment ions). Evidently, fragment ions possess a mass that is lower than the mass of the intact analyte molecules; for this reason they do not allow any strong conclusion concerning the molecular mass of an analyte. Fragment ions, however, are still of great benefit because they are often formed along comprehensible chemical paths and thus may deliver detailed information about structural units contained in a molecule.

A special class of ions represent cationic or anionic molecular groups that are charged components of salts and are transferred during the ionization process directly into the gas phase and the spectrometer devoid of their counter ions. Like fragment ions, such ions do not allow any conclusions about the mass of the effectively investigated compounds, because the mass of the counter ions remains fully hidden when the measurements are performed in one polarity mode only. An imminent problem with such ions is also the fact that they can get easily confused with adduct ions (e.g., with $[M + H]^+$ ions), which could lead to serious misinterpretations of the spectra.

The Term "Mass" in Mass Spectrometry

Molecular Mass (M_R) Most elements occur in nature as mixtures of different isotopes in well-known, defined compositions (polyisotopic elements). The tabulated *atomic masses* of the elements that are commonly used to calculate the *molar masses* (also *molecular masses* or *molecular weights*) of a compound correspond to the weighted average of the masses of all these isotopes. These "atomic" masses (and hence also the derived molecular masses) therefore do not refer to the masses of individual atomic or molecular species but to the average mass of several different particles! MS, on the other hand, separates these different species and delivers for a given structure an MS "signal" consisting of several peaks representing the several isotopomers of the ionized compound. For this reason, not the atomic masses found in the periodic table of the elements or

elsewhere are of use for MS but rather the individual masses of the isotopes the particles are composed of. The term "mass" and its use in MS thus need to be clarified.

Mass Number (A) The *mass number* (also *atomic mass number* or *nucleon number*) is the total number of nucleons (protons and neutrons) in an atomic nucleus. It is an integer and it corresponds to the relative atomic mass rounded to the next integer. The different isotopes of an element are thus characterized by different mass numbers (e.g., ^{12}C: A = 12; ^{13}C: A = 13) and the same mass number can characterize species of different compositions (e.g., A = 17 for $^{13}C^1H_4$, $^{16}O^1H$, and $^{14}N^1H_3$).

Nominal Mass We are used to estimate the molecular mass of a compound by addition of the tabulated atomic masses rounded to the next integer. We are aware that the result of this procedure is not very precise, but we also know that it suffices for many purposes. The same is true for the *nominal mass*. The nominal mass of an element is defined as the mass number of its most abundant isotope in the natural isotopic mixture. It often corresponds to the mass number of the lowest-mass isotope of this element, as it is the case with the most common elements used in organic chemistry: C, H, N, O, S, Si, F, Cl, Br, and I. Typical counterexamples for this rule are elements such as Ar, Fe, Se, Pd, or Pt (see **Table 4.1**). The nominal mass of an ion or a molecule is finally the sum of the nominal masses of the elements contained in the particle of scrutiny. In the spectra of small molecules, the signals most often are labeled with the nominal masses of the respective ions. For larger compounds or for compounds that contain isotopes with typical mass defects, it is more appropriate to use the *monoisotopic masses* instead, rounded to the first decimal if the measurements were not done under high-accuracy conditions. We use in this book exclusively isotopic or monoisotopic masses, which correspond to nominal masses for small molecules if given without decimals.

Monoisotopic Mass (M) The *monoisotopic mass* of an element is the exact mass of the most abundant isotope of the element. It is in most cases quite close to the nominal mass but not identical with it (e.g., ^1H: 1.007825 u or ^{14}N: 14.003074 u). An exception is ^{12}C, which has a mass of 12.000000 u. The monoisotopic mass of a molecular entity is consequently the sum of the monoisotopic masses of the elements that compose the particle, plus (or minus) the masses of the electrons that are added (or removed) to result in the charge of the particle. Thus, it is the accurate mass of the structural unit that comprises solely the highest abundant isotopes of the elements it is composed of. It is very often the ion of lowest m/z value within a group of isotopomers. The corresponding ions are called *monoisotopic ions*, the MS signal *monoisotopic signal*, or *monoisotopic peak*. For more simple routine applications, it is sufficient to specify the masses to the full integer (= nominal mass; for M_R up to ca. 500 u) or to one decimal only (for M_R of ca. 500–2,000 u). Exact

Table 4.1 Atomic masses of some principal elements, their natural isotopes with natural abundance and accurate masses, and their classification[1]

Element	Standard atomic mass (u)[*]	Isotopes	Atomic number A (u)	Isotopic mass (u)[**]	Isotopic composition (%)[***]	Normalized isotopic composition (%)	Classification
H	1.01	^{1}H	1	1.007825	99.99	100	X
		^{2}H = D	2	2.014101	0.01	0.01	**X+1**
Li	6.94	^{6}Li	6	6.015123	7.59	8.21	X−1
		^{7}Li	7	7.016003	92.41	100	X
B	10.81	^{10}B	10	10.012937	19.9	28.44	**X−1**
		^{11}B	11	11.009305	80.1	100	X
C	12.01	^{12}C	12	12.000000	98.93	100	X
		^{13}C	13	13.003355	1.07	1.08	X+1
N	14.01	^{14}N	14	14.003074	99.64	100	X
		^{15}N	15	15.000109	0.36	0.37	**X+1**
O	16.00	^{16}O	16	15.994915	99.76	100	X
		^{17}O	17	16.999132	0.04	0.04	X+1
		^{18}O	18	17.999160	0.20	0.21	**X+2**
F	19.00	^{19}F	19	18.998403	100	100	X
Na	22.99	^{23}Na	23	22.989769	100	100	X
Si	28.09	^{28}Si	28	27.976927	92.22	100	X
		^{29}Si	29	28.976495	4.69	5.08	X+1
		^{30}Si	30	29.973770	3.09	3.35	**X+2**
P	30.97	^{31}P	31	30.973762	100	100	X
S	32.06	^{32}S	32	31.972071	94.99	100	X
		^{33}S	33	32.971459	0.75	0.79	X+1
		^{34}S	34	33.967867	4.25	4.47	**X+2**
		^{36}S	36	35.967081	0.01	0.01	X+4
Cl	35.45	^{35}Cl	35	34.968853	75.76	100	X
		^{37}Cl	37	36.965903	24.24	32.00	X+2
Ar	39.95	^{36}Ar	36	35.967545	0.33	0.34	X−4
		^{38}Ar	38	37.962732	0.06	0.06	X−2
		^{40}Ar	40	39.962383	99.60	100	X
Fe	55.85	^{54}Fe	54	53.939609	5.85	6.37	**X−2**
		^{56}Fe	56	55.934936	91.75	100	X
		^{57}Fe	57	56.935393	2.12	2.31	X+1
		^{58}Fe	58	57.933274	0.28	0.31	X+2
Se	78.97	^{74}Se	74	73.922476	0.86	1.79	X−6
		^{76}Se	76	75.919214	9.23	18.89	X−4
		^{77}Se	77	76.919214	7.60	15.38	X−3
		^{78}Se	78	77.917309	23.69	47.91	X−2
		^{80}Se	80	79.916522	49.80	100	X
		^{82}Se	82	81.916700	8.82	17.60	X+2
Br	79.90	^{79}Br	79	78.918338	50.69	100	X
		^{81}Br	81	80.916290	49.31	97.28	**X+2**
Pd	106.42	^{102}Pd	102	101.90560	1.02	3.73	X−4
		^{104}Pd	104	103.904031	11.14	40.76	X−2
		^{105}Pd	105	104.905080	22.33	81.71	X−1
		^{106}Pd	106	105.903480	27.33	100	X
		^{108}Pd	108	107.903892	26.46	96.82	X+2
		^{110}Pd	110	109.905172	11.72	42.88	X+4
I	126.90	^{127}I	127	126.90447	100	100	X
Pt	195.09	^{190}Pt	190	189.95993	0.01	0.04	X−5
		^{192}Pt	192	191.96104	0.78	2.31	X−3
		^{194}Pt	194	193.962681	32.86	97.44	X−2
		^{195}Pt	195	194.964792	33.78	100	X−1
		^{196}Pt	196	195.964952	25.21	74.61	X
		^{198}Pt	198	197.96789	7.36	21.17	X+1

Mass of the electron: 0.0005485799 u.

X = most abundant isotope. Its mass is used for the calculation of the monoisotopic mass of an ion. The "X + n" classification is denoted relative to the most abundant isotope. In bold: characteristic isotope that is indicative for the respective element.

[*]rounded to two decimals.

[**]given with six decimals (or, if less, according to IUPAC values).

[***]does not always add up to 100% due to rounding errors.

[1]https://www.isotopesmatter.com/applets/IPTEI/IPTEI.html (May 2019)

masses, calculated and measured to four to five decimal digits, are important in high-accuracy HR-MS, which is discussed in Section 4.4.2.

Isotopic Mass The *isotopic mass* is the exact mass of a specific particle composed of defined isotopes of the participating elements. It refers to a structural unit for which all the isotopes in the chemical formula are explicitly declared.

Most Abundant Mass (M_{ma}) The mass that belongs to the most intense peak within the isotopic distribution of a particle is called the *most abundant mass*. This mass is of particular importance for large molecules for which the signal intensities of the monoisotopic species become very low. It almost corresponds in such cases, after subtraction of the mass of the ionizing particle, to the molecular mass (M_R) of the compounds. MS signals of large molecules are thus best described with this mass (see below and **Fig. 4.9**). However, the most abundant mass is also important for the description of small molecules for which the monoisotopic signal does not correspond to the most abundant ion. As an example, the spectrum of the $[M + Na]^+$ ions of the sugar derivative **5** is shown in **Fig. 4.5**. Due to the three chlorine atoms present in the compound, the most abundant ion is registered at $m/z = 804.1$, while the monoisotopic peak is the one at $m/z = 802.1$.

Location of the "Mass" in a Spectrum The calculated and measured spectra shown in **Fig. 4.3** to **Fig. 4.5** shall illustrate the meaning of the different terms of "mass" used in MS.

If the most abundant isotopes of all the elements of a sample molecule are the lowest-mass isotopes, then the monoisotopic mass corresponds to the mass of the ions registered at the lowest m/z value of the several isotopic peaks belonging to the same ion structure. The respective ions are called monoisotopic ions, their signal monoisotopic peak, and their mass monoisotopic mass. An example for such a compound is 1,2-diiodo-4,5-dinitrobenzene (**3**), for which the calculated spectrum of the $M^{+\bullet}$ ions is shown in **Fig. 4.3**. The monoisotopic $M^{+\bullet}$ ions of the compound are registered at $m/z = 419.8$ (labeled in green). The signal arises exclusively from species with the isotopic composition of $^{12}C_6{}^{1}H_2{}^{127}I_2{}^{14}N_2{}^{16}O_4$—thus acknowledging the term *monoisotopic*. The other signals found at $m/z = 420.8$ and 421.8 are called isotopic signals, the respective ions isotopic ions, and the masses isotopic masses. They stem from isotopomeric ions of the same mass number, which are not resolved in ordinary MS. The signal at $m/z = 420.8$ is thus not the signal of a single type of species but the response to species of different isotopic compositions such as $[^{12}C_5{}^{13}C_1{}^{1}H_2{}^{127}I_2{}^{14}N_2{}^{16}O_4]^{+\bullet}$, $[^{12}C_6{}^{1}H_2{}^{127}I_2{}^{14}N^{15}N^{16}O_4]^{+\bullet}$, etc. It must be mentioned here that the terms *isotopic mass* and *isotopic ion* have slightly different meanings in low-resolution MS (as above) and in HR-MS: in HR-MS, the two ions mentioned above are registered as individual isotopic ions with distinct

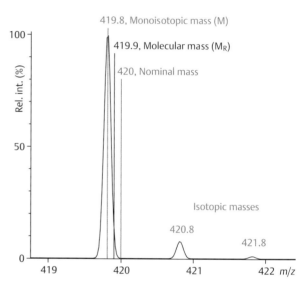

3

$C_6H_2I_2N_2O_4$, M = 419.8

Fig. 4.3 Calculated spectrum of the $M^{+\bullet}$ ion of 1,2-diiodo-4,5-dinitrobenzene (**3**) with labeling of the peaks with the respective exact isotopic masses (*green*). For comparison, the positions of the molecular mass (M_R, *red*) and the nominal mass (*blue*) of the compound are also shown.

isotopic masses (420.81320 u for $[^{12}C_5{}^{13}C_1{}^{1}H_2{}^{127}I_2{}^{14}N_2{}^{16}O_4]^{+\bullet}$ and 420.80688 u for $[^{12}C_6{}^{1}H_2{}^{127}I_2{}^{14}N^{15}N^{16}O_4]^{+\bullet}$), while in ordinary MS, all the isotopic ions that share the same rounded mass are collectively called isotopic ions.

The positions of the molecular mass (M_R) and the nominal mass of **3** are marked in red and blue, respectively, in the spectrum shown in **Fig. 4.3**. Already for a small molecule such as **3**, the differences of the three masses are quite distinct. They become even larger when the molecules grow in size. The nominal mass of a hypothetical $M^{+\bullet}$ ion of a peptide of the formula $C_{1000}H_{1500}N_{280}O_{290}S_{15}$, e.g., is 22,540 u, its monoisotopic mass (M) 22,550.7 u, and the molecular mass (M_R) of the peptide 22,565.6 g mol^{-1}. The most abundant mass (M_{ma}) of this compound would be 22,564.7 u—thus, rather close to M_R (see below and **Fig. 4.9**).

The monoisotopic peak is not always the peak registered at the lowest m/z value of a signal ensemble. If a species contains an element whose most abundant isotope is not corresponding to the one of lowest mass, then the monoisotopic peak of the species is accordingly shifted to a higher mass. This is the case, e.g., for ions that contain rhenium. The most abundant isotope of

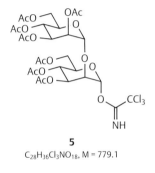

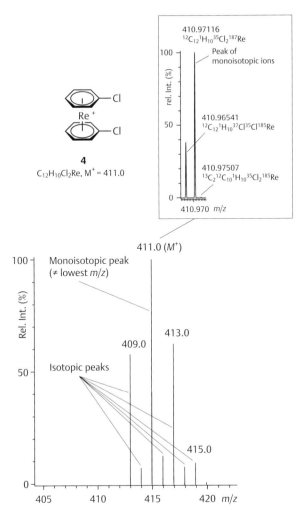

4

$C_{12}H_{10}Cl_2Re$, $M^+ = 411.0$

410.97116
$^{12}C_{12}{}^1H_{10}{}^{35}Cl_2{}^{187}Re$
Peak of monoisotopic ions

410.96541
$^{12}C_{12}{}^1H_{10}{}^{37}Cl^{35}Cl^{185}Re$

410.97507
$^{13}C_2{}^{12}C_{10}{}^1H_{10}{}^{35}Cl_2{}^{185}Re$

rel. Int. (%)

410.970 m/z

411.0 (M⁺)

Monoisotopic peak
(≠ lowest m/z)

409.0   413.0

Isotopic peaks

415.0

Rel. Int. (%)

405   410   415   420 m/z

Fig. 4.4 ESI-MS of the rhenium derivative **4** (sample from R. Alberto, University of Zurich) with its monoisotopic peak at $m/z = 410.97$ (M^+).

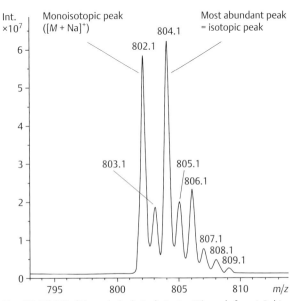

5

$C_{28}H_{36}Cl_3NO_{18}$, M = 779.1

Fig. 4.5 ESI-MS of the carbohydrate derivative **5** (sample from J. Robinson, University of Zurich). The monoisotopic peak detected for the [M + Na]⁺ ion at $m/z = 802.1$ shows a lower intensity than that of the isotopic peak registered at $m/z = 804.1$, which is due to the triple chlorination of the compound.

rhenium is ^{187}Re with a natural abundance of 62.6%. This isotope is accompanied by ^{135}Re (37.4%). Consequently, the monoisotopic peak for the M^+ ions of the rhenium derivative **4** is not the peak registered at $m/z = 409$ but the one disclosed at $m/z = 411$ (**Fig. 4.4**). Contradictory to its name *monoisotopic*, the respective peak is in fact not the response of a single monoisotopic type of species but rather the overlay of nonresolved signals of different isotopomeric ions. If the signal at $m/z = 411$ is resolved (upper spectrum in **Fig. 4.4**), three peaks can be recognized, and only the response at $m/z = 410.97116$ is due to the monoisotopic ions [$^{12}C_{12}{}^1H_{10}{}^{35}Cl_2{}^{187}Re$]⁺. The peaks at $m/z = 410.96541$ (38%) and $m/z = 410.97507$ (1%) are due to the isobaric ions [$^{12}C_{12}{}^1H_{10}{}^{37}Cl^{35}Cl^{185}Re$]⁺ and [$^{13}C_2{}^{12}C_{10}{}^1H_{10}{}^{35}Cl_2{}^{185}Re$]⁺, respectively.

Isotope Distribution

As already mentioned above, most chemical elements occur naturally as several stable isotopes (see **Table 4.1** for a number of selected elements and **Table 4.19**, p. 421 for a complete list). Elements that are composed of more than two isotopes are called *polyisotopic elements*, those that comprise merely two isotopes *diisotopic elements*, and the few that occur in a single isotopic version only *monoisotopic elements*.

The natural isotopic composition of an element is a characteristic feature and it propagates statistically in larger atomic assemblies. This leads to characteristic isotopic distributions (isotopic patterns) for ion signals, depending on the respective elemental compositions of the species. Since MS maps isotopic compositions, the isotopic pattern of an MS signal already reveals structural information about the analyte. Isotopic distributions can easily be estimated for small molecules or calculated by computers and then compared with measured data. Of particular relevance are elements that have for themselves already very typical isotopic characteristics and thus readily reveal their presence within a structure.

Examples of monoisotopic elements, also called X elements, that often can be recognized upon close scrutiny of the spectra are, e.g., fluorine (^{19}F), sodium (^{23}Na), phosphorus (^{31}P), and, in particular, iodine (^{127}I). Ions that contain these components show isotopic peaks that are usually characteristically lower in intensity than those that would be expected for ions of other organic molecules with similar mass. An example is shown with the MS spectra of 2-fluoro-1-iodonaphthalene (**6**, M = 272) and the halogen-free triaryl compound **7** (M = 274) in **Fig. 4.6**. The peak for the X + 1 isotope of the molecular ion $M^{+\cdot}$ of **6** is registered at m/z = 273.0 with an intensity of 11.3 rel.% (spectrum **a**), which is markedly lower than the intensity of the X + 1 isotopic peak of the $[M + H]^+$ ion of compound **7**, which is found at m/z = 276.2 with 20.4 rel.% (spectrum **b**). These intensities can be compared with the estimated relative intensity of the X + 1 peak of a pure hydrocarbon of the mass 275 u, which is ca. 22% (p. 340, isotope distribution). The markedly lower intensity of the X + 1 peak of the $M^{+\cdot}$ ion of **6** indicates that a considerable portion of the mass of the compound is due to monoisotopic elements that displace a rather large number of carbons (an X + 1 element).

Particularly characteristic and lucid isotopic distributions are provided by diisotopic elements. These can further be subclassified as X + 1, X + 2, and X – 1 elements, thus as elements for which the most conspicuous isotopic peaks are detected at 1 or 2 mass units higher or 1 mass unit lower, respectively, than the monoisotopic peaks.

The elements of highest importance in organic chemistry, carbon (^{12}C, ^{13}C) and hydrogen (^{1}H, ^{2}H≡D), belong to the X + 1 elements, but also nitrogen (^{14}N, ^{15}N). Due to the very small fraction of deuterium present in natural hydrogen (0.015%), this element is usually treated as monoisotopic (an X element). This approximation is still sufficiently accurate even for larger structures with 100 and more H atoms, where the first isotopic peaks are mainly dominated by other X + 1 elements, mostly by carbon (1.1% ^{13}C) and nitrogen (0.366% ^{15}N).

The best known X + 2 elements are the halogens chlorine (^{35}C, ^{37}C) and bromine (^{79}Br, ^{81}Br), which are encountered on a regular basis in organic molecules and which show very striking and characteristic ion patterns. But copper (^{63}Cu, ^{65}Cu), silver (^{107}Ag, ^{109}Ag), and other elements also belong to this category. Oxygen

(^{16}O, (^{17}O), ^{18}O), silicon (^{27}Si, (^{28}Si), ^{29}Si), and sulfur (^{32}S, (^{33}S), ^{34}S, (^{36}S)), i.e., polyisotopic elements, are for practical reasons often counted among the X + 2 elements too; oxygen likewise can also be regarded as an X element due to low content of ^{17}O (0.038%) and ^{18}O (0.20%) in the natural isotopic mixture. Ions that contain typical X + 2 elements such as chlorine or bromine show conspicuous and characteristic isotopic patterns in their mass spectra with two or more isotopic peaks of rather high intensity (relative to the monoisotopic peak) that are separated in steps of 2 mass units from each other. This pattern often allows the identification of X + 2 elements easily. Some examples of ion distributions of halogen-containing species can be found in **Table 4.16** (p. 414).

Lithium (^{6}Li, ^{7}Li) and boron (^{10}B, ^{11}B) are X – 1 elements and iron (^{54}Fe, ^{56}Fe, (^{57}Fe), (^{58}Fe)) can be regarded as an X – 2 element. Ions that contain elements of these types show prominent isotopic peaks at m/z values corresponding to lower masses than those of the monoisotopic ions. The lithium adduct of compound **8**, $[M + Li]^+$, shows the signal of the X – 1 ions at m/z = 484.2, while the monoisotopic peak is registered at m/z = 485.2 (**Fig. 4.7**, spectrum **a**).

Most elements of the periodic table are polyisotopic with three and more isotopic components, and rarely the major isotopomer is the one of lowest mass. Thus, analogous to compounds with X – 1 elements, the monoisotopic peak for compounds with such elements usually does not correspond to the signal recorded at the lowest m/z value. This is for example the case for the Pd complex **9**, for which the ESI-MS is shown in **Fig. 4.7** (spectrum **b**). The monoisotopic peak for the $[M – Cl]^+$ ions is found at m/z = 689.3, while a number of isotopic peaks are registered within the range of m/z = 685.3 to 695.3. The broad isotopic distribution of a signal readily indicates the presence of a polyisotopic element, and the number of peaks together with their relative intensities often allow the direct deduction of the involved element. The interpretation of such ion signals may however become rather complex when mono- and diisotopic elements are also present in an ion species, which is usually the case.

Isotopic patterns become complex and sometimes confusing not only when polyisotopic elements are involved but also when large molecular assemblies are analyzed, e.g., natural polymers such as nucleic acids, proteins, and polysaccharides. The isotopic distribution of the ions becomes broader and broader with increasing size of an organic molecule, adopting a Gauss-type form for large compounds. This is illustrated with the isotopic distribution of C_n with increasing n (**Fig. 4.8**, calculated isotopic distributions for hypothetical species C_n^+): While the spectrum of C_1^+ displays directly the relative abundance of the two isotopes ^{12}C and ^{13}C, the intensity of the X + 1 peaks increases with increasing number of C atoms. It already reaches 21.6% for C_{20}^+, and the X + 2 peak is recognized clearly for this species as

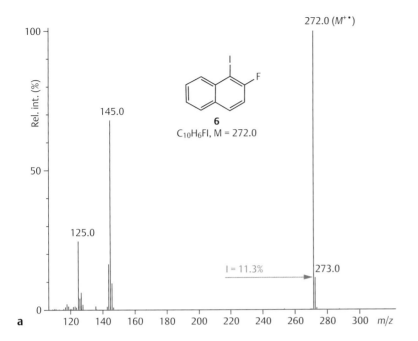

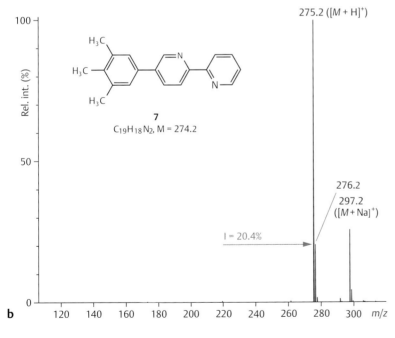

Fig. 4.6 (a) EI-MS of compound **6** ($M^{+\bullet}$ at $m/z = 272.0$) with an isotopic X + 1 peak of lower intensity due to the presence of the monoisotopic elements F and I in comparison to **(b)** the ESI-MS of compound **7** ($[M + H]^+$ at $m/z = 275.2$ and $[M + Na]^+$ at $m/z = 297.2$). (sample from J. Siegel, University of Zurich) with "normal" intensities of the two X + 1 isotopic peaks).

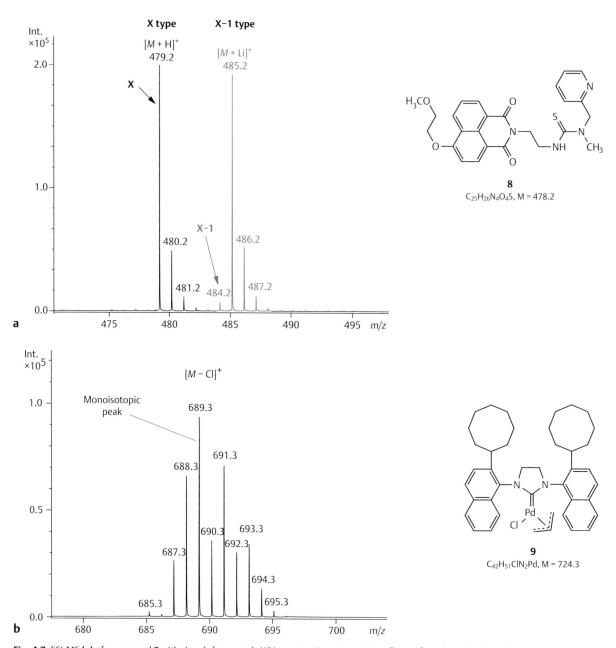

Fig. 4.7 ESI-MS **(a)** of compound **8** with signals for a purely X/X + n type ion composition ([M + H]⁺, m/z = 479.2) and for an ion that contains Li as an X − 1 element ([M + Li]⁺, m/z = 485.2) (sample from N. Finney, University of Zurich) and **(b)** of the complex **9** that reveals in the [M − Cl]⁺ ion (m/z = 689.3, peak of the monoisotopic ion) the presence of the polyisotopic element Pd by a broad isotopic peak distribution (sample from R. Dorta, University of Zurich).

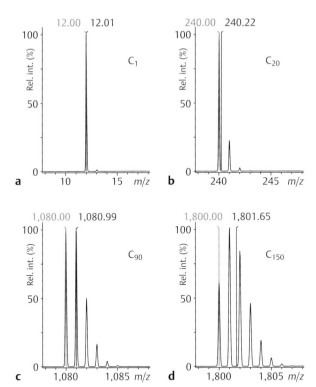

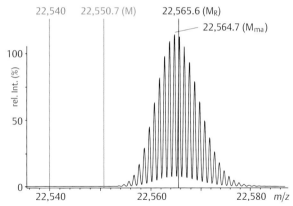

Fig. 4.9 Calculated isotopic distribution for the $M^{+\bullet}$ ion of a hypothetical biopolymer (peptide) of the formula $C_{1,000}H_{1,500}N_{280}O_{290}S_{15}$ (*blue*: nominal mass; *green*: monoisotopic mass (M); *red*: position of the molecular mass (M_R), which is close to the most abundant mass (M_{ma}, *black*)).

Fig. 4.8 Influence of the number of C atoms on the isotopic distribution of larger molecular structures by means of the calculated signals for the $M^{+\bullet}$ ions of hypothetical C_n compounds (*green*: monoisotopic mass; *red*: position of the molecular mass (M_R)).

well (2.2%). With even larger numbers of C atoms, peaks for X + 3, X + 4, etc. ions are also found. The X + 1 peak shows almost the same intensity as the monoisotopic peak for C_{90}^+, and for an even larger n, e.g., for C_{150}^+, the monoisotopic peak does no longer represent the base peak, and isotopic peaks dominate the spectrum. The monoisotopic peaks become marginal with ions of larger compounds and are no longer observed with very large structures.

This is particularly experienced in the analysis of biopolymers, e.g., peptides. For a hypothetical peptide ion with the formula $[C_{1,000}H_{1,500}N_{280}S_{15}]^{+\bullet}$ (M_R = 22,565.6 u, labeled red in the simulated spectrum shown in **Fig. 4.9**), the monoisotopic peak, expected at m/z = 22,550.7 (green), is too low in intensity to be detected. This is typical for ions of macromolecular assemblies. Thus, the signals of such species can no longer be described with their monoisotopic peaks. The signals of the most abundant ions can sensibly be used instead. Such ions are registered at m/z values that are close to the values of the molecular masses of the parent species, which themselves do not represent masses of MS-detectable ions. In the case of our hypothetical

peptide ion with a molecular mass M_R of 22,565.6 u (red), the most abundant ion (M_{ma}) would be registered at m/z = 22,564.7 (black). To illustrate once more the difference between nominal mass and monoisotopic mass, the position of the nominal mass of the peptide ion is also shown in the spectrum (blue). This example shows impressively how important it is to use accurate rather than nominal masses for the calculation of MS signals of larger compounds (above ca. 1,000 u), and how large errors may become otherwise.

Multiply Charged Ions

Multiply charged ions are often formed when mild ionization methods are used and polar molecules are analyzed. Such ions can generally be recognized in two ways.

Neighboring Isotopic Peaks. Signals of ions that are relatively low in mass are usually well resolved in MS. For such signals, the distance between neighboring isotopic peaks reveals directly the charge of the respective observed species. Because ions are separated in the MS according to their m/z value, the distance between isotopic peaks in m/z units is $\Delta m/z$ and thus dependent on the charge of the ions. Since Δm for neighboring isotopic peaks is usually 1 (mass unit), $\Delta m/z$ is 1/2 for doubly charged ions, 1/3 for triply, $1/z$ for z-fold charged ions. For example, three groups of ions are detected in the ESI-MS of the polyheterocyclic compound **10** (**Fig. 4.10**): at m/z = 1,001.4, 501.2, and 334.5. If the regions of the three signals are magnified and the isotopic patterns are scrutinized, it is recognized that the isotopic peaks belonging to the monoisotopic signal at m/z = 1,001.4 are separated by one m/z unit ($\Delta m/z$ = 1), corresponding to a charge of z = 1. The respective ions are the $[M + H]^+$ ions, as can

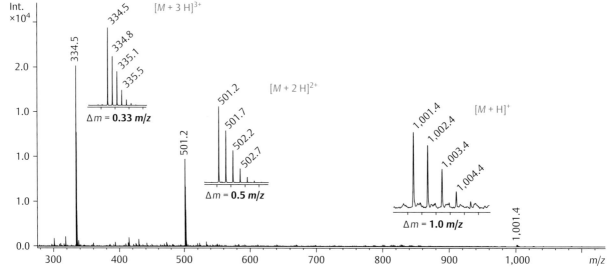

10

$C_{60}H_{64}N_4S_5$, M = 1,000.4

Fig. 4.10 ESI-MS of compound **10** with signals for ions of the types $[M + H]^+$, $[M + 2\,H]^{2+}$, and $[M + 3\,H]^{3+}$ (sample from N. Finney, University of Zurich).

readily be verified by calculation. The isotopic peaks proximate to $m/z = 501.2$ and 334.5 are separated by 1/2 and 1/3 m/z units, corresponding to ion charges of $z = 2$ and 3, respectively, the responsive ions being $[M + 2\,H]^{2+}$ and $[M + 3\,H]^{3+}$.

The monoisotopic mass (M) of a neutral molecular species giving rise to multiply charged ions can be calculated from the observed monoisotopic m/z value of any of the charged ions, provided the way of ion formation is known (e.g., by protonation as in the case of compound **10**). The respective formula is:

$$M = z\cdot m_{obs} - n\cdot m_a$$

where z = ion charge, m_{obs} = m/z value of the observed ion, n = number of the attached ionizing particle(s), and m_a = their mass. The monoisotopic mass of **10**, for which three signals for the ionized molecules are observed, is thus calculated as:

$[M + H]^+$ ($m/z = 1,001.4$): M = 1·1,001.4 − 1·1.0 = 1,000.4 u
$[M + 2\,H]^+$ ($m/z = 501.2$): M = 2·501.2 − 2·1.0 = 1,000.4 u
$[M + 3\,H]^+$ ($m/z = 334.5$): M = 3·334.5 − 3·1.0 = 1,000.5 u

For reasons of clarity, the unit "u" is usually not appended when spectra are labeled or ions are characterized in schemes.

Multiply Charged Ions. Multiply charged ions are particularly easily formed with polar molecules that are ionized with mild ionization methods. In such cases, the isotopic peaks of an ion are often no longer resolved, and the method described above cannot be applied anymore for the determination of the ionic charges. On the other hand, spectra of such compounds usually show a row of ions of different charges. This phenomenon does not only allow the recognition of multiply charged ions but also the determination of the ionic charges from the distance of two neighboring ion signals according to the formula:

$$z_2 = \frac{m_1 - m_a}{m_1 - m_2}$$

where z_2 = charge of the ion with its signal at $m/z = m_2$, m_1 = m/z value of the signal to the right of m_2 (charge = $z_2 - 1$), and m_a = mass of the attached ionizing particle(s).

For example, the charge of the ion registered at m/z = 848.5 of horse myoglobin (**Fig. 4.11**) with m_1 = 893.1, m_2 = 848.5, and m_a = 1 is calculated to be z_2 = 20+:

$$z_2 = \frac{893.1 - 1.0}{893.1 - 848.5} = 20$$

The mass of the related uncharged molecular assembly, which corresponds to the most abundant mass (M_{ma}) and therefore approximately to the molecular mass of the analyte (M_R) (**Fig. 4.9**), can then be determined as described above. This yields a mass of 16,950 u for horse myoglobin, calculated with the data of the signal registered at m/z = 848.5:

$$[M + 20\,H]^+ \ (m/z = 848.5):$$
$$M = 20 \cdot 848.5 - 20 \cdot 1.0 = 16,950\ \text{u}$$

Calculations done with different signals of a signal ensemble usually deliver slightly different values for the respective ion masses, which is mostly due to the limited accuracy of the instruments. MS instruments thus calculate molecular masses not only with a single signal pair but across all registered signals according to an algorithm that is specific to the spectrometer. The result is then revealed in the form of a *deconvoluted spectrum*, which shows the simulated peak for the most abundant (hypothetical) molecular ion $M^{+\cdot}$, displayed with a signal width that reflects the mass uncertainty. The analysis of horse myoglobin shows the signal at m/z = 16,952 ± 2 u in the deconvoluted spectrum (**Fig. 4.11**).

On a first glance, synthetic polymers display similar pictures of ion distributions as multiply charged ions of macromolecules (**Fig. 4.12**). However, their spectra represent mappings of polymer mixtures with signals of polymer components that have the same level of ionization (the same charge) but different sizes due to different numbers of monomer units. Such spectra are easily recognized and readily distinguished from MS of differently charged macromolecules by the equidistance of the several signals. The difference in m/z between two adjacent signals corresponds to the mass of the monomeric units the polymer is composed of. For polymer **11**, this mass is 58 u and corresponds to propylene oxide (C_3H_6O).

Description of a Spectrum

Routine Spectra (spectra with m/z values given with up to two significant decimals). Unlike the description of other analytical data, the report of a mass spectrum does not include the enlisting of all registered signals. It is sufficient to report just the monoisotopic peaks because the isotopic peaks—their presence and distribution—are inherently connected to a given particle (chemical formula). The monoisotopic peaks are described with their respective m/z values (given in the appropriate precision), together with the signal intensities and, if possible and sensible, the ion interpretations. The description of the EI-MS of butan-2-one (**27**, **Fig. 4.43**, p. 344) would then be:

EI-MS (70 eV, m/z (%)): 72 (25, $M^{+\cdot}$), 57 (8, $[M - \text{Me}]^+$), 43 (100, $[M - \text{Et}]^+$), 29 (17, $[\text{Et}]^+$), 27 (8).

Isotopic peaks are described, as an exception to the rule given above, when very characteristic diagnostic isotopic patterns are observed as, e.g., with halogen derivatives. The signal of the protonated 1-chlorobutan-2-one (C_4H_7ClO, M = 106) that would be obtained by chemical ionization could either be described as m/z = 107 (100, $[M + H]^+$) or as m/z = 107/109 (100/32, $[M + H]^+$). The presence of the Cl atom is certainly much better acknowledged with the latter description.

Spectra of Large Molecular Structures. As discussed above, it is not appropriate to describe mass spectra of very large molecular structures with their monoisotopic peaks, since these signals are usually very low in intensity if observable at all. The most

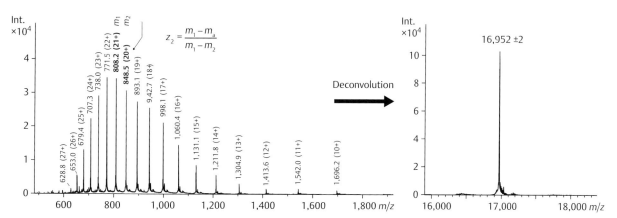

Fig. 4.11 ESI-MS of horse myoglobin: original spectrum and spectrum after deconvolution (calculation of M^+). The calculation of the charge of the ions with m/z = 848.5 (*red signal*) is indicated in the left spectrum.

Mass Spectrometry

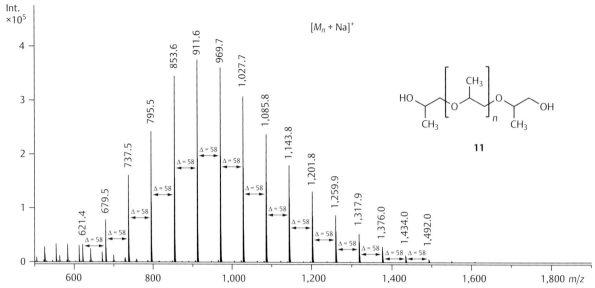

Fig. 4.12 ESI-MS of polypropylene glycol 1,000 (PPG 1,000, **11**).

abundant ion is used instead. This would be m/z = 22,564.7 for the hypothetical species of the formula $C_{1,000}H_{1,500}N_{280}S_{15}$ (simulated ion signal in **Fig. 4.9**), and the spectrum could be described as:

$$\text{ESI-MS (MeOH + 0.1\% HCO}_2\text{H): 22,564.7 ([}M + \text{H]}^+, M_{ma}).$$

The mass of the (hypothetical) most abundant $M^{+\bullet}$ ion (M_{ma}), as revealed by the deconvoluted spectrum, is used most sensibly for the description of the mass spectrum of a macromolecular compound for which only an assembly of multiply charged ions is registered. This mass corresponds largely to the molecular mass of the neutral analyte molecule (M_R) as already discussed above. The declaration of the "deconvoluted" m/z value should be supplemented with the indication of the measurement error and the type of ions that were used for the deconvolution. The spectrum of horse myoglobin (**Fig. 4.11**) could thus be described as:

$$\text{ESI-MS (MeOH): 19,952 ± 2 (deconvoluted from}$$
$$\text{signals of the type [}M + 27 \text{ H]}^{27+} \text{ to [}M + 10 \text{ H]}^{10+}).$$

High-Resolution Mass Spectra (spectra with m/z values given with four to five significant decimals). The m/z values obtained with HR-MS are listed with the number of significant decimals, appropriate for the performed measurement, instrument, and related error limits. The experimental data are usually complemented with calculated ion masses to allow rapid comparisons and plausibility assessments. It is absolutely imperative when HR-MS data are described that not only the resolution of the used instrument is declared in the report but also the mass

accuracy of the acquisition according to the instruments' calibration to account for the significance of the measured values. This information is usually given in the section *General* of an *Experimental Part* of a report, where also the instrumentation is described. Such a statement could read for an HR-ESI-MS as follows:

High-resolution electrospray ionization mass spectra (HR-ESI-MS): *QExactive* (*Thermo Fisher Scientific*, Bremen, Germany); .....; mass ranges 80 to 1,200, 133 to 2,000, or 200 to 3,000 u; resolution (full width at half-maximum) 70,000; mass calibration < 2 ppm accuracy for m/z 130.06619 to 1,621.96509 in (+)-ESI and for m/z 265.14790 to 1,779.96528 in (−)-ESI with *Pierce* ESI calibration solutions (*Thermo Fisher Scientific*, Rockford, United States); lock masses: ubiquitous erucamide (m/z 338.34174, (+)-ESI) and palmitic acid (m/z 255.23295, (−)-ESI).

For the description of HR-MS signals, two cases have to be distinguished.

Case A: the peak of lowest mass within a signal set is always the response of an isotopically pure species, provided the investigated molecular assembly is sufficiently small to allow the detection of the full signal ensemble. In the absence of X − n elements, this peak is the monoisotopic peak, otherwise it is an isotopic peak. If these lowest-mass signals are detected, the spectrum can be described by directly listing the respective m/z values, seconded with the values calculated for the respective isotopic ions. Addition of the mass deviations in ppm facilitates the scrutiny of the results. The spectrum of the $[M + H]^+$ ion of a compound of the formula $C_{16}H_{19}N_3O_4$ (M_R = 317.35 g mol^{-1}) can be described as:

> HR-ESI-MS (MeOH): 318.14476 ($C_{16}H_{20}N_3O_4^+$; [M + H]$^+$; calc. 318.14483; Δm = −0.2 ppm).

If the lowest-mass peak does not correspond to the monoisotopic ion as above, it is necessary to explicitly specify the exact isotopic composition of the chosen ion (see below).

Case B: if no signal of an isotopically pure ion is available for the determination of an accurate mass, then the measurement must be performed with an isotopic peak that is the overlap of signals of isotopomeric species. Usually, the most intense signal (the signal of the most abundant ions) is used for this purpose. This situation occurs when the peak of lowest mass is no longer detected in the mass spectrum, for instance due to the large size of the analyte molecule (**Fig. 4.9**) and/or the presence of X − n elements, such as Fe or Pd.

Since the signal to be scrutinized is no longer the signal of a single type of species but of a mixture of isotopomers, its spectrometrically determined m/z position cannot be compared directly with a calculated isotopic mass. The shape of the signal gets markedly altered by the admixture of isotopomeric species, and as a result the signal maximum does not correspond to the mass of a real particle. This in fact no longer represents a problem when newer mass spectrometers are used. Their software is able to precisely calculate the shape and the maximum of a peak to be expected from an ion mixture of a given composition, taking the resolution of the instrument into account. The respective calculated peak maximum can then be taken as the reference for comparison with experimental data. The software calculates additionally the overall isotopic pattern of the signal too, which is then compared with the effectively measured spectrum and ranked (Section 4.4.2, p. 341). Such a software-assisted spectral interpretation and ranking is more precise and more reliable than the direct comparison of a measured mass number with a calculated value (case A), and is for this reason often also applied in cases where the signal of the isotopically pure, lowest mass ions is obtained.

Both possibilities mentioned above could be applied to describe the HR-MS of the M^+ ion of compound **4** (**Fig. 4.4**, p. 295). Either the peak registered at m/z = 409.0, derived from the isotopically uniform ion containing ^{185}Re (case A), or the peak found at m/z = 411.0, which is the response of the monoisotopic ion together with other isotopic ions (case B), is used. Principally, any other peak of the signal could be used likewise.

The HR-MS description of the M^+ ion signal of **4** by means of the peak at m/z = 409.0 is done as described above for case A. Because the ions registered at m/z = 409.0 are not the

monoisotopic ions, the isotopic composition of the ions under scrutiny is specified as:

> HR-ESI-MS (MeOH): 408.96836 ($C_{12}H_{10}Cl_2^{185}$Re; M^+; calc. 408.96753; Δm = +2.0 ppm).

In the HR-MS description of a peak derived from a mixture of isotopic ions, the experimentally determined mass number is reported together with the chemical formula of the respective ion, the exact mass that was calculated by the MS software, and the mass deviations in ppm provided by the software as well. The spectral report for the M^+ signal of compound **4** by the use of the peak of the most abundant (and monoisotopic) ion at m/z = 449.1 could thus look like:

> HR-ESI-MS (MeOH): 410.96959 ($C_{12}H_{10}Cl_2$Re$^+$; M^+; calc. 410.96856; Δm = +2.5 ppm).

Tandem Mass Spectra (MS/MS, MS^n). The description of a tandem mass spectrum follows the same lines as that of an ordinary spectrum with the only difference that the precursor ion (P^+) or the cascade of the precursor ions ($P_1^+ > P_2^+ > ...$) and the method of collision activation (collision gas, collision energy E_c, also called fragmentation amplitude) have to be added. An appropriate description of the atmospheric pressure chemical ionization (APCI)-MS/MS for which a precursor ion P^+ was selected at m/z = 452 and activated for collision induced dissociation (CID) by collision with Ar would be:

> APCI-MS/MS (Ar; 452 (P^+, E_c = 1.2): 452 (15, P^+), 435 (69, [P − NH$_3$]$^+$), 396 (32, [P − C$_4$H$_8$]$^+$), 352 (100, [P − C$_4$H$_8$ − CO$_2$]$^+$).

For an MS^3 the respective description would be:

> APCI-MS^3 (Ar; 452 (E_c = 1.0) > 352 (P^+, E_c = 1.2)): 352 (15, P^+), 335 (69, [P − NH$_3$]$^+$), 296 (32, [P − C$_4$H$_8$]$^+$), 252 (100, [P − C$_4$H$_8$ − CO$_2$]$^+$).

4.3 Instrumental Aspects

4.3.1 Sample Introduction (Injection) and Ion Types

Since only charged particles that are in the gas phase can be analyzed by MS, two technical problems have to be solved: the analytes have to be transferred into the gas phase (sample inlet) and need to be ionized (ion formation).

Sample Inlet

It is technically a challenging task to transfer a sample from ambient pressure into the high vacuum (down to 10^{-9} Pa) without interruption of the vacuum and without decomposition of the analyte. This is particularly so because the probes might be composed of very diverse substances such as neutral

Mass Spectrometry

compounds, salts, complexes, or other aggregates. The samples may in addition be available in different forms: as pure compounds ready to be analyzed (solids, liquids, or gases), dissolved in a solvent, or embedded in a matrix.

Two different inlet systems have been developed for sample injection: the *direct inlet* that directly admits the sample via a vacuum lock into the instrument (used with EI, CI, and matrix-assisted laser desorption/ionization (MALDI)) and the *direct infusion* which continuously introduces the sample through a small opening or a capillary (used in particular with ESI and other spray methods, but also with EI and CI; see below). The continuous sample delivery with the direct infusion allows the coupling of this inlet system with chromatographic methods, which became the state of the art in many applications.

Direct Inlet. Direct inlet is used for crystalline, lacquer-like, viscous, or fluid samples. The material to be analyzed is either placed onto a support or filled into a metal crucible, which are then positioned directly into the ion source of the mass spectrometer via a vacuum lock chamber.

Vaporizable compounds are usually brought into the mass spectrometer in a crucible (mostly of Al or glass; internal diameter: 1 mm). The loaded crucible is fixed onto the tip of a heatable probe, which is inserted without heating into a lock chamber. After evacuation of the chamber, the still cooled tip of the probe is brought into the ion source, where it is slowly heated until the sample vaporizes. The tip of the probe can also be cooled to enable the measurement of highly volatile samples that would otherwise evaporate too rapidly in the hot ion source or to quickly re-cool an overheated sample.

Compounds that are difficult to evaporate or cannot be vaporized at all can be placed onto a small metal plate that is introduced into the vacuumed ion source of the mass spectrometer as described above. However, the molecules are then not transferred into the gas phase by heating of the sample but rather by its bombardment with particles (e.g., Ar atoms) or with a laser beam. This bombardment not only results in vaporization of the probe but leads also to the ionization of the analyte molecules. The type of the used radiation determines the classification of the so-called desorption method.

Direct Infusion. Direct infusion is particularly suited for samples that are already dissolved in a solvent or are gaseous. It is principally a two-step process. In the classical case, an internal reservoir is filled with the sample, which is then transferred into the mass spectrometer through a gas inlet. Liquids are either directly injected with a microsyringe through a septum into the previously evacuated reservoir, or they are vaporized into it out of a separate receptacle. Gaseous samples are brought into the reservoir by means of a small vessel with a break seal. Transfer of the analytes from the reservoir into the ion source takes place via an orifice, which is a capillary of defined dimension with a valve. The reservoir itself is heatable (max. 150°C at permanent use) to secure continuous vaporization and controlled delivery of the analyte.

A gas chromatograph (GC) can be installed in the instrument rather than a reservoir. In this case, samples are introduced directly from the GC capillary into the ion source. Samples in solution can for their part be sprayed or vaporized at ambient pressure into a spray chamber upon which concurrent ionization occurs (see ESI or APCI, p. 313). The ions are then directly transferred from the spraying chamber into the analyzer of the instrument through a capillary and a skimmer that ensures maximal ion transmission without collapsing the vacuum in the analyzer.

Direct infusion is best suited for the investigation of very small sample amounts. These inlet systems allow a simple coupling of an ion source to chromatographic systems such as GC, liquid chromatography (LC; high-performance liquid chromatography (HPLC) or ultra-high performance liquid chromatography (UHPLC)), or capillary electrophoresis (CE). Such coupled systems are widely used these days because they allow in one analytical procedure the separation of analytes (by retention or mass) and obtaining mass spectrometric information about components within mixtures (Section 4.3.6). In addition, they enhance the sensitivity of the measurements.

Required Sample Amount. The amount of sample needed for an MS analysis is in general very small, but it is dependent on the compound class (functional groups, size of the molecules, etc.) and on the instrumental setup, primarily on the used sample inlet and the ionization method. Usually, 0.001 to 0.1 mg of a sample is sufficient for a routine measurement; a little more, if more elaborate studies should be undertaken. If MS is used in combination with chromatographic methods (GC or HPLC), even amounts as low as 10^{-6} to 10^{-12} g—depending on the substance class and the ionization method—may still deliver good results. It is to mention at this place that the amounts of the samples indicated above must be directly available and may not be distributed as thin films on the surface of a vial!

Ion Types

Neutral molecules must be converted to ionized species after they were transferred into the gas phase. Only this way they can be separated according to their m/z values. The process of ionization takes place in the so-called ion source, and several different ionization processes are available.

Removal of an Electron. A radical cation with the charge 1+ is formed when an electron is removed from an uncharged molecule (**Scheme 4.1**). Such a particle has the same nominal mass as the starting molecule, and it is called molecular ion ($M^{+\bullet}$). Removal of an electron is the process that is operative in EI and is mainly observed if this ionization method is applied. EI-MS is particularly used for the analysis of small and rather nonpolar compounds, and it usually delivers highly energetic molecular ions that readily undergo heavy fragmentation.

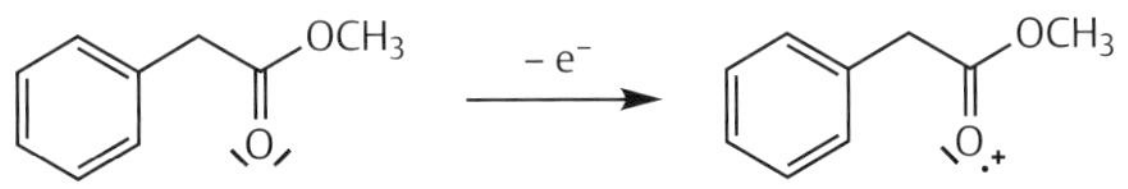

12
(M = 150)

12$^{+\bullet}$
(m/z = 150, $M^{+\bullet}$)

Scheme 4.1 Ionization by removal of an electron.

Addition of an Electron. The addition of an electron to a charge-neutral molecule generates a molecular ion ($M^{-\bullet}$) with the charge 1− (**Scheme 4.2**). This type of ionization is of minor importance for MS and can only be used for molecules of high electron affinity. Examples are halogenated or nitrated aromatics that can readily be ionized this way in an APCI ion source.

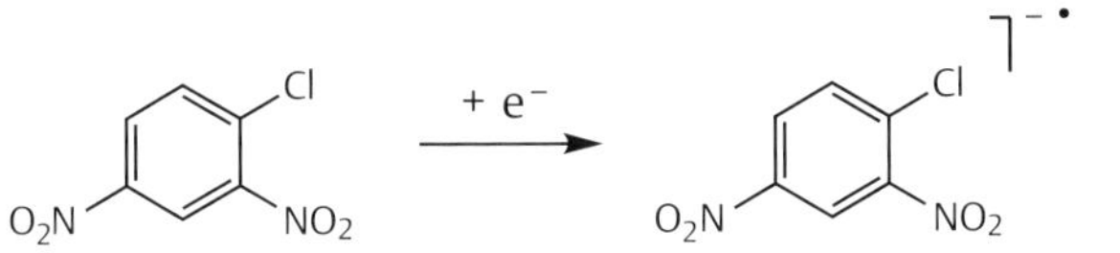

13
(M = 202)

13$^{-\bullet}$
(m/z = 202, $M^{-\bullet}$)

Scheme 4.2 Ionization by addition of an electron.

Protonation. Neutral molecules can be converted to charged particles by mono or multiple protonation, delivering cations with the charge $n+$, with n = number of attached protons (**Scheme 4.3**). The nominal mass of such ions (in u) is naturally larger by the number of the attached protons as compared to the mass of the neutral molecules. Protonation occurs preferentially at basic positions in a molecule, e.g., at the N atoms of amines or at the O atoms of diverse functional groups. The basicities of the functional groups give a good idea of where the charge(s) within a molecular framework will mainly be located after ionization. Methods such as CI, ESI, APCI, and MALDI (in positive mode) deliver mainly protonated species.

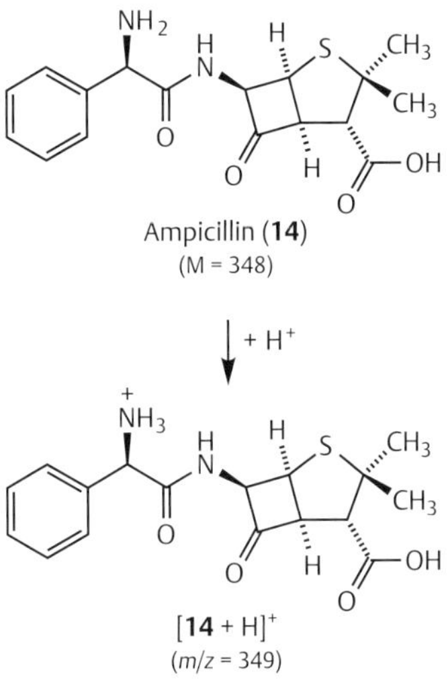

Ampicillin (**14**)
(M = 348)

$+ H^+$

[**14** + H]$^+$
(m/z = 349)

Scheme 4.3 Ionization by protonation.

Deprotonation. Anions can be formed from uncharged molecules by the abstraction of one or more protons (**Scheme 4.4**). These anions possess the charge $n-$, with n = number of removed protons, and have a nominal mass that is smaller by $\Delta m = -n$ than the mass of the nonionized molecules. Ionization of molecules by deprotonation is particularly successfully used with acidic compounds such as carboxylic acids, phosphoric acids, and phenols. The respective ionization methods are again CI, ESI, APCI, and MALDI, and the instrument needs to be operated in the negative mode.

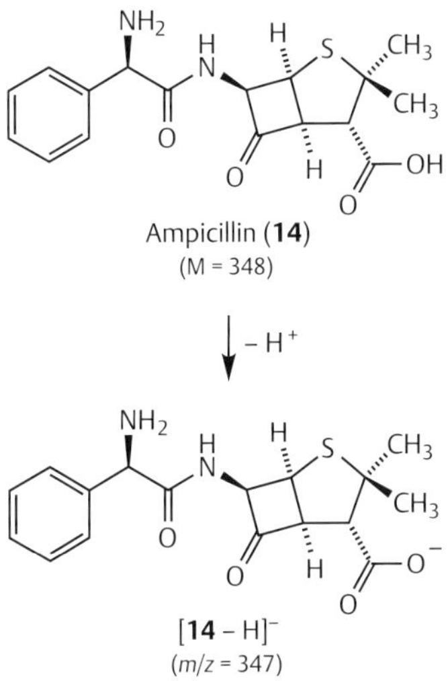

Ampicillin (**14**)
(M = 348)

$- H^+$

[**14** − H]$^-$
(m/z = 347)

Scheme 4.4 Ionization by deprotonation.

Addition of a Cation or an Anion. A great number of weakly polar compounds such as esters or amides are hardly protonated or deprotonated when soft ionization methods are used. Alternatively, cations or anions can be bound through noncovalent interactions to such molecules (**Scheme 4.5**). The most commonly observed ions of this type are adducts with Na^+ ($+23$ u), K^+ ($+39$ u), NH_4^+ ($+18$ u), I^- ($+127$ u), or $CF_3CO_2^-$ ($+113$ u). They are often observed by the use of soft ionization methods such as ESI and MALDI, and their abundance can be increased by the addition of respective salts to the analyte. The mass differences between adduct ions and the respective analyte molecules are sometimes rather large, which needs to be considered during the interpretation of spectra that might show signals of such species.

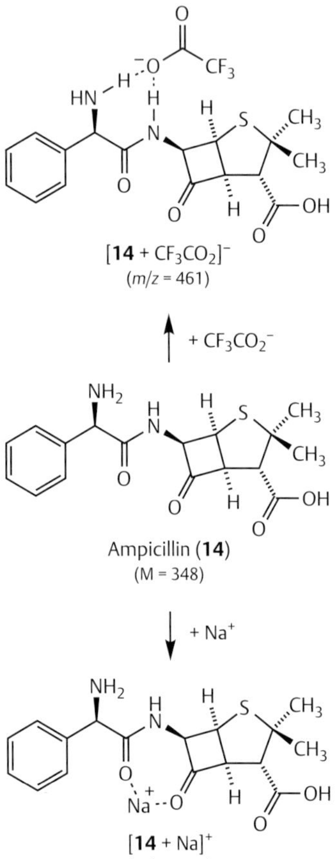

$[14 + CF_3CO_2]^-$
($m/z = 461$)

↑ $+ CF_3CO_2^-$

Ampicillin (**14**)
(M = 348)

↓ $+ Na^+$

$[14 + Na]^+$
($m/z = 371$)

Scheme 4.5 Ionization by addition of an anion ($CF_3CO_2^-$) or a cation (Na^+).

Dissociation and Direct Transfer of Ions into the Gas Phase. Salts, acids, and bases dissociate in aqueous or other media and form charged species that can be transferred directly from the

solution into the gas phase and analyzed by MS (**Scheme 4.6**). Since the ion optics of a spectrometer allows operation only in one mode of polarity at a given time (positive or negative mode), cations and anions that are formed together by dissociation cannot be analyzed simultaneously. Several measurements with alternating polarities are required to detect both types of ions. Certain instruments, however, are able to do such alternating acquisitions within milliseconds in one analytical run.

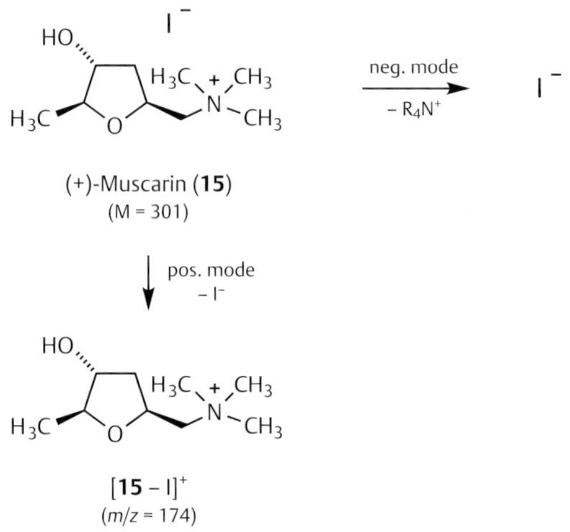

(+)-Muscarin (**15**)
(M = 301)

↓ pos. mode
$- I^-$

$[15 - I]^+$
($m/z = 174$)

Scheme 4.6 Ionization by dissociation and transfer of the ions into the gas phase.

4.3.2 Ionization Methods

The ionization methods used in MS can be sorted into three categories: the vaporization methods, the desorption methods, and the spray methods.

Vaporization Methods: Electron Ionization, Chemical Ionization

With the vaporization methods, samples—in most cases organic compounds—are vaporized by heating in the vacuum, transferred as gases into the ion source, and ionized there. This process requires that the analytes are fairly stable and possess a reasonable volatility. Thus, the methods are suited for rather small molecules with not too reactive functional groups and not for macromolecules or molecules of very high polarity that hardly can be evaporated without decomposition. The melting behavior of a compound can be taken as a good measure to predict the compatibility of a compound with vaporization methods: the melting point of a compound to be analyzed should not exceed 300°C, and the melting should proceed without decomposition (no pyrolytic decomposition). The most important vaporization methods are EI and CI.

Electron Ionization. EI is the oldest ionization method used in MS. With this method, the ions are generated as follows: from the inlet systems (gas or direct inlet), a fine beam of molecules streams into the ion source where it intersects perpendicularly with a beam of electrons, formed between a hot filament (cathode) and an anode (**Fig. 4.13**). The energy of the electrons can be varied between 10 and 300 eV by adjusting the potential between the cathode and the anode. An electron energy of 70 eV is usually employed for standard measurements. Only in special cases, when fragmentation needs to be suppressed, electrons of lower energy between 12 and 15 eV are used. The interaction of the energetic electrons with paired electrons of the neutral molecules in a near-miss path generates positively charged radical ions ($M^{+\cdot}$, molecular ions) by electron ejection according to the equation:

$$M + e^- \rightarrow M^{+\cdot} + 2\,e^-$$

Molecules with strongly electronegative elements or groups can be ionized by electron capture. They are bombarded in this case with thermal electrons of just a few electronvolts and form negatively charged radical ions ($M^{-\cdot}$):

$$M + e^- \rightarrow M^{-\cdot}$$

Which atom or which bond loses an electron upon electron bombardment and, thus, at which position the molecule gets ionized is dependent on the exact structure of the analyte. It is usually one of the easiest-to-remove n-electrons of free electron pairs of heteroatoms (ionization potential of functionalities with N and O atoms ≈ 8–9 eV) that is expelled. More difficult is the removal of π-electrons from double bonds and aromatic systems (ionization potential of alkenes and aromatics ≈ 8.8–9.5 eV), and even harder is the removal of σ-electrons from single bonds (ionization potential ≈ 10.6 eV). These three types of electron removal are shown exemplary with compound **16** (M = 124) in **Scheme 4.7**. It is easily understood that the exact structures of the radical cations—and, thus, the knowledge of their tendency to be formed—are of paramount importance for the way the ions undergo subsequent fragmentations.

The sample molecules absorb some of the kinetic energy of the electrons when bombarded—ca. 10 to 12 eV. With the typical ionization potential of 10 ± 3 eV of organic molecules, an excess of energy (ca. 3–5 eV) remains on the formed ions, and this energy is dissipated by fragmentation. See Section 4.4.3 (p. 343) for typical fragmentation reactions and spectra.

The nonionized particles are removed from the ionization chamber by high vacuum pumps to avoid scattering and electrical discharges in the ion source (vacuum of 10^{-3}–10^{-4} Pa). The molecular and fragment ions that have been generated, on the other hand, are led out of this chamber by means of a voltage

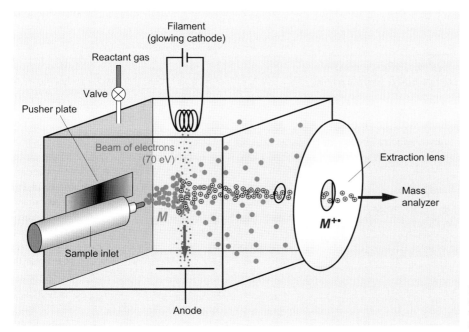

Fig. 4.13 Ionization chamber for electron ionization (reactant gas valve closed) and for chemical ionization (reactant gas valve opened).

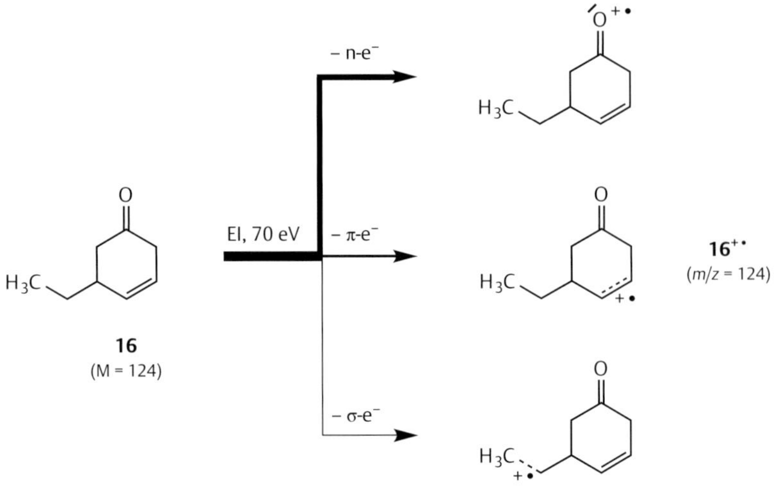

Scheme 4.7 Possible formations of radical cations by electron bombardment of ketone **16**. The ease to remove an electron decreases in the order of n- > π- > σ-electron.

that is applied between a pusher plate and a first extraction lens. After the ions have left the ionization chamber, they are accelerated, focused, and finally released through an exit slit into the analyzer. The acceleration of the ions is achieved by applying an acceleration potential that varies with the type of instrument from 2 to 10 kV, and the final speed of the ions is reached at the exit slit. Separation of the ions according to their m/z values follows in the analyzer.

It is easy to see that the setup of an EI source does not allow the simultaneous analysis of cations (positive mode) and anions (negative mode), because the escorting and acceleration voltages must be of opposite polarities for the two types of ions.

The importance of EI decreases markedly when the molecular masses of the analytes exceed ca. 500 u. Molecules of this size are not only problematic with regard to their unaffected vaporization, but compounds of this type also often undergo such heavy fragmentation that almost no structural information can be acquired from their spectra any longer. For this reason, CI, which is much milder, was developed as an alternative.

Chemical Ionization. The instrumental setup for CI is principally the same as that for EI (**Fig. 4.13**). Ionization of the neutral molecules in the ionization chamber, however, is not achieved by electron bombardment but by chemical reaction with cationic particles that are formed locally from reactant gases. To form the ionizing species, the ionizing chamber is filled with a reactant gas by opening the respective valve, and the gas is then ionized by electron bombardment (EI process) as described above. Collision of the ionized reactant gas species with analyte molecules leads to the transfer of an ionized particle and, thus, to ionization of the sample

molecules. The transfer of the ionized analyte species from the ionization chamber to the analyzer takes place as described above.

CH_4, isobutane, and NH_3 are the most commonly used reactant gases, and H_2, H_2O, and ROH are used less frequently. Ionization occurs usually by protonation of the neutral molecules with the formation of ions of the type $[M + H]^+$, in certain cases by the addition of ionized reactant gas molecules with the formation of species such as $[M + C_2H_5]^+$ or $[M + NH_4]^+$ (with CH_4 and NH_3 as reactant gases, respectively). It is also possible to generate negatively charged ions of the type $[M − H]^-$. Such ions are either formed by deprotonation (particularly suited for acidic compounds) by the use of a basic reactant gas. Occasionally, molecular ions of the type M^- are also observed. They are formed by the capture of electrons that have lost most of their kinetic energy due to impacts with reactant gas (this process, in fact, corresponds to EI in the negative mode).

Looking at the ionization process in more detail, the reactant gas (ca. 1 kPa) is ionized and activated for chemical reaction by EI with highly energetic electrons of 150 to 200 eV. In the case of methane, $CH_4^{+\cdot}$ is formed as an initial cation, which reacts with an additional methane molecule to form the primary protonating species CH_5^+:

$$EI \text{ part: } CH_4 + e^-_{200\,eV} \rightarrow CH_4^{+\cdot} + 2\,e^-$$
$$CI \text{ part: } CH_4^{+\cdot} + CH_4 \rightarrow CH_5^+ + CH_3^{\cdot}$$
$$M + CH_5^+ \rightarrow [M + H]^+ + CH_4$$

A number of other ions such as $C_2H_5^+$, $C_3H_5^+$, or CH_3^+ are formed in addition to CH_5^+ with CH_4 as the reactant gas, and all of them can act as proton sources. However, they alternatively can also form adduct ions. Analogously to CH_5^+ from CH_4, the reactive ions $C_3H_7^+/C_4H_9^+$ from isobutane, NH_4^+ from NH_3, and H_3O^+/ROH_2^+ from H_2O and ROH are formed.

The recalescence of the proton transfer reactions decreases in correlation with the acidity of the reactant species in the following order:

$$H_3^+ > CH_5^+ > C_3H_7^+ > H_3O^+ > C_4H_9^+ > NH_4^+$$

This means that the ionization with H_3^+ (hard) to NH_4^+ (soft) proceeds more and more gently. Thus, it can be controlled with the choice of the reactant gas how much excess energy is transferred onto an ionized molecule and, hence, how much fragmentation will be induced. The CI-MS fragmentation behavior of cinnamic acid derivative **17** (M = 219) ionized with (a) CH_4 (hard), (b) isobutane (medium hard), and (c) NH_3 (soft) as the reactant gases is shown with the three spectra in **Fig. 4.14**. While in spectrum

(a) the response of a fragment ion with $m/z = 131$ dominates as the base peak over the signal of the ionized molecule $[M + H]^+$, the respective peak is observed in spectrum (b), obtained upon CI with isobutane, with low intensity only. By the use of NH_3 as the reactant gas, almost no fragmentation is observed.

Spectrum (a) shows in addition to the signals for $[M + H]^+$ at $m/z = 220$, $[M + H - H_2O]^+$ at $m/z = 202$, and $[M + H - H_2N(CH_2)_4OH]^+$ at $m/z = 131$, signals for further adduct ions ($[M + C_2H_5]^+$ at $m/z = 248$ and $[M + C_3H_5]^+$ at $m/z = 260$). Adduct signals are quite common for CI, in particular NH_4^+ adducts by the use of NH_3 as the reactant gas. The respective signal is not observed, however, in the case of compound **17**.

By the use of deuterated reactant gases, acidic H atoms of the analyte can be exchanged, which delivers additional structural information.

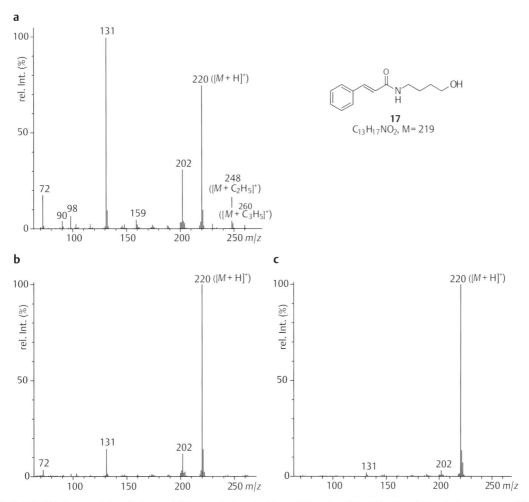

Fig. 4.14 CI-MS of N-cinnamoyl-4-aminobutanol (**17**; sample from M. Hesse, University of Zurich) with **(a)** CH_4, **(b)** isobutane, and **(c)** NH_3 as the reactant gas.

A special case of CI is the *direct CI*, which can be used to acquire also CI-MS of compounds of low volatility. In this case, a sample is first loaded in solution onto a wire loop made of Pt, Re, or W, which is fixed onto the tip of a direct inlet probe. The probe is inserted into the ionization chamber after the solvent has evaporated, and the thin substance coating on the wire is vaporized within seconds by rapid heating. The molecules liberated this way are then ionized as usual. The rapid heating results in less destructive vaporization and allows the investigation of molecules with masses of up to 1,000 u, which would not be amenable to conventional CI-MS.

Prime Desorption Method: MALDI

MS instruments were almost exclusively equipped with EI or CI sources until the 1980s. These sources, however, allow only the analysis of small and rather nonpolar molecules that can be vaporized without decomposition. They are not suited for the investigation of larger and more highly functionalized compounds, in particular not for the study of macromolecular assemblies such as synthetic and biological polymers (peptides, proteins, nucleic acids, and polysaccharides) that increasingly gained importance. No structural information can be gathered for such compounds by MS because neither ions that contain the complete structure of the analyte molecules nor spectra that show clear and meaningful fragmentation patterns are accessible. Thus, not even the mass of the sample molecules can be determined, let alone functional groups or structural entities.

One possibility to solve these problems is offered by the desorption methods that allow the direct and mild transfer of ions from the surface of a sample to the gas phase. The energy needed for ionization and liberation of the molecules from the molecular association can be supplied, e.g., by heating or irradiation with light or particles. The mass range for ionization with desorption methods is largely unlimited; the limitations for MS measurements are rather given by the analyzer and detection systems. The desorption methods have been successfully used for oligopeptides (smaller proteins), oligonucleotides, oligosaccharides, synthetic polymers, and dendrimers. Their application might become problematic with very large molecules because desorption usually leads to singly charged species that are more difficult to detect when the *m/z* values become larger.

Matrix-Assisted Laser Desorption/Ionization (MALDI). MALDI is a relatively new method that was developed in the late 1980s for the analysis of synthetic and biological polymers. It is a very efficient and sensitive ionization method and currently the most popular desorption method used in MS. The method allows the nondestructive measurement of molecules with masses of up to 1,000,000 u and beyond, the usual mass

region being 300 to 100,000 u. A measurement cannot be done over the full mass range, however: specific subranges have to be chosen such as 300 to 5,000 u for smaller, 1,000 to 20,000 u for medium-sized, and 10,000 to 100,000 u for larger molecules.

The MALDI method is very mild and can even transfer noncovalently bonded molecular aggregates into the gas phase. **Fig. 4.15** shows a typical example for a MALDI spectrum of a peptide ($C_{197}H_{339}N_{53}O_{52}S$; $M_R = 4,314.2$). This spectrum reveals in addition to ions of the types $[M + H]^+$ ($m/z = 4,314$), $[M + Na]^+$ ($m/z = 4,336$), and $[M + 2\,H]^{2+}$ ($m/z = 2,157.5$) the formation of aggregate ions of the type $[2\,M + H]^+$ ($m/z = 8,627$).

The energy needed for ion formation and desorption is delivered in MALDI by a laser beam. Because many sample molecules do not possess a chromophore that absorbs light of the laser's wavelength, the analytes are usually embedded into a suitable matrix (ratio of matrix to analyte ≈ 1,000:1). Radiation of such a mixture with energy-rich laser pulses is assumed to photoionize the matrix molecules, upon which proton exchange with the sample molecules becomes possible (formation of $[M + H]^+$ in the positive mode and of $[M - H]^-$ in the negative mode). Analytes with very low proton affinities such as lipids, neutral oligosaccharides, or many synthetic peptides get ionized by formation of cation adducts with Na^+ or K^+. The matrix also accepts the energy that is needed for the ion desorption. A rapidly expanding plume with a hot plasma is formed upon irradiation, and the generated ions are pulled off by means of an extraction plate and are transferred to the mass analyzer (**Fig. 4.16**).

Because the ions are generated by laser pulses, they arise in small distinct ion packages, with most of the ions being singly charged. The formation of such ion packages is ideally suited to be combined with a time-of-flight (TOF) analyzer that requires a "pulsed" ion supply and can separate ions over a wide mass range with an almost infinite *m/z* scale (Section 4.3.3).

The sample preparation is of utmost importance and has a great influence on the quality of a MALDI spectrum. In particular, the matrix needs to be chosen with care to fit the chemical nature of the analyte. It should absorb the major portion of the irradiated ultraviolet (UV) light and protect the sample molecules during their evaporation against the destructive power of the laser beam. At the same time, the matrix must still guarantee an efficient ionization of the analyte by protonation or deprotonation. A compilation of the most common matrix materials, together with the respective applications, is found in **Table 4.2**.

It is also possible to directly ionize samples by laser irradiation without their embedding into a matrix if the analyte possesses a chromophore that absorbs the laser light (λ_{max} at 330–360 nm for conventional lasers). They get directly excited and ionized by the incoming light through photoionization. This technique is

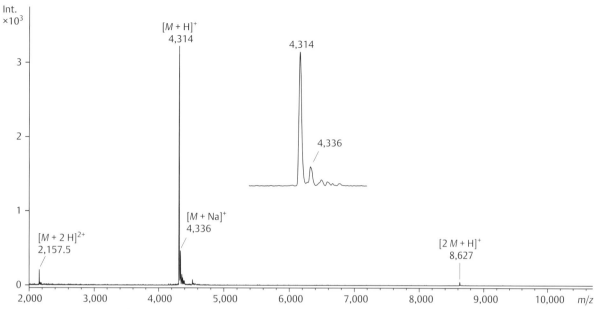

Fig. 4.15 MALDI spectrum of a peptide ($C_{197}H_{339}N_{53}O_{52}S$, $M_R = 4{,}314.2$, $M_{ma} = 4{,}314$) acquired by excitation with an N_2 laser within a matrix of sinapic acid.

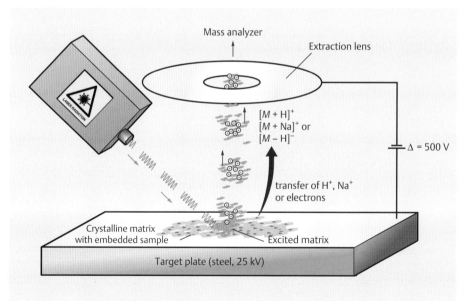

Fig. 4.16 Ionization process with MALDI.

Mass Spectrometry

particularly used for the analysis of insoluble substances and of polymers that are directly brought as solids onto the target plate. As a prerequisite, of course, the compounds need to be able to vaporize and ionize in the excited state without decomposition.

Target plates with nanostructured porous silicon surfaces (DIOS) or nanotubes of metal oxides or nitrides (NALDI) are available for some special applications. The samples are brought directly onto the target plates where they are adsorbed and ready for investigation without the need for a matrix. Such systems permit the direct laser desorption of small molecules (<1,000 u) from biological samples and even allow the direct quantification of metabolized compounds, e.g., within a blood sample.

The MALDI method can also be combined with thin layer chromatography (TLC). The surface of a TLC plate, sprayed after its

Table 4.2 Most commonly used matrix substances for MALDI-MS with an N_2- or Smart-Beam-YAG laser at 337 or 353 nm, respectively

Main application	Matrix	Structure
Peptides	α-Cyano-4-hydroxycinnamic acid (HCCA)	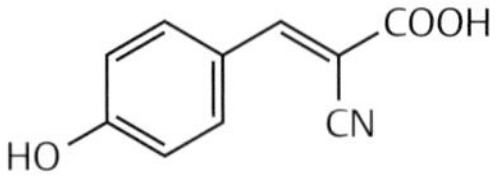
Proteins, peptides	Sinapinic acid (SA)	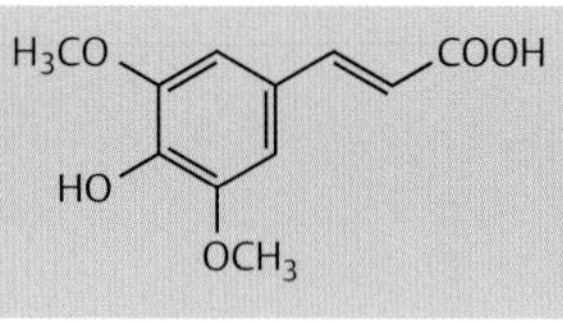
Water-soluble lipids, proteins, non-covalent complexes	6-Aza-2-thiothymine (ATT)	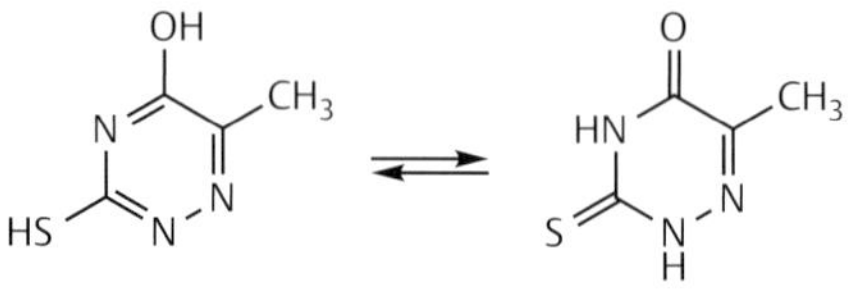
Peptides, small proteins, oligosaccharides	2,5-Dihydroxybenzoic acid (DHB)	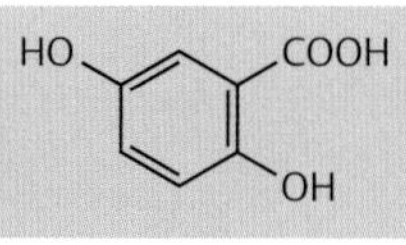
Small oligonucleotides	3-Hydroxypicolinic acid (3-HPA)	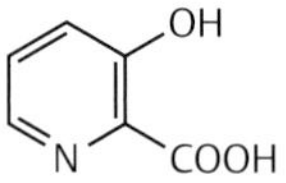
Apolar, synthetic polymers, polyaromatic hydrocarbons	3-Indolacrylic acid (IAA)	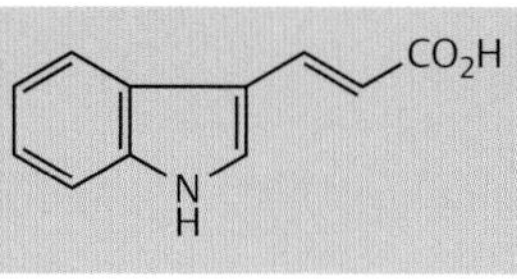
Polyaromatic hydrocarbons, synthetic polymers, lipids	1,8,9-Trihydroxyanthracene (Dithranol)	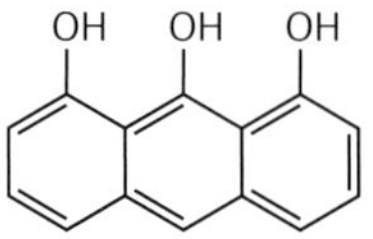
Lipids, polyaromatic hydrocarbons	trans-2-[3-(4-tert-Butylphenyl)-2-methylprop-2-enyliden] malononitrile (DCTB)	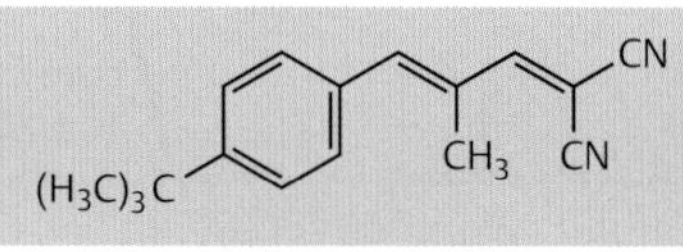
Lipids (neg. mode)	5-Chloro-2-mercaptobenzothiazole (CMBT)	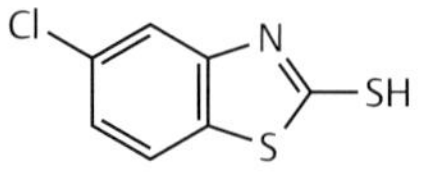

development with a matrix compound, can be scanned, and a location-dependent set of spectra can be obtained. This method is particularly valuable in the analysis of lipid derivatives.

Additional Desorption Methods

Secondary Ion Mass Spectrometry (SIMS). SIMS is primarily used in the analysis of surfaces. Positively or negatively charged ions are generated from a sample that was placed onto the surface of a metal plate (e.g., Ag) by its bombardment with a beam of highly energetic ions such as Cs^+, $Ar^{+\cdot}$, or Ga^+ at 2 to 20 keV. Aside from the molecular ions $M^{+\cdot}$ and $M^{-\cdot}$, $[M + H]^+$, $[M - H]^-$, $[M + Na]^+$, $[M + Ag]^+$, and fragment ions may also be formed. The metal ions originate from the surface of the metal plate or from impurities (Na^+ is ubiquitous). SIMS is well suited for the investigation of organic compounds that are either nonvolatile or only slightly volatile such as ammonium salts, peptides, or oligosaccharides. The method can also be used for the detection and identification of contaminants of surfaces or as an ion microscope to generate three-dimensional (3D) pictures of surfaces or their inner structures, respectively.

Fast Atom Bombardment (FAB). FAB is also used for the ionization of organic molecules that are difficult or impossible to vaporize, e.g., organic acids, polypeptides, oligosaccharides, or oligonucleotides. Ionization is achieved in this case by bombardment of a thin sample layer with a continuous beam of fast and charge-neutral atoms (in most cases Ar^0). The sample is dissolved in a matrix substance (usually glycerol, thioglycerol, or 3-nitrobenzyl alcohol) and brought into the ion source on a flattened copper tip. Impact of the fast atoms does usually lead not only to desorption of sample molecules and their conversion to adduct and fragment ions, but also to desorption and ionization of the matrix molecules (e.g., formation of $[glycerol_n]^+$). This can become rather bothersome when smaller molecules (M < 400 u) are analyzed. FAB usually leads to $[M + H]^+$, $[M + Na]^+$, and $[M - H]^-$ ions that are analyzed in the positive or negative mode of the instrument, respectively.

Field Desorption (FD) and Field Ionization (FI). FD and FI were developed for the analysis of large nonpolar molecules. They are demanding techniques that have largely been replaced by more convenient desorption and spray methods such as MALDI, ESI, and APCI. The principle of FD is based on the desorption of ions from a heated filament under the influence of strong electric fields. The solution of an almost nonvolatile sample is loaded onto the emitting wire by the use of a capillary. To obtain a high efficiency of the desorption process, the emitting wire is activated by exposure at high temperature (ca. 1,200°C) to the gas of benzonitrile. This forms fine needles (whiskers) of elemental carbon that surround the wire like a blanket of felt. The emitter is then loaded with a dissolved analyte, and the analyte ions are desorbed from the whiskers by the application of high voltages to extraction plates and pulling orifices. The ions that are formed are of the types $M^{+\cdot}$, $[M + H]^+$, $[M + Na]^+$ (positive mode), and $M^{-\cdot}$, $[M - H]^-$ (negative mode).

FI is closely related to FD. The analytes are ionized analogously to FD, but the sample is loaded to the activated emitter by vaporization rather than by application of a sample solution.

Prime Spray Methods: ESI and APCI

It probably was one of the most challenging tasks for MS engineers to develop methods that allow to generate ions directly out of solutions and to transfer these ions to a mass analyzer. The major problem in this endeavor is to selectively ionize an analyte that is surrounded by a massive excess of solvent and to transfer

the analyte ions from ambient pressure to the high vacuum of the mass analyzer (10^{-5}–10^{-9} mbar). If the coupling of an HPLC system with MS is considered, where the chromatography is run with a flow of 1 mL min^{-1}, the vaporization of the solvent alone would result in a gas flow of 300 to 2,000 L s^{-1}. Such a gas flow can hardly be managed by even a sophisticated pump system. The ions, thus, have to be generated directly from solution at atmospheric pressure and have then to be transferred into the analyzer portion of the MS instrument by electric fields.

Several spray methods, the so-called *atmospheric pressure ionization* methods, that directly form ions from solution at atmospheric pressure had been developed in the 1980s. The development started with the launch of the *thermospray ionization*, which, however, never asserted itself. ESI and APCI followed shortly afterwards, and ESI is currently by far the most widespread spray method. A more recent method is the *atmospheric pressure photoionization* (APPI) that is particularly well suited to ionize also nonpolar compounds.

The use of protic and polar solvents (e.g., H_2O, MeOH, *i*-PrOH, and MeCN) in combination with volatile buffers (e.g., HCO_2H, $MeCO_2H$, CF_3CO_2H, NH_4OAc, and NH_3) is essential to ensure efficient ion formation with spray methods. Usually, spray methods generate ions that encompass the full original structure of the analyte molecules by protonation, deprotonation, or often also by addition of a metal cation. The analytes need to be well soluble in the solvent systems, even though only very low (μM) sample concentrations are typically required. Spray methods allow an easy coupling of a mass spectrometer with liquid chromatographic devices (HPLC, UHPLC, and CE) and, thus, the immediate acquisition of MS spectra of compounds directly after their chromatographic separation. This facilitates the identification, quantification, and structural elucidation of substances directly out of mixtures (Section 4.3.6).

Electrospray Ionization (ESI). ESI had originally been developed for the mass spectrometric investigation of biological macromolecules. However, it currently finds also broad appreciation in the MS of small polar compounds and, because of its very mild character, also for the study of metal complexes and noncovalent interactions.

The method is technically simple, robust, and very sensitive (ca. 1,000–10,000 times more sensitive than CI). For ion formation, a solution of the sample is sprayed with a flow rate of ca. 1 to 20 μL min^{-1} through a fine capillary (inner $\varnothing \approx 100$ μm) of metal or quartz glass (fused silica) into a chamber. A voltage of 2 to 5 kV is set between the capillary and a cathode, which acts as a spray shield and which represents the passage from the spray chamber to the vacuumed region of the mass spectrometer (**Fig. 4.17**).

Mass Spectrometry

The liquid leaves the capillary in a Taylor cone (A), and a fine jet of charged droplets is ejected from its apex toward the counter electrode. The charge density on the surface of these droplets increases upon evaporation of the solvent until the electrostatic repulsion becomes so large that the droplets disintegrate explosively into smaller droplets (Coulomb explosion, B). The process of solvent evaporation and Coulomb explosion is repeated until the droplets are small enough to expulse ions from their surface by desorption (C). The generated ions are released from the spray chamber through a small aperture in the spray shield, focused, and guided by a capillary to the analyzer of the mass spectrometer (**Fig. 4.18**).

When ESI-MS is coupled with HPLC, the flow rate of the sample solution administered to the ion source is ca. 100 to 500 µL min⁻¹, rather than 1 to 20 µL min⁻¹ mentioned above. To evaporate such a large amount of solvent, it is necessary to assist the process by the use of a spray and drying gas (N_2). An example of an ESI ion source is displayed schematically in **Fig. 4.18**. This ion source is of an *off-axis* design, which means that the sample solution is not sprayed perpendicularly toward the spray shield. This construction is the current standard. It raises the tolerance for high flow rates and contaminations, which reduces losses of sensitivity.

Nanoelectrospray ionization (nano-ESI) is used for the analysis of smallest sample amounts or very highly diluted samples. The principle of ionization in this method is the same as in ESI. The sample solution, however, is sprayed into the ion chamber by a special sprayer at a flow rate of merely 10 nL min⁻¹ to 1 µL min⁻¹. By the use of a very fine capillary, a reduced flow of N_2, and the positioning of the sprayer into the immediate proximity of the entry of the mass spectrometer, the sensitivity of the measurement can be enhanced as compared to conventional ESI by a factor of up to 100. In the ideal case, over 60% of the analyte molecules can be ionized and transferred to the mass analyzer by this method (compared to 10–20% in ESI).

ESI and nano-ESI are particularly well suited to investigate polar compounds that readily get ionized by acid–base reactions or by aggregation with ions in solution. Neutral or basic polar compounds are usually protonated or ionized by bonding to cations and, hence, are detected as cations in the positive mode of the spectrometer. Small molecules of M_R < ca. 1,000 u are usually converted to singly charged ions of the type $[M + X]^+$, while larger molecules lead to multiply charged species of the type $[M + n\,X]^{n+}$, with n = up to 100 (see, e.g., the ESI-MS of horse myoglobin in **Fig. 4.11**, p. 301, which exhibits signals of ions with charges of 10+ to 27+).

Basic compounds are preferentially ionized by protonation(s). With decreasing basicity of the analytes, however, signals due to protonated molecules decrease in intensity in favor of signals deriving from Na^+ adducts. It is possible, too, that protons as well as alkali cations are attached in varied ratios to analyte molecules, which evidently enhances the complexity of the data.

A typical ESI-MS, obtained from the cyclic tetradecapeptide **18** (M_R = 1,805.2), is shown in **Fig. 4.19**. Signals of six different types of adduct ions are found in the spectrum: three of charge 1+ ($[M + H]^+$, $[M + Na]^+$, $[M + K]^+$) and three of charge 2+ ($[M + 2\,H]^{2+}$, $[M + H + Na]^{2+}$, $[M + 2\,Na]^{2+}$). Sometimes, such spectra can be simplified by the addition of an acid, NaI, or by measurement in the negative mode of the instrument.

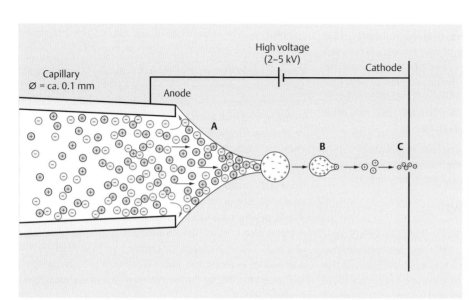

Fig. 4.17 Electrospray ionization (ESI): **(A)** the Raleigh limit is reached at the tip of the Taylor cone and droplets are formed. **(B)** Evaporation of the solvent leads to Coulomb explosion and to the formation of smaller droplets. **(C)** Ions are desorbed from the smallest droplets and transferred through an aperture in the cathode to the mass analyzer.

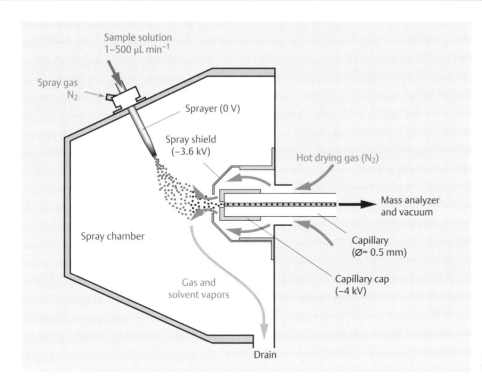

Fig. 4.18 Construction of an ESI ion source.

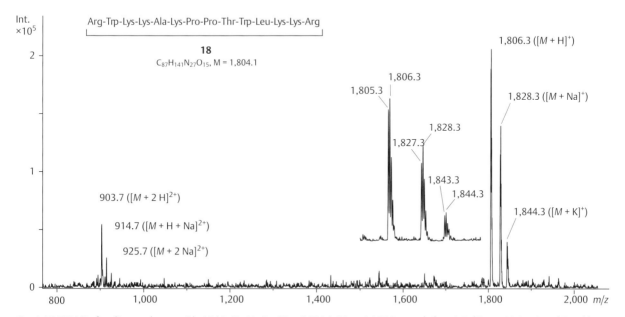

Fig. 4.19 ESI-MS of cyclic tetradecapeptide **18** ($C_{87}H_{141}N_{27}O_{15}$, M = 1,804.1, M_R = 1,805.2; sample from J. Robinson, University of Zurich) as a characteristic example for a peptide spectrum. The signals are labeled with the most abundant masses (M_{ma}).

Mass Spectrometry

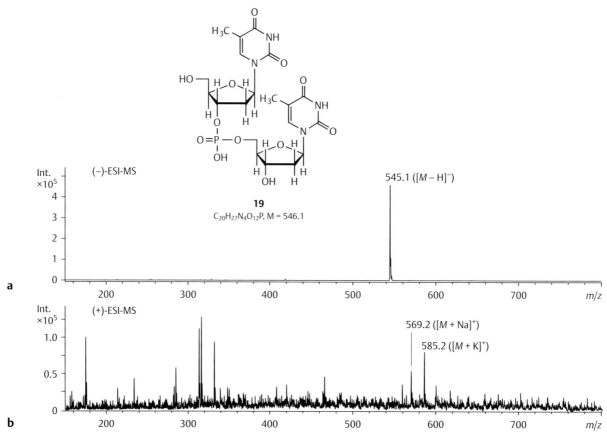

Fig. 4.20 Influence of the mode of measurement on the quality of ESI-MS. Spectrum of dinucleotide **19** measured **(a)** in the negative mode (−)-ESI-MS and **(b)** in the positive mode (+)-ESI-MS. The two spectra were obtained from the same sample of **19**, dissolved in MeCN/H$_2$O 1:1 + 0.1% HCO$_2$H.

Acidic compounds, even if they possess functional groups that could be protonated, are particularly well observed by their measurement in the negative mode. This is nicely demonstrated with the example of dinucleotide **19** (M = 546.1), for which two ESI-MS spectra, measured in the negative (a) and positive (b) mode, are shown in **Fig. 4.20**. It is readily perceived that the quality of spectrum (a) with its sole signal at m/z = 545.1 for the [M − H]$^-$ ions and almost no recognizable background noise is of much higher quality than that of spectrum (b) that shows in addition to the signals at m/z = 569.2 and 585.2 for the adduct ions [M + Na]$^+$ and [M + K]$^+$ a number of unassigned signals and heavy noise.

Atmospheric Pressure Chemical Ionization (APCI). The APCI method was developed already in the 1970s for the analysis of relatively nonpolar analytes directly from solutions. Similar to the procedure used for ESI, a dissolved sample is directly submitted to the ion source. The sample solution exits a heated

capillary together with a spraying gas (N$_2$) at atmospheric pressure as a steam jet. In contrast to ESI, no voltage is applied to the capillary. The mixture of N$_2$ and analyte vapor is rather ionized by means of a discharge needle, which finally forms the ionized species (**Fig. 4.21**).

The mechanism of ion formation in APCI is complex and can be described in a simplified manner as follows. In a first step, primary ions are formed from the N$_2$ spray gas at the tip of the discharge needle. These ions react with the vaporized solvent to form the effectively ionizing species such as, e.g., H$_3$O$^+$, OH$^-$, or OH$^·$ from H$_2$O. Ionization of the sample molecules takes place subsequently, usually by proton transfer reactions or, as an exception, by the addition of a cation or an anion. Basic analytes, thus, typically form cations of the type [M + H]$^+$ (positive mode) and acidic compounds, accordingly, ions of the type [M − H]$^-$ (negative mode). Analytes such as aromatic nitro compounds that are easily reduced,

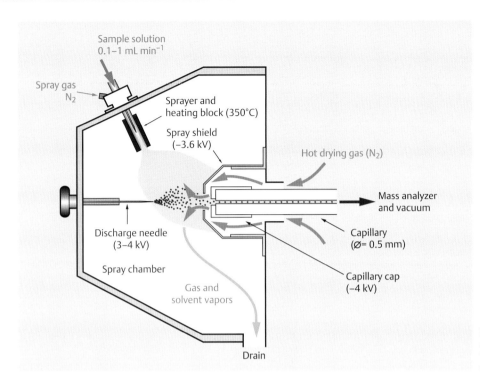

Fig. 4.21 Construction of an APCI ion source. Vapor is indicated with the *blue zone* and ions are represented as *red spots*.

on the other hand, are often transferred to radical anions $M^{-\bullet}$ by electron transfer.

In contrast to ESI, APCI allows also the efficient ionization of rather nonpolar compounds. Additionally, the formation of Na^+ adducts is suppressed, which is an advantage because the accuracy of quantitative measurements can often be significantly enhanced by this. On the other hand, a sample might get heated up to 400°C, which can lead to thermal decomposition of labile compounds. **Fig. 4.22** shows as an example the APCI-MS of polyaromatic compound **20** (M = 1,190.4). A strong signal of a single adduct ion of the type $[M + H]^+$ can be identified at m/z = 1,191.4 in this spectrum—a respective signal is hardly found in the ESI-MS of the compound. However, molecule **20** evidently loses readily N_2 from the triazole ring (signal at m/z = 1,163.4), which most probably is due to the heating of the sample at vaporization rather than due to an MS fragmentation.

ESI and APCI ion sources are very similar in construction (**Figs. 4.18** and **4.21**), and the switch from one method to the other is thus usually rather easy and straightforward. There are even combined ion sources available that allow the alternating measurement of ESI- and APCI-MS within one analytical run.

APCI, like ESI, is often used in combination with HPLC. The respective parameters are comparable to those discussed for ESI, except for the flow rates that can be slightly higher (0.1–1 mL min⁻¹). Some specific APCI sources were developed to be coupled to GC.

Additional Spray Methods

Several additional spray methods have been developed over the recent years with the goal to broaden the field of MS applications. The major objectives have been to increase the ionization efficiency (and, thus, to increase the sensitivity of the MS methods), to allow the direct investigation of analytes within complex sample materials without the need for laborious sample preparations, and to detect and quantify traces of metallic and nonmetallic elements at very low concentrations. Three examples are concisely presented below.

Inductive Coupled Plasma Ionization (ICP). ICP was developed and is used for the quantitative analysis of more than 80 elements and their isotopes from metal complexes or organic molecules. Samples can be analyzed directly from acidic

Mass Spectrometry

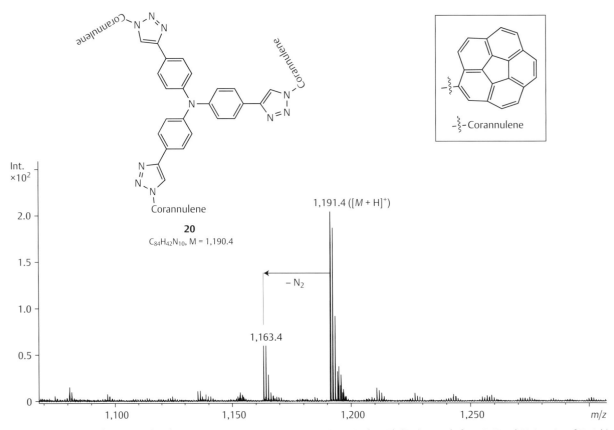

Fig. 4.22 APCI-MS of compound **20** ($C_{84}H_{42}N_{10}$, M = 1,190.4, M_R = 1,191.3, MeOH/CH_2Cl_2 (3:1); sample from J. Siegel, University of Zurich) that readily loses N_2 during ionization.

solutions, or they are separated into their components by chromatography (GC or HPLC) and administered then to the ion source. The eluate meets in the ICP source a torch of >10,000°C that is formed with an inert gas such as Ar. The sample species disintegrates within this hot torch to their elemental constituents and form positively charged atomic ions that are finally separated and quantified by MS.

Desorption Electrospray Ionization (DESI). DESI is a technique that is used for the analysis of surfaces. Analyte molecules are extracted and ionized upon spraying charged solvent droplets with high velocity onto a sample plate. This method allows the direct analysis of analytes from TLC plates, on organic tissues, from blood plasma, or from pills.

Atmospheric Pressure Photoionization (APPI). APPI is used as a complementary method to ESI or APCI for the analysis of lipophilic, unsaturated, slightly polar compounds such as some vitamins. Sample solutions are sprayed directly into an ionization chamber, where ion formation is effected by irradiation with a strong krypton UV lamp. Two mechanisms of ion formation

are possible. The sample molecules can be excited directly by the UV radiation, which leads to radical cations $M^{+\bullet}$, or they are ionized indirectly when no suitable chromophore is available in the molecule. In this case, a doping compound (e.g., acetone or toluene) that gets ionized by the UV irradiation and acts as a protonating species (as in APCI) needs to be added to the solvent.

4.3.3   Mass Analyzers

Properties of Mass Analyzers

The separation and detection of the ions that were formed in the ion source are realized in the mass analyzer of an MS instrument. Several mass analyzers are available. They can be classified and characterized according to their working principle and construction as well as their technical characteristics such as *mass range*, *scan rate*, *mass accuracy*, and *mass resolution*. These properties are of fundamental significance for the appropriate problem-specific choice of the MS method as well as for the interpretation of the obtained data. The terms are concisely explained as follows.

Mass Range. The mass range of an analyzer is the range of m/z values within which an instrument can separate and detect ions (e.g., 50–2,000 m/z). The mass ranges covered by some modern analyzers meanwhile exceed 100,000 m/z, but the range is highly dependent on the type of instrument and should be carefully adapted to the analytical problem that needs to be solved (see below). It must be mentioned that the mass range of a specific measurement (the measuring range) is usually smaller than the mass range specified for the instrument. Limitations exist with regard to the ion formation, the ion optics, or the used scan rate, and the instrument's parameters have to be adjusted to optimize a measurement in the mass range of interest.

Scan Rate. The scan rate corresponds to the rate with which the mass range of an analyzer can be examined. It is specified, depending on the type of instrument, either in "$m/z\ s^{-1}$" (e.g., 2,000 $m/z\ s^{-1}$) or "s decade^{-1}" (the time in seconds to scan the mass range of a decimal power, e.g., 10–100 or 100–1,000 m/z). The scan rate is of particular importance for combinations of MS with chromatographic methods, where the sample composition continuously changes with time.

Mass Accuracy. The mass accuracy of an analyzer is given by the ratio $\Delta m/m$, with m = mass of the investigated ion and Δm = mass difference between the experimental mass of the ion and its effective (calculated) mass. The accuracy of an instrument is usually determined and optimized prior to high-resolution measurements by means of reference compounds.

Mass Resolution. There are two definitions used in MS for the resolution R: the *10% valley* definition and the *full width at half maximum* (*FWHM*) definition.

The 10% valley definition is explained with the representation in **Fig. 4.23** (left): two signals are regarded as resolved according to this definition if the signal intensity in the valley between the two signals is ≤10% of the intensity of the signal maxima. With

such a small overlap, the peak maxima of the two signals are not noticeably affected. The resolution R of the instrument corresponds then to the ratio $m/\Delta m$, the ratio of the mass number m of a peak to the difference of the mass number to a neighboring peak of equal intensity that overlaps with 10% of the peak intensities. This is equal to the ratio of the m value of a peak to its peak width Δm at 5% intensity.

The formula $R = m/\Delta m$ allows rapid evaluation of the resolution an instrument must provide to be able to separate two signals of ions with given masses. A resolution of 4,000 is, for example, required to separate the signals of isotopic ions of a compound with M = 4,000 (m = 4,000; Δm = 1). This resolution is the minimal standard for almost all current MS instruments. Significantly higher resolutions are needed, though, for the resolved detection of isobaric ions (ions of the same nominal mass but different exact masses due to different elemental compositions), where Δm becomes very small. For the separation of anions with the chemical formulas $C_{22}H_{21}O_{12}^-$ (M = 477.10385) and $C_{17}H_{21}HN_2O_{10}S_2^-$ (M = 477.06431), for example, the required resolution of the MS would approximately be 12,000 (m = 477 and Δm = 0.03954).

Mass spectrometers are delivered with a specified minimal resolution that not necessarily corresponds to the effectively provided resolution for a given experiment. The effective resolution can be determined, however, with the 10% valley definition either by finding two equally intense signals that overlap with 10% intensity (which is not realistic) or by the scrutiny of a single signal and determination of its signal width at 5% intensity. For the latter, a noise-free spectrum with a symmetric signal would be required, which is difficult to find as well. Less problematic with regard to noise and signal symmetry is the peak width at 50% intensity (FWHM). For this reason, this value is usually used to characterize the resolution of an instrument. In this case, the resolution is also defined as $R = m/\Delta m$, with Δm = FWHM (**Fig. 4.23**, right). This resolution is larger by a factor of approximately 2 as compared to the resolution of the 10% valley definition. As can be seen in **Fig. 4.23**, two signals that overlap at their 50% intensity (blue lines) can still be distinguished as individual peaks (black line); they are, thus, resolved. However, the observed peak maxima do no longer correspond to the effective m/z values of the underlying ions. This effect is even more pronounced in cases of overlapping signals of different intensities, where the peak of the minor component may not even be recognized as a separate peak but rather as a shoulder.

The performance of analyzers has been improved in parallel to the development of the spraying methods with regard to all four characteristics mentioned above, and more achievements are certainly to come.

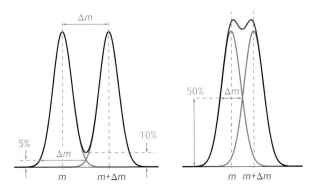

Fig. 4.23 Visualization of the resolution $R = m/\Delta m$ of a mass analyzer according to the 10% valley definition (*left*) and full width at half-maximum (FWHM) definition (*right*).

Classification of Analyzers. Mass analyzers can be classified according to their operation modes into three categories:

continuously working *ion beam analyzers*, pulsed *ion trap*, and pulsed *TOF analyzers*. The combination of several different analyzers within one instrument is also possible (hybrid instruments). The most important analyzer systems are concisely introduced below—arranged according to the three main categories mentioned above.

Ion Beam Mass Analyzers (Sector Field and Quadrupole Analyzers)

For ion beam analyzers, ions that are formed in the ion source are accelerated by means of an electric field to a defined kinetic energy and subjected to the mass analyzer in a continuous ion beam. Ions differing in mass and velocity are then separated by deflection of the ion beams in electrostatic and/or magnetic fields.

Sector Field Mass Analyzer. Sector field mass analyzers are the oldest analyzer systems known in MS. They consist of electrostatic and magnetic sectors in which the ions are differently deflected from their trajectories due to different *m/z* values or momentums. Current instruments possess in most of the cases one electrostatic and one magnetic field sector and are called accordingly double-focusing mass spectrometers (variations with three to five sectors are also available, but less common). The best performing instruments have a mass range of up to 6,000 u and allow—depending on the operating mode (see below)—scan rates of up to 0.1 s per decade, high resolutions exceeding 60,000 (10% valley definition), and high accuracies of down to ca. 1 ppm deviation from the theoretical values.

The schematically shown apparatus in **Fig. 4.24** corresponds to a double-focusing sector field analyzer with Nier–Johnson geometry, where the ions entering from the ion source first pass an electrostatic (*E*) and then a magnetic (*B*) sector.[b] The electrostatic analyzer acts as an energy dispersive device, focusing particles of identical energy and reducing the kinetic energy distribution of an ion beam that was generated by acceleration of the ions formed in the ion source by means of a strong electric field *U* (2–10 kV). This focusing is of particular importance when EI is used as the ionization method, where the translation energies of the particles would be too dispersed for high-accuracy measurements because of charge–space effects. The electrostatic analyzer, thus, is in fact not a mass analyzer but rather a device that sorts the ions according to their kinetic energies.

The effective mass analysis, the separation of the ions by their *m/z* values, occurs in the magnetic sector of the instrument. Ions of lower mass are deflected to a greater extent from a straight trajectory by the action of a strong magnetic field (approximately 1 T) than ions of higher mass and the same charge. Thus, ions of different masses fly along bent trajectories with different radii, according to the following equations:

$$m/z = \frac{r_m^2 \cdot B^2}{2 \cdot U} \text{ or } r_m^2 = \frac{2 \cdot U}{B^2} \cdot (m/z)$$

---

[b] The reversed arrangement, the inverse Nier–Johnson geometry with an upstream magnetic and a downstream electrostatic field sector, is possible as well and does not differ in its mode of action (function principle).

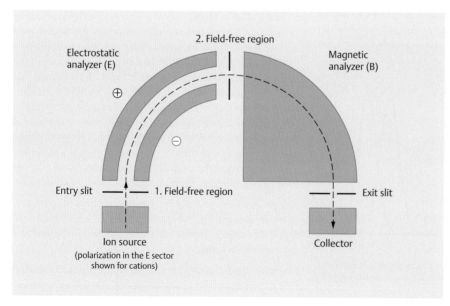

Fig. 4.24 Schematic representation of a double-focusing sector field analyzer with *E–B* configuration (Nier–Johnson geometry).

The deflection radius r_m thus depends on the m/z value of the particle, the acceleration voltage U, and the magnetic field strength B. The equations given above directly allow the recognition of some device-specific peculiarities with regard to the ion detection.

In the simplest mode of operation of a double-focusing mass spectrometer, the strength of the magnetic field is varied (magnet scan) while the acceleration potential U and the potential in the electrostatic sector are kept constant. This effects that ions of different m/z values leave the analyzer through the exit slit at different magnetic field strengths. High scan rates can be realized with this mode of operation, but the measurements cannot reach the maximal sensitivities, accuracies, and resolutions provided by the instrument. Magnetic scan allows the measurement of routine spectra (EI and CI) and the linkage to chromatographic methods (e.g., GC).

If the instrument is operated in the *electric scan* mode, then the magnetic field strength B in the magnetic sector is kept constant. The acceleration potential U is varied instead, and the voltage in the electrostatic sector is correlated with the acceleration potential. Higher resolutions and accuracies can be attained with this mode of operation, but the resolution of the measurement is also not the best in this case.

A sector field mass analyzer allows the so-called *peak matching*, which is a technique for the most accurate determination of ion masses. In the peak matching mode, two components—a known reference compound and an unknown sample compound of similar mass—are ionized and measured simultaneously. The measurement is done at a constant B field and by the synchronized variation of the acceleration potential U and the electrostatic field strength in sector E. In contrast to the electric scan mode described above, the potential U is not varied to cover the full mass range of a spectrum. It is rather modified alternatingly within a small range just to exactly match the two signals. The additional voltage ΔU needed to match the two signals can be determined with high precision and, since the mass of the reference ion m_r is well known, the mass of the unknown component can m_x be determined with high accuracy according to

$$m_x = m_r \cdot \frac{U_r}{U_r + \Delta U}$$

The peak matching mode is usually the mode of operation for high-accuracy measurements that allow the determination of elemental compositions of ions. The high resolution of the analyzer (above 60,000 (10% valley definition)) often allows also the separation and characterization of isobaric ions—ions with the same mass number but different chemical formulas.

Due to the field-free region between the ion source and the analyzer, metastable ions also can be observed with a sector field instrument. A special acquisition mode allows the detection of the broad transition signals that can be very helpful for the elucidation of mass spectrometric fragmentation mechanisms.

Quadrupole Mass Analyzer (Quad). Most of the mass spectrometers in use are equipped with a quadrupole mass analyzer. The respective instruments are easy to handle, compact, and cost-effective. The Quad can be operated in combination with most of the ion sources and also coupled to chromatographic methods due to their high scan rates (5,000 $m/z\ s^{-1}$). In contrast to other analyzers, however, they are characterized by rather low mass ranges (up to ca. 4,000 m/z) and low-resolution powers (up to ca. 3,000).

A quadrupole mass analyzer consists essentially of four hyperbolically shaped metal rods that act in pairs as electrodes (**Fig. 4.25**). The opposite electrodes are held at the same potential composed of a direct voltage (DC voltage) U overlaid with an alternating voltage (AC voltage) $+V_0 \cdot \cos \omega t$ or $-V_0 \cdot \cos \omega t$, respectively. This generates an electric field within the cavity of the four rods that leads to stable oscillating trajectories along the z-axis for ions of specific m/z values. Thus, ions with these m/z values pass the analyzer unaffected (solid line) and reach the detector while ions of different masses get quenched by collision and discharge at the metal rods (dashed lines).

It is not possible to analyze several ions of different masses at the same time with a Quad analyzer as can readily be understood on the basis of the instrument's operating principle. The voltages applied to the electrodes can, however, be varied within µs, which allows a very rapid scanning of the mass range and almost continuous detection of ions with different mass numbers.

The combined operation of three quadrupole analyzers in succession (*triple quadrupole* system; QqQ) offers the possibility of a number of specific tandem mass spectrometric experiments that are beneficial for the solution of problems such as the

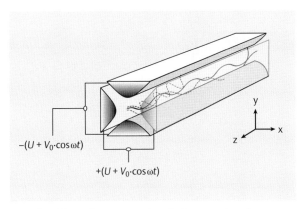

Fig. 4.25 Schematic representation of a linear quadrupole.

quantification of analytes in complex matrices or the structural elucidation of unknown compounds (Section 4.3.5, p. 326).

Ion Trap Mass Analyzers (QIT, FT-ICR, and Orbitrap)

Ion trap mass analyzers are operated in a pulsed fashion. Ions are accumulated in a first step and subsequently separated and detected as ion packages. Three fundamentally different major types of ion traps are currently in use: *quadrupole ion traps* (QITs), *Fourier-transform ion cyclotron resonance* (FT-ICR) analyzers, and *orbitraps*. The three analyzer types differ profoundly with regard to operating principle, scan rate, mass accuracy, and mass resolution.

Quadrupole Ion Traps. Two major types of QITs are operational: the older and widely spread cylindrical QIT (also called *Paul trap*) and the more modern linear QIT. QITs allow very high scan rates of up to 52,000 m/z s^{-1} and reach in the full scan mode sensitivities that are much higher than those of conventional quadrupole analyzers. The mass ranges of the instruments extend to approximately 6,000 m/z, and the resolutions are relatively high with up to 20,000 (FWHM). The analyzers are particularly well suited also for MS^n measurements in which up to 10 cascades of ion isolation with subsequent activation and fragmentation can be realized (MS^{10}).

The cylindrical ion trap is described here in some more detail. It consists of a central ring electrode and two end-cap electrodes (**Fig. 4.26**), which keep the ions by means of time-variable and three-dimensional electric fields on specific orbits that are m/z-dependent. This allows the specific accumulation (isolation and storage) of ions for a short time.

The amplitude of the alternating radio frequency (RF) field applied to the end-cap electrodes is then successively enhanced to acquire a mass spectrum. The accumulated ions are sequentially expelled according to their m/z value from the ion trap and transferred to the detector. With scan rates of up to 52,000 m/z s^{-1}, up to 20 spectra s^{-1} can be acquired over the full mass range of the analyzer in the alternating polarity mode. This makes the QIT analyzer ideal to be combined with chromatographic methods.

QIT cells are charged with a light "buffer" gas, usually He (0.133 Pa). The trapped ions get decelerated (cooled) by collision with this gas and are forced to the center of the ion trap. The m/z scan then starts for all ions from the same origin, which leads to the enhancement of accuracy and sensitivity of the measurement. The background gas can be used at the same time as collision gas for CID in tandem MS (MS^n, tandem-in-time MS, p. 328).

QIT analyzers are characterized by a higher sensitivity (over a high mass range and in the MS^n mode) as compared to conventional quadrupole analyzers because they collect and analyze all ions of interest at the same time. A considerable disadvantage of QITs is, however, their *low mass cut-off*. For a given RF voltage at the ring electrode that is optimized for a specific (ideal) ion mass, only ions with masses larger than >1/3 of this ideal mass can be collected and analyzed. The sensitivity of QIT analyzers, thus, is not constant over the full mass range, and the RF voltage must be adjusted, depending on the samples, to ensure that no signals get missed.

Fourier-Transform Ion Cyclotron Resonance Mass Analyzer. FT-ICR instruments are at present the most powerful mass analyzers. They impress with highest sensitivities with a detection limit of ca. 10 ions in the trap, highest resolutions of >10,000,000, and incredible mass accuracies in the sub-ppb range.

FT-ICR analyzers consist of small cylindrical cells (ICP cells) that are embedded into strong homogenous magnetic fields B along the z-axis (**Fig. 4.27**). The magnetic fields are generated by superconducting magnets and are typically of the magnitude of 7 to 15 T. The ionized molecules are introduced into the ICR cells and kept trapped there by electric fields. A strong magnetic field forces the ions to move along circular orbits perpendicular to this field. Thereby, the rotation frequency of the particles, the cyclotron frequency f_c, is dependent solely on the m/z value of the ions and not on the thermal energies of the particles. This allows the correlation of the mass numbers of ions directly with the cyclotron frequencies according to the following equation:

$$f_c \propto \frac{1}{m/z} = \text{constant at a constant field } B$$

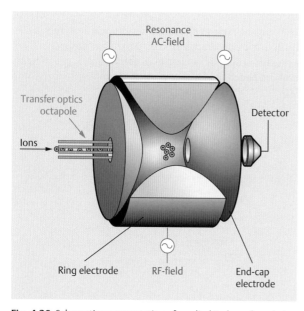

Fig. 4.26 Schematic representation of a cylindrical quadrupole ion trap (QIT).

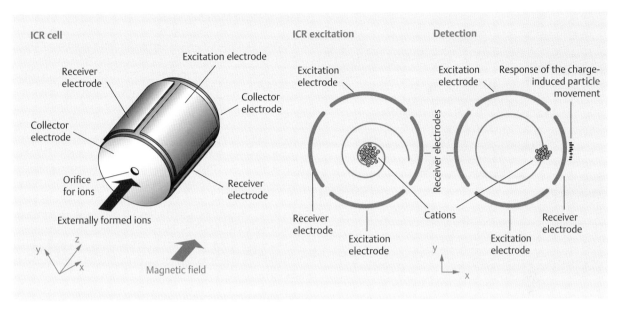

Fig. 4.27 Schematic representation of an ICR cell.

While the cyclotron frequencies (or the angular velocities) of the ions of the same mass numbers are constant (at a given B field) and independent of the velocities of the species, the radii of the circular orbits are a function of the thermal energy of the particles. By irradiation of an electric AC pulse in the RF region along the x-axis (excitation electrode), all ions trapped in the ICP cell are simultaneously accelerated and forced onto orbits with larger radii. The ions then pass the receiver electrode (y-axis) in close proximity, inducing an RF signal (an *image current*). This signal is registered as the overlay of all ion signals. It is amplified and transferred from the time to the frequency domain by Fourier transformation and converted to m/z values. The frequency amplitudes of the individual signals correspond to the ion abundances.

The resolution of an FT-ICR instrument depends strongly on the available time for the observation of the image currents. Ions can circulate within an ICR cell for several seconds if it is used with a very good vacuum (super-high vacuum of $p < 10^{-8}$ Pa). Such observation periods, eternities in the time scale of MS, result in mass resolutions in the region of 10^6 and above. The possibility to perform exact frequency analyses and the independence of the mass analysis of the initial velocities of the particles lead also to very high mass accuracies in the ppb range.

The extremely high resolution and high mass accuracy of the FT-ICR instrument allow the resolution of isotopic fine structures of naturally occurring isotopes within the investigated particles, which leads to shorter hit lists in the analysis of the HR-MS data (see Section 4.4.2).

Orbitrap Mass Analyzer. Orbitrap mass analyzers are the result of the latest technical developments in MS. They deliver mass spectra over a mass range of up to ca. 6,000 m/z at high resolution (up to ca. 450,000) and high mass accuracy (<3 ppm with external and <1 ppm with internal calibration). No superconducting magnets are required for this type of analyzers, which makes them more compact and less expensive than FT-ICR instruments. The combination of Orbitrap analyzers with chromatographic or other mass spectrometric methods is also possible.

The construction of an Orbitrap analyzer is based on a central spindle-shaped electrode that is embraced by a split outer electrode (**Fig. 4.28**). The ions are introduced in a tangential direction into the electric field between the two electrodes, and they remain trapped in the cavity due to balanced attraction to the central electrode and centrifugal forces. The ions cycle around the central electrode and oscillate along the z-axis in a frequency that is solely dependent on the m/z value of the ions:

$$f_z \propto \frac{1}{\sqrt{m/z}}$$

These ion oscillations induce in the outer electrodes, which also serve as detector plates, RF signals (*image currents*) that can be converted analogous to FT-ICR to m/z values. The principle of an Orbitrap analyzer is thus very similar to that of

Mass Spectrometry

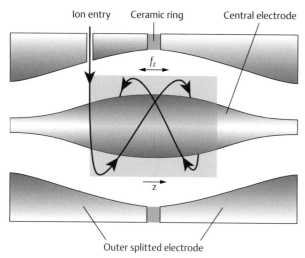

Ion entry   Ceramic ring   Central electrode

f_z

z

Outer splitted electrode

Fig. 4.28 Schematic representation of an Orbitrap mass analyzer.

an FT-ICR analyzer with the substantial difference that no RF pulses (resonance) or superconducting magnets with elaborate cooling are needed.

Commercial Orbitrap instruments use an upstream-connected curved linear ion trap, the so-called C-trap, for ion injection. This C-trap collects the ions, cools them, and releases them in bunched packages into the Orbitrap.

A broad spectrum of MS and MS^n experiments at high resolution, accuracy, and sensitivity becomes possible when Orbitraps are combined with linear ion traps and/or collision cells.

Time-of-Flight Mass Analyzers

TOF mass analyzers separate ions on the basis of different flight times. Ions of different masses have, after their acceleration by an electrostatic field, different velocities and will therefore need different times to pass a given distance. TOF analyzers are the ideal instruments to separate and study ions that are distributed over a large mass range and for ions that are generated or provided in a "pulsed form", such as by MALDI or by an ion-mobility device. TOF analysis is very sensitive because practically all ions that are released from the ion source into the analyzer are detected. The mass range is practically unlimited (above 100,000 m/z) and, due to the high scan rate, up to 500 spectra s^{-1} can be measured in the full scan mode. The instruments were initially characterized by rather low resolutions and accuracies, but these problems have been resolved for the most part by now (resolutions up to ca. 80,000 and accuracies of ≤1 ppm).

The schematic representation in **Fig. 4.29** shows the construction of a typical MALDI-TOF mass spectrometer. In the shown case, bunched ions packages are formed by pulsed laser excitation on the MALDI plate—if other ion sources are used, then the ions to be analyzed have to be brought into a pulsed package form by other means. This can be done for sources that generate ions as continuous ion beams by the use of a cooling cell, ending with a gate and a transfer lens that collects, stores, and releases ions as packages on demand.

The time measurement is triggered in MALDI-TOF by the ionizing laser pulse that is irradiated onto the sample plate. During

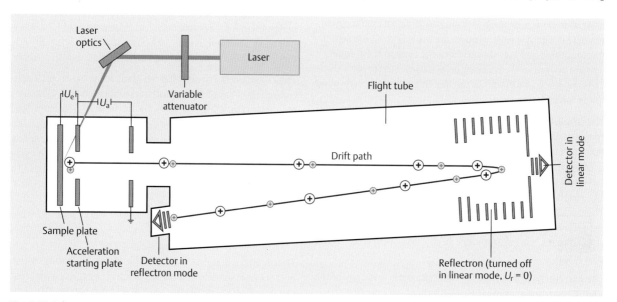

Fig. 4.29 Schematic representation of a MALDI-TOF mass spectrometer.

the desorption process, the sample plate is kept field-free, but immediately after the ions are generated, they are transferred to the aperture of the acceleration starting plate by applying a short (ca. 100 ns) positive or negative extraction voltage U_e. The compact ion packages accumulated this way are then accelerated toward the flight tube by a potential U_a (5–20 kV). At the junction to the flight tube, all ions should have in the ideal case the same kinetic energy and, thus, different mass-dependent velocities. The ions of different speeds finally traverse the field-free flight tube (length of up to ca. 3 m) to the detector where they are registered after different times. The dependence of the flight time on m/z of an observed particle is described with the following relation:

$$t_{TOF} \propto \sqrt{m/z}$$

which shows that lower-mass ions travel faster through the tube than ions with higher masses. The flight times (t_{TOF}) are finally converted to m/z values by calculation.

If the flight times are determined after the ions have directly crossed the flight tube, the respective mode is called linear TOF. This corresponds to the original measuring mode of a TOF instrument. The resolution in this measuring mode is, however, rather low due to the finite duration of the ionization pulse (time distribution), the volume within which the ions are formed (space distribution), and the energy distribution of the ions. This problem can be addressed with several measures at the stage of the ionization but also with the analyzer.

Differences in kinetic energies of the particles can be compensated by reflecting the ions at one end of the flight tube in the so-called reflectron and detecting them with a second detector after they have passed the tube back. Because ions of higher energy (and, thus, higher velocity) penetrate the reflectron more deeply before they are reflected and sent back to the other end

of the flight tube, they have to pass a longer distance to the detector, which compensates for the speed differences and leads to more compact ion packages. This and also the doubling of the overall flight distance achieved with this arrangement lead to distinctively enhanced resolutions (up to 80,000 (FWHM)) as well as higher mass accuracies (up to <1 ppm). The reflectron mode is optimal for the analysis of molecules that are stable and not too large (up to ca. 20,000 u); ions that decompose between the ion source and reflectron are lost for the measurement. Linear TOF is thus used (by switching off the reflectron; $U_r = 0$) when maximal sensitivity is demanded, and the reflectron mode ($U_r \neq 0$) when high resolution and high accuracy are required.

4.3.4 Detectors

Destructive Detector Systems

The ions generated in a mass spectrometer are, except for FT-ICR and Orbitrap instruments, detected by destructive methods. They collide after their exit from the mass analyzer with a collector unit. This can be a Faraday cup that registers the ion stream directly as an electric current. Because the incoming particles lead to very small signals only, an amplification device is often used though. By collision of the ions with a high-voltage dynode or a scintillator plate, electrons or photons are formed as secondary particles. These can be multiplied and eventually deliver signals that are strong enough to be digitalized and transmitted to a computer. In the several detector systems currently in use, the most widespread secondary particle amplifiers are *secondary electron multipliers* (**Fig. 4.30**), *photomultipliers*, and *microchannel plates* (a special version of a photomultiplier).

Two important aspects have to be mentioned: spectra are never generated without any noise signals, and ion responses usually depend on the ion masses.

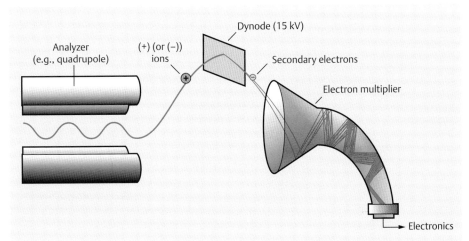

Dynode (15 kV)

Analyzer (e.g., quadrupole)

(+) (or (−)) ions

Secondary electrons

Electron multiplier

Electronics

Fig. 4.30 Transmitting dynode secondary electron multiplier as an example of the most commonly used detector in quadrupole and ion trap instruments.

Mass Spectrometry

Noise in a mass spectrum originates on the one hand from signals derived from metastable ions or residual gas that are registered stochastically over the full mass range of the spectrometer. On the other hand, it also comes from the electronics of the instrument that has to register and amplify very small currents. The interpretation of MS data and, in particular, the interpretation of signals of low intensities must therefore be done with appropriate care.

The fact that the ion response in MS detectors is mass-dependent and not proportional to the effective ion population is of particular importance and relevance for quantitative MS analyses. The effect is especially pronounced with MALDI-TOF-MS, where the measurements are done over very large mass ranges and where the very heavy ions can often no longer lead to the formation of secondary particles upon their impact onto a dynode or a scintillator. Careful calibrations with external or internal standards are thus always required for quantitative MS analyses.

Fourier-Transform Detector Systems

FT detector systems are inherently connected to FT-ICR and Orbitrap analyzers. They are, in contrast to the methods described above, nondestructive. The ions are not quenched and destroyed in an FT detector but rather used to induce alternating currents with frequencies that stand in direct correlation with the mass numbers of the ions. The detected wave signals are converted by Fourier transformation from the time domain to the frequency domain and finally to m/z values. Since the signal amplitudes are directly proportional to the ion populations, FT detectors are the devices of choice for the accurate determination of isotope ratios. However, FT-MS instruments do not allow relative quantifications of ions with very different masses because ion-trap instruments suffer from a mass-dependent ion discrimination (low mass cut-off). This discrimination is not solely an effect originating in the detector; other instrumental units and instrumental properties can affect ion formation and ion stability and, hence, ion populations as well (e.g., ionization methods, ion transfer electronics and orifices, excitation electronics, and others). Careful evaluation of the methods and calibrations are thus also indispensable with FT detectors.

4.3.5   Tandem Mass Spectrometry

Tandem MS is a technique that allows the investigation of selected ions by several consecutively performed mass spectrometric experiments. To this purpose, ions of a specific m/z value are selected first, and then the so-called precursor ions (e.g., P^+) are submitted to a subsequent MS experiment. They are usually excited and brought to fragmentation, and the resulting fragment ions F_1^+, F_2^+, etc. are separated and detected by MS.

Depending on the instrumental setup, the MS cascade can be extended over several cycles by selection of fragment ions and their use as new precursor ions until no species remain to be studied (MS^n; n = number of MS cycles).

Several methods are available for the excitation of precursor ions such as the collision with inert gases (CID), the collision with slow electrons (*electron-capture dissociation*), the exposure to photons (*photodissociation*), or ion–ion reactions that occur with excited electron transfer (*electron-transfer dissociation*). In organic chemistry, almost exclusively CID is used in tandem MS, and mainly three types of MS/MS instruments are employed: triple quadrupole, quadrupole–TOF, and ion-trap devices. With the first two instrument types, the ions are isolated, excited, and analyzed in three spatially separated regions of the apparatus. This type of MS/MS is called *tandem-in-space*. In contrasts to this, ion isolation, excitation, and analysis occur in ion-trap instruments in the same region of the machine but separated in time, which is called *tandem-in-time*. Tandem-in-space and tandem-in-time are fundamentally different, and the two methods consequently allow a number of important alternative experiments (see below).

Tandem-in-Space

Experimental Principle. The experimental setup required for performing tandem-in-space experiments consists of (at least) three instrumental elements arranged in sequence. A first mass analyzer is used to select the precursor ions, an excitation chamber to induce fragmentation, and a second mass analyzer to separate the fragment ions (**Fig. 4.31**). All types of mass analyzers can principally be used in MS/MS instruments, and the two analyzers of an instrument can be of the same type (e.g., in triple quadrupole [Q1q2Q3]) or of different types (e.g., Quad-TOF or linear-QIT-Orbitrap instruments). The use of different analyzers in an instrument has benefits from the individual strength of the components regarding, e.g., speed of measurements or quality of MS/MS spectra.

The instrumental setup shown in **Fig. 4.31** is that of a triple quadrupole instrument (Q1q2Q3) with a collision cell for CID experiments. Precursor ions of an arbitrarily selectable mass can be sorted out with the first analyzer (Q1) and accelerated toward the collision cell. This cell is a special quadrupole (q2) that is filled with an inert gas such as Ar or N_2 at a pressure of ca. 0.133 Pa. The precursor ions that enter the cell with high translational energy collide with the collision gas, which leads to their excitation and fragmentation (CID). The excitation energy can be set by variation of the acceleration potential applied between Q1 and q2. The collision energy is usually optimized in a way that as many fragment ions as possible are formed and maximal structural information can be gained. The fragment

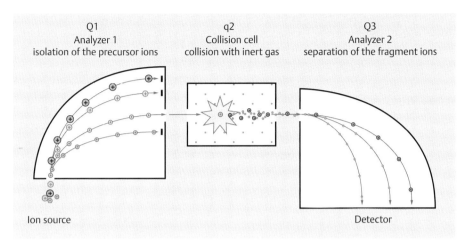

Q1
Analyzer 1
isolation of the precursor ions

q2
Collision cell
collision with inert gas

Q3
Analyzer 2
separation of the fragment ions

Ion source

Detector

Fig. 4.31 Schematic representation of the instrumental setup for tandem-in-space mass spectrometry.

ions formed in the collision cell are finally transferred to the second quadrupole analyzer (Q3), where they are separated and brought to detection.

Other MS/MS Scan Modes. In addition to the straightforward MS/MS experiment described above, delivering a spectrum of product ions for a selected precursor ion and, for this reason, is called *product-ion scan*, several further experiments are possible with triple quadrupole instruments. Such MS/MS experiments can contribute to the solution of very specific problems, and they can improve selectivity and/or increase sensitivity of particular measurements.

The *precursor-ion scan* allows the detection of all precursor ions that deliver ions of a specific m/z value. The first analyzer scans the full mass range in this operating mode, while the second analyzer is set to a single m/z value, the m/z value of the product (fragment) ions of interest with m_F. The output of this experiment is a spectrum that shows signals for all precursor ions that share a common product ion of the mass m_F. Not the precursor ions with the mass m_P themselves but the product ions with m_F are in fact detected.

The precursor-ion scan can be used to selectively identify molecules or ions that are structurally related to each other. This can be relevant, e.g., for the investigation of analyte mixtures or for the elucidation of MS fragmentation pathways (Section 4.4.8, p. 383). For example, all esters of phthalic acid show a significant signal at $m/z = 149$ (**Scheme 4.31**, **Fig. 4.73**, p. 362f). Thus, the fragment ions of this mass can be used to identify any signal in an ordinary mass spectrum that is related to phthalic esters. It is not valid, however, to conclude that all signals that are recorded in a precursor-ion scan are in fact structurally related. An ion of $m/z = 149$ can principally

be formed from many structurally very diverse precursor species.

Another possibility with tandem-in-space instruments is the simultaneous scan of the two analyzers over the full mass range with a defined mass offset ($\Delta m/z$). In this mode, signals are registered (at the m/z value of the precursor ion) only when the precursor and product ions differ in mass by Δm. This technique is called *neutral-loss scan* because it selectively delivers signals for precursor ions that lose the neutral fragments of the same specified mass.

This method allows—as the precursor-ion scan—the identification of related compounds and ions. In this case, however, focus is given to the common neutral fragments that are lost during MS fragmentations rather than to the shared structural units that remain after a fragmentation reaction. For instance, a neutral-loss scan with $\Delta m/z = 162$ could be used to selectively identify glycosides within a mixture of other compounds because the glycosides lose monosaccharide units of $C_6H_{10}O_5$ ($m = 162$ u).

The highest sensitivity is observed in MS when analyzer 1 as well as analyzer 2 are fixed to defined mass numbers, m_1/z and m_2/z. By this mode of operation, signals are only recorded for precursor ions of the mass m_1 that lead directly or via reaction cascades to product ions of the mass m_2. The technique is called *single-reaction monitoring* or *selected-reaction monitoring* or, when several distinct mass pairs m_1/m_2, m_3/m_4, etc. are selected, *multiple-reaction monitoring*.

Because of its high sensitivity and selectivity, the multiple-reaction monitoring mode is the method of choice to be used for quantifications of analytes, particularly for analytes that are incorporated into complex matrices such as biological samples

Mass Spectrometry

(e.g., urine, cell extracts, etc.). An important prerequisite for the method is, however, that the analyte and its fragmentation behavior are known.

Tandem-in-Time

The working principle of tandem-in-time MS is shown schematically with the example of a QIT instrument in **Fig. 4.32**. In the "normal" QIT-MS mode of the instrument, the ions obtained from the ion source are accumulated in the ion trap (accumulation, step 1) and subsequently released by means of electric RF fields to the detector according to their m/z values (p. 322).

Instead of emptying the trap, precursor ions of selected m/z values can also be retained by appropriate choice of the RF fields (isolation, step 2). These ions can be accelerated by the application of an RF field at the end-cap electrodes of the ion trap, which leads to collisions with the buffer gas that transfer enough energy to the ions (excitation, step 3) that they undergo fragmentation (step 4). The fragment ions are then cooled and concentrated in the center of the ion trap (relaxation, step 5) and finally expelled to the detector to deliver an MS^2 spectrum (detection, step 6). Alternatively, a new precursor ion can be selected from the fragment ions (isolation, step 2), and the process of

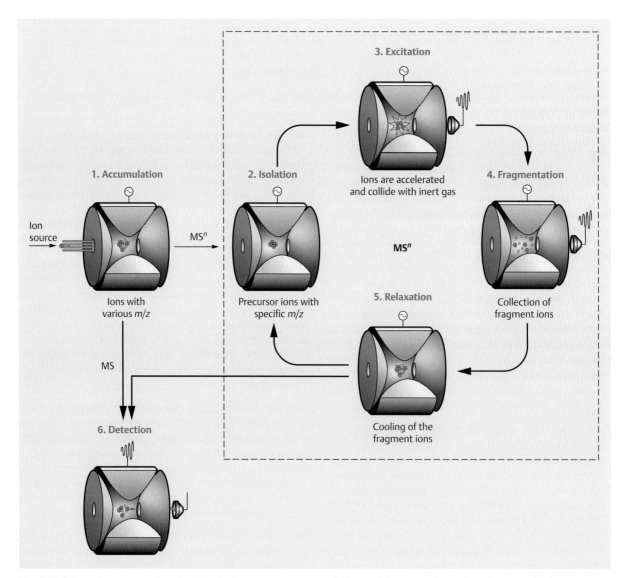

Fig. 4.32 Schematic representation of tandem-in-time mass spectrometry (MS^n, $n \geq 2$) by means of a quadrupole ion trap (QIT).

excitation, fragmentation, and relaxation (steps 3–5) is repeated to deliver an MS^3 spectrum. Provided the ion population is large enough, ions of up to the 10th fragmentation generation (MS^{10}) can be generated by repeated selection of ions and their fragmentation.

Information gained by MS^n experiments often allow the detailed elucidation of mass spectrometric fragmentation paths and mechanisms. It must be mentioned at this place, however, that the *low mass cut-off* already described for "ordinary" QIT measurements is also effective for MS^n experiments. Thus, ions with $m/z < 1/3$ of the m/z value of the precursor ions are not detected by the use of standard settings, which might represent a crucial limitation of the method for MS investigations where ions of higher and lower mass have to be observed at the same time.

Diversity of Tandem Mass Spectra

Diversity. The term MS^n is used in a rather broad way and stands for any experiment that generates fragment ions separated spatially or in time. However, despite the universal term, MS^n delivers quite different data depending on the instrument used and the measuring conditions employed. This is due to the fact that the formation of fragment ions is very dependent on the collision energy, the collision gas, and the type and construction of the collision cell.

The measuring conditions are usually optimized for MS/MS experiments in a way that the spectrum shows as many signals for different fragment ions as possible—while the signal for the precursor ion is still visible (in the ideal case with an intensity of ca. 30 rel. %). The collision energy has to be consequently adjusted substance-specifically for each measurement anew to attain optimal instrumental conditions. Modern instruments can follow preprogrammed energy ramps (called, e.g., *Smart-Frag* or *Pulsed Q Dissociation*), which increases the number of detected signals and improves the signal distribution within the observed mass range.

Because of the high diversity regarding available tandem mass spectrometers and used measuring conditions, the comparison of MS/MS data is possible only with limitations and shall be done sensibly in consideration of the effectively used parameters such as energy, type of collision gas, and instrumentation. The problematic reproducibility of MS/MS spectra is also the reason why international collections of reference tandem mass spectra are still much less extensive than databases with EI mass spectra. The diversity of spectra acquired at diverse collision energies and the sophistication of computer-aided spectral comparisons, however, are increasing, which also leads to growing MS/MS databases.

Example of an MS/MS Spectrum. The peculiarities of a tandem MS in comparison with an ordinary MS are illustrated with the typical MS^2 spectrum shown in **Fig. 4.33**. The upper spectrum in **Fig. 4.33** shows the normal spectrum obtained with amino acid derivative **21** (M = 295.1) as the sample compound, which was ionized by ESI and separated with a QIT analyzer. For the tandem mass spectrum shown below, the ions with $m/z = 318.1$ ([**21** + Na]$^+$, **Scheme 4.8**) were selected and brought to collision with He. The ions of the type [**21** + Na]$^+$ were isolated monoisotopically, within a mass range of ±0.6 m/z, which means that the [**21** + Na]$^+$ adduct ions with ^{37}Cl isotopes are expelled from the MS/MS measurement as are all other ions of the spectrum such as those registered at $m/z = 403.2, 340.1, 282.2,$ or 262.0.

The MS/MS presents itself as a clear almost noise-free spectrum with just a few lines that can easily be interpreted (**Scheme 4.8**). It is noteworthy that no isotopic patterns for the several signals can be recognized any longer, which is a consequence of the monoisotopic isolation of the precursor ion. This is advantageous on the one hand because signals of isotopomers can be distinguished from signals of alternative particles of the same mass. On the other hand, some information is lost because certain elements that show characteristic isotopic distributions can no longer be recognized in the signals of fragment ions. For instance, the two fragment ions **a** ($m/z = 262.0$) and **b** ($m/z = 218.0$) both contain a Cl atom, which, however, is not manifested with the typical X + 2 signals that are observed for Cl derivatives in ordinary MS. The presence of the element Cl in fragment **a**, however, can still be substantiated by MS because the ESI-MS shows the signal of these fragment ions too, with all the isotopic peaks to be expected for a Cl-containing particle. This is not the case for fragment **b**, which is only formed by CID. In this case, the presence of the halogen must be verified by other means, for instance by choosing a broader mass range at the isolation of the precursor ion (±3 m/z), by MS/MS of the isotopic peak, or by high accuracy HR-MS.

4.3.6 Coupling of Mass Spectrometry with Chromatographic Methods

Mass spectrometers are popular detectors for GCs, liquid chromatographs (LC, HPLC, or UHPLC), or CE because of their high sensitivity that allows the detection of even very small sample amounts. Above all, MS detectors additionally deliver structure-related information to the individual chromatographic fractions, allowing in the ideal case identification, structural elucidation, and quantification of analytes out of mixtures without their physical separation. The combination of chromatographic methods with mass spectrometers also allows the acquisition of less ambiguous MS data. It guarantees

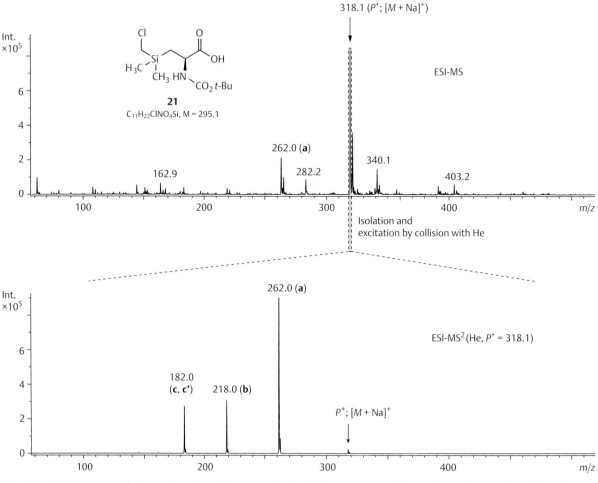

Fig. 4.33 ESI-MS² spectrum of amino acid derivative **21**, measured with a QIT instrument and with He as the collision gas (sample from S. Bienz, University of Zurich).

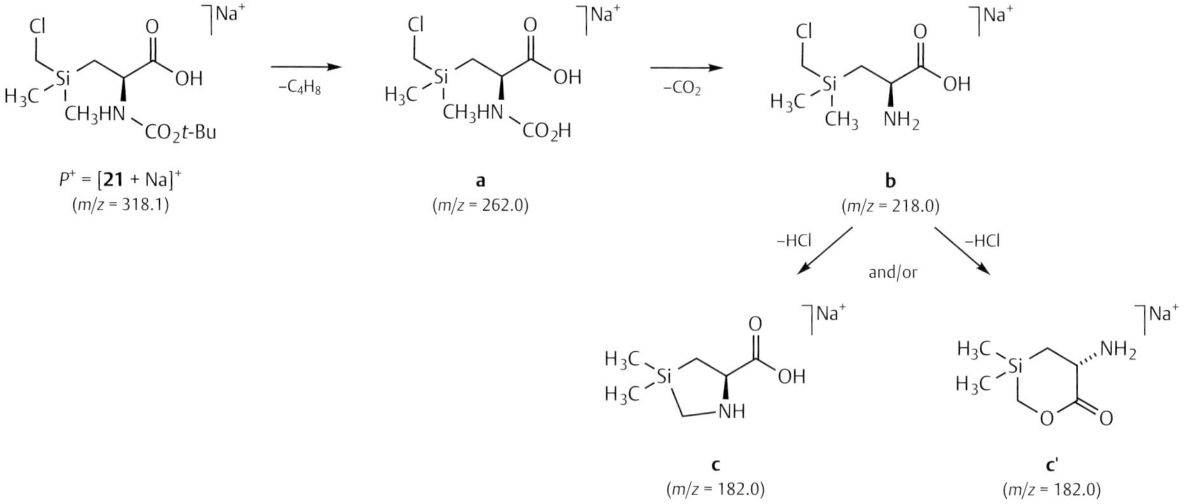

Scheme 4.8 Proposed fragmentation path for the precursor ion [**21** + Na]⁺ (= P⁺) upon CID.

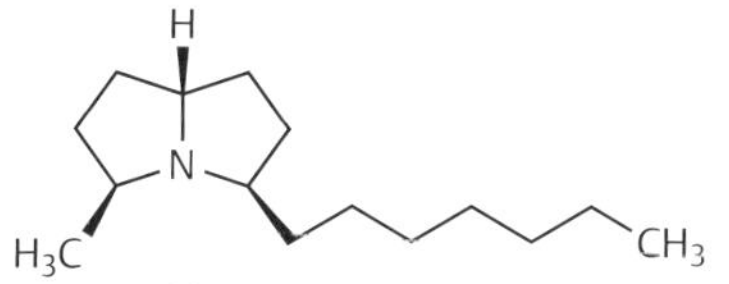

22
$C_{15}H_{29}N$, M = 223.2

Fig. 4.34 Pyrrolizidine alkaloid **22**.

a certain level of purity of the analyte that is brought for the MS analysis and usually enhances the sensitivity of an MS measurement because the individual analytes, derived, e.g., from complex samples or mixtures, can be investigated in very small sample volumes (down to below 1 µL). Naturally, analyte mixtures can also be physically separated into fractions, which subsequently can be investigated by MS. This usually requires quite some efforts and is avoided if ever possible. In some cases, however, a classical analyte separation cannot be avoided, for instance when sample components have to be investigated for biological activities or when additional analytical data are needed for structural elucidations, e.g., nuclear magnetic resonance (NMR) spectra.

Gas Chromatography–Mass Spectrometry

GC-MS is nowadays a standard analytical method in organic analytical chemistry. GC capillary columns (ca. 15–30 m) are easily combined with mass spectrometers. The end of the capillary located in a thermostatically controlled chamber (T = up to ca. 350°C) can be inserted directly into the ion source of the mass spectrometer. Sometimes, a T-piece is inserted in between the GC column and the MS entry, which not only allows the regulation of the amount of carrier gas introduced into the instrument but also splitting off a portion of the sample to be detected independently, for instance with a flame ionization detector (FID; **Fig. 4.35**).

The components that are separated by chromatography leave the column individually and can then directly (on-line) be studied one after the other by EI- or CI-MS. The connecting transfer line between the GC and the mass spectrometer must be kept as short as possible to avoid remixing of the separated components, and it must be heated to ca. 20°C above the highest temperature used in GC to prevent partial recondensation of the analytes.

The compounds separated by a GC column usually elute with a peak width of ca. 1 s, which is an important time factor with respect to the acquisition of the chromatograms and the mass spectra. A sufficiently high scan rate is required to obtain a reasonably resolved peak that can be sensibly integrated in the chromatogram. A scan rate of 7 to 10 acquisition points per peak

is the accepted minimal requirement. High scan rates are hence the prerequisite for an instrument that is intended to be used for quantitative measurements. Scan rates, however, cannot be increased indefinitely, since the sensitivity of a measurement decreases with increasing scan rates and the amount of data increases excessively too.

To illustrate a GC-MS experiment, the separation and analysis of a mixture of four compounds is shown in **Fig. 4.35**. From the original sample, it was only known that the major component **B** should be compound **22** (M = 223.2, **Fig. 4.34**). The upper chromatogram in **Fig. 4.35** was detected by FID and shows five signals: a signal for the solvent and four additional peaks for the four components **A**, **B**, **C**, and **D** of the mixture. The FID, however, does not reveal any additional information about the chemical nature of the detected components.

This is different for the EI-MS chromatogram (lower chromatogram in **Fig. 4.35**). It was recorded by the continuous reading-out of the *total ion current* (TIC) registered at the mass detector. The resultant chromatogram itself is very comparable to that of the FID-detected version, revealing the four peaks for **A**, **B**, **C**, and **D** in comparable intensities.[c] Only the solvent peak is missing, which is due to the fact that the solvent was not measured to avoid saturation of the mass detector. In contrast to the FID-detected chromatogram, the chromatographic peaks of the GC-MS contain additional hidden information. When a GC-MS spectrum is measured, full-scan EI-MS spectra are in fact continuously acquired at given time intervals, and these spectra are summarized to the TICs recorded in the chromatogram. Since the individual full-scan spectra can be retrieved for the several data points of the chromatogram, structural information about the detected analytes can be obtained. The EI-MS spectra for the several peaks corresponding to the components **A**, **B**, **C**, and **D** are shown in the lower part of **Fig. 4.35**.

Liquid Chromatography–Mass Spectrometry

The coupling of LC with MS was established in the 1990s together with the development of ESI and APCI ion sources. The LC-MS method has the great advantage that mass spectra of nonvolatile, large, polar, and/or labile analyte components can be individually and directly acquired from mixtures. For a long time, the problem with the LC-MS coupling had been that the high flow rates of HPLC (>1 mL min^{-1}) were incompatible with the high vacuums required for MS. The interface between LC and MS, however, no longer represents a problem. On the one hand, HPLC systems have been developed that allow reproducible gradient elution at flow rates of <100 nL min^{-1}, and,

---

[c] FID- and MS-detected chromatograms do not always resemble so closely as in this case because the responses of compounds in FID and MS often differ markedly.

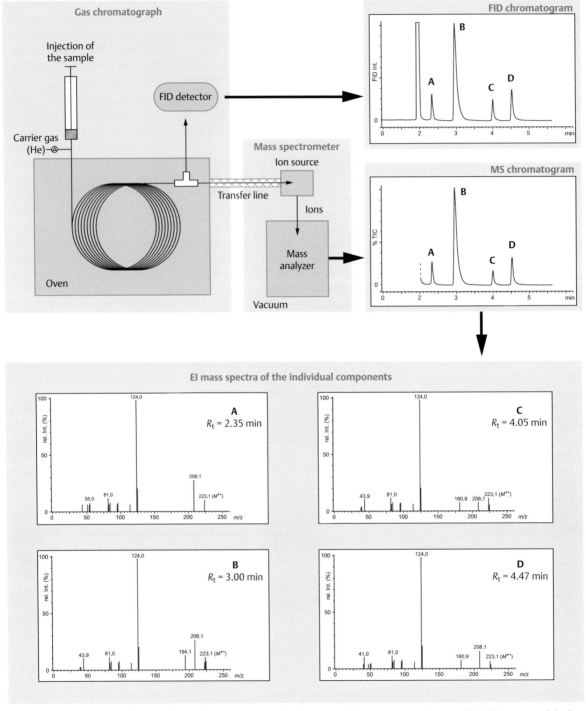

Fig. 4.35 Schematic representation of a GC-MS system, equipped with an optional flame ionization detector (FID). The spectra of the four diastereoisomeric components **A**, **B** (major component), **C**, and **D** of a mixture of pyrrolizidine derivatives were individually acquired and can be compared.

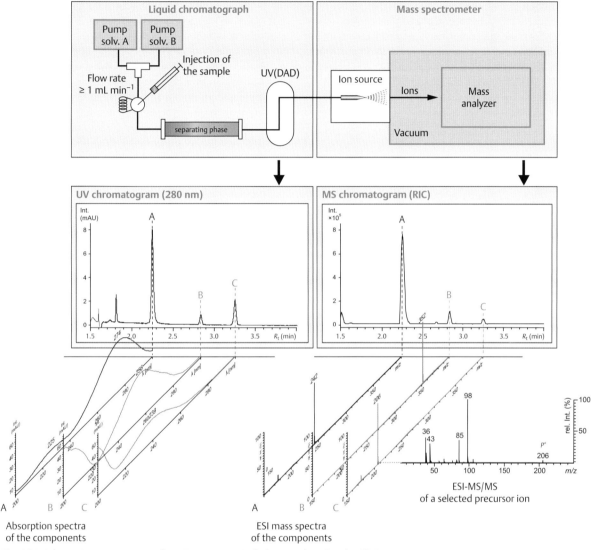

Fig. 4.36 Schematic representation of an LC-MS system, including a UV(DAD) and MS/MS.

on the other hand, new generations of ion sources and very powerful turbomolecular pumps can cope with flow rates of >500 µL min⁻¹.

The LC-MS combination is particularly well suited when reversed phase columns are used in the chromatography because polar solvents are advantageous for the ionization with spray methods. Solvent systems with H_2O as the weakly eluting and MeCN or MeOH as the strongly eluting components, admixed with volatile additives, are commonly used. The separation usually is done on packed columns of 50 to 300 mm length with particle sizes of 3 to 5 µm. The separation power of such systems is, in comparison to GC or CE, significantly lower, which

usually leads to rather broad chromatographic peaks (up to min). However, these shortcomings have also been reduced with the development of UHPLC and new stationary phases with particle sizes of <2 µm.

While GC-MS can be combined with an FID detector, LC-MS is frequently supplemented with a UV detector, inserted in between the column and the mass spectrometer as shown in **Fig. 4.36**. The combination of LC-UV-MS has the advantage that three methods that all require just smallest amounts of sample material are linked together to complement each other optimally. Two types of UV detectors are commonly used: spectrometers register a UV chromatogram either by recording the absorption at a specific, selectable wavelength (*variable wavelength detector*) or by the use of a *UV diode array detector* (UV(DAD)) that records complete UV spectra throughout the chromatographic run. The advantage of UV(DAD) is obvious: the method delivers additional analytical data also related to the individual components of a mixture that can be used—as the MS information—for the characterization, identification, or structural elucidation of the analytes.

The schematic representation of an HPLC-UV(DAD)-ESI-MS instrument with a respective data output is shown in **Fig. 4.36**. The UV and MS chromatograms recorded in the LC-UV(DAD)-MS combination differ distinctly more from each other than the FID and MS chromatograms obtained using the GC-FID-MS combination. This is due to the fact that a prerequisite for UV detection is the presence of a chromophoric group in the analyte, while MS detection is less demanding with regard to specific functionalities that have to be present. It must be mentioned at this point that the MS hardware of an LC-MS instrumentation is the same as used in an ordinary MS instrument. This means that most of the MS experiments that are possible without the LC coupling are also feasible with an LC-MS instrument, e.g., MS/MS experiments. The analytical information that can be obtained by an LC-MS/MS experiment is represented in **Fig. 4.46** with the data output shown at the bottom part of the graphic. For each of the chromatographic peaks, a UV absorption spectrum (chromophore), an MS (molecular mass, maybe chemical formula), and an MS/MS spectrum (fragmentation rules) are available. The latter is shown for reasons of clarity just for fraction C but could of course be retrieved for all fractions.

Capillary Zone Electrophoresis–Mass Spectrometry

Certain mixtures of compounds can also be separated by *capillary zone electrophoresis* (CE or CZE). For this type of separation, pKa, charge, size, and form of the molecules are of significance. The method is characterized by a particularly low demand of analyte in terms of amount (fmol to even amol), which makes its combination with the very sensitive MS method obvious. The method is very robust and allows the investigation of complex samples, such as beer, shampoo, blood, or urine, without pretreatment. It is particularly suited for the separation of small ionic species, peptides, and proteins. Neutral molecules can be separated by *micellar electrokinetic chromatography*, a modification of CE.

For a CE analysis, some nL of a sample are injected into a capillary that is filled with a buffer solution. A high voltage of some kV is applied to the capillary, which results in an electroosmotic flow directed toward the MS instrument. Ionic analytes

move—depending on their size and charge state—more or less rapidly in the same or the reversed direction through the capillary. The elution time of the individual species is finally a function of their charges, viscosities, and sizes, and of the rate of the electroosmotic flow. Because the electroosmotic flow is usually higher than the velocities of the opposite migrating analyte ions, both cations and anions can be separated at the same time with CE and brought for analysis into an MS instrument in one analytical run. The polarity of the instrument (positive or negative measuring mode) has to be naturally adjusted appropriately.

CE is an extremely efficient method that yields highest separation performances. It reacts on subtle structural differences, and even racemates can sometimes be separated into the antipodes by the use of chiral additives.

Data Analysis

A GC-, LC-, or CE-MS generates a large amount of data that are handled by computers. The chromatograms are obtained by plotting ion currents that had been acquired for the many individual MS measurements versus the measuring time. Usually, the TIC or the *reconstructed ion chromatogram* is used, which principally leads to the registration of all analytes. An alternative would be to use just the intensities of the base peaks of the individual spectra (*base peak chromatogram*, BPC). Such a selective processing of the MS data leads to chromatograms that are characterized by a higher signal-to-noise ratio because the background signals are ignored and, hence, the sensitivity for the signal of interest is increased. As an illustration for this, the TIC (a) and BPC (b) traces of a plant extract of *Arabidopsis thaliana* are shown in **Fig. 4.37**. The two chromatograms deliver virtually the same information, as can easily be recognized. The BPC, however, shows much less noise and a baseline at nearly 0% intensity.

It is also possible to create a chromatogram by registering solely the signal intensities at selected m/z values. This is called an *extracted ion chromatogram* (EIC or XIC). The chromatograms selectively show peaks for analytes of the same masses only. Examples are the three EICs of the same extract presented in **Fig. 4.37**: the EIC processed with ion signals within the rather broad mass window of $m/z = 477.1 \pm 0.2$ (c) and the two (HR)-EICs detected with ions within the narrow mass windows of $m/z = 477.10385 \pm 0.01$ (d) and $m/z = 477.06431 \pm 0.01$ (e). Of the many signals detected in the TIC or BPC, just three peaks remain in the EIC (c)—all signals of compounds delivering ions with $m/z \neq 477$ are masked out. The three signals correspond to compounds **23** to **25** that share the same nominal mass (M = 478.1; **Fig. 4.38**). As can be seen with the chromatograms (d) and (e), HR-(EIC) even allows the detection of isobaric compounds separately. Chromatogram (d) reveals only the signal for the [*M* − H]⁻ ions (*m/z* = 477.10385) of compound **23** (C₂₂H₂₂O₁₂), while chromatogram (e) discloses the two peaks for the [*M* − H]⁻ ions (*m/z* = 477.06431) of the isomeric compounds **24** and **25** (C₁₇H₂₂N₂O₁₀S₂). The latter ions obviously cannot be separated by mass spectrometric means. However, as the corresponding compounds show different retention times in HPLC, individual mass spectra for **24** and **25** can be acquired, including specific MS^n.

In summary, because each point in an MS chromatogram corresponds to an MS scan, it is possible to retrieve at each point a mass spectrum and, provided the instrument allows it, to select a precursor ion for additional MS^n experiments. Thus, a single chromatographic run provides a wealth of information and data. Depending on the scan rates and the duration of the analyses, hundreds to thousands of mass spectra might be acquired, and the evaluation of the individual spectra allow mass and structural analyses of the several components of a mixture. EI-MS and MS/MS data are best analyzed by computer-aided comparisons with spectra of MS libraries (Section 4.4.6).

4.3.7 Ion-Mobility Mass Spectrometry

Over the last few years, ion-mobility separation has been implemented to mass spectrometers providing a new separation

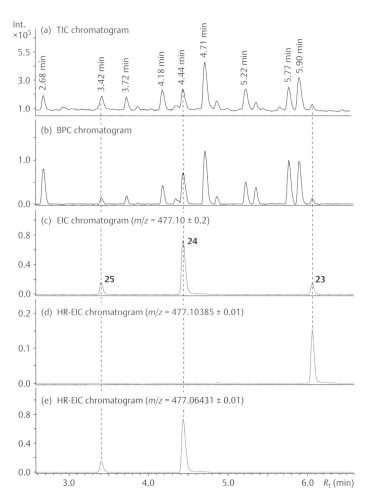

Fig. 4.37 Comparison of TIC-, BPC-, EIC-, and two HR-EIC-traces of the HPLC-ESI-MS analysis (negative mode) of an extract of *Arabidopsis thaliana* (thale cress, mouse-ear cress).

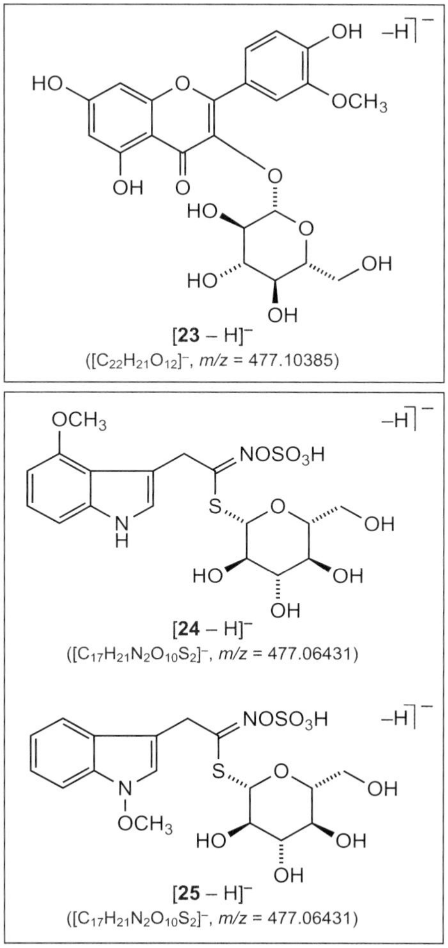

Fig. 4.38 Constituents of *Arabidopsis thaliana* (thale cress, mouse-ear cress).

dimension and also additional structural data related to the analytes. The technique is based on the mobility of ions passing through an inert carrier buffer gas. Its principle is shown in **Fig. 4.39** with the schematic representation of an ion-mobility device of the latest generation of a trapped ion-mobility TOF mass spectrometer (TIMS-TOF from Bruker).

In this instrument, ion packages are generated from a continuous ion beam in an ion-trapping section of the ion-mobility interface (section 1, ion accumulation). These packages are then released in pulsed form into the mobility separation unit and pushed through the drift tube (section 2) by means of the carrier gas (typically N_2 at a pressure of ca. 1 mbar). While pushed in one direction, the ions are retained at the same time by means of an electric field. Since the electrostatic retention forces are the same for particles of the same charge state and the pushing forces by the carrier gas are dependent on the collisional cross-sections of the ions—larger for ions with larger collisional cross-sections and smaller for ions with smaller collisional cross-sections—ions of the same charge state are separated in the drift tube according to their size and shape.

The ion-mobility separation is very efficient and allows the rapid separation (in ms) of very similar ions such as regioisomeric, stereoisomeric, or isobaric ions that cannot be separated by mass spectrometric means. It also reduces false-positive or -negative MS responses in the analyses of biological samples, e.g., by suppressing chemical noise through separation of the ions of interest from matrix interferences. Last but not least, the ion-mobility separation includes at the same time the measurement of the collision cross-sections (CCS) of the ions, which are related to the structures of the species and can—within certain boundaries—be calculated and compared.

4.3.8   Selection of the Appropriate Method

A given compound can be analyzed in most cases by several ionization methods and different types of mass analyzers. However, the type of the analytical problem that needs be solved is decisive for the choice of the most appropriate procedure. It makes a difference whether a sample is a pure analyte or a mixture of compounds or whether the ionized molecule is hard to fragment or dissociates easily. It needs to be known whether fragmentation and information to be deduced therefrom is welcome or not, whether combinations with chromatographic methods are required or not, or whether exact masses

with calculation of probable chemical formulae are needed (demanded) or not.

Not all methods are equally well suited to analyze all analytes and to solve all types of problems. Although there exist MS instruments that can be equipped and used with different ion sources, it is often not trivial to shift from one configuration to another (exceptions are the switches from ESI to APCI or from EI to CI and *vice versa*). This problem is often avoided by using several instruments that work permanently in dedicated configurations.

For the choice of the method, it is best to start with the selection of the proper ionization technique. The selection of the mass analyzer is of secondary priority—if the elected ionization method even leaves space for a further choice.

Ion Generation

The success of an MS analysis depends primarily on the proper choice of the appropriate ionization mode. MS analyses are impossible if the ions of interest are not formed or cannot be transferred to the mass analyzer. The decisive criteria for the choice of the ionization method are given by the nature of the analyte—foremost its polarity, volatility, solubility (in different solvents), and the required mass range—but also by the type of problem to be solved. Essential parameters are also the purity and composition of a sample (pure compound, compound mixture, mixture enriched with buffers, etc.), its thermal stability, and the presence of impurities such as salts or biological matrices (e.g., proteins or lipids). Additionally, the aspired information content of the analysis may never be neglected.

The several application areas of the individual ionization techniques as well as their advantages and disadvantages have already been described in the respective parts of Section 4.3.2. **Table 4.3** and the graphic in **Fig. 4.40** summarize this information concisely according to the properties of the analytes to give the user a fast and lucid decision-making support.

Mass Analyzer

The choice of the proper acquisition method comprises also the selection of an appropriate mass analyzer that can solve the given problem. It also applies here that different characteristics of an instrument are requested for addressing different

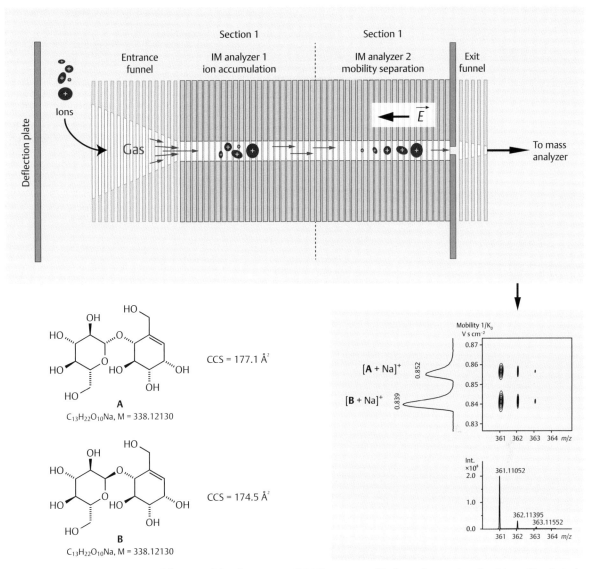

Fig. 4.39 Schematic representation of the ion-mobility device in a TIMS-TOF instrument of Bruker and separation of a mixture of two isobaric disaccharides **A** and **B** using ion-mobility. The two carbohydrates are baseline separated as their $[M + Na]^+$ adduct ions with an ion-mobility resolving power of >100 and are individually detected with a high-resolution quadrupole TOF mass analyzer.

problems. A mass analyzer needs, e.g., a high mass accuracy and mass resolution if MS should be used as a tool to determine the elemental compositions of ions. This is, on the other hand, not necessary for the analysis of polymers where the capability to measure over a wide mass range is of greater importance. High scan rates are required if chromatographic methods or ion-mobility separation are coupled to the MS; ease of handling and

robustness of the instrument should be decisive if the machine should be run in an open-access mode or for the analysis of compounds within complex matrices such as in biological samples. **Table 4.4** summarizes the most important properties of the mass analyzers that have already been described in more detail in Section 4.3.3. This overview should help you to efficiently find the appropriate instrumental setup to tackle your problems.

Table 4.3 Overview of the characteristics and the field of application of the most widespread ion sources used in organic MS laboratories

	Vaporization methods		Desorption methods	Spray methods	
	EI	CI	MALDI	ESI	APCI
Main characteristics of the sample	Vaporizable without decomposition	Vaporizable without decomposition	Polar to medium polar with M ≥ ca. 500	Soluble in polar solvents such as H_2O, CH_3OH, or CH_3CN (to a limited extend also DMF, DMSO, $CHCl_3$, or CH_2Cl_2); compatible with volatile buffers (conc. <0.1% or <10 mM, resp.)	Soluble in polar to midpolar solvents such as CH_3OH, CH_3CN, $CHCl_3$, or CH_2Cl_2 (to a limited extend H_2O, toluene, acetone, or ethers); compatible with volatile buffers (conc. <0.1% or <10 mM, resp.); easy to protonate or deprotonate or strongly electron deficient
Classes of compounds (i.a.)	Alkanes, halogenides, aromatics, esters, ketones, nitriles, nitro compounds, amines	Alcohols, ketones, esters, amines, lipids up to M = 1'000	Proteins, peptides, polymers, polycyclic aromatic hydrocarbons (PAH), polar lipids	Peptides, proteins, polymers, polar lipids, saccharides, organometallic compounds, complexes	Nitro compounds, alcohols, esters, amines, steroids
Possible decomposition of the sample	Thermal	Thermal	Thermal and photolytic (modestly)	Only as an exception	Thermal (modestly)
Suitable molecular mass (ca.)	2 – 1,000	18 – 1,500	300 →100,000	18 →100,000; (formation of M^{n+} ions)	18 – 2,000
m/z range (ca.)	1 – 1,600	10 – 1,600	100 →100,000 (not simultaneously over the full mass range)	10 – 4,000	10 – 2,000
Minimal required quantity per analysis	Picomol	Picomol	Femtomol	Femto- to picomol (1'000 more sensitive with nano-ESI)	Picomol
Common ionized molecule	$M^{+\cdot}$	$[M + H]^+$, $[M + C_4H_9]^+$ (with isobutane), $[M + NH_4]^+$ (with NH_3), $[M - H]^-$	$[M + H]^+$, $[M + Na]^+$, $[M + K]^+$, $[M - H]^-$, $[2M + H]^+$, $[2M + Na]^+$	$[M + H]^+$, $[M + Na]^+$, $[M + K]^+$, $[M + nH]^{n+}$, $[M - H]^-$, $[M - nH]^{n-}$, $[2M + H]^+$, $[2M + Na]^+$	$[M + H]^+$, $[M - H]^-$, $M^{-\cdot}$, $[M + solv. + H]^+$
Frequent on-line-coupling	GC	GC		HPLC, CE	HPLC, GC
Comments	Positive ions only; analysis of mixtures only in combination with GC; $M^{+\cdot}$ sometimes missing	Structure elucidation by variation of the reactand gas (H/D-exchange); analysis of mixtures only in combination with GC	Interference of matrix background signals with the measurement of small molecules (M <1,000); tolerates rather high salt concentrations; quantifications are challenging	Ideal for ionic substances, metal complexes, and the investigation of noncovalent interactions; polymers give complex spectra because of the formation of M^{n+}-ions	Method of choice for HPLC-MS of relatively apolar substances; a transfer line is required to connect with GC

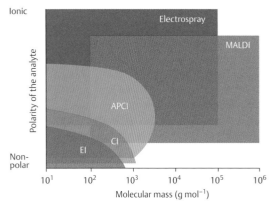

Fig. 4.40 Selection of the different ionization methods in the space of analytes' molecular mass and polarity.

4.4 Interpretation of Spectra and Structural Elucidation

MS is alongside NMR spectroscopy and single-crystal X-ray analysis probably the most important structural analytical method used in organic chemistry these days. It particularly impresses with its high sensitivity and the associated low sample demand. The deduction of structural information from mass spectra is, however, not trivial and requires a high level of expertise and experience.

To use MS efficiently and effectively for structural elucidations, a gradual approach is advised. Mass spectra are best analyzed in a first step broad-brushed and only later in close detail. The following sections are arranged according to such a line of working.

4.4.1 Preparation for the Interpretation

MS provides, depending on the method used, a large amount of data that are best subjected to a preliminary assessment. A first inspection of MS spectra provides often already advanced information regarding elemental composition and structural elements of the analytes, which facilitates the detailed evaluation and interpretation of the complete data. The following questions should be addressed in the course of an initial MS interpretation:

- Is the signal of the molecular ion or of an alternative ion that reflects the unaltered molecular composition of the analyte molecules recognizable and explainable by means of reasonable molecular eliminations or adduct formations?
- Is it possible to recognize or exclude chemical elements as constituents of the analytes by means of the registered MS signals (mass defect, isotopic distributions, and nitrogen rule)?
- Is the sample contaminated with impurities that could hamper the interpretation of the MS (or other analytical) data?

Double Bond Equivalents

It is rather straightforward and easy to postulate molecular formulas of analytes with the aid of the masses of molecular ions or other ions representing the intact original structures of the sample molecules; HR-MS often provides this information even directly and with high reliability! If the molecular formula of a compound or of an ion is known, the following simple formula

Table 4.4 Overview of the characteristics and the field of application of the most widespread mass analyzers used in organic MS laboratories

	Sector instrument	Quadrupole	Quadrupole ion trap	Orbitrap	FT-ICR	Time-of-flight (TOF)
Compatible ionization methods	EI, CI, (ESI)	EI, CI, ESI, APCI	EI, CI, ESI, APCI, (MALDI)	ESI, APCI, (MALDI)	ESI, APCI, MALDI	MALDI, EI, CI, ESI, APCI
Common mass range	10 – 6,000	10 – 2,000	50 – 6,000	50 – 4,000	100 – 10,000	10 – >300,000
Resolution (FWHM)	60,000	4,000	4,000–20,000	140,000	>1,000,000	4,000 – 80,000
Accuracy	2 – 5 ppm	<100 ppm	<100 ppm	2 ppm	<1 ppm	1 – 200 ppm
Scan rate	Low	Medium to high	Medium to high	Low to medium	Low to medium	High
MS/MS	Rare	MS^2 (with triple-quadrupoles)	MS^n ($n \leq 11$)	MS^n (with linear ion traps)	MS^2 (with quadrupole or ECD)	MS^2 (with quadrupole/TOF or TOF/TOF)
Handling	Demanding	Easy	Easy	Medium	Demanding	Easy to medium
Strength	• Resolution, • accurate quantification	• Quantification in MS^2 mode with highest sensitivity and precision, • alternating (+/−)-switching	• Sensitivity, • scan rate, • alternating (+/−)-switching, • MS/MS with ETD as option	• Resolution, • sensitivity, • MS/MS with ETD as option	• Resolution, • accuracy, • MS/MS with ECD as option	• Scan rate (currently highest), • accuracy, • sensitivity
Weakness	• Scan rate at expense of resolution	• Low accuracy, • small mass range	• Low accuracy, • 1/3 mass cut-off	• Scan rate at expense of resolution	• Scan rate at expense of resolution, • price (superconducting magnet)	• No alternating (+/−)-switching

Mass Spectrometry

delivers readily the number of unsaturations or rings (*double bond equivalents*, DBEs) contained in the structure of a charge-neutral species of the general formula $C_cH_hN_nO_oX_x$ (X stands for a halogen):

$$DBE = \frac{(2c + 2) - (h + x) + n}{2}$$

The number of O atoms does not appear in this formula! The formula can be extended to further elements whereby the elements of equal valences are combined. The general formula is:

$$DBE = \frac{[2 \cdot (\text{tetravalent} + 2)] - (\text{monovalent}) + (\text{trivalent})}{2}$$

The calculations of DBEs with the formulas above deliver integers for radical ions and half-integers for nonradical ions. The effective DBEs are the values that are obtained by rounding up or rounding down to the next integer. It must be rounded down for carbenium ions; for all other cations such as oxonium, immonium, or ammonium ions, rounding up is required. For anions that are formed by deprotonations (may be just formally), it is rounded down to the next integer; for all other anions it is rounded up (e.g., $[BMe_4]^-$).

Nitrogen Rule

Radical ions with odd mass numbers possess an odd number of N atoms (N_1, N_3, N_5, etc.) and those with even mass numbers either no or an even number of N atoms (N_0, N_2, N_4, etc.). Accordingly, nonradical ions with odd mass numbers contain no or an even number of N atoms and nonradical ions with even mass numbers have an odd number of N atoms. Thus, if the type of ion is known, which is mostly the case for molecular ions or for adduct ions, then the detected m/z values give directly evidence for an odd or an even number of N atoms in the species. For instance, an $M^{+\cdot}$ ion of m/z = 315 would contain an odd number of N atoms as would an $[M + H]^+$ ion of m/z = 316.

On the other hand, the nitrogen rule can also be used to determine whether an observed ion with known chemical formula is a radical or not. For instance, a species with the chemical formula $C_{10}H_{12}N$ (even number of H atoms, thus, even mass) is a nonradical cationic species ($[C_{10}H_{12}N]^+$), while a respective oxygen derivative with the formula $C_{10}H_{12}O$ (even mass, not containing nitrogen) is a radical cation ($[C_{10}H_{12}O]^{+\cdot}$).

Isotope Distribution

Certain elements can be readily recognized or excluded as components of MS-detected species by inspection of the isotopic distribution of the respective signals (Section 4.2.2, p. 295; **Tables 4.16** and **4.19**, pp. 414 and 421). This is particularly the

case for the X + 2 elements Cl and Br that cannot be missed as constituents in molecular frameworks with their natural isotopic distribution of ca. 3:1 ($^{35}Cl/^{37}Cl$) and 1:1 ($^{79}Br/^{81}Br$). Also, other elements reveal clearly their presence upon closer scrutiny of the spectra. The elements S (relatively high percentage of the X + 2 isotope) and Si (relatively high percentage of X + 1 and X + 2 isotopes) require a deeper analysis of the isotopic pattern. Their presence can, however, be readily excluded if the expected X + 2 isotopic peaks are missing.

The intensities of the X + 1 and X + 2 peaks of a hydrocarbon with a mass M can be roughly estimated as $I_{(X + 1)} \approx$ $(M/14) \cdot 1.1\%$ and $I_{(X + 2)} \approx (M/14) \cdot 0.012\%$. If the isotopic distribution of a signal markedly deviates from these values, it is appropriate to search for elements that could be responsible for this. Monoisotopic elements such as F, I, or P lead to X + 1 peaks that are too weak, and X + 2 elements such as S and Si lead to X + 2 peaks that are too intense.

Mass Defect

Heavy elements show a pronounced mass defect that can be recognized in the exact mass of a particle. The presence of the element I (mass defect $\Delta m \approx -0.1$ u) or the accumulation of S ($\Delta m \approx -0.03$ u), Se ($\Delta m \approx -0.07$ u), P ($\Delta m \approx -0.03$ u), or other heavy elements leads to isotopic masses that are markedly smaller than those that would be expected on the basis of the isotopic masses of compounds composed only of C, H, N, and O. Due to the mass defect of P, nucleic acids are readily distinguished from proteins.

Rational Loss of Fragments

Mass spectrometric fragmentations correspond to chemical reactions that follow common chemical principles. Mass differences between ions that are interrelated by fragmentation should therefore be chemically explainable. The compilation in **Table 4.15** (p. 406) lists frequently observed and, thus, reasonable fragment losses that can be very helpful for the interpretation of mass spectra. Mass differences that cannot be explained directly by sensible chemical transformation are most often due to two fragmentation reactions that proceed in parallel. A mass difference of Δm = 14 u between two ions, e.g., corresponds hardly to the loss of a CH_2 group but is rather the result of the loss of two homologous fragments from a common precursor ion, for instance the loss Me and Et.

The transition of ions with even mass numbers to ions with even or odd mass numbers (or *vice versa*) is also important with regard to the identification of rational fragmentations. In neutral processes that involve the elimination of nonradical fragments, the mass numbers of precursor and fragment ions remain either even or odd. The transition of mass numbers from even to odd (or *vice versa*) indicates the loss of a radical species. In EI-MS, where the ionization process usually produces radical cations, the energy excess of the $M^{+\cdot}$ ions is generally depleted early by the loss of radical species. Subsequent fragmentations are normally neutral processes. Fragmentations that occur after CI or upon collision activation (CID) are usually neutral processes; radical reactions are very rare and limited to special functional groups only.

4.4.2   Structural Information from HR-MS

The knowledge of the chemical formula of an analyte and its fragments can be of decisive importance for the characterization of a sample compound, in particular of an unknown sample compound. Powerful mass analyzers that work with high resolutions and high accuracies (<5 ppm) can directly provide chemical formulas by scrutiny of the MS signals. In this regard, the mass accuracy of an instrument is more important than its resolution.

In contrast to elemental analysis by combustion, where the sample purity is of central importance, HR-MS can propose chemical formulas also directly for components of mixtures. A prerequisite, however, is that the ions of the several components can selectively and independently be detected. From an MS point of view, this is only problematic for isobaric or isomeric ions. The resolution needs to be very high to allow the mass separation of ions of the same mass numbers but different chemical formulas, and the separation of isomeric ions is not possible by mass spectrometric means alone. These problems may be solved only in combination with chromatographic methods and/or with ion mobility.

Because fragment ions are also accurately registered and analyzed, HR-MS delivers information that goes far beyond the chemical formula of an analyte and, thus, far beyond the information that can be acquired by a combustion analysis. The knowledge of the elemental composition of fragment ions is important, e.g., to recognize functional groups within partial structures of a molecule or also in connection with the identification of fragmentation reactions and the elucidation of their mechanisms (Section 4.4.8, p. 383).

HR-MS can principally be measured by the use of any ionization method and with either mode of polarity (detection of cations or anions). The accuracy of the instruments is by now that high that the masses in HR spectra can be given with five significant decimals. This high accuracy, however, is only given if the instrument was carefully calibrated. Calibrations can be done externally (prior to an analysis) or internally, when analyses of highest accuracies are demanded. Mixtures of known compounds that deliver signals over the full mass range of the intended acquisition are used as references to calibrate the mass scale at different positions in the spectrum. The *peak matching* technique is often used with sector field instruments for the most accurate determination of individual masses. In this case, an internal reference signal with a mass offset of maximal 10% with regard to the mass of interest is used (Section 4.3.3, p. 321).

HR-MS data are usually evaluated by the use of a dedicated software, which calculates possible elemental compositions for a given mass and compares the respective theoretical masses with the effectively measured value within the mass range given by the accuracy of the instrument. The length of the list of possible elemental compositions depends on the size of the molecules. It grows with increasing mass of the analyte. For instance, the data analysis of the signal corresponding to the $[M + H]^+$ ion of compound **26**, registered at m/z_{exp} = 504.22591 by ESI-MS (**Fig. 4.41**), affords a list of 3,240 possible ion compositions if a mass accuracy of 5 ppm is assumed and the elemental composition is not restricted with regard to the number of C, H, N, and O atoms, and restricted only slightly with respect to the presence of other elements. The marginal restriction in our case was defined as $C_nH_nN_nO_nP_{0-2}Na_{0-1}S_{0-2}Si_{0-1}F_{0-6}$. This example shows that a reasonable restriction with regard to the elemental composition of an analyte is necessary to reduce the hit list to a digestible size. The introduction of restraints, however, also comprises always the risk that the correct hit is filtered out. For this reason, restrictions have to be introduced considerately and expertly.

If it is considered that the monitored ion, formed by ESI, should have an even number of electrons such as, e.g., ions of the type $[M + H]^+$ or $[M + Na]^+$, the number of proposed formulas is already reduced to 1,553. It must be mentioned at this point that the lists generated this way do not allow any prioritizations according to probabilities. They merely show all chemical formulas of ions with masses that lie within the range $\{m/z_{exp} \pm$ mass accuracy$\}$, whereby the ions that possess masses closer to the experimental mass can only be regarded to a very limited extent as more probable candidates for the sample molecules.

The number of hits can further be reduced by narrowing the range $\{m/z_{exp} \pm$ mass accuracy$\}$ by the use of an instrument of higher accuracy, but the hit list may still be too large to draw sensible conclusions. For the ion with m/z_{exp} = 504.22591 described above, still 612 chemical formulas are offered when the accuracy of the measurement is enhanced from 5 to 2 ppm.

Modern HR-MS software tools fortunately use more elaborate algorithms than just the direct comparison of calculated with experimental masses as described above. If the nitrogen rule is taken into account and also the fact that the number of H atoms cannot be arbitrarily large (the number of DBE must be ≥ –0.5), then the program still delivers 211 proposals. If the operator specifies with chemical intelligence a reasonable H/C ratio, in

Mass Spectrometry

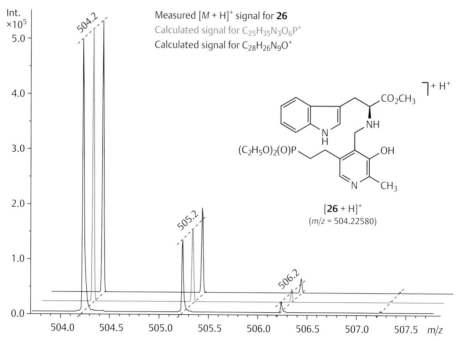

Fig. 4.41 Measured signal for the $[M + H]^+$ ions ($C_{25}H_{35}N_3O_6P^+$) of compound **26** (*black*) together with the calculated spectra for ions of the formulas $C_{25}H_{35}N_3O_6P$ (correct formula, score 100, *green*) and $C_{28}H_{26}N_9O$ (incorrect formula, score 51.7, *red*) for comparison (sample from S. Bienz, University of Zurich).

our case ≤3, to exclude suggestions with oversized N and too small C contents, the number of hits is reduced to 156. This number can even be reduced more if the minimal and maximal numbers of elemental appearances in the chemical formula are reasonably narrowed down. In our case, 97 hits are obtained if the presence of a minimum of 17 C atoms is assumed, which is the minimal number of C atoms that was estimated by the software on the basis of the isotopic peak distribution. If more information about the analyte were available, e.g., from infrared, UV-Vis, and/or NMR spectroscopy, even more restraints could be defined, which would further reduce the number of hits. The lists, however, would remain in most cases still too large.

The search for reasonable ion formulas can, however, be further refined on the basis of MS information that has not been fully considered so far. More recent HR-MS software actually does not only consider the mass of a monoisotopic signal (and the defined restrictive filters), but also the respective isotopic distribution of the overall signal, which is diagnostic for the elemental composition of the observed ion. Inclusion of these elements into the evaluation not only reduces the number of sensible hits, but also allows the allocation of probability factors (scores) to them and to sort them in a ranking list. The score is calculated, *inter alia*, from the mass deviation ($\Delta m = m_{calc} - m_{exp}$) and, with higher weighting, a factor that results from the comparison of the experimental and the theoretical isotopic distributions (mSigma or mFit, depending

on the instrument used. The fit is the better, the smaller the value of the factors is.) The score list can be reduced by defining a minimal quality of the isotopic fit. Setting the maximal allowed mSigma value to <60 returns only three ($m/z = 504.22591$) and seven entries ($m/z = 526.20715$) for compound **26** (**Fig. 4.42**).

Fig. 4.41 shows the experimentally obtained $[M + H]^+$ signal of compound **26** ($C_{25}H_{34}N_3O_6P$, $M = 503.3$) in black ($m/z_{monoisotopic} = 504.22591$) together with two calculated signals for $[C_{25}H_{35}N_3O_6P_1]^+$ ($m/z = 504.22580$, green, $\Delta m = -0.22$ ppm, correct formula) and for $[C_{28}H_{26}N_9O_1]^+$ ($m/z = 504.22548$, red, $\Delta m = -0.85$ ppm, alternative chemical formula). While the mass deviations of the observed signal from the calculated values for the two alternatively proposed ions lie within ±2 ppm, which would still be well compatible with both ion types; the mSigma values calculated for the two proposals (15.1 and 37.5, respectively) allow a clear preference of the correct "green" spectrum (score = 100.0) over the "red" spectrum (score = 51.7). The differences of the isotopic distributions for the proposed signals are indeed clearly recognizable in the lineup of the spectra in **Fig. 4.41**, but it would be difficult and arduous to determine and assess such subtle differences manually and individually in the wealth of many proposals of chemical formulas.

The example of the HR-MS measurement of compound **26** that results the "correct" chemical formula for the $[M + H]^+$ ion with a calculated score of 100 appears to be rather unambiguous.

HR-ESI Report

Meas. m/z	#	Formula	Score	m/z	err [mDa]	err [ppm]	mSigma	rdb	e⁻ Conf	N-Rule
504.22591	1	C 25 H 35 N 3 O 6 P	100.00	504.22580	-0.12	-0.23	15.1	10.5	even	ok
	2	C 28 H 26 N 9 O	51.70	504.22548	-0.43	-0.86	37.5	20.5	even	ok
	3	C 30 H 31 N 3 Na O 3	54.83	504.22576	-0.15	-0.30	39.9	16.5	even	ok
526.20715	1	C 22 H 33 N 5 O 8 P	93.98	526.20613	-1.03	-1.95	14.5	9.5	even	ok
	2	C 15 H 33 N 7 Na O 12	98.25	526.20794	0.79	1.50	18.7	2.5	even	ok
	3	C 23 H 29 N 9 O 4 P	100.00	526.20746	0.31	0.59	28.4	14.5	even	ok
	4	C 22 H 26 N 13 Na P	76.67	526.20640	-0.76	-1.44	30.4	16.5	even	ok
	5	C 25 H 34 N 3 Na O 6 P	82.46	526.20774	0.59	1.12	30.9	10.5	even	ok
	6	C 28 H 32 N O 9	82.64	526.20716	0.01	0.01	40.8	13.5	even	ok
	7	C 28 H 25 N 9 Na O	53.16	526.20743	0.28	0.52	52.0	20.5	even	ok

Fig. 4.42 Typical HR-ESI report obtained for the measurement of compound **26**.

In practice, however, it is not too rare that the score for a "correct" chemical formula is <100 and, thus, a "wrong" formula is higher prioritized. Score lists have therefore to be handled very critically and with appropriate care. If, for example, the result of the analysis of the ion registered at m/z = 526.20715 for the [M + Na]⁺ ion of **26** is considered in the HR-MS report (**Fig. 4.42**), the "correct" chemical formula $C_{25}H_{34}N_3NaO_6P$ for the registered species is found at the fifth position only, with a score of 82.46. Evidently, the deduction of the correct chemical formula of compound **26** would not have been possible if only the signal of the [M + Na]⁺ ion would have been available.

The HR-MS report reproduced in **Fig. 4.42** illustrates well the value of the presence of signals for several types of ions in a spectrum, which can be—in addition to the most common [M + H]⁺ ions—adducts with NH_4^+, Li⁺, Na⁺, or K⁺. A mass difference of 22 u between two ions such as those registered at m/z = 504.2 and 526.2 indicates with highest probability the presence of ions that differ by the exchange of a proton by a sodium ion, e.g. [M + H]⁺ and [M + Na]⁺. This structural difference should also find its manifestation in the listing of the HR-MS report. In fact, two pairs of proposals reflecting this structural variation are found in HR-MS compilation for **26**, namely the pairs of the chemical formulas $C_{25}H_{35}N_3O_6P/C_{25}H_{34}N_3NaO_6P$ (green) and $C_{28}H_{26}N_9O/$ $C_{28}H_{25}N_9NaO$ (red). Only these proposals have to be further considered as potential chemical formulas for the observed ions; all others can be excluded. With the scores of 100.00/82.46 (green) and 51.70/53.16 (red) for the two alternatives, a safe allocation of the "green" chemical formulas to the respective ions is possible.

Because mass differences within a spectrum can be determined with highest precision—with a higher accuracy than the instrumental specifications suggested—the chemical formulas of the released molecular portions are usually revealed with high reliability. Modern software tools such as *Smart Formula*

3D can include the composition of such molecular losses into their evaluations of the signals of precursor ions, enhancing the reliability of the resulting proposals.

Despite the help of increasingly more accurate MS instruments and more and more elaborate software, HR-MS also finds its limits. The number of possible chemical formulas increases exponentially with increasing mass of the investigated particles, so that it becomes very difficult to obtain reasonably short hit lists (<5 hits) for compounds with M > 1,000 u, despite data reduction and software help. The probability to filter off reasonable proposals or even the correct molecular formula of an ion by the application of constraints increases with increasing mass of the analyte as well.

4.4.3 Fragmentation Reactions in EI-MS

In EI-MS, molecules are bombarded in the gas phase with energetic electrons, converting the neutral species by removal of an electron to molecular ions $M^{+\bullet}$. The energy of the ionizing electrons, usually 70 eV, is distinctly higher than the ionization energy of an organic molecule (10 ± 3 eV). Thus, the bombardment of a molecule does not only lead to its ionization, but it transfers also enough energy onto the particle that it can undergo fragmentation. The minimal energy required for a molecule to undergo fragmentation is the dissociation energy of C–C, C–H, C–O, or other bonds, thus, ca. 3 to 5 eV. Since the energy transfer during EI usually markedly exceeds this minimal energy, EI is denoted as a "hard" ionization method that generally leads to strong fragmentation.

The ionization of molecules and their subsequent fragmentation do not proceed just stochastically. For the most part, they follow rules that can be chemically conceived and are the basis for EI-MS interpretations. The following sections describe the

Mass Spectrometry

most important and most frequently observed fragmentation reactions of organic molecules by means of concrete examples.

α-Cleavage

In functionalized molecules, the bonds in the α-position to heteroatoms X (X = N, O, S, etc.) are cleaved preferentially. The electron bombardment in the EI process is supposed to predominantly drive out an electron of the lone pair of a heteroatom, leading to a radical cation of the type **A** (**Scheme 4.9**). This radical cation would then undergo fragmentation by cleavage of the bond in the α-position to the X atom, forming cation **B** in which the ensuing charge is stabilized by the heteroatom. In addition to the charged species, the charge-neutral (and thus undetected) radical species **C** is formed. The fragmentation is thus mechanistically related to the Norrish type 1 reaction, which is alternatively also called α-cleavage in photochemistry.

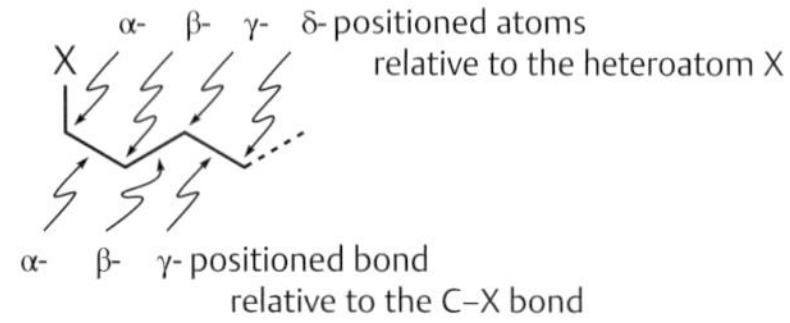

α- β- γ- δ- positioned atoms
relative to the heteroatom X

α- β- γ- positioned bond
relative to the C–X bond

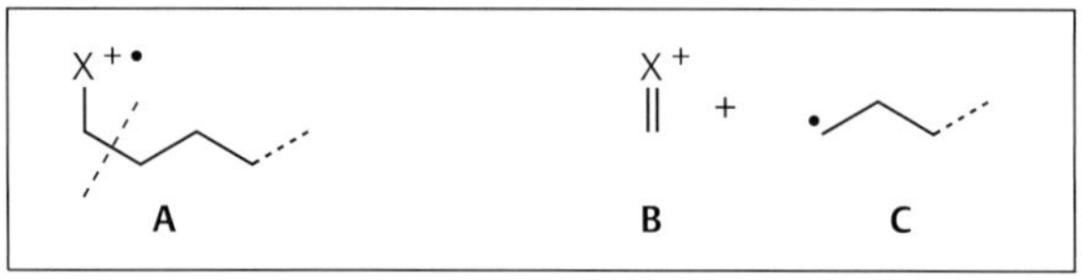

Scheme 4.9 α-Cleavage: general course of the fragmentation reaction.

Aside from a very few exceptions, α-cleavage can occur only once in a decomposition chain (sequential fragmentation reactions). This is because the homolytic cleavage in a cation that has been created by the α-cleavage of a radical cation would require too much energy.

The mass spectrum of butan-2-one (**27**, M = 72) is displayed in **Fig. 4.43**. Two characteristic fragment ions are detected for this compound at $m/z = 43$ and 57. Their masses differ from that of the molecular ion ($M^{+\cdot}$, $m/z = 72$) by 29 and 15 u, respectively, which means that the corresponding fragment ions have been formed by the loss of the radicals Et$^\cdot$ and Me$^\cdot$, respectively, from the molecular ion. The loss of Et$^\cdot$ could alternatively also be conceived as the sequential loss of the radical Me$^\cdot$ and a CH$_2$ portion. However, the loss of a CH$_2$ fragment from an ion is extremely rare, if observed at all. Therefore, the two-step process can be excluded from our considerations.

The formation of fragment ions from butan-2-one is shown in **Schemes 4.10** and **4.11**, and **Fig. 4.44**. To explain the commonly

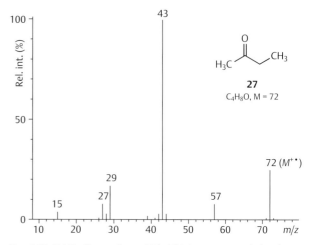

Fig. 4.43 EI-MS of butan-2-one (**27**). (This image is provided with permission of NIST [NIST 2005]).

used notations for writing mass spectrometric decomposition reactions, several possibilities are discussed in detail with this example. Under electron bombardment, one electron is ejected from the neutral molecule **27**, and the positively charged molecular ion **27**$^{+\cdot}$ is formed. This ion is registered on the m/z scale (mass per charge) at 72 ($M^{+\cdot}$, $m/z = 72$). By writing *formula* $^{+\cdot}$ it is meant that no assumption is made about the location of the charge within the molecular ion (**Scheme 4.10**). Because both fragment ions **a** and **b** result from a species with the localization of the charge and the radical at the O atom, the respective notations shown in **Schemes 4.10** and **4.11** for **27**$^{+\cdot}$ are used. The location of the charge and radical to the heteroatom (O$^{+\cdot}$) indicates that the species deriving by the removal of an electron from a lone pair of the oxygen is regarded as the starting point for drawing the fragmentation mechanism.

Regarding the structure of **27**$^{+\cdot}$ in **Scheme 4.10** or **4.11**, two bonds are present that are α-positioned to the O atom.[d] By pairing the unpaired electron on the O atom with an electron of either of the two single bonds at the carbonyl C atom, the two oxonium ions **a** or **b** are formed, which are individually represented with the signals at $m/z = 57$ and 43. The second electron of the σ-bond remains with the alkyl fragments, in this case the Me (creating Me$^\cdot$) or the Et (creating Et$^\cdot$) group.[e] The formed radicals are uncharged and thus not recorded by the mass spectrometer. The fragment ions are labeled with lower

---

[d] The bond is positioned α to the O atom and not to the C=O group!
[e] The shift of a single electron is indicated by the use of a one-sided arrow (→; fish hook) and the shift of an electron pair by a double-sided arrow (→). In principle, the shift of each electron should be indicated individually by a fish hook as shown in Scheme 4.10. We prefer however to use the abbreviated notation shown in Scheme 4.11, which is unequivocal as well.

case letters (**a**–**z**) when they are referred to in the text, and the respective numbering starts with each new scheme anew. It is quite useful to provide the mass number (*m/z* value) of the particle in parentheses under the symbol of the fragment ion. It sometimes is also useful to indicate the neutral fragments with their mass (in particular with heavier fragments). This is then done as depicted in **Scheme 4.10** for Me˙ (15 u) or Et˙ (29 u). The ions registered at *m/z* = 29 and 15 in the spectrum are *nota bene* not formed from the isobaric radicals released by α-cleavage. They originate from the ions **a** (*m/z* = 57) and **b** (*m/z* = 43), respectively, by the loss of CO (p. 363f).

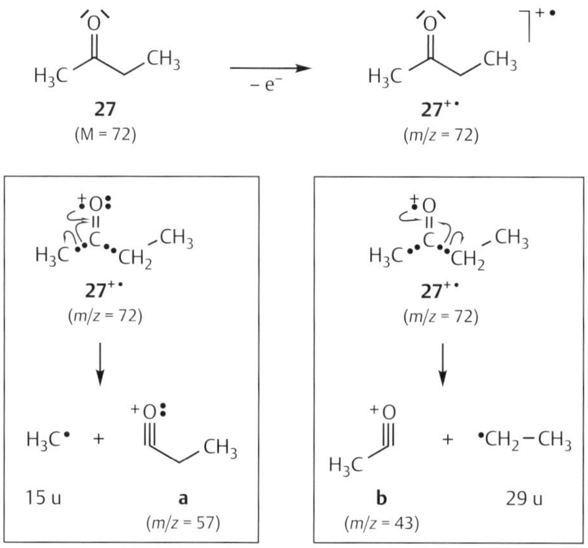

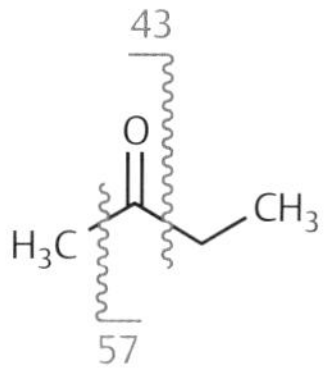

Fig. 4.44 Shorthand way of indicating the main fragmentation of butan-2-one (**27**) by α-cleavage with indication of the respective masses of the fragment ions.

The representation given in **Fig. 4.44** should be chosen if solely the main fragment ions formed by α-cleavage should be indicated without mechanistic explanation. Suitably modified schemes can also be used to indicate other fragmentation reactions.

Unsymmetrically substituted compounds of the general structure **28** (**Table 4.5**) with homologous groups R^1 and R^2 usually undergo, as a general rule, α-cleavage in a way that the heavier substituent gets predominantly lost. The analogous behavior is observed with compounds of the type **29** (**Table 4.6**).

While the α-cleavage of aliphatic compounds leads directly to the formation of fragment ions, the corresponding alicyclic compounds yield only isomeric molecular ions. Cyclohexanone (**30**; M = 98) illustrates such a case. The base peak of the spectrum (**Fig. 4.45**) is found at *m/z* = 55, and it was shown by labeling experiments that the respective ions are formed along a reaction cascade as shown in **Scheme 4.12**.

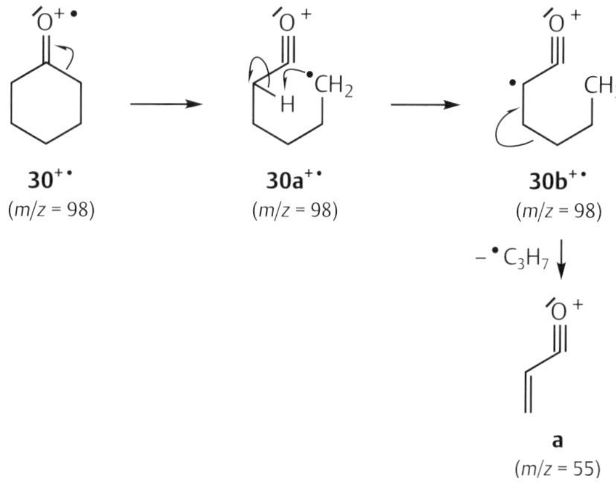

Scheme 4.10 Detailed notation for writing the main fragmentation of butan-2-one (**27**).[f]

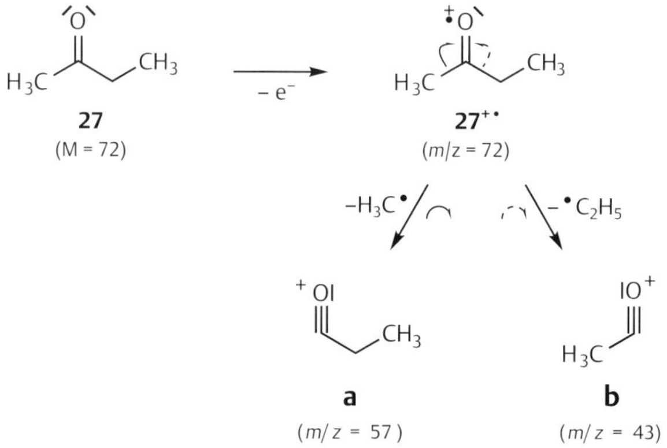

Scheme 4.12 Decomposition of the molecular ion of cyclohexanone (**30**) leading to fragment ion **a** recorded as the base peak in the EI-MS (Fig. 4.45).

Scheme 4.11 Shortened notation for Scheme 4.10.

Thus, molecular ion **30**⁺˙ leads initially to the isomeric molecular ion **30a**⁺˙ by α-cleavage. The highly reactive primary radical in this ion captures, via a six-membered cyclic transition state, an H atom from the activated α-position to the CO group and forms the subsequent isomeric molecular ion **30b**⁺˙. The radical of this ion is resonance

---

[f] The structures of the fragment ions are written such that the geometry of the original molecular ion is retained. As a result, the geometry of the fragment ions may sometimes be incorrectly represented.

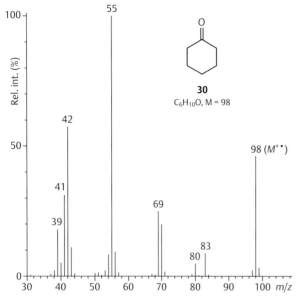

30
$C_6H_{10}O$, M = 98

98 ($M^{+\bullet}$)

Fig. 4.45 EI-MS of cyclohexanone (**30**). (This image is provided with permission of NIST [NIST 2005]).

stabilized and consequently energetically more favored than its precursor. A subsequent radical cleavage reaction, analogous to the α-cleavage, results in the formation of a propyl radical and the ion **a** ($m/z = 55$), in which the two multiple bonds are conjugated.

Substituted derivatives of cyclohexanone show an analogous fragmentation behavior. Depending upon the type of substituent and the site of substitution, alkyl derivatives of cyclohexanone may also show a signal for ions of the type **a** in addition to signals of ions homologous to **a** (**Scheme 4.13**). If, for example, a methyl group is located at the 4-position (compound **31**), the molecular ions are indeed recorded at $m/z = 112$ (+14 u as compared to **30**), but a signal still appears at $m/z = 55$, at the same mass number as the base peak found for **30**. This signal stands for both compounds for ions **a**, as can easily be conceived with the schematically indicated scissions in **Scheme 4.13**.

The situation is different for the 2- and 3-methylated cyclohexanones **32** and **33**. These lead not only to ions of the type **a** but also to those of the types **b** and **c**, respectively, both being registered at $m/z = 69$ (= 55 + 14 for methyl derivatives). The two types of ions (**a/b** or **a/c**) arise because the initiating α-cleavage

Table 4.5 Signal intensities of fragment ions that were formed by α-cleavage starting from compounds of the type **28**

	Compound	R¹	R²	a [M – R¹]⁺ rel. Int. (%)	b [M – R²]⁺ rel. Int. (%)
	Ketones				
	Butan-2-one	CH_3	C_2H_5	6	100
	Hexan-2-one	CH_3	C_4H_9	4	100
	Hexan-3-one	C_2H_5	C_3H_7	61	100
	Octan-4-one	C_3H_7	C_4H_9	75	100
	Decan-4-one	C_3H_7	C_6H_{13}	66	100
	sec. Alcohols				
	Butan-2-ol	CH_3	C_2H_5	19	100
	Pentan-2-ol	CH_3	C_3H_7	6	100
	Hexan-2-ol	CH_3	C_4H_9	5	100
	Hexan-3-ol	C_2H_5	C_3H_7	41	100
	sec. Thiols				
	Butane-2-thiol	CH_3	C_2H_5	5	100
	Pentane-2-thiol	CH_3	C_3H_7	2	100
	Amines				
	Butan-2-amine	CH_3	C_2H_5	11	100

Table 4.6 Signal intensities of fragment ions that were formed by α-cleavage starting from compounds of the type **29**

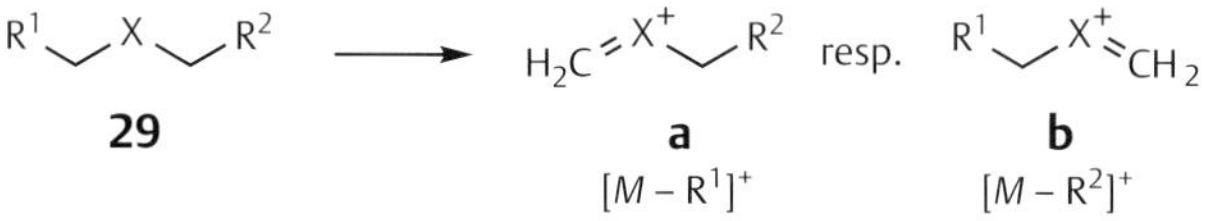

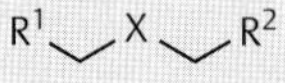

$R^1 \diagup X \diagup R^2 \longrightarrow H_2C= X^+ \diagup R^2$ resp. $R^1 \diagup X^+ = CH_2$

	29	**a**	**b**
		$[M-R^1]^+$	$[M-R^2]^+$

$R^1 \diagup X \diagup R^2$	**Compound**	**R¹**	**R²**	**a [M − R¹]⁺** rel. Int. (%)	**b [M − R²]⁺** rel. Int. (%)
$R^1 \diagup O \diagup R^2$	**Ethers**				
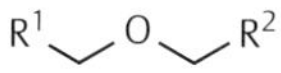	Butyl ethyl ether	CH_3	C_3H_7	2	100
	Butyl propyl ether	C_2H_5	C_3H_7	54	100
$R^1 \diagdown \underset{N}{\overset{H}{|}} \diagup R^2$	**Amines**				
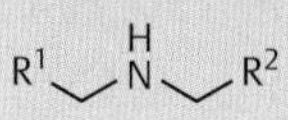	Ethyl(propyl)amine	CH_3	C_2H_5	10	100
	Butyl(ethyl)amine	C_2H_5	C_3H_7	43	100

can occur likewise to the *right* as well as to the *left* side of the molecules. Based on the knowledge of the fragmentation mechanism discussed above, the enonium ions to be formed as fragments of **32** and **33**, as well as those of compounds with other substitution patterns, can readily been predicted.

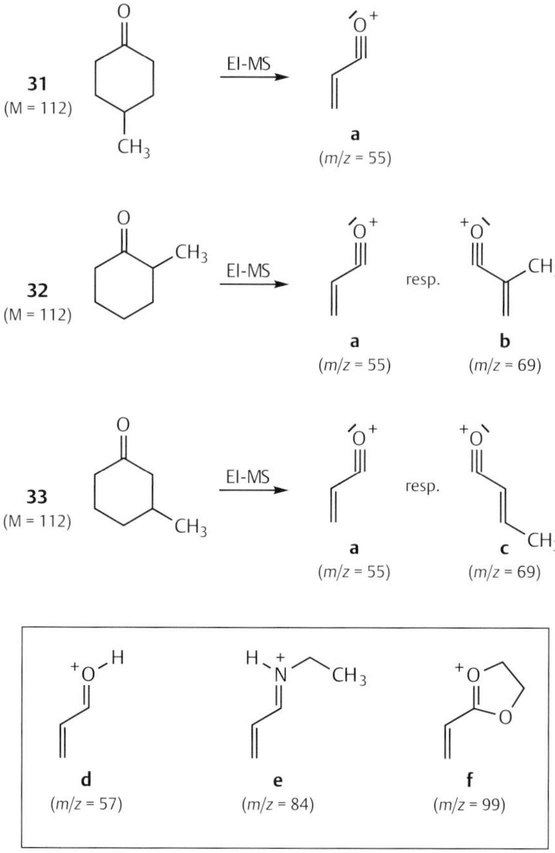

Other 5- and 6-membered alicyclic functionalized compounds decompose along similar paths as the cyclohexanone derivatives described above and deliver ions of analogous structures, such as **a**, **b**, and **c**. Cyclohexanol (**34**; M = 100), for example, leads to ions **d** (**Scheme 4.13**). Their signal is registered at $m/z =$ 57 (**Fig. 4.46**), thus, shifted by two mass units as compared to the respective ketone, which corresponds to the reduced form of the compound. Ions **e** with $m/z = 84$ are detected in the EI-MS of N-ethylcyclohexylamine (**35**; M = 127, **Fig. 4.47**) at $m/z = 84$, demonstrating that the fragmentation path is quite general and not restricted to oxygen derivatives. Very typical and prominent is also the formation of ions of the type **f** ($m/z = 99$) for ethylene acetals of cyclohexanone derivatives. The respective signal for

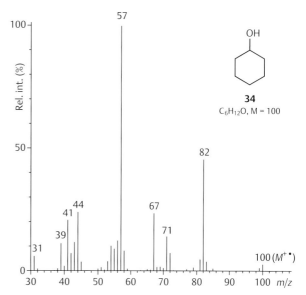

Fig. 4.46 EI-MS of cyclohexanol (**34**). (This image is provided with permission of NIST [NIST 2005]).

Scheme 4.13 MS fragmentations of substituted cyclohexanone and functionalized cyclohexane derivatives.

Mass Spectrometry

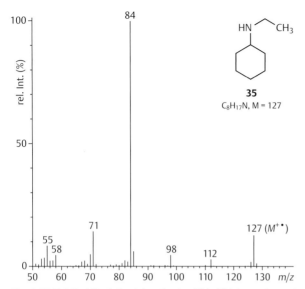

35
$C_8H_{17}N$, M = 127

Fig. 4.47 EI-MS of *N*-ethylcyclohexylamine (**35**). (This image is provided with permission of NIST [NIST 2005]).

the steroid derivative **36** (M = 318, **Fig. 4.48**) represents the base peak in the spectrum.

This example demonstrates well how strongly the presence of an ethylene acetal dominates the fragmentation behavior of a compound. If, namely, just a carbonyl group is incorporated into a larger alicyclic assemblage instead of an acetal, fragment ions that are formed by reaction cascades initiated by α-cleavage adjacent to the carbonyl group are usually detected in low abundance only. For the 5α-androstan-3-one derivative **36**, however, ions that arise from α-cleavage to the acetal O atoms dominate

the spectrum. Not only the signal at *m/z* = 99 stems from such ions, but also the signal at *m/z* = 125 (**Fig. 4.48**).

The primary fragmentation reaction for **36**$^{+\cdot}$ is the α-cleavage, which is controlled by the ethylene acetal residue. In contrast to an ethylene acetal of cyclohexanone, the two α-bonds adjacent to the functional group (C(2)–C(3) and C(3)–C(4)) are not equivalent, because the "cyclohexane ring" is substituted at C(5) and C(10). As a result, it is apparent that **36**$^{+\cdot}$ can undergo two different α-cleavages, as it was also found for the ions of **32** and **33** (see above). Both these α-cleavages are in fact operative, and the respective reaction paths that have been confirmed by D-labeling experiments are shown in **Scheme 4.14**.

Cleavage of the α-positioned C(3)–C(4) bond produces the isomeric ion **a**, which is comparable to ion **30a**$^{+\cdot}$ obtained from cyclohexanone (**Scheme 4.12**).[g] The resonance-stabilized and still isomeric ion **b** is then generated by the intramolecular transfer of an H atom from the C(2)H$_2$ group via a six-membered cyclic transition state. Scission of the C(1)–C(10) bond finally leads to ion **c**, which gives rise to the base peak at *m/z* = 99. In a similar way, the second α-positioned bond C(2)–C(3) is cleaved. This delivers the isomeric ions **d** and, after H transfer as described above, **e**. Cleavage of the C(5)–C(10) bond of **e**, however, does not result in the loss of a radical species as with **b**, but further forms an isomeric ion **f**. The tertiary radical of this species then accepts an H atom from C(6)H$_2$, again via a six-membered transition state, and leads to ion **g**, which is

---

[g] The α-cleavages of the bonds C(2)–C(3) and C(3)–C(4) are indicated in Scheme 4.14 with 3∤4 and 2∤3, respectively. This represents a further alternative to the notations given in Scheme 4.11.

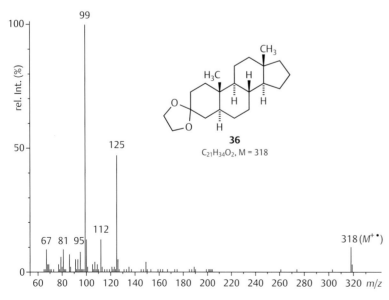

36
$C_{21}H_{34}O_2$, M = 318

Fig. 4.48 EI-MS of 5α-androstan-3-one ethylene acetal (**36**). (This image is provided with permission of NIST [NIST 2005]).

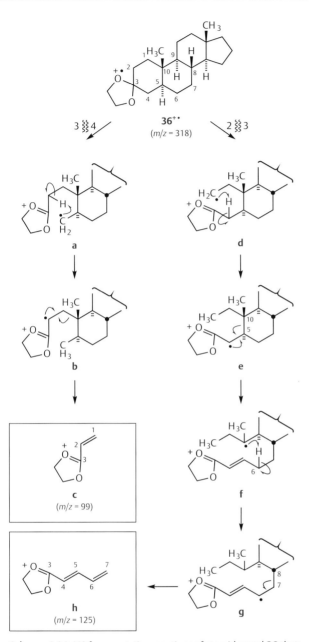

$3 \not\S\S 4$ **36$^{+\bullet}$** $2 \not\S\S 3$
($m/z = 318$)

a

b

c
($m/z = 99$)

d

e

f

g

h
($m/z = 125$)

Scheme 4.14 MS fragmentation reactions of steroid acetal **36** that lead to the spectrum in **Fig. 4.48**.

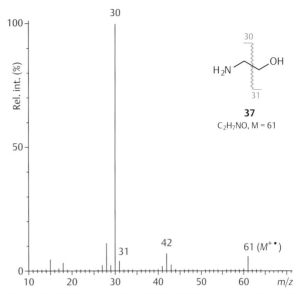

Fig. 4.49 EI-MS 2-aminoethanol (**37**). (This image is provided with permission of NIST [NIST 2005]).

37
C_2H_7NO, M = 61

As already mentioned, the ethylene acetal group is able to effect decomposition reactions that start with α-cleavage to a much greater extent than the carbonyl group. The fact that ketones, secondary alcohols , and amines induce α-cleavage to different degrees can readily be seen also by comparing the spectra in **Fig. 4.45** to **Fig. 4.47**. With compounds of the type X–CH$_2$–CH$_2$–Y, where X and Y stand for different functional groups, it is possible to observe directly the effect of such functional groups on the α-cleavage. The spectrum of 2-aminoethanol (**37**; M = 61, **Fig. 4.49**) can be used as an example. Scission of the C–C bond by α-cleavage starting with the molecular ions with the radical cation located at the NH$_2$ or the OH group, respectively, yields the ions with $m/z = 30$ (α to NH$_2$) and $m/z = 31$ (α to OH). As can be seen in **Fig. 4.49**, the intensity of the signal at $m/z = 30$, corresponding to the N-containing fragment ion, far exceeds that of the signal at $m/z = 31$. Hence, it follows that the NH$_2$ group is significantly more charge-stabilizing and therefore directs the fragmentation more strongly than the OH group.

Table 4.7 lists some experimental values that can be used to compare individual functional groups with each other with respect to the relative abundances of the ions that arise from direct α-cleavage. Although the numeric values should not be overrated, they still give a relative order of effectiveness of substituents for inducing α-cleavage in similar structural environments. While the distinct inducing effect of an ethylene acetal was already mentioned, it is also worthwhile to point out the very strong inducing effect of N,N-dialkylamino groups. Such groups are frequently found in alkaloids, and the related fragmentation reactions often permit the localization of such groups within a

still isomeric to the original molecular ion. Only now, this ion releases a major part of the steroid framework by scission of the C(7)–C(8) bond and delivers ion **h**, possessing three conjugated double bonds. The respective signal is found at $m/z = 125$. The two signals for the ions **c** and **h** dominate the spectrum of **36** and underline the particular role of the acetal group for the fragmentation of organic compounds.

Mass Spectrometry

Table 4.7 Reference values for the relative power of substituents to stabilize a positive charge in the α-position and, hence, to promote α-cleavage

Functional group	Reference value	Functional group	Reference value
–COOH	1	–I	109
–Cl / –OH	8	–SCH$_3$	114
–Br	13	–NHCOCH$_3$	128
–COOCH$_3$	20	–NH$_2$	990
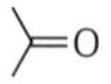 ⟩=O	43	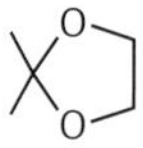 (1,3-dioxolane)	1,600
–OCH$_3$	100	–N(CH$_3$)$_2$	2,100

molecule or the determination of the position of substituents that are close to them.

In summary, it can be said that α-cleavage is the most important primary mass spectrometric fragmentation process. It can even be reinforced by modification of functional groups within a molecule, e.g., by conversion of an alcohol into a ketone, an acetal, or an amine. This allows obtaining valuable structural information of compounds, as will be seen later again with examples used to discuss other fragmentation reactions.

Benzyl, Allyl, and Propargyl Cleavage

Analogous to heteroatoms, aromatic rings, delocalized double bond systems, and isolated double or triple bonds also initiate α-cleavages: benzylic or allylic and propargylic bonds are particularly prone to be cleaved.

Benzyl Cleavage. Fig. 4.50 shows the EI mass spectrum of ethylbenzene (**38**; M = 106) with its base peak registered at m/z = 91. This signal is the response of ions formed from the molecular ion **38**$^{+\bullet}$ by cleavage of the benzylic C–C bond with the loss of a methyl radical (**Scheme 4.15**).

The high intensity of the signal at m/z = 91 indicates that the corresponding ions are very stable. This cannot be explained solely on the basis of the resonance stabilization of a benzyl cation **a**, but suggests the participation of the very stable aromatic tropylium ion **b**. It was proven that this ion in fact represents the central species, for instance, by the course of its subsequent fragmentation to ion **c** (m/z = 65) under loss of C$_2$H$_2$. While all C and H atoms in the symmetric tropylium ion (**b**) are equivalent to each other, three different types of C atoms (CH$_2$, CH, and C) and two types of H atoms (CH$_2$ and CH) are present in ion **a**; the five CH groups not being equivalent as well (*ortho*, *meta*, and *para*). For ^{13}C- or D-labeled compounds **38**, it would thus be expected to find a dependence of the labeling degree of ion **c** on the original labeling

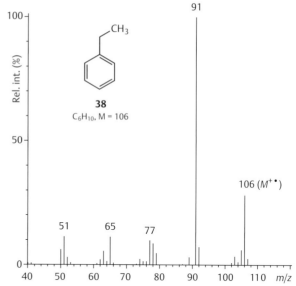

Fig. 4.50 EI-MS of ethylbenzene (**38**). (This image is provided with permission of NIST [NIST 2005]).

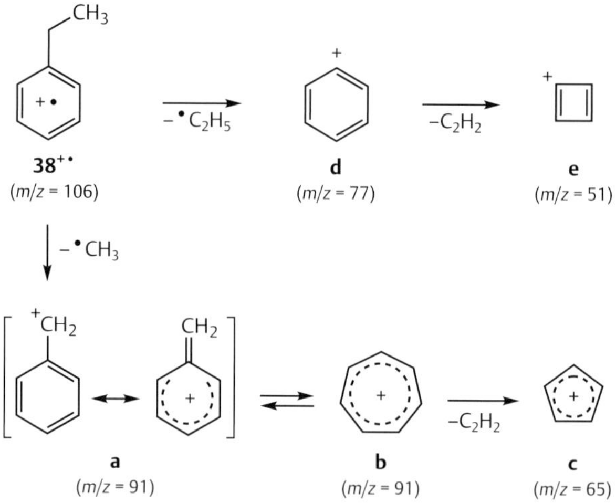

Scheme 4.15 MS fragmentation reactions of ethylbenzene (**38**) that lead to the spectrum shown in **Fig. 4.50**.

positions in **38** if ion **a** would be its precursor, whereas this would not be the case for **b**. Investigation of several alkylbenzenes confirmed the equivalence of the C atoms in the ion registered at m/z = 91, which favors the assumption that ion **b** is involved.

An abundant signal at m/z = 91 allows, as a reverse conclusion and strong argument, the suggestion of the presence of a benzyl residue in a compound of unknown structure. Weak signals at this mass number, however, are less characteristic because the highly stable tropylium ion can also be formed by more complex rearrangements.

The spectrum of **38** does not only show the dominance of benzylic cleavage in an alkyl-substituted benzene but also shows that the phenyl cleavage (formation of **d**, $m/z = 77$) is significantly less favored. Like the tropylium ion, ion **d** also loses acetylene (ion **e**, $m/z = 51$), and the pairs of ions, **b**/**c** ($m/z = 91/65$) and **d**/**e** ($m/z = 77/51$), are typical and diagnostic for monosubstituted alkyl-aromatics. The appearances of signal pairs at $m/z = 91/65$ and $77/51$ are thus strong indicators, but not proof, for the presence of monosubstituted alkylbenzenes. The same ions are also found, for example, with *o*-, *m*-, and *p*-xylene, which *a priori* would not be expected and which indicates that possibly a different or an additional mechanism is operative in the formation of the respective ions. In this context, it is interesting to compare the spectra of benzyl chloride (**39**; M = 126, **Fig. 4.51**) and those of the mono-chlorotoluenes, representatively shown with the spectrum of *p*-chlorotoluene (**40**; M = 126) in **Fig. 4.52**. All spectra of these compounds look virtually the same, aside from small differences in the relative abundances.

In line with the above discussion, the spectrum of benzyl chloride appears as expected because a Cl atom is more easily removed from the benzylic position than an H atom. This would lead to the respective benzylic cation **a** and, after rearrangement, to the tropylium ion (**b**) as observed with the signal at $m/z = 91$ (**Scheme 4.16**). The spectrum of *p*-chlorotoluene, however, is surprising. The Cl atom is directly attached to the benzene ring and, as a result, the tendency for it to be ejected by formation of an ion $[M − Cl]^+$ is small. It would rather be assumed that a chloro-substituted benzyl cation **c** (or the respective tropylium ion, $m/z = 125$) would be formed by the loss of an H· from the Me

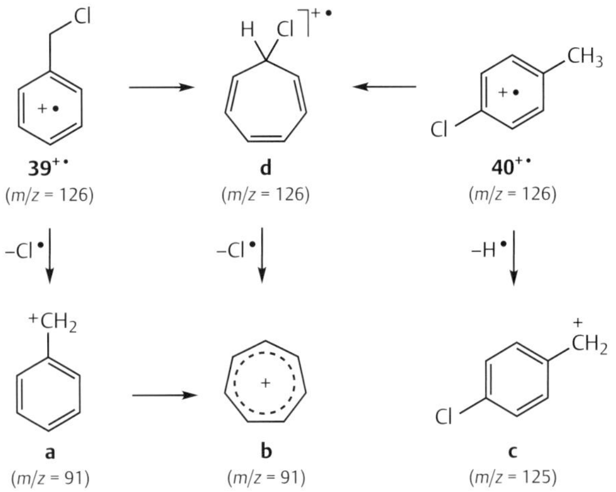

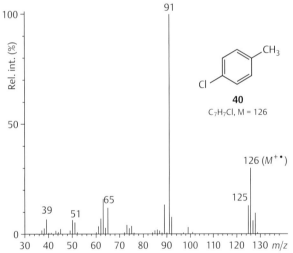

Scheme 4.16 Potential fragmentation path for the formation of the tropylium ion **b** from the molecular ions of benzyl chloride (**39**) and *p*-chlorotoluene (**40**).

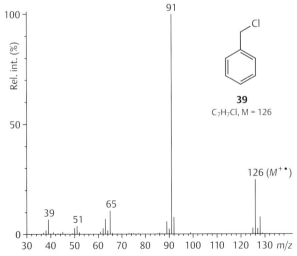

Fig. 4.51 EI-MS of benzyl chloride (**39**). (This image is provided with permission of NIST [NIST 2005]).

Fig. 4.52 EI-MS of *p*-chlorotoluene (**40**). (This image is provided with permission of NIST [NIST 2005]).

group of **40**⁺· by benzyl cleavage. However, this ion is detected with low abundance only, relative to the tropylium ion at $m/z = 91$, which corresponds to the expulsion of a Cl·. Because the two spectra of **39** and **40** are almost the same, it must be assumed that **40**⁺· rearranges to **39**⁺· or, more likely, that the two $M^{+·}$ ions isomerize to a common species, e.g., ion **d**, before they lose a Cl· (**Scheme 4.16**).

The EI mass spectrum of 1-chloro-4-ethylbenzene (**41**; M = 140, **Fig. 4.53**) indicates that both processes, direct benzyl cleavage and isomerization of the $M^{+·}$ ion, most possibly are operative

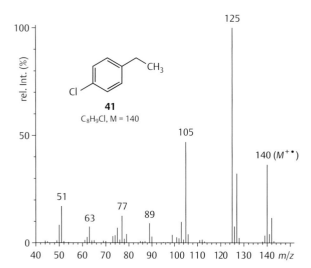

Fig. 4.53 EI-MS of 1-chloro-4-ethylbenzene (**41**). (This image is provided with permission of NIST [NIST 2005]).

in parallel. For compound **41**, the signal for $[M - Me]^+$, which corresponds to the result of the direct benzyl cleavage and loss of a methyl radical, is registered with high intensity at $m/z = 125$, but also a highly prominent signal is found at $m/z = 105$, which stands for ions of the type $[M - Cl]^+$, formed after initial rearrangement of the $M^{+\cdot}$ ion.

Allyl and Propargyl Cleavage. Allyl and propargyl cleavages are less pronounced than benzyl cleavages because the stabilization through the formation of the resulting allyl or propargyl cation is smaller. Nevertheless, such cleavages are still characteristic, and the ions formed by this process may dominate a spectrum, in particular when no other functional groups are present in a molecule.

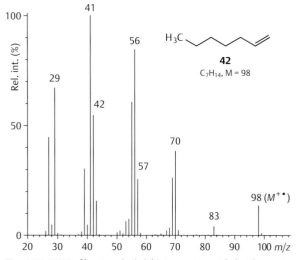

Fig. 4.54 EI-MS of heptene (**42**). (This image is provided with permission of NIST [NIST 2005]).

Figs. 4.54 and **4.55** show the mass spectra of heptene (**42**; M = 98) and 4-methylhexene (**15**; M = 98), respectively. The base peaks are found in both spectra at $m/z = 41$, derived from ions **a** (**Scheme 4.17**). Ion **a** is predominantly formed from the respective $M^{+\cdot}$ ions **42**$^{+\cdot}$ and **43**$^{+\cdot}$ by allyl cleavage because the allylic bond is the weakest bond within the two molecular assemblies and the allylic cation is the most stable cation that can be formed from the compounds by C–C scission. The complementary molecular portions, which are formed by allyl cleavage as well, are registered at $m/z = 57$. The intensity of the respective signal for compound **42** is distinctly lower (27 rel.%) as compared to the intensity of the signal due to the ions **a** ($m/z = 41$; 100 rel.%). For compound **43**, on the other hand, the intensities of the signals at $m/z = 41$ (100 rel.%) and $m/z = 57$ (95 rel.%) are comparable. The higher intensity of the signal at $m/z = 57$ for

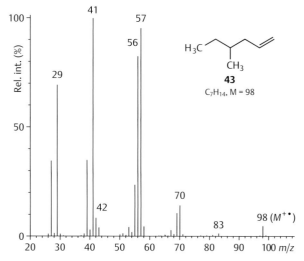

Fig. 4.55 EI-MS of 4-methylhexene (**43**). (This image is provided with permission of NIST [NIST 2005]).

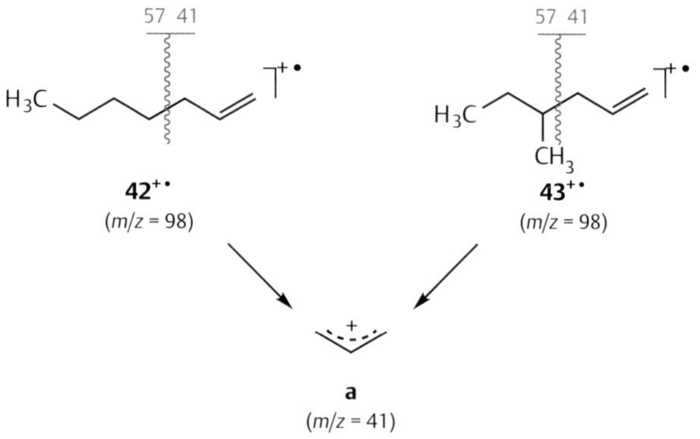

Scheme 4.17 Allyl cleavage with heptene (**42**) and 4-methylhexene (**43**).

compound **43** can be explained with the higher stability of the formed secondary carbocation (as compared to the primary carbocation formed with **42**).

A number of additional signals are found in the two spectra of the compounds **42** and **43**. The signals recorded at m/z = 42 and 56 can be explained with ions formed by McLafferty rearrangement, a common fragmentation reaction that is discussed later. Other signals, such as, e.g., those registered at m/z = 69 and 70, can reasonably be explained only by the shift of the double bond in the $M^{+\cdot}$ ion prior to decomposition. Double-bond shifts and other competing side reactions are quite common in EI-MS for alkenes, which complicates the interpretation of the spectra.

As a summary and conclusion of the examples discussed above, it can be stated that allylic bonds (the same applies analogously to propargylic bonds), in comparison to nonactivated ordinary C–C bonds, are preferentially cleaved in alkenes. The charge stabilization realized with the formation of allylic/propargylic cations, however, is not suited to reliably estimate signal intensities. Because a number of competing reactions are also possible with activated alkenes, above all double bond shifts that impede the recognition and/or the localization of double bonds within a molecule, the allyl cleavage is not a fragmentation reaction that allows very reliable conclusions to be made. With known compounds, some of the observed signals can be readily explained. Reverse conclusions, the proposal of structural units from the EI-MS of an alkene, however, are not possible. For the structural elucidation of unknown compounds, it is often better and safer to "fix" the positions of the C=C bonds of an alkene by derivatization and to analyze the derivatives by MS rather than the original sample. Suitable derivatives are, for example, acetonides of the corresponding diols or, for alkynes, carbonyl derivatives that are obtained by the addition of water to the C≡C bond.

Cleavage of "Nonactivated" Bonds

Compounds without bonds that are activated by heteroatoms (α-cleavage), aryl groups (benzyl cleavage), or C=C or C≡C bonds (allyl/propargyl) essentially undergo fragmentations with stochastic bond scissions. They nevertheless can deliver patterns with a recognition value in EI-MS. **Fig. 4.56** shows the mass spectrum of hexadecane (**44**; M = 226). This spectrum is typical of an unbranched, straight-chain hydrocarbon.

For such compounds, the most abundant signals are usually located in the region corresponding to fragments with three to five C atoms; the base peak most often corresponding to $[C_3H_7]^+$ (m/z = 57). The abundance of the homologous ions decreases almost asymptotically with the increasing number of C atoms.

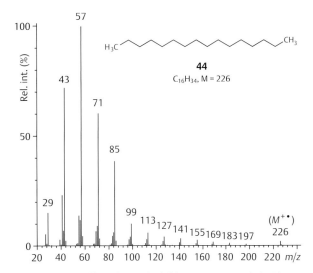

Fig. 4.56 EI-MS of hexadecane (**44**). (This image is provided with permission of NIST [NIST 2005]).

The $[M – Me]^+$ ions are usually not recorded, but the signal due to the molecular ions is always found and allows the determination of the molecular mass of the analyte. The $M^{+\cdot}$ signal becomes even more intense for larger molecules than **44**. The general profile of the spectrum (the connection lines between the highest peaks within each group of signals) is typical, and it is useful to memorize this profile, because hydrocarbons are often found as impurities in samples.

If a spectrum shows the profile of a hydrocarbon as discussed above but reveals signals that stand out from this uniform picture, then these additional signals must have a meaning for the structural analysis. The uniformity of the hydrocarbon spectra is indeed essentially the result of the fact that the cleavage of each of their internal C–C bonds results in the formation of species of the same type: a primary carbocation and a primary radical. This is no longer the case for branched hydrocarbons. In addition to primary carbocations and radicals, secondary and even tertiary carbocations and radicals can be formed as well. Since such higher substituted carbocations and carbon radicals are more stable, they are more readily formed, and the respective signals stand clearly above the general profile of the spectrum. This effect can be easily recognized in the spectrum of 7-propyltridecane (**45**; M = 226, **Fig. 4.57**), where the signals at m/z = 183/43 and 141/85 are particularly high in intensity.

While the spectrum of **45** still allows a straightforward analysis, the spectra of hydrocarbons with nonsymmetrically attached branchings and the spectra of multiply branched hydrocarbons

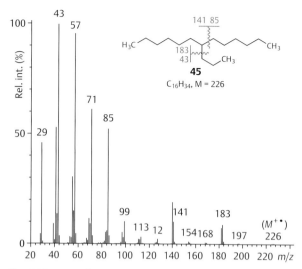

Fig. 4.57 EI-MS of 7-propyltridecane (**45**). (This image is provided with permission of NIST [NIST 2005]).

may become confusingly complex. But despite potential difficulties in the interpretation of the spectra, the mass spectrometric analysis of higher hydrocarbons is still of particular importance because no other analytical methods, with the exception of ^{13}C NMR spectroscopy, are available to be used for structural elucidations of such compounds.

The spectral profile of hydrocarbons may also become apparent for functionalized compounds that possess hydrocarbon portions. For example, the larger a hydrocarbon portion of a

monofunctionalized compound becomes, the more its spectrum looks like the spectrum of a pure hydrocarbon. E.g., the spectrum of hexadecanol (**46**; M = 242; **Fig. 4.58**) shows the typical profile of a linear hydrocarbon, overlaid with the characteristic signals for primary alcohols at m/z = 224 ([$M − H_2O$]$^{+•}$), m/z = 196 ([$M − EtOH$]$^{+•}$, for long-chained primary alcohols, **Table 4.15**, p. 408 and 411), and m/z = 31 ([CH_2OH]$^+$, product of α-cleavage, **Table 4.14**, p. 395). Contrary to the situation with pure hydrocarbons and as found in the spectrum of alcohol **46**, the molecular ions of such compounds are very often no longer detected.

The decomposition of aliphatic halogen hydrocarbons is determined only to a very small extent by α-cleavage adjacent to the halogen atom. For this reason, these compounds are discussed in this section too. Only fluorohydrocarbons show reliably well recognizable signals for ions that have arisen through α-cleavage, e.g., the signal at m/z = 33 ([CH_2F]$^+$, ion **a**, **Fig. 4.59**) in the spectrum of fluoroheptane (**47**; M = 118, **Fig. 4.60**). Other halogenated compounds show such signals only occasionally.

Particularly characteristic is the behavior of 1-chloro- and 1-bromohydrocarbons with at least five linearly arranged methylene groups. They usually form the five-membered cyclic chloronium and bromonium ions **b** and **c** (**Fig. 4.59**), exhibiting in their signals the isotopic ratios characteristic of the respective halogen (**Table 4.16**, p. 414). Iodoalkanes, on the other hand, decompose often by scission of the C–halogen bond and formation of I$^+$ (**d**, m/z = 126.9); a reaction that is less prominent for other halogen alkanes. The signal for the I$^+$ ion can readily be identified in the spectra, even in low-resolution mode, due to the distinct mass defect of the halogen (−0.1 u). To illustrate these points, the mass spectra of the four 1-halogenoheptanes **47–50** are given in **Fig. 4.60** to **4.63**.

Retro-Diels–Alder Reaction

Six-membered cyclic systems that contain a double bond can dissociate into two fragments, an ene and a diene component, by a concerted de-cyclization reaction, the *retro*-Diels–Alder reaction (RDA reaction, **Scheme 4.18**). The diene component is usually favored as the charge carrier; however, the ene part is

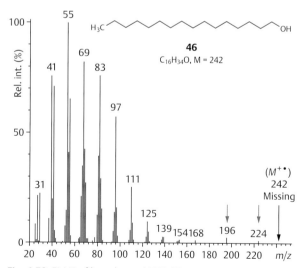

Fig. 4.58 EI-MS of hexadecanol (**46**). (This image is provided with permission of NIST [NIST 2005]).

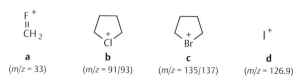

Fig. 4.59 Dominant ions derived from 1-fluorohydrocarbons, from 1-chloro- and 1-bromohydrocarbons with minimal five linearly arranged methylene groups, and from iodohydrocarbons.

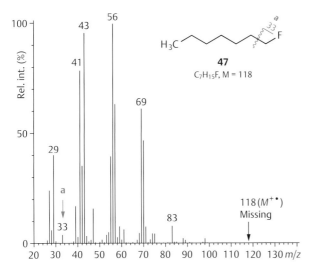

Fig. 4.60 EI-MS of fluoroheptane (**47**). (This image is provided with permission of NIST [NIST 2005]).

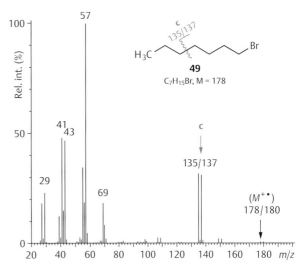

Fig. 4.62 EI-MS of bromoheptane (**49**). (This image is provided with permission of NIST [NIST 2005]).

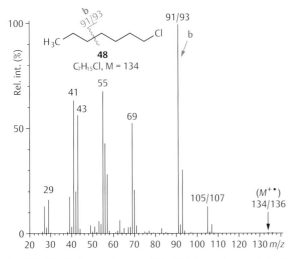

Fig. 4.61 EI-MS of chloroheptane (**48**). (This image is provided with permission of NIST [NIST 2005]).

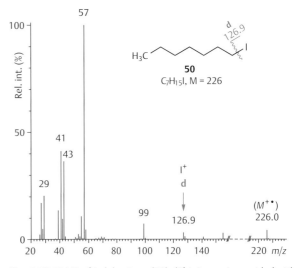

Fig. 4.63 EI-MS of iodoheptane (**50**). (This image is provided with permission of NIST [NIST 2005]).

also frequently observed in the mass spectra. The cyclohexene ring does not need to contain any heteroatoms, but it may possess one or more of them. It could also be part of a larger ring system. The RDA reaction can take place in molecular ions as well as in fragment ions in which the double bond in the ring may have been formed only as a result of another fragmentation reaction (e.g., through an α-cleavage). The RDA reaction is a neutral process in which no radical is cleaved off. This means that a radical cation precursor delivers again a radical cation product and a nonradical precursor delivers a nonradical product.

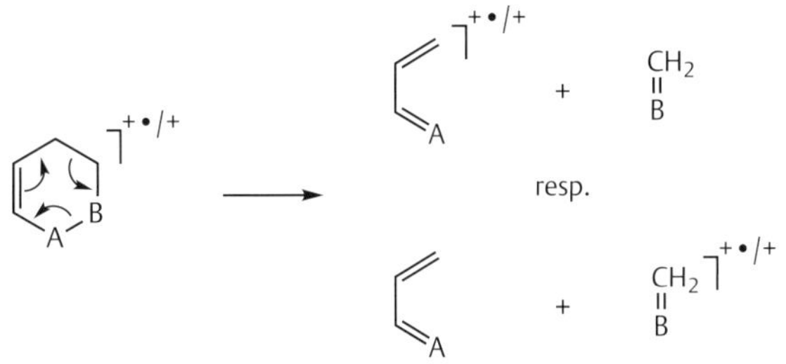

Scheme 4.18 *Retro*-Diels–Alder reaction with cyclohexene derivatives.

The RDA reaction shall be reviewed in more detail by means of the mass spectrum of 1,2,3,4-tetrahydrocarbazole (**51**; M = 171,

Fig. 4.64). The most intense signal in the spectrum of **51** is found at $m/z = 143$. It is the response of a fragment ion of the formula $[C_{10}H_9N]^{+\cdot}$ that was formed from the molecular ion $M^{+\cdot}$ ($m/z = 171$) by the loss of the $C(2)H_2C(3)H_2$ portion (**Scheme 4.19**). This was established by HR-MS and by investigations with isotopically labeled analogs. It thus was excluded that the ion with $m/z = 143$ is of the formula $[C_{11}H_{11}]^+$, which in principle could be proposed to be formed by the loss of H and HCN from the molecular ion, or that the loss of C_2H_4 originates from a different portion of the molecule.

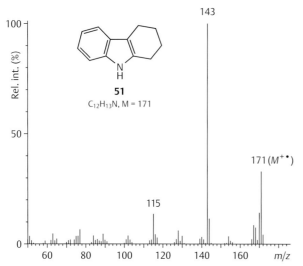

Fig. 4.64 EI-MS of 1,2,3,4-tetrahydrocarbazole (**51**). (This image is provided with permission of NIST [NIST 2005]).

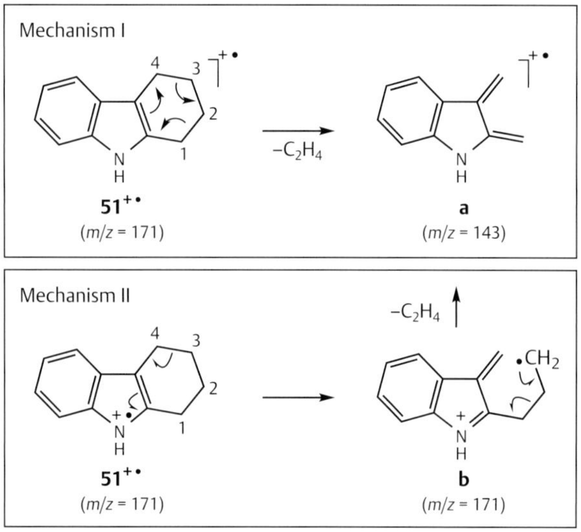

Structure **a** (**Scheme 4.19**) was finally assigned to the ion responsible for the signal at $m/z = 143$, because it is chemically the most plausible: the Diels–Alder reaction is a well-known reversible thermally allowed electrocyclic reaction, which suggests that the concerted RDA reaction (mechanism I) would be a rather obvious process to take place in excited cyclohexene systems. However, an alternative, stepwise reaction cannot be excluded on the basis of the available data. It is principally possible that fragment **a** is formed by a cascade of a vinylogous α-cleavage (formation of **b**) followed by the scission of the C(1)–C(2) bond by radical elimination (mechanism II). Such a mechanism is in fact conceivable for 1,2,3,4-tetrahydrocarbazole **51** but less for other structures such as those shown in **Fig. 4.65**.

The RDA reaction was observed with a number of other systems, for example with 1,2,3,4-tetrahydro-β-carboline (**52**), 1,2,3,4-tetrahydroisoquinoline (**53**), 1,2,3,4-tetrahydronaphthalene (**54**), 4H-chromen-4-one (**55**), isochroman-3-one/-3-imine (**56**), or 2,3,4,6,9,9a-hexahydro-1H-quinolizine (**57**). The RDA reaction delivers for all of these compounds the ions responsible for the base peaks in the spectra: at $m/z = 143$ for compound **52**, at $m/z = 120$ for compound **55**, at $m/z = 104$ for the compounds **53**, **54**, and **56**, and at $m/z = 83$ for compound **57**.

Many organic natural products contain the types of ring systems shown below, and the elucidation of their structures is achieved frequently with the help of the mass spectrometric RDA reaction. Particular examples are indole alkaloids (with **51** and **52** as partial structures), tetrahydroisoquinoline alkaloids (with **53** as partial structure), and many natural products that belong to the flavonoid family (e.g., flavones, isoflavones, and rotenoids with **55** as a partial structure).

Scheme 4.19 Mechanistic possibilities for the formation of the RDA fragment **a** of compound **51**.

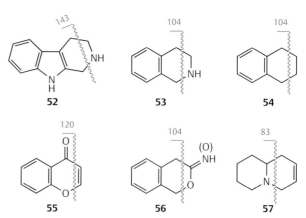

Fig. 4.65 Some examples of structural units that show strong fragmentation by RDA reaction.

The mass spectrum of 5,7-dihydroxy- 4'-methoxyisoflavone (biochanin A, **58**; M = 284) is reproduced in **Fig. 4.66**. The molecular ion is cleaved by an RDA reaction in ring C into two parts, of which each can carry the charge (**Scheme 4.20**). Thus, **58**$^{+\cdot}$ can form either the charged diene component **a** with m/z = 152 or the ene component **b** with m/z = 132. The latter dominates over fragment **a**. Fragment **b** can additionally lose in a subsequent step a methyl radical, leading to ions that are registered at

m/z = 117. It can be concluded from the masses of the observed fragment ions **a** and **b** that ring A of compound **58** (the mass of the unsubstituted ion of type **a** would be 120 u) is substituted with two OH groups (+32 u) and that ring B (the mass of the unsubstituted ion **b** would be 102 u) carries a MeO group (+30 u) or an OH (+16 u) and a Me group (+14), which cannot be distinguished by MS.

It can be very helpful for the structural elucidation of an unknown compound to be able to assign certain substituents to either the diene or the ene fragment, e.g., the two OH groups to **a** and a MeO group or a Me and an OH group to **b** as for compound **58**. The MS method, however, does often not allow us to figure out the exact locations of these substituents within the structures of the diene and ene fragments. For example, the positions of the two OH groups at ring A or of the MeO group at ring B of **58** remain undisclosed by the data obtained with the MS spectrum shown in **Fig. 4.66**. Evidently, more analytical data are needed for their determination.

McLafferty Rearrangement

The McLafferty rearrangement, also known as β-cleavage with hydrogen shift, is related to the ester pyrolysis, the Tschugaev reaction, the ene reaction (thermal reactions), or the Norrish type II reaction (photochemical radical reaction). In this reaction, an H atom is transferred via a six-membered cyclic transition state from the γ-position of an alkene or a heteroene to the terminal doubly bonded atom. Simultaneously, a migration of the double bond occurs and a neutral fragment containing the β- and γ-positioned atoms is ejected. The McLafferty rearrangement is a neutral process like the RDA reaction. It can proceed either in a concerted fashion, written with one- or two-electron shifts (mechanisms Ia and Ib), or in a stepwise fashion (mechanism II; **Scheme 4.21**). Radical as well as nonradical cations are possible starting ions for the reaction, and the required acceptor double bond can be either already present in the starting molecule or formed by another fragmentation reaction (e.g., by α-cleavage). Groups that can give rise to a McLafferty rearrangement are among others C=O (carboxylic acids, esters, aldehydes, ketones, amides, lactams, lactones), C=N (azomethines or Schiff bases, hydrazones, oximes, semicarbazones), S=O (sulfonic acid esters), and C=C (alkyl arenes, alkyl heterocycles, benzyl ethers, olefins).

The McLafferty rearrangement shall be illustrated by means of two esters. Methyl butanoate (**59**; M = 102, **Fig. 4.67**) loses ethylene by McLafferty rearrangement and forms

Fig. 4.66 EI-MS of biochanin A (**58**). (This image is provided with permission of NIST [NIST 2005]).

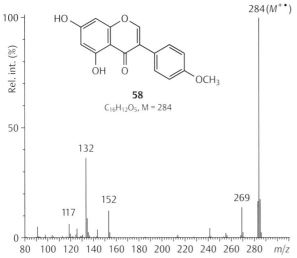

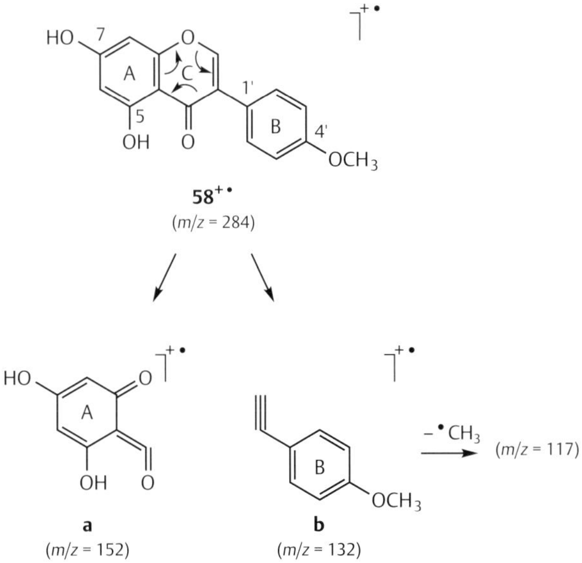

Scheme 4.20 Mass spectrometric fragmentation of biochanin A (**58**).

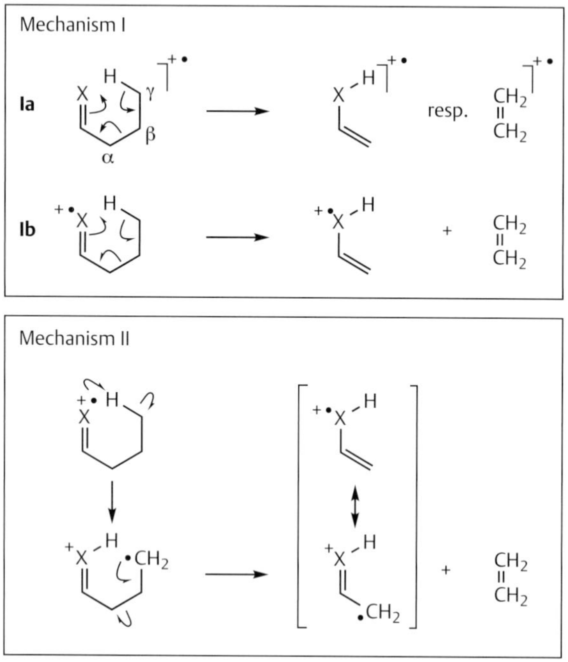

Mechanism I

Ia

resp.

Ib

Scheme 4.21 Mechanisms of the McLafferty rearrangement.

Mechanism II

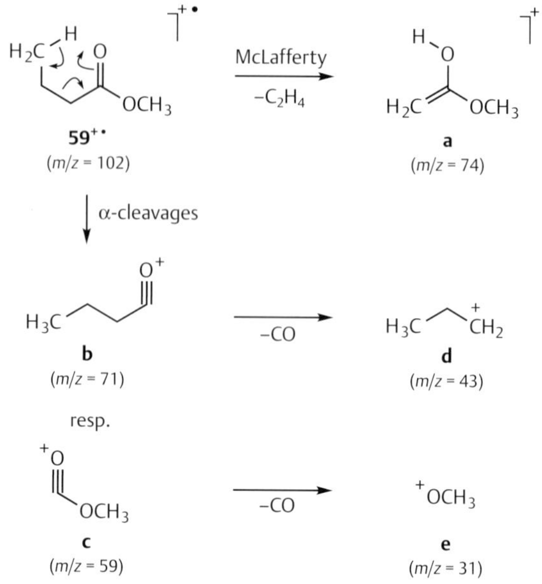

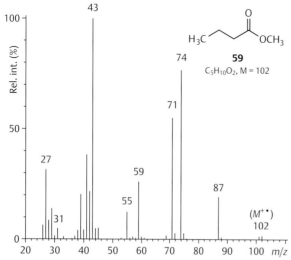

Scheme 4.22 Mass spectrometric fragmentation of methyl butanoate (**59**).

ion **a**, which is prominently recorded in the spectrum at $m/z = 74$ (**Scheme 4.22**). The signals at $m/z = 71$ and 59 can be explained with fragment ions **b** and **c**, both formed by α-cleavage, and the signals found at $m/z = 43$ and 31 by the ions **d** and **e**, the products of CO loss from the oxonium ions

b and **c**, respectively (p. 363). The signal at $m/z = 87$, finally, was shown to be the response of an ion that was formed by scission of the σ-bond in γ-position to the carboxyl group (γ-cleavage), which is slightly favored over the cleavage of other "nonactivated" σ-bonds.

The mass spectrometric behavior of compound **59** is characteristic of esters of aliphatic acids and, in fact, also of aliphatic acids and amides. The signals at $m/z = 74$ (McLafferty rearrangement) and $m/z = 59$ (α-cleavage) are thus characteristic of all methyl esters of unbranched aliphatic acids. The respective signals for ethyl esters are found at $m/z = 88$ ($+14$) and $m/z = 73$ ($+14$), for propyl esters at $m/z = 102$ ($+28$) and $m/z = 87$ ($+28$), etc. The corresponding ions formed from carboxylic acids are detected at $m/z = 60$ (McLafferty rearrangement) and $m/z = 45$ (α-cleavage), and those from N-unsubstituted amides at $m/z = 59$ and 44. The signals of γ-cleavage products, found at $m/z = 87$ for **59**, increases in intensity for derivatives of higher fatty acids.

Esters of higher alcohols and related N-substituted amides undergo McLafferty rearrangement also toward the sides of the substituents at the O and N atoms. In addition, they usually also show signals of ions that correspond to the protonated carboxylic acids or the protonated dealkylated amides. These ions are the products of a McLafferty rearrangement with additional H transfer. The behavior of such compounds is exemplified with the spectrum of butyl benzoate (**60**; M = 178, **Fig. 4.68**). The

Fig. 4.67 EI-MS of methyl butanoate (**59**). (This image is provided with permission of NIST [NIST 2005]).

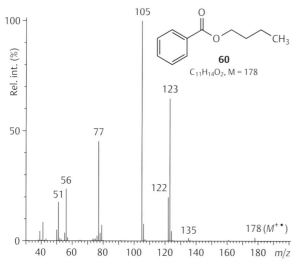

Fig. 4.68 EI-MS of butyl benzoate (**60**). (This image is provided with permission of NIST [NIST 2005]).

spectrum of **60** shows most of the signals that are expected on the basis of the fragmentation reactions discussed above. The signals of the ions **a** and **b** derived from α-cleavages are recorded at $m/z = 135$ and 105 (**Scheme 4.23**). Not observed, however, is a signal at $m/z = 101$ that would account for ion **c**, expected as a product of the α-cleavage of the CO–Ph bond. The signals at $m/z = 122$ and 56 are due to the ions **d** and **e** derived from the molecular ion **60**$^{+\bullet}$ by the ordinary McLafferty rearrangement. The intense signal at $m/z = 123$, finally, corresponds to protonated benzoic acid, ion **f**, formed by the abovementioned McLafferty reaction with additional H transfer. An analogous reaction is also observed with *N*-alkylamides, for example with compound **64** (**Fig. 4.72** and **Scheme 4.30**). The mechanism of the formation of ion **f** can be understood as a radical-induced γ-H abstraction followed by an onium reaction as shown in **Scheme 4.24**. The onium reaction is described in the next section.

The description of the spectrum of **60** would not be complete, if the rather conspicuous signals at $m/z = 77$ and 55 would not be explained. They correspond to the ions **g** and **h** that are formed from ion **b** ($m/z = 105$) by the loss of CO (**g**, p. 363) and subsequent expulsion of C_2H_2 (**h**) as discussed previously. We would like to mention at this point as well that the signal at $m/z = 105$, together with the signals at $m/z = 77$ and 51, is very typical and diagnostic for benzoyl derivatives such as benzoic esters, benzoic amides, and phenyl ketones. Corresponding signal cascades are also found for substituted benzoyl derivatives.

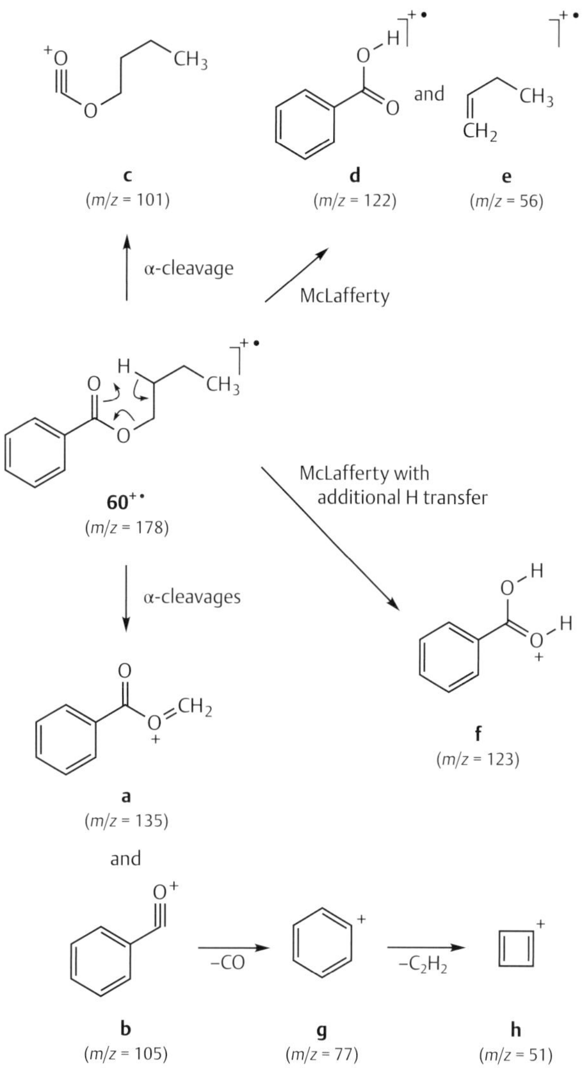

Scheme 4.23 Fragmentation of butyl benzoate (**60**).

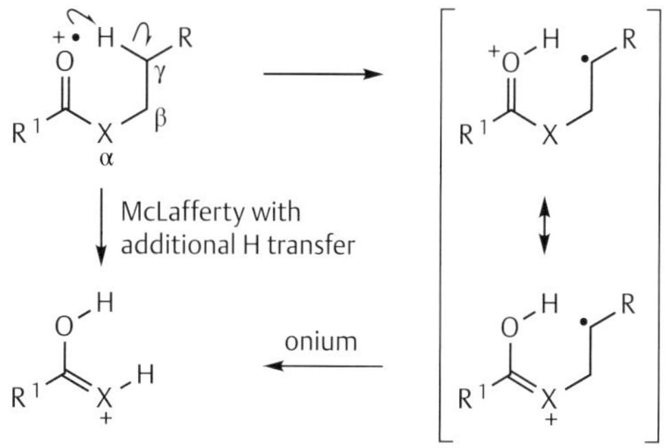

Scheme 4.24 Mechanism of the McLafferty rearrangement with additional H transfer.

Mass Spectrometry

A McLafferty rearrangement that occurs under the influence of an aromatic C=C bond is represented, for example, by the signal at $m/z = 92$ for ion **a** (**Scheme 4.25**) in the spectrum of butylbenzene (**61**; M = 134, **Fig. 4.69**). The source of the remaining prominent signals at $m/z = 91, 77, 65$, and 51 has been explained previously (**Fig. 4.50** and **Scheme 4.15**, p. 350); the signal at $m/z = 105$ is the response to ions formed by γ-cleavage.

The McLafferty rearrangement is also a common fragmentation reaction observed with simple alkenes. For example, the signals found at $m/z = 42$ and 56 in the spectra of the previously discussed compounds **42** and **43** (**Figs. 4.54** and **4.55**, p. 352) are due to the ions **a**, **b**, and **c**, formed from the respective molecular ions **42**$^{+\bullet}$ and **43**$^{+\bullet}$ by McLafferty rearrangements (**Scheme 4.26**).

McLafferty rearrangements that start from fragments rather than from molecular ions are observed too. They are not discussed here separately, but examples are given in the next section with amine **62** (**Fig. 4.70** and **Scheme 4.28**) and ether **63**

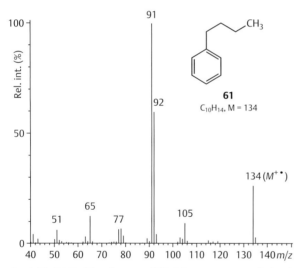

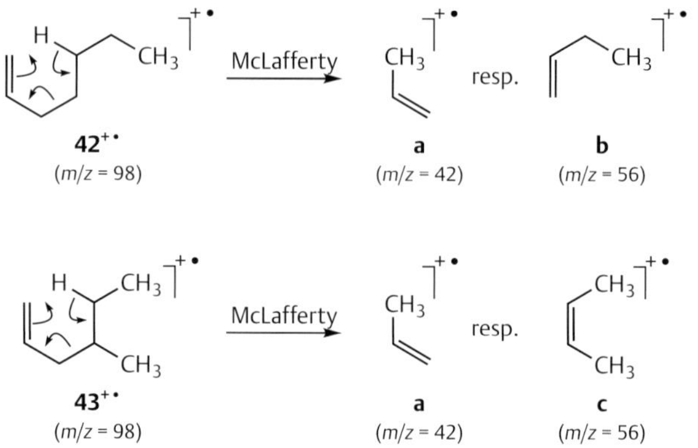

Scheme 4.26 Formation of ions **a**, **b**, and **c** from the alkenes **42** and **43** by McLafferty rearrangement (for spectra, see **Figs. 4.54** and **4.55**).

(**Fig. 4.71** and **Scheme 4.29**). Rare but also known are McLafferty rearrangements in which a γ-alkyl or another group in the γ-position is rearranged instead of a γ-H atom.

Onium Reaction

The onium reaction is a common fragmentation process that is often observed with cationic fragment ions in which a heteroatom carries the formal charge. This is the case, e.g., with oxonium, ammonium, phosphonium, or sulfonium fragment ions. In the onium reaction, an alkyl group with a chain length of minimal two carbons is replaced at the heteroatom of the onium species **a** by an H atom under formation of ion **b** (**Scheme 4.27**). It is a neutral process that is mechanistically not well understood. Since the source of the transferred H atom cannot be specified as it was shown with labeling experiments, the generalized notation as shown in **Scheme 4.27** can be used to describe the decay of ion **a**.

N-Isopropyl-*N*-methylbutanamine (**62**; M = 129, **Fig. 4.70** and **Scheme 4.28**) is discussed as a typical example of a compound that shows a prominent onium reaction in its mass spectrometric behavior. The compound loses a Me˙ or a C_3H_7˙ fragment upon ionization by electron bombardment and α-cleavage, delivering the ammonium ions **a** or **b**, respectively, recorded as signals at $m/z = 114$ and 86. The double bond in ion **a** has a γ-positioned H atom that allows the species to undergo McLafferty rearrangement and formation of ion **c** ($m/z = 72$). Such a γ-H atom is not available in the alternatively formed α-cleavage product **b** and, thus, McLafferty rearrangement is not possible with this ion. Both ions **a** and **b**, and in fact also ion **c**, can undergo the onium reaction under elimination of an *N*-alkyl residue and simultaneous H transfer. Thus, elimination of C_4H_8 (56 u) from **a** gives ion **d** with $m/z = 58$, while the removal of C_3H_6 (42 u) from **b** or of C_2H_4 (28 u) from **c** produces ion **e** with $m/z = 44$.

Fig. 4.69 EI-MS of butylbenzene (**61**). (This image is provided with permission of NIST [NIST 2005]).

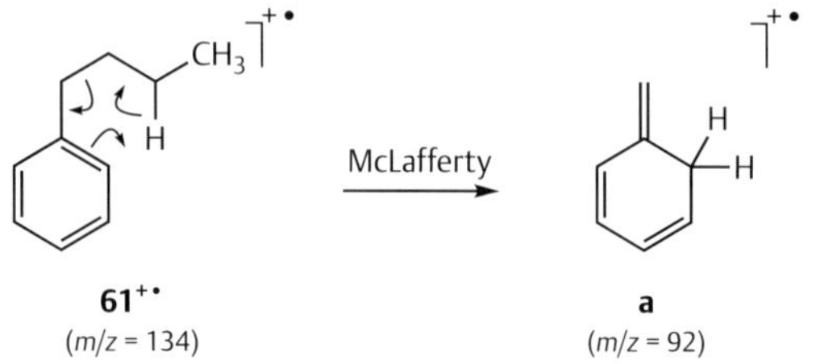

Scheme 4.25 McLafferty rearrangement with butylbenzene (**61**).

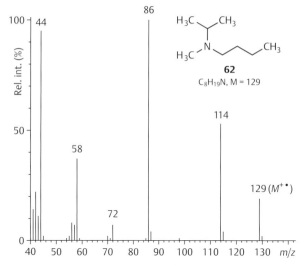

Fig. 4.70 EI-MS of *N*-isopropyl-*N*-methylbutylamine (**62**). (This image is provided with permission of NIST [NIST 2005]).

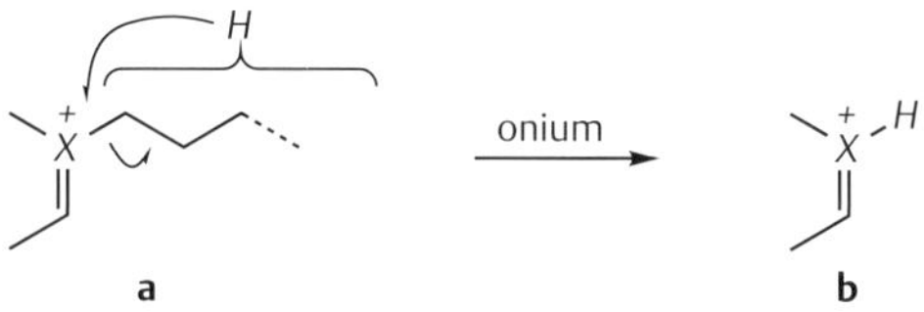

Similar decomposition sequences can be observed with ethers or thioethers as it is shown for butylethylether (**63**; M = 102, **Fig. 4.71**) in **Scheme 4.29**.

In addition to alkyl substituents (other than Me), acyl residues can also undergo the onium reaction. In the mass spectrum of *N*-butylacetamide (*N*-acetylbutanamine, **64**; M = 115, **Fig. 4.72**), the signal at m/z = 30 represents the base peak. It is the response to ions of the type **b**, which are formed by the cascade of α-cleavage of **64**$^{+\cdot}$ to ion **a** and onium reaction with the loss of ketene (**Scheme 4.30**). The ion **a** and also the ion **c**, formed by the alternative α-cleavage, are both represented in the spectrum of **64** with their signals at m/z = 72 and 43 too. It is noteworthy to mention here that the signal at m/z = 43 is a general indication of the presence of an acetyl compound (ketone, *O*- or *N*-acetyl derivative), and that *N*-alkylated acetamides are characterized by a signal registered at m/z = 60, derived from ion **d** that is formed by the McLafferty rearrangement with additional H shift. A base peak signal at m/z = 30, on the other hand, is not indicative of an amide derivative. The same signal is found as the base peak for any primary *n*-alkylamine.

Other *N*-acyl compounds behave similarly to **64**. It is remarkable, however, that acyl residues that do not possess any sp³-bonded H atoms can also be cleaved off from an onium species through an onium reaction (i.e., with an H migration!). Such residues are, e.g., benzoyl or benzenesulfonyl groups.

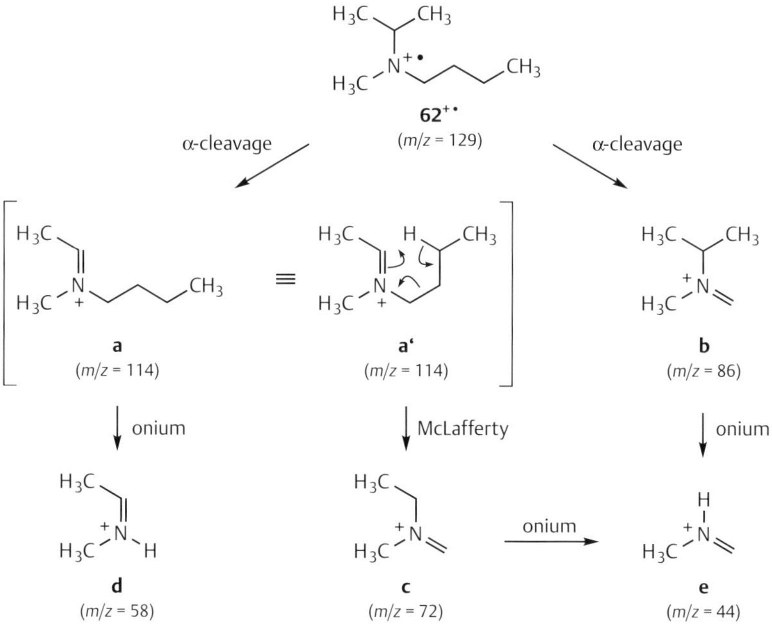

Scheme 4.27 Schematic representation of the onium reaction.

Scheme 4.28 Fragmentation of *N*-isopropyl-*N*-methylbutylamine (**62**).

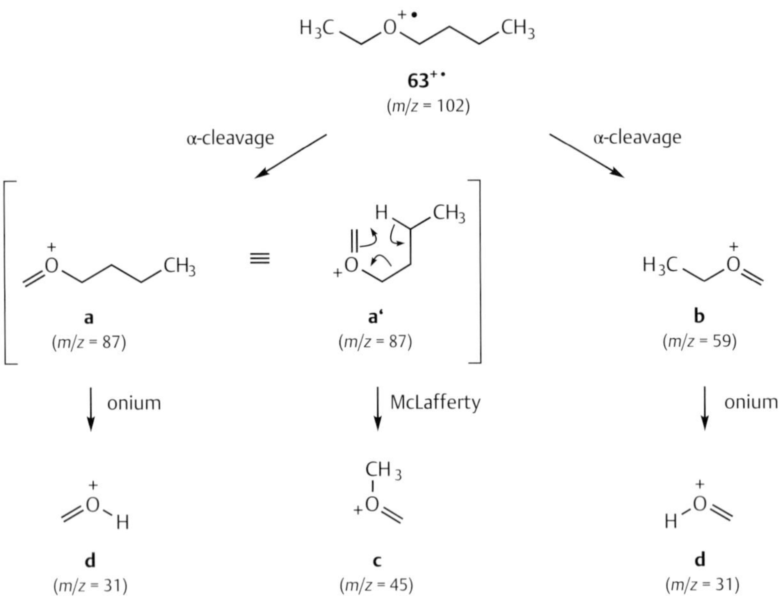

Scheme 4.29 Fragmentation of butylethylether (**63**).

The most abundant ion registered in the spectra of dialkyl esters of phthalic acid—common plasticizers that are often observed in EI-MS as contaminants—is also the result of an onium reaction. The ion **c**, giving rise to the base peak at $m/z = 149$ in the spectrum of diethyl phthalate (**65**; M = 222, **Fig. 4.73**), is produced by a fragmentation cascade that is initiated by the α-cleavage of a C–O bond under ejection of an ethoxy residue (**Scheme 4.31**). The resultant ion **a** ($m/z = 177$) may cyclize due to the proximity of the neighboring ethoxycarbonyl group to the isomeric ion **b**, which then can undergo the onium reaction to form ion **c** ($m/z = 149$).

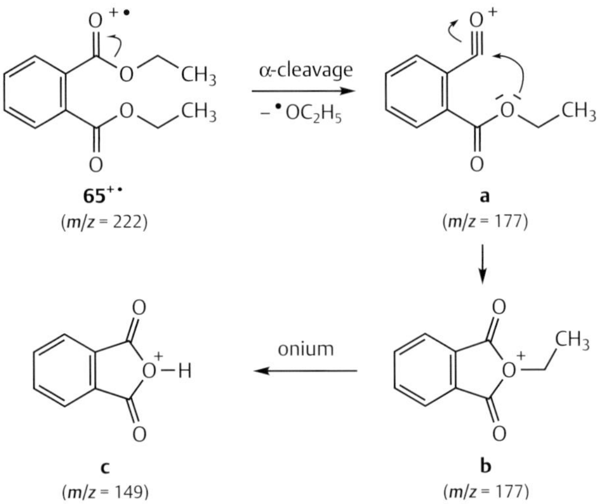

Scheme 4.31 Fragmentation of diethyl phthalate (**65**).

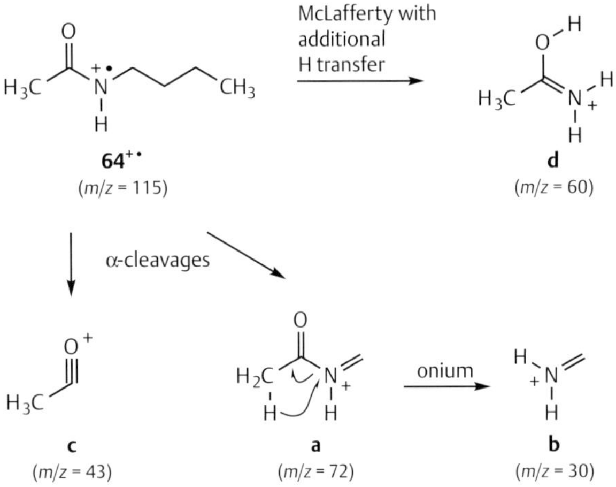

Scheme 4.30 Fragmentation of N-butylacetamide (**64**).

o-Disubstituted benzene derivatives show often a special mass spectrometric behavior when compared with the m- and p-isomers. This effect is known as the *ortho* effect and is discussed in Section 4.4.5, p. 373.

In some cases, the ejection of acyl residues that are bound to O or N atoms occurs directly from molecular ions. This behavior is in particular observed with acyloxybenzenes and N,N-diacyl-alkanamines, which can form the corresponding, rather stable phenol or N-monoacylalkanamine radical cations. It must be

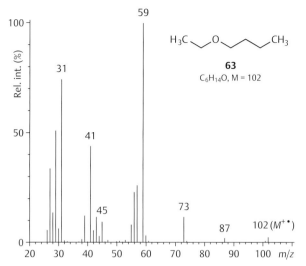

Fig. 4.71 EI-MS of butylethylether (**63**). (This image is provided with permission of NIST [NIST 2005]).

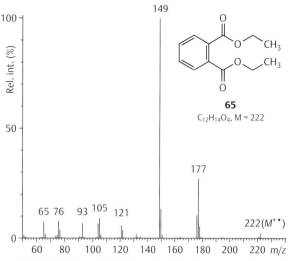

Fig. 4.73 EI-MS of diethyl phthalate (**65**). (This image is provided with permission of NIST [NIST 2005]).

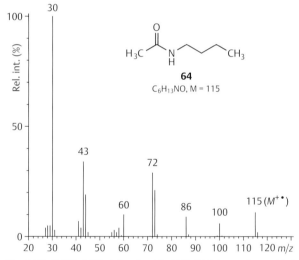

Fig. 4.72 EI-MS of N-butylacetamide (**64**). (This image is provided with permission of NIST [NIST 2005]).

added here that the respective ions could also be the molecular ions of artifacts. The compounds mentioned above are very easily hydrolyzed to the respective phenols or N-acylalkylamines that very easily might find their way into the analysis.

Loss of CO

Cyclic, highly unsaturated compounds and ions that have been formed by α-cleavage adjacent to a carbonyl or carboxyl group have the tendency to eject CO (28 u). For the behavior of the latter, see the spectra and fragmentation schemes given for the compounds **59** and **60** (p. 358 and p. 359). If several CO

groups are present in a molecule, they can be eliminated one after another.

The mass spectrum of tropone (**66**; M = 106, **Fig. 4.74**) documents clearly the fragmentation reaction with the loss of CO (28 u), which is favored in highly unsaturated cyclic systems. The ions corresponding to CO loss from **66**$^{+\bullet}$ are detected at m/z = 78 and give rise to the base peak of the spectrum. The remaining part of the spectrum strikingly resembles that of benzene, which indicates that the ion resulting from the loss of CO might in fact be the molecular ion of benzene.

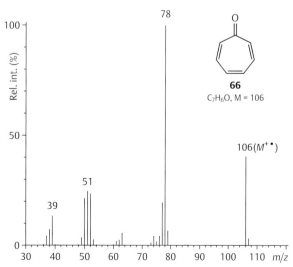

Fig. 4.74 EI-MS of tropone (**66**). (This image is provided with permission of NIST [NIST 2005]).

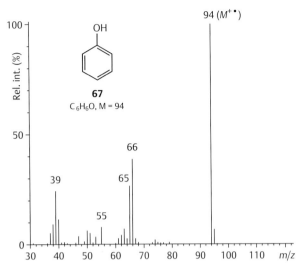

Fig. 4.75 EI-MS of phenol (**67**). (This image is provided with permission of NIST [NIST 2005]).

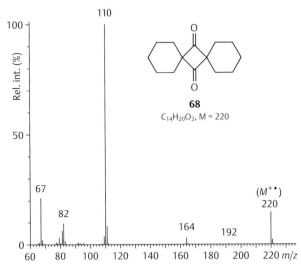

Fig. 4.76 EI-MS of dispiro[5.1.5.1]tetradecan-7,14-dione (**68**). (This image is provided with permission of NIST [NIST 2005]).

Compounds whose enol form in solution is more abundant than the keto form can also readily lose CO in the mass spectrometer. Phenols are typical representatives for this class of substances.

Phenol itself (**67**; M = 94) gives the spectrum shown in **Fig. 4.75**. The most abundant fragment ion is recorded at $m/z = 66$. It obviously corresponds to an ion that was formed by CO loss (–28 u) from the molecular ion **67**$^{+\bullet}$ ($m/z = 94$). It most probably arose from the keto form **a** of the molecular ion and can be described with structure **b** (**Scheme 4.32**). This structure is supported by the observation of the signal at $m/z = 65$, which is assigned to the cyclopentadienyl cation **c**. The formation of this ion can be explained with of the loss of a highly energetic H atom from **b**, a reaction that is usually not preferred in

EI-MS but is possible in this special case because the resultant resonance-stabilized species is particularly low in energy.

Dispiro[5.1.5.1]tetradecane-7,14-dione (**68**; M = 220, **Fig. 4.76**) is an example of a compound that can lose two CO groups in EI-MS. The base peak of its spectrum is found at $m/z = 110$, which is half the value of the molecular ion. However, the peak is not due to a doubly charged molecular ion M^{++} as it could be suspected: there are no isotopic peaks at half m/z values present, which would be the case for doubly charged species. The responsible ion for the base peak is **a**, formed by a *retro*-[2+2]-cycloaddition (**Scheme 4.33**). This ion can lose CO and most probably delivers cyclopentene ion **b**, which is recorded at $m/z = 82$. Alternatively, the molecular ion **68**$^{+\bullet}$ loses stepwise or in a single step two CO moieties, forming ions **c** and **d**, which show their signals at $m/z = 192$ and 164. Analogous to compound **68**, other diketones, e.g., anthraquinone, may lose both CO groups upon EI-MS.

Tabular Summary

The α-cleavage with its variants of the benzyl, allyl, and propargyl cleavages, together with the McLafferty rearrangement, the onium reaction and, because of the ubiquity of the C=O group, also the CO loss are the key fragmentation reactions in EI-MS. The RDA reaction requires specific structural features and thus is observed less frequently, and the cleavage of "nonactivated" C–C bonds is of importance only for compounds for which no other fragmentation reactions are dominant. The several EI fragmentation reactions mentioned above are concisely summarized in **Table 4.8**.

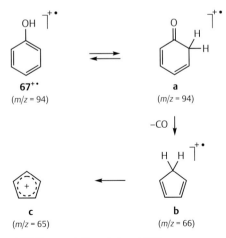

Scheme 4.32 Fragmentation of phenol (**67**).

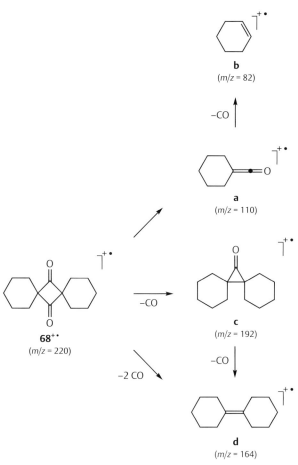

b
(m/z = 82)

–CO

a
(m/z = 110)

–CO

c
(m/z = 192)

68$^{+\bullet}$
(m/z = 220)

–CO

–2 CO

d
(m/z = 164)

Scheme 4.33 Fragmentation of dispiro[5.1.5.1]tetradecan-7,14-dione (**68**).

The range of possible EI fragmentation reactions, however, is not exhausted with the listing of the major fragmentation reactions compiled in **Table 4.8**. Certain functional groups or a particular arrangement of atoms can sometimes cause special fragmentation processes. Some examples of reactions that have not been discussed here are the loss of water and ammonia, S_Ni reactions, and reactions with neighboring group participation (Section 4.4.5, p. 373).

4.4.4 Collision-Induced Dissociation

If mild ionization methods are used in MS, the spectra very often show signals of ionized molecules only and none of fragments. This is advantageous, on the one hand, because the mass of a molecule can be determined much more safely. On the other hand, this is a drawback because no information can be gained from fragment ions that might be characteristic or

diagnostic of a given structure. To compensate for this loss of information, ions can be selected by a mass filter and brought to fragmentation by collision activation (CID). CID is not restricted to MS with mild ionization methods, but can be combined with any MS method, hence also with EI-MS that truly does not lack fragmentation behavior.

The ions that are selected for CID are called *precursor ions* and denoted as P^+. If the process is repeated with precursor ions of different masses, these ions are symbolized with P_1^+, P_2^+, etc. The selection of ions and their activation/fragmentation is called MS/MS (or MS^2). Experiments for which fragment ions of MS/MS experiments are successively selected and used for consecutive CID are called MS^n, with n = number of MS cascades. It is important to note at this point that it is of no relevance for the fragmentation behavior of a precursor ion on which stage of an MS cascade it was selected. What is relevant is solely the amount of energy that is transferred onto the particle upon collision activation (collision energy E_c or fragmentation amplitude).

Important Fragmentation Reactions

Fragmentation reactions upon CID and the respective mechanisms have not been investigated comprehensively. For this reason, many of the fragment ions formed by CID are just regarded as "reasonable" with respect to rational expulsions of molecular portions from a base structure. However, because the precursor ions that are selected for CID are in most cases nonradical species such as adduct ions or fragments thereof, their reactions can often be explained by common transformations of ionic compounds also known from classical chemistry. Particularly frequent are elimination processes, intramolecular substitution and condensation reactions, and also the formation of acylium ions from carboxylic acid derivatives. A typical example of an ESI-MS^2 is shown with the spectrum of coumaroylputrescine (**69**; M = 234, **Fig. 4.77**).

Fragmentation paths that could lead to ions **a** to **d** detected in the CID spectrum of the precursor ion P^+ are shown in **Scheme 4.34**. The ions registered at m/z = 72 (**a**) and m/z = 218 (**d**) are probably formed by substitutive expulsion of the amide portion or of NH$_3$ from P^+, respectively. Protonation at the carboxyl group of P^+ needs to be assumed to explain the former reaction. The formation of acylium ions is characteristic of carboxylic acid derivatives, and the signal of acylium ion **c** is accordingly found at m/z = 147 for amide **69**. The reaction to **c** corresponds with respect to its result to the α-cleavage that was discussed in EI-MS, but the two processes are chemically barley related.

Substitution reactions and the formation of acylium ions are often accompanied by H transfers, producing charged particles from "neutral" fragments by transprotonation. For instance, ion

Mass Spectrometry

Table 4.8 Summary of the major fragmentation reactions observed in EI-MS

Type	Precursor ion	Repetition of the same reaction type	Example
α-Cleavage radical process *requirement* radical cation with localization of the radical on a hetero-atom (N, O, S, rarely halogen)	Molecular ion and radical fragment ions	No	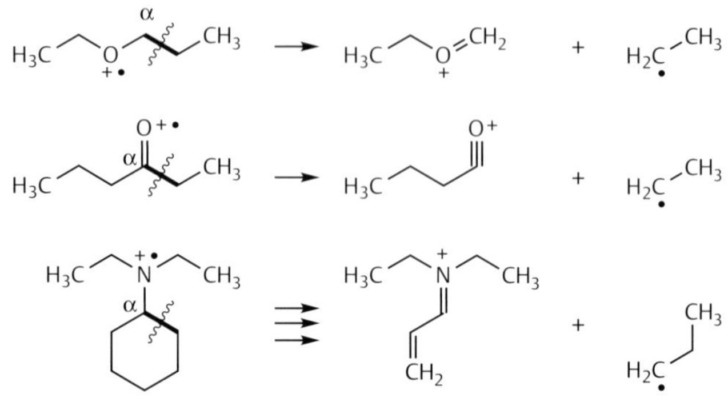
Benzyl, allyl, and propargyl cleavage radical process *requirement* benzyl-, allyl-, or propargyl-bonded substituent	Molecular ion and radical fragment ions	No	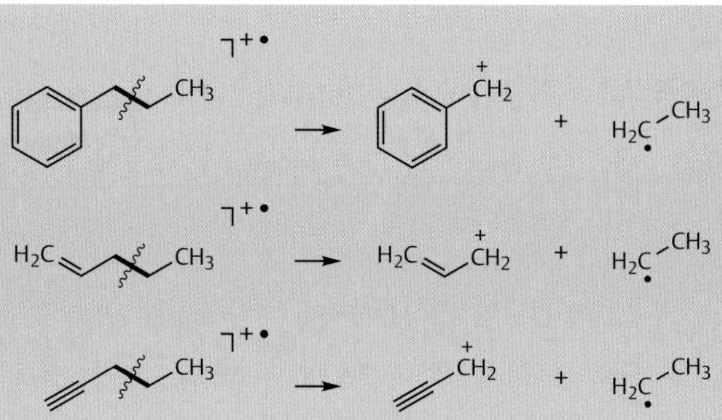
Retro-**Diels–Alder reaction (RDA)** neutral process *requirement* 6-membered alicyclic or heterocyclic ring with at least one double bond	Molecular ion and fragment ions	Yes	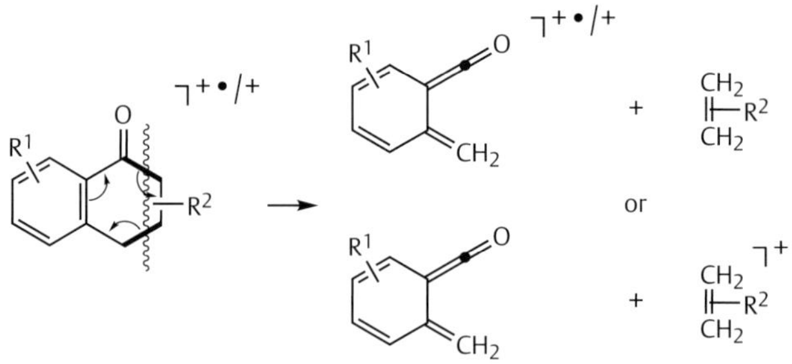

(Continued)

Table 4.8 (Continued)

Type	Precursor ion	Repetition of the same reaction type	Example
McLafferty rearrangement Neutral process *requirement* an H-atom in γ-position to a double bond	Molecular ion and fragment ions	Yes	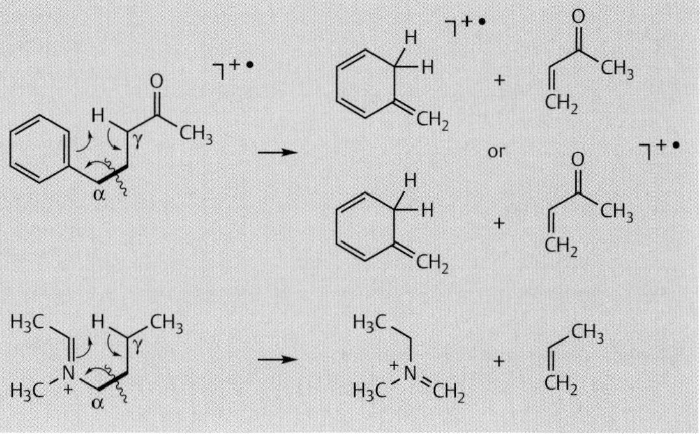
Onium reaction Neutral process *requirement* alkyl substituent (except methyl) bonded to a heteroatom of an onium species such as immonium, oxonium), etc.	Fragment ion	Yes	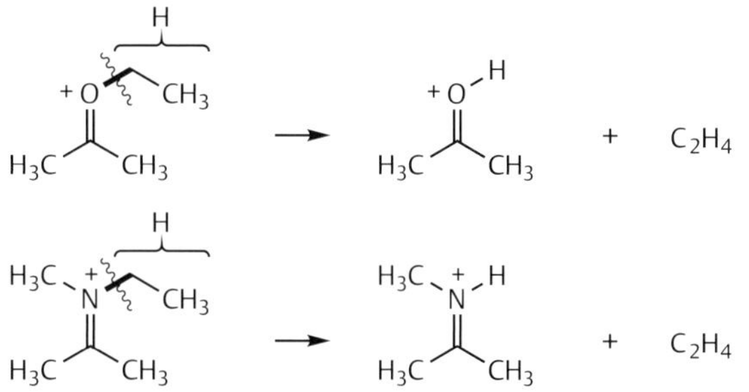
Loss of CO Neutral process *requirement* cyclic carbonyl derivatives (ketones, chinones), phenols, metal carbonyls, carbonyl-containing fragment ions (e.g., acylium ions from α-cleavage of carbonyl and carboxyl derivatives)	Molecular ion and fragment ions	Yes	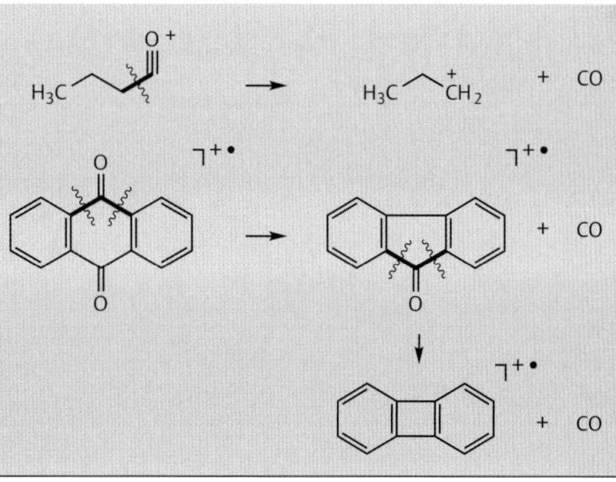

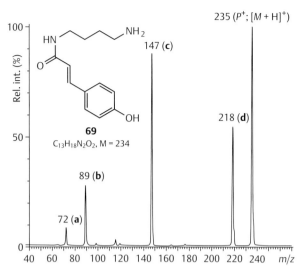

Fig. 4.77 ESI-MS² (Ar; $P^+ = [M + H]^+ = 235$, $E_c = 15$ eV) of couma-roylputrescine (**69**).

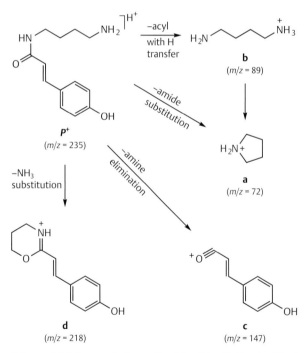

Scheme 4.34 Fragmentation paths of the $[M + H]^+$ ion ($= P^+$) of coumaroylputrescine (**69**) upon CID with Ar.

b ($m/z = 89$) is formed from P^+ by expulsion of 1,4-diaminobutane (loss of fragment **c**), which is immediately cationized by protonation. The process of expulsion of a neutral fragment and its cationization by transprotonation is of particular importance

for peptides. It explains the formation of the diagnostic ions of the types **c″** and **y″** (**Scheme 4.35**, p. 369).

In addition to the reactions discussed above, further neutral processes can take place in CID. We are already familiar with some of them from EI-MS: the RDA reaction, the McLafferty rearrangement, and the CO loss are also common fragmentation reactions observed in CID.

Peptides

The CID of peptides follows the principles discussed above. Because the same functional group is repeated along the oligo- or polymer chain in such compounds, the associated fragmentations can be put into rules and used for the mass spectrometric sequencing of peptides (the determination of their amino acid sequences).

The sequencing of peptides involves mainly six series of product ions. These series are divided into two subsets on the basis of the peptide terminus contained in the ions (**Scheme 4.35**). The ions of the types **a**, **b**, and **c″** include the N-terminus of a peptide, while the ions of the types **x**, **y″**, and **z** hold the C-terminus (ion nomenclature according to Biemann[2] and to Roepstorff and Fohlman[3]). The double quotation marks with the ions **c″** and **y″** refer to the two additional H atoms as compared to the scission lines drawn in the structures. The **b** and **y** type ion series are the most important for the sequencing of a peptide because they are usually quite complete and give rise to distinct signals.

The mass difference between neighboring signals within the same ion series corresponds to the mass of the amino acid residue (amino acid: –18 u; trunk mass) as can easily be recognized by means of the representation shown in **Scheme 4.36**. The residual masses for the 20 proteinogenic amino acids are compiled in **Table 4.9**. The N- and C-terminal amino acids of a peptide can be identified by the masses of the a_1 and the $y_1″$ (mass of the respective amino acid: +1 u) ions. For this reason, the masses of the amino acids and the m/z values for the a_1 ions are also listed in the table.

In the ideal case, the full primary structure of a peptide can be disclosed by means of the several ion series, starting with the amino acid at the N- or C-terminus and going from one amino acid to another amino acid. However, the problem is usually more complex when a real spectrum needs to be interpreted. There exist, on the one hand, amino acids that have the same mass (Leu and Ile) or very similar masses (Gln and Lys). This often leads for larger molecules such as peptides to signals that remain unresolved (or can only be resolved by HR-MS and/or ion mobility-MS). On the other hand, the ion series produced by CID of peptides are usually not complete, and additional ions derived from internal fragmentations (not corresponding

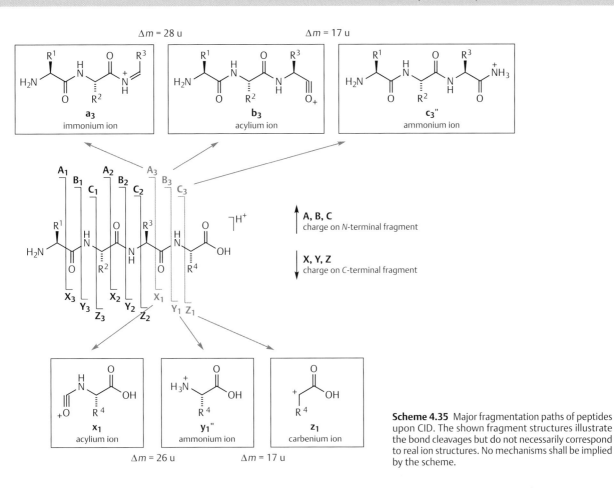

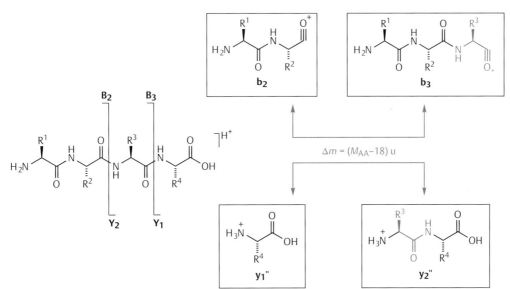

Scheme 4.35 Major fragmentation paths of peptides upon CID. The shown fragment structures illustrate the bond cleavages but do not necessarily correspond to real ion structures. No mechanisms shall be implied by the scheme.

Scheme 4.36 Mass difference between two ions of a series, shown with the **b₂/b₃** and **y₁″/y₂″** ion pairs of a generic tetrapeptide structure.

Table 4.9 Monoisotopic masses, trunk masses, and m/z values of the $\mathbf{a}_1$ fragments of the 20 proteinogenic amino acids[4]

Amino acid	Abbreviation	Code	Mass (neutral, u)	Trunk mass (u)	$\mathbf{a}_1$-Signal position (m/z)
Glycine	Gly	G	75.03203	57.02146	30.03383 (–)*
Alanine	Ala	A	89.04768	71.03711	44.04948 (–)
Serine	Ser	S	105.04259	87.03203	60.04439 (+)
Proline	Pro	P	115.06333	97.05276	70.06513 (++)
Valine	Val	V	117.07898	99.06841	72.08078 (++)
Threonine	Thr	T	119.05824	101.04768	74.06004 (+)
Cysteine	Cys	C	121.01975	103.00918	76.02155 (–)
Leucine	Leu	L	131.09463	113.08406	86.09643 (++)
Isoleucine	Ile	I	131.09463	113.08406	86.09643 (++)
Asparagine	Asn	N	132.05349	114.04293	87.05529 (–)
Aspartic acid	Asp	D	133.03751	115.02694	88.03930 (–)
Glutamine	Gln	Q	146.06914	128.05858	101.07094 (+)
Lysine	Lys	K	146.10553	128.09496	101.10732 (–) 84.08132 (–NH_3; +)
Glutaminic acid	Glu	E	147.05316	129.04259	102.05495 (–)
Methionine	Met	M	149.05105	131.04049	104.05285 (–)
Histidine	His	H	155.06948	137.05891	110.07127 (++)
Phenylalanine	Phe	F	165.07898	147.06841	120.08078 (++)
Arginine	Arg	R	174.11168	156.10111	129.11347 (–)
Tyrosine	Tyr	Y	181.07389	163.06333	136.07569 (++)
Tryptophane	Trp	W	204.08988	186.07931	159.09167 (++) 130.06523 (–CH_2NH; ++)

* Signal intensity of type $\mathbf{a}_1$ ions: (–) low, (+) medium, (++) high.

to Biemann fragments) are found as well. This substantially complicates the spectral interpretation, but fortunately the analysis of CID spectra of peptides is efficiently supported by powerful computer software. Modern sequencing software reliably allows the elucidation of primary structures of peptides without much effort.

The interpreted CID mass spectrum of the pentapeptide Phe-Leu-Val-Ala-Gly (**70**; M = 505.3) is shown as a simple example for the sequencing of a peptide in **Fig. 4.78**. The singly charged ionized molecule $[M + H]^+$ was chosen as the precursor ion P^+ in this case. Multiply charged ions, however, could likewise be chosen as precursor ions. They usually deliver more complete ion series and more lucid spectra. The fragmentation paths and the structures of the relevant species can be gathered from **Fig. 4.79**.

The sequencing of a peptide is still possible even when several ion sequences are incomplete, as can be seen with the example of peptide **70**. In this case, ions of the types $\mathbf{c}''$, $\mathbf{x}$, and $\mathbf{z}$ are missing as well as the ions $\mathbf{a}_5$, $\mathbf{b}_1$, and $\mathbf{y}_1''$. The information found in the spectrum, however, is redundant. The signals recorded at $m/z = 171.1$ ($\mathbf{y}_3'\mathbf{b}_4$), $m/z = 213.2$ ($\mathbf{y}_4'\mathbf{b}_3$), and $m/z = 284.2$ ($\mathbf{y}_4''\mathbf{b}_4$),

due to ions that were formed by internal fragmentations, contribute to this redundancy.

Nucleic Acids and Saccharides

Other natural polymers such as nucleic acids and polysaccharides (polymers in general) also follow distinct fragmentation paths that can be used for structural elucidations. Fragmentation rules have been compiled for CID MS of nucleic acids and polysaccharides, analogous to those discussed above for the peptides.

The analysis of nucleic acids is done in the negative ionization mode, and a multiply charged ion, formed by deprotonation of the analyte molecule, is usually selected for CID. Because the charge of the precursor ion will get distributed onto the fragments, series of ions with a lower charge than the precursor ion will be formed. This leads to signals that are detected at higher m/z values than that of the precursor ion.[h] The fragmentation

---

[h]This is in fact also the case with the CID of peptides if multiply charged ions are selected there.

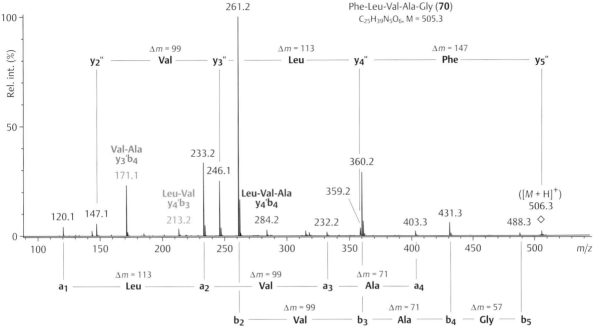

Fig. 4.78 ESI-MS^2 (Ar; P^+ = 506.3) of pentapeptide **70** with signal interpretation.

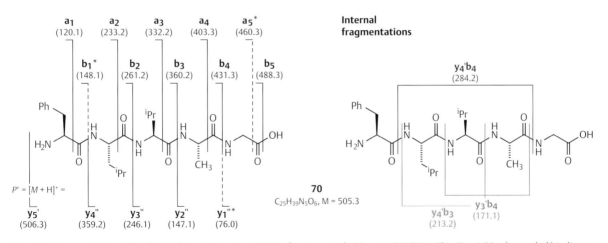

Fig. 4.79 Fragmentation of the $[M + H]^+$ ion (P^+; m/z = 506.3) of pentapeptide **70** upon CID (ESI-MS^2 in **Fig. 4.78**; the symbol* indicates that the respective fragments are not observed).

of nucleic acids can be described with the bond cleavages indicated in **Fig. 4.80** with a generic oligonucleotide. The most important fragments, which can also be used for the sequencing of the compounds, are the ions of the types $[a_n - B_n]$ and w_n (nomenclature according to McLuckey et al[5]). The mass differences Δm between two neighboring ions within an ion series are characteristic of the nucleic bases (and the sugars) that are localized in between the bond scissions. In the case shown in **Fig. 4.80**, the mass difference Δm (B$_2$) for the nucleobase B$_2$ is,

for example, found between w_3 and w_2. The Δm (B) values for the different nucleobases and the respective nucleic acids (DNA or RNA) are compiled in **Table 4.10**.

Polysaccharides decompose upon collision activation along the patterns indicated in **Fig. 4.81** (nomenclature according to Domon and Costello[6]). CID of saccharides, however, rarely delivers complete structures because the sugar units, contrary to amino acids or nucleobases, do not systematically differ in mass. Branchings and, if reference monosaccharide units (with

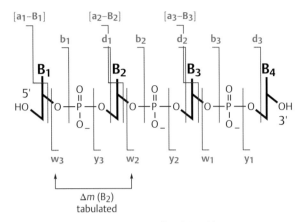

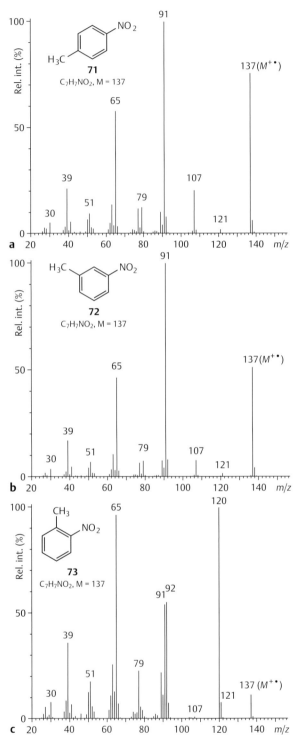

Fig. 4.80 Fragmentation pattern of nucleic acids.

Table 4.10 Mass differences Δm (B) between two neighboring ions of an ion series obtained by CID of nucleic acids

Base (B)	Code	DNA Δm (B) (u)	RNA Δm (B) (u)
Adenine	A	313.05761	329.05252
Guanine	G	329.05252	345.04744
Cytosine	C	289.04637	305.04129
Thymine	T	304.04604	(320.04096)
Uracil	U	(290.03039)	306.02530

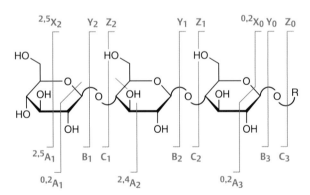

Fig. 4.81 Fragmentation pattern of saccharides.

characteristic masses) are available, also partial sequences of polysaccharides may be determined by the use of MS^n.

4.4.5 Neighboring Group Participation and Stereoisomerism

Neighboring Groups

The mass spectrometric decomposition of di- and polyfunctionalized compounds is primarily characterized by two

Fig. 4.82 EI-MS of **(a)** *p*-nitrotoluene (**71**), **(b)** *m*-nitrotoluene (**72**), and **(c)** *o*-nitrotoluene (**73**). (These images are provided with permission of NIST [NIST 2005]).

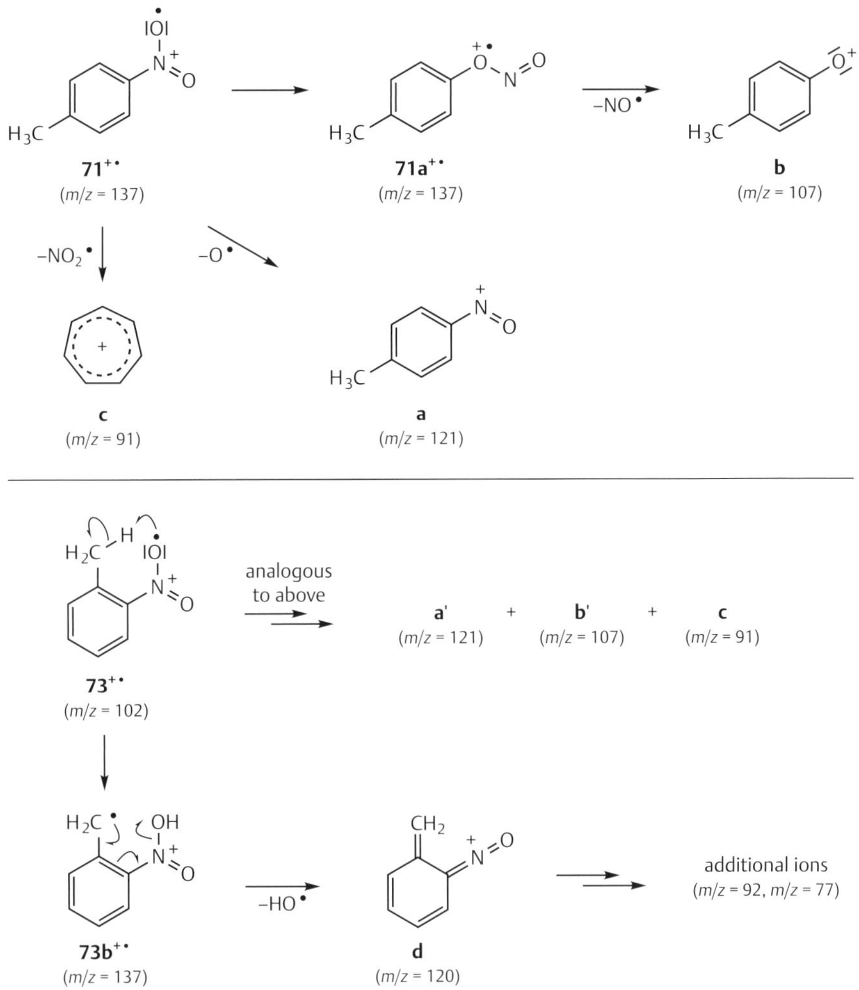

Scheme 4.37 Fragmentation of p-nitrotoluene (**71**) and o-nitrotoluene (**73**) in EI-MS.

different types of processes. The formation of fragment ions is determined on the one hand by the separate fragmentation of the molecules due to the individual functional groups but also, on the other hand, by new fragmentation processes that are the result of interactions between functional groups of a compound. Such reactions have been found for many α,ω-difunctionalized alkanes, e.g., for esters of ω-hydroxy, ω-methoxy, ω-oxo, and ω-amino carboxylic acids, for acetals of ω-hydroxy ketones, for α,ω-diaminoalkanes, N,N-diacyl-α,ω-diaminoalkanes, and ω-aminophenylalkanes. The distance between the functional groups, their spatial proximity, plays a crucial role in fragmentation reactions occurring with neighboring group participation. Reactions that can proceed via cyclic transition states or reactions that result in ring formations are particularly favored. An example for this phenomenon is given with the spectra of the three nitrotoluene derivatives **71** to **73** (M = 137, **Fig. 4.82**).

The mass spectrometric differentiation of p-, m-, and o-derivatives of benzene is often impossible because the three respective spectra are usually very similar. If, however, functional groups are present that can react intramolecularly and hence lead to specific fragmentations, the spectra of the o-isomers differ significantly from those of the p- and m-isomers. This is called the *ortho*-effect.

The situation for the *ortho*-effect to become operative is given for o-nitrotoluene (**73**). As can be seen in **Fig. 4.82**, the spectra of the p- and m-isomers **71** and **72** of nitrotoluene are virtually identical, which does not allow any assignments to be made. However, the spectrum of o-nitrotoluene (**73**) is clearly different due to fragmentations proceeding with neighboring group participation.

The spectra of the p- and m-isomers **71** and **72** exhibit the signals that would be expected for the nitroaromatic compounds.

The fragment ions **a** ($[M - O]^+$) and **b** ($[M - NO]^+$ that typically are found for nitro compounds are registered at $m/z = 121$ and 107, respectively, and the signal at $m/z = 91$ demonstrates the formation of the tropylium ion **c** ($[M - NO_2]^+$). The respective reactions are exemplarily delineated for compound **71** in **Scheme 4.37**.

The signals characteristic of the nitrotoluenes **71** and **72** are also found for the *o*-isomer **73**. However, several further signals are observed for this compound—above all the very intense signal at $m/z = 120$, which represents the base peak of the spectrum. This signal corresponds to an ion of type **d** that is formed by the loss of OH⋅ from the molecular ion in the form of **73b**⁺⋅ according to **Scheme 4.37**. The spectrum shows also a number of further signals with significant intensities that additionally substantiate the uniqueness of the *ortho* compound. The signals recorded at $m/z = 92$ and 65 are due to ions that derive from **d** by loss of CO ($m/z = 92$) and subsequent loss of HCN ($m/z = 65$). They are explained with rather complex reactions that include rearrangements.

Neighboring group participation is not only observed for EI fragmentation but also for fragmentation due to collision activation. For example, the CID spectra of polyamine derivatives are dominated by signals that arise from fragment ions that are explained by intramolecular substitution reactions via five- or six-membered cyclic transition structures.[7] As an example, the precursor ion $[M + H]^+$ selected at $m/z = 417$ for CID in the ESI-MS/MS of the polyamine spider toxin **74** (M = 416) delivers a prominent signal at $m/z = 343$ (20%) for ions of the type **a**, while the respective signal for ion **a'** from the isomeric compound **75** (M = 416) is not found at all. The prominent formation of **a** can be explained with the loss of diaminopropane by intramolecular nucleophilic substitution via an advantageous five-membered cyclic transition structure as shown in **Scheme 4.38**. The corresponding reaction to ion **a'** would need to pass through an unfavorable four-membered cyclic structure, and is thus not observed.[7,8] This pattern applies also for other signals observed in the MS/MS of the two and many other related compounds.

Stereoisomers

Stereoisomers are individual and distinct compounds and, thus, show usually different physical and chemical properties. For optical antipodes (and, hence, also for racemates), this is however only true in a "chiral" environment, which is not given in MS. Mass spectra of enantiomers are for this reason identical.

The situation is different for diastereoisomers. Such compounds are not isometric and, thus, can (but do not need to) show different mass spectra. The spectra are different when the different spatial arrangements give preference to distinct fragmentation reactions for the two isomeric compounds. If the fragmentation processes take place outside the sphere of influence of the stereogenic elements contained in the compounds, no differences in the spectra can be expected.

The spectra of *E,Z*-isomers are often not identical. In most of the cases, however, they deviate from each other just in the relative intensities of the signals. Only in rare cases, very different spectra are obtained for such compounds. This is demonstrated with the spectra of the two isomeric compounds maleic acid (**76**; M = 116) and fumaric acid (**77**; M = 116) shown in **Fig. 4.83**.

Decarboxylation and formation of ions of the type $[M - CO_2]^{+\cdot}$ ($m/z = 72$) evidently plays a major role in the fragmentation behavior of *Z*-configured maleic acid (**76**), while this reaction is negligible for the *E*-configured counterpart **77**. On the other hand, loss of water ($[M - H_2O]^{+\cdot}$, $m/z = 98$) and decarbonylation ($[M - CO]^{+\cdot}$, $m/z = 88$) are important processes for **77** but less or not for **76**. While the differences of the spectra of the two compounds are evident, the interpretation of the different behaviors of the two isomeric compounds is not trivial. It is not *a priori* perceptive why the loss of water (formation of ions with $m/z = 98$) should be more prominent for fumaric as compared to maleic acid. Water loss would intuitively (and based on the wet-chemical behaviors of the compounds) rather be expected for the *Z*-configured maleic acid, where intramolecular substitution and cyclization to the anhydride are possible.

Not only *E/Z*-isomers of compounds but also diastereoisomers that differ in the relative configurations of stereogenic centers can deliver different mass spectra. The mass spectra can be more or less distinct, which depends on the type of compound, the possible fragmentation reactions (and their prerequisites), and the spatial arrangements of the substituents (that might allow neighboring interactions of functional groups or lead to conformational restraints).

For example, the two diastereoisomeric compounds *cis*-cyclohexane-1,4-diol (**78**; M = 116) and *trans*-cyclohexane-1,4-diol (**79**; M = 116) deliver all in all very similar spectra, except for the signal that corresponds to ions of water loss $[M - H_2O]^{+\cdot}$ ($m/z = 98$) (**Fig. 4.84**). The more prominent loss of water from **79**⁺⋅ (100 rel.%) as compared to **78**⁺⋅ (50 rel.%) can be ascribed to a transannular 1,4-elimination, which is possible in the boat conformation of **79**⁺⋅ as indicated in **Fig. 4.85**. Ion **78**⁺⋅ cannot form an analogous conformation that would allow the same reaction.

In summary, MS is not the method of choice to solve stereochemical problems. In some cases, differences in the spectra of stereoisomers can be recognized. They arise in most cases

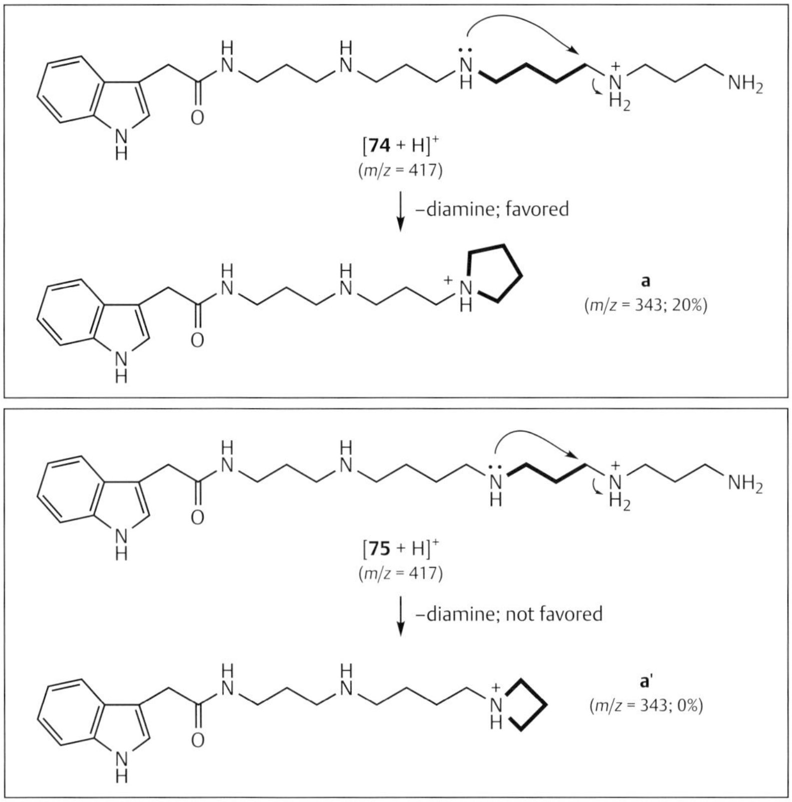

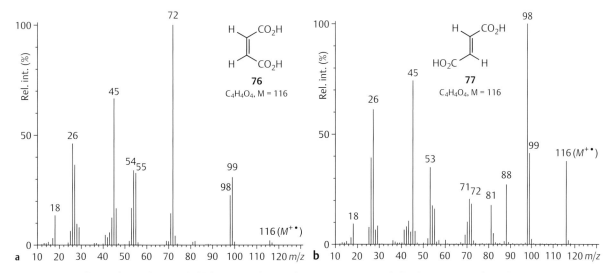

Scheme 4.38 Preferred fragmentation of poly-amine derivatives by intramolecular substitution via a five-membered transition state.[7,8]

Fig. 4.83 EI-MS of **(a)** maleic acid (**76**) and **(b)** fumaric acid (**77**). (These images are provided with permission of NIST [NIST 2005]).

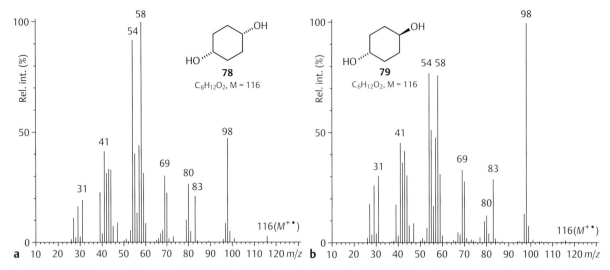

Fig. 4.84 EI-MS of **(a)** *cis*-cyclohexane-1,4-diol (**78**) and **(b)** *trans*-cyclohexane-1,4-diol (**79**). (These images are provided with permission of NIST [NIST 2005]).

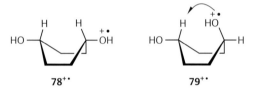

Fig. 4.85 Structural predestination for transannular 1,4-elimination of H$_2$O for *trans*-cyclohexane-1,4-diol (**79**) as compared to *cis*-cyclohexane-1,4-diol (**78**).

from neighboring group participations, and their interpretation is rather ambiguous and, thus, of little help for structural assignments.

4.4.6   Spectral Libraries

Mass spectra are produced in a digital format (mass number versus relative abundance) that can be stored electronically and compiled to spectral libraries, which can be consulted for spectral comparison. It is sensible to store only mass spectra of structurally known compounds of highest purity in a spectral library, and the libraries should be partitioned into sections according to the applied method. An EI-MS of a given compound is, for example, hardly comparable with the CID-MS of the same compound, and different CID-MS measured under varied conditions (type of instrument, collision gas, collision energy, etc.) may result in distinctively different spectra as well.

Extensive libraries were established in particular for EI-MS. EI-MS was the routine method of organic MS for a long time

and, thus, many spectra for many organic compounds were accessible. Additionally, EI-MS (contrary to CID-MS) is rather insensitive to the used instrument and, thus, the spectra are well reproducible, which is a prerequisite if spectral comparisons are envisaged. The major use of spectral databases is the identification of known compounds in samples of unknown provenience by spectral comparisons. It is possible, e.g., that a computer directly compares measured EI spectra obtained by GC-MS of multicomponent mixtures with spectra of a database and delivers potential sample structures in a hit list. However, the database matching can only be successful for known analytes for which the spectra are filed in the database. Databases are less beneficial for the analysis of unknown compounds.

Databases are continuously gaining in importance because they become increasingly important to master the ever-expanding data volume. Many different spectral libraries that contain several hundreds of thousands of EI and other MS spectra are accessible free of charge or through commercial channels. The algorithms for the electronic comparison of spectra become progressively more elaborate as well. Despite all the positive developments, however, a critical evaluation of the computer-generated output by analytically trained personnel still remains indispensable.

Since the algorithms for the software-generated hit lists are virtually inaccessible and nontransparent, there are some general aspects to be considered when dealing with spectral comparisons.

The reproducibility of mass spectra is limited, even for EI-MS. A mass spectrum is dependent not only on the ionization method and the type of instrument used, but also on many factors related to the sample itself (e.g., its purity) and to instrumental parameters such as the electronic optimization of the ion transfer or other rather unfathomable issues. A given sample can deliver distinctively different spectra even if it is analyzed on the same instrument under apparently the same conditions so that different analytes could be suspected for the different measurements. The range of variation in the spectral appearance is particularly high for MS/MS spectra, and related spectral databases are of limited value for this reason.

Due to the limited reproducibility of the MS spectra, increasingly spectral sets related to a specific compound, obtained by measurements under varied conditions, are acquired and stored in the databases. All these spectra should then be considered in the matching process of a spectral comparison. A spectral database, however, contains to some extent data of fantastical unique compounds also that hardly will be re-encountered as unknowns but delivered spectra which can be similar to those of new samples. Another problem with spectral databases is their incompleteness. Even very simple compounds that are well known for a long time might not be included there. Thus, finding no match never means that the respective compound is not known.

All these problems, combined with the occurrence of a slightly falsified spectrum due to impurities, make the unequivocal identification of compounds by matching their spectra with reference spectra of known compounds difficult. The result of the search for possible analyte substances on the basis of a slightly contaminated sample of coumarin (**80**; M = 146) illustrates this problem. The EI-MS (70 eV) of the sample was measured on a Finnigan-MAT 95 instrument, and for spectral matching the *NIST 2005 MS Search Program* was used. The respective list of the 10 best-matching compounds **80** to **89** is shown in **Fig. 4.86**. As can be seen, the 10 compounds are structurally very diverse, and it becomes obvious with this example that the interpretation of the computer output still needs chemical intelligence and know-how of the user.

The computer support in the spectral interpretation should, however, not be underestimated. There are algorithms incorporated into the software that are quite effective and produce prioritized hit lists with the structures delivering the most similar spectra placed on the top. Two different matching factors are used for the prioritization of hit results: the SI factor (*similarity index*, also called *match* or *fit*) and the RSI factor (*relative similarity index*, also called *Rmatch* or *Rfit*). The SI factor is a measure for the direct match of the measured spectrum with a spectrum stored in the database. The RSI factor is an inverse matching factor that quantifies the level of conformity of two spectra solely considering the signals that appear in both of them (ignoring the signals that are unique for one or the other

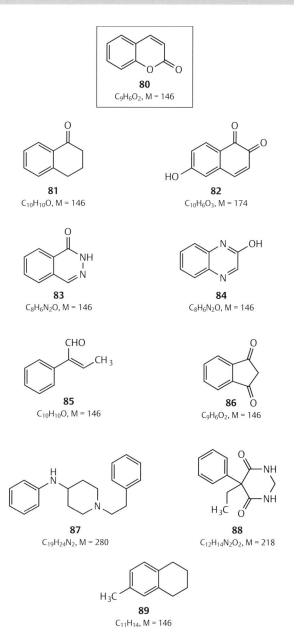

Fig. 4.86 Structures of the first 10 hits of the list that was obtained by the *NIST 2005 MS Search Program* upon comparison of the EI-MS spectrum of a slightly contaminated sample of coumarin (**80**) with database spectra. The compounds are arranged in the order of decreasing priorities as determined by the program.

spectrum). The *NIST 2005 MS Search Program*, e.g., calculates from the two matching factors, with a stronger weighing of the SI factor, probability values (*Prob*) that are used for the ranking of the hits.

Mass Spectrometry

Hit	SI	RSI	Prob	Name	Library Name
1	931	933	86.24	Benzophenone	mainlib
2	837	845	7.06	1,2,4-Trioxolane, 3,3,5-triphenyl-	mainlib
3	830	887	5.41	1,2,4,5-Tetroxane, 3,3,6,6-tetraphenyl-	mainlib
4	743	763	0.47	2,5-Dimethyl-7,7-diphenyl-3-aza-4,6-dioxabicyclo[3.2.0]hept-2-ene	mainlib
5	712	739	0.13	Azobenzene	mainlib

Fig. 4.87 The first four hits of the list (computer output) that was obtained by the *NIST 2005 MS Search Program* upon comparison of the GC/EI-MS spectrum of a sample of benzophenone (**90**) with database spectra.

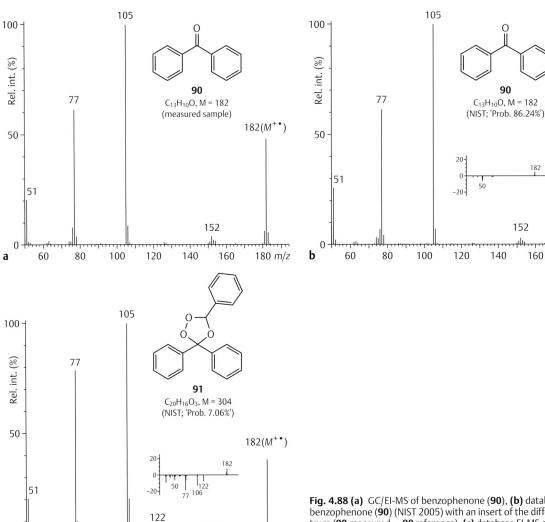

Fig. 4.88 (a) GC/EI-MS of benzophenone (**90**), **(b)** database EI-MS of benzophenone (**90**) (NIST 2005) with an insert of the differential spectrum (**90** measured − **90** reference), **(c)** database EI-MS of ozonide **91** (NIST 2005) with an insert of the differential spectrum (**90** measured − **91** reference). (These images are provided with permission of NIST [NIST 2005]).

The hit list (only the first four of an overall of 67 hits) obtained for the search of matching spectra to benzophenone (**90**; M = 182) is displayed in **Fig. 4.87**. The measurement with an authentic sample was performed by GC/EI-MS to ensure that the obtained spectrum derives from a pure analyte. **Fig. 4.88** shows (a) the measured spectrum of **90** and the spectra for the first two hits from the database: (b) the reference spectrum of **90**, and (c) the reference spectrum of the peroxy compound **91** (M = 304).

It is hardly comprehensible to understand why the program delivers a hit probability of 86% for the correct structure **90** and only 7% for compound **91** in light of the minor differences in the three spectra that can hardly be recognized upon visual comparisons. The spectra's deviations, however, become evident when the difference spectra (database spectrum – measured spectrum) are considered (see the corresponding inserts in (b) and (c)).

The hit rate in spectral comparisons is usually high. Nevertheless, it is important to always keep in mind, as already mentioned, that MS spectra can never be measured with 100% reproducibility, that a database can contain erroneous information, and that the number of spectra filed in a database is limited.

It would certainly be desirable to have access to a personal database that is customized with regard to the local instrumentation and needs. Only this would ensure the most efficient and reliable comparison of spectra by computational means. To establish such a database, however, is very laborious and time-consuming and only worthwhile when the database is used intensively and over a long time. This is the case when specific substances have to be repetitively analyzed and identified, e.g., in forensics, doping, fragrance, or environmental analytics.

4.4.7   Special Techniques

Derivatization

Not all compounds can be analyzed with equal ease with all MS methods. EI- and CI-MS of highly polar compounds, for example, are hardly accessible because the analytes frequently cannot be vaporized without decomposition as required for these methods. Derivatization of the original samples is often helpful in such cases. Very polar groups such as OH, R_2NH/RNH_2, or CO_2H can be converted into less polar groups by silylation (TMS or TBDMS), acylation, or methylation. Even though this leads to compounds with higher molecular masses, the derivatives are usually more volatile and less reactive. Typical examples are polyhydroxy

compounds such as sugars, which do not deliver EI-MS because they rather decompose than vaporize upon heating. This is the case also with iditol (**92**, M = 182), a hexitol that is formed and enriched in the human body as a result of metabolic disorders. The compound can be made indirectly accessible to EI-MS, however, by its conversion into its persilylated derivative **93** (M = 614, **Scheme 4.39**). The hexatrimethylsilyl derivative **93** not only can be evaporated without decomposition, but it also survives gas chromatographic separations and, thus, allows rapid analysis by GC-MS. **Fig. 4.89** shows as an illustration the GC/EI-MS result of the analysis of a sample of **93**.

Derivatization is not only used to convert polar and reactive analytes into more volatile and inert compounds but often also to improve chromatographic behaviors. Diethanolamine (**94**; M = 105), for example, can readily be analyzed by EI-MS but

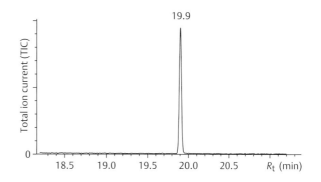

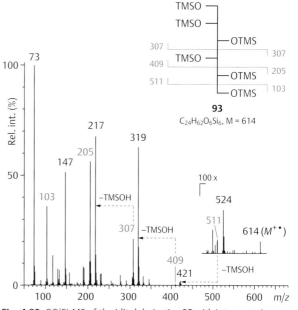

Fig. 4.89 GC/EI-MS of the iditol derivative **93** with interpretation.

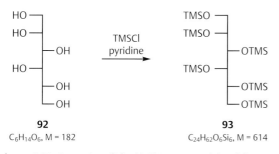

Scheme 4.39 Conversion of iditol (**92**) into its persilylated derivative **93** to enhance sample volatility and sample stability.

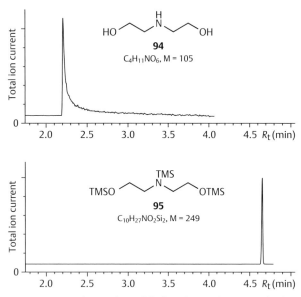

Fig. 4.90 EI-MS-detected GC of diethanolamine (2,2'-iminodietha-nol, **94**, upper chromatogram) and its silylated derivative **95** (lower chromatogram).

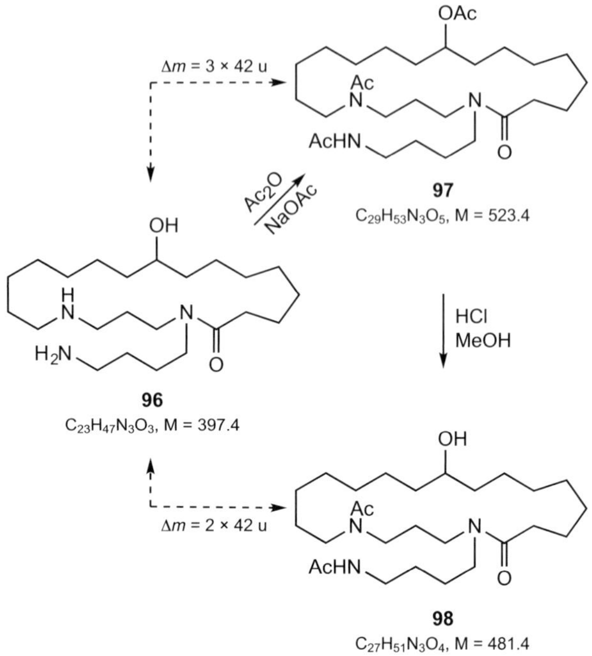

Scheme 4.40 Conversion of inandenine-13-ol (**96**) to its peracetylated derivative **97**, and the N-acetylated derivative **98**.[9]

shows a pronounced tailing in GC (**Fig. 4.90**, upper chromatogram), which can become problematic in analyses of mixtures. Silylation of diaminoalcohol **95** affords the tris-trimethylsilylated derivative **95** (M = 249), which not only delivers a well interpretable EI-MS but also a sharp and optimally shaped signal in the GC analysis (lower chromatogram in **Fig. 4.90**).

Besides the possibilities to enhance volatility and inertness of a compound or to modify its chromatographic behavior in favor of simplified analytics, derivatization also offers possibilities to specifically identify functional groups within a molecule. The acylation of hydroxy and amino groups can be used for the determination of the number of such units in structural elucidations of unknown compounds, in particular of natural products. If, for example, an alkaloid is treated with Ac_2O in the presence of a base, complete acetylation of all primary and secondary amino groups and of all hydroxy groups is affected. The mass difference between the original alkaloid and the derivative reveals the number of "active" functional groups contained in the molecule. Subsequent selective cleavage of the O-acetates by transesterification with hydrochloric methanol allows the differentiation of alcohols and amines.

Acetylation was used as described above in the structural elucidation of the alkaloid inandenine-13-ol (**96**; M = 397.4, **Scheme 4.40**).[9] Peracetylation of the compound delivered a product with a mass of M = 523.4, which later was assigned to **97**. The mass difference of +126 u (= 3 × 42 u) allowed the conclusion that the original natural product contains three NH and/or OH groups. Acidic methanolysis of **97** afforded a product with

one less acetyl group (**98**; M = 481.4, Δm = +84 u), indicating that one of the three acetylated groups of **96** was an alcohol and the other two amines.

Other derivatization reactions such as esterifications, acetalizations (for carbonyl compounds or diols), and formation of oximes or hydrazones (for carbonyl compounds) could also be used for the identification of functional groups analogous to acetylation (Table 4.11). The more specific a reaction is, the more significant is the information that can be obtained.

Table 4.11 Common derivatizations used for mass spectrometry

Functional group	Derivative	Mass increase per derivatization (u)
–OH	–OTMS	72
	–OTBDMS	114
	–OAc	42
–CO₂H	–CO₂CH₃	14
	–CO₂TMS	72
	–CO₂TBDMS	114
–NH₂, –NH–	–N(H)TMS	72
	–N(H)TBDMS	114
	–N(H)Ac	42
	–N(H)COCF₃	96
	–N–CHN(CH₃)₂	55
C=O	C=N–OCH₃	29
	C=N–OTMS	87
	C=N–NHPh	102

The specific complexation of functional groups can be regarded as a special case of a derivatization. It is known, for example, that hydroxamic acids and phenols form stable complexes with Fe^{3+}, while other functional groups are inert toward this metal ion. Thus, compounds that contain hydroxamic acid functionalities or phenolic groups can readily be identified by ESI-MS if iron (III) salts are admixed to the samples prior to their analysis. Polyaromatic hydrocarbons that selectively form $[M + Ag]^+$ adducts with silver ions are further examples for this type of derivatization.

H/D Exchange

A special case of derivatization is the exchange of acidic H by D atoms. Such a H/D exchange can provide valuable structural information not only in ^{1}H NMR spectroscopy but also in MS. The information gained is similar, but to a great extent complementary to that obtained by some of the derivatizations described above. Acetylation, e.g., can distinguish between acidic H atoms of amines, alcohols, and carboxylic acids, which H/D exchange cannot. In contrast, H/D exchange allows the differentiation between substitution levels of amines (primary, secondary, and tertiary), which acetylation does not.

Technically seen, H/D exchange is limited to ionization methods that generate the charged particles directly from solutions (e.g., ESI and APCI). With vaporization methods, previously H/D exchanged samples usually undergo partial (or even complete) re-exchange (D to H) before they reach the ionization chamber of the instrument. In the case of spray methods, H/D exchange can be achieved very easily *in situ*. The sample is simply dissolved in (or chromatographed with) deuterated rather than the normally used solvents. The number of exchangeable H atoms can then be directly established from the mass shift Δm of the adduct ions, taking the type of ionization into account. If the positive mode of an instrument is used and the ionization proceeds by protonation, the number of exchangeable H atoms is $n = \Delta m - 1$. In the negative mode with deprotonation as the ionization process, $n = \Delta m + 1$. If neither a proton is attached nor removed, $n = \Delta m$. The latter is the case in the $(+)$-ESI-MS of hesperidin (**99**; M = 610.2, **Fig. 4.91**), where the $[M + Na]^+$ ions of the undeuterated sample are detected at $m/z = 633.2$ and those of the H/D exchanged probe at $m/z = 641.2$. The mass difference of $\Delta m = 8$ stands for the eight exchangeable OH atoms of the compound.

The elucidation of the polyamine backbone of the spider toxin LF487A (**100**; M = 487.3, **Fig. 4.92**), a component of the complex venom mixture of *Larinioides folium*, exemplifies the use of H/D exchange in structural analysis.[10] The thorough analysis of the MS data acquired for LF487A by HPLC-ESI-MS/MS left open the question of the source of the signals found at $m/z = 329$ (**b**)

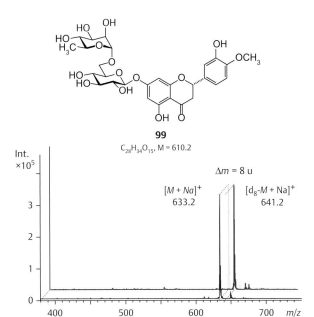

99

$C_{28}H_{34}O_{15}$, M = 610.2

Fig. 4.91 ESI-MS of hesperidin (**99**) measured with nondeuterated and deuterated solvent (H_2O/MeOH 1:1).

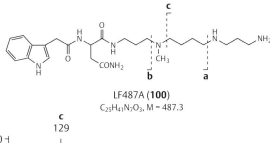

LF487A (**100**)

$C_{25}H_{41}N_7O_3$, M = 487.3

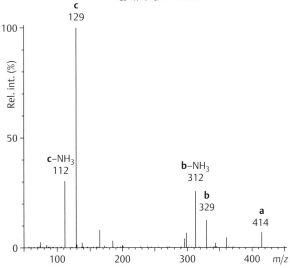

Fig. 4.92 ESI-MS^2 (Ar; $P^+ = [M + H]^+ = 488$) of LF487A (**100**).

Mass Spectrometry

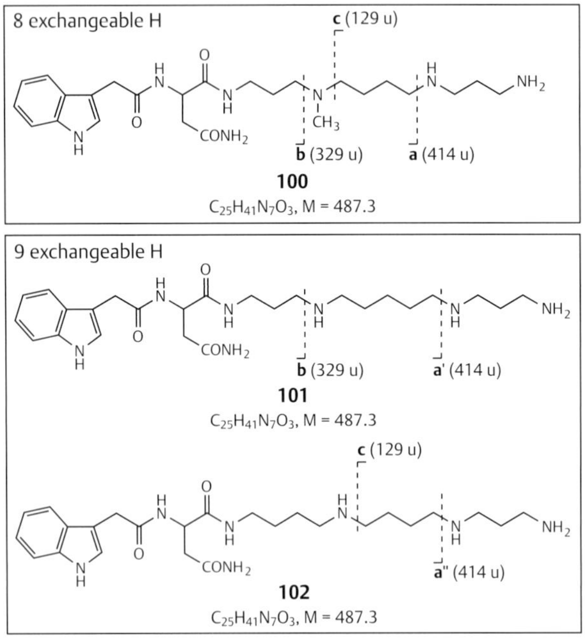

Fig. 4.93 Potential sources of the signals found at $m/z = 329$ and 129 in the spectrum of **Fig. 4.92:** either fragments **b** and **c** derive from compound **100** or fragment **b** stems from **101** and fragment **c** from **102**, the two compounds measured as a mixture (co-eluting).

and $m/z = 129$ (**c**). Two possibilities stood under discussion (**Fig. 4.93**): the signals could either be the responses to the fragment ions **b** and **c** that are both formed from the N-methylated derivative **100** or they could be the signals of the same ions that are, however, formed from two different compounds, ion **b** from compound **101** (M = 487.3) and ion **c** from compound **102** (M = 487.3), that accidently are co-eluting in the HPLC. It was shown by *on-column* H/D exchange, upon use of D_2O and d-TFA as acidic components of the mobile phase in the chromatography, that the analyte molecules possess eight exchangeable H atoms. This excludes the possibility that the investigated sample could be a mixture of the two compounds **101** and **102** and supports the proposed structure **100** for LF487A.

Degree of Isotopic Labeling

There are a number of questions that might find their answers with the help of isotopically labeled compounds. Such questions arise, e.g., during the elucidation of MS fragmentation mechanisms (p. 384) or of reaction mechanisms in general, and also biological questions such as "what is the origin and the fate of natural products?" can be addressed with the investigation of labeled compounds. Apart from the position of the labeling, a very important criterion in such investigations is often also the degree of isotopic labeling found in the products.

MS is generally the method of choice to determine the degree of isotopic labeling of a compound (while NMR is more suited for the determination of labeling positions, in particular if labeling with D and ^{13}C atoms is used). The analysis is usually done by the direct comparison of the spectra of the labeled and the unlabeled samples. Several considerations and measures are however required to make sure that reasonable results are obtained. First of all, it must be ensured that the MS spectra of both samples, the labeled and the unlabeled, are free of disturbing external signals in the region of interest. This means in practice that analyses are performed preferably with chemically uniform compounds or with samples that only have impurities with signals in spectral regions that are not considered in the analyses. That the purities of the samples are sufficient is usually secured by independent analytical methods such as m.p., TLC, GC, HPLC, NMR, and/or combustion analysis. Chemical uniformity of the analytes may also be obtained by the use of coupled MS methods such as GC-MS or HPLC-MS. It is also advantageous to determine prior to the MS investigations the labeling positions within a sample structure. It is often possible to estimate the degree of labeling during this process, and, hence, a later misinterpretation of the data becomes less likely.

For the effective determination of the labeling degree of a compound, the signal of the molecular ion or of an adduct ion is used. This signal must be of suitable intensity and should not be overlaid, in the ideal case, with signals of fragments. For example, molecular ions of samples that lose $H^{\bullet}$ or H_2 are not suited for this reason. If the signal intensity is too low or if fragment ions interfere with the analysis, which is often the case with EI-MS, it is advisable to change the ionization method. To finally determine the labeling degree, samples of labeled and unlabeled analytes have to be repetitively measured and the averaged spectra compared. It is important that the unlabeled sample is also measured and that its averaged spectrum rather than the theoretical isotopic distribution is used as the reference. The spectra should be measured one after the other by always using the same analytical conditions to exclude errors and to ensure a most reliable result. Thorough care should be taken so that memory effects are excluded, i.e. a measured sample is completely removed from the instrument before the next sample is processed.

The data evaluation of a labeling experiment is explained with the following example: compound **A** with the chemical formula $C_9H_{11}NO$ (M = 149) was synthesized by using a ^{13}C-labeled reagent in the isotopically enriched form (**A***, $^{12}C_8{}^{13}C_1H_{11}NO$; M = 150). The EI-MS spectra of the unlabeled sample **A** and the labeled sample **A*** were measured five times each, and the obtained averaged isotopic distributions for the $M^{+\bullet}$ ions are:

Unlabeled sample **A**:

EI-MS ($M^{+\cdot}$): 149 (100.00), 150 (10.59).

Labeled sample **A***:

EI-MS ($M^{+\cdot}$): 149 (10.94), 150 (100.00), 151 (11.11).

The easiest way to determine the labeling degree of **A*** is by comparing the intensities of the monoisotopic peaks of **A** and **A*** in the sample of the labeled compound. The intensity of the monoisotopic peak of **A** (m/z = 149) can directly be extracted from the spectrum: it is 10.94%. The intensity of the monoisotopic peak of **A*** (m/z = 150), on the other hand, needs to be calculated because the signal at m/z = 150 is a superposition of the monoisotopic peak of **A*** and the M + 1 isotopic peaks of **A**. The intensity of the M + 1 isotopic peaks of **A** (m/z = 150), however, can readily be determined by using the information acquired for the unlabeled sample **A**: it is 10.59% of the monoisotopic peak. This corresponds for **A** in the ^{13}C-enriched sample to 10.59% of 10.94% or 1.56%. The intensity of the monoisotopic peak of **A*** in the labeled sample is thus 98.44% (100.00%–1.56%), and the ratio **A**/**A*** follows as 10.94/98.44. Hence, 90% of the sample molecules in the probe are ^{13}C-labeled.

The evaluation for multiply labeled compounds naturally becomes more complex.

4.4.8 Elucidation of Fragmentation Mechanisms

MS offers a number of possibilities to elucidate fragmentation mechanisms. The most expedient methods are MS^n and linked scan methods, combined with high accuracy MS, because they are readily available, fast, and very informative, already answering in many cases the posed questions. Alternatively, labeling experiments might be considered. Since labeling of a compound is usually a laborious and time-consuming task, however, labeling experiments are generally used only if direct MS/MS experiments fail to solve a problem.

MS^n and Linked Scan Methods

Mass spectrometers that have at least two mass analyzers and an activation cell offer the possibility of tandem MS (Section 4.3.5, p. 326). This type of complex instrumentation allows us to tackle specific questions regarding MS fragmentation mechanisms and to figure out, for instance, which fragment ions are derived from a chosen precursor ion. More complex (collision induced) dissociations can be addressed with MS^n experiments, wherein the information content increases with increasing number of MS generations.

Further questions such as "which are the precursor ions that lead to a specific fragment ion?" or "which ions lose a fragment

with a given mass?" can be solved with linked scan methods on triple quadrupole instruments.

Product Ion Scan. The standard MS/MS experiment uses the product ion scan method. It provides directly the information about which precursor ion P^+ gives rise to which product fragment ions F_1^+, F_2^+, F_3^+, etc. Additional information about an ion can be obtained with an in-source CID experiment. Increase of the ion acceleration during ionization results in heavier collisions of the ions with residual gas on their way to the analyzer, resulting in the formation of fragment ions.

More specific information for clearly defined precursor ions can be obtained with a QIT mass spectrometer. If fragment ions are selected as precursor ions in a second or a subsequent generation of an MS^n experiment, the product ion scan allows the identification of fragmentation cascades. For example, the result of the following two scans

$$MS^2: M^{+\cdot} \rightarrow F_1^+, F_2^+, F_3^+, F_4^+, F_5^+, ... \text{ and}$$
$$MS^3: F_1^+ \rightarrow F_3^+, F_5^+, ...$$

shows that the product ions F_3^+ and F_5^+ detected in the MS^2 of $M^{+\cdot}$ can be (but not necessarily need to be) formed from fragment ions F_1^+ and that F_2^+, on the other hand, cannot be derived from these ions. The MS^2 and MS^3 results thus suggest, but do not prove, that a fragmentation cascade $M^{+\cdot} \rightarrow F_1^+ \rightarrow F_3^+, F_5^+$ is operative, and they exclude a fragmentation path, $M^{+\cdot} \rightarrow F_1^+ \rightarrow F_2^+$. By scanning of more product ions or by using MS^n ($n > 2$), fragmentation paths can be followed in more detail—at least along the line of masses. If in the line of HR-MS experiments also accurate masses and, hence, chemical formulas of fragment ions and/or of lost molecular moieties are available, very often structures can be directly assigned to fragments. This facilitates the determination of fragmentation paths and fragmentation mechanisms even more.

Precursor Ion Scan. In some cases, it is desirable to clarify which precursor ions lead to specific product ions (with a particular mass m_F). The precursor ion scan is the method of choice to solve this problem. In this experiment, the full mass range of the ions formed in the ion source and transferred to the activation cell is scanned in the first analyzer, and the second analyzer is set to detect only the ions of the specified value $m/z = m_F$. This way, the detector registers a signal for each precursor ion at $m/z = m_P$ by actually counting its derived ions at $m/z = m_F$. The spectrum thus shows signals for all precursor ions that can (or do) lead to decomposition—directly or within a fragmentation cascade—under formation of the fragment of the specified mass. The respective information can be of high relevance for the elucidation of fragmentation paths too. The combination of product ion scan and precursor ion scan can possibly also solve the question that was left open above, whether particular

Mass Spectrometry

fragment ions ($F_3{}^+$ and $F_5{}^+$ in the example above) are derived from several ions or from a single ion only.

Neutral Loss Scan. For the determination of fragmentation mechanisms, but also for the solution of other problems, it might be of interest to identify fragments (or compounds) that lose the same neutral fragments. Such an identification is done with a neutral loss scan. In this case, the two analyzers of the MS/MS arrangement are set to scan in parallel, the detecting analyzer being synchronized to record ions with an m/z value that is lower than the m/z value of the currently selected precursor ions by the mass of the neutral loss fragment. For example, all compounds in a mixture that show a loss of CO_2 would be registered when $\Delta m/z = 44$ would be chosen for the parallel scanning.

Labeling Experiments

The specific labeling of functional groups (or their surroundings) by isotopes or by derivatization is a common technique used in spectroscopic, kinetic, bioorganic, or mechanistic investigations. Most frequently ^{2}H (D), ^{13}C, ^{15}N, and ^{18}O are used for isotopic labeling; for bioorganic studies, radioactive isotopes such as ^{3}H (T), ^{14}C, ^{18}F, and ^{131}I are also employed. Many reagents and compounds containing these isotopes are commercially available such that almost any compound with any specific labeling pattern can principally be accessed by synthesis. These

syntheses can be more or less tricky, depending on the complexity of the structures to be obtained. By the very nature of organic synthesis, it usually is more intricate to synthesize C- or N- than H- or O-labeled compounds. However, we do not want to discuss here in more detail how isotopically labeled compounds may be manufactured.

Isotopically labeled compounds can be used in MS to establish the positions of functional groups within the molecular assembly of the observed ions. The fate of labeled structural units during fragmentation processes is of particular interest in investigations regarding the elucidation of mass spectral fragmentation mechanisms. Because the synthesis of labeled compounds is very laborious and expensive, the planning of labeling experiments must be done very carefully. An example of the successful use of isotopic labeling for the elucidation of a fragmentation cascade is given with the ^{15}N-labeled derivatives **103a** to **103c** of the polyamine spider toxin IndAc3334 (**103**; M = 416, **Fig. 4.94**).[8]

It was recognized in connection with previous investigations of polyamine spider toxins that compound **103** (M = 416) delivers unexpectedly fragment signals at $m/z = 115$ and 98 upon CID-ESI-MS. These signals are characteristic of polyamine compounds with a PA33 terminus—they correspond to fragments of the type **a** that subsequently lose NH_3 to form ions **b** (**Scheme 4.41**)—but compound **103** does not possess such a PA33 terminus.

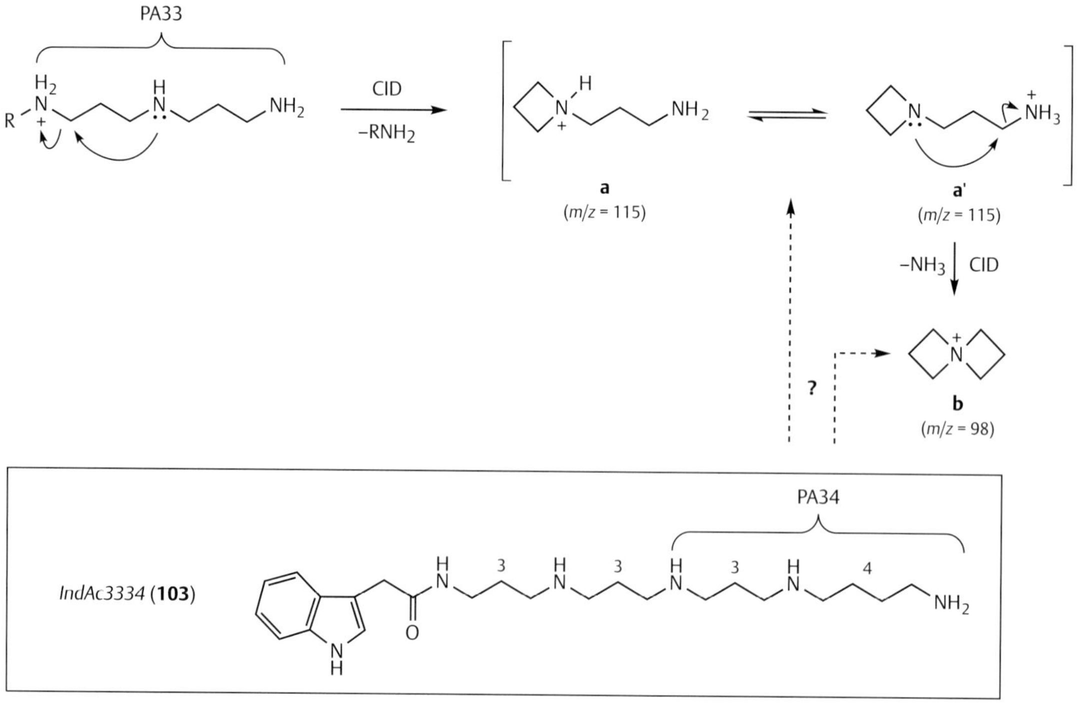

Scheme 4.41 Mass spectrometric fragmentation of the terminal PA33 moiety of an acyl polyamine.

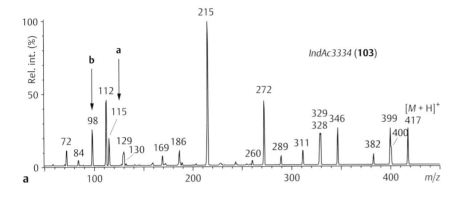

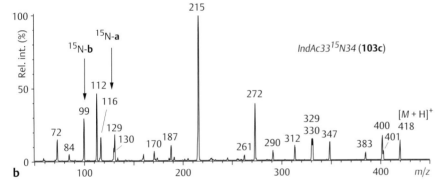

Fig. 4.94 ESI-MS² of the $[M + H]^+$ precursor ions **(a)** of **103** $(P^+ = 417)$ and **(b)** of **103c** $(P^+ = 418)$.[8]

Because the structural elucidation of polyamine spider toxins crucially relies on MS fragmentations, it was important to understand the formation of the fragment ions **a** and **b** as well as the structural prerequisites needed in the precursor molecules. To uncover the molecular portion within compound **103** that is contained in the unusual fragment ions, the three ¹⁵N-lableled derivatives (**103a–103c**) were specifically synthesized and analyzed by MS/MS. It turned out that only compound **103c** delivered ¹⁵N-labeled fragments of the types **a** ($m/z = 116$) and **b** ($m/z = 99$), as recognized in the spectrum of the compound (**Fig. 4.94**). This observation allowed making the conclusion that the highlighted portion of **103c** represents the structural basis for these ions (**Scheme 4.42**). The proposed fragmentation path from the ionized molecule [**103c** + H]⁺ to the ions **a** and **b**, with ion **c** (observed at $m/z = 347$) postulated as an intermediate, is shown in **Scheme 4.42**.

Alternative fragmentation paths that would involve transamidation or transalkylation reactions could be excluded by the labeling experiments.

4.5 Sample Preparation

The success of an MS analysis is strongly dependent on the quality of the studied sample, its proper preparation for the measurement, and the appropriateness of the used method and the technical procedures. It is rather easy and usually does not require special precautions if a pure and homogenous substance is used for analysis. This is particularly true if already some information about the structure and physical properties of the sample molecules is available. However, MS analyses are also possible with mixtures and with completely unknown

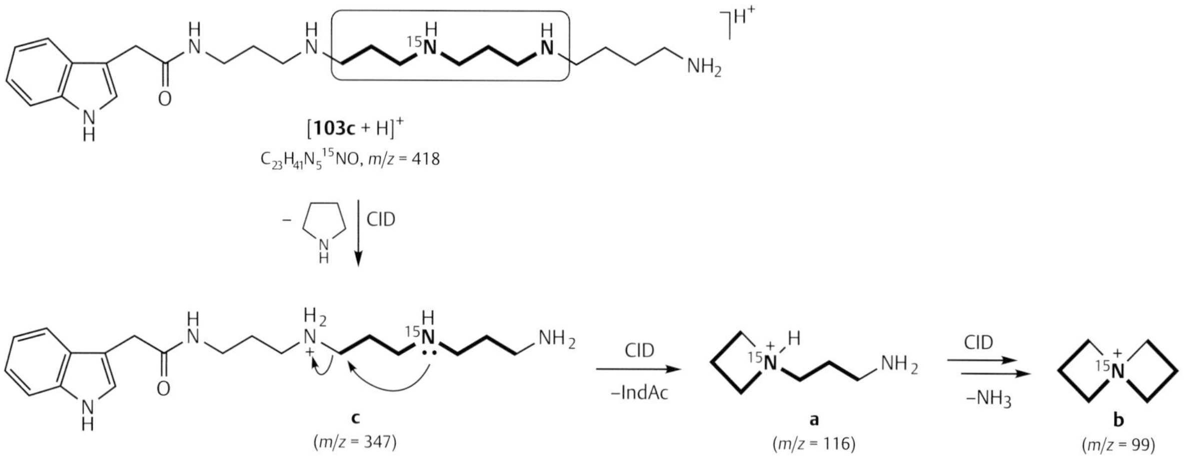

Scheme 4.42 Mechanistic proposal for the formation of the fragment ions **a** and **b** starting from [**103c** + H]$^+$ by CID.[8]

compounds for which, depending on the envisaged MS method, certain quality criteria need to be fulfilled.

4.5.1 Purification, Preparation, and Enrichment

Most organic samples can be purified and prepared for MS analyses by extraction, distillation, crystallization, and/or chromatography. These classical purification techniques can also be used in microscales, allowing rapid and simple access to samples appropriate for MS. Among these, microextraction has become particularly popular as a mini work-up procedure that delivers reliably and reproducibly MS samples.

More and more frequently, however, the classical procedures are not able to provide suitable MS samples. Many reasons can be responsible for this: the sample amount could be too low for classical purification (≤1 mg), the concentration of the analytes might be too low in the sample, the analytes can be contained in an extremely complex matrix (e.g., a biological matrix such as blood), or the sample might contain extremely large amounts of salts, e.g., nonvolatile buffers after chromatographic separations. For such cases, micro work-up procedures are available that may efficiently provide better samples. The techniques presented below are based on commercially available tools.

Precipitation and Filtration

Buffers, proteins, or other high-molecular compounds that obstruct proper MS measurements may often be precipitated by the addition of an organic solvent such as MeCN and removed by filtration. It will be appropriate in most cases to use centrifugation to attain proper sedimentation, and the sample is then collected by decanting. Sedimentation and decanting can be omitted for very small sample amounts by the use of

fine-pored microfilters (0.22 μm). Two respective systems are commercially available: microfilter funnels, through which the solutions are sucked by means of centrifugation, and microfilter extensions for syringes, through which the solutions are pressed. Several different filter materials are in use, such as modified cellulose, polyvinylidene fluoride (PVDF), or, if very aggressive or purely organic solvents are used, polytetrafluoroethene (PTFA, teflon). Solutions that are obtained this way can usually directly be employed for GC-MS, LC-MS, or ESI-MS.

Size-Exclusion Chromatography

If compounds of particular molecular sizes are hampering MS measurements, they can be selectively removed from a sample by *size-exclusion chromatography* (SEC). The procedure is from a technical point of view the same as described above for microfiltration. The filter materials, however, are characterized by mass cut-offs (*nominal molecular weight limits*) that can be chosen to be 1, 3, 5, 10, 30, 50, or 100 kDa. Filtration through such materials retains the molecules that exceed the cut-off mass on the filter and allows the smaller molecules to pass. Commercial filter cartridges are usually designed for sample volumes of 0.5 mL.

Simple filtration through such cartridges, hence, removes macromolecular contaminants from smaller analyte molecules, which is of relevance, e.g., in the analytics of metabolites, which arise in solutions together with plasma proteins. Size-exclusion cartridges can also be used to desalinate and accumulate probes of macromolecules. The material retained on the cartridge can be rinsed with solvents and subsequently backwashed. SEC nowadays is often preferred over dialysis to free (LC)-MS samples of salts because the procedure is simple and fast, and it delivers the samples in higher concentrated forms.

Solid-Phase Extraction

An additional technique to desalinate samples and in particular also to obtain more concentrated sample solutions is *solid-phase extraction*. Reversed-phase or cation exchange resins are used instead of size exclusion material in this case, which offers different selectivity criteria for the analyte substances and contaminants as compared to SEC.

Micro-Solid Phase Extraction. Micropipette tips filled with up to 1 mg of adsorbent resins are commercially available for micro-solid phase extractions. Analyte substances are collected on the resin by repetitive sucking up and blowing out of the analyte solution. The adsorbed material can then be freed from salts or ionic detergents by washing and rinsing, and finally liberated from the adsorbent material by elution with a small volume (in the μL region) of a suitable solvent. Such samples can then be directly used for ESI-, MALDI-, or GC/LC-MS measurements.

Solid-Phase Extractions with Larger Amounts of Material. Solid-phase extraction is also suited to process larger volumes of sample solutions. Cartridges that allow the convenient handling of up to 1 L of sample solutions are commercially available. The use of such large cartridges not only leads to larger amounts of the analytes but the process of concentrating the analytes becomes also more effective (concentration enhancements of up to 1,000-fold) and more reproducible. This results also in higher reliabilities of subsequent MS investigations, which is of particular importance for quantifications.

Table 4.12 summarizes the most important and most frequently encountered problems to be solved in sample preparation together with the respective recommended processing methods. The effect of proper sample preparation on the quality

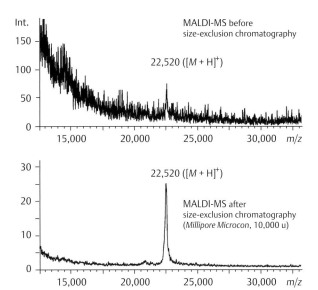

Fig. 4.95 Effect of size-exclusion chromatography (*Millipore Microcon*, 10,000 u) of a protein sample (M_R = 22,520; sample from J. Robinson and O. Zerbe, University of Zurich) that was contaminated with 0.02% of a detergent on the quality of the MALDI-MS.

of an MS spectrum of a protein sample is shown with two spectra in **Fig. 4.95**.

4.5.2 Sample Submission and Declaration of Sample Properties

Sample Properties

It is of major importance for the analyst working on the MS instruments to be comprehensively informed about the sample

Table 4.12 Problems with MS samples and their resolution

Approach	Source of problems				Associated effort
	Suspension	Contains salts/buffers/ detergents	Contains macromolecules (M >2,000)	Low concentration	
Precipitation/filtration	–	±	+	–	High
Centrifugation/decantation	+	–	–	–	Low
Microfiltration (0.22 μm)	++	–	–	–	Very low
Size-exclusion chromatography	–	+	++	+ (for macromolecules)	Very low
Dialysis	±	++	++	+ (for proteins)	High
Micro solid phase extraction	+	++	+	+	Very low
Solid phase extraction	–	++	+	++	High
Liquid/liquid extraction (micro-preparation)	–	±	+	±	Low
++ appropriate, + suited, ± suited in some cases, – inappropriate					

Mass Spectrometry

to be analyzed and the problem to be solved. Only this way, the proper method can be chosen and the appropriate experiments can be performed. Thus, a sample should not just be submitted to a service center but the submission should be accompanied with as much detailed information about the sample as possible. Each additional piece of information can lead to a more reliable and more meaningful analytical result.

As a standard, the melting point and, if possible, also boiling point of a pure sample compound should be reported upon its MS submission. This information allows the estimation of the volatility and the thermal stability of the compound to be measured. Also important for MS technicians is information regarding solubility of the material in different solvents. This information is needed not only to choose the appropriate MS method such as GC-MS, HPLC-MS, or MALDI-MS, but also to estimate the ease of ionization of the sample molecules under different conditions (EI, ESI, APCI, MALDI, (+)-mode, (−)-mode). It is also mandatory to declare all that is known or suspected about the composition and the structure of the sample compound (chemical formula, functional groups, molecular weight, etc.). This information also facilitates the proper choice of the MS method, and it narrows the measuring range, which guarantees efficient analytics.

Amount and Form of the Sample

The amount of sample and its proper form needed for an MS measurement are primarily dependent on the ionization method to be used. For applications with vaporization methods, the samples are needed as pure compounds (without solvents) in amounts that are "visible" (a few mg is sufficient).

Probes can also be accepted as solutions in appropriate solvents when spray or desorption methods are used. The amount of sample effectively needed with these methods is very low. Amounts of a few pmol (MALDI) or nmol (ESI/APCI) are usually sufficient for sensible analyses. The concentrations of the solutions should be in the range of nM to µM, and the respective information needs to be provided upon submission of the sample. It is usually better to submit samples of higher concentration because they can be easily diluted prior to the measurements if required. Concentrating a solution is more laborious.

4.5.3   Specific Preparations for the Measurement

Vaporization Methods

Sample preparation for vaporization methods consists essentially of the transfer of ca. 0.1 to 1 mg of a solvent-free sample into a small glass or metal crucible, which is then fixed onto the tip of a heatable probe that is inserted into a lock chamber (the ion source). If highly volatile samples are used, the crucible

needs to be covered with a cap to prevent premature evaporation of the analyte.

The crucible is loaded only in special cases with a sample solution rather than with solvent-free material. The solvent to be used in these cases must be easily vaporizable such as diethyl ether or ethyl acetate. Chlorinated solvents such as chloroform or dichloromethane, which are sometimes known to strongly adhere to the analytes, are not suitable. After evaporation of the solvent, the measurement can be performed as described above.

Desorption Methods

It is already sufficient to have 0.5 to 2 µL of an analyte solution which is concentrated in the low µM region (pmol µL^{-1}) to obtain good and reliable data with MALDI-MS. This is, however, only possible if an appropriate sample preparation on the MALDI plate is performed. It is particularly important to ensure that the matrix is crystallized as perfect as possible in order to obtain the most homogeneous analyte distribution. Since MALDI plates are with increasing frequency prepared within research units outside of the MS laboratories, the two most important methods for their preparation are briefly introduced below.

Dried-Droplet Method. The easiest and most direct method of MALDI sample preparation is the *dried-droplet* method. This method can be applied for all types of samples with all types of matrix substances (**Table 4.2**, p. 312). To prepare a MALDI plate, the analyte solution (µM) is mixed with the matrix solution (saturated, ca. 10 g L^{-1}) in a ratio of 1:1, and 1 µL of the resultant solution is pipetted onto the sample carrier. Crystallization of the matrix/analyte mixture then takes place upon evaporation of the solvent.

Double-Layer Method. The *double-layer* method is regarded as the method of choice for the sample preparation of probes with proteins. However, it can also be used for peptides or other compounds. In this method, a concentrated matrix solution is initially applied onto the sample carrier to provide a thin matrix layer as a support for the actual sample. Sinapic acid (SA) is advantageously used because it crystalizes evenly and usually results in a homogeneous and consistently thick crystalline layer. Onto this bedrock, a 1:1 mixture of the analyte solution (1–100 µM) and the matrix solution (saturated) is applied, and crystallization of this mixture is again effected by leaving the solvents to evaporate. Aqueous, slightly acidic solvents can be used for the sample preparation, but it needs be considered that the solvent shall not dissolve the thin matrix support layer.

The advantage of the double-layer method over the dried-droplet method is that the crystallization of the sample/matrix mixture that is applied onto an already crystalline support proceeds in a more controlled and uniform way and that contaminants such as buffers or detergents could be removed more easily from the surface of the probe.

Handling of Contaminations with Buffers and Detergents. The MALDI method is generally rather tolerant with regard to contamination of the samples with buffers or detergents. Too high concentrations of such contaminants, however, influence the crystallization behavior of the matrix substances and, thus, affect the quality of the MALDI-MS data. Small amounts of impurities can still be washed off from the crystal surface immediately after the sample preparation with buffer solutions (e.g., H_2O + 0.1% TFA). The maximal concentrations of common sample contaminants that still allow decent sample preparations for MALDI-MS are compiled in **Table 4.13**. If larger amounts of impurities are present in a sample, a prior sample treatment as described in Section 4.5.1 (p. 386f) is required.

Table 4.13 Estimated maximal concentrations of sample additives or contaminants tolerated in MALDI sample preparation[11]

Additive	Maximal tolerated concentrations
Urea, guanidine, dithiothreitol	0.5 M
Tris, phosphate	50 mM
Sodium dodecylsulfate (SDS)	0.01%
Detergent (except SDS)	0.1%
Glycerol	1%
Alkali metal ions (e.g., Na^+)	0.5 M

Spray Methods

The samples to be analyzed by the use of spray methods are supplied to the MS in solution. This requires naturally that the analytes are sufficiently soluble in the selected solvent and that the solvent and its additives are appropriate for the chosen MS method. Usually, amounts of 100 to 200 µL of analyte solutions (a few µM, max. 100 nmol mL^{-1}) are required for a measurement; for nano-ESI, even less than 100 µL is sufficient. The type of solvents and the type of additives that are suited to be used for sample admission depend on the exact ionization method used.

ESI Samples. ESI samples are usually measured from solutions of polar and protic solvents, e.g., from H_2O, MeOH, i-PrOH, MeCN, and mixtures thereof. For less polar analytes, mixtures with larger portions of nonpolar solvents can be used, e.g., MeOH/DMSO (max. 10%), MeOH/DMF (max. 10%), MeOH/CHCl$_3$ (max. 70%), MeOH/CH$_2$Cl$_2$ (max. 50%), or MeOH/THF (max. 30%). In these cases, however, the operating conditions of the ion source have to be adjusted to guarantee the

formation of stable sprays (higher capillary voltage and lower gas flow).

It is assumed that the ions observed in ESI-MS had already been preformed in the sample solutions. Thus, acids or bases are often added to the sample solutions to increase the concentrations of cationized or anionized molecules. Volatile buffers such as HCO_2H, $MeCO_2H$, CF_3CO_2H, NH_4OAc, Et_3N, and NH_4OH in concentrations of <0.1% or <10 mM are commonly used. Ion formation and ion response can also be enhanced for slightly basic or slightly acidic molecules with SH or OH groups (e.g., sugars), amide, ester, or ether functional groups by the addition of alkali metal salts such as NaOAc, NaOTf, and NaI (or the respective Li^+ or K^+ salts; <0.1 mM). Adduct ions of the type [M + Alkali]$^+$ are then prominently formed.

APCI Samples. APCI-MS is typically used in combination with HPLC at flow rates above 100 µL min^{-1}, and the same solvents are compatible for APCI as those mentioned above for ESI. The sensitivity of the measurements is generally higher if the solvents are less polar and if they contain some nonpolar components such as CH_2Cl_2 or Et_2O. The addition of (exclusively volatile!) buffers is also supportive of the ionization with APCI, but not the addition of alkali salts, which usually leads to almost full signal suppression. This is due to the nature of the ionization process in APCI, which almost exclusively comprises proton transfer reactions with uncharged sample molecules in the gas phase.

4.6   Artifacts

A very important and not too rare problem with MS analyses is the occurrence of artifact signals. Such signals may have different causes, and three major types of artifact signals can be distinguished. These are (1) signals that arise without any connection to the sample molecules due to the electronics of the MS instrument, (2) signals that are responses of samples that were investigated previously (memory effect), and (3) signals that can be ascribed to artifact structures that were formed from the studied analytes.

4.6.1   Memory Effect

If residues of substances remain after a measurement within the ion source of a mass spectrometer (e.g., due to condensation in regions of lower temperature) and if signals related to these substances are found in the spectrum of a subsequent measurement, then this is called *memory effect*. A memory effect is also observed when instrumentation that was used for handling samples (during reactions, work-up, sample preparation, etc.) are contaminated with "old" analytes and, thus, are re-subjected for measurements.

Mass Spectrometry

The presence of undesired sample residues in the MS instrument can be discovered by the measurement of background spectra prior to a real measurement. "Old" analytes that might have been transferred into the MS with the sample, however, cannot be recognized this way. To recognize such memory effects, high awareness and attention of the instrument's operator and, in particular also, of the sample supplier is required. This is particularly the case when the measurements were performed at different times.

Memory effects that are due to the residual sample in the MS instrument can be eliminated by baking out the ion source or by its mechanical cleaning. To avoid memory effects due to the resubmission of "old" analytes, absolute care should be taken to work with meticulously clean equipment.

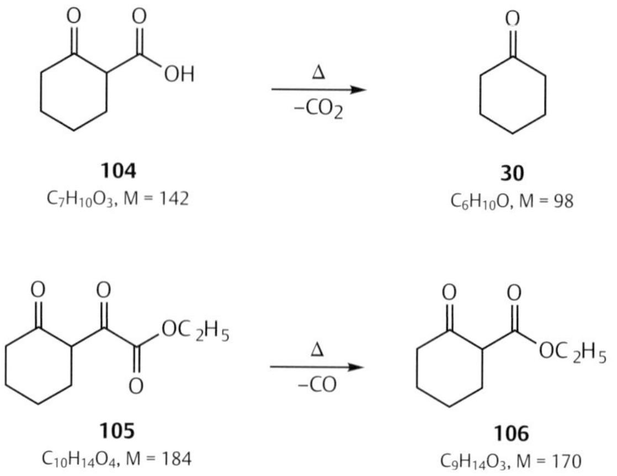

4.6.2 Formation of Artifacts in the Ion Source

The mass spectrum of a sample that forms artifacts in the MS instrument during vaporization or ionization does no longer correspond to the effective spectrum of the investigated compound but rather to the spectrum of one or more decomposition products, maybe in the mixture with the original analyte. Such artifacts can impede the interpretation of MS data or even render it impossible and lead to erroneous conclusions and incorrect structural assignments.

Artifact formation in the MS instrument is most common with the application of vaporization methods. The samples are usually heated with these methods, which can initiate thermal reactions. Thermal reactions, however, can also occur by the use of spray or desorption methods! Acid- or base-initiated reactions are more common with these ionization methods, because the ion formation starts usually from an acidic or basic environment.

Artifacts that are formed upon vaporization or ionization of a probe (heating, concentrating) usually correspond to reaction products that could also be expected as the result of "common" reactions known from synthetic organic chemistry. Some important and frequently encountered examples are introduced below.

Loss of Small Fragments

Loss of CO_2 (Decarboxylation). β-Oxocarboxylic acids are well known to readily undergo decarboxylation. However, decarboxylation is not only observed with such compounds but also with aromatic and other compounds that possess several carboxyl groups. This is particularly the case for larger molecules for which intensive heating is required for vaporization.

Decarboxylation is such a prominent reaction of 2-oxocyclohexane carboxylic acid (**104**; M = 142, **Scheme 4.43**, spectrum not

104
$C_7H_{10}O_3$, M = 142

30
$C_6H_{10}O$, M = 98

105
$C_{10}H_{14}O_4$, M = 184

106
$C_9H_{14}O_3$, M = 170

Scheme 4.43 Examples of artifact formation by thermal decarboxylation or decarbonylation, respectively.

shown) that the EI-MS of this compound is hardly distinct from the spectrum of cyclohexanone (**30**; M = 98, EI-MS: **Fig. 4.45**, p. 346). Solely an additional signal at $m/z = 44$ in the spectrum of **104** accounts for the formed $CO_2^{+\cdot}$ ions.

Loss of CO (Decarbonylation). α-Oxocarboxylic acids are, similarly to β-oxocarboxylic acids, rather labile compounds. They willingly lose CO, sometimes already upon distillation. It is not astonishing, thus, that loss of CO is also a prominent reaction that can be observed when such an analyte is transferred to the ion source. A striking example represents glyoxylic acid ester **105** (M = 198, **Scheme 4.43**). Its EI-MS corresponds, apart from a low-intensity signal at $m/z = 28$ for $CO^{+\cdot}$ ions, exactly to the spectrum of β-ketoester **106** (M = 170).

Loss of $MeCO_2H$ and Other Carboxylic Acids. Esters of aliphatic carboxylic acids can lose the respective carboxylic acid by ester pyrolysis. This reaction occurs particularly easily if the double bond that is formed during the reaction ends up in conjugation to a double bond that is already present in the starting molecule. As an example, the behavior of the indole alkaloid O-acetylhervin (**107**; M = 426, **Scheme 4.44**) is shown. Compound **107** revealed at the beginning of the measurement (EI-MS, direct inlet at 250°C) two prominent signals at $m/z = 426$ and $m/z = 366$ for the $M^{+\cdot}$ and the [M − AcOH]$^+$ ions in a ratio of 72:100. Later, after 3 minutes, the signal at $m/z = 426$ had completely disappeared, and the signal at $m/z = 366$ remained as the signal of highest mass number. The formation of [M − AcOH]$^+$ can principally be explained by the McLafferty rearrangement of the molecular ion $M^{+\cdot}$ (p. 358). Alternatively, ester pyrolysis of **107** in the sample crucible and submission of compound **108** (M = 366) to the MS would account for the same ion. The time dependence of the spectrum allows making the conclusion that

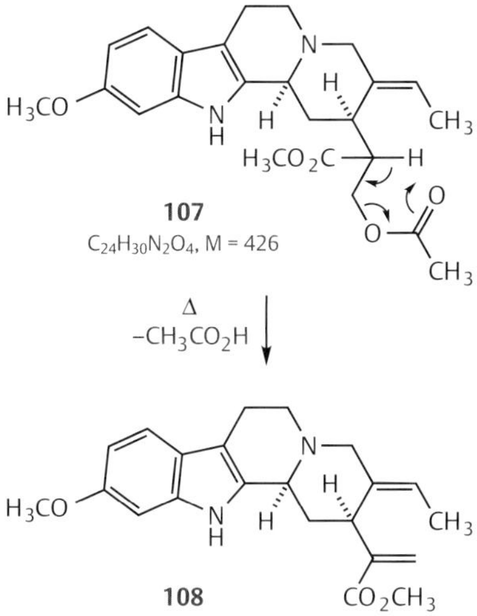

107
$C_{24}H_{30}N_2O_4$, M = 426

Δ
$-CH_3CO_2H$

108
$C_{22}H_{26}N_2O_3$, M = 366

Scheme 4.44 Example for artifact formation by ester pyrolysis.

the McLafferty rearrangement cannot be the only way, the ions with m/z = 366 are formed, but that artifact formation must be involved.

Loss of HX (H_2O, NH_3, HCl, etc.). There are many examples that document the loss of H_2O, NH_3, and hydrohalogenic acids in MS or prior to MS analyses. These reactions are very common for alcohols, amines, and alkyl halogenides but not limited to these functionalities. Loss of H_2O is in particular also observed with certain *N*-oxides (after rearrangement) or amides (lactams) that can undergo condensation with a transannular amino group to form amidines.

Retro-Reactions

***Retro*-Aldol Reaction.** β-Hydroxycarbonyl and β-hydroxycarboxyl derivatives that possess the general structural unit **109** (**Scheme 4.45**) can decompose thermally with the loss of CH_2O (or a corresponding neutral portion) to compounds of the type **110**. Examples of this kind were observed, for example, with basic natural products (alkaloids).

***Retro*-Diels–Alder Reaction.** The RDA reaction is a very frequently observed reaction with alicyclic or heterocyclic monounsaturated six-membered ring systems. Because it also proceeds

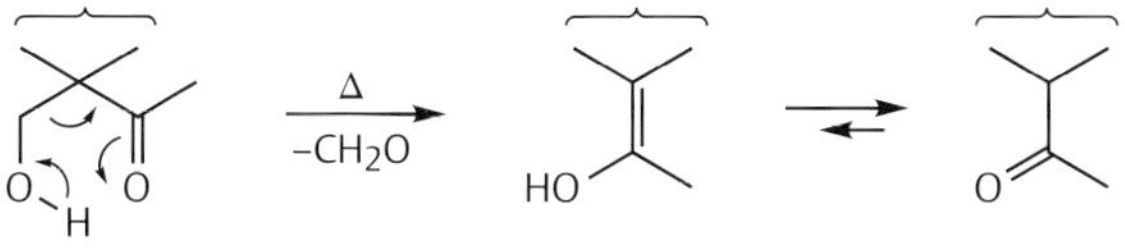

109 **110**

Scheme 4.45 Structural requirement for a *retro*-aldol reaction.

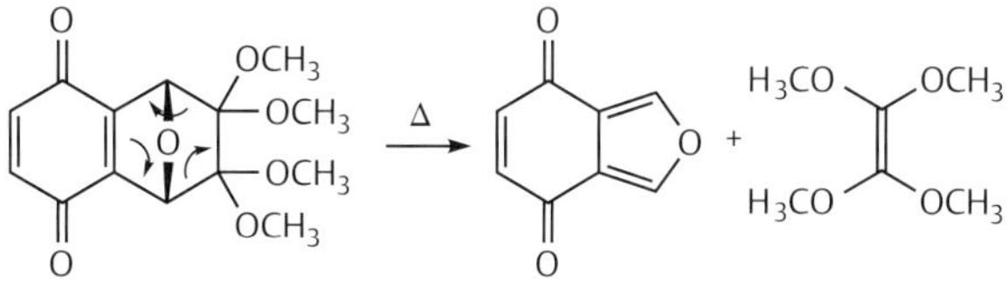

111
$C_{14}H_{16}O_7$, M = 296

112
$C_6H_4O_3$, M = 148

113
$C_6H_{12}O_4$, M = 148

Scheme 4.46 Example for artifact formation by *retro*-Diels–Alder reaction.

prominently as an MS fragmentation, the distinction of artifact formation (decomposition prior to the introduction of the sample to the ion source) and mass spectrometric fragmentation is sometimes difficult. Indications for (at least partial) artifact formation are time-dependent spectra (see above) or spectra that are dependent on the temperature of the sample entry. Quinone derivative **111** (M = 296, **Scheme 4.46**), for example, shows in the EI-MS obtained upon direct inlet of the compound at 170°C the signal for the $M^{+\bullet}$ ion at m/z = 296. This signal is no longer observed when the sample is introduced into the instrument via a gas inlet at 200°C. A spectrum that corresponds to the overlay of the spectra of the portions **112** (M = 148) and **113** (M = 148) is found for the latter case.

Isomerization

Of particular importance, because very difficult to perceive in structural elucidations, are isomerization reactions that lead to very different new functionalities and new structural units in a molecule. An example for a compound that shows such a behavior is given with the *ortho*-substituted phenol derivative **114** (M = 160, **Fig. 4.96**). It is evident that the EI-MS of the compound is in striking contradiction to the structure of the sample molecule. While the loss of a methyl radical ($[M-Me]^+$, m/z = 145) from the molecular ion could still be explained, the very prominent elimination of an ethyl radical ($[M-Et]^+$, m/z = 131) is not conceivable with the structure of an $M^{+\bullet}$ of **114** and the common MS fragmentation reactions. The signal at m/z = 131 is, however, readily explained if it is assumed that compound **114** rearranges to **115** prior to the MS analysis (**Scheme 4.47**)—a reaction that was performed also on a preparative

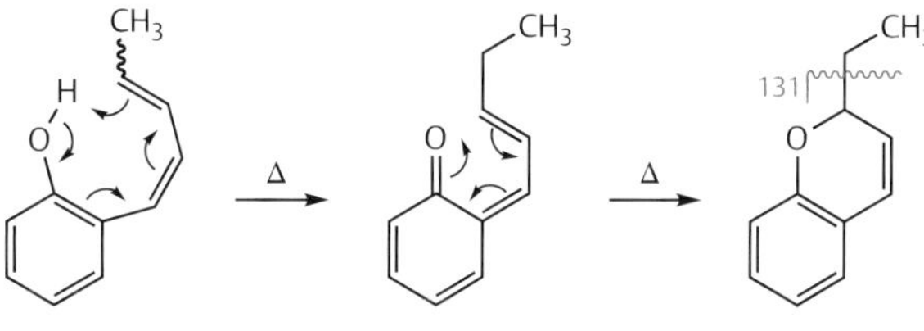

114
$C_{11}H_{12}O$, M = 160

115
$C_{11}H_{12}O$, M = 160

Scheme 4.47 Example of artifact formation with subsequent loss of an unexcepted fragment.

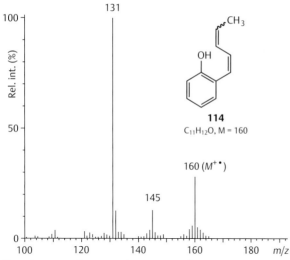

Fig. 4.96 EI-MS of compound **114** or its artifact **115**, respectively.

scale. The loss of Et· corresponds then to the product of a simple α-cleavage. The spectrum of a synthetic sample of **115** was in fact virtually identical with that of compound **114**.

Disproportionation, Dehydration, and Hydration

It can happen with the investigation of certain compound classes that molecular signals of ions of hydrogenation and dehydrogenation products of the analytes are observed rather than those of the expected molecular ions themselves. For example, dihydroquinoxaline derivative **116** (M = 118) shows solely signals for the molecular ions of quinoxaline derivative **117** (M = 116) and tetrahydroquinoxaline derivative **118** (M = 120) in its EI-MS (**Scheme 4.48**).

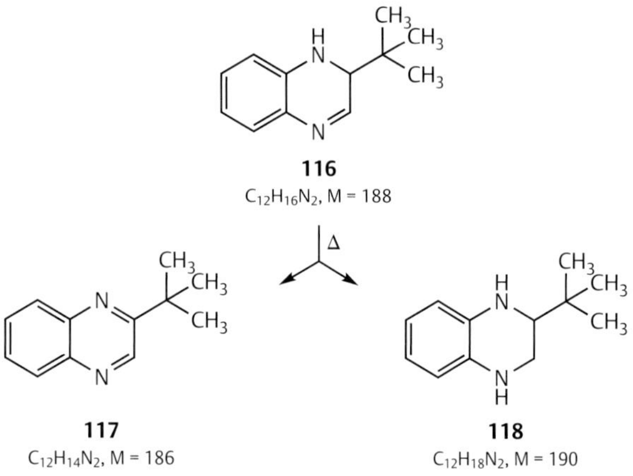

116
$C_{12}H_{16}N_2$, M = 188

117
$C_{12}H_{14}N_2$, M = 186

118
$C_{12}H_{18}N_2$, M = 190

Scheme 4.48 Example for artifact formation by disproportionation.

The occurrence of signals that would correspond to $[M + 2]^+$ ions is typical for quinones. They have, however, to be ascribed to the signals of the respective hydroquinones that were shown to be formed within the inlet and ion source of the mass spectrometer. On the other hand, oxidation of hydroquinones to quinones in the inlet system of an MS also can be observed, which leads to ions with the mass corresponding to $[M - 2]^+$.

Pyrolysis of Quaternary Nitrogen Compounds

Quaternary nitrogen compounds cannot be evaporated without decomposition in most cases. Thus, upon application of vaporization methods, the volatile decomposition products rather than the original sample compounds are observed in MS. These are usually products of substitution and/or elimination reactions, often with the counterions acting as nucleophiles (prominent with iodides, rare with fluorides; demethylations are particularly favored) or as bases. The respective possible reaction paths for a generic ammonium species **119** is shown in **Scheme 4.49**.

Alkylation and Transalkylation

It is possible that an alkyl group (Me or, more rarely, Et) is transferred in a thermal reaction from one functional group to another. Esters, aryl ethers, or ammonium groups are possible alkyl donors, acids, phenols, or primary or secondary (or even tertiary) amines acceptors. The transalkylations can occur inter- as well as intramolecularly. In the case of an intermolecular transmethylation, ions that correspond by mass to ions of the type $[M + 14]^{+\cdot}$ and $[M' - 14]^{+\cdot}$ are formed, deriving from the two species reacting with each other. Thus, after the methyl transfer (15 u), a proton must be transferred back to the donor molecule. Instead of a proton, an alkyl group can also be transferred back to the donor molecule. An example for artifact formation by transalkylation is shown with the reaction of tetrahydroquinoline derivative **120** (M = 147) that reacts with ester **121** (M = 142) to amine **122** (M = 161) and carboxylic acid **123** (M = 128) as shown in **Scheme 4.50**.

Deoxygenation of Oxidized Nitrogen Functional Groups

Nitro, nitroso, oxime, N-oxide, and N-hydroxy derivatives are occasionally deoxygenated upon APCI. A mechanistic explanation for this reaction is pending, but it is known for the deoxygenation of N-hydroxy compounds that the transformation is neither a mass spectrometric fragmentation nor a reaction that occurs prior to the sample inlet. The reaction, thus, must proceed within the ion source. It was shown that the extent of the reaction is strongly dependent on the pH of the analyte solution, which was used as a means to identify specifically the presence of N-hydroxylamines within a molecule.[12]

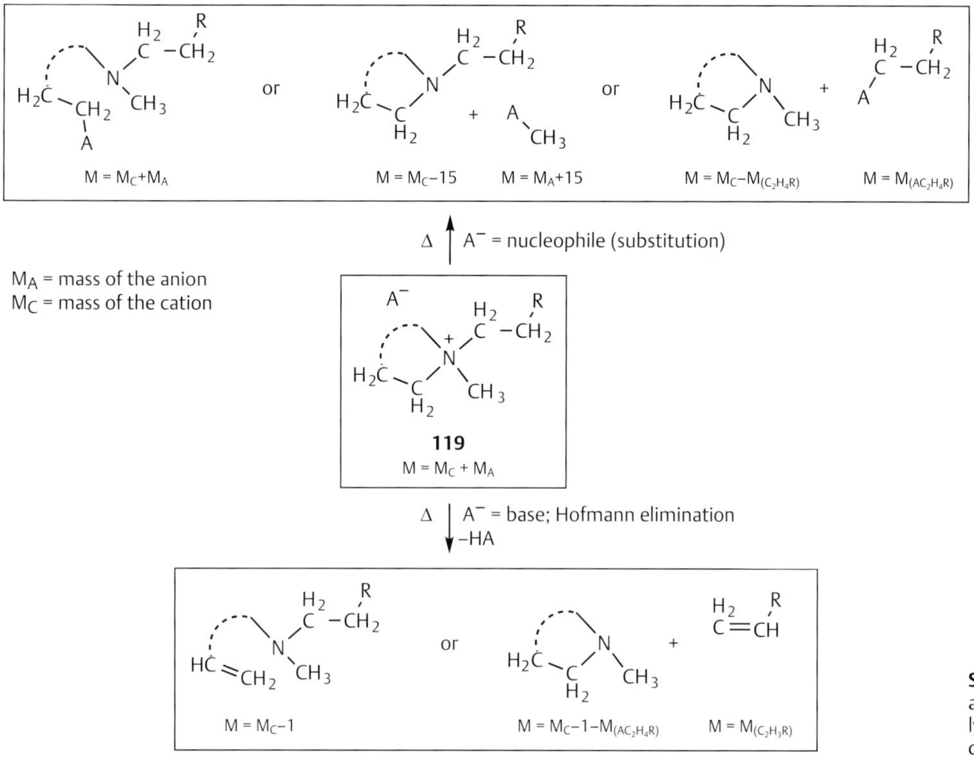

Scheme 4.49 Example for artifact formation by pyrolysis of quaternary nitrogen compounds.

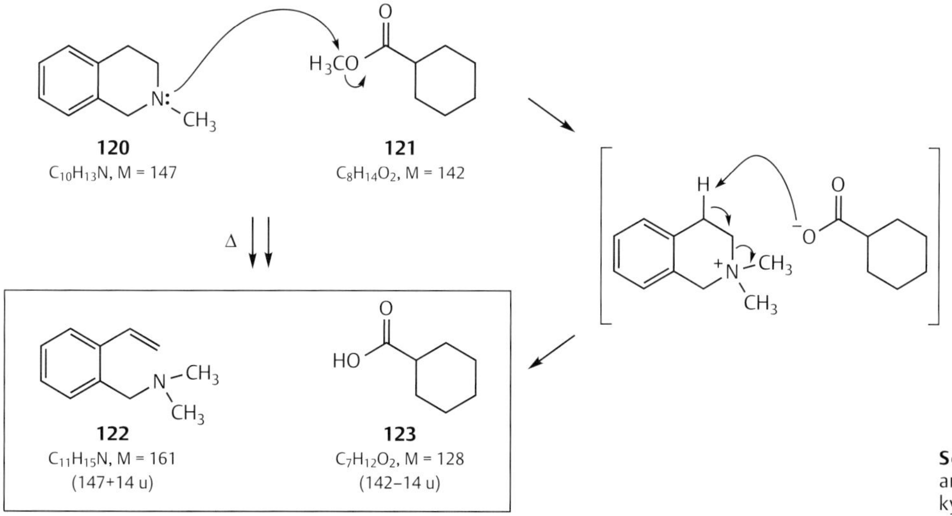

M_A = mass of the anion
M_C = mass of the cation

Scheme 4.50 Example for artifact formation by transalkylation.

4.6.3 Identification of Artifacts

Artifact Signals from the Electronics of the Instrument. Artifact signals that stem from the electronics of the instrument are usually readily recognized. Such signals are typically very sharp and isolated lines (spikes), for which the isotopic signals are missing and which are not registered at full mass numbers (the first decimals often do not correspond to reasonable mass defects).

Investigation of the Thermal Stability of a Sample. Most artifacts are formed during the transfer of the probe into the instrument, and they are products of "ordinary" thermal decomposition reactions of the sample compounds. The stability of a compound upon its distillation or sublimation in high vacuum (at $p \leq 0.1$ Pa) can be taken as a measure for the probability of the compound to decompose on its way into the mass spectrometer. There is no reason to believe that a sample is thermally decomposed upon its introduction into a mass spectrometer, if its distillate or sublimate can be unaffectedly collected. If, however, decomposition during distillation is observed, the same type of decomposition is also possible when an analyte is transferred to the mass spectrometer. Indication for thermal lability of a probe can also be the observation of temperature- and/or time-dependent spectra (see above).

Acquisition of a Different Type of Mass Spectrum. Mass spectrometric fragmentations can also be distinguished from artifact formation that occurs outside of the MS instrument by the acquisition of MS spectra with different methods. If, e.g., a spectrum is acquired by the use of a milder ionization method, less MS fragmentation products should be observed, but the results of thermal reactions occurring before or during the sample inlet remain the same (as long as the conditions of the sample inlet are not changed as well).

Investigation of the Fragmentation Pattern. If the mass spectrum of a sample is obviously in contradiction to the structure (maybe only to the proposed structure) of the sample molecule (e.g., as observed with compound **114**, **Fig. 4.96**, p. 391f), or if mass differences are observed that can only hardly or not at all be explained by reasonable fragmentation reactions (e.g., $\Delta m = -14$, -16, or -20 u), then the samples most probably are mixtures that either were present from the beginning or were formed by artifact formation during the sample inlet.

4.6.4   Prevention of Artifact Formation

It is most important to know the reason for artifact observation to be able to prevent it. It might be sufficient in some cases just to purify the compounds by recrystallization, filtration, or distillation to effect stabilization of the material (removal of catalysts). In most cases, however, the functional groups contained in a molecule are responsible for the artifact formation and, thus, the problem is inherently connected to the structure of the molecule and cannot be solved directly. Using more mild ionization methods and entry systems that transfer less energy onto the analytes is always a possibility, that is worthwhile to be tested. If the ionization and MS method is to be retained, derivatization might be used to stabilize a compound and to enhance its volatility (Section 4.4.7, p. 379).

4.7   Tables to the Mass Spectrometry

4.7.1   Frequently Detected Ions

Table 4.14, p. 395–405.

4.7.2   Frequently Detected Mass Differences

Table 4.15, p. 406–413.

4.7.3   Isotope Distributions of Halogenated Compounds

Table 4.16, p. 414.

4.7.4   Solvents and Frequent Impurities

Table 4.17, p. 415–417

Table 4.18, p. 418–420

4.7.5   Isotopes of Naturally Occurring Elements

Table 4.19, p. 421–428.

4.7.1 Frequently Detected Ions

Table 4.14 Frequently detected ions in mass spectrometry and their mass spectrometric origin

Mass (m/z)	Ion	Proposed structure	Origin/comment
1–10	–		Are not registered under normal conditions
14	CH_2^+		Unspecific
15	CH_3^+		Unspecific
16	$CH_4^{+\bullet}$, NH_2^+, $O^{+\bullet}$		Unspecific
17	NH_3^+, OH^+		Unspecific
18	$H_2O^{+\bullet}$		Unspecific
19	F^+		F derivatives
20	$HF^{+\bullet}$		F derivatives
26	$C_2H_2^{+\bullet}$		Unspecific
	CN^+		Nitriles
27	$C_2H_3^+$		Unspecific
	$HCN^{+\bullet}$		Nitriles
28	$C_2H_4^{+\bullet}$	$\bullet$	Unspecific
	$CO^{+\bullet}$		Carbonyl and carboxyl derivatives
	N_2^+		N_2 from air, azides
29	$C_2H_5^+$		Ethyl derivatives
	CHO^+		Aldehydes
30	$C_2H_6^{+\bullet}$		Unspecific
	CH_4N^+	$H_2C=NH_2^+$	Secondary acylamides, primary amines, amines (onium reaction)
	NO^+		Nitrosamines
31	CH_3O^+	$H_2C=OH^+$	Primary alcohols, ethers (onium reaction), methyl esters, CH_3OH (solvent)
32	$O_2^{+\bullet}$		Peroxides, O_2 from air
	S^+		S derivatives
	$CH_4O^{+\bullet}$		CH_3OH (solvent)
33	CH_2F^+		Aliphatic F derivatives
	HS^+		S derivatives
34	$H_2S^{+\bullet}$		S derivatives
35	$^{35}Cl^+$		Cl derivatives, quaternary ammonium chlorides, hydrochlorides in combination with $m/z = 37$ (characteristic isotopic distribution for Cl)
36	$H^{35}Cl^{+\bullet}$		Cl derivatives, quaternary ammonium chlorides, hydrochlorides in combination with $m/z = 38$ (characteristic isotopic distribution for Cl)
39	$C_3H_3^+$		Alkynes, occasionally aromatic compounds
40	$Ar^{+\bullet}$		From air
41	$C_3H_5^+$	$H_2C=CH-CH_2^+$	Allyl derivatives
	$C_2H_3N^{+\bullet}$	$H_2C=C=NH^{+\bullet}$	Oximes
		$H_3C=CN^{+\bullet}$	C-Methyl-N-heterocycles, CH_3CN (solvent)
42	$C_3H_6^{+\bullet}$		Unspecific
	$C_2H_4N^+$	$H_2C=N^+=CH_2$	Cyclic amines
	$C_2H_2O^{+\bullet}$		Acetyl derivatives
43	$C_3H_7^+$		Propyl derivatives
	$C_2H_5N^{+\bullet}$	$H_2C=NH^+-CH_2^\bullet$	Cyclic amines
	$C_2H_3O^+$		Acetyl derivatives, methyl ketones
	$CHNO^{+\bullet}$	$HN=C=O^{+\bullet}$	Urethanes

(Continued)

Mass Spectrometry

Table 4.14 Frequently detected ions in mass spectrometry and their mass spectrometric origin (continued)

Mass (m/z)	Ion	Proposed structure	Origin/comment
44	$C_2H_6N^{+\bullet}$	$H_2C=NH^+-CH_3$	Acyl amides, amines
	$C_2H_4O^{+\bullet}$	$H_2C=CH-OH^{+\bullet}$	Cyclobutanol derivatives, aldehydes, vinyl ethers
	CH_2NO^+	$O=C=NH_2^+$	N-Monosubstituted carboxylic acid amides
	$CO_2^{+\bullet}$		Unspecific (from carboxylic acids, carbonates, or CO_2 from solvent, air)
45	$C_2H_5O^+$	$H_3C-CH=OH^+$	Alkan-2-ol derivatives
		$H_2C=O^+-CH_3$	Methyl ethers, ethyl ethers
		$H_5C_2-O^+$	Ethyl ethers, ethyl esters
	$CHO_2^{+\bullet}$	$COOH^{+\bullet}$	Carboxylic acids
	CHS^+	$HC≡S^+$	Disulphides, aromatic and other unsaturated S heterocycles
46	$C_2H_6O^{+\bullet}$		C_2H_5OH (solvent)
	$CH_2S^{+\bullet}$		Thioethers
47	CH_3S^+	$H_2C=SH^+$	Thioethers, primary thiols
49	$CH_2{}^{35}Cl^+$		Cl derivatives in combination with m/z = 51 (characteristic isotopic distribution for Cl)
50	$C_4H_2^{+\bullet}$		Ortho-disubstituted phenylcarbonyl derivatives (from m/z = 76)
	$CH_3{}^{35}Cl^{+\bullet}$		Quaternary methylammonium chlorides in combination with m/z = 52 (characteristic isotopic distribution for Cl)
51	$C_4H_3^+$		Aromatic compounds (from m/z = 77)
52	$C_4H_4^{+\bullet}$		Aromatic compounds (from m/z = 78)
55	$C_4H_7^+$		Unspecific
	$C_3H_3O^+$	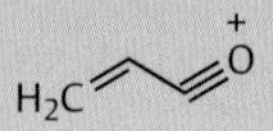	Cyclopentanone and cyclohexanone derivatives
56	$C_4H_8^{+\bullet}$		Alkenes, alkenoles, butyl esters
	$C_3H_6N^+$	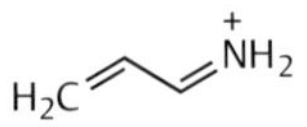	Cyclopentyl and cyclohexylamino derivatives
	$C_2H_2NO^+$	$H_2C=N^+=C=O$	Isocyanates
57	$C_4H_9^+$		Alkanes, Boc derivatives
	$C_3H_7N^{+\bullet}$	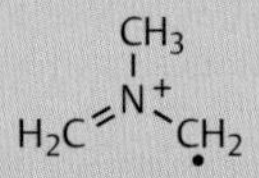	Cyclic amines
	$C_3H_5O^+$	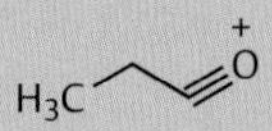	Ethyl ketones, propanoic acid derivatives
		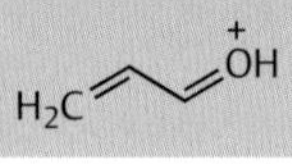	Cyclopentanol and cyclohexanol derivatives
58	$C_3H_8N^+$	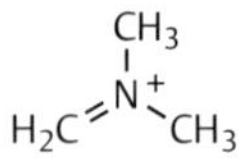	N,N-Dimethylated amines
	$C_3H_6O^{+\bullet}$	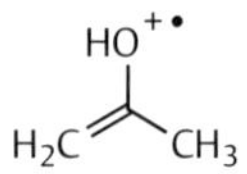	Methyl alkyl ketones
		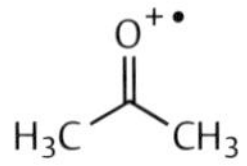	Acetone (solvent)

Table 4.14 (Continued)

Mass (m/z)	Ion	Proposed structure	Origin/comment
59	$C_3H_7O^+$	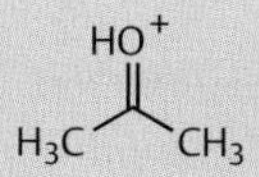 HO$^+$ / H$_3$C—C—CH$_3$	2-Methylalkan-2-ol and Boc derivatives
		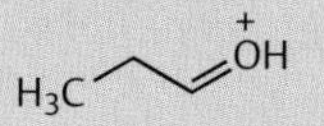 H$_3$C—CH=$\overset{+}{O}$H	Alkan-3-ol derivatives, propyl ethers
		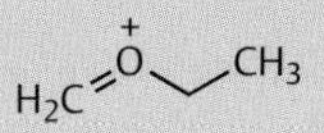 H$_2$C=$\overset{+}{O}$—CH$_3$	Ethyl ethers
	$C_2H_5NO^{+\bullet}$	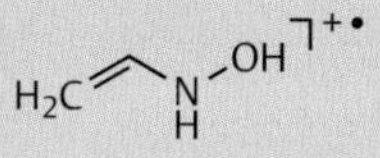 [H$_2$C=N(H)—OH]$^{+\bullet}$	Oximes
		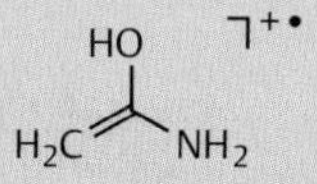 [HO—C(=CH$_2$)—NH$_2$]$^{+\bullet}$	N-Unsubstituted amides
	$C_2H_3O_2^{+\bullet}$	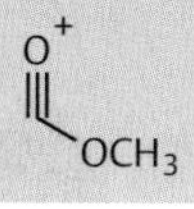 O$\overset{+}{=}$C—OCH$_3$	Methyl esters
60	$C_3H_8O^{+\bullet}$	e.g. H$_7$C$_3$—OH$^{+\bullet}$	Propanols (solvents)
	$C_2H_4O_2^{+\bullet}$	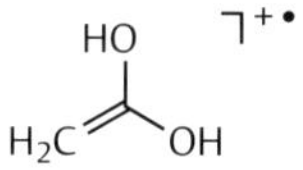 [HO—C(=CH$_2$)—OH]$^{+\bullet}$	Aliphatic carboxylic acids
		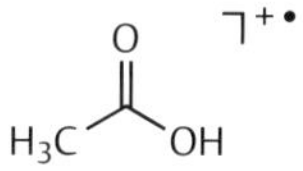 [H$_3$C—C(=O)—OH]$^{+\bullet}$	Acetic acid (solvent)
61	$C_2H_5S^+$	H$_3$C—CH=SH$^+$ H$_2$C=S$^+$—CH$_3$ H$_5$C$_2$—S$^+$	Alkane-2-thiol derivatives Methyl thioethers, ethyl thioethers Ethyl thioethers, ethyl thiolesters
63	$CH_3O_3^+$	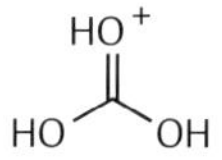 HO$^+$ / HO—C—OH	Carbonic acid dialkylesters
	CH_3OS^+	H$_3$C—S$^+$=O	Alkyl sulfoxides
65	$C_5H_5^+$	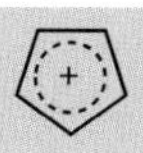 (cyclopentadienyl cation)	Alkylated aromatic compounds (from m/z = 91), N-heterocycles, aromatic amines (from m/z = 92)
66	$H_2S_2^{+\bullet}$	HSSH$^{+\bullet}$	Disulfides
69	$C_4H_5O^+$	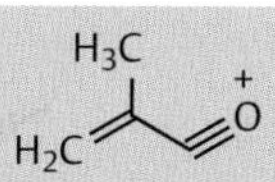 H$_3$C—C(=CH$_2$)—CH=$\overset{+}{O}$	2-Methylcyclopenanone or hexanone derivatives
		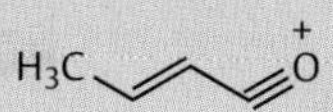 H$_3$C—CH=CH—CH=$\overset{+}{O}$	3-Methylcyclopenanone or hexanone derivatives
	CF_3^+		Trifluoromethyl or trifluoroacetyl derivatives, polyfluoro hydrocarbons

(Continued)

Table 4.14 Frequently detected ions in mass spectrometry and their mass spectrometric origin (continued)

Mass (m/z)	Ion	Proposed structure	Origin/comment
70	$C_4H_8N^+$	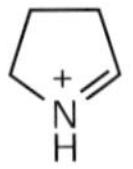	2-Substituted pyrrolidine derivatives
71	$C_5H_{11}^+$		Alkanes
	$C_4H_7O^+$	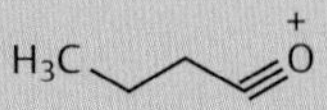	Propyl ketones, butyric acid derivatives
		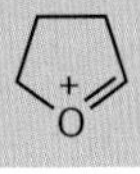	2-Substituted tetrahydrofurane derivatives
72	$C_4H_{10}N^+$	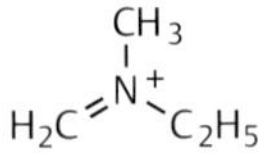	Amines
	$C_4H_8O^{+•}$	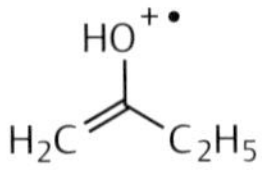	Ethyl alkyl ketones
	$C_2H_2NS^+$	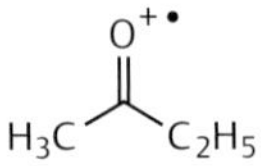	Butan-2-one (solvent), analogous ion of butanal
		$H_2C=N^+=C=S$	Isothiocyanates
73	$C_4H_9O^+$	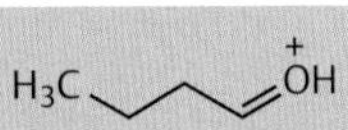	Alkan-4-ol derivatives, butyl ethers
		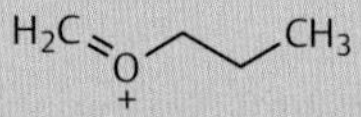	Propyl ethers
		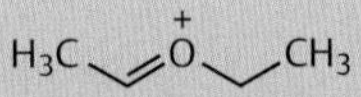	Diethyl ether (solvent)
	$C_3H_7NO^{+•}$	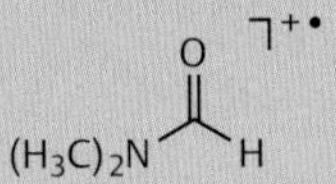	Dimethylformamide (solvent)
	$C_3H_9Si^+$	$(H_3C)_3Si^+$	TMS and TBDMS derivatives
	$C_3H_5O_2^+$	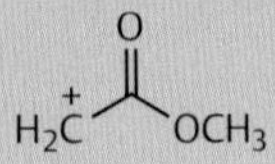	Methyl esters
		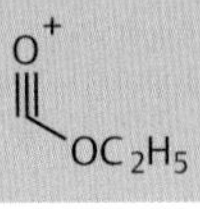	Ethyl esters
74	$C_3H_6O_2^{+•}$	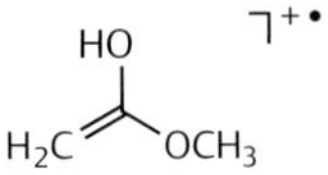	Methyl esters
	$C_4H_{10}O^{+•}$	$H_5C_2OC_2H_5^{+•}$	Diethyl ether (solvent)

Table 4.14 (Continued)

Mass (m/z)	Ion	Proposed structure	Origin/comment
75	$C_3H_7O_2^+$	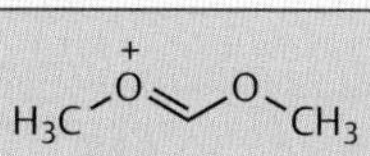	Dimethyl acetals
	$C_3H_7S^+$	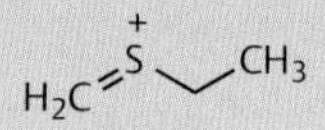	Ethyl alkyl sulfides
	$C_2H_7OSi^+$	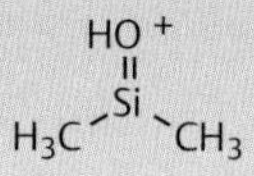	O-TMS, O-TBDMS derivatives and other silyl ethers
76	$C_6H_4^{+\bullet}$	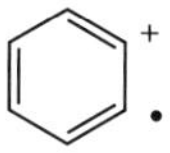	*Ortho*-disubstituted phenylcarbonyl derivatives (forms m/z = 50)
	$CS_2^{+\bullet}$		CS_2 (solvent)
77	$C_6H_5^+$	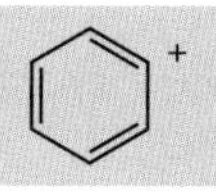	Phenyl derivatives (forms m/z = 51)
78	$C_6H_6^{+\bullet}$	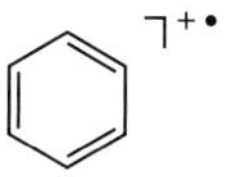	Phenyl derivatives (forms m/z = 52), benzene (solvent)
79	$C_5H_5N^{+\bullet}$	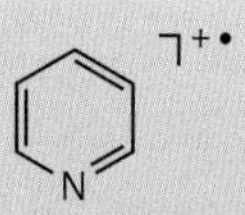	Pyridine (solvent)
	$^{79}Br^+$		Br derivatives, quaternary ammonium bromides, hydrobromides in combination with m/z = 81 (characteristic isotopic distribution of Br)
80	$C_5H_6N^+$	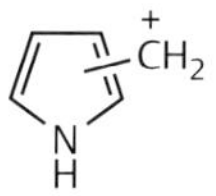	C-Alkylpyrrol derivatives
		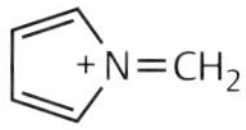	N-Alkylpyrrol derivatives
		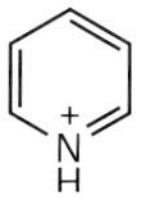	Pyridine derivatives
	$H^{79}Br^{+\bullet}$		Br derivatives, quaternary ammonium bromides, hydrobromides in combination with m/z = 82 (characteristic isotopic distribution of Br)
81	$C_5H_5O^+$	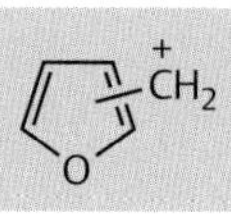	Alkylfurane derivatives
82	$C_5H_8N^+$	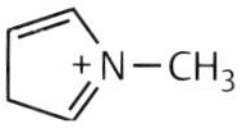	Dihydropyrrole derivatives

(Continued)

Table 4.14 Frequently detected ions in mass spectrometry and their mass spectrometric origin (continued)

Mass (m/z)	Ion	Proposed structure	Origin/comment
83	$CH^{35}Cl_2^+$		$CHCl_3$ (solvent) in combination with $m/z = 85, 87$ (characteristic isotopic distribution for Cl_2)
84	$C_5H_{10}N^+$	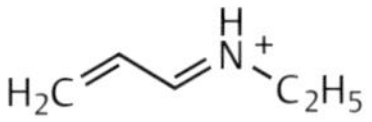	N-Ethylcyclopentyl and N-ethylcyclohexyl amine derivatives (forms $m/z = 56$)
		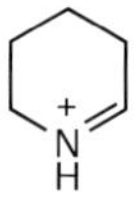	2-Substituted piperidine derivatives
		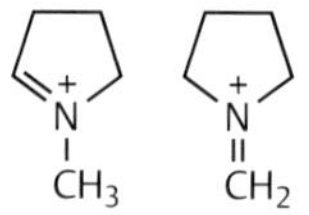	Pyrrolidine derivatives
	$CH_2^{35}Cl_2^{+\bullet}$		CH_2Cl_2 (solvent) in combination with $m/z = 86, 88$ (characteristic isotopic distribution for Cl_2)
85	$C_6H_{13}^+$		Alkanes
	$C_5H_9O^+$	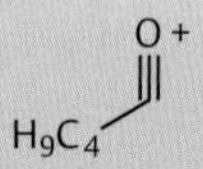	Butyl ketones, pentanoic acid derivatives
		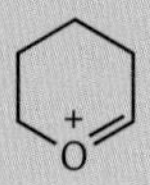	2-Substituted pyran derivatives
	$C_4H_5O_2^+$	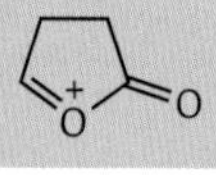	4-Substituted γ-lactones
86	$C_6H_{14}^{+\bullet}$		Hexane (solvent)
87	$C_5H_{11}O^+$	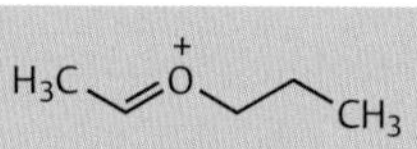	Ethers (other isomeric structures are also possible)
	$C_4H_7O_2^+$	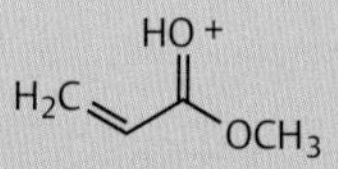	Methyl esters
		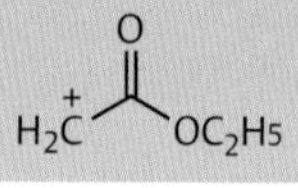	Ethyl esters
88	$C_4H_8O_2^{+\bullet}$		Dioxane, ethyl acetate (solvents)
90	$C_4H_{10}O_2^{+\bullet}$		Glycol dimethyl ether (solvent)
91	$C_7H_7^+$	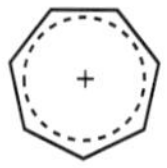	Benzyl derivatives (forms $m/z = 65$)
	$C_4H_8^{35}Cl^+$		1-Chloroalkane derivatives in combination with $m/z = 93$ (characteristic isotopic distribution for Cl)

Table 4.14 (Continued)

Mass (m/z)	Ion	Proposed structure	Origin/comment
92	$C_7H_8^{+\bullet}$	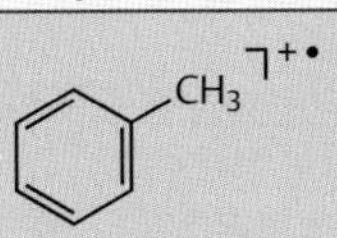	Toluene (solvent)
		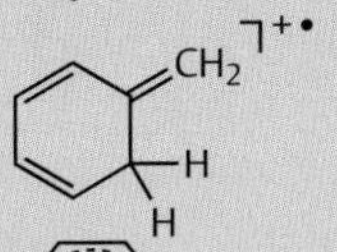	Alkylbenzene derivatives
	$C_6H_6N^+$	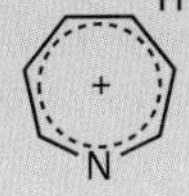	Alkylpyridine derivatives
93	$CH_2^{79}Br^+$		Alkyl bromides in combination with $m/z = 95$ (characteristic isotopic distribution for Br)
94	$C_6H_6O^{+\bullet}$	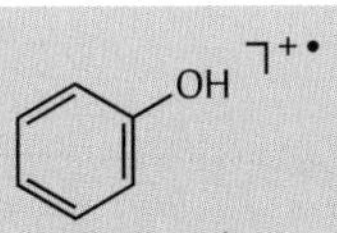	Phenol, phenyl ethers, aromatic sulfones
	$C_5H_4NO^+$	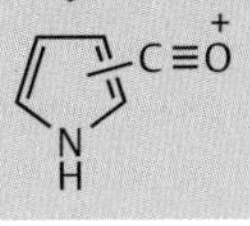	Pyrrolecarbonyl derivatives
95	$C_5H_3O_2^+$	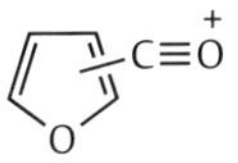	Furylcarbonyl derivatives
97	$C_5H_5S^+$	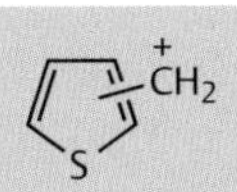	Alkylthiophene derivatives
	$C_2F_3O^+$	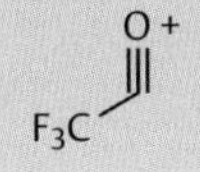	Trifluoroacetyl derivatives
98	$C_6H_{12}N^+$	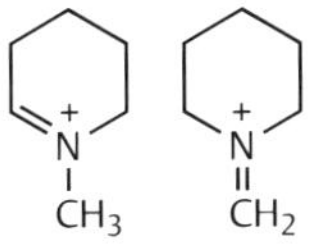	Piperidine derivatives
99	$C_5H_7O_2^+$	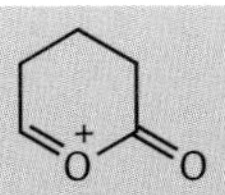	5-Substituted Δ-lactones
		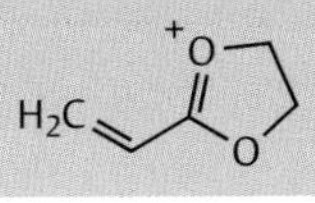	Ethylene acetals
101	$C_5H_9O_2^+$	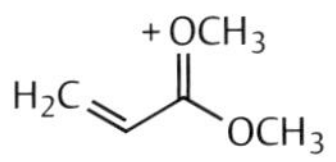	Dimethyl acetals
102	$C_6H_{14}O^{+\bullet}$		Alcohols, ethers (e.g., diisopropyl ether) (solvents)

Mass Spectrometry

(Continued)

Table 4.14 Frequently detected ions in mass spectrometry and their mass spectrometric origin (continued)

Mass (m/z)	Ion	Proposed structure	Origin/comment
103	$C_8H_7^+$	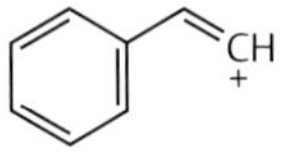	Cinnamic acid derivatives (from $m/z = 131$, forms $m/z = 77$)
	$C_4H_{11}OSi^+$	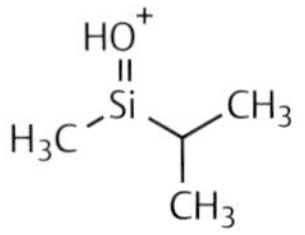	Methyl(isopropy)silyl ethers
104	$C_8H_8^{+\bullet}$	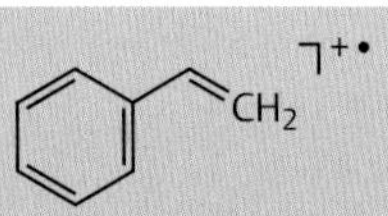	Phenethyl derivatives
		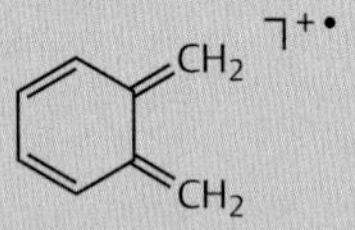	1,2,3,4-Tetrahydro-(hetero)-naphthalene derivatives (RDA), ortho-methyldiphenylmethane derivatives
	$C_7H_4O^{+\bullet}$	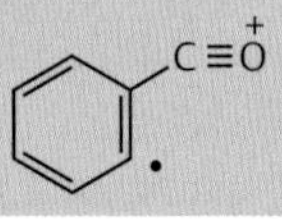	2-Substituted benzoic acid derivatives
105	$C_8H_9^+$	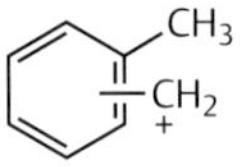	Alkyl toluenes
	$C_7H_5O^+$	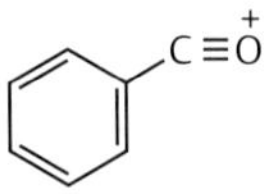	Phenylcarbonyl derivatives (e.g., from $m/z = 123$, 122, forms $m/z = 77$)
	$C_6H_5N_2^+$	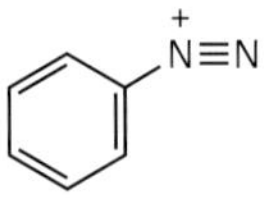	Aromatic azo derivatives
107	$C_7H_7O^+$	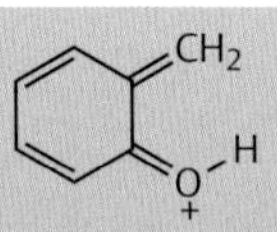	Benzylic phenol derivatives
		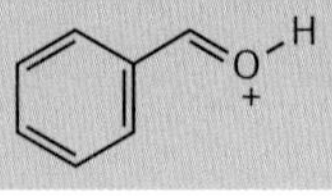	Benzyl ethers, (Cbz-) Z-derivatives
108	$C_7H_8O^{+\bullet}$	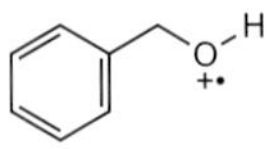	Benzylalcohol, benzyl ethers, (Cbz-) Z-derivatives
110	$C_7H_{12}N^+$	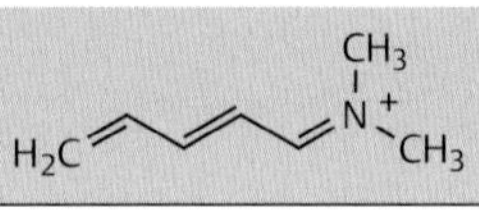	Dimethylamino steroids, e.g.

Table 4.14 (Continued)

Mass (m/z)	Ion	Proposed structure	Origin/comment
111	$C_5H_3OS^+$	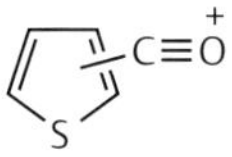	Thiophenecarbonyl derivatives
117	$C^{35}Cl_3^+$		CHCl$_3$ (solvent) in combination with $m/z = 119, 121$ (characteristic isotopic distribution for Cl$_3$)
118	$CH^{35}Cl_3^{+\bullet}$		CHCl$_3$ (solvent) in combination with $m/z = 120, 122$ (characteristic isotopic distribution for Cl$_3$)
119	$C_8H_7O^+$	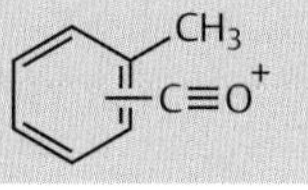	(Methylphenyl)carbonyl derivatives
120	$C_7H_4O_2^{+\bullet}$	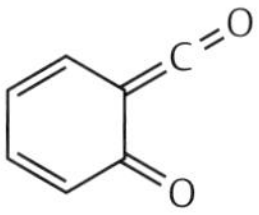	4H-Chromen-4-one derivatives
121	$C_8H_9O^+$	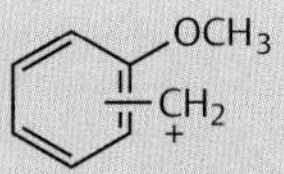	(Methoxyphenyl)alkyl derivatives
	$C_7H_5O_2^+$	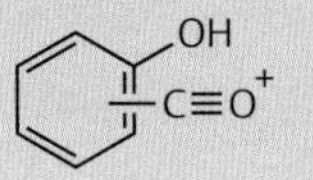	(Hydroxyphenyl)carbonyl derivatives
122	$C_7H_6O_2^{+\bullet}$	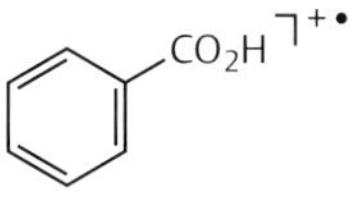	Benzoic acid, benzoic acid esters (forms $m/z = 105, 77$)
123	$C_7H_7O_2^+$	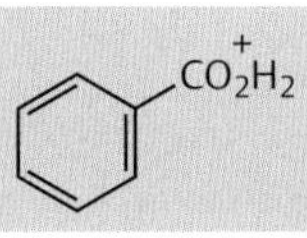	Benzoic acid, benzoic acid esters (forms $m/z = 105, 77$)
125	$C_7H_9O_2^+$	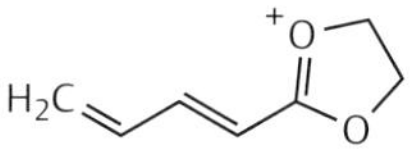	Ethylene acetals
127	$C_7H_{11}O_2^+$	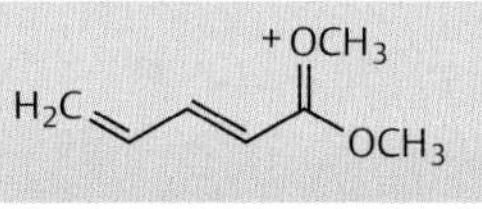	Dimethyl acetales
	I^+		I derivatives, quaternary ammonium iodides, hydroiodides
128	$HI^{+\bullet}$		I derivatives, quaternary ammonium iodides, hydroiodides
130	$C_9H_8N^+$	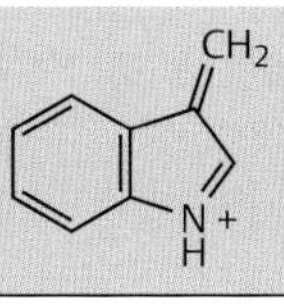	Indole and indoline derivatives, characteristic for indole alkaloids (in combination with $m/z = 144$)

(Continued)

Table 4.14 Frequently detected ions in mass spectrometry and their mass spectrometric origin (continued)

Mass (m/z)	Ion	Proposed structure	Origin/comment
131	$C_9H_7O^+$	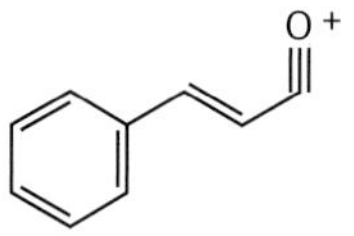	Cinnamic acid derivatives (forms $m/z = 103, 77$)
	$C_5H_7S_2^+$	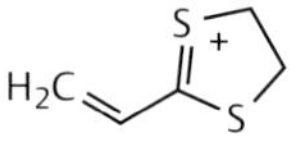	Ethylendithio acetals
	$C_6H_{15}OSi^+$:	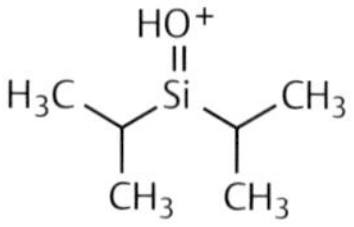	O-TIPS derivatives
135	$C_4H_8{}^{79}Br^+$	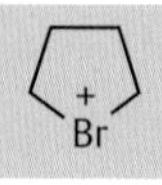	1-Bromoalkane derivatives in combination with $m/z = 137$ (characteristic isotopic distribution for Br)
139	$C_7H_7OS^+$	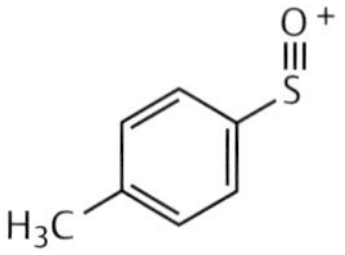	Tosyl derivatives (in combination with $m/z = 155$, forms $m/z = 91$)
142	$CH_3I^{+\bullet}$		Quaternary methylammonium iodides
144	$C_{10}H_{10}N^+$	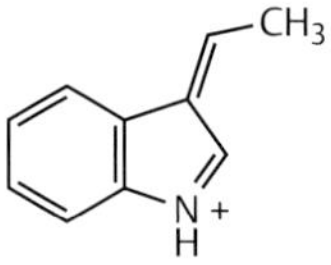	Indole and indoline derivatives, typical for indole alkaloids (in combination with $m/z = 143, 130$)
147	$C_5H_{15}SiO^+$	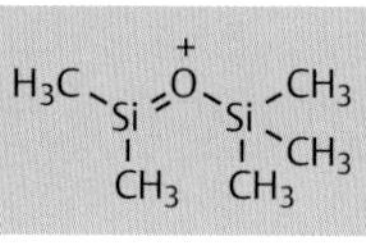	Poly-TMS derivatives, siloxanes (in combination with $m/z = 73$)
149	$C_9H_9O_2^+$	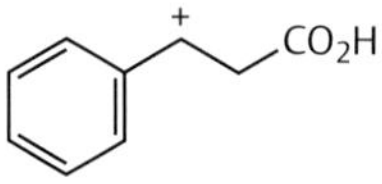	β-Substituted dihydrocinnamic acid derivatives
	$C_8H_5O_3^+$:	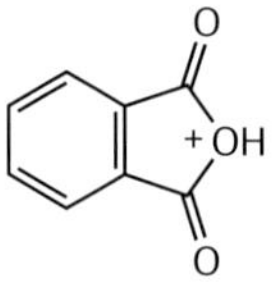	Phthalic acid esters (plasticizers)
155	$C_7H_7O_2S^+$	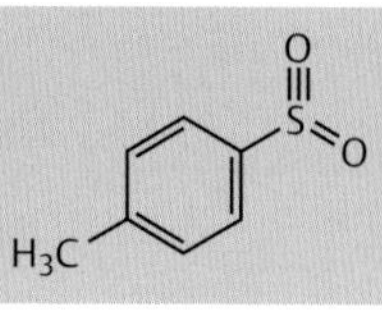	Tosyl derivatives (in combination with $m/z = 139$, forms $m/z = 91$)
156	$C_2H_5I^{+\bullet}$		Quaternary alkylammonium iodides

Table 4.14 (Continued)

Mass (m/z)	Ion	Proposed structure	Origin/comment
157	$C_7H_9S_2^+$	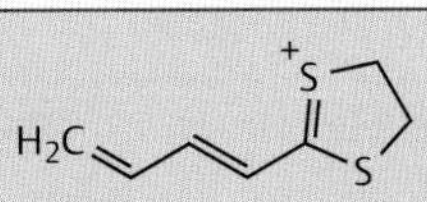	Ethylenedithio acetals
	$C_9H_{21}Si^+$	$(i\text{-Pr})_3Si^+$	TIPS derivatives
160	$C_9H_6NO_2^+$	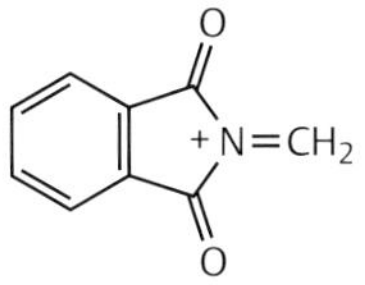	N-Alkylphthalimides
164	$C_2{}^{35}Cl_4^{+\bullet}$		Tetrachloroethylene (solvent) in combination with m/z = 166, 168, 170 (characteristic isotopic distribution for Cl_4)
165	$C_{13}H_9^+$	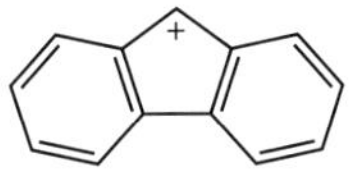	Fmoc derivatives
178	$C_{14}H_{10}^{+\bullet}$	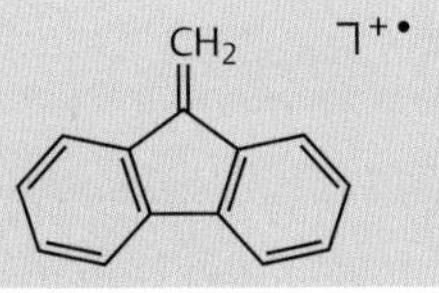	Fmoc derivatives
179	$C_6H_{18}N_3OP^{+\bullet}$	$PO(N(CH_3)_2)_3^{+\bullet}$	Hexamethylphosphoric acid triamide (HMPA; solvent additive)
196	$C_{14}H_{12}O^{+\bullet}$	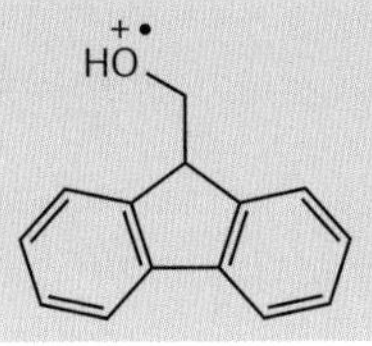	Fmoc derivatives
205	$C_{14}H_{21}O^+$	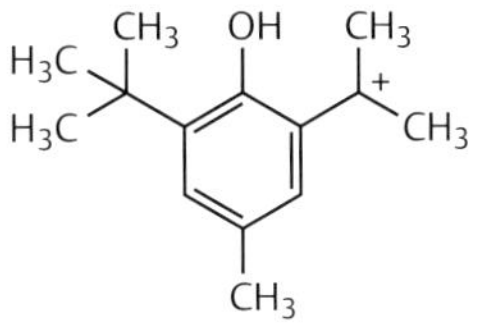	2,3-Di(tert-butyl)-4-methylphenol (ether stabilizer)
220	$C_{15}H_{24}O^{+\bullet}$	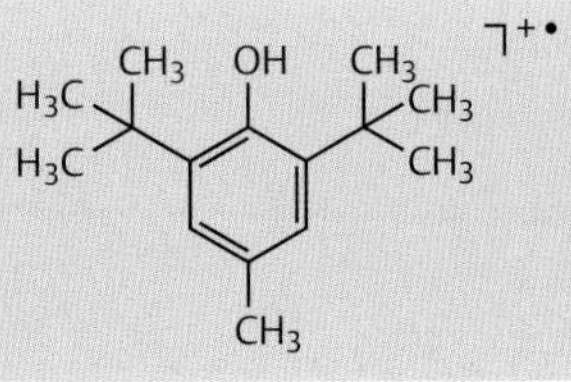	2,3-Di(tert-butyl)-4-methylphenol (ether stabilizer)
256	$S_8^{+\bullet}$		Elemental sulfur (forms also m/z = 224, 192, 160, ... , 32)

4.7.2 Frequently Detected Mass Differences

Table 4.15 Mass differences resulting from synthetic and mass spectrometric transformations

Mass difference (Δm in u)	Chemical transformation	MS transformations	Source / Comment
± 0		$[M + NH_4 - H_2O]^+$	CI-MS with NH_3 as reactant gas
± 0	 (methyl ester → tertiary alcohol)		"+ CH_4 – O"
+1		+ H^+	Amines, nitriles (EI); protonation in positive mode
– 1		– $H^{\bullet}$	Amines, primary and secondary alcohols, ethers, nitriles, some aromatic derivatives (EI)
		– H^+	Deprotonation in negative mode
± 1	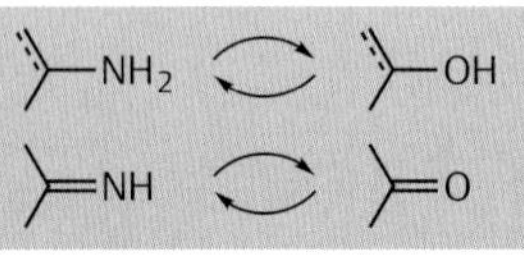 (–NH₂ ⇌ –OH; =NH ⇌ =O)		
+ 2		+ H_2	Quinones
– 2		– H_2	Hydroquinones
		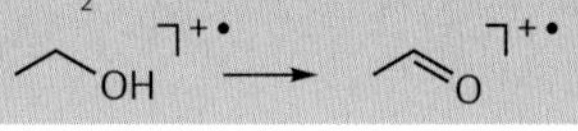 (alcohol → aldehyde/ketone radical cation)	Saturated primary alcohols (EI)
± 2	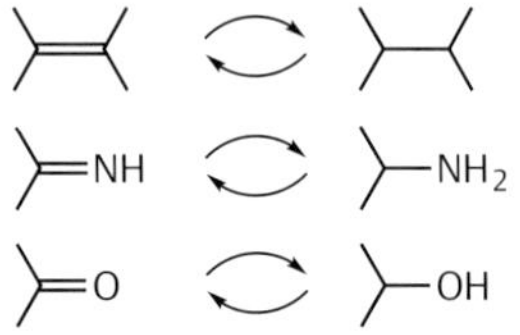 (=CR₂ ⇌ –CR₃; =NH ⇌ –NH₂; =O ⇌ –OH)		
– 3		– $H_3 (- H_2 - H^{\bullet})$ e. g.: 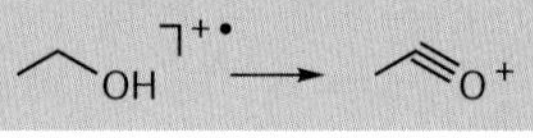 (alcohol radical cation → acylium)	Saturated primary alcohols (EI)
± 4	(–C≡C– ⇌ alkane; –C≡N ⇌ –NH₂; lactone ⇌ diol; –OCH₃ ⇌ –Cl; ester –OCH₃ ⇌ acyl –Cl)		Isotopic pattern!
			Isotopic pattern!

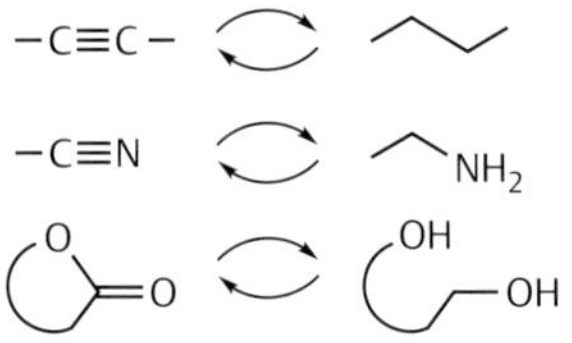

Table 4.15 (Continued)

Mass difference (Δm in u)	Chemical transformation	MS transformations	Source / Comment
± 6		$[M + H]^+ \rightleftarrows [M + Li]^+$	Exchange of the cationizing particle in ESI/MALDI
+ 7		$+ Li^+$	Li$^+$ addition (positive mode) in ESI/MALDI
+ 11		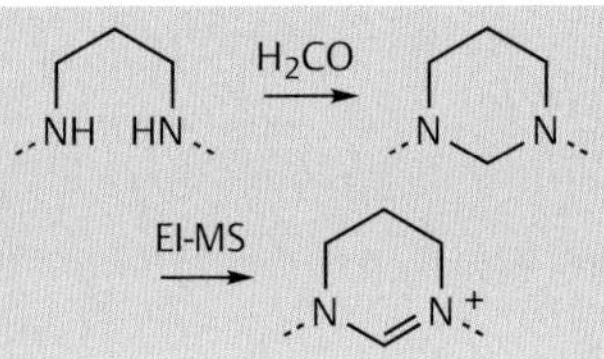	Typical for 1,3-diamines when using MeOH as solvent
± 12	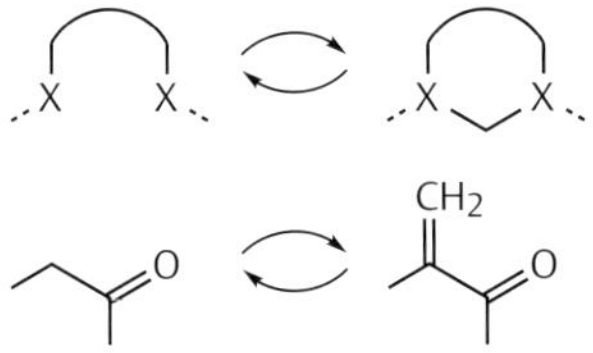		
± 14	homologs (± CH$_2$) 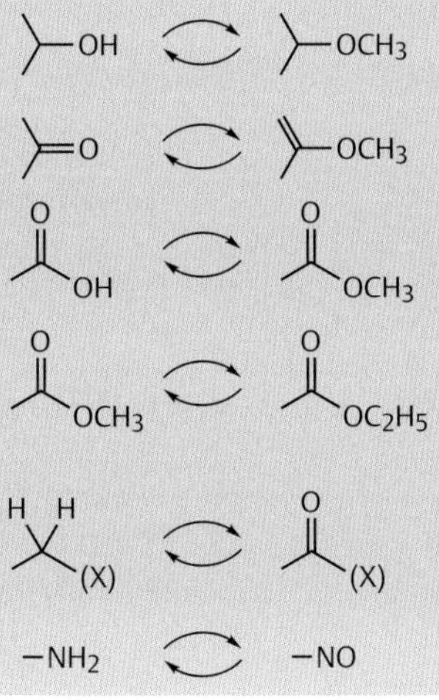		
– 15		$- CH_3{}^\bullet$	Unspecific (EI)
– 16		$- NH_2{}^\bullet$	Primary amides, amines, sulfonamides (EI)
		$- O$	diaryl sulfoxides, nitro derivatives, N-oxides, sulfones (EI)

(Continued)

Table 4.15 Mass differences resulting from synthetic and mass spectrometric transformations (continued)

Mass difference (Δm in u)	Chemical transformation	MS transformations	Source / Comment
± 16		$[M + Li]^+ \rightleftarrows [M + Na]^+$ $[M + Na]^+ \rightleftarrows [M + K]^+$	Exchange of the cationizing particle in ESI/MALDI
	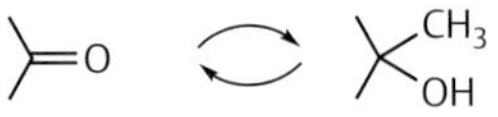		Reductive methylation
	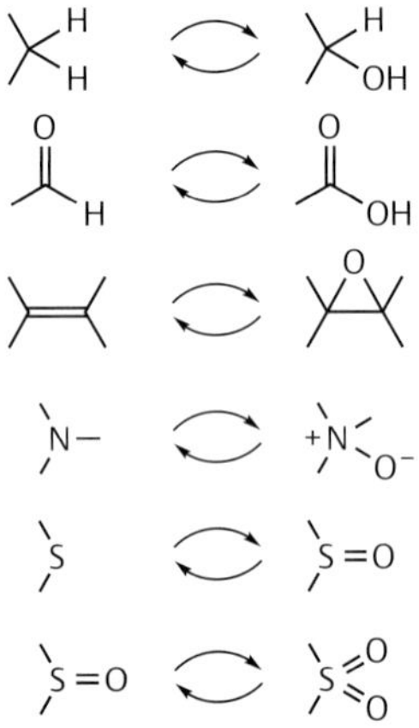		Oxidations
− 17		− NH_3 − $OH^{\bullet}$	Amines, ammonium salts Alcohols, carboxylic acids, N-oxides, oximes, sulfoxides, generally O derivatives (EI)
± 17	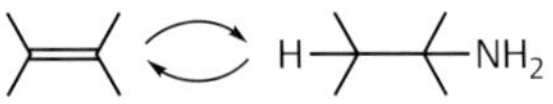	$[M + H]^+ \rightleftarrows [M + NH_4]^+$	Exchange of the cationizing particle in CI-MS with NH_3 as reactant gas
− 18		− H_2O	Aldehydes, alcohols, ethers, carboxylic acids, lactones, N-oxides, generally O derivatives
± 18		± NH_4^+	NH_4^+ addition (positive mode; CI (NH_3)/ESI/MALDI) or loss of NH_4^+ (negative mode; NH_4 salts)
	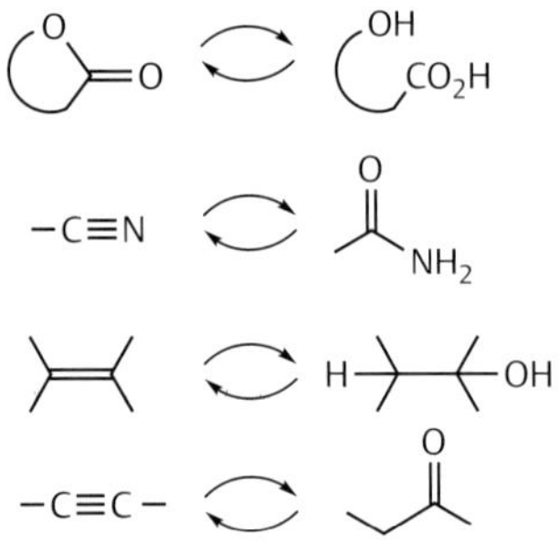		Addition of H_2O, hydrolysis
			Isotopic pattern!

Table 4.15 (Continued)

Mass difference (Δm in u)	Chemical transformation	MS transformations	Source / Comment
– 19		– F˙	F derivatives
		– H₂O – H˙	Aldehydes, alcohols, ethers, carboxylic acids, lactones, N-oxides, generally O-derivatives
± 19	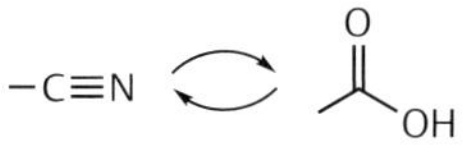		
– 20		– HF	F derivatives
± 20			
± 22		$[M + H]^+ \rightleftarrows [M + Na]^+$	Exchange of the cationizing particle in ESI/ MALDI
± 23		± Na⁺	Na⁺ addition (positive mode) or loss of Na⁺ (negative mode, Na salts) in ESI/ MALDI
– 26		– C₂H₂	Aromatic hydrocarbons
		– CN˙	Aromatic nitriles
– 27		– C₂H₃˙	Terminal vinyl derivatives
		– HCN	Aromatic amines, aromatic N-heterocycles
± 27	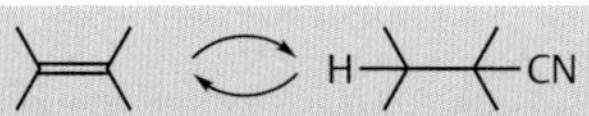 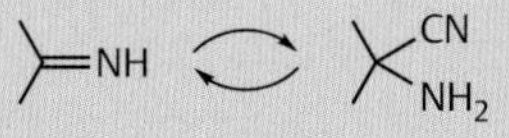 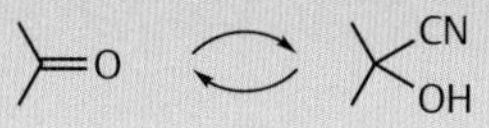		
– 28		– C₂H₄	McLafferty rearrangement (e.g., ethyl ester), RDA reaction (e.g., 1-tetralone derivatives), ethyl onium species (onium reaction) (EI)
		– CO	Aldehydes, quinones, O-heterocycles, lactams, unsaturated lactones, phenols, co loss after α-cleavage of carbonyl derivatives (EI)
± 28	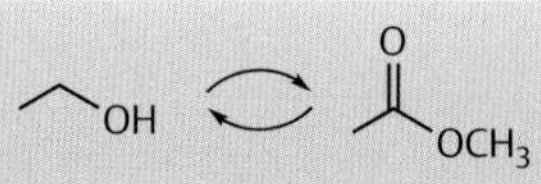 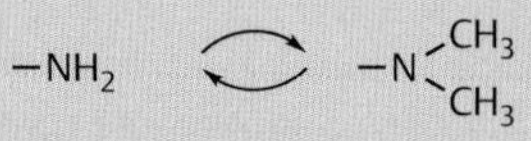   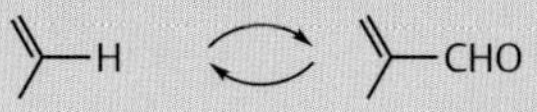 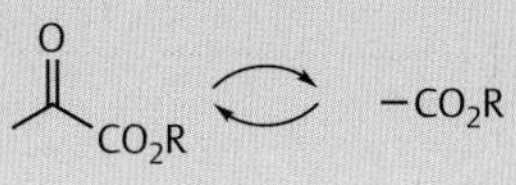		

Mass Spectrometry

Table 4.15 Mass differences resulting from synthetic and mass spectrometric transformations (continued)

Mass difference (Δm in u)	Chemical transformation	MS transformations	Source / Comment
– 29		– C$_2$H$_5$$^{\bullet}$	Ethyl derivatives (EI)
		– CHO$^{\bullet}$	Aromatic aldehydes, aromatic methoxy derivatives (EI)
– 30		– CH$_2$O	Cyclic ethers, aromatic methoxy derivatives (EI)
		– NO	Nitro derivatives (EI)
± 30	$-NH_2 \rightleftarrows -NO_2$		
– 31		– CH$_3$O$^{\bullet}$	Primary alcohols, methyl ethers, methyl esters (EI)
– 32		– CH$_3$OH	Methyl esters (EI), methoxy derivatives
		– S	S derivatives (EI)
± 32	$\diagup\!\!=\!\!\diagdown \rightleftarrows H\!-\!\diagup\!\!\diagdown\!-\!OCH_3$	$[M + Li]^+ \rightleftarrows [M + K]^+$	Exchange of the cationizing particle in ESI/MALDI
– 33		– CH$_3$$^{\bullet}$ – H$_2$O	Methyl esters (EI)
		– HS$^{\bullet}$	Isothiocyanates (EI), thiols
– 34		– H$_2$S	Thiols
± 34	$\diagup\!\!=\!\!\diagdown \rightleftarrows H\!-\!\diagup\!\!\diagdown\!-\!SH$		
	$\diagdown\!\!-\!H \rightleftarrows \diagdown\!\!-\!Cl$		Isotopic pattern!
– 35		– Cl$^{\bullet}$	Cl derivatives (EI)
– 36		– HCl	Cl derivatives
± 36	$\diagup\!\!=\!\!\diagdown \rightleftarrows H\!-\!\diagup\!\!\diagdown\!-\!Cl$		Isotopic pattern!
± 38		$[M + H]^+ \rightleftarrows [M + K]^+$	Exchange of the cationizing particle in ESI/MALDI
± 39		± K$^+$	K$^+$ addition (positive mode) or loss of K$^+$ (negative mode, K salts) in ESI/MALDI
– 41		– C$_3$H$_5$$^{\bullet}$	Alicycles
		– CH$_3$CN	Aromatic N-heterocycles
– 42		– C$_3$H$_6$	McLafferty (butylcarbonyl derivatives, propyl esters, propyl amides, etc.), onium reaction (e.g., O- and N-propyl derivatives), aromatic propyl ethers
		– C$_2$H$_2$O	α,β-Unsaturated cyclohexanones, 2-tetralones (RDA), derivatives of acetic acid

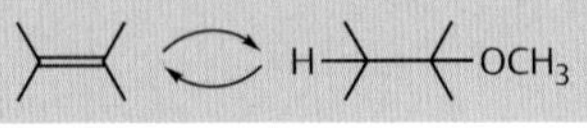

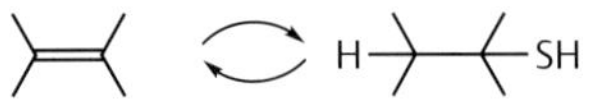

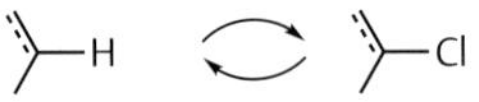

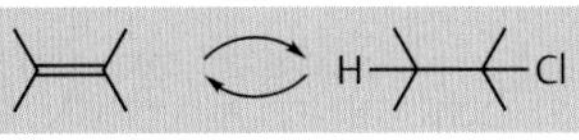

Table 4.15 (Continued)

Mass difference (Δm in u)	Chemical transformation	MS transformations	Source / Comment
± 42	  —OH ⇌ —OAc —NH$_2$ ⇌ —NHAc		Acetyl derivatives (common derivatization for GC-MS analyses)
– 43		– C$_3$H$_7$$^{•}$ – C$_2$H$_3$O$^{•}$: – CH$_3$CO$^{•}$ – CH$_3$$^{•}$ – CO	Propyl and isopropyl derivatives Acetyl derivatives, aldehydes Aromatic methyl esters
– 44		– C$_2$H$_4$O – CO$_2$	Aldehydes (McLafferty) Anhydrides, carboxylic acids, carbonic Acid ester
± 44	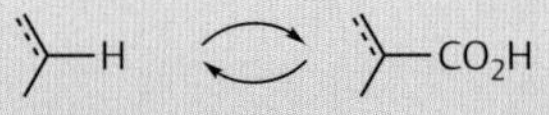 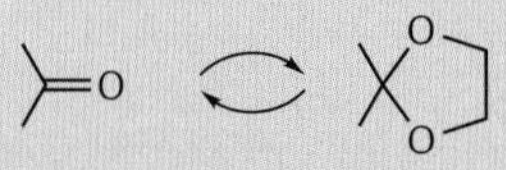  —H ⇌ —CO$_2$H =O ⇌ (1,3-dioxolane) —OH ⇌ —OMOM		MOM = methoxymethyl
– 45		– C$_2$H$_7$N – C$_2$H$_5$O$^{•}$ – CHO$_2$$^{•}$	e.g., *N,N*-Dimethylamino derivatives *O*-Ethyl derivatives Carboxylic acids, lactones
– 46		– C$_2$H$_6$O: – C$_2$H$_5$OH – C$_2$H$_4$ – H$_2$O – NO$_2$$^{•}$	Long chain primary alcohols, ethyl ethers Ethyl esters Nitro derivatives
± 46	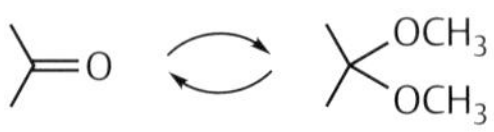 =O ⇌ C(OCH$_3$)(OCH$_3$)		
– 48		– SO	Diarylsulfoxides
– 55		– C$_4$H$_7$$^{•}$	Aromatic butyl ester, alicycles
– 56		– C$_4$H$_8$ – C$_2$O$_2$	Butyl esters, Boc derivatives Diketone, unsaturated lactones
± 58	=O ⇌ (1,3-dioxane)		
– 59		– CO$_2$CH$_3$$^{•}$	Methyl esters
– 60		– CH$_3$CO$_2$H	O-Acetyl derivatives
– 72		– C$_2$O$_3$	Aromatic anhydrides
± 72	 —OH ⇌ —OTMS		TMS = trimethylsilyl (common derivatization for GC-MS analyses)

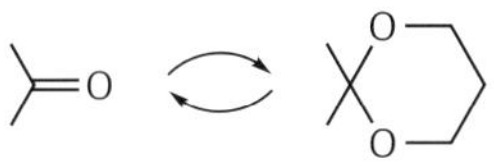

(Continued)

Table 4.15 Mass differences resulting from synthetic and mass spectrometric transformations (continued)

Mass difference (Δm in u)	Chemical transformation	MS transformations	Source / Comment
−73		$- C_3H_5O_2{}^{\bullet}: - CH_2CO_2CH_3{}^{\bullet}$ $- CO_2C_2H_5{}^{\bullet}$	Methyl esters Ethyl esters
± 76	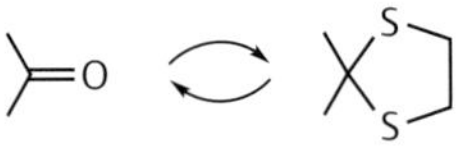 >=O ⇌ >C(S–CH₂CH₂–S) (1,3-dithiolane)		
− 77		$- C_6H_5{}^{\bullet}$	Phenyl derivatives
± 78	>−H ⇌ >−Br		Isotopic pattern!
	>−OH ⇌ >−OMes		Mes = methane sulfonyl
− 79		$- Br^{\bullet}$	Br derivatives
− 80		$- HBr$	Br derivatives
± 80	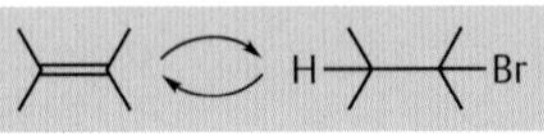 >=< ⇌ H−>−<−Br		Isotopic pattern!
± 84	 >−OH ⇌ >−OTHP		THP = tetrahydropyranyl
± 88	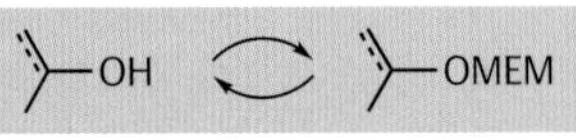 >−OH ⇌ >−OMEM		MEM = (2-methoxyethoxy)methyl
± 90	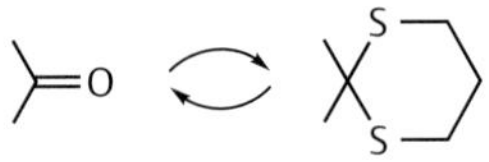  >=O ⇌ >C(S–CH₂CH₂CH₂–S) (1,3-dithiane) >CH₂ (H,H) ⇌ >C(S–CH₂CH₂–S)		
− 91		$- C_7H_7{}^{\bullet}$	Benzyl derivatives
± 92	 >−OH ⇌ >−OTBDMS		TBDMS = *tert*-butyldimethylsilyl (common derivatization for GC-MS analyses)
± 96	  >−OH ⇌ >−OCOCF₃		Trifluoroacetates (common derivatization for GC-MS analyses)
	>−NH₂ ⇌ >−NHCOCF₃		
± 100	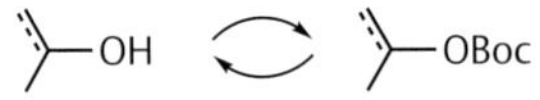  >−OH ⇌ >−OBoc		Boc = *tert*-butoxycarbonyl
	>−NH₂ ⇌ >−NHBoc		

Table 4.15 (Continued)

Mass difference (Δm in u)	Chemical transformation	MS transformations	Source / Comment
± 104	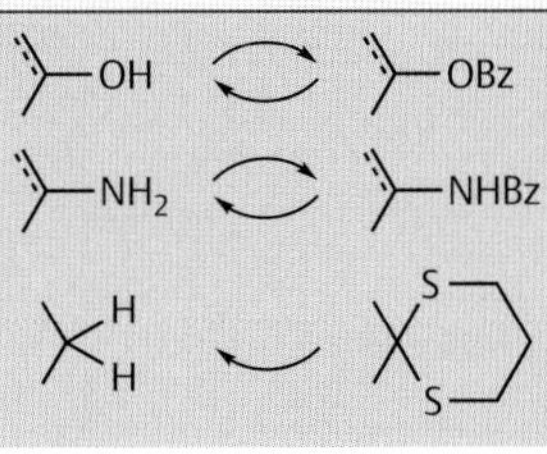 >—OH ⇌ >—OBz >—NH$_2$ ⇌ >—NHBz (dithiane)		Bz = benzoyl
± 120	 >—OH ⇌ >—OSEM		SEM = (2-trimethylsilylethoxy)methyl
± 126	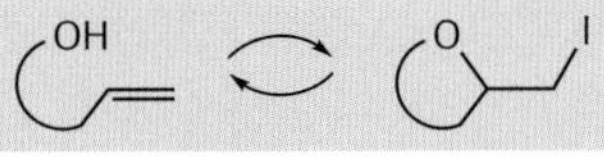 (OH) ⇌ (O—I)		
− 127		− I˙	I derivatives
− 128		− HI	I derivatives
± 128	 >=< ⇌ H—>—<—I		
± 130	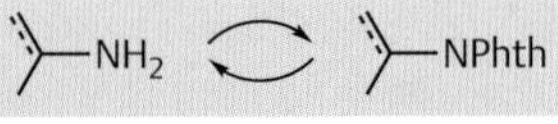 >—NH$_2$ ⇌ >—NPhth		Phth = phthaloyl
± 134	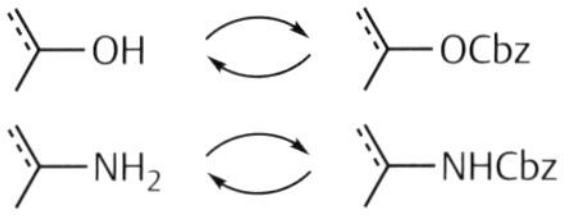 >—OH ⇌ >—OCbz >—NH$_2$ ⇌ >—NHCbz		Cbz (Z) = benzyloxycarbonyl
± 136	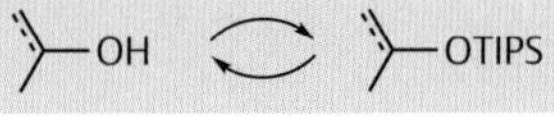 >—OH ⇌ >—OTIPS		TIPS = triisopropylsilyl
± 154	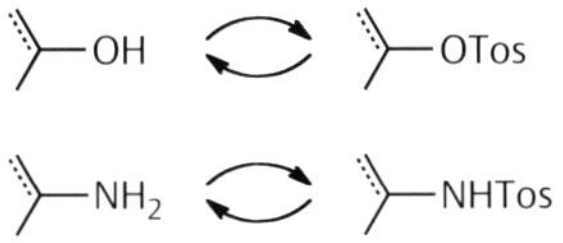 >—OH ⇌ >—OTos >—NH$_2$ ⇌ >—NHTos		Tos = *para*-toluene sulfonyl
− 155		− C$_7$H$_7$O$_2$S˙	*para*-Toluene sulfonyl (Tos) derivatives (EI)
± 222	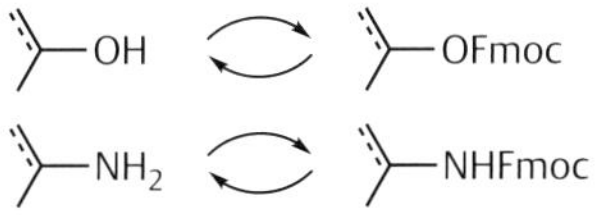 >—OH ⇌ >—OFmoc >—NH$_2$ ⇌ >—NHFmoc		Fmoc = fluorenylmethoxycarbonyl
± 228	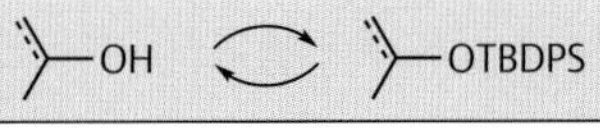 >—OH ⇌ >—OTBDPS		TBDPS = *tert*-butyldiphenylsilyl

4.7.3 Isotope Distributions of Halogenated Compounds

Table 4.16 Isotopic patterns for compounds containing Cl and/or Br atoms

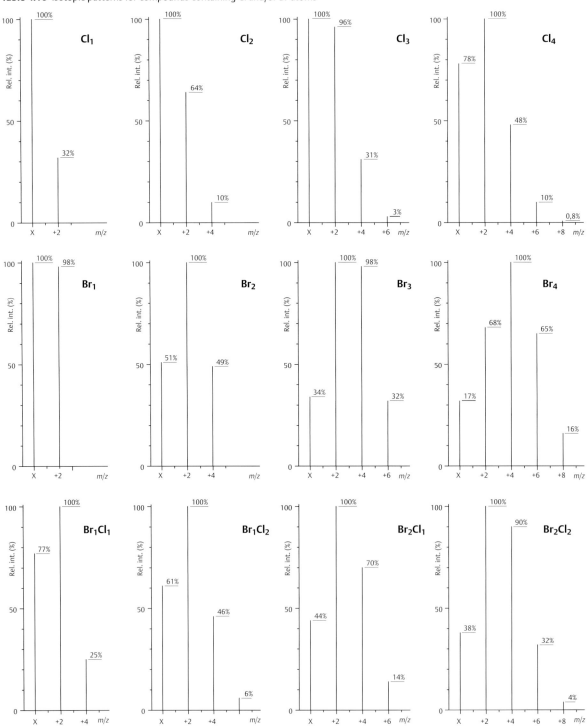

4.7.4 Solvents and Frequent Impurities

Table 4.17 EI-MS of common solvents, additives, and contaminants

Solvents

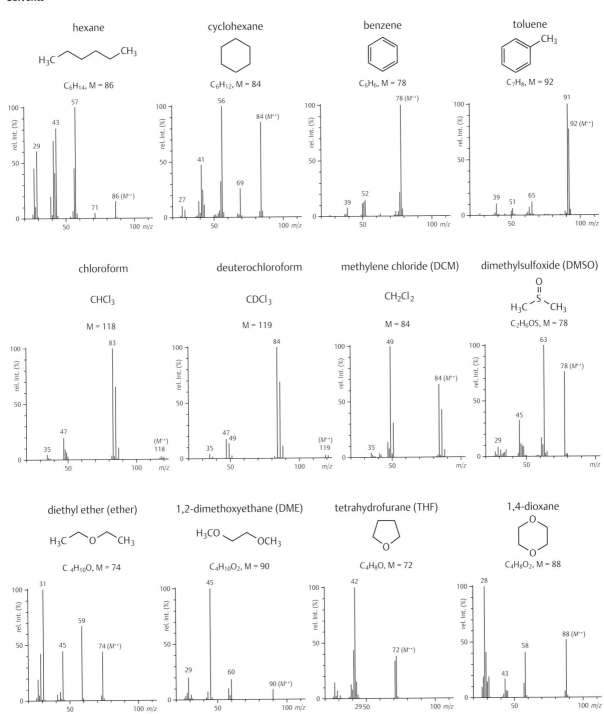

(Continued)

Table 4.17 EI-MS of common solvents, additives, and contaminants (Continued)
Solvents

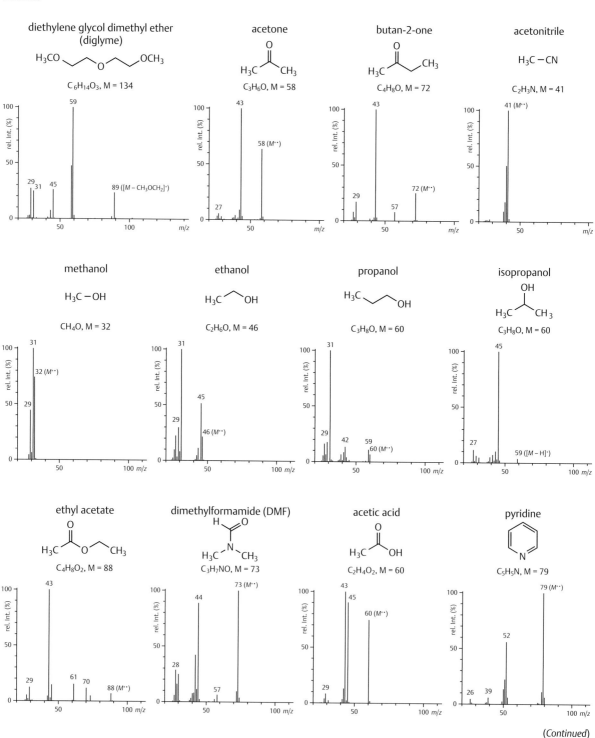

(Continued)

Table 4.17 (Continued)

Additives and contaminants

hexamethylphosphoric acid triamide
(hexametapol, HMPA)

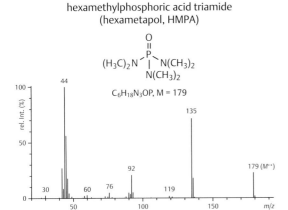

$C_6H_{18}N_3OP$, M = 179

2,6-di-tert-butyl-methylphenol
(jonol, stabilizer in ether)

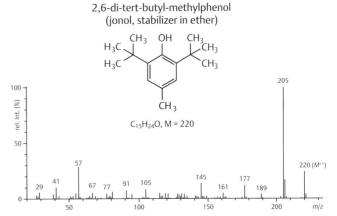

$C_{15}H_{24}O$, M = 220

phthalic acid dibutyl ester (plasticizer)

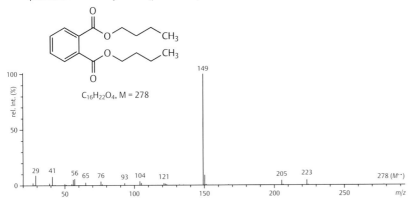

$C_{16}H_{22}O_4$, M = 278

Table 4.18 Frequently observed signals of solvent residues and contaminants in mass spectrometry[13]

Cation or Radical cation [m/z]	Anion [m/z]	Formula	Source
73.04680		$C_3H_9Si^+$	Cation from stationary phase (GC column bleeding)
78.01339		$C_2H_6OS^{+\bullet}$	Dimethylsulfoxide (DMSO)
82.94498		$CHCl_2^+$	$CHCl_3$
83.95126		$CDCl_2^+$	$CDCl_3$ (fragment ion)
101.00316		$C_2H_6NaOS^+$	Dimethylsulfoxide (DMSO)
102.12773		$C_6H_{16}N^+$	Triethylamine
107.04082		$C_2D_6NaOS^+$	d_6-Dimethylsulfoxide (d_6-DMSO)
	112.98559	$C_2F_3O_2^-$	Trifluoroacetic acid (TFA)
117.91384		$CHCl_3^{+\bullet}$	$CHCl_3$
118.92011		$CDCl_3^{+\bullet}$	$CDCl_3$
122.08117		$C_4H_{12}NO_3^+$	Tris(hydroxymethyl)aminomethane (TRIS)
123.09167		$C_7H_{11}N_2^+$	Dimethylaminopyridine (DMAP)
130.15903		$C_8H_{20}N^+$	Diisopropylethylamine (DIPEA)
136.98208		$C_2HF_3NaO_2^+$	Trifluoroacetic acid (TFA)
144.17468		$C_9H_{22}N^+$	Tripropylamine (TPA)
147.06560		$C_5H_{15}Si_2O^+$	Cation from stationary phase (GC column bleeding)
149.02332		$C_8H_5O_3^+$	Fragment ion from phthalates (plasticizer)
150.12773		$C_{10}H_{15}N^+$	Phenyldiethyl amine
153.13862		$C_9H_{17}N_2^+$	Diazabicycloundecene (DBU)
	153.01933	$C_7H_5O_4^-$	2,5-Dihydroxybenzoic acid (DHB, MALDI matrix)
155.03389		$C_7H_7O_4^+$	2,5-Dihydroxybenzoic acid (DHB, MALDI matrix)
157.03515		$C_4H_{13}O_2S_2^+$	Dimethylsulfoxide (DMSO)
158.96403		$C_2F_3Na_2O_2^+$	Trifluoroacetic acid (TFA)
169.11047		$C_4HD_{12}O_2S_2^+$	d_6-Dimethylsulfoxide (d_6-DMSO)
179.01709		$C_4H_{12}NaO_2S_2^+$	Dimethylsulfoxide (DMSO)
186.22163		$C_{12}H_{28}N^+$	Tributylamine (TBA)
	188.03532	$C_{10}H_6NO_3^-$	α-Cyano-4-hydroxycinnamic acid (HCCA, MALDI matrix)
190.04987		$C_{10}H_8NO_3^+$	α-Cyano-4-hydroxycinnamic acid (HCCA, MALDI matrix)
191.09241		$C_4D_{12}NaO_2S_2^+$	d_6-Dimethylsulfoxide (d_6-DMSO)
195.06519		$C_{10}H_{11}O_4^+$	Dimethyl phthalate
205.15869		$C_{14}H_{21}O^{+\bullet}$	2,6-Di-*tert*-butyl-4-methylphenol (stabilizer in ether)
207.03235		$C_5H_{15}O_3Si_3^+$	Cation from stationary phase (GC column bleeding)

Table 4.18 (Continued)

Cation or Radical cation [m/z]	Anion [m/z]	Formula	Source
214.08963		$C_{10}H_{16}NO_2S^+$	Butylbenzene sulfonamide (NBBS, plasticizer)
220.18217		$C_{15}H_{24}O^{+\bullet}$	2,6-Di-*tert*-butyl-4-methylphenol (stabilizer in ether)
221.08439		$C_7H_{21}Si_3O_2^+$	Cation from stationary phase (GC column bleeding)
	223.06120	$C_{11}H_{11}O_5^-$	Sinapinic acid (SA, MALDI matrix)
225.07575		$C_{11}H_{13}O_5^+$	Sinapinic acid (SA, MALDI matrix)
225.19614		$C_{13}H_{25}N_2O^+$	Dicyclohexylurea (DCU)
	226.97845	$C_4HF_6O_4^-$	Trifluoroacetic acid (TFA)
236.07157		$C_{10}H_{15}NNaO_2S^+$	Butylbenzene sulfonamide (NBBS, plasticizer)
242.28423		$C_{16}H_{36}N^+$	Tetrabutylammonium hydroxide (TBAOH)
243.11683		$C_{19}H_{15}^+$	Fragment ion from the trityl protecting group
	248.96039	$C_4F_6NaO_4^-$	Trifluoroacetic acid (TFA)
257.03103		$C_6H_{18}NaO_3S_3^+$	Dimethylsulfoxide (DMSO)
	262.97601	$C_5F_9O_2^-$	Nonafluoropentanoic acid (NFPA)
267.17197		$C_{12}H_{28}O_4P^+$	Tributylphosphate (plasticizer)
273.12739		$C_{20}H_{17}O^+$	Fragment ion from monomethoxytrityl protecting group (MMT)
278.08550		$C_{18}H_{15}OP^{+\bullet}$	Triphenylphosphine oxide
278.15126		$C_{16}H_{22}O_4^{+\bullet}$	Dibutyl phthalate (plasticizer)
279.09333		$C_{18}H_{16}OP^+$	Triphenylphosphine oxide
279.15909		$C_{16}H_{23}O_4^+$	Dibutyl phthalate (plasticizer)
281.05114		$C_7H_{21}O_4Si_4^+$	Cation from stationary phase (GC column bleeding)
282.27914		$C_{18}H_{36}NO^+$	Oleamide (endogenous substance, lubricants in polyethylene)
284.29479		$C_{18}H_{38}NO^+$	Stearamide (lubricants in polyethylene)
	291.08339	$C_{10}H_{15}N_2O_8^-$	Ethylendiaminetetraacetic acid (EDTA)
293.09794		$C_{10}H_{17}N_2O_8^+$	Ethylendiaminetetraacetic acid (EDTA)
295.10318		$C_9H_{27}O_3Si_4^+$	Cation from stationary phase (GC column bleeding)
301.14103		$C_{16}H_{22}NaO_4^+$	Dibutyl phthalate (plasticizer)
303.13796		$C_{21}H_{19}O_2^+$	Fragment ion from dimethoxytrityl protecting group (DMT)
304.26109		$C_{18}H_{35}NNaO^+$	Oleamide (endogenous substance, lubricants in polyethylene)
	307.04594	$C_{14}H_{11}O_8^-$	2,5-Dihydroxybenzoic acid (DHB, MALDI matrix)
309.06049		$C_{14}H_{13}O_8^+$	2,5-Dihydroxybenzoic acid (DHB, MALDI matrix)

Mass Spectrometry

(Continued)

Table 4.18 Frequently observed signals of solvent residues and contaminants in mass spectrometry[13] (continued)

Cation or Radical cation [m/z]	Anion [m/z]	Formula	Source
315.07989		$C_{10}H_{16}N_2NaO_8^+$	Ethylendiaminetetracetic acid (EDTA)
338.34174		$C_{22}H_{44}NO^+$	Erucamide (lubricants in polyethylene)
355.06995		$C_9H_{27}O_5Si_5^+$	Cation from stationary phase (GC column bleeding)
360.32369		$C_{22}H_{43}NNaO^+$	Erucamide (lubricants in polyethylene)
369.12197		$C_{11}H_{33}O_4Si_5^+$	Cation from stationary phase (GC column bleeding)
371.10123		$C_{10}H_{31}O_5Si_5^+$	Decamethylcyclopentasiloxane (deodorants, hair spray)
	377.07791	$C_{20}H_{13}N_2O_6^-$	α-Cyano-4-hydroxycinnamic acid (HCCA, MALDI matrix)
379.09246		$C_{20}H_{15}N_2O_6^+$	α-Cyano-4-hydroxycinnamic acid (HCCA, MALDI matrix)
391.28429		$C_{24}H_{39}O_4^+$	Dioctyl phthalate (plasticizer)
	412.96643	$C_8F_{15}O_2^-$	Perfluorooctanoic acid (PFOA)
413.26623		$C_{24}H_{38}NaO_4^+$	Dioctyl phthalate (plasticizer)
429.08872		$C_{11}H_{33}O_6Si_6^+$	Cation from stationary phase (GC column bleeding)
445.12002		$C_{12}H_{37}O_6Si_6^+$	Dodecamethylcyclohexasiloxane (deodorant, hair spray)
	447.12967	$C_{22}H_{23}O_{10}^-$	Sinapinic acid (SA, MALDI matrix)
449.14422		$C_{22}H_{25}O_{10}^+$	Sinapinic acid (SA, MALDI matrix)
449.38500		$C_{26}H_{49}N_4O_2^+$	Dicyclohexylurea (DCU, plasticizer)
519.13882		$C_{14}H_{43}O_7Si_7^+$	Tetradecamethylcycloheptasiloxane
803.54324		$C_{48}H_{76}NaO_8^+$	Dioctyl phthalate (plasticizer)
Characteristic, repetitive signal mass differences (Δm in u)			
44		$-[CH_2\text{-}CH_2]_n-$	Polyethylene glycol (PEG), triton X-100, nonidet, e.g.
58		$-[CH\text{-}CH_3\text{-}CH_2]_n-$	Polypropylene glycol (PPG)
68		$(HCOONa)_n$	Sodium formiate cluster (calibration)
71		$-[CH_2\text{-}CH(CONH_2)]_n-$	Polyacrylamide
82		$(H_3CCOONa)_n$	Sodium acetate clusters (calibration)
149.89		$(NaI)_n$	Sodium iodide clusters
165.87		$(KI)_n$	Potassium iodide clusters

4.7.5   Isotopes of Naturally Occurring Elements

Table 4.19 The natural elements, sorted according to element symbols, their atomic numbers (AN), the mass numbers (A) of their natural isotopes with the accurate isotopic masses, isotopic compositions, and normalized isotopic compositions, and the elements' standard atomic masses[1]

Element	Standard atomic mass (u)*	Atomic number A (u)	Isotopic mass (u)	Isotopic composition (%)*	Normalized isotopic composition (%)**
Ag	107.87	107	106.90509	51.84	100
		109	108.904755	48.16	92.91
Al	26.98	27	26.981539	100	100
Ar	39.95	36	35.967545	0.34	0.34
		38	37.962732	0.06	0.06
		40	39.962383	99.60	100
As	74.92	75	74.921595	100	100
Au	196.97	197	196.966569	100	100
B	10.81	10	10.012937	19.90	24.84
		11	11.009305	80.10	100
Ba	137.33	130	129.90632	0.11	0.15
		132	131.905061	0.10	0.14
		134	133.904508	2.42	3.37
		135	134.905688	6.59	9.19
		136	135.904576	7.85	10.95
		137	136.905827	11.23	15.67
		138	137.905247	71.70	100
Be	9.01	9	9.012183	100	100
Bi	208.98	209	208.98040	100	100
Br	79.90	79	78.918338	50.69	100
		81	80.916290	49.31	97.28
C	12.01	12	12.000000	98.93	100
		13	13.003355	1.07	1.08
Ca	40.08	40	39.962591	96.94	100
		42	41.958618	0.65	0.67
		43	42.958766	0.14	0.14
		44	43.955482	2.09	2.15
		46	45.95369	<0.01	<0.01
		48	47.952523	0.19	0.19
Cd	112.41	106	105.906460	1.25	4.35
		108	107.904183	0.89	3.10
		110	109.903007	12.47	43.47
		111	110.904183	12.80	44.55
		112	111.902763	24.11	83.99
		113	112.904408	12.23	42.53
		114	113.903365	28.75	100
		116	115.904763	7.51	26.07
Ce	140.12	136	135.907129	0.19	0.21
		138	137.90599	0.25	0.28

(Continued)

Mass Spectrometry

Table 4.19 The natural elements and their isotopes (continued)

Element	Standard atomic mass (u)*	Atomic number A (u)	Isotopic mass (u)	Isotopic composition (%)*	Normalized isotopic composition (%)**
		140	139.90544	88.45	100
		142	141.90925	11.11	12.57
Cl	35.45	35	34.968853	75.76	100
		37	36.965903	24.24	32.00
Co	58.93	59	58.933194	100	100
Cr	52.00	50	49.946042	4.35	5.19
		52	51.940506	83.79	100
		53	52.940648	9.50	11.34
		54	53.938879	2.36	2.82
Cs	132.91	133	132.905452	100	100
Cu	63.55	63	62.929598	69.15	100
		65	64.927790	30.85	44.61
Dy	162.50	156	155.92428	0.06	0.20
		158	157.92442	0.10	0.34
		160	159.92520	2.33	8.24
		161	160.92694	18.89	66.84
		162	161.92681	25.48	90.15
		163	162.92874	24.90	88.10
		164	163.92918	28.26	100
Er	167.26	162	161.92879	0.14	0.41
		164	163.92921	1.60	4.78
		166	165.93030	33.50	100
		167	166.93205	22.87	68.26
		168	167.93238	26.98	80.52
		170	169.93547	14.91	44.50
Eu	151.96	151	150.91986	47.81	91.61
		153	152.92124	52.19	100
F	19.00	19	18.998403	100	100
Fe	55.85	54	53.939609	5.85	6.37
		56	55.934936	91.75	100
		57	56.935393	2.12	2.31
		58	57.933274	0.28	0.31
Ga	69.72	69	68.925574	60.11	100
		71	70.924703	39.89	66.37
Gd	157.25	152	151.91980	0.20	0.81
		154	153.92087	2.18	8.78
		155	154.92263	14.80	59.58
		156	155.92213	20.47	82.41
		157	156.92397	15.65	63.00
		158	157.92411	24.84	100
		160	159.92706	21.86	88.00

Table 4.19 (Continued)

Element	Standard atomic mass (u)*	Atomic number A (u)	Isotopic mass (u)	Isotopic composition (%)*	Normalized isotopic composition (%)**
Ge	72.63	70	69.924249	20.52	55.50
		72	71.922076	27.45	74.37
		73	72.923459	7.76	21.13
		74	73.921178	36.52	100
		76	75.921403	7.75	21.32
H/D	1.01	1	1.007825	99.99	100
		2	2.014102	0.01	0.01
He	4.00	3	3.016029	<0.01	<0.01
		4	4.002603	>99.99	100
Hf	178.49	174	173.94005	0.16	0.46
		176	175.94141	5.26	14.99
		177	176.94323	18.60	53.02
		178	177.94371	27.28	77.77
		179	178.94582	13.62	38.83
		180	179.94656	35.08	100
Hg	200.59	196	195.965830	0.15	0.50
		198	197.966769	10.04	33.39
		199	198.968281	16.94	56.50
		200	199.968327	23.14	77.36
		201	200.970303	13.17	44.14
		202	201.970643	29.74	100
		204	203.973494	6.82	23.01
Ho	164.93	165	164.93033	100	100
I	126.90	127	126.90447	100	100
In	114.82	113	112.904062	4.28	4.48
		115	114.903879	95.72	100
Ir	192.22	191	190.96059	37.30	59.49
		193	192.96292	62.70	100
K	39.10	39	38.963706	93.26	100
		40	39.963998	0.01	0.01
		41	40.961825	6.73	7.22
Kr	83.80	78	77.920365	0.36	0.62
		80	79.916378	2.29	4.01
		82	81.913483	11.59	20.34
		83	82.914127	11.50	20.18
		84	83.911478	56.99	100
		86	85.910611	17.28	30.32
La	138.91	138	137.90712	0.09	0.09
		139	138.90636	99.91	100
Li	6.94	6	6.015123	7.59	8.21
		7	7.016003	92.41	100

Mass Spectrometry

(Continued)

Table 4.19 The natural elements and their isotopes (continued)

Element	Standard atomic mass (u)*	Atomic number A (u)	Isotopic mass (u)	Isotopic composition (%)*	Normalized isotopic composition (%)**
Lu	174.97	175	174.94078	97.40	100
		176	175.94269	2.60	2.66
Mg	24.31	24	23.985042	78.99	100
		25	24.985837	10.00	12.66
		26	25.982593	11.01	13.94
Mn	54.94	55	54.938044	100	100
Mo	95.95	92	91.906808	14.65	61.06
		94	93.905085	9.19	38.16
		95	94.905839	15.87	65.73
		96	95.904676	16.67	68.95
		97	96.906018	9.58	39.52
		98	97.905405	24.29	100
		100	99.907472	9.74	39.98
N	14.01	14	14.003074	99.64	100
		15	15.000109	0.36	0.37
Na	22.99	23	22.989769	100	100
Nb	92.91	93	92.90637	100	100
Nd	144.24	142	141.90773	27.15	100
		143	142.90982	12.17	44.85
		144	143.91009	23.80	87.50
		145	144.91258	8.29	30.51
		146	145.91312	17.19	63.24
		148	147.91690	5.76	20.96
		150	149.92090	5.64	20.59
Ne	20.18	20	19.992440	90.48	100
		21	20.993847	0.27	0.30
		22	21.991385	9.25	10.22
Ni	58.69	58	57.935342	68.08	100
		60	59.930786	26.22	38.52
		61	60.931056	1.14	1.67
		62	61.928345	3.63	5.34
		64	63.927967	0.93	1.36
O	16.00	16	15.994915	99.76	100
		17	16.999132	0.04	0.04
		18	17.999160	0.20	0.21
Os	190.23	184	183.952489	0.02	0.05
		186	185.95384	1.59	3.90
		187	186.95575	1.96	4.81
		188	187.95584	13.24	32.47
		189	188.95814	16.15	39.60

Table 4.19 (Continued)

Element	Standard atomic mass (u)*	Atomic number A (u)	Isotopic mass (u)	Isotopic composition (%)*	Normalized isotopic composition (%)**
		190	189.95844	26.26	64.39
		192	191.96148	40.78	100
P	30.97	31	30.973762	100	100
Pb	207.20	204	203.973044	1.40	2.67
		206	205.974466	24.10	45.99
		207	206.975897	22.10	42.176
		208	207.976653	52.40	100
Pd	106.42	102	101.90560	1.02	3.73
		104	103.904031	11.14	40.76
		105	104.905080	22.33	81.71
		106	105.903480	27.33	100
		108	107.903892	26.46	96.82
		110	109.905172	11.72	42.88
Pr	140.91	141	140.90766	100	100
Pt	195.08	190	189.95993	0.01	0.04
		192	191.96104	0.78	2.31
		194	193.962681	32.86	97.44
		195	194.964792	33.78	100
		196	195.964952	25.21	74.61
		198	197.96789	7.36	21.17
Rb	85.47	85	84.911790	72.17	100
		87	86.909181	27.83	38.56
Re	186.21	185	184.952955	37.40	59.74
		187	186.95575	62.60	100
Rh	102.91	103	102.90550	100	100
Ru	101.07	96	95.907590	5.54	17.56
		98	97.90529	1.87	5.93
		99	98.905934	12.76	40.44
		100	99.904214	12.60	39.94
		101	100.905577	17.06	54.07
		102	101.904344	31.55	100
		104	103.90543	18.62	59.02
S	32.06	32	31.972071	94.99	100
		33	32.971459	0.75	0.79
		34	33.967867	4.25	4.47
		36	35.967081	0.01	0.01
Sb	121.76	121	120.90381	57.21	100
		123	122.90421	42.79	74.79
Sc	44.96	45	44.955908	100	100

(Continued)

Table 4.19 The natural elements and their isotopes (continued)

Element	Standard atomic mass (u)*	Atomic number A (u)	Isotopic mass (u)	Isotopic composition (%)*	Normalized isotopic composition (%)**
Se	78.97	74	73.922476	0.86	1.79
		76	75.919214	9.23	18.89
		77	76.919214	7.60	15.38
		78	77.917309	23.69	47.91
		80	79.916522	49.80	100
		82	81.916700	8.82	17.60
Si	28.09	28	27.976927	92.22	100
		29	28.976495	4.69	5.08
		30	29.973770	3.09	3.35
Sm	150.36	144	143.91201	3.08	11.48
		147	146.91490	15.00	56.04
		148	147.91483	11.25	42.02
		149	148.91719	13.82	51.66
		150	149.91728	7.37	27.59
		152	151.91974	26.74	100
		154	153.92222	22.74	85.05
Sn	118.71	112	111.904824	0.97	2.98
		114	113.902783	0.66	2.03
		115	114.903345	0.34	1.04
		116	115.901743	14.54	44.63
		117	116.902954	7.68	23.57
		118	117.901607	24.22	74.34
		119	118.903311	8.59	26.37
		120	119.902202	32.58	100
		122	121.903440	4.63	14.21
		124	123.905277	5.79	17.77
Sr	87.62	84	83.913419	0.56	0.68
		86	85.909261	9.86	11.94
		87	86.908878	7.00	8.48
		88	87.905613	82.58	100
Ta	180.95	180	179.94746	0.01	0.01
		181	180.94800	99.99	100
Tb	158.93	159	158.92535	100	100
Te	127.60	120	119.90406	0.09	0.26
		122	121.90304	2.55	7.48
		123	122.90427	0.89	2.61
		124	123.90282	4.74	13.91
		125	124.90443	7.07	20.75
		126	125.90331	18.84	55.28
		128	127.904461	31.74	93.13
		130	129.906223	34.08	100

Table 4.19 (Continued)

Element	Standard atomic mass (u)*	Atomic number A (u)	Isotopic mass (u)	Isotopic composition (%)*	Normalized isotopic composition (%)**
Th	232.04	230	230.03313	0.02	0.02
	232.04	232	232.03806	99.98	99.98
Ti	47.87	46	45.952628	8.25	11.19
		47	46.951759	7.44	10.09
		48	47.947942	73.72	100
		49	48.947866	5.41	7.34
		50	49.944787	5.18	7.03
Tl	204.38	203	202.972345	29.52	41.88
		205	204.974428	70.48	100
Tm	168.93	169	168.93422	100	100
U	238.03	234	234.04095	0.01	0.01
		235	235.04393	0.72	0.73
		238	238.05079	99.27	100
V	50.94	50	49.947156	0.25	0.25
		51	50.943957	99.75	100
W	183.84	180	179.946710	0.12	0.39
		182	181.948204	26.50	86.49
		183	182.950223	14.31	46.70
		184	183.950931	30.64	100
		186	185.954360	28.43	92.79
Xe	131.29	124	123.90589	0.10	0.35
		126	125.90430	0.09	0.33
		128	127.903531	1.91	7.10
		129	128.904781	26.40	98.11
		130	129.903509	4.07	15.13
		131	130.905084	21.23	78.91
		132	131.904155	26.91	100
		134	133.905395	10.44	38.78
		136	135.907214	8.86	32.92
Y	88.91	89	88.90584	100	100
Yb	173.05	168	167.93389	0.13	0.41
		170	169.93477	3.02	9.55
		171	170.93633	14.22	44.86
		172	171.93639	21.75	68.58
		173	172.93822	16.10	50.68
		174	173.93887	31.90	100
		176	175.94258	12.89	40.09
Zn	65.38	64	63.929142	49.17	100
		66	65.926034	27.73	57.96
		67	66.927128	4.04	8.50
		68	67.924845	18.45	39.41
		70	69.92532	0.61	1.31

Table 4.19 The natural elements and their isotopes (continued)

Element	Standard atomic mass (u)*	Atomic number A (u)	Isotopic mass (u)	Isotopic composition (%)*	Normalized isotopic composition (%)**
Zr	91.22	90	89.90470	51.45	100
		91	90.90564	11.22	21.81
		92	91.90503	17.15	33.33
		94	93.90631	17.38	33.78
		96	95.90827	2.80	5.44

Addendum: Mass of the electron: 0.0005485799 u

*rounded to two decimals.
**relative to most abundant isotope (100%), rounded to two decimals.
[1] https://www.isotopesmatter.com/applets/IPTEI/IPTEI.html (May 2019).

4.8   Literature

1. Coursey JS, Schwab DJ, Tsai JJ, Dragoset RA. Atomic weight and isotopic compositions with relative atomic masses. Available at: https://www.nist.gov/pml/atomic-weights-and-isotopic-compositions-relative-atomic-masses. Accessed March 2019

2. Biemann K. Appendix 5. Nomenclature for peptide fragment ions (positive ions). Methods Enzymol 1990;193:886–887

3. Roepstorff P, Fohlman J. Proposal for a common nomenclature for sequence ions in mass spectra of peptides. Biomed Mass Spectrom 1984;11:601

4. Biemann K. Appendix 6. Mass values for amino acid residues in peptides. Methods Enzymol 1990;193:888

5. McLuckey SA, Van Berkel GJ, Glish GL. Tandem mass spectrometry of small, multiply charged oligonucleotides. J Am Soc Mass Spectrom 1992;3:60–70

6. Domon B, Costello CE. A systematic nomenclature for carbohydrate fragmentations in Fab-MS/MS spectra of glycoconjugates. Glycoconj J 1988;5:397–409

7. Tzouros M, Chesnov S, Bigler L, Bienz S. A template approach for the characterization of linear polyamines and derivatives in spider venom. Eur J Mass Spectrom (Chichester) 2013;19:57–69 PubMed

8. Tzouros M, Manov N, Bienz S, Bigler L. Tandem mass spectrometric investigation of acylpolyamines of spider venoms and their ^{15}N-labeled derivatives. J Am Soc Mass Spectrom 2004;15(11):1636–1643

9. Doll MKH, Guggisberg A, Hesse M. Spermidine alkaloids type inandenine from *Oncinotis tenuiloba*. Phytochemistry 1995;39:689–694

10. Eichenberger S, Bigler L, Bienz S. Structure elucidation of polyamine toxins in the venom of the spider *Larinioides folium*. Chimia 2007;61:161–164

11. Klinter M, Schermann NE. Protein sequencing and identification using tandem mass spectrometry. New York: John Wiley & Sons; 2000

12. Eichenberger S, Méret M, Bienz S, Bigler L. Decomposition of *N*-hydroxylated compounds during atmospheric pressure chemical ionization. J Mass Spectrom 2010;45:190–197

13. Keller BO, Sui J, Young AB, Whittal RM. Interferences and contaminants encountered in modern mass spectrometry. Anal Chim Acta 2008;627:71–81

Stefan Bienz, Laurent Bigler

5

Handling of Spectra and Analytical Data: Practical Examples

Examples

5 Handling of Spectra and Analytical Data: Practical Examples

5.1 Introduction

Several analytical methods, enriched with compilations of data that shall facilitate their application, were introduced separately in independent chapters. Analytical work, however, does usually not mean that the diverse data acquired by individual analytical methods are scrutinized and interpreted independently. It rather means that all analytical information about a sample, derived from several analytical methods, shall find combined and simultaneous attention. It is very rare that the data of a single specific analytical method can be fully interpreted in all details without taking additional information from other methods into account, and it is equally rare that an isolated spectroscopic method can provide enough data that allow the conclusive deduction and characterization of the structure of a sample molecule. It is usually the interplay of several spectroscopic and analytical methods, each having its own strengths and weaknesses, which allows efficient, reliable, and comprehensive analytics. Accordingly, chemists and chemical analysts need to be skilled in understanding this interplay—they should be able to let their eyes drift from one dataset to another, and they should be able to know strategies that enable them to find efficient and safe ways to solve dependably their analytical problems.

In the next few sections we intend to present some possible approaches to attack analytical challenges. By means of practical examples, several analytical issues are covered: Section 5.2 deals with the characterization and analytical description of compounds for which the structures are secured, Section 5.3 discusses the structural elucidation and scrutiny of products that arise from reactions of known (or allegedly known) components, and Section 5.4 presents possible tactics and methods to reveal the structures of fully unknown compounds solely on the basis of analytical data. The examples used in this book are of rather complex nature, to give us the opportunity to discuss the diverse aspects and problems of analysis and interpretation of spectra in detail. To provide the readers with opportunities to train their analytical skills "from scratch", however, we offer electronically some additional examples, starting with very simple problems and continuing with tasks of increasing complexity (https://thieme.de/HMZ).

5.2 Characterization of Compounds

It may be regarded as trivial to describe the analytical data of a compound. At closer look, however, several crucial questions arise. (1) Which data shall be considered and described when a compound needs to be characterized? What is minimally necessary, what is desirable, and what is superfluous? (2) Which data shall be interpreted into which depths? Which signals need to be explained and assigned and for which signals the report of their presence is sufficient? (3) How shall the data be described and presented? Is it suitable to show them within a text flow, or is it more appropriate to summarize them in tables or show them just as graphical sketches? Is it maybe even necessary to show them in the form of the original spectra? Certainly, the questions will have to be answered differently depending on the exact problem and its context. The way data presentation will finally be demanded, however, has its consequences on the scope of the required data acquisition and should, thus, be considered early in an investigation.

We fundamentally believe that the following analytical data should be acquired and described for any new compound:

- Color, aggregation state, and viscosity or crystalline appearance.
- Melting interval, boiling interval (or conditions for distillation), and maybe refractive index.
- Specific optical rotation (if enantiomerically enriched).
- Infrared (IR) and ultraviolet/visible light (UV/Vis) spectra (the latter only when the compound possesses a chromophoric group).
- ^{1}H and ^{13}C NMR spectra (with the respective multiplicities derived from DEPT 90, DEPT 135, or heteronuclear single quantum coherence (HSQC) spectra).
- Mass spectrometry (MS) spectrum (measured with an appropriate ionization method), if possible complemented with high-resolution MS (HR-MS) data.
- Combustion analysis (elemental analysis).

For known compounds, a characterization that confirms the equality of the investigated sample with a trustworthy reference compound would in principle suffice. However, it might become beneficial to have access to the breadth of a full analytical dataset of a compound if a more extensive research is performed in a field where related compounds are of common interest. Often it is thus worthwhile to acquire more data than currently needed.

The question about the required depth and completeness of a spectral interpretation needs to be answered in the respective context. In most cases of spectral analyses, the expected and required signals for a proposed structure are readily found. If this is unquestionably sufficient to be certain about a structural assignment, it is not necessary to go into all details of spectral interpretation. It is not essential to be able to assign all signals to the exact respective structural units. For instance, the fingerprint region of an IR spectrum is definitely diagnostic for a given compound, but its detailed interpretation usually does not make much sense. Likewise, if all expected nuclear magnetic resonance (NMR) signals for a compound are found but cannot be individually assigned to the respective nuclei, additional efforts to enforce such assignments can be omitted, when the assignments are not decisive to prove the structure. Spectral interpretation is then confined to those declarations that are possible and appropriate on the basis of the given data: diagnostic signals are reported with the respective assignments and the remaining signals may be interpreted in a summary manner. In the IR spectrum of compound **1** (**Fig. 5.1** and Example 1), for example, merely the signals at 1,697s ($v_{C=O}$) (**Table 2.10**, p. 62) and 1,635m ($v_{C=C}$) cm^{-1} (**Table 2.13**, p. 65) that are typical for α,β-unsaturated carbonyl compounds are interpreted; the signals at 2,927s and 2,858m cm^{-1} or those at <1,500 cm^{-1} in the fingerprint region are just listed as a reference. Most of the NMR signals corresponding to the several CH$_2$ groups of compound **1**, likewise, are just interpreted in a summary manner. It is not possible to perform an unequivocal assignment of all these signals to the respective nuclei on the basis of simple NMR experiments. Asking for additional experiments for this purpose appears to be unnecessary and not sensible.

The situation is somewhat different when new and more complex compounds need to be described and in particular also when stereochemical aspects are involved as well (see compound **2**, **Fig. 5.1**, and Example 2). In such cases, signal assignments should be done as completely as ever possible. Maybe additional data from, for instance, correlation spectroscopy (COSY), nuclear Overhauser effect spectroscopy (NOESY), HSQC,

heteronuclear multiple-bond correlation (HMBC), and/or other spectra will be needed to allow full interpretation of the simple one-dimensional NMR (1D-NMR) spectra. The acquisition of a rather broad set of analytical data might even be advisable when dealing with known compounds. Particularly when working with compounds of rather complex structures, it is always sensible to carefully scrutinize the signal assignments reported in the literature. If these are not unequivocally supported by strong experimental evidence, it is recommended to secure them with own data. Often enough errors in the literature could be detected and corrected by such a process.

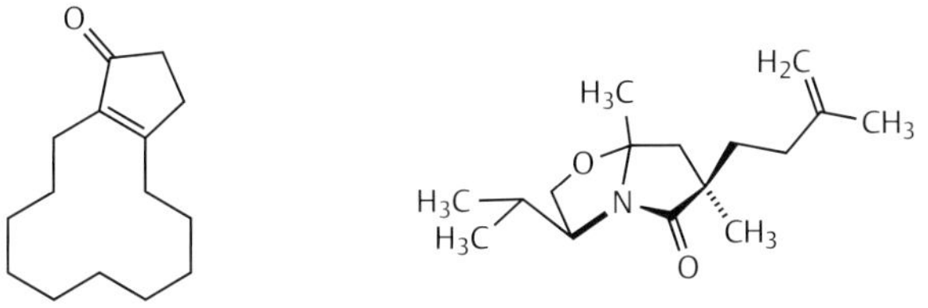

When all the required analytical data are acquired, they are conveniently described in a way that creates the highest recognition value. This means, for instance, that for IR and MS spectra not only those signals are listed that are diagnostic for functional groups or structurally identified ions but rather all signals (above a threshold) together with information about their intensities (s, m, and w for strong, middle, and weak for IR absorptions and the values of rel.%-intensities for MS signals). In the same spirit, NMR signals with well-structured peak patterns should be described in recognizable detail such as "t-like m," "symm. m with eight lines," "A-portion of ABX, four lines," or likewise. Unfortunately, such signals are progressively described as m only, which contains no recognition value, probably for the reason of convenience.

Analytical data are usually described in the text form. This is also possible for more complex datasets, for instance, when the results of two-dimensional NMR (2D-NMR) experiments have to be described. In such cases, however, it is often more appropriate to display the data graphically or in a table form. Respective presentations can be found with the examples shown below. If more complex spectroscopic problems are discussed and/or the interpretation of analytical data is the major point of focus, it is also reasonable to provide the original spectra. This is the case, for instance, when structural elucidations of natural products are described. Many scientific journals nowadays ask as a standard for representations of original spectra in the *Supplementary Material*.

Example 1

Bicyclo[10.3.0]pentadec-1(12)-en-13-one (**1**) was synthesized with a reliable procedure.[1] It delivered the data and spectra shown below. These data shall be described as completely as possible, recognizing that signal assignments shall be done solely on the basis of the given data (thus, without referring to information found in the literature for **1**). Which NMR signals can be assigned unequivocally to the respective nuclei in compound **1** and which can not? Which assignments are crucial to explicitly confirm the structure of **1** and which are irrelevant? If it would be demanded to assign all NMR signals to the respective nuclei, which additional experiments would you perform?

Fig. 5.1 Structures used to exemplify the description of spectral data.

Examples

Example 1

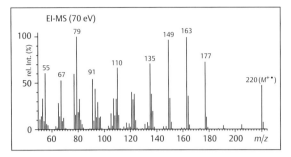

EI-MS (70 eV)

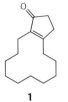

colorless oil
b.p. 162–163 °C (4.7 mbar)
n_D 1.5012

$C_{15}H_{24}O$ (220.36):   C 81.76, H 10.98
found: C 81.83, H 11.16

1

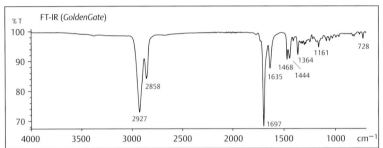

FT-IR (*GoldenGate*)

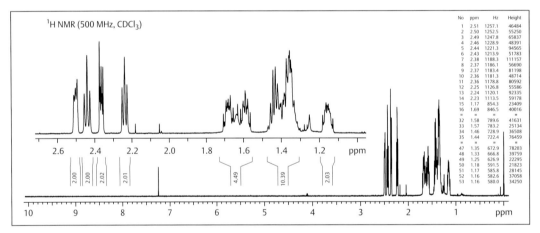

¹H NMR (500 MHz, CDCl₃)

No	ppm	Hz	Height
1	2.51	1257.1	46484
2	2.50	1252.5	55250
3	2.49	1247.8	65837
4	2.46	1228.9	48391
5	2.44	1221.3	94565
6	2.43	1213.9	51783
7	2.38	1188.3	111157
8	2.37	1186.1	56690
9	2.37	1183.4	81198
10	2.36	1181.3	48714
11	2.36	1178.8	80592
12	2.25	1126.8	55586
13	2.24	1120.1	92335
14	2.23	1113.5	59178
15	1.17	854.3	23409
16	1.69	846.5	40016
...	...	...	...
32	1.58	789.6	41631
33	1.57	783.2	25134
34	1.46	728.9	36508
35	1.44	722.4	76459
...	...	...	...
47	1.35	672.9	78203
48	1.33	666.8	39759
49	1.25	626.9	22295
50	1.18	591.5	21823
51	1.17	585.8	28145
52	1.16	582.6	37058
53	1.16	580.0	34250

¹³C{¹H} NMR (125 MHz, CDCl₃)

DEPT 90

DEPT 135

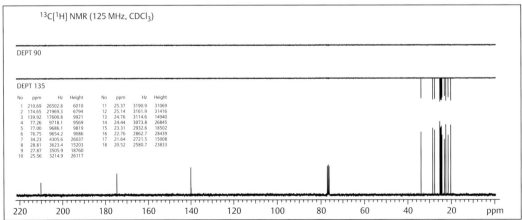

No	ppm	Hz	Height	No	ppm	Hz	Height
1	210.69	26502.8	6010	11	25.37	3190.9	31069
2	174.65	21969.3	6794	12	25.14	3161.9	31416
3	139.92	17600.8	9921	13	24.76	3114.6	14940
4	77.26	9718.1	9569	14	24.44	3073.8	26845
5	77.00	9686.1	9819	15	23.31	2932.6	18502
6	76.75	9654.2	9086	16	22.76	2862.7	28439
7	34.23	4305.6	26037	17	21.64	2721.5	15008
8	28.81	3623.4	15203	18	20.52	2580.7	23833
9	27.87	3505.0	18760				
10	25.56	3214.9	26117				

Solution and Discussion of Example 1

In a first instance, the experimental data are compared with the expected values (**Table 5.1**).

The compilation in **Table 5.1** shows that all the expected diagnostic signals (with their signal positions, intensities, and fine structures) are found in the four spectra. In the IR spectrum, only the absorptions registered at 1,697 (**Table 2.10**, p. 63) and 1,635 cm^{-1} (**Table 2.13**, p. 65), characteristic for α,β-unsaturated carbonyl compounds, are structurally relevant. Thus, these signals are reported together with the respective interpretations, while all other signals are simply listed. The bands observed at 2,957 and 2,858 cm^{-1} could indeed also be assigned, namely to the symmetric CH valence vibration of CH$_2$ groups (**Table 2.1**, p. 59). However, since CH$_2$ groups are almost ubiquitous for organic compounds, these signals are not diagnostic and are not further interpreted. A more thorough interpretation of the fingerprint region of the IR spectrum is not appropriate for the same reason. For compound **1** it is also not essential that all the signals in the aliphatic region of the NMR spectra are (or

can be) assigned to the respective nuclei of the molecule: the structure of compound **1** is still sufficiently secured without such a detailed interpretation. In the description of the spectra, the respective signals are globally interpreted as corresponding to "CH$_2$" groups. This is different from the description of signals that can be assigned unequivocally, which has to be more specific (see below). For the description of an electron ionization-MS (EI-MS) spectrum, it is sufficient to label only the obvious signals with their interpretations: for molecular ions and for ions that derive from common fragmentation reactions. For compound **1**, solely the signal at $m/z = 220$, corresponding to the $M^{+\cdot}$ ions, is such a signal. It certainly would be possible to propose fragments that could be responsible for other signals of the spectrum. However, without more in-depth MS investigations, their assignment would be pure speculation. Accordingly, the description of the analytical date of compound **1** could be:

Colorless oil, b.p. 162–163°C (4.7 mbar), n_D^{23} 1.5012. IR (film): 2,957s, 2,858m, 1,697s ($v_{(C=O)}$, enone), 1,635m ($v_{(C=O)}$, enone), 1,468m, 1,444m, 1,364m, 1,161w, 728w. ^{1}H NMR (500 MHz, CDCl$_3$): 2.51–2.49 (m, CH$_2$, cyclopentenone); 2.44 (t, $J = 7.5$,

Table 5.1 Comparison of expected and experimental analytical data for compound **1**

Method	Expected		Interpretation	Found	
IR	1,685–1,665s		$v_{(C=O)}$, enone	1,697s	
	1,640–1,590s		$v_{(C=C)}$, enone	1,635m	
	2,960–2,850s (2–3 bands)		v_{CH}, aliph. CH$_2$	2,957s, 2,858m	
	fingerprint:				
	1,470–1,430m		δ_{CH}, symm. CH$_2$	1,468m, 1,444m, 1,364m, 1,161w	
	$\approx$720w		δ_{CH}, rock. CH$_2$	728w	
^{1}H NMR	2.6–2.2	(2m, AA'XX', 4 H)	CH$_2$ α to CO	2.38–2.36	(m, 2 H)
			CH$_2$ β to CO cyclopentenone	2.51–2.49	(m, 2 H)
	2.5–2.2	(2t, 2 × 2 H)	all. CH$_2$ γ to CO	2.44	(t, $J = 7.5$, 2 H)
			all. CH$_2$ β to CO	2.24	(t, $J = 6.7$, 2 H)
	1.9–1.2	(m, 16 H)	8 aliph. CH$_2$	1.71–1.57	(m, 4 H)
				1.46–1.33	(m, 10 H)
				1.19–1.13	(m, 2 H)
^{13}C NMR	208	(s)	CO	210.7	(s)
	170	(s)	COC=C	174.7	(s)
	138	(s)	COC=C	139.9	(s)
	34	(t)	CH$_2$, α to CO	34.2	(t)
	28–20	(11t)	11 CH$_2$, cycle	28.8, 27.9, 25.6, 25.4, 25.1, 24.8, 24.4, 23.3, 22.8, 21.6, 20.5 (11t)	
EI-MS	220		$M^{+\cdot}$	220 (47)	
				177 (72), 164 (36), 163 (99), 150 (34), 149 (96), 137 (19), 136 (38), 135 (71), 123 (38), 122 (32), 121 (40), 111 (15), 110 (67), 109 (33), 108 (15), 107 (33), 105 (17), 95 (29), 93 (43), 91 (54), 81 (32), 80 (19), 79 (100), 78 (16), 77 (60), 67 (53), 65 (29), 55 (61), 53 (33)	

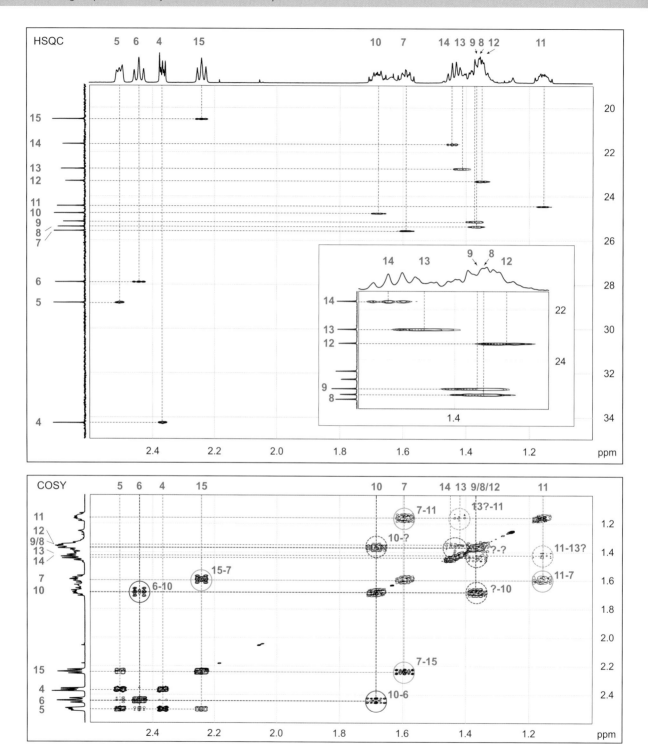

Fig. 5.2 HSQC (with expanded section) and COSY spectrum of compound **1**. The ^{13}C signals are numbered from low field to high field, and the ^{1}H signals are numbered according to the HSQC correlations.

all. CH$_2$, γ to CO); 2.38–2.36 (*m*, CH$_2$, cyclopentenone); 2.24 (*t*, *J* = 6.7, all. CH$_2$, β to CO); 1.71–1.57 (*m*, 2 CH$_2$); 1.46–1.33 (*m*, 5 CH$_2$); 1.19–1.13 (*m*, CH$_2$). ^{13}C NMR (125.7 MHz, CDCl$_3$): 210.7 (*s*, CO); 174.7 (*s*, COC=C); 139.9 (*s*, COC=C); 34.2 (*t*, CH$_2$, α to CO); 28.8, 27.9, 25.6, 25.4, 25.1, 24.8, 24.4, 23.3, 22.8, 21.6, 20.5 (11*t*, 11 CH$_2$, macrocycle). EI-MS: 220 (47, $M^{+\cdot}$), 177 (72), 164 (36), 163 (99), 150 (34), 149 (96), 137 (19), 136 (38), 135 (71), 123 (38), 122 (32), 121 (40), 111 (15), 110 (67), 109 (33), 108 (15), 107 (33), 105 (17), 95 (29), 93 (43), 91 (54), 81 (32), 80 (19), 79 (100), 78 (16), 77 (60), 67 (53), 65 (29), 55 (61), 53 (33). Anal. calc. for C$_{15}$H$_{24}$O (220.36): C 81.76, H 10.98; found: C 81.83, H 11.16.

Several approaches would be possible if it would be required to assign all the ^{1}H and ^{13}C NMR signals to the several CH$_2$ groups of the macrocycle. The most common method starts with the correlation of the directly connected ^{1}H and ^{13}C nuclei through an HSQC spectrum. In an ideal case, the ^{1}H spin systems can subsequently be perambulated by the identification of neighboring ^{1}H groups with a COSY spectrum. The interconnection of the separate spin systems, finally, can be achieved by means of HMBC correlations—maybe supplemented with information from NOESY/rotating-frame Overhauser spectroscopy (ROESY) spectra.

This procedure, however, is not successful in the case of compound **1**. While the correlation of the signals of directly bound ^{1}H and ^{13}C nuclei by the HSQC spectrum is still possible (**Fig. 5.2**), the complete assignment of the several neighboring CH$_2$ groups by means of the COSY spectrum is not feasible: starting from the two allylic methylene groups of the large ring, the connections C(15)–C(7)–C(11) (green) and C(6)–C(10) (red) can still be readily followed (**Fig. 5.2**), but additional correlations cannot be done unequivocally because the signals of the remaining protons overlap too strongly. Incredible natural-abundance double-quantum transfer experiment (INADEQUATE, p. 218) spectroscopy could provide the solution to this problem! However, because it relies upon the coupling of neighboring ^{13}C nuclei, this method is quite insensitive. It requires rather large amounts of sample, needs long measurement times, and is therefore not popular. The HSQC-total correlated spectroscopy (HSQC-TOCSY, p. 213) experiment can often be used as an alternative.

In fact, the complete assignment of the remaining NMR signals (**Fig. 5.3**) was finally possible by means of the HSQC-TOCSY spectrum of **1** (**Fig. 5.4**). It is not trivial to assign the correlations in this spectrum as well, but it is still possible to proceed from CH$_2$ group to CH$_2$ group of the complete spin system

due to the better resolution of the signals in the ^{13}C spectrum. The assignment can be completed with the correlation signals marked with green or red for the left- or right-hand portions of the molecule.

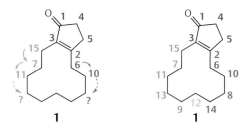

Fig. 5.3 Signal assignments by means of the COSY spectrum (*left*) and by interpretation of the HSQC-TOCSY spectrum (*right*).

Example 2

Compound **2** was synthesized from the corresponding α-mono-substituted bicyclic lactam by alkylation of its enolate with methyl iodide (**Scheme 5.1**).[2] The connectivities shown with structure **2** can be regarded as safe, likewise the absolute configuration of the stereogenic center of the aminoalcohol portion in the oxazolidine substructure, which is derived from L-valine. The configurations of the remaining stereogenic centers were postulated on the basis of "chemical experience."

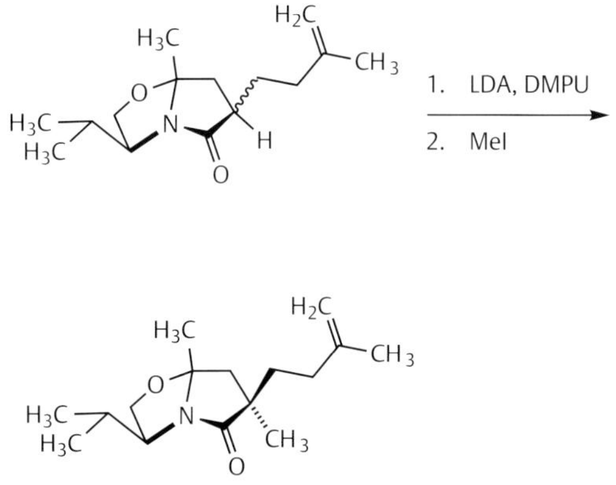

Scheme 5.1 Reaction scheme for the synthesis of compound **2**.

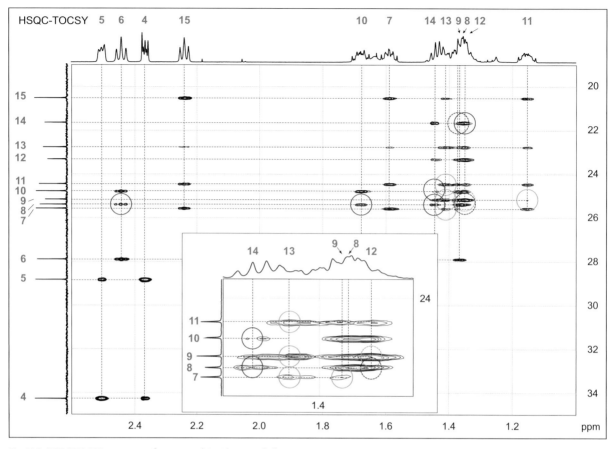

Fig. 5.4 HSQC-TOCSY spectrum of compound **1** with expanded section.

The standard analytical data of compound **2** are shown below. The spectra shall be interpreted and described as comprehensively as possible. In particular, the spatial structure of **2** shall be substantiated by the analytical data.

Solution and Discussion of Example 2

As exercised with the previous example, the experimental analytical data of compound **2** are compared with the data expected for the compound. To facilitate the assignments of the NMR signals, the C atoms of compounds with known connectivities are numbered—either according to the nomenclature rules of IUPAC or generically. We use for our discussion the generic numbering shown in **Table 5.2**.

As can be seen from the compilation in **Table 5.2**, all expected signals are found in the acquired spectra. Most of them are readily allocated to the respective structural units as well, except some—for instance the ^{13}C NMR signals of the five CH_3 groups—for which the assignments are not possible with the given data but also not required to secure the structure of the analyte. The chemical shifts of the NMR signals correspond in most cases quite well with the calculated values, except for some, where the correlation is less satisfactory. The latter, however, cannot be taken as an indication for an incorrect structural assignment because the increment system used by the ChemDraw program just delivers rudimentary estimates of chemical shifts. It is often not very reliable—particularly not in cases of rigid cyclic structures.

Example 2

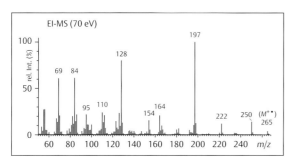

EI-MS (70 eV)

colorless oil
n_D 1.4784
$[\alpha]_D$ +16.7 (c = 0.92, CHCl$_3$)

$C_{16}H_{27}NO_2$ (265.40):   C 72.41, H 10.25, N 5.28
found: C 72.47, H 10.36, N 5.13

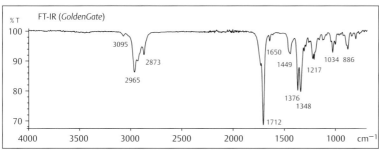

FT-IR (*GoldenGate*)

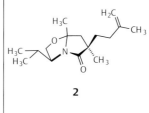

2

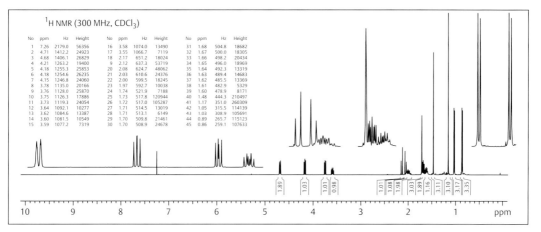

^{1}H NMR (300 MHz, CDCl$_3$)

No	ppm	Hz	Height	No	ppm	Hz	Height	No	ppm	Hz	Height
1	7.26	2179.0	56356	16	3.58	1074.0	13490	31	1.68	504.8	18682
2	4.71	1412.2	24923	17	3.55	1066.7	7119	32	1.67	500.0	18305
3	4.68	1406.1	26829	18	2.17	651.2	18024	33	1.66	498.2	20434
4	4.21	1263.2	19400	9	2.12	637.3	53719	34	1.65	496.0	18969
5	4.18	1255.3	25853	20	2.08	624.7	48062	35	1.64	492.3	13319
6	4.18	1254.6	26235	21	2.03	610.6	24376	36	1.63	489.4	14683
7	4.15	1246.8	24060	22	2.00	599.5	18245	37	1.62	485.5	13369
8	3.78	1135.0	20166	23	1.97	592.7	10038	38	1.61	482.9	5329
9	3.76	1128.0	25870	24	1.74	521.9	7188	39	1.60	478.9	8171
10	3.75	1126.3	17886	25	1.73	517.8	120944	40	1.48	444.3	210497
11	3.73	1119.3	24054	26	1.72	517.0	105287	41	1.17	351.0	260309
12	3.64	1092.1	10277	27	1.71	514.5	13019	42	1.05	315.5	114139
13	3.62	1084.6	13387	28	1.71	513.1	6149	43	1.03	308.9	105691
14	3.60	1081.5	10549	29	1.70	509.8	21461	44	0.89	265.7	115123
15	3.59	1077.2	7319	30	1.70	508.9	24678	45	0.86	259.1	107633

1.89 1.03 1.01 0.98   1.01 1.08 1.98 3.03 1.89 1.16 3.11 3.10 3.17 3.35

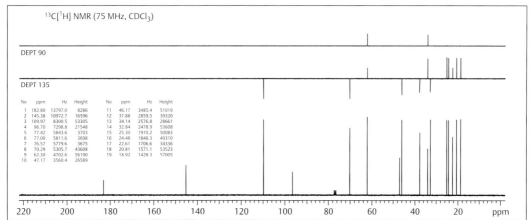

^{13}C[^{1}H] NMR (75 MHz, CDCl$_3$)

DEPT 90

DEPT 135

No	ppm	Hz	Height	No	ppm	Hz	Height
1	182.80	13797.0	8286	11	46.17	3485.4	51019
2	145.38	10972.7	16596	12	37.88	2859.5	39320
3	109.97	8300.5	53305	13	34.14	2576.8	28661
4	96.70	7298.8	21548	14	32.84	2478.9	53608
5	77.42	5843.6	3703	15	25.30	1910.2	50083
6	77.00	5811.6	3698	16	24.48	1848.3	49310
7	76.57	5779.6	3675	17	22.61	1706.6	34336
8	70.29	5305.7	43608	18	20.81	1571.1	53523
9	62.30	4702.6	56190	19	18.92	1428.3	57005
10	47.17	3560.4	26589				

Table 5.2 Comparison of expected and experimental analytical data for compound **2**. The chemical shifts of the ^1H and ^{13}C signals were predicted with the program ChemDraw Professional (version 18)

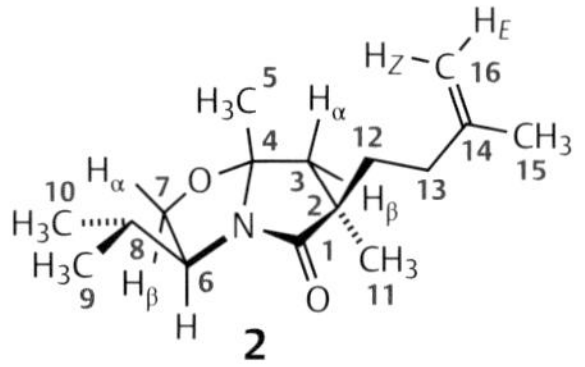

Method	Expected		Interpretation	Found	
IR	3,095–3,075*m*		$\nu_{(=CH_2)}$	3,095*w*	
	2,960–2,850*s* (2–3 bands)		$\nu_{(CH)}$, aliph. CH_2	2,965*s*, 2,873*m*	
	1,717*s*		$\nu_{(C=O)}$, lactam (5-membered)	1,712*s*	
	1,680–1,620*v*		$\nu_{(C=C)}$	1,650*w*	
	fingerprint:				
	1,470–1,430*m*		$\delta_{(CH)}$, symm. CH_2 and CH_3	1,449*m*	
				1,376*s*, 1,348*s*, 1,217*m*, 1,034*m*, 886*m*	
^1H NMR	4.72, 4.61	(2*d*, 2 × 1 H)	$C(16)H_2$	4.71 4.68	(br.s, 1 H) (br.s, 1 H)
	3.69, 3.44	(2*dd*, 2 × 1 H)	$C(7)H_2$	4.18 3.75	(*dd*, *J* = 8.6, 7.9, 1 H) (*dd*, *J* = 8.6, 7.0, 1 H)
	3.60	(*ddd*, 1 H)	C(6)H	3.59	(*dt*-like *m*, "*J*" = 10.5, 7.4, 1 H)
	2.22	(*m*, 1 H)	C(8)H	1.65–1.61	(*m*, 1 H)^a
	2.08 1.82	(2*m*, 2 × 1 H)	$C(3)H_2$	2.14 2.06	(A of AB, *J* = 14.0, 1 H) (B of AB, *J* = 14.0, 1 H)
	1.94	(*m*, 2 H)	$C(13)H_2$	2.03–1.95	(*m*, 2 H)
	1.79	(*s*, 3 H)	$C(15)H_3$	1.73	(br.s, 3 H)
	1.49	(*m*, 2 H)	$C(12)H_2$	1.71–1.66	(*m*, 2 H)^a
	1.50	(*s*, 3 H)	$C(5)H_3$	1.48	(*s*, 3 H)
	1.27	(*s*, 3 H)	$C(11)H_3$	1.17	(*s*, 3 H)
	0.88	(2*d*, 6 H)	$C(9)H_3$, $C(10)H_3$	1.04 0.87	(*d*, *J* = 6.6, 3 H) (*d*, *J* = 6.6, 3 H)
^{13}C NMR	176.1	(*s*)	C(1)	182.8	(*s*)
	145.8	(*s*)	C(14)	145.4	(*s*)
	110.6	(*t*)	$C(16)H_2$	110.0	(*t*)
	89.4	(*s*)	C(4)	96.7	(*s*)
	69.5	(*d*)	C(6)H	62.3	(*d*)
	66.9	(*t*)	$C(7)H_2$	70.3	(*t*)
	42.4	(*s*)	C(2)	47.2	(*s*)
	42.0	(*t*)	$C(3)H_2$	46.2	(*t*)
	36.7	(*t*)	$C(12)H_2$	37.9	(*t*)
	32.4	(*d*)	C(8)H	34.1	(*d*)
	33.8	(*t*)	$C(13)H_2$	32.8	(*t*)
	25.9–19.9	(5*q*)	5 CH_3	25.3 24.5 22.6 20.8 18.9	(*q*)^a (*q*)^a (*q*)^a (*q*)^a (*q*)^a
EI-MS			$M^{+\cdot}$	265 (4)	
			$[M - CH_3]^+$	250 (13)	
			$[M - CH_3 - CO]^+$	222 (12)	
			$[M - isoprene]^{+\cdot}$	197 (100)	
				166 (10), 164 (21), 154 (17), 128 (80), 126 (23), 123 (16), 112 (21), 111 (15), 110 (24), 100 (11), 97 (11), 96 (11), 95 (22), 86 (22), 85 (16), 84 (61), 83 (12), 82 (20), 70 (21), 69 (61), 68 (15), 67 (18), 56 (28), 55 (28).	

^{a}Assignment not safe.

The question that arises now is whether the conformity of the experimental and the expected data suffices to prove the proposed structure of the molecule in question. Can the acquired analytical data in fact substantiate the relative configurations assigned to the several stereogenic centers of compound **2**? This question has to be clearly negated! The IR, NMR, and MS spectra of the several stereoisomeric forms of **2** would be expected to be very similar, thus they cannot vouch specifically for any of the several stereoisomers. Additional information is needed to reveal and prove safely the stereochemical arrangements.

Such information is accessible from NOESY spectra (p. 170). The correlation signals highlighted in the NOESY spectra shown in **Fig. 5.5** allow the detailed interpretation of the 1D-¹H NMR spectrum. The signals encircled in red correspond to nuclei that are α-oriented, i.e., located above the bicyclic system, and the signals labeled in green are derived from protons that are

β-positioned. It can readily be deduced from these signals that the structure shown for compound **2** is stereochemically correct. For instance, the signals labeled in red along the highlighted axis at $\delta = 1.48$ reveal clearly that C(5)H$_3$, C(8)H, C(12)H$_2$, C(3)H$_\alpha$, and C(7)H$_\alpha$ are positioned on the same face of the bicyclic system. The signals marked in green along the axis at $\delta = 1.14$, on the other hand, show that C(11)H$_3$ and C(3)H$_\beta$ are located on the other side of this system.

Even though not required for the structural proof of compound **2**, the NOESY spectrum also allows to precisely assign the two doublets for the CH$_3$ groups of the isopropyl moiety and the two broadened singlets derived from the vinylic protons of the C(16)H$_2$ group: while the protons of C(9)H$_3$ solely show a contact to C(6)H, the protons of C(10)H$_3$ reveal also their proximity to C(7)H$_2$. The correlation signals of the vinylic protons, finally, show the spatial proximities of C(16)H$_E$ to C(15)H$_3$ and of C(16)H$_Z$ to C(12)H$_2$ and C(13)H$_2$, respectively.

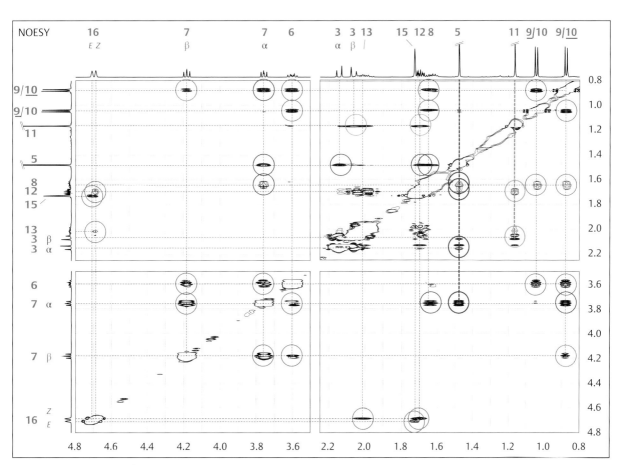

Fig. 5.5 NOESY spectrum of compound **2**.

5.3 Structure Elucidation of Allegedly Known Compounds and of Products Arising from Syntheses

The structure elucidation of compounds that arise from syntheses differs fundamentally from that of completely unknown samples. After all, the starting materials used in a chemical transformation are known, and their structural frameworks usually remain to a great extent unchanged in a reaction. Thus, reaction products contain in most cases structural units that are already known from the starting materials, and this often allows a rather simple data interpretation. Positions in a structure that got modified during a reaction are usually readily recognized by simple comparison of the spectra of starting materials and products. With some common sense and chemical intelligence, it is then possible to readily propose reasonable structures and to do straightforward signal assignments.

Let us consider the ¹H and ¹³C spectra shown in **Fig. 5.6** as an example: it is not trivial to decide solely on the basis of these two spectra which of the compounds **3** or **4** would be the analyte. The two compounds would be expected to give rise to very similar spectra with signals of the same multiplicities and comparable chemical shifts.

However, it becomes immediately evident that structure **3** correctly represents the analyte when it is known that the compound arose from the reaction of ethyl 4-hydroxybenzoate with methyl iodide in the presence of a base. The formation of compound **4** under such conditions would not be possible from a chemical point of view.

While the knowledge of starting materials, reagents, and the reactions to be expected can be very rewarding for the structural elucidation of a synthesis product and the related data interpretation, the associated anticipations might also lead to biased structural ideas that could become problematic. It is not too rare that a reaction does not lead to the desired or expected product. This is usually not a real problem when the structures of the anticipated and the obtained products differ significantly, and thus the analytical data unambiguously reveal that an unexpected product had formed. However, if the structures of an anticipated and the effectively formed product differ just slightly, erroneous interpretations might arise due to the just small differences in the analytical data that are expected for the

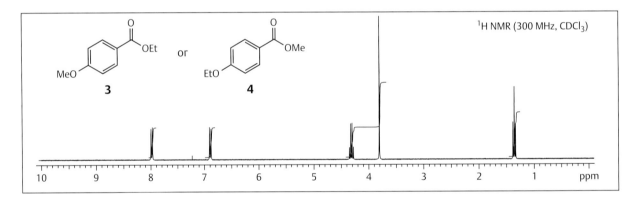

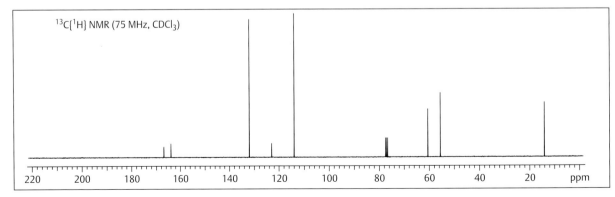

Fig. 5.6 ¹H and ¹³C NMR spectra of one of the two compounds **3** or **4**.

two alternatives. It is therefore always imperative to scrutinize analytical data with the greatest care and diligence, also when presumably known compounds such as reaction products are considered. It is in any case essential to identify all chemically possible structural variations of a product, to include these into analytical considerations, and to constantly question preliminary interpretations. This is in particular true when experimental and theoretical data diverge without a perceptive reason. In such cases, or certainly also when a dataset does not allow an unequivocal structural assignment, the acquisition of complementary analytical data is compulsory.

Example 3

Treatment of D-fructopyranose with 2,2-dimethoxypropene and acetone under acidic and dehydrating conditions delivers according to literature the bisacetonide **5** (**Scheme 5.2**).[3] A student of a chemical laboratory class followed the published procedure and obtained a product that delivered analytical data that are in fact compatible with the desired structure. However, the synthetic step to be performed subsequently, the oxidation of the secondary alcohol to the respective ketone, remained unsuccessful: oxidation occurred but a different product was formed. Thus, the question arose whether the student synthesized the correct compound. Which chemically conceivable alternative products could be formed by the reaction described above and how could a corresponding structure be analytically proven?

Solution and Discussion of Example 3

If the analytical data of Example 3 are interpreted on their own and not compared directly with the published data for compound **5**, there is no powerful argument to be found that would raise doubt that the desired product was really formed. All the signals to be expected for the fructopyranose framework are found in the ^{1}H and the ^{13}C NMR spectra below, as well as the signals that are required as proof for the formation of the two acetonides. The assignments of the ^{1}H signals are possible on the basis of their chemical shifts and the multiplicities to be expected for the several proton positions. Solely the signal structure of the AB system of the protons C(1)H$_2$ appears

somewhat unusual, because there exists no conceivable reason for a line broadening in the signal of the A portion of this system. Even though the ^{13}C signals cannot be assigned to their individual nuclei on the basis of the given data—except for C(2)—this appears not crucial because the spectrum shows all the signals that are expected for structure **5** with the required multiplicities in the appropriate spectral regions.

However, if the unexpected reactivity of the analyte is taken seriously and the formation of product **5** is questioned, it is readily recognized that D-fructopyranose could principally lead to two alternative bis-acetonides with β-configuration at the anomeric center, compounds **6** and **7**, and an additional three isomers, compounds **8**, **9**, and **10**, which are α-configured (**Fig. 5.7**). These six compounds are closely related and all of them could deliver the data of Example 3. Additional information is thus needed to allow an unequivocal structural assignment.

The ^{1}H NMR signals of the sample molecules can readily be correlated to the respective ^{13}C signals by means of an HSQC spectrum. This, however, does not directly solve the problem, and the respective spectrum is thus not shown here. The ^{1}H and ^{13}C NMR data allow only together with the HMBC spectrum, supplemented and supported by information gained from the NOESY spectrum, an unambiguous structural assignment. **Fig. 5.8** shows the HMBC correlations of the several nuclei of the two acetonides. They reveal that A, D, and F on the one hand and B, C, and E on the other hand belong to the same structural unit, respectively. A single HMBC signal, the cross-peak for C(5)H→C(A), indicates that the acetonide unit A-D-F corresponds to a 4,5-isopropylidene group. This is corroborated by the NOESY spectrum (see below). Thus, the structures **7** and **10** with their C(5)OH group can be excluded for the sample molecule. The assignment of the two acetonide groups cannot be finalized since no additional HMBC signal was found that would correlate any of the remaining ^{1}H signals of the fructose skeleton with C(A) or particularly with C(B).

Only the analysis of the NOESY spectrum allows finally the revelation of the complete structure of the sample molecule. The compound that delivered the data presented in Example 3 must be the bis-acetonide **6**, and its conformation must be predominantly the one shown in **Fig. 5.9**.

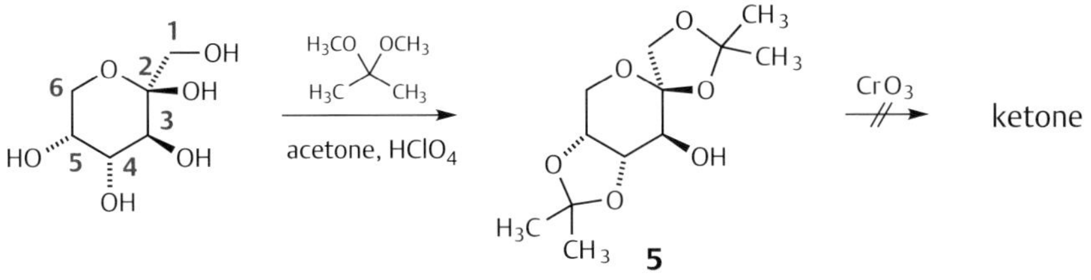

Scheme 5.2 Synthesis of 1,2:4,5-di-*O*-isopropylidene-β-D-fructopyranose (**5**).

Examples

Example 3

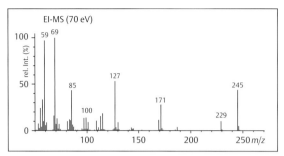

EI-MS (70 eV)

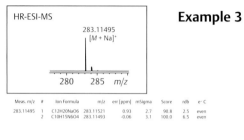

HR-ESI-MS

283.11495
$[M + Na]^+$

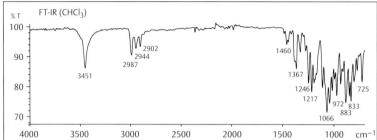

Meas. m/z	#	Ion Formula	m/z	err [ppm]	mSigma	Score	rdb	e⁻ C
283.11495	1	C12H20NaO6	283.11521	0.93	2.7	90.8	2.5	even
	2	C10H15N6O4	283.11493	-0.06	3.1	100.0	6.5	even

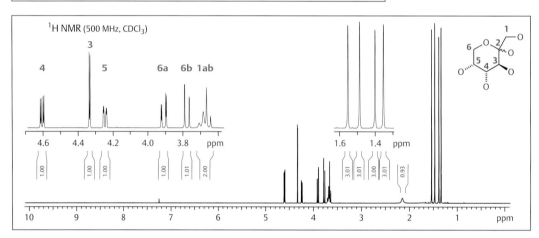

FT-IR (CHCl₃)

colorless needles
m.p. 96–97 °C (acetone/pentane 1:1)
$[\alpha]_D^{25}$ –38.1 (c = 1.7, acetone)

$C_{12}H_{20}O_6$ (260.29): C 55.37, H 7.75
found: C 55.42, H 7.83

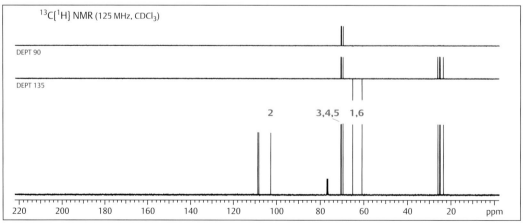

¹H NMR (500 MHz, CDCl₃)

¹³C[¹H] NMR (125 MHz, CDCl₃)

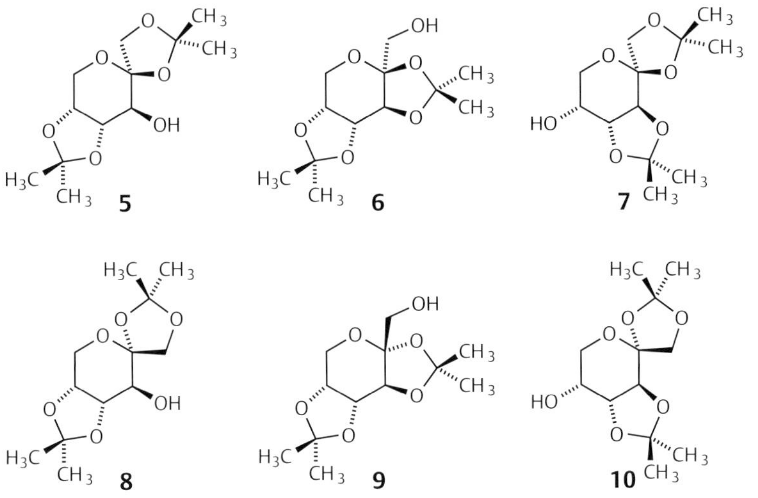

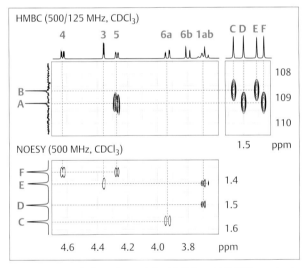

Fig. 5.7 Possible structures for the reaction product used as the sample in Example 3.

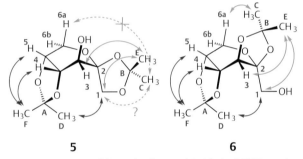

Fig. 5.9 Structures of the molecules **5** and **6** with the NOESY correlations that are (or would be) relevant for the structural proof indicated with *red* and *green* double arrows.

Fig. 5.8 Expanded sections of the HMBC and NOESY spectra to Example 3.

to C(1)H is not observed. This not only proves that compound **6** represents the sample molecule analyzed in Example 3 but also explains why the attempted CrO_3 oxidation of the student's experiment failed.

Example 4

In the course of the synthesis of $(+)\text{-}\Delta^{9(12)}$-capnellene, compound **2** (see Example 2) was ozonized and the resulting ketone **11** subsequently treated with ethylene glycol and an acid at a *Dean–Stark* water trap (**Scheme 5.3**).[2] In a first attempt, the mixture was heated to reflux for the prolonged period of a weekend. Under these conditions, the reaction delivered just a little of the desired acetal **12**, and an unknown compound **13** was formed as the main product instead. The structure of **13** shall be deduced from the analytical data shown below.

The NOESY contacts of the CH_3 group F to C(4)H and C(5)H confirm the assignment of the A-D-F unit to be a 4,5-isopropylidene group, and the correlation of the CH_3 group D with C(1)H$_2$ shows clearly that these three groups are positioned on the same face of the hexahydropyran structure. Hence, the two remaining β-anomers **8** and **9** can be excluded as the sample molecules. The NOESY cross-peaks for CH_3 (C)↔C(6)H$_a$ and for CH_3 (E)↔C(3)H, finally, are only compatible with structure **6** and not with structure **5**. While no NOE can be expected for an interaction CH_3 (C)↔C(6)H$_a$ in structure **5**, the obvious contact of CH_3 (C)

Examples

Example 4

colorless oil
$[\alpha]_D^{25} -75.0$ (c = 1.3, CHCl$_3$)

combustion analysis
found: C 72.18, H 9.33, N 5.41

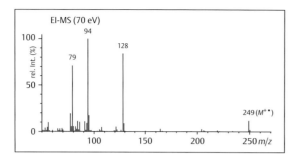

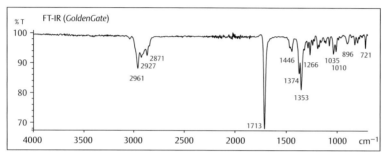

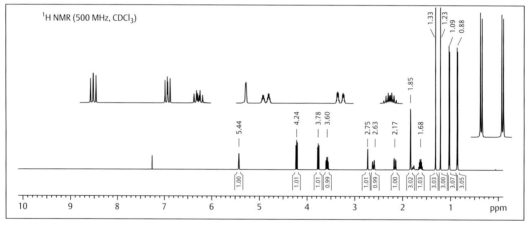

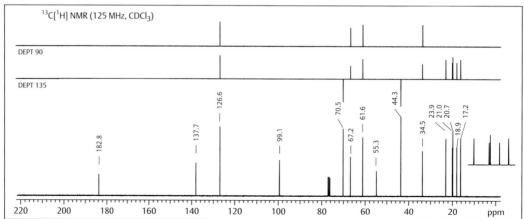

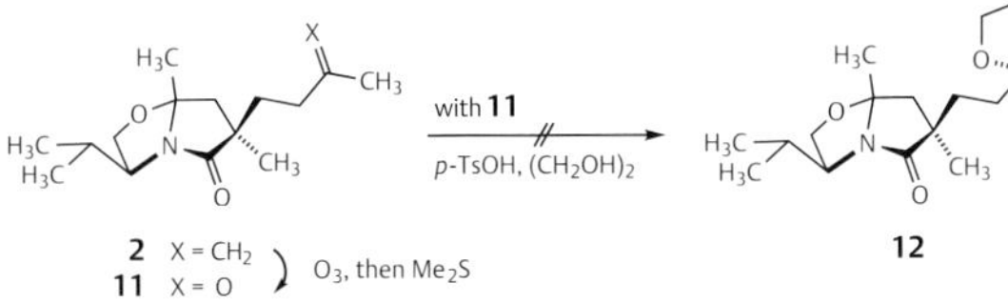

Scheme 5.3 Planned synthesis of acetal **12** starting from compound **2**.

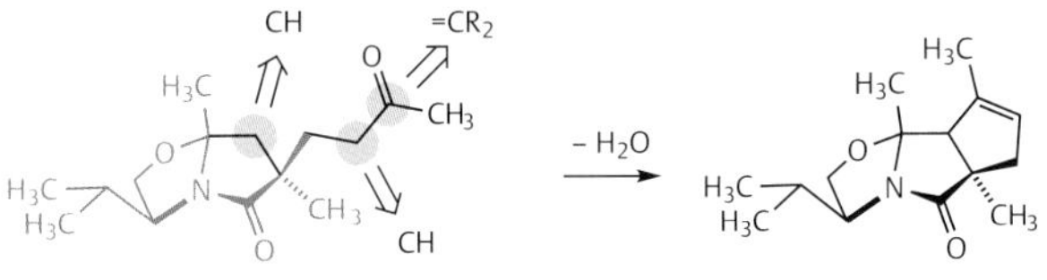

Scheme 5.4 Formation of compound **13** by intramolecular condensation of the ketone unit of compound **11**.

Solution and Discussion of Example 4

When the analytical data of compound **13** are compared with the data of compound **2** (Example 2), it is readily recognized that the better part of the original bicyclic lactam portion is retained in the product: the typical signals for the protons of the oxazolidine moiety are found in the ^{1}H NMR spectrum at $\delta = 4.24$, 3.78, and 3.60, together with the two doublets at $\delta = 1.09$ and 0.80 and the multiplet at $\delta = 1.68$ for the isopropyl group. The IR spectrum—in combination with the ^{13}C NMR spectrum—reveals additionally that compound **13** still contains a γ-lactam unit ($\nu_{C=O} = 1,713$ cm^{-1}, **Table 2.10**, p. 64; $\delta_{C=O} = 182.8$, **Table 3.33**, p. 185) and also a quaternary C atom in the α-position to the respective carboxyl group ($\delta = 99.1$). The compound still possesses five CH$_3$ groups as well, which are readily recognized with the respective signals in the ^{1}H and ^{13}C NMR spectra. One of these CH$_3$ groups is most probably vinylic as suggested by the chemical shift and the broadening of its signal ($\delta = 1.85$) in the ^{1}H NMR spectrum.

In contrast to the starting material **11**, the ^{13}C NMR spectrum of product **13** reveals just two (rather than three) CH$_2$ groups and four (rather than two) CH fragments. The protons of the CH$_2$ group that does not belong to the oxazolidine unit show in the ^{1}H NMR spectrum the signals of an AM(X$_n$) system at $\delta = 2.63$ and 2.17, and the protons of the two new CH fragments give rise to broadened singlets at $\delta = 5.44$ (vinylic H) and 2.75. Instead of the signal for the carbonyl C atom of the starting material **11**, we can find for **13** a signal at $\delta = 126.6$ in the ^{13}C NMR, which can be attributed to a quaternary olefinic C atom. Finally, the EI-MS, in combination with the elemental analysis, reveals C$_{15}$H$_{23}$NO$_2$ as the chemical formula of compound **13**, which means that compound **11** lost H$_2$O during its transformation into the unknown product.

Scheme 5.4 shows those structural units of the starting material **11** highlighted in blue that are found unchanged in product **13**. The units that got modified during the reaction, and thus were involved in the transformation, are labeled in orange. If this

and all aspects described above are taken into account, solely the structure shown in **Scheme 5.4** remains feasible for compound **13**. It can chemically be explained by an intramolecular condensation of the side-chain ketone with an intermediary formed enamide.[4]

Example 5

In the attempt to allylate 2-nitrocyclohexanone (**14**) by a Pd(0)-catalyzed reaction with carbonate ester **15** (**Schema 5.5**),[5] not the desired nitroketoester **16** but a compound **17** with an intensive and characteristic smell of lovage (*Levisticum officinale*) was formed. The reaction was performed, contrary to well-established literature procedures, in the presence of CH$_3$ONa in CH$_3$OH, probably to assure full deprotonation of the starting nitroketone. The quest is now to deduce the structure of the unexpected product **17** from its analytical data and to explain its formation.

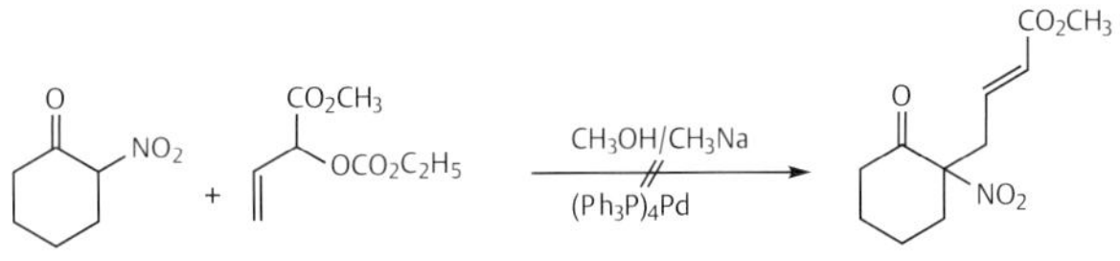

Scheme 5.5 Attempt of the Pd(0)-catalyzed allylation of 2-nitrocyclohexanone (**14**) with allylic carbonate **15**.

Solution and Discussion of Example 5

Already the superficial review of the analytical data of compound **17** revealed that the reaction in discussion proceeded fundamentally different than planned. In contrast to Example 4, where also an unexpected reaction was observed but the basic structure of the starting material was found to be unmodified in the product, no trace of the starting nitroketone **14** can be recognized in the spectra of **17**. The ^{13}C NMR spectra show the signal for merely a single CH_2 group, and the signals at m/z = 200 for the $M^{\cdot+}$ ions in the EI-MS and at m/z = 223 for the $[M + Na]^+$ ions in the electrospray ionization-MS (ESI-MS), respectively, indicate that the analyte most probably is nitrogen-free (nitrogen rule; p. 340). This raises the strong suspicion that compound **17** does not derive from a reaction of the presumed starting material **14** but rather from the allylation reagent **15**. However, since no direct relationship of **17** with **15** is recognizable, the structure of **17** first needs to be deduced as far as possible without taking starting materials and reagents used for its preparation into account.

It usually turns out to be very helpful to first determine for a completely unknown analyte its chemical formula and its functional groups. HR-ESI-MS reveals readily $C_9H_{12}O_5$ as the formula of **17**, which can likewise be deduced from the masses of the molecular ions $M^{\cdot+}$ (m/z = 200, EI-MS) or the ionized molecules $[M + H]^+$ and $[M + Na]^+$ (m/z = 201 and 223, ESI-MS) together with the elemental analysis (see Example 7, p. 452, for such a calculation). The compound, thus, possesses four double-bond equivalents (DBEs; p. 339).

The most important and obvious functional groups of **17** can be deduced from the IR spectrum in combination with the NMR spectra. The double band registered at 1,751 and 1,732 cm^{-1} most probably stands for two C=O stretching vibrations. With the relatively high frequencies, they could originate from esters or, in the case of the vibration at 1,751 cm^{-1}, maybe also from a γ-lactone (**Table 2.10**, p. 63). The strong IR bands at 1,100 and 1,200 cm^{-1}, typical for C–O valence vibrations of esters (**Table 2.19**, p. 67), and the two singlets observed in the ^{13}C NMR spectrum at δ = 168.8 and 168.9 (**Table 3.33**, p. 185) would support this proposal. The broad IR signal at 3,337 cm^{-1} indicates furthermore the presence of an OH group with a rather acidic or hydrogen-bonded H atom (**Table 2.4**, p. 59). The respective proton signal is found in the ^1H NMR spectrum at δ = 5.59 as a broad singlet.

The NMR spectra supplement the picture as follows. Two singlets and a triplet for three CH_3 groups and two multiplets, each integrating for one H atom, account in the ^1H NMR spectrum for the remaining protons. The singlet at δ = 3.79 would be typical for a CH_3 group of a methyl ester, and the singlet at δ = 1.93 could be assigned to a vinylic CH_3 group (**Table 3.13**, p. 139). The remaining three signals, the two multiplets centered at δ = 2.29 and 1.94 and the triplet at δ = 0.90 (J = 7.4 Hz), represent the response of an AMX_3 spin system. Since the compound does not possess any CH fragment, evidenced by the DEPT 90 spectrum, the AM-portion of this spin system must stem from diastereotopic protons of a CH_2 group.

Thus, compound **17** possesses an ethyl group and at least one stereogenic unit.

The ^{13}C NMR spectra support all the conclusions drawn above. In addition to the signals of the two carboxyl C atoms already mentioned, the spectra show the signals for three CH_3 groups (δ = 53.1 (OCH$_3$), 8.8, and 7.0) and for a CH_2 group (δ = 27.0), which is compatible with the information deduced from the ^1H NMR spectrum. The ^{13}C NMR spectrum reveals also three additional singlets that allow the completion of the chemical formula. Two of these signals, registered at δ = 138.0 and 128.9, can be attributed to quaternary olefinic C atoms and the final one, found at δ = 88.0, fits best to a quaternary sp^3-hybridized C atom that would be connected to an O atom.

The six structural units identified so far are shown in **Fig. 5.10**. If exotic structures are excluded, only the four structural variations **A** to **D** can be assembled from these six portions. Since we know the reaction conditions that led to compound **17**, we can exclude the two structures **C** and **D** by using our chemical intellect. While the two compounds **A** and **B** can be understood retrosynthetically as condensation products of two molecules of 2-oxobutanoic ester, which is closely related to compound **15**, no retrosynthesis to comparably likely synthons is possible for the compounds **C** and **D**. Thus, the remaining question is: Which of the structures **A** or **B** correctly represent compound **17**?

Answering this question is not trivial. The analytical data do not allow exclusion of one or the other structure nor do they offer an argument to prefer one over the other. Thus, additional information is needed.

Such information might arise from a NOESY and/or an HMBC spectrum. A contact of the OCH$_3$ group with the proximate CH_2 group in compound **A** could be expected to deliver a correlation signal in its NOESY spectrum, while a signal for the contact of the OCH$_3$ with the neighboring vinylic CH_3 group would be possible for compound **B**. In the HMBC spectrum, a common correlation of the protons of the OCH$_3$ and the CH_2 groups with one of the two carboxyl C atoms would provide clarification in favor of structure **A**, while correlation of the two groups to different carboxyl C atoms would vouch for **B** as the correct structure.

Both spectra were measured, but they did not provide the desired answer. No NOE correlation whatsoever was observed for the OCH$_3$ group, and the information offered by the HMBC spectrum was not conclusive. Although some correlations of ^1H NMR signals in the region of δ = 1.98 to 1.92 with the two ^{13}C NMR signals of the carboxyl C atoms at δ = 168.9 and 168.8 were found, the unequivocal assignment of the respective ^1H and ^{13}C nuclei and thus of the

Example 5

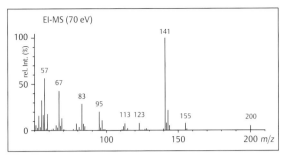

EI-MS (70 eV)

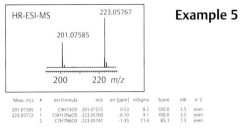

HR-ESI-MS

Meas. m/z	#	Ion Formula	m/z	err [ppm]	mSigma	Score	rdb	e⁻ C
201.07585	1	C9H13O5	201.07575	0.53	8.2	100.0	3.5	even
223.05772	1	C9H12NaO5	223.05769	-0.10	9.1	100.0	3.5	even
	2	C7H7N6O3	223.05741	-1.35	11.4	85.1	7.5	even

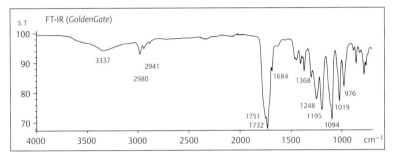

FT-IR (*GoldenGate*)

colorless crystals
m.p. 84–85 °C (Et₂O)

combustion analysis
found: C 53.59, H 5.88

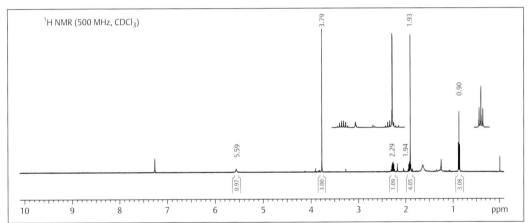

¹H NMR (500 MHz, CDCl₃)

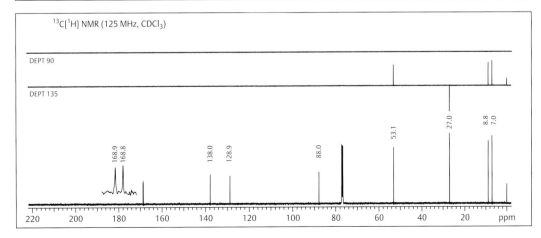

¹³C[¹H] NMR (125 MHz, CDCl₃)

Examples

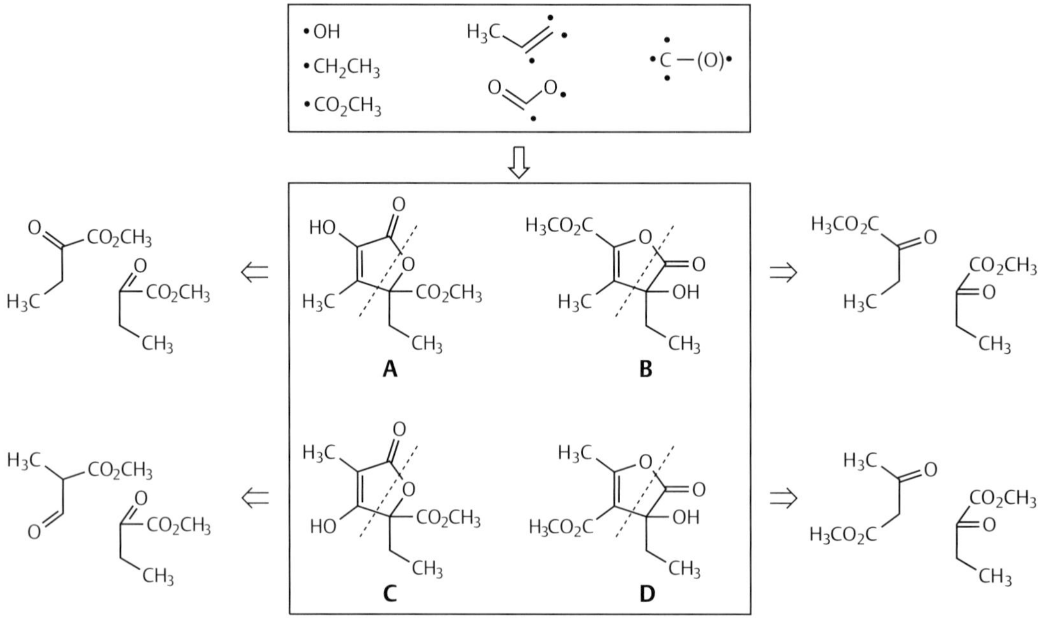

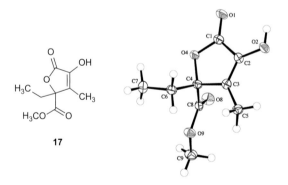

Fig. 5.10 Structural units that were deduced from the analytical data of compound **17**, the four structures **A–D** that can be reasonably assembled from them, and the conceivable retrosynthetic dissections of the latter.

actually correlating nuclei was not possible. The ^{1}H signals of the CH_2 and the vinylic CH_3 group overlap too strongly, and also the ^{13}C signals of the carboxyl C atoms are barely resolved (data not shown).

Fig. 5.11 Structure of compound **17** together with the Ortep plot of the single-crystal X-ray analysis.

X-ray analysis with a single crystal finally showed that compound **17** possesses structure **A** (**Fig. 5.11**). With this unambiguous clarification, compound **17** and some analogs were synthesized on purpose.[6]

5.4   Structure Elucidation of Completely Unknown Compounds

Some of the problems associated with the structure elucidation of fully unknown compounds were already addressed during the discussion of Example 5 above. Since in such cases no (or at best just a little preliminary) information about a compound is available, a two-step process is proposed to approach its structure. It is recommended to acquire in a first step by broad-brush analysis of all the available data a quick and general overview of the problem. It should be the goal at this stage to deduce the chemical formula of the compound as well as its most important and distinct functional groups and structural units. Focus has to be given just to the very obvious and unambiguous, not to produce prejudices that might impede the subsequent process. This sets a reliable basis for the detailed interpretation of the analytical data that has to follow and very often opens starting points for structural assignments. The following examples shall present some possible approaches and strategies that could allow the efficient disclosure of previously unknown structures.

Example 6

The sample of a compound **18** was found in the legacy of a PhD student. The structure of this compound shall be elucidated on the basis of the analytical data.

Solution and Discussion of Example 6

We would like to show with the rather simple Example 6 an organized and step-wise procedure that can be followed to elucidate the structure of an unknown compound.

At first the mass spectra—maybe in combination with an elemental analysis—shall be consulted to determine the chemical formula of an analyte. In the case of Example 6, compound **18** shows in the EI-MS the signal for the particle of highest mass at $m/z = 238$. This is the signal for the $M^{+\cdot}$ ion as can be readily recognized from the HR-ESI-MS. This spectrum reveals three signals at $m/z = 193$, 239, and 261. The latter two are characterized by a mass difference of 22 u, which is typically found for pairs of ionized molecules of the type $[M + H]^+$ and $[M + Na]^+$. Their m/z values confirm the EI-MS assignment of the $M^{+\cdot}$ peak. The HR-MS software proposes for the ionized molecules $[M + H]^+$ and $[M + Na]^+$ elemental compositions that substantiate the chemical formula $C_{13}H_{18}O_4$ for the sample molecule **18**. The signals at $m/z = 193$ found in both MS spectra correspond to identical fragment ions that arose from the $M^{+\cdot}$ or $[M + H]^+$ ions by the loss of an EtO$^{\cdot}$ or an EtOH portion, respectively. Since the MS data are conclusive enough, an elemental analysis of **18** is not required. The number of DBEs of **18** is 5, as can be calculated from the chemical formula (p. 339).

The IR spectrum of **18** exhibits a strong double absorption at 1,721 and 1,693 cm^{-1}, which can stand for two C=O groups (**Table 2.10**, p. 62ff), maybe also for an enol or phenol ether (**Table 2.13**, p. 65). The band at 1,251 cm^{-1} in the IR spectrum (**Table 2.19**, p. 67) and the singlet at $\delta = 165.9$ in the ^{13}C NMR spectrum (**Table 3.33**, p. 185) suggest the presence of an ester. The three signals at $\delta = 7.78$, 6.42, and 6.41 in the ^{1}H NMR

spectrum (**Table 3.24**, p. 149) and the six signals in the region of $\delta = 166$ to 100 in the ^{13}C NMR spectrum (**Table 3.43**, p. 194) additionally reveal clearly that compound **18** is a triple-substituted benzene derivative. The aromatic substructure together with the ester functionality fully account for the 5 DBEs of the compound.

To facilitate the detailed interpretation of the NMR spectra, it is advised to label their signals suitably. We recommend to number the ^{13}C NMR signals consecutively from the left to the right (from low to high field) and to add the multiplicities that are gained from the DEPT spectra. If an HSQC spectrum is available, the ^{1}H signals shall obtain the same numbers as the correlating ^{13}C signals, otherwise—as here in Example 6—they are independently labeled in alphabetic order (**Fig. 5.12**). Such a labeling of the spectra does usually not represent a problem, except when signals strongly overlap and consequently are difficult to differentiate. Such a case is given, e.g., in the ^{1}H NMR spectrum of compound **18**. The spectrum shows at $\delta = 4.01$ a line pattern that could be mistaken for a sextet if the spectrum would be examined just superficially. The observed lines, however, belong to two quartets that partially overlap. This can be recognized by the relative line intensities, which are not correct for a sextet, and also by the integral that accounts for four protons, which cannot be reasonably explained upon consideration of all NMR spectra. That the signal pattern at $\delta = 4.01$ is in fact the overlap of two quartets is nicely shown with the inserts in the spectrum

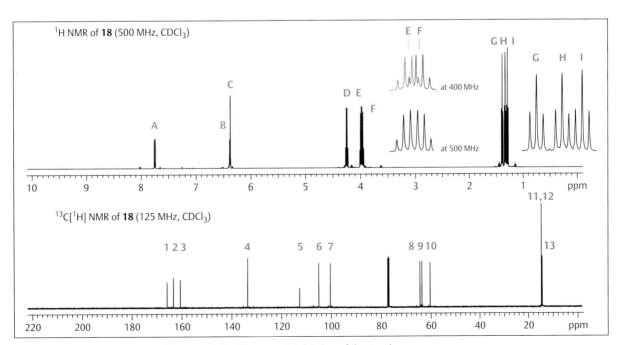

Fig. 5.12 ^{13}C and ^{1}H NMR spectra of compound **18** with appropriate labeling of the signals.

Example 6

colorless oil
b.p. 120 °C (1.3 mbar)

EI-MS (70 eV)

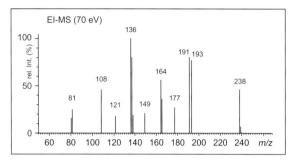

HR-ESI-MS

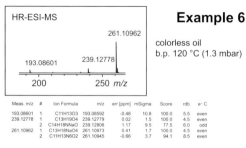

Meas. m/z	#	Ion Formula	m/z	err [ppm]	mSigma	Score	rdb	e⁻ C
193.08601	1	C11H13O3	193.08592	-0.48	10.8	100.0	5.5	even
239.12778	1	C13H19O4	239.12779	0.02	1.5	100.0	4.5	even
	2	C14H18NaO	239.12806	1.17	9.5	77.5	6.0	odd
261.10962	1	C13H18NaO4	261.10973	0.41	1.7	100.0	4.5	even
	2	C11H13N6O2	261.10945	-0.66	3.7	94.1	8.5	even

FT-IR (*GoldenGate*)

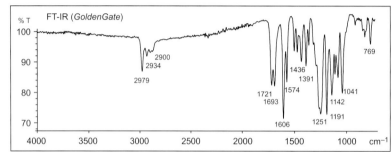

UV-Vis $c = 5.19 \cdot 10^{-3}$ g L^{-1} (EtOH)

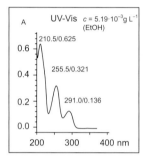

¹H NMR (500 MHz, CDCl₃)

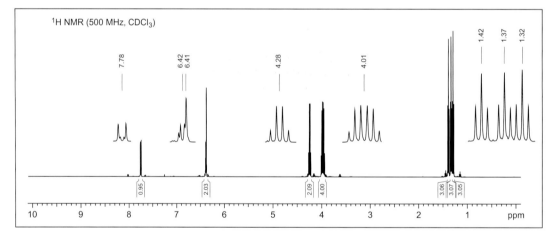

¹³C{¹H} NMR (125 MHz, CDCl₃)

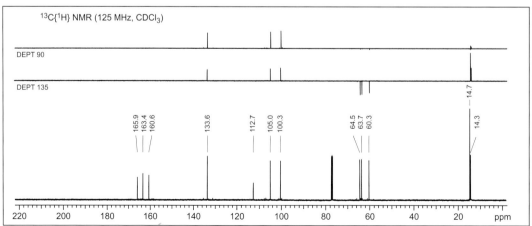

of **Fig. 5.12**. These inserts show the respective spectral region measured once at 500 MHz and once at 400 MHz.

If the ^{1}H NMR spectrum is scrutinized more profoundly, it is recognized that the signals A, B, and C show the pattern of an AXY spin system. The protons A and B must be positioned *ortho* in the aromatic system because A delivers a *d*-like *m* with "*J*" ≈ 8.5 Hz that finds its counterpart in signal B. Because signal C appears almost as a singlet, the respective proton can neither be positioned *ortho* to A nor to B. Thus, compound **18** must be a 1,2,4-trisubstituted benzene derivative. The respective ^{13}C NMR signals are found at δ = 163.4, 160.6, and 112.7 (three singlets) and at δ = 133.6, 105.0, and 100.3 (three doublets).

The ^{1}H NMR spectrum also provides some information about the substituents attached to the aromatic moiety. The three triplets G, H, and I, each integrating for three protons, correspond to three CH$_3$ groups that are each connected to a CH$_2$ group. The proton signals of the latter are found as the three quartets D, E, and F at δ = 4.28, 4.02, and 4.00. The high chemical shifts of the methylene signals suggest that the disclosed three ethyl groups have to be connected to O atoms. The ^{13}C NMR spectrum shows signals at δ = 64.5, 63.7, and 60.3 (three triplets) and at δ = 14.7 and 14.3 (two quartets, of which the first stands for two nuclei that absorb isochronously by accident), which are compatible with the three ethoxy groups suggested above. It finally becomes evident that the three substituents of the aromatic moiety must be two of these ethoxy groups and one carboethoxy unit. Taking the aromatic portion and the three ethoxy groups into account, just a C=O group is missing to complete the chemical formula of **18**, and the IR and the ^{13}C NMR spectra already revealed the presence of an ester group. The presence of a carboethoxy group is also supported by the signals observed at *m/z*

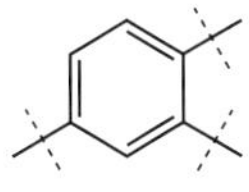

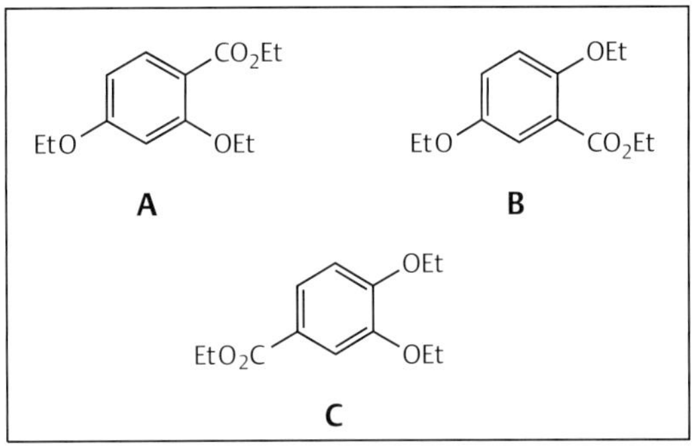

Fig. 5.13 Possible structures of compound **18** derived from direct spectroscopic evidence.

= 193 in the two MS spectra. They can be readily explained by oxonium ions that are typically formed from $M^{+\cdot}$ ions of ethyl esters by the loss of EtO$^{\cdot}$ by α-cleavage or from [*M* + H]$^+$ ions by the loss of EtOH. Thus, only the structures **A**, **B**, and **C** shown in **Fig. 5.13** can be proposed for compound **18**.

It is not possible to deduce the exact structure of compound **18** directly from the given spectral data. However, structure **A** can be safely assigned to compound **18** if the estimated chemical shifts of the signals of the aromatic protons and carbons are considered for the three alternative structures **A**, **B**, and **C**. Only the δ-values estimated for the signals of **A**, e.g., by increment calculations (**Table 5.3**), are well enough in agreement with the

Table 5.3 Expected chemical shifts of the ^{1}H and ^{13}C NMR signals of the aromatic structures **A–C** and experimental values determined for compound **18**

	Structure A		Structure B		Structure C		Sample 18	
	δ calc.[a]	δ calc.[b]	δ calc.[a]	δ calc.[b]	δ calc.[a]	δ calc.[b]	δ exp.	Signal
C(a)	112.3	107.4	123.0	116.4	122.3	123.3	112.7	C(5)
C(b)	164.3	165.2	156.0	151.7	155.0	149.3	163.4	C(2)
C(c)	161.0	161.7	149.5	152.7	149.2	144.8	160.6	C(3)
C(d)	131.1	131.3	119.3	118.9	122.1	122.3	133.6	C(4)
C(e)	105.9	105.4	116.1	115.4	113.6	115.4	105.0	C(6)
C(f)	101.5	98.5	114.9	114.4	111.4	114.4	100.3	C(7)
H(d)	8.01	7.77	7.20	6.91	7.62	7.44	7.78	H(A)
H(e)	6.72	6.48	7.51	7.41	7.44	7.41	6.42	H(B)
H(f)	6.66	6.45	7.05	6.81	7.09	6.81	6.41	H(C)

[a]Calculated with *ChemDraw Pro 18*.
[b]Calculated with the increments provided in this book (**Table 3.24**, p. 149, and **Table 3.43**, p. 194).

Examples

experimental data obtained for **18**. The congruence of the data is so distinct that the ^{13}C signals can be assigned readily and with high reliability as shown in the compilation of **Table 5.3**. This is not the case for the ^{1}H NMR signals. The observed signal distribution still matches best with structure **A**, but the assignment of the two signals at $\delta = 6.42$ and 6.41 to the nuclei H(e) and H(f), respectively, as indicated in the table, cannot be done on the basis of the chemical-shift estimations alone. However, it can be deduced from the signal multiplicities.

Example 7

Sometimes, the structural elucidation of even relatively small molecules is not trivial. A high school student was studying some antidepressant drugs and their metabolites for her final graduation project. She synthesized compound **19**, which she used as a reference substance in her investigation, and the analytical data of this material are shown below. Can you determine the structure of **19** on the basis of the analytical data alone, without any information about its synthesis?

Solution and Discussion of Example 7

First, the chemical formula of compound **19** shall be deduced from the given information. The ESI-MS suggests approximately 323 g mol^{-1} for the molecular mass of the compound, if it is assumed that the signal of highest mass, recorded at $m/z = 324$, is the response to $[M + H]^+$ ions (alternatively, also $[M + Na]^+$ ions have to be considered). With the experimentally determined proportions of 66.93% C, 6.15% H, and 4.88% N, the partial formula with respect to these elements can be deduced as $C_{18}H_{20}N$ (nominal 250 u):

$$\text{Number of C: } \frac{323 \times 0.6693}{12} = 18.0$$

$$\text{Number of H: } \frac{323 \times 0.0615}{1} = 19.8$$

$$\text{Number of N: } \frac{323 \times 0.0488}{14} = 1.12$$

Thus, 73 u remain to complete the molecular mass. These 73 u, however, cannot be accounted for by just adding, as is often the case, some O atoms to the formula. The compound evidently comprises yet at least one other type of heteroatom. The nature and number of the remaining elements is revealed by the HR-MS spectrum (**Chapter 4.4.2**, p. 341ff). The MS software proposes for the ions registered at $m/z = 324.15653$ six chemical formulas. However, the numbers of C atoms of only two of them are consistent with the elemental analysis, namely those of proposals 2 and 3. These two comprise very different chemical formulas—one with F atoms and the other without—and both are comparable with respect to accuracy and mSigma

value. Solely the content of hydrogen, which is somewhat too high for proposal 3, lets proposal 2 to appear a little more likely.

The NMR spectra can be used to assess whether or not the prioritized formula $C_{18}H_{20}F_3NO$ for compound **19** is correct. If compound **19** would in fact contain F atoms, these would unavoidably be attached to C atoms and, because ^{19}F nuclei possess a nuclear spin, would show ^{19}F–^{13}C coupling. Indeed, the ^{13}C NMR spectrum of **19** shows some signals that are split due to coupling with F nuclei, thus, proving that $C_{18}H_{20}F_3NO$ is the correct chemical formula of the compound. It can be recognized in the spectral expansion shown in **Fig. 5.14** that the low-intensity lines registered in the region of $\delta = 121$ to 128 represent in fact two quartets: signal 7 at $\delta = 124.4$ with $J = 271$ Hz (highlighted in green) and signal 8 at $\delta = 122.7$ with $J = 33$ Hz (highlighted in red). At closer inspection also signal 5 turns out to be a quartet ($\delta = 126.7, J = 4$ Hz). The three ^{13}C NMR signals, thus, indicate the presence of a CF_3 group (signal 7) that is attached to an aromatic moiety (signals 8 and 5; **Table 3.48**, p. 198).

It is worthwhile to mention and to emphasize at this point how highly valuable the knowledge of the chemical formula of an unknown compound is to successfully initiate its structural elucidation. If the chemical formula of the sample compound **19** would not be known and the ^{13}C signals in its NMR spectrum were to be numbered consecutively, the problem would arise how to deal with the low-intensity signals in the region of $\delta = 121$ to 128. One might be inclined to ignore these signals as deriving from impurities, which could direct you onto a wrong track and, in the worst case, to erroneous deductions.

The preliminary considerations are concluded with the cursory interpretation of the IR spectrum. Only a few characteristic bands are observed there. The two signals at 1,614 and 1,517 cm^{-1} can account for the $\nu_{(C=C)}$ of the aromatic group that already had been identified by NMR in connection with the CF_3 group. The strong absorption to be expected for a CF_3 group in the region of 1,365 to 1,120 cm^{-1} could be present. However, it is not possible to definitely assign it to one of the several bands observed in the respective region of the spectrum.

After having acquired a rough overview of the major structural features of compound **19**, the NMR spectra of the analyte are scrutinized in detail to do the actual structural elucidation. The ^{1}H NMR spectrum of the sample shows signals in three regions. Four signals (A–D, Fig. **5.14**), integrating for an overall of nine protons, are found in the region of $\delta \approx 7.0$ to 7.5. They can be the responses of the protons of two aromatic groups: A/D ($2d$ with $J = 8.5$ Hz, each 2 H) for the protons of a *para*-disubstituted aromatic moiety that possesses an electron-withdrawing as well as an electron-donating substituent and B/C (m, 4 + 1 H) for the protons of a phenyl group connected to a substituent that neither exhibits distinct donor nor acceptor properties

Example 7

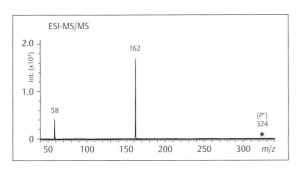

ESI-MS/MS

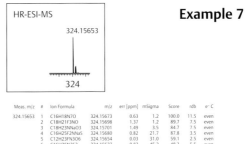

HR-ESI-MS

324.15653

324

Meas. m/z	#	Ion Formula	m/z	err [ppm]	mSigma	Score	rdb	e⁻ C
324.15653	1	C16H18N7O	324.15673	0.63	1.2	100.0	11.5	even
	2	C18H21F3NO	324.15698	1.37	1.2	89.7	7.5	even
	3	C18H23NNaO3	324.15701	1.49	3.5	84.7	7.5	even
	4	C16H25F2NNaS	324.15680	0.82	21.7	87.8	3.5	even
	5	C12H25F3N3O6	324.15654	0.03	31.0	59.1	2.5	even
	6	C16H26N3S2	324.15627	-0.82	45.2	48.3	5.5	even

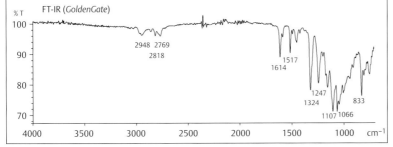

FT-IR (*GoldenGate*)

colorless oil
R_f = 0.3 (CH₂Cl₂/MeOH 10:1)

combustion analysis
found: C 66.93, H 6.15, N 4.88

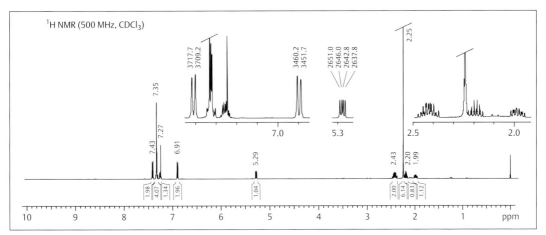

¹H NMR (500 MHz, CDCl₃)

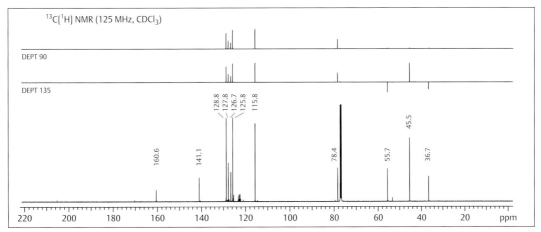

¹³C{¹H} NMR (125 MHz, CDCl₃)

DEPT 90

DEPT 135

Examples

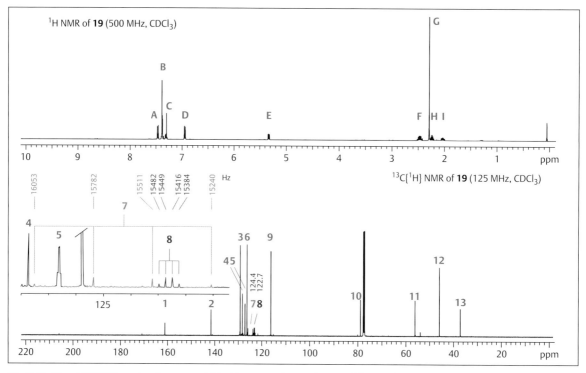

Fig. 5.14 ¹H and ¹³C NMR spectra of compound **19** with consecutive numbering of the signals.

(**Table 3.24**, p. 149). This assessment is corroborated by the ¹³C NMR spectrum, which exhibits in the aromatic region (**Table 3.43**, p. 194) the signals 1 and 2 for two quaternary C atoms, the signals 3, 4, 5, 6, and 9 for five CH groups, and signal 8 for a further quaternary C atom that is bound to a CF₃ group. The signals 3, 6, and 9 show twice the intensity of signal 4 and, thus, stand each for two symmetry-related C atoms. Signal 5 is split into an incompletely resolved quartet, most probably due to coupling with the ¹⁹F nuclei of an *ortho*-positioned CF₃ group, and presumably is also the response of two chemically equivalent C atoms. In addition, signal 1 can readily be assigned to an O-substituted aromatic C atom due to its chemical shift of δ = 160.6. It evidently belongs to the *para*-disubstituted aromatic moiety that also holds the two quite differently shielded protons H(A) and H(D). Thus, the two partial structures **A** and **B** (**Fig. 5.15**) can be recognized by now.

The second region of the ¹H NMR spectrum exhibits a single signal E at δ = 5.29, a doublet × doublet with *J* = 8.2 and 5.0 Hz, integrating for 1 H. This signal belongs to a strongly deshielded proton of a CH fragment, with the respective carbon nucleus absorbing at δ = 78.4 in the ¹³C NMR spectrum (signal 10). The chemical shifts of the ¹H as well as of the ¹³C signal indicate that the CH group is benzylic and O-substituted. It represents obviously a stereogenic center, and this renders the protons of the connected CH₂ diastereotopic and accounts well for the

Fig. 5.15 Structural elements and their assembly to compound **19**.

observed multiplicity of signal E. The proton signals of the CH₂ group connected to C(E)H are found within the third region of the ¹H NMR spectrum with the multiplets H and I at δ = 2.20

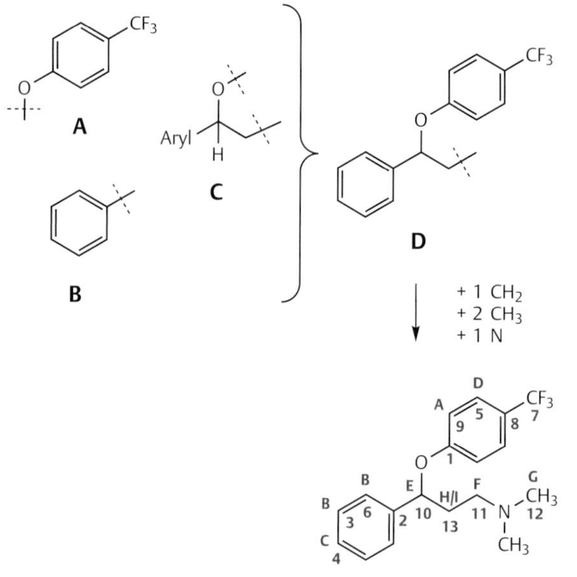

and 1.99. The corresponding methylene ^{13}C NMR signal is either signal 11 or 13. The scrutinized data thus reveal the partial structure **C**, which can be assembled with the previously revealed fragments **A** and **B** to the moiety **D**.

To finalize the structure of compound **19**, the remaining protons and carbons correlated to the signals F/11 or F/13 (a CH_2 group) and G/12 (two chemically equivalent CH_3 groups) and an N atom need to be accommodated. The only possible way to link all these groups to the previously deduced fragment **D** is to elongate the latter with a $CH_2N(CH_3)_2$ group, for which the signal at $m/z = 58$ in the MS/MS might have been indicative. The resulting structure of compound **19** is shown in **Fig. 5.15**; the assignment of the several signals was done on the basis of correlation tables and incremental calculations.

Example 8

A methanolic extract of the root bark of *Capparis decidua* (Forssk.) Edgew. was separated chromatographically and delivered—among others—a compound **20**.[7] The product was taken up in MeOH, treated with MeOH/HCl, and the solvent was evaporated to deliver the sample that was used to acquire the analytical data given below. Compound **20** shows on the thin layer chromatogram upon staining with an iodoplatinate solution (Schlittler reagent[8]) a dark bluish-brown and with a $FeCl_3$ solution a reddish-brown coloring, which indicates the presence of an amine and a phenol/enol, respectively. The structure of compound **20** shall be deduced on the basis of the provided analytical data.

Solution and Discussion of Example 8

The initial analysis of the data gives the following picture: compound **20** possesses the chemical formula $C_{25}H_{29}N_3O_4$, which can be readily deduced from the HR-ESI-MS in combination with the overall inspection of the remaining data. The compound is an amine (reaction with iodoplatinate) that has to be present as a hydrochloride salt due to the treatment with MeOH/HCl. The respective N–H valence vibrations are found in the IR spectrum at 2,742 and 2,661 cm^{-1} (**Table 2.6**, p. 60). The spectrum additionally suggests the presence of N-monosubstituted amides with the bands at 3,413, 3,232, and 3,065 cm^{-1}, which are typical absorptions for amide N–H groups (**Table 2.5**, p. 60), and at 1,651 and 1,550 cm^{-1} for C=O valence vibrations (amides I and II; **Table 2.10**, p. 64). The presence of amides is supported by the ^{13}C NMR spectrum. It shows two signals at $\delta \approx 165$ for two carboxyl C atoms that are shifted to higher field, probably due to conjugation of the C=O groups with an alkene or aromatic moiety (**Table 3.33**, p. 185). Seven additional signals for methylene C atoms are found in the aliphatic region of the ^{13}C NMR spectrum ($\delta = 22$–46) and seven singlets and seven doublets (two with doubled intensities) in the aromatic

and unsaturated region—broadly distributed over the range of $\delta = 110$ to 156. Thus, the ^{13}C NMR spectrum well accounts with its 23 signals for the 25 C atoms of the compound. Likewise, the ^{1}H NMR spectrum is fully in agreement with the data discussed so far. All the proton signals that have to be expected on the basis of the ^{13}C signals described above are found in the spectrum (one being covered by the signal of H_2O), and the spectrum also reveals four additional signals that evidently stand for five protons bonded to heteroatoms. This is confirmed in the HSQC spectrum.

The spectra above are labeled according to the rules that were already proposed in the discussion of Example 6. The ^{13}C signals were numbered from the left to the right, and the ^{1}H signals obtained the respective numbers of the carbons correlating with them in the HSQC spectrum. The four remaining ^{1}H NMR signals that do not show HSQC correlation with any of the ^{13}C signals are labeled with A, B, C, and D. The respective protons are evidently bonded to heteroatoms.

After this initial labeling procedure, it is advisable to collect all the NMR information from the several 2D-NMR spectra and summarize them within a table. This was done for compound **20**, and the information is found in **Table 5.4**. It is crucial when doing such a compilation that all signals and correlations—also ambiguous ones—are enlisted in the table. Of course, ambiguities have to be emphasized appropriately. The actual interpretation may start only after all the data are collected and suitably arranged.

The detailed data interpretation is best initiated by the identification of the individual ^{1}H spin systems and the deduction of the related structural units. For this purpose, first and foremost information from the COSY (or, when available also the TOCSY) spectrum is used. In the case of compound **20**, the analysis of the COSY spectrum reveals readily that the compound possesses six spin systems for the C-bound protons.

Two AX systems can be identified for H(6)/H(14) and H(7)/H(12), respectively. They belong to the protons of two 1,2-disubstituted alkene units, which can be deduced on the basis of the chemical shifts of the ^{1}H and of the related ^{13}C signals. Both double bonds must be (E)-configured to account for the large vicinal coupling constants of $J = 15.5$ and 15.7 Hz, respectively. The HMBC spectrum reveals then that the two alkene units are each connected to a carboxyl group and an aromatic quaternary carbon atom: the C(6)/C(14) unit with C(2)=O and C(10), and the C(7)/C(12) unit with C(1)=O and C(8).

The signals of the four aromatic protons H(9)$_2$/H(13)$_2$ reveal an (AX)$_2$ or an AA'XX' spin system. They most probably are the responses of the protons of a *para*-disubstituted aromatic system, because only such a system provides the required local symmetry. This interpretation is corroborated by the HMBC spectrum, which shows that the aromatic portion is completed

Example 8

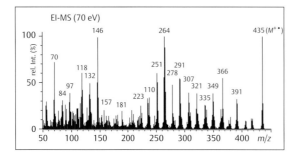

EI-MS (70 eV)

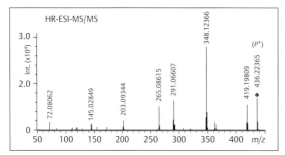

HR-ESI-MS/MS

colorless, amorphous solid, obtained
from MeOH/HCl by evaporation of the solvent
m.p. 246–247 °C (decomp.)

HR-ESI-MS/MS Report

Meas. m/z	#	Ion Formula	m/z	err [ppm]	mSigma	Score	rdb	e⁻ C
265.08615	1	C17H13O3	265.08592	-0.86	6.1	100.0	11.5	even
291.06607	1	C18H11O4	291.06519	-3.04	94.7	96.4	13.5	even
	2	C19H7N4	291.06552	1.55	98.6	100.0	18.5	even
348.12366	1	C21H18NO4	348.12303	-1.81	8.7	100.0	13.5	even
	2	C22H14N5	348.12437	2.03	22.1	73.5	18.5	even
419.19809	1	C25H27N2O4	419.19653	-3.70	15.1	57.6	13.5	even
	2	C26H23N6	419.19787	-0.51	23.8	100.0	18.5	even
436.22365	1	C25H30N3O4	436.22308	-1.30	12.6	100.0	12.5	even
	2	C26H26N7	436.22442	1.77	23.9	71.0	17.5	even

FT-IR (*GoldenGate*)

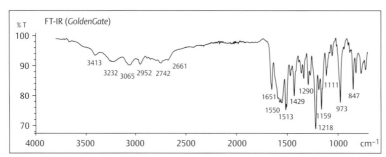

UV-Vis

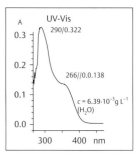

¹H NMR (600 MHz, DMSO-d₆)

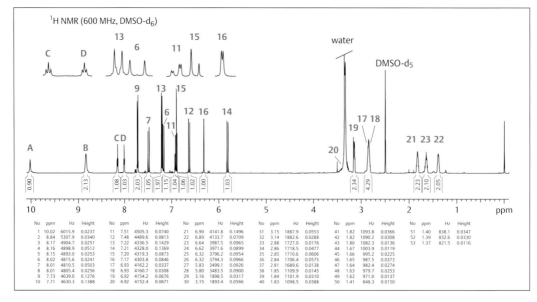

No	ppm	Hz	Height	No	ppm	Hz	Height	No	ppm	Hz	Height	No	ppm	Hz	Height	No	ppm	Hz	Height	No	ppm	Hz	Height
1	10.02	6015.9	0.0237	11	7.51	4505.3	0.0740	21	6.90	4141.8	0.1496	31	3.15	1887.9	0.0553	41	1.82	1093.8	0.0366	51	1.40	838.1	0.0347
2	8.84	5307.9	0.0340	12	7.48	4489.6	0.0813	22	6.89	4133.7	0.0709	32	3.14	1882.6	0.0288	42	1.82	1090.2	0.0308	52	1.39	832.6	0.0330
3	8.17	4904.7	0.0251	13	7.22	4336.5	0.1429	23	6.64	3987.5	0.0965	33	2.88	1727.0	0.0176	43	1.80	1082.3	0.0136	53	1.37	821.5	0.0116
4	8.16	4898.9	0.0512	14	7.21	4328.0	0.1369	24	6.62	3971.6	0.0899	34	2.86	1718.5	0.0477	44	1.67	1003.9	0.0119				
5	8.15	4893.0	0.0253	15	7.20	4319.3	0.0873	25	6.32	3796.2	0.0954	35	2.85	1710.6	0.0606	45	1.66	995.2	0.0225				
6	8.02	4815.6	0.0241	16	7.17	4303.8	0.0846	26	6.32	3794.3	0.0966	36	2.84	1706.4	0.0575	46	1.65	987.5	0.0373				
7	8.01	4810.5	0.0503	17	6.93	4162.2	0.0337	27	5.83	3499.1	0.0926	37	2.81	1689.6	0.0138	47	1.64	982.4	0.0274				
8	8.01	4805.4	0.0256	18	6.93	4160.7	0.0308	28	5.80	3483.5	0.0900	38	1.85	1109.9	0.0145	48	1.63	979.7	0.0253				
9	7.73	4639.0	0.1276	19	6.92	4154.2	0.0676	29	3.16	1898.5	0.0317	39	1.84	1101.9	0.0310	49	1.62	971.0	0.0137				
10	7.71	4630.3	0.1388	20	6.92	4152.4	0.0671	30	3.15	1893.4	0.0566	40	1.83	1098.5	0.0388	50	1.41	848.3	0.0150				

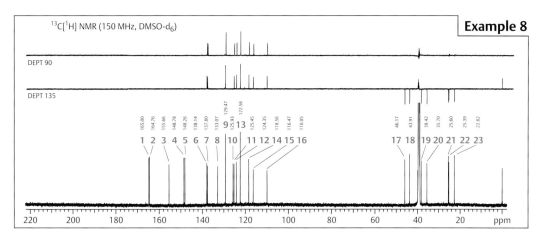

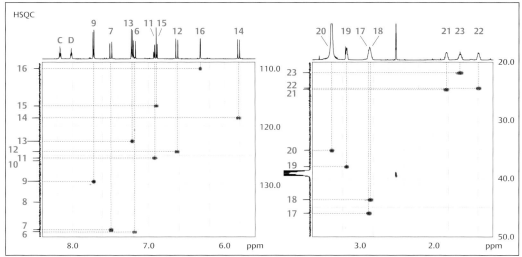

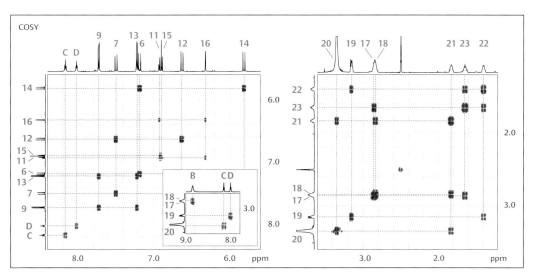

Example 8

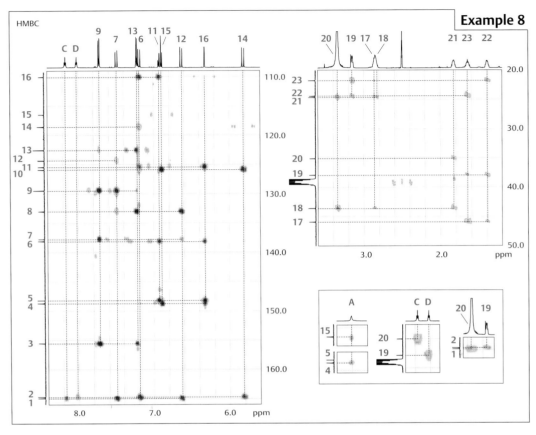

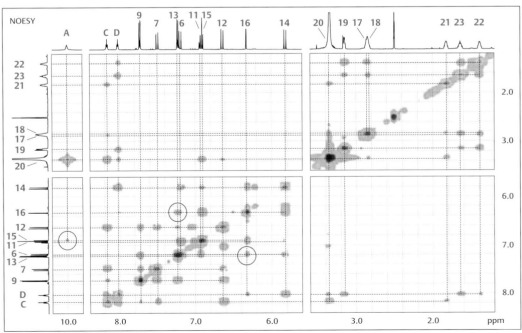

Table 5.4 Compilation of the NMR data and 2D-NMR correlations of compound **20**

No.	^{13}C	HMBC (C→H)	^{1}H		HMBC (H→C)	COSY	NOESY
1	165.0 (s)	7, 12, 20, C	—	—	—	—	—
2	164.8 (s)	6, 14, 19, D	—	—	—	—	—
3	155.6 (s)	9, 13	—	—	—	—	—
4	148.8 (s)	15, 16?, A	—	—	—	—	—
5	148.3 (s)	11, 16	—	—	—	—	—
6	138.1 (d)	11, 16	7.19	(d, J = 15.5 Hz, 1 H)	2, 10?, 11, 14, 16	14	11, 14, 16?
7	137.8 (d)	9, 12w	7.50	(d, J = 15.7 Hz, 1 H)	1, 8, 9, 12	12	9, 12
8	133.1 (s)	7, 12, 13	—	—	—	—	—
9	129.5 (d)	7, 9, 13w	7.72	7.72 (d, J = 8.7 Hz, 2 H)	3, 7, 9, 13	13	7, 12, 13, Cw
10	125.9 (s)	6?, 11?, 14, 15	—	—	—	—	—
11	125.5 (d)	6, 16	6.93	(dd, J = 8.3, 1.9 Hz, 1 H)	5, 6, 10?, 16	15, 16w	6, 14?
12	124.4 (d)	7	6.63	(d, J = 15.9 Hz, 1 H)	1, 7w, 8	7	7, 9, 20w?, C
13	122.6 (d)	9, 13	7.22	(d, J = 8.5 Hz, 2 H)	3, 8, 9w, 13	9	9, 12w?, 14?, 16
14	118.6 (d)	6	5.82	(d, J = 15.7 Hz, 1 H)	2, 10	6	6, 11, 16, D
15	116.5 (d)	A	6.90	(d, J = 8.1 Hz, 1 H)	4, 10	11w	14w, 16, (20), A
16	110.1 (d)	6, 11	6.32	(d, J = 1.9 Hz, 1 H)	4?, 5, 6, 11	11w	6, (11, 15), 13, 14
17	46.2 (t)	22, 23	2.88–2.81	(m, 2 H) high-field portion	18, 22w, 23w	23, B	22, B
18	43.9 (t)	17, 20, 21	2.88–2.81	(m, 2 H) low-field portion	21	21, B	23?, B
19	38.4 (t)	22, 23w, D	3.16–3.14	(m, 2 H)	2, 22, 23	22, D	22, 23, D
20	35.7 (t)	21, C	≈3.35	(m, 2 H)	1, 18, 21	21, C	21, C, Dw
21	25.6 (t)	18w, 20	1.85–1.80	(m, 2 H)	18, 20	18, 20	18w, (20), C
22	25.4 (t)	17w, 19, 23	1.41–1.37	(m, 2 H)	17, 19, 23	19, 23	(17, 18), 19, D
23	22.8 (t)	17w, 19, 22	1.67–1.62	(m, 2 H)	17, 19w, 22	22, 17	(17, 18), 19, 20, D
A	—	—	10.02	(br. s, 1 H)	4, 15	—	15
B	—	—	8.84	(br. s, 2 H)	—	17, 18	(17, 18)
C	—	—	8.16	(t, J = 5.9 Hz, 1 H)	1, 20	20	7, 9w, 12, (17, 18)w, (20), 21, D
D	—	—	8.01	(t, J = 5.1 Hz, 1 H)	2, 19	19	14, 19, (20), 22, 23, C

with the two quaternary sp^2 C atoms C(3) and C(8). The two structural units deduced from the spin systems H(7)/H(12) and H(9)$_2$/H(13)$_2$ can finally be connected to the first larger partial structure of compound **20**, the cinnamic acid moiety **A** (**Fig. 5.16**).

The signals for H(11), H(15), and H(16), forming an ABX spin system, are found in the aromatic region of the ^{1}H NMR spectrum, too. The signal patterns correspond to the proton responses of a 1,3,4-trisubstituted benzene unit. The doublet of H(16) with the small coupling constant of J = 1.9 Hz stands for the isolated proton, and the AB portion of the spectrum for the neighboring H(11) and H(13), for which the *ortho* coupling of H(11) with H(15) and the *meta* coupling of H(11) with H(16) are recognizable. The HMBC spectrum reveals finally that

Examples

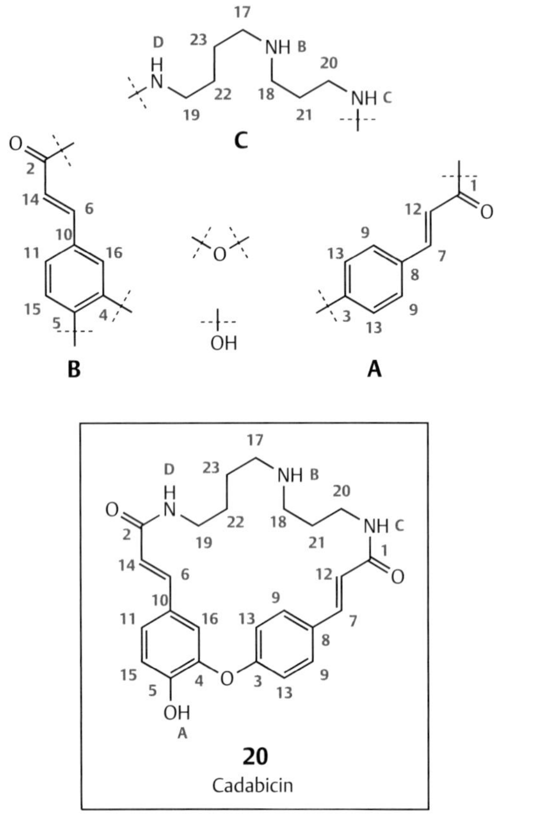

Fig. 5.16 Partial structures that were deduced from the NMR spectra and complete structure of compound **20**.

the three quaternary C atoms C(4), C(5), and C(10) belong to the same aromatic carbon framework. This—together with the C(6)/C(14)/C(2) unit identified above, which is connected to C(10)—results in fragment **B** shown in **Fig. 5.16**. The individual carbon atoms can be readily assigned on the basis of the ^3J correlations of C(5)→H(11)/H(16) and H(15)→C(4)/C(10).

Two further ^1H spin systems, found in the aliphatic region of the COSY spectrum, remain to be interpreted. They belong to a butylene moiety C(17)–C(23)–C(22)–C(19) and a propylene moiety C(18)–C(21)–C(20), respectively. The protons of their terminal C(17)H$_2$ and C(18)H$_2$ groups correlate with signal B, which can be explained by a bonding of the two methylene groups to the protonated N atom that still has to be accommodated in the structure. The protons of the two other terminal methylene groups C(19)H$_2$ and C(20)H$_2$ show correlations to the triplet signals C and D, respectively, which themselves could arise from amide N–H protons. Thus, moiety **C**, a α,ω-diacylated spermidine, can be suggested as the third partial structure of compound **20** (shown in **Fig. 5.16** without protonation).

Having progressed that far, only two O atoms and one H atom have still to be accounted for to complete the chemical formula of compound **20**. The reaction of **20** with FeCl$_3$ suggests the presence of a phenol in the compound. This group has to be localized at the aromatic group of the cinnamic acid moiety **B**, since otherwise the remaining O atom cannot find its place in the molecule. But with the OH group attached to moiety **B**, the final O atom can close the open valences of the two structural units **A** and **B** in the form of a diaryl ether bridge.

There are two final questions to be answered to complete the structure of compound **20**. (1) How is the spermidine moiety **C** connected to the two cinnamic acid groups **A** and **B**? (2) At which of the two carbons C(4) and C(5) the OH group and the diaryl ether linkage have to be placed?

The HMBC spectrum answers both of these questions. The correlations of the signals C(1)→H(C)/H(20) and C(2)→H(D)/H(19) reveal unmistakably that the C$_3$ unit of the spermidine moiety is connected to the carboxyl group of **A** and the C$_4$ unit to the carboxyl group of **B**. The correlations of the phenolic proton signal H(A) with the signals of C(4) and, in particular, of C(15) prove for their part that the OH group cannot be connected to C(4) and therefore must be connected to C(5). C(4) is thus connected to the O atom that forms the diaryl ether bridge. Both the positioning of the OH at C(5) and the diaryl ether linkage at C(4) are supported by the NOESY spectrum: it shows a contact of H(A) with H(15) and a contact of H(13) with H(16). All these interpretations and deductions finally lead to the structure shown in **Fig. 5.16** for compound **20**, which is known in the literature as cadabicin.[9]

Example 9

Spider venoms are complex mixtures of diverse compounds that are obviously accessible in small amounts only. Structural elucidations of natural components of spider venoms can therefore only rely on data that are derived from very sensitive analytical methods, and experience has shown that it is also often necessary to have access to secured preknowledge. It is known for the polyamine spider toxins of *Larinioides folium* that many of these compounds have the general structure shown in **Fig. 5.17**: they possess a linear α,ω-diaminoazaalkane scaffold of the type **A** as their basic structure that is connected at one end to an aromatic carboxylic acid **B** of the row **21** to **29** *via* an asparagine linkage. The polyamine portion usually consists of several diaminopropane subunits and a diaminobutane, diaminopentane, or diaminohexane group, and some of the N atoms within the polyamine chain are occasionally methylated.

(U)HPLC-ESI-MS/MS has been established as the first choice method for the structural elucidation of such toxins (p. 331). The coupling of two most efficient separation methods—the

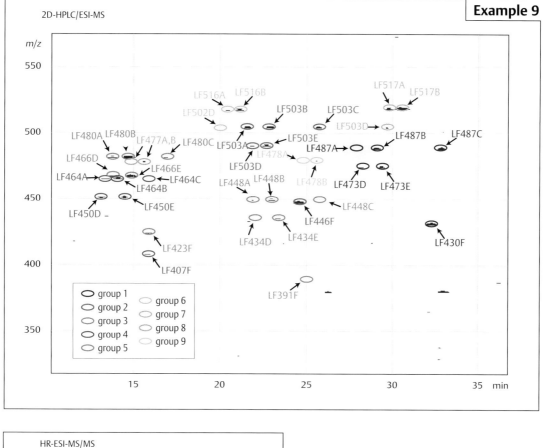

Example 9

2D-HPLC/ESI-MS

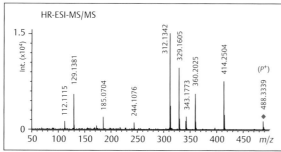

HR-ESI-MS/MS

UV(DAD) (H$_2$O): λ_{max} 193, 216, 219, 279, 287

The amino acid analysis of an UHPLC fraction, collected at 27–28 min, revealed the presence of asparagine and/or aspartic acid.

HR-ESI-MS/MS Report

Meas. m/z	Ion Formula	m/z	err [ppm]	rdb	exch. H*
112.1115	C7H14N	112.1121	-4.79	1.5	0
129.1381	C7H17N2	129.1386	-4.43	0.5	3
185.0704	C11H9ON2	185.0709	-3.17	8.5	1
244.1076	C13H14N3O2	244.1081	-2.04	8.5	4
312.1342	C17H18N3O3	312.1343	-0.09	10.5	2
329.1605	C17H21N4O3	329.1608	-0.85	9.5	5
343.1773	C18H23N4O3	343.1765	2.55	9.5	4
360.2025	C18H26N5O3	360.2030	-1.45	8.5	7
414.2504	C22H32N5O3	414.2500	1.17	9.5	5
488.3339	C25H42N7O3	488.3344	1.02	8.5	9

*exch. H = exchangeable H, determined by MS/MS with deuterated solvents.

Examples

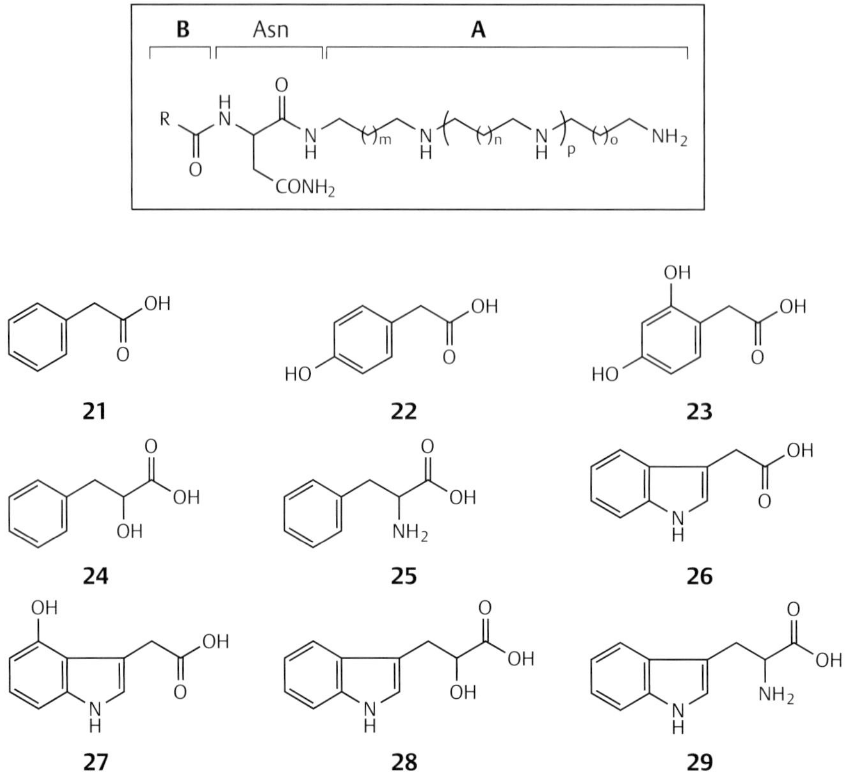

Fig. 5.17 General structure and structural portions of the polyamine toxins from the venom of the spider *Larinioides folium*.

chromatographic separation by (U)HPLC and the mass separation by MS—enables the analysis of smallest amounts of toxins directly from a natural, mostly very complex venom sample. For instance, the method allows the isolation of the ionized compound **LF487A** from the venom of *L. folium* (see the yellow-marked signal in the 2D-HPLC/ESI-MS shown in the collected data of example 9) and to analyze it by MS/MS. The respective MS/MS data, together with some additionally acquired information, the knowledge of the general structural features of such toxins, and the knowledge of the MS/MS behavior of related analogous compounds (p. 384ff), allowed the deduction of the correct structure of the compound (www.venoms.ch).[10]

The task of this Example is to elaborate a structural proposal for **LF487A**, solely on the basis of the information that was also available during the original analysis of the native venom.

Solution and Discussion of Example 9

To start with, we assume that **LF487A** in fact possesses the general structure of the toxins usually found in *L. folium*. This

hypothesis is supported by the amino acid analysis that revealed asparagine or aspartic acid as a constituent of the compound. Aspartic acid, however, can readily be excluded as a component of the structure: the exact mass of the P^+ ion ($= [M + H]^+$, $m/z = 488.3339$) in the MS suggests the chemical formula $C_{25}H_{41}N_7O_3$ for the toxin, which means that the compound does not contain enough oxygen to accommodate aspartic acid as a linker (one of the O atoms needs to be part of the group **B**). The chemical formula of **LF487A** also indicates the nature of group **B**. The formula $C_{25}H_{41}N_7O_3$ corresponds to a compound with nine DBEs (p. 339), which means that seven DBEs have to be allocated to **B** (overall nine DBEs minus two DBEs for asparagine) and that **B**, thus, could be the derivative of one of the indole-containing carboxylic acids **26** to **29**. Acids **27** and **28**, however, can immediately be eliminated as structural components because they would include an additional O atom, which would not be compatible with the chemical formula of the toxin.

Two general structures result for **LF487A** when the two carboxylic acids **26** and **29** are considered as possible components: **26–Asn–A¹** and **29–Asn–A²** (**Fig. 5.18**), and the respective polyamine portions A¹ and A² would have the chemical formulas

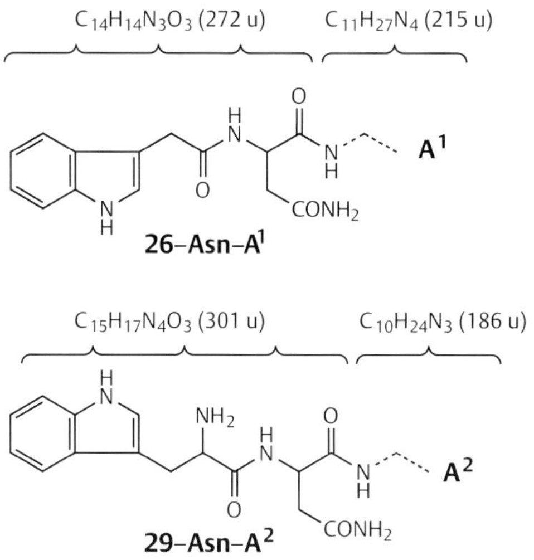

$C_{14}H_{14}N_3O_3$ (272 u) $C_{11}H_{27}N_4$ (215 u)

26–Asn–A¹

$C_{15}H_{17}N_4O_3$ (301 u) $C_{10}H_{24}N_3$ (186 u)

29–Asn–A²

Fig. 5.18 Possible partial structures of **LF487A**.

$C_{11}H_{27}N_4$ and $C_{10}H_{24}N_3$. The former could correspond to tetramines with two propylene groups and one pentylene unit (3/3/5, 3/5/3, 3/3/5) or *N*-methylated tetramines with two propylene groups and one butylene unit (3/3/4, 3/4/3, 4/3/3) in the main framework. The latter polyamine component could principally be an unprecedented triamine with 3/7, 4/6, or 5/5 methylene units in between the N atoms, or, which would come closest to previously found frameworks, a trimethylated spermidine group (3/4).

To determine the ultimate structure of **LF487A**, it is finally necessary to get to the bottom of the composition and the structure of its polyamine framework. The appropriate tool to attain this task is MS/MS. It is known from many investigations that acylpolyamines undergo well-defined and predictable fragmentation reactions upon collision activation (**Fig. 5.19**).[11] From all

these reactions, those that involve intramolecular substitutions *via* five- or six-membered cyclic transition states are usually preferred and deliver dominant signals. Thus, for the generic structure shown in **Fig. 5.19**, for instance, intense signals would be expected for the fragments $\mathbf{a_1}$ and $\mathbf{a_4}$, as well as for $\mathbf{z_1}$ and $\mathbf{t_1}$, all deriving from reactions around the C_4 unit highlighted with bold bonds.

The search for the **a**-type fragments in the MS/MS of **LF487A** is focused on ions with high degrees of unsaturation. These denote the presence of the intact head group consisting of the chromophoric acyl moiety and asparagine. Such ions are found at $m/z = 329$ ($\mathbf{a_1}$) and $m/z = 414$ ($\mathbf{a_2}$), corresponding to particles $[P - C_8H_{21}N_3]^+$ and $[P - C_3H_{10}N_2]^+$, respectively (**Fig. 5.20**). Hence, **LF487A** must represent a derivative of a tetramine, which would be compatible with a structure **26–Asn–A¹** but not with the structure **29–Asn–A²**. Signal $\mathbf{a_2}$ indicates the loss of a terminal diaminopropane group from the precursor ions P^+ and signal $\mathbf{a_1}$ the loss of an additional $C_5H_{11}N$ unit. The two signals allow to conclude that the polyamine terminus of **LF487A** is either a homospermidyl (-NH(CH$_2$)$_5$NH(CH$_2$)$_3$NH$_2$) or an *N*-methylated spermidyl group (-N(CH$_3$)(CH$_2$)$_4$NH(CH$_2$)$_3$NH$_2$). The pair of signals at $m/z = 129/112$ for the $\mathbf{z_2}/\mathbf{y_2}$ ions $[C_7H_{17}N_2]^+/[C_7H_{14}N]^+$ finally vouches unequivocally for the latter. The presence of an *N*-methylated compound is additionally supported by the number of exchangeable protons found for the P^+ ions; there are nine rather than ten that would be expected for the alternative structure.

The complete structure of **LF487A** finally results from the assembly of the head group **26–Asn** and the tail portion "*N*-methylspermidine," which have to be interconnected by a propylene unit—the partial structure that is needed to complete the chemical formula of the compound. The structure of **LF487A** proposed this way[10] is shown in **Fig. 5.20**, where also the fragments observed in the MS/MS are indicated.

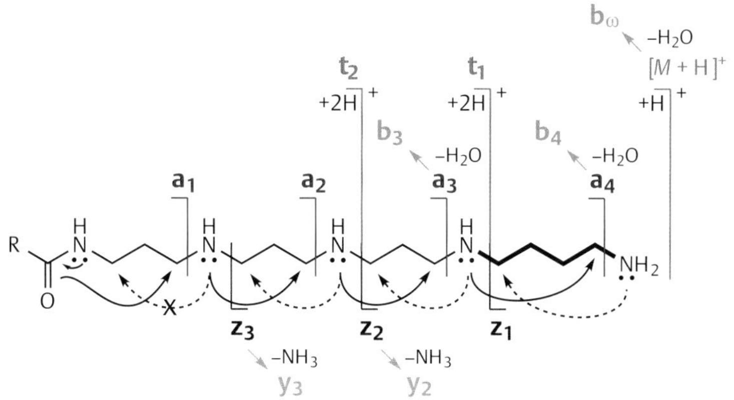

Fig. 5.19 Common fragmentation reactions observed with acylpolyamines.

Examples

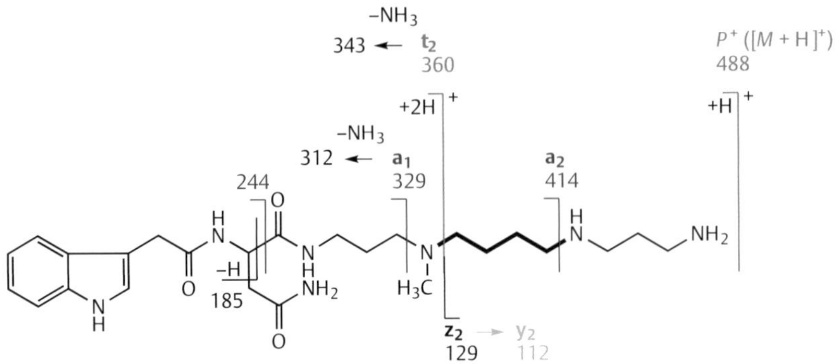

Fig. 5.20 Structural proposal for **LF487A** on the basis of MS/MS data, and indication of the MS/MS fragments found for the compound.

Example 10

Toxin **LF487A** was synthesized to prove unequivocally its structure and also the appropriateness of the methods that were used for its structural elucidation.[12] The equivalence of the synthetic compound and the natural toxin was ensured by HPLC-MS/MS. The task of Example 10 is to deduce the structure of **LF487A** on the basis of the extended analytical data (Examples 9 and 10) alone, ignoring additional information such as the source of the natural product, known analogs, and synthetic precursors.

Solution and Discussion of Example 10

The structure proposed for **LF487A** in Example 9 is fully consistent with all the data obtained for the toxin upon the UHPLC-HR-ESI-MS/MS analysis of the natural venom. The deduction, however, substantially relies on structural assumptions that have not been proven but were accepted for the reason of structural analogy. For instance, the MS/MS can neither prove definitely that indole acetic acid is a constituent of the toxin nor can it be used to determine where a potential CH_2COX group would be positioned at the aromatic moiety. MS/MS is also not suited to prove the linear nature of the polyamine backbone—the C framework could in principle be branched, and the same fragment pattern would be expected. Nevertheless, taking all information and knowledge into account—including biogenetic considerations—the structure of **LF487A** can be regarded in all probability as correct. Still, the desire for a complete and unequivocal structural elucidation remains for any analytical chemist, and such an elucidation of **LF487A** is only possible by means of the extended set of analytical data presented below (Example 10).

We start our widened considerations with a summary of the secured insights gathered with Example 9: **LF487A** is a compound with the chemical formula $C_{25}H_{41}N_7O_3$. It contains asparagine or aspartic acid as a constituent and, recognized by the nine DBEs and the UV-Vis spectrum, an extended region of unsaturation.

The first analysis of the ^{13}C NMR spectrum reveals that all 25 C atoms are accounted for with individual signals. In fact, the spectrum exposes an overall of 27 signals, but the two signals found at $\delta = 171.4$ (d) and 49.3 (q) have to be assigned to formic acid and methanol, respectively. Methanol had been added as a reference to the NMR solvent and formic acid was a constituent of the solvent mixture used in the HPLC purification of the sample.

The signals in the several NMR spectra shown above are labeled according to the rules already applied previously: the ^{13}C NMR signals are consecutively numbered from low to high field, and the ^{1}H NMR signals obtained the respective numbers of the correlating signals in the HSQC spectrum. The ^{13}C signals are distributed over three regions of the spectrum. Three signals ($3s$) are found at $\delta \approx 175$ in the region of the absorption of carboxylic C atoms (**Table 3.33**, p. 185), eight ($3s$ and $5d$) in the region of sp²-hybridized carbons (aromatic and other unsaturated C atoms, **Tables 3.41** and **3.43**, p. 192 and 193), and the remaining 14 ($1d$, $12t$, and $1q$) in the aliphatic region. The respective proton signals are found in the ^{1}H NMR spectrum in the aromatic range at $\delta = 7.50–7.24$ (five distinct signals for one proton each), at $\delta = 4.55$ ($1dd$ for a methine proton), at $\delta = 3.77$ ($1s$ for a methylene group), and within the regions of $\delta \approx 3.3–2.7$ ($7 \times CH_2$ and $1 \times CH_3$) and $\delta \approx 2.0–1.6$ ($4 \times CH_2$). The ^{13}C and ^{1}H NMR signals are compiled in **Table 5.5** together with the correlation information of the HMBC (C→H and H→C) and COSY spectra.

Again, it is a reasonable approach to perform the NMR interpretation stepwise. The signals in the aromatic/olefinic regions of the spectra may represent a possible entry point. The

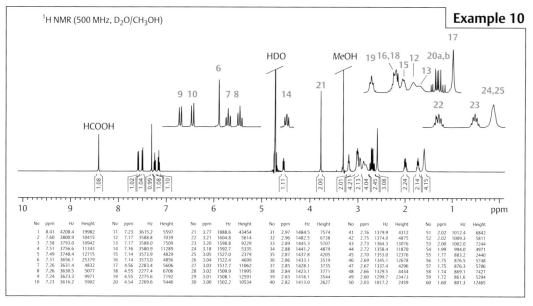

¹H NMR (500 MHz, D₂O/CH₃OH)

Example 10

No	ppm	Hz	Height		No	ppm	Hz	Height		No	ppm	Hz	Height		No	ppm	Hz	Height		No	ppm	Hz	Height		No	ppm	Hz	Height
1	8.41	4208.4	19982		11	7.23	3615.2	5597		21	3.77	1888.6	43454		31	2.97	1484.5	7574		41	2.76	1379.9	4312		51	2.02	1012.4	6842
2	7.60	3800.9	10415		12	7.17	3588.8	7039		22	3.21	1604.8	5614		32	2.96	1482.5	6738		42	2.75	1374.0	4875		52	2.02	1009.3	5811
3	7.58	3793.0	10942		13	7.17	3588.0	7509		23	3.20	1598.8	9229		33	2.89	1445.3	5707		43	2.73	1364.3	13076		53	2.00	1002.0	7244
4	7.51	3756.6	11343		14	7.16	3580.9	11289		24	3.18	1592.7	5335		34	2.88	1441.2	4879		44	2.72	1358.4	11870		54	1.99	994.0	4971
5	7.49	3748.4	12715		15	7.14	3573.9	4829		25	3.05	1527.0	2379		35	2.87	1437.8	4205		45	2.70	1353.0	12376		55	1.77	883.2	2440
6	7.31	3656.1	25379		16	7.14	3573.0	4856		26	3.04	1522.4	4699		36	2.86	1433.1	3519		46	2.69	1345.1	12678		56	1.75	876.5	5748
7	7.26	3631.4	4832		17	4.56	2283.4	5606		27	3.03	1517.7	11062		37	2.85	1428.1	3735		47	2.67	1337.4	4296		57	1.75	876.3	5706
8	7.26	3630.5	5077		18	4.55	2277.4	6706		28	3.02	1509.9	11995		38	2.84	1423.1	3771		48	2.66	1329.5	4434		58	1.74	869.1	7427
9	7.24	3623.3	9971		19	4.55	2275.6	7192		29	3.01	1508.1	12591		39	2.83	1418.1	3544		49	2.60	1299.7	23473		59	1.72	861.6	5294
10	7.23	3616.2	5902		20	4.54	2269.6	5446		30	3.00	1502.2	10534		40	2.82	1413.0	2627		50	2.03	1017.2	2459		60	1.60	801.3	12465

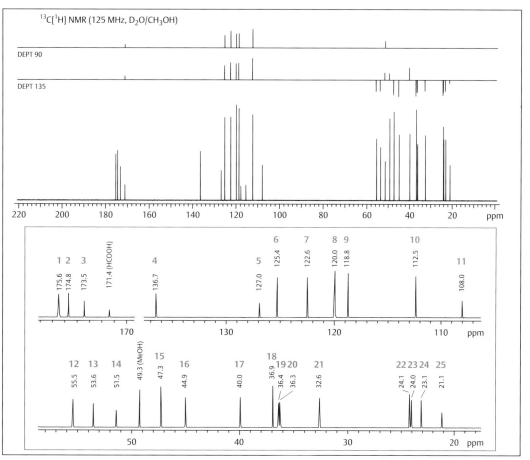

¹³C[¹H] NMR (125 MHz, D₂O/CH₃OH)

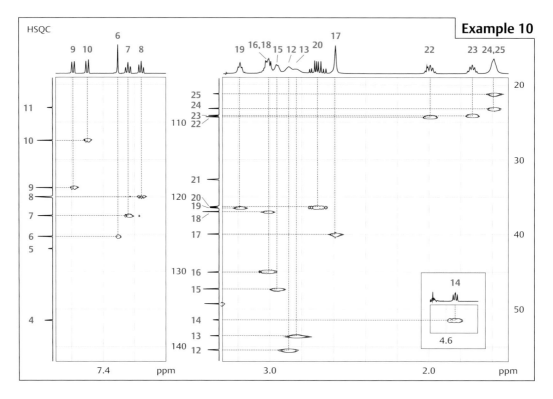

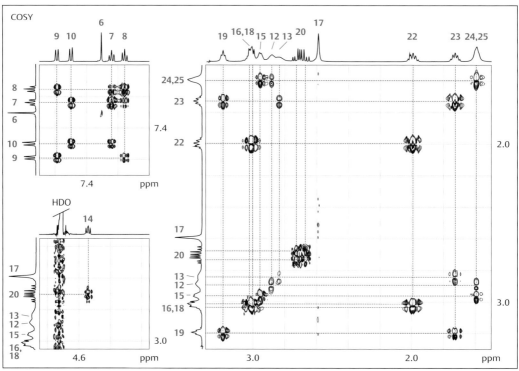

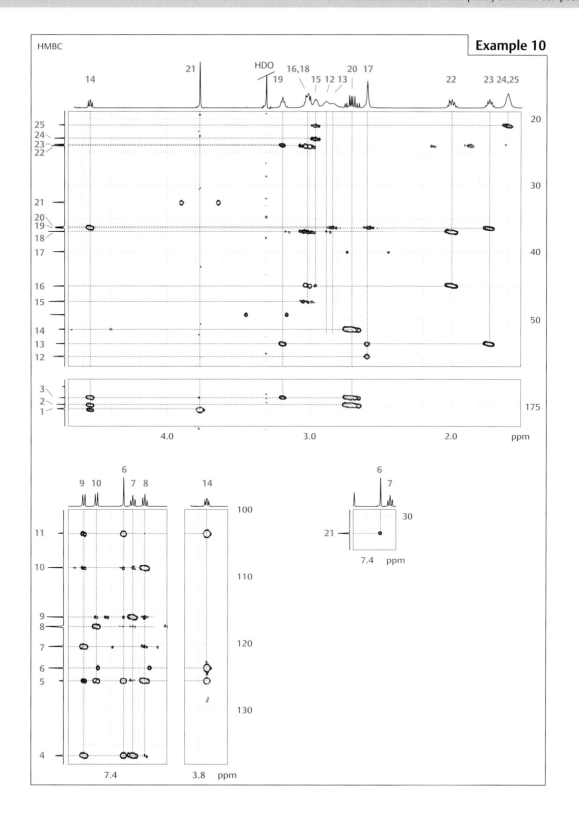

Table 5.5 Compilation of the NMR data and the 2D-NMR correlations found for **LF487A**

No.	^{13}C	HMBC (C→H)	^1H			HMBC (H→C)	COSY
1	175.6 (s)	14, 21	—	—		—	—
2	174.8 (s)	14, 20	—	—		—	—
3	173.5 (s)	14, 19, 20	—	—		—	—
4	136.7 (s)	6, 7, 9	—	—		—	—
5	127.0 (s)	6, 8, 9, 10, 21	—	—		—	—
6	125.4 (d)	21	7.31	(s, 1 H)		4, 5, 10w, 11	—
7	122.6 (d)	8, 9	7.24	(td, J = 7.6, 0.9 Hz, 1 H)		4, 9	8, 10
8	120.0 (d)	10	7.16	(td, J = 7.5, 0.9 Hz, 1 H)		5, 7w, 9w, 10	7, 9
9	118.8 (d)	7, 8	7.59	(d, J = 8.0 Hz, 1 H)		4, 5, 7, 10w, 11	8
10	112.5 (d)	8	7.50	(d, J = 8.1 Hz, 1 H)		5, 6w, 8, 9w	7
11	108.0 (s)	6, 9, 21,	–	–		—	—
12	55.5 (t)	17	2.89	(br. s, 2 H)		—	24/25
13	53.6 (t)	17, 19, 23	2.85	(br. s, 2 H)		19	23
14	51.5 (d)	20	4.55	(dd, J = 7.8, 6.0 Hz, 1 H)		1, 2, 3, 20	20
15	47.3 (t)	16/18	2.97	(br. d-like m, 2 H)		16, 24, 25	24/25
16	44.9 (t)	15w, 18, 22	3.05–2.99	(m, 2 H, low-field portion)		18, 22/23	22
17	40.0 (q)	—	2.60	(s, 3 H)		12, 13	—
18	36.9 (t)	16, 22	3.05–2.99	(m, 2 H, high-field portion)		15w, 16, 22/23	22
19	36.4 (t)	13, 23	3.20	(br. t, J = 6.0 Hz, 2 H)		3, 13, 22/23	23
20	36.3 (t)	14	2.73 / 2.68	(A of ABX, J = 15.6, 6.0 Hz, 1 H) (B of ABX, J = 15.6, 7.9 Hz, 1 H)		2, 3, 14 / 2, 3, 14	14 / 14
21	32.6 (t)	6	3.77	(s, 2 H)		1, 5, 6, 11	—
22	24.1 (t)	16/18	2.00	(quint, J = 7.7 Hz, 2 H)		16, 18	16/18
23	24.0 (t)	19	1.73	(quint, J = 7.2 Hz, 2 H)		13, 19	13, 19
24	23.1 (t)	15	1.60	(br. s, 2 H, isochron. with H(25))		25	12, 15
25	21.1 (t)	15, 24	1.60	(br. s, 2 H, isochron. with H(24))		—	12, 15

COSY spectrum shows that the ^1H NMR signals 7 to 10 found in this region belong to the same spin system: H(7) couples with H(8) and H(10), and H(8) couples with H(7) and H(9). The signals of H(7) and H(8) are triplets and those of H(9) and H(19) doublets. This constellation corresponds to a conjugated triene C(9)H–C(8)H–C(7)H–C(10)H or, more probably, to an ortho-disubstituted benzene with C(9)H and C(10)H being the peripheral units. In fact, a benzene system can be readily completed with C(4) and C(5). The two C atoms show strong HMBC correlations with H(7)/H(9) and H(8)/H(10), respectively. Taking into account that ^3J HMBC correlations in aromatic systems are prominent, this is consistent with the assignment shown for the benzene part of the indolic group **E** in **Fig. 5.21**. The HMBC correlation H(9)→C(11), also considered to be due to ^3J coupling, suggests that C(11) is connected to C(5), and the correlations

H(6)→C(5)/C(4)/C(11) imply that the system is completed with C(6) and an N(H) to an indole group (**Fig. 5.21**). In fact, all the chemical shifts of the ^1H and ^{13}C signals in question as well as all the respective HMBC correlations that have not been discussed so far are in full agreement with an indole partial structure (p. 241) and with the signal assignments indicated in **Fig. 5.21**. Incidentally, the HMBC spectrum reveals with the H(21)→C(1)/ C(5)/C(6)/C(11) correlations too that C(11) is connected to C(21) and C(21) to C(1). Thus, the first partial structure of **LF497A**, the indole acetyl group **E**, is secured by spectral proof.

As a second partial structure, the asparagine-diyl unit **F** can be easily recognized. This substructure is already known from the amino acid analysis, and the respective NMR signals are readily found: the signal at δ = 4.55 (dd, H(14)) corresponds to the doublet × doublet that is expected at this chemical shift for

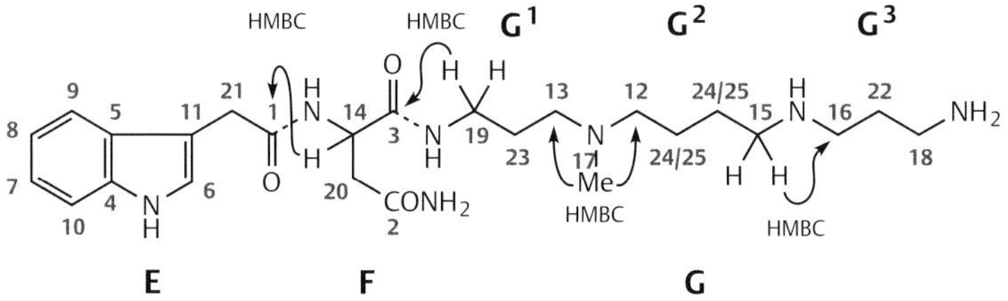

Fig. 5.21 Structure of **LF487A** and its partial structures that can be deduced from the NMR spectra.

the α-methine proton of asparagine, and the signals of the two diastereotopic protons H(20a/b), belonging to the same spin system as H(14), are found as an AB portion of an ABX system at $\delta = 2.73$ and 2.68. The HMBC correlations C(2)→H(14)/H(20) and C(3)→H(14)/H(20), finally, show that the atoms C(2) and C(3) also belong to the asparagine unit.

The HMBC correlation H(14)→C(1) shows that the units **E** and **F** are linked to each other via the α-N atom of the amino acid, and the HMBC correlation C(3)→H(19) indicates that C(3) is connected to the rest of the molecule (substructure **G**). The latter, however, cannot be proven on the basis of the given data because the HMBC correlations of the two carboxyl C atoms with the protons H(14) and H(20) are of equal significance and, thus, do not allow a definite assignment. This would require additional information to be obtained from, for instance, a NOESY or a ROESY spectrum.

The structure of the polyamine portion **G** that is connected to the units **E** and **F** results again from the combined interpretation of the COSY and the HMBC spectra. Starting with C(19)H$_2$, recognized above to be connected to C(3) via NH by an amide bond, the first spin system of the polyamine portion **G¹** can be walked through in the COSY spectrum: H(19)→H(23)→H(13). Unfortunately, no HMBC correlations are found for H(13) or C(13) with nuclei of portion **G²** that would indicate how the structure continues. However, the *N*–C(17)H$_3$ group is helpful in this respect. Indeed, the methyl protons C(17)H$_3$ show distinct HMBC correlations to both C(13) as well as C(12), proving that these two carbons have to be connected to each other via the methyl-bearing N atom. The COSY spectrum finally reveals that H(24)/H(25)—the two groups cannot be distinguished in the ¹H NMR spectrum—and H(15) belong to the same spin system, by which the subportion **G²** is completed.

According to the COSY spectrum, the last subunit **G³** of the polyamine portion is composed of C(16)H$_2$, C(22)H$_2$, and C(18)H$_2$. The protons H(16) and H(18) absorb isochronously and are, thus, not distinguished in the ¹H NMR spectrum.

However, the assignment of the C atoms of the unit **G³** as shown in **Fig. 5.21** is possible on the basis of the HMBC correlation H(15)→C(16).

With the final assignment of unit **G³**, the structure of **LF487A** is completely determined and secured. The analysis of the complete set of spectra, thus, revealed the same structure as the one that was deduced for the natural product from the MS/MS data alone.

5.5   Literature

1. Biemann K, Büchi G, Walker BH. The structure and synthesis of muscopyridine. J Am Chem Soc 1957;79:5558–5564
2. Meyers AI, Bienz S. Asymmetric total synthesis of (+)-$\Delta^{9(12)}$-capnellene. J Org Chem 1990;55:791–798
3. Perali RS, Mandava S, Bandi R. A convenient synthesis of L-ribose from D-fructose. Tetrahedron 2011;67:4031–4035
4. Bienz S, Busacca C, Meyers AI. The ambiphilic nature of *N*-acyliminium ion enamide tautomers — a novel annulation to enantiomerically pure polycyclic frameworks. J Am Chem Soc 1989;111:1905–1907
5. Ognyanov VI, Hesse M. Palladium-catalyzed synthesis of 2-allyl-2-nitrocycloalkanones. Synthesis 1985;6/7:645–647
6. Stach H, Huggenberg W, Hesse M. Synthesis of 2-hydroxy-3-methyl-2-hexen-4-olid. Helv Chim Acta 1987;70:369–374
7. Forster Y, Ghaffar A, Bienz S. A new view on the codonocarpine type alkaloids of *Capparis decidua*. Phytochemistry 2016;128:50–59
8. Schlittler E, Hohl J. Über die Alkaloide aus *Strychnos melinoniana* Baillon. Helv Chim Acta 1952;35:29–45
9. Ahmad VU, Amber AUR, Arif S, Chen MHM, Clardy J. Cadabicine, an alkaloid from *Cadaba farinosa*. Phytochemistry 1985;24:2709–2711
10. Eichenberger S, Bigler L, Bienz S. Structure elucidation of polyamine toxins in the venom of the spider *Larinioides folium*. Chimia 2007;61:161–164
11. Tzouros M, Chesnov S, Bigler L, Bienz S. A template approach for the characterization of linear polyamines and derivatives in spider venom. Eur J Mass Spectrom 2013;19(1):57–69
12. Pauli D, Bienz S. Regioselective solid-phase synthesis of *N*-mono-hydroxylated and *N*-mono-methylated acylpolyamine spider toxins using an 2-(ortho-nitrophenyl)ethanal-modified resin. Org Biomol Chem 2015;13:4473–4485

Examples

Index

Subject Index

Places where terms are explained in more detail are highlighted in semi-bold.

Index

Specific Compound Index and Classes of Compounds

References to displayed spectra are highlighted in semi-bold print. Indications of substance class are highlighted in italics. NMR: Single Compounds discussed in the text and Classes of Compounds described in tables.

Index